Energy, Environment, and Sustainability

Series Editor

Avinash Kumar Agarwal, Department of Mechanical Engineering, Indian Institute of Technology Kanpur, Kanpur, Uttar Pradesh, India

AIMS AND SCOPE

This books series publishes cutting edge monographs and professional books focused on all aspects of energy and environmental sustainability, especially as it relates to energy concerns. The Series is published in partnership with the International Society for Energy, Environment, and Sustainability. The books in these series are edited or authored by top researchers and professional across the globe. The series aims at publishing state-of-the-art research and development in areas including, but not limited to:

- Renewable Energy
- Alternative Fuels
- Engines and Locomotives
- Combustion and Propulsion
- Fossil Fuels
- Carbon Capture
- Control and Automation for Energy
- Environmental Pollution
- Waste Management
- Transportation Sustainability

Review Process

The proposal for each volume is reviewed by the main editor and/or the advisory board. The chapters in each volume are individually reviewed single blind by expert reviewers (at least four reviews per chapter) and the main editor.

Ethics Statement for this series can be found in the Springer standard guidelines here https://www.springer.com/us/authors-editors/journal-author/journal-author-helpdesk/before-you-start/before-you-start/1330#c14214

More information about this series at https://link.springer.com/bookseries/15901

Swatantra P. Singh · Avinash Kumar Agarwal ·
Kamlesh Kumar · Simant Kumar Srivastav
Editors

Metal Nanocomposites for Energy and Environmental Applications

Springer

Editors
Swatantra P. Singh
Department of Environmental Science and Engineering (ESED)
Indian Institute of Technology Bombay
Mumbai, Maharashtra, India

Avinash Kumar Agarwal
Department of Mechanical Engineering
Indian Institute of Technology Kanpur
Kanpur, Uttar Pradesh, India

Kamlesh Kumar
Department of Chemistry
Institute of Science, Banaras Hindu University (BHU)
Varanasi, Uttar Pradesh, India

Simant Kumar Srivastav
University Department of Chemistry
Lalit Narayan Mithila University
Darbhanga, Bihar, India

ISSN 2522-8366 ISSN 2522-8374 (electronic)
Energy, Environment, and Sustainability
ISBN 978-981-16-8598-9 ISBN 978-981-16-8599-6 (eBook)
https://doi.org/10.1007/978-981-16-8599-6

This Springer imprint is published by the registered company Springer Nature Singapore Pte Ltd.
The registered company address is: 152 Beach Road, #21-01/04 Gateway East, Singapore 189721, Singapore

Preface

The unsustainable growth of the world resulted in a polluted ecosystem and with high energy demand. In recent years, nanomaterials and nanocomposites have shown their potentials for environmental remediation. Different nanomaterials (CBNs), such as metal, metal oxide-based nanomaterials, carbon-based nanomaterials and other nanocomposites have been used widely since the last decades. These nanomaterials and nanocomposites have shown their usefulness for energy and environmental remediation through adsorption, reactive oxygen species generation through electro- and photocatalytic processes.

The International Society for Energy, Environment and Sustainability (ISEES) was founded at the Indian Institute of Technology Kanpur (IIT Kanpur), India, in January 2014 to spread knowledge/ awareness and catalyze research activities in the fields of Energy, Environment, Sustainability, and Combustion. Society's goal is to contribute to the development of clean, affordable, and secure energy resources and a sustainable environment for society and spread knowledge in the areas mentioned above, and create awareness about the environmental challenges the world is facing today. The unique way adopted by ISEES was to break the conventional silos of specializations (Engineering, science, environment, agriculture, biotechnology, materials, fuels, etc.) to tackle the problems related to energy, environment, and sustainability in a holistic manner. This is quite evident by the participation of experts from all fields to resolve these issues. The ISEES is involved in various activities such as conducting workshops, seminars, conferences, etc., in the domains of its interests. Society also recognizes the outstanding works of young scientists, professionals, and engineers for their contributions in these fields by conferring them awards under various categories.

Fifth International Conference on 'Sustainable Energy and Environmental Challenges' (V-SEEC) was organized under the auspices of ISEES from December 19-21, 2020, in virtual mode due to restrictions on travel because of the ongoing Covid-19 pandemic situation. This conference provided a platform for discussions between eminent scientists and engineers from various countries, including India, Spain, Austria, Bangladesh, Mexico, USA, Malaysia, China, UK, Netherlands, Germany, Israel, and Saudi Arabia. At this conference, eminent international speakers presented their views on energy, combustion, emissions, and alternative energy resources for sustainable development and a cleaner environment. The conference presented two high voltage plenary talks by Dr. VK Saraswat, Honorable Member, NITI Ayog, on 'Technologies for Energy Security and Sustainability' and Prof. Sandeep Verma, Secretary, SERB, on 'New and Equitable R&D Funding Opportunities at SERB.'

The conference included nine technical sessions on topics related to energy and environmental sustainability. Each session had 6-7 eminent scientists from all over the world, who shared their opinion and discussed the trends for the future. The technical sessions in the conference included Emerging Contaminants: Monitoring and Degradation Challenges; Advanced Engine Technologies and Alternative Transportation Fuels; Future Fuels for Sustainable Transport; Sustainable Bioprocessing for Biofuel/ Non-biofuel Production by Carbon Emission Reduction; Future of Solar Energy; Desalination and Wastewater Treatment by Membrane Technology; Biotechnology in Sustainable Development; Emerging Solutions for Environmental Applications' and Challenges and Opportunities for Electric Vehicle Adoption. 500+ participants and speakers from all over the world attended this three days conference. The conference concluded with a high voltage panel discussion on 'Challenges and Opportunities for Electric Vehicle Adoption,' where the panelists were Prof. Gautam Kalghatgi (University of Oxford), Prof. Ashok Jhunjhunwala (IIT Madras), Dr. Kelly Senecal (Convergent Science), Dr. Amir Abdul Manan (Saudi Aramco) and Dr. Sayan Biswas (University of Minnesota, USA). Prof. Avinash K Agarwal, ISEES, moderated the panel discussion. This conference laid out the roadmap for technology development, opportunities, and challenges in Energy, Environment, and Sustainability domain. All these topics are very relevant for the country and the world in the present context. We acknowledge the support received from various agencies and organizations for the successful conduct of the Fifth ISEES conference V-SEEC, where these books germinated. We want to acknowledge SERB (Special thanks to Dr. Sandeep Verma, Secretary) and our publishing partner Springer (Special thanks to Ms. Swati Mehershi).

The editors would like to express their sincere gratitude to large number of authors from all over the world for submitting their high-quality work in a timely manner and revising it appropriately at a short notice. We would like express our special thanks to Dr. Ravinder Singh, Dr. Uttamam Mukherjee, Dr. Atul Dhar, Dr. Sandeep Singh Patel, Dr. Sumit Pramanik, Dr. Bhupesh Mishra and Dr. Ritesh Chourasia, who reviewed various chapters of this monograph and provided their valuable suggestions to improve the manuscripts.

In recent years, nanomaterials and nanocomposites have shown their potentials for energy and environmental applications. The unsustainable growth of the world economy leads to air, water and soil pollution. Additionally, high energy demand for economic growth is essential. This book focused on some recent development of the metal nanomaterials and nanocomposites for energy and environmental applications such as pollution control in water, air and soil pollution. The chapters have been incorporated carbon-based, metal-based and metal-organic framework-based nanomaterials and nanocomposites for emerging contaminants (pharmaceuticals and microplastics) and other traditional pollutants remediation along with energy storage, sensing of air and water pollutants and carbon capture and storage (CCS). We hope that the book would be of great interest to the professionals, postgraduate students involved in energy and environmental science and engineering research.

Mumbai, India — Swatantra P. Singh
Darbhanga, India — Simant Kumar Srivastav
Varanasi, India — Kamlesh Kumar
Kanpur, India — Avinash Kumar Agarwal

Contents

Editors and Contributors

About the Editors

Dr. Swatantra P. Singh is an environmental engineer with experience in membrane fabrication, environmental nanotechnology, fate, and transport of pollutants and emerging contaminants in the environment. Currently, he is an assistant professor in the Environmental Science and Engineering Department at Indian Institute of Technology (IIT) Bombay, India. He has developed a key technology to fabricate the membranes for water purification and print graphene in situ in a single step. He has four US patents (two granted and two provisional) on membrane and laser-based graphene fabrication techniques. He has authored 19 journal articles, one book and three book chapters. He recently won the INAE Young Engineer Award (2020) and ISEES Young Scientists Award (2020).

Dr. Simant Kumar Srivastav is an assistant professor in University Department of Chemistry at Lalit Narayan Mithila University, Darbhanga, Bihar, India. He obtained his Ph.D. degree from Indian Institute of Technology (IIT) Kanpur in 2014. He worked as post-doctoral fellow at Hebrew University of Jerusalem Israel and National Postdoctoral Fellow at IIT Delhi. He is a material chemist with a vast experience in the synthesis and characterizations of multifunctional materials. He has over 10 years of research experience in the area of material chemistry. He has developed a facile method for the synthesis of phase pure metal oxide nanomaterials. He has authored around 15 international scientific journals and 3 book chapters. In LNMU, he is leading the Advanced Multifunctional Materials (AMFM) research group. Dr. Srivastav is currently supervising 4 doctoral and 5 Master's students.

Dr. Kamlesh Kumar is currently an assistant professor in the Department of Chemistry, Institute of Science, BHU (India). He obtained his Ph.D. degree from Indian Institute of Technology (IIT) Kanpur in 2012. He worked as postdoctoral fellow at Technion-Israel Institute of Technology, Israel followed by Lund University, Sweden and University of Johannesburg, South Africa. His research interest is mainly focused on inorganic and organometallic chemistry as well as catalysis. His research group is developing novel inorganic materials as catalysts for water splitting as well as organic transformation reactions. Dr. Kumar has published 17 research articles in peer-reviewed international journals and is the author of 4 book chapters. He has supervised 9 Master's students and is currently supervising 4 doctoral scholars.

Prof. Avinash Kumar Agarwal joined IIT Kanpur in 2001. He worked at the Engine Research Center, UW@Madison, the USA as a Post-Doctoral Fellow (1999–2001). His interests are IC engines, combustion, alternate and conventional fuels, lubricating oil tribology, optical diagnostics, laser ignition, HCCI, emissions, and particulate control, 1D and 3D Simulations of engine processes, and large-bore engines. Prof. Agarwal has published 435+ peer-reviewed international journal and conference papers, 70 edited books, 92 books chapters, and 12200+ Scopus and 19000+ Google Scholar citations. He is the associate principal editor of FUEL. He has edited *Handbook of Combustion* (5 Volumes; 3168 pages), published by Wiley VCH, Germany. Prof. Agarwal is a Fellow of SAE (2012), Fellow of ASME (2013), Fellow of ISEES (2015), Fellow of INAE (2015), Fellow of NASI (2018), Fellow of Royal Society of Chemistry (2018), and a Fellow of American Association of Advancement in Science (2020). He is the recipient of several prestigious awards such as Clarivate Analytics India Citation Award-2017 in Engineering and Technology, NASI-Reliance Industries Platinum Jubilee Award-2012; INAE Silver Jubilee Young Engineer Award-2012; Dr. C. V. Raman Young Teachers Award: 2011; SAE Ralph R. Teetor Educational Award-2008; INSA Young Scientist Award-2007; UICT Young Scientist Award-2007; INAE Young Engineer Award-2005. Prof. Agarwal received Prestigious CSIR Shanti Swarup Bhatnagar Award-2016 in Engineering Sciences. Prof. Agarwal is conferred upon Sir J. C. Bose National Fellowship (2019) by SERB for his outstanding contributions. Prof. Agarwal was a highly cited researcher (2018) and was in the top ten HCR from India among 4000 HCR researchers globally in 22 fields of inquiry.

Contributors

A. A. Abdala Chemical Engineering Program, Texas A&M University at Qatar, Doha, Qatar

Avinash Kumar Agarwal Department of Mechanical Engineering, Indian Institute of Technology Kanpur, Kanpur, India

Piyush Avasthi School of Engineering, Indian Institute of Technology Mandi, Mandi, Himachal Pradesh, India

Viswanath Balakrishnan School of Engineering, Indian Institute of Technology Mandi, Mandi, Himachal Pradesh, India

Sweta Balchandani Department of Chemical Engineering, Pandit Deendayal Energy University, Gandhinagar, Gujarat, India

Devidas S. Bhagat Department of Forensic Chemistry and Toxicology, Government Institute of Forensic Science, Aurangabad, Maharashtra, India

P. R. Bhilkar Department of Chemistry, Seth Kesarimal Porwal College of Arts, Science and Commerce, Kamptee, India

G. S. Bhusari Research and Development Division, Apple Chemie India Private Limited, Nagpur, India

Narendra Bihari University Department of Physics, Lalit Narayan Mithila University, Darbhanga, India

Gurvinder S. Bumbrah Department of Chemistry, Biochemistry and Forensic Science, Amity School of Applied Sciences, Amity University, Haryana, 122413 India

Debrup Chakraborty School of Mechanical and Aerospace Engineering, Queen's University Belfast, Belfast, Northern Ireland, UK

R. G. Chaudhary Department of Chemistry, Seth Kesarimal Porwal College of Arts, Science and Commerce, Kamptee, India

Jyoti Roy Choudhuri BMS Institute of Technology and Management, Bengaluru, Karnataka, India

Nitesh K. Chourasia School of Physics, Indian Institute of Science Education and Research, Thiruvananthapuram, Kerala, India

Ritesh Kumar Chourasia University Department of Physics, Lalit Narayan Mithila University, Darbhanga, India

Subrata Das Department of Chemistry, National Institute of Technology Patna, Patna, India

Khemchand Dewangan Department of Chemistry, Indira Gandhi National Tribal University, Amarkantak, Madhya Pradesh, India

Anirban Dey Department of Chemical Engineering, Pandit Deendayal Energy University, Gandhinagar, Gujarat, India

S. G. Dhanushkumar Kumaraguru College of Technology, Coimbatore, Tamil Nadu, India

Atul Dhar School of Engineering, Indian Institute of Technology Mandi, Mandi, Himachal Pradesh, India

A. Geethakarthi Kumaraguru College of Technology, Coimbatore, Tamil Nadu, India

K. Giftlin Devapriya Kumaraguru College of Technology, Coimbatore, Tamil Nadu, India

Pranshu Kumar Gupta Department of Chemistry, Institute of Science, Banaras Hindu University, Varanasi, India

Wasudeo B. Gurnule Department of Chemistry, Kamla Nehru Mahavidyalaya, Nagpur, Maharashtra, India

Hansa Department of Materials Science and Engineering, Gachon University, Seongnam-si, Gyeonggi-do, South Korea

Kritika Jashrapuria Environmental Science and Engineering Department (ESED), Indian Institute of Technology Bombay, Mumbai, India

Anima Johari DST, Science and Engineering Research Board, New Delhi, India

Pradip Kalbar Interdisciplinary Program in Climate Studies, Centre for Urban Science and Engineering, Indian Institute of Technology Bombay, Mumbai, India

J. S. Khan Department of Chemistry, Seth Kesarimal Porwal College of Arts, Science and Commerce, Kamptee, India

Manisha Khandelwal Department of Chemistry, University College of Science, Mohanlal Sukhadia University, Udaipur, Rajasthan, India

Rama Kanwar Khangarot Department of Chemistry, University College of Science, Mohanlal Sukhadia University, Udaipur, Rajasthan, India

TaeYoung Kim Department of Materials Science and Engineering, Gachon University, Seongnam-si, Gyeonggi-do, South Korea

Mitil M. Koli Environmental Science and Engineering Department (ESED), Indian Institute of Technology Bombay, Mumbai, India

Kamlesh Kumar Department of Chemistry, Institute of Science, Banaras Hindu University, Varanasi, India

Manish Kumar Experimental Research Laboratory, Department of Physics, A.R.S.D. College, University of Delhi, New Delhi, India

Pramanand Kumar Department of Chemistry, National Institute of Technology Patna, Patna, India

Chandra Shekhar Kushwaha Department of Polymer Science, Bhaskaracharya College of Applied Sciences, University of Delhi, Delhi, India

Agasthiyaraj Lakshmanan Applied NanoPhysics Laboratory, Department of Physics and Nanotechnology, SRM Institute of Science and Technology, Kattankulathur, India

D. T. Masram Department of Chemistry, Delhi University, Delhi, India

B. Mirudhula Kumaraguru College of Technology, Coimbatore, Tamil Nadu, India

L. Monisha Kumaraguru College of Technology, Coimbatore, Tamil Nadu, India

Uttama Mukherjee Department of Chemistry and Centre for Energy Science, Indian Institute of Science Education and Research, Pune, Pune, Maharashtra, India

Fenil Patel Department of Chemical Engineering, Pandit Deendayal Energy University, Gandhinagar, Gujarat, India

A. K. Potbhare Department of Chemistry, Seth Kesarimal Porwal College of Arts, Science and Commerce, Kamptee, India

Sumit Pramanik Functional and Biomaterials Engineering Lab, Department of Mechanical Engineering, Faculty of Engineering and Technology, SRM Institute of Science and Technology, Tamil Nadu, Kattankulathur, KancheepuramChennai, India

Abhishek Rai Department of Chemistry, Faculty of Science, L.N. Mithila University, Darbhanga, India

Pawan Kumar Sada Department of Chemistry, Faculty of Science, L.N. Mithila University, Darbhanga, India

Shalini Sahani Department of Materials Science and Engineering, Gachon University, Seongnam-si, Gyeonggi-do, South Korea

Bharti Saini Department of Chemical Engineering, School of Technology, Pandit Deendayal Energy University, Gandhinagar, Gujarat, India

S. Sanjaikabilan Kumaraguru College of Technology, Coimbatore, Tamil Nadu, India

Debabrata Sarkar Applied NanoPhysics Laboratory, Department of Physics and Nanotechnology, SRM Institute of Science and Technology, Kattankulathur, India

Pooja P. Sarngan Applied NanoPhysics Laboratory, Department of Physics and Nanotechnology, SRM Institute of Science and Technology, Kattankulathur, India

Nilesh Satpute Department of Chemistry, Indira Gandhi National Tribal University, Amarkantak, Madhya Pradesh, India

Miraj Savani Department of Chemical Engineering, The Maharaja Sayajirao University of Baroda, Vadodara, Gujarat, India

Nisha Saxena Department of Chemistry, M.R.M. College, A Constituent Unit of L.N. Mithila University, Darbhanga, India

Rishabh Sharma Interdisciplinary Program in Climate Studies, Indian Institute of Technology Bombay, Mumbai, India

T. S. Shrirame Department of Chemistry, Seth Kesarimal Porwal College of Arts, Science and Commerce, Kamptee, India

Kamlesh Shrivas School of Studies in Chemistry, Pt. Ravishankar Shukla University, Raipur, Chhattisgarh, India

Saroj Kr Shukla Department of Polymer Science, Bhaskaracharya College of Applied Sciences, University of Delhi, Delhi, India

Ajay Kumar Singh Cenral Revenues Control Laboratory, New Delhi, India

Pratibha Singh Department of Polymer Science, Bhaskaracharya College of Applied Sciences, University of Delhi, Delhi, India

Ravindra Singh Department of Chemistry, Maharani Shri Jaya Government Post-Graduate College, Bharatpur, Rajasthan, India

Ritika Singh Department of Chemistry, Indira Gandhi National Tribal University, Amarkantak, Madhya Pradesh, India

Swatantra P. Singh Interdisciplinary Program in Climate Studies, Department of Environmental Science and Engineering (ESED), Centre for Research in Nanotechnology & Science (CRNTS), Indian Institute of Technology Bombay, Maharashtra, India

Vikas Kumar Sonu Department of Chemistry, Faculty of Science, L.N. Mithila University, Darbhanga, India

Abhay Nanda Srivastva Department of Chemistry, A Constituent Unit of B.R.A. Bihar University, Nitishwar Mahavidyalaya, Muzaffarpur, India

Ankita Srivastava Department of Physics, Faculty of Science and Humanities, Darbhanga College of Engineering, Darbhanga, India

Simant Kumar Srivastav University Department of Chemistry, L. N. Mithila University, Darbhanga, Bihar, India

Umang Sutariya Department of Chemical Engineering, Pandit Deendayal Energy University, Gandhinagar, Gujarat, India

M. S. Umekar Department of Chemistry, Seth Kesarimal Porwal College of Arts, Science and Commerce, Kamptee, India

Ravi Vaghasia Department of Chemical Engineering, School of Technology, Pandit Deendayal Energy University, Gandhinagar, Gujarat, India

Sandeep Yadav School of Engineering, Indian Institute of Technology Mandi, Mandi, Himachal Pradesh, India

Part I
Nanocomposites for Energy Applications

Chapter 1
Metal Nanocomposites for Energy and Environmental Applications

Swatantra P. Singh, Simant Kumar Srivastav, Kamlesh Kumar, and Avinash Kumar Agarwal

Abstract In recent years, metal nanocomposites have shown their potentials for energy and environmental applications. The unsustainable growth of the world economy leading to air, water, and soil pollution. Additionally, high energy demand for economic growth is essential. This book focused on some recent development of metal nanomaterials and nanocomposites for energy and environmental application such as pollution control in water, air, and soil pollution, carbon capture and storage (CCS), and sensing applications. The chapters have been incorporated carbon-based, metal-based and metal–organic framework-based nanomaterials and nanocomposites for emerging contaminants (pharmaceuticals and microplastics) and other traditional pollutants remediation along with energy storage, sensing of air and water polutents and CCS. We hope that the book would be of great interest to the professionals, post-graduate students involved in energy and environmental science and engineering research.

Keywords Metal and metal oxide nanomaterials · Nanocomposites · Environmental remediation · Carbon capture and storage (CCS) · Sensors

---

S. P. Singh (✉)
Department of Environmental Science and Engineering (ESED), Indian Institute of Technology Bombay, Maharashtra 400076, India
e-mail: swatantra@iitb.ac.in

Interdisciplinary Program in Climate Studies, Indian Institute of Technology Bombay, Mumbai 400076, India

Centre for Research in Nanotechnology & Science (CRNTS), Indian Institute of Technology Bombay, Mumbai 400076, India

S. K. Srivastav
University Department of Chemistry, L. N. Mithila University Darbhanga, Bihar 846006, India

K. Kumar
Department of Chemistry, Institute of Science, Banaras Hindu University, Varanasi 221005, India

A. K. Agarwal
Department of Mechanical Engineering, Indian Institute of Technology Kanpur, Kanpur 208016, India

S. P. Singh et al. (eds.), *Metal Nanocomposites for Energy and Environmental Applications*, Energy, Environment, and Sustainability,
https://doi.org/10.1007/978-981-16-8599-6_1

1.1 Introduction

The importance of sustainable energy sources and a clean environment is the key to the better health of humans and the ecosystem. The health problems from the consumption of polluted air, water, and food, and also global warming due to CO_2 emissions are a significant cause of human misery and the extinction of many other species from the Earth. The advancements in nanoscience and nanotechnology, especially in materials science, improve the ability of CCS, sensing of pollutants, measurement, degradation, and degradation of pollutants have increased significantly over the years. Metal-based nanomaterials and nanocomposites, based on carbon nanomaterials (CNMs), metal–organic framework, and their derivatives have strongly impacted the field of nanotechnology due to its physical, electronic, and chemical properties. These nanocomposites have shown their potential to be prominent adsorbents for CCS, pollutants, electrochemical degradation and conversion, and sensing applications.

In Chap. 2, the authors have discussed the lightweight metallic nanocomposites in energy Applications. They have provided the importance of composite materials like, metal matrix composites (MMCs) due to their enhanced high strength to weight ratio and lightweight properties. The authors have summarised the different manufacturing, processing, and characterization techniques of various MMCs comprising metals or alloys matrix and metallic oxides, nitride, borides, carbide, and their subgroups in energy application with the importance of these composites in the near future.

Chapter 3 deals with the metal oxides as an interesting candidate for electrochemical energy devices owing to their significant theoretical capacity and promising material as an electrode. The combination of metal oxides with carbon-based nanostructures gives rise to new nanocomposites with the improvement in structural and functional properties to serve the purpose of the electrochemical storage application. This chapter deals with the combination of metal oxides with one-, two-, or three-dimensional (1D, 2D, or 3D) carbon materials and discusses the various characteristic (physical, electrical, chemical, and structural properties) features of such unique composites and their application in the field of electrochemical storage.

Chapter 4 has provided the role of metal nanocomposites for electrode fabrication and it role for CCS. This chapter has provided details about the electrochemical catalysts based on metal nanocomposites for an effective CO_2 conversion to other high-energy molecules for future energy generations. Furthermore, Chap. 5 has discussed the recent advancement in the area of metal nanocomposites for CO_2 adsorption. The high adsorption by metal nanocomposites reported due to large surface area, plenty of nanopores, enhanced reactivity, and porosity.

Chapter 6 has discussed the technologies such as Carbon Capture and Storage (CCS) and Carbon Capture and Utilization (CCU) have become inevitable to achieve negative CO_2 emissions required to attain sustainable climate goals. This chapter discusses the different available methods for carbon capture, including absorption, membrane-based techniques, adsorption techniques, cryogenic methods, and biological methods that include microbial and algal techniques. Furthermore, this also focuses on converting captured CO_2 it into graphene, which exhibited a wide range of applications, including environmental remediation application. Chapter 7 provided the details about metal–organic framework (MOFs) as an adsorbent material for CO_2 capture due to MOFs highly crystalline nature with large surface areas and good selectivity. This chapter summarized the synthesis, performance, and regeneration of MOFs. Furthermore, Chap. 8 has demonstrated that the Carbon capture and sequestration (CCS) technology can be used as a potential tool for the significant reduction of CO_2 and this involves the capture of CO_2 from large sources like fossil fuel stations and transportation of CO_2 to the deep underground storage sites. This chapter highlights the use of amine-based nano-fluids in carbon capture, which exhibited more absorption efficiency than conventional solvents.

Chapter 9 talks about water splitting as a potential process of clean energy source which produces hydrogen and oxygen. Gurnule et al. described the advances in the synthesis of nano-composite materials and their application as efficient electro-catalyst in water splitting reactions to generate hydrogen and oxygen. Chapter 10 deals with the development of semiconductor and polymer-metal composite nanofiber for photocatalysis applications for dyes degradation and water splitting for H_2 production. Furthermore, this also discussed the latest development of nanocomposite membranes for the air filter and face mask because in the present Coronavirus disease 2019 (COVID-19) pandemic situation, disease transmission prevention is a major task that can be controlled by using a good face mask and air filter. Chapter 11 talks about the development of novel nano-composites that can effectively act towards ecological remediation to overcome the detrimental ecological impacts. In this chapter, the author discussed the various popular and effective synthesis roots of nano-composites useful in environmental remediation. Chapter 12 provided the details for designing green materials with biodegradable polymers and metal nanocomposites, and their utilization for different devices such as the development of electrochemical cells and batteries for energy storage, electrochromic devices for conservation of energy, and gaseous capturing devices for the safety of the environment.

Chapter 13 has provided a picture of nanocomposites application for energy and environmental applications. They have summarized several alkalines and alkali earth metal-based nanocomposites that are used in basic heterogeneous catalysis for cleaner fuel production along with transition metal/metal oxide-based nanocomposites (TMMONs) as photocatalysts. Chapter 14 summarizes the graphene-based polymer nano-composites application for the removal of organic pollutants. This chapter summarizes the most recent studies on the modification of graphene with polymers and their applications in the photolytic degradation of organic pollutants (OPs). Similarly, Chap. 15 has discussed functionalized graphene-based materials for the selective removal and detection of heavy metals. The selective removal of heavy metal ions by any Gr/GO-based nanomaterials does not require utterly demineralized water and maintaining water quality and human health.

In recent years, metal–organic frameworks (MOFs) have been employed in technologies to bring an inspiring breakthrough for wastewater treatment. This is because of unique characteristics of MOFs materials accountable for the wastewater treatment like easily synthesizable, various size cavities with different-different functional group, surface functional groups, various functionality where host–guest interaction takes place and high surface area, which responsible for high absorption capacity. Chapter 16 describes the synthesis of MOF-based nano-materials and their application towards the elimination of heavy metal ions from wastewater as well as energy storage devices. Chapter 17 provided the importance of metal nanocomposites for groundwater pollutant removal. This chapter summarized typical synthesis methods, removal mechanism and chemistry, adsorption capabilities, and influence of pH on adsorption of these contaminants by different nanocomposites.

Chapter 18 describes the fundamental principle, design strategies, and preparation protocols of metal nanocomposites. Furthermore, the sensing methods developed to determine harmful and hazardous metal ions, chemicals, and bio-molecules presence in the environmental, aquatic, and other related samples are also discussed. Similarly, Chap. 19 has talked about nanocomposites for ultra-fast, and selective multi-sensor activity over narrow detection limits are of an urgent demand for national, homeland, and environment safety. The fabrication of these biosensors endorses their progress in environmental and material aspects. This chapter deals with recent developments made in the field of nanocomposite-sensors, their synthesis, the mechanism for their sensing action, and environmental application involving mainly optical sensors. Furthermore, Chap. 20 describes the basic principle of sensing techniques (optical, electrical, and mechanical) have been covered along with the synthesis of transition metal and conducting polymer nanocomposites with their noble properties for sensing different environmental parameters and gases like humidity, ammonia, and other gaseous pollutants. Chapter 21 has talked about copper-based composites for sensing applications due to their low cost, easy availability, and high electrical conductivity. This chapter describes the synthesis technologies of copper-based polymer nanocomposites (CuPNCs) and their different applications as sensors in gas, chemical, and humidity sensing.

The last chapter of the book has demonstrated stacked stainless steel mesh with iron oxide nanostructures as a substrate for NOx emission control of diesel engines. The high conversion efficiency was achieved with 24 stacks which indicate stacking impotence of the SS mesh. This work explores the feasibility of low-cost iron oxides as a catalyst support material for Selective catalytic reduction (SCR) of NOx in diesel engines by making it dense, effective, and cheap. Overall, this book offers a state of the art literature for researchers and academicians working on nanocomposites applications in the field of energy and environment.

Chapter 2
Lightweight Metallic Nanocomposites in Energy Applications

Debrup Chakraborty and Sumit Pramanik

Abstract The demand from consumers leads to the contemporary design of cars with lightweight components. The renewable energy industry is focused on the design of an optimized shape of the components to increase electricity production and deter the usage of fossil fuels. Composite materials like, metal matrix composites (MMCs) are preferred over the use of pure metals and alloys due to their enhanced high strength to weight ratio and lightweight properties. Here, the different manufacturing, processing, and characterization techniques of various metal matrix nanocomposites comprising metals or alloys matrix and metallic oxides, nitride, borides, carbide, and their subgroups in energy-related applications have been elucidated. Various phenomena related to the improvement of the strength of nanocomposites have been discussed. The strength of composites depends on the base metal or reinforced components such as glass fiber and carbon fiber, and other reinforcement agents. Different salient properties related to energy applications and their testing procedures for the MMCs nanomaterials have been highlighted. The advantages of the composites and their future potential in energy applications are described thoroughly in this chapter.

Keywords Composite · MMC · Energy · Nanoparticle · Specific strength · Lightweight

---

D. Chakraborty
School of Mechanical and Aerospace Engineering, Queen's University Belfast, Ashby Building, Stranmillis Road, Belfast BT9 5AH, Northern Ireland, UK

S. Pramanik (✉)
Functional and Biomaterials Engineering Lab, Department of Mechanical Engineering, Faculty of Engineering and Technology, SRM Institute of Science and Technology, Tamil Nadu, Kattankulathur, KancheepuramChennai 603203, India
e-mail: sumitprs@srmist.edu.in

S. P. Singh et al. (eds.), *Metal Nanocomposites for Energy and Environmental Applications*, Energy, Environment, and Sustainability,
https://doi.org/10.1007/978-981-16-8599-6_2

2.1 Introduction

In the past few decades, scientists and industrialists are emphasizing on increased performance and lightweight properties of materials in various applications such as automobile, aerospace, power generation, and so on (Hunt and Miracle 2001; Miracle 2001). Metal matrix composites (MMCs) due to their high specific strength (i.e., strength to weight ratio) and stiffness compared to traditional metals and alloys are considered as the most suitable in applications where weight saving is of utmost importance. Automobile engine and brake components are mostly made up of MMCs because of their better wear properties and high specific strength compared to conventional metals (Purant et al. 2020). Special features such as high yield strength, greater stiffness, and good electrical conductivity have become a necessity in every modern engineering application. Renewable energy systems most commonly solar and wind energy resources due to their carbon neutral emissions during power generation are replacing fossil fuels energy sources (Liu et al. 2019). Energy storage systems such as rechargeable batteries, supercapacitors, and fuel cells are crucial in modern engineering utility for their clean energy production and zero carbon particulate emissions (Kulova 2013). Lithium ion batteries (LIBs) possess high working voltage and are playing a pivotal role as a large-scale electrochemical energy storage device for their applications in the latest transportation, power-grids, and electronic devices (Kim et al. 2020a). The improvement in the cathode and anode materials in LIBs leads to greater voltage and capacity. Oxide materials such as nickel oxide (NiO_2), manganese oxide (Mn_2O_4), and titania (TiO_2) are commonly used in the cathode of LIBs to improve their energy density. Several other transition metal oxides namely stannous oxide (SnO_2), copper oxides (CuO/Cu_2O), and zinc tungstate ($ZnWO_4$) have been incorporated as cathode resulting in improved energy storage capacity. Carbonaceous allotropes such as graphite, graphene, multi-walled carbon nanotubes (MWCNTs), etc. when used as anodes materials have produced high specific capacity in terms of electrochemical performance (Divakaran et al. 2021; Goriparti et al. 2014; Jamil et al. 2021). Metal fluorides were used generally in lithium batteries as a cathode component (Amatucci and Pereira 2007; Andre et al. 2015; Cabana et al. 2010). However, the huge changes in volume during the combined lithiation-delithiation process led to a reduction in its capacity and overall performance. So, carbon metal fluoride nanocomposite due to its superior electrochemical performance has made an edge over the use of conventional metal fluorides (Helen et al. 2020). With high cyclic performance and stability, vanadium pentoxide-gold (V_2O_5-Au) nanocomposite film is a prominent candidate for cathode material in lithium-ion micro batteries (Hongliang et al. 2021). MMCs also possess high fatigue life and high impact strength that creates an edge for its application in automobile parts in terms of enhanced durability in long-term utilization.

In this context, aluminium matrix composites (AMCs) gained a huge relevance preferably in automobile and aerospace industries due to their good ductility, high strength, and low wear resistance (Kala et al. 2014). Hence, the focus of material researchers is on the improvement of the tensile strength, tribological features, and better corrosion resistance of AMCs. The incorporation of various reinforcement agents namely alumina (Al_2O_3) and silicon carbide (SiC) improvised the material properties and mechanical performance of the AMCs (Liang et al. 2013; Ravi Kumar et al. 2017). In particular, the abrasion resistant characteristics were found to be increased for Al/Al_2O_3 and Al/SiC AMCs. Sometimes, dispersion of hard particulates such as titanium boride (TiB_2), silicon carbide (SiC), tungsten carbide (WC), titanium carbide (TiC), and so on in Al-matrix had improved the wear resistance of the AMCs. Special AMC designed by 3 M Multinational Conglomerate Company is trying to replace the traditional cast iron for a new generation brake caliper (*3 M Brake Caliper*). Rear brake rotors comprising Al/SiC MMC had gained huge attention in the Partnership for New Generation Vehicle Program (PNGV) as specified by General Motors, Chrysler Plymouth Power. These brake rotors significantly decreased the weight up to 60% compared to the use of cast iron. AMCs possess a low thermal coefficient of expansion and as a result, can withstand high mechanical and thermal loads thus increasing the overall performance in engine applications (Prasad and Asthana 2004). Among the various manufacturing techniques stir-casting, squeeze-casting pressure die-casting, and powder metallurgy techniques are most commonly used for the production of AMCs. A comparative study showed that the stiffness and tensile strength of Al-SiC MMC were greater for powder metallurgy technique whereas compressive strength was larger for stir casting fabrication. The optimum blending and composition are the most important process parameters that affect the mechanical and tribological characteristics of the AMCs besides selecting the best manufacturing method for getting the enhanced performance of the MMCs (Mhaske and Shirsat 2021).

Carbonaceous nano reinforcements namely MWCNTs are a good alternative to metal oxides and their particulates because of their lubricious nature. A slight variation of MWCNTs in Al-matrix leads to a fine microstructure and improved hardness and thermal conductivity (Jayaraj et al. 2020). The tendency of corrosion in the composite gets reduced as compared to the base metal (Samuel Ratna Kumar et al. 2017). Recently, graphene made a huge relevance in electronics manufacturing (Vallés et al. 2008; Liang and Zhi 2009; Li et al. 2019; Zhang and Pang 2021) and energy storage tools (Robinson et al. 2008; Fowler et al. 2009). In comparison to polymer and ceramic matrix composites, graphene reinforced metal matrix composite (GRMMC) outperformed them in terms of its ability to disperse in the matrix easily. It also has a larger specific surface area, larger strength, and stiffness, environmentally less perilous than other carbon allotropes (Prashantha Kumar and Anthony Xavior 2014). Metals and their oxides mixed with graphene are getting a spotlight in energy storage applications (Zheng et al. 2017, 2018; Xue et al. 2019). Besides, graphene-based metal composites are used in manufacturing nanocomposites for their application as supercapacitor electrodes (Singh et al. 2011; Krishnan et al. 2012). Moreover, the cost of graphene is low and it possesses high electrochemical stability (He et al. 2018). MMC is also playing a key role in renewable energy generation such as in solar cells. Advanced solar cells are gaining huge relevance in space applications due to their capability of becoming thinner leading to the weight reduction of the overall device. The probability of fracture in the solar cell device increases due to excessive mechanical stress, causing electrical loss (Kntges et al. 2011). To avoid this crack propagation from the tip, MMC film structure is developed with Ag as the matrix layer and MWCNTs spray deposited over it improving the electrical conductivity alongside maintaining the mechanical properties (Abudayyeh et al. 2015, 2019; Cox et al. 2016; Wilt et al. 2017).

On the other side, increased demand for low-cost reinforcing agents in MMCs is growing for their necessities as the usage of by-products from industries and agro-wastes. These are known as green composites produced by reinforcing fly-ash, maize stalk waste, leftover eggshells, and so on (Aigbodion and Nwigbo 2012; Bose et al. 2019). Suitable chemical or thermal treatments might be needed for its effectiveness as a reinforcing agent. The treatments for checking the chemical composition of the ceramic phases procured as industrial and agro-wastes are of significant importance to future research. Green composite is a positive indication towards sustainable energy usage and reducing the consumption of all the fossil fuels (Heidarzadeh et al. 2021).

2.2 Types of Metal Matrix Composites

Based on the matrix materials, composite materials can be broadly divided into three categories such as polymer matrix composites (PMCs) where polymers act as binders, ceramic matrix composites (CMCs) where ceramics act as binders, and metal matrix composites (MMCs) where metals act as binders. Further classification of MMCs depicted in Fig. 2.1 can be done depending upon the reinforcing agents used in the metal matrix. The reinforcing agents are mainly of three types such as particulates, fibrous, and laminates. Thus, based on the density of the MMCs, they

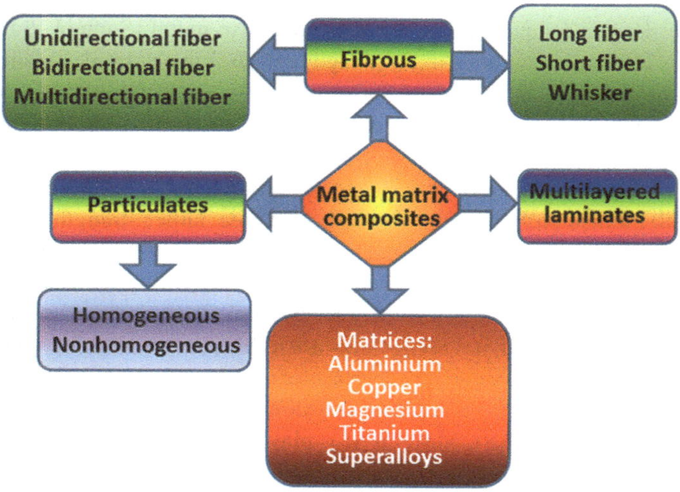

Fig. 2.1 Classification of metal matrix composite materials

can be used in different energy related applications. Some advanced nanocomposites used in various energy related purposes are mentioned in the following subsections.

2.2.1 Particle-Strengthened MMC

Particle is a kind of particulate that has a small aspect ratio, i.e., length to diameter ratio, near to 1. It is the most widely used MMC. There are different shapes of particles are available but in most instances, nearly equiaxed size are used. The combination of particle matters especially in the form of oxides, carbides, nitride, and borides as the metal matrix forms a MMC. MMCs can be named as per the metal or alloy used as matrix material e.g., for aluminium (Al) matrix composites are commonly regarded as AMCs. In MMCs, the usual ceramic particle reinforcing agents preferred are Al_2O_3, SiO_2, SiC_p, TiC, TiB_2, BN, and so on because of their high strength, modulus, and hardness compared to the metal matrix. The ceramic strengthened MMCs possess high tensile strength and good wear resistance ability. Besides, they have low density and are lightweight (Karvanis et al. 2016; Pramod et al. 2018).

2.2.2 Fiber-Reinforced MMC

Fiber is a type of particulate that has a very high aspect ratio, more than 100. There are several types of fibrous materials such as SiC_f, carbon fiber (CF), carbon nanotube (CNT), and so on used as reinforcements in the metal matrices. The mechanical as well as functional properties can be improved using this fiber reinforcement in the MMC. Heat sink material made up of aluminium and copper reinforced with carbon fibers (CFs) offers a huge advantage. Due to high wear resistance, CFs are greatly used for bearings and brake components. The addition of over 30% short CFs in the metal matrix increases the coefficient of friction (COF). CFs are resistant to extreme temperature and possess a high modulus of elasticity, tensile strength, and better electrical conductivity. The carbon nanotubes (CNTs) have greater stiffness and elastic modules close to 1 TPa than the CFs (Yu et al. 2000; Bakshi et al. 2009; Shirvanimoghaddam et al. 2017). The hardness of MMC is found to increase to a huge extent with a 2.5% weight fraction of multi-walled CNTs (MWCNTs) in the base aluminium alloy. The inclusion of 1.5 wt.% MWCNTs in the A356 aluminium alloy enhanced the ultimate tensile strength and yield strength of the A356/MWCNT nanocomposite by 50–60% (Elshalakany et al. 2014). In this context, the 2D allotrope of carbon named graphene (Gr) is regarded as a more suitable reinforcement (usually known as GrMMC) than CNTs as it contains a large surface area and mixes easily with the metal matrix. These have good electrical and thermal conductivity alongside high yield strength and extremely high elastic modulus (Chen et al. 2020).

2.2.3 Whisker-Reinforced MMC

The whiskers are kind of small needle like particles of typical range in size from 500 nm to 10 μm in diameter and a length of few micrometers to a few centimeters. Generally, whiskers have very high strength. In this regard, silicon carbide whisker (SiC_w) had shown very high strength and Young's modulus. The high oriented structure of whiskers provides good mechanical characteristics particularly high strength to weight ratio, which is very important for energy consumption in automobile and aerospace applications. Other properties namely optical, dielectric, and unique magnetic properties (Wong et al. 1997; Boehlert et al. 2009) enable them to be a notable reinforcement agent in comparison to hard particulate oxides, borides, carbides, and carbonaceous fibres. The common whisker materials used as reinforcement are SiC_w, $Al_{18}B_4O_{33w}$, and Al_2O_{3w}. At a high aspect ratio, the whiskers may tend to agglomerate and lead to breakage (Wang et al. 2019b). Whiskers can transfer loads and bridge cracks thus enhancing the fracture toughness of the MMCs. A well bonded mixture of $Mg_xAl_{(1-x)}B_2$ whiskers with $MgAl_2O_4$ nanoparticles dispersed in aluminium matrix enhanced the tensile strength and hardness of the MMCs, but the elongation at break decreased (Wang et al. 2018). On the other hand, whiskers such as $MgAlB4_w$ based AMC have improved ductility due to the perfectly aligned direction of the whiskers (Li et al. 2020). The whisker MgB_2O_{5w} reinforced in a metal matrix was found to improve the friction and wear properties of the MMCs. Some commonly used MMCs formed with Al alloy and reinforcement agent MgB_2O_{5w} with a coating of ZnO and CuO are $Mg_2B_2O_{5w}$/6061Al, ZnO/$Mg_2B_2O_{5w}$/6061Al, and CuO/$Mg_2B_2O_{5w}$/6061Al (Jin et al. 2014).

2.2.4 Laminated MMC

The laminated structure is known to have used for improving the mechanical properties of the MMCs. Generally, the laminated MMCs are prepared by hot pressing by rolling of alternately stacked metal matrix foils and ceramic fabrics (Fan et al. 2017) or metallic lamina (Wang et al. 2016). It exhibits a superior strength-ductility synergy compared to the bulk MMC. Some recent investigations have reported that laminated structure can effectively be used to inhibit the strain localization during the plastic deformation. Hence, the laminated MMC shows a superior strength-ductility combination (Ma et al. 2016). Fan et al. (2017) reported that the Ti foil-(SiC_p/Al) laminated MMCs has significant improvements in the mechanical properties compared to SiC_p/Al bulk MMCs. It is resulted from the effect of interface on local stress/strain transfer behaviors. They showed that the crack initiation and propagation of SiC_p/Al layers can be strongly suppressed, and the mechanical properties can also be further improved by controlling the formation of interfacial phases (Fan et al. 2017).

2.2.5 *Green MMC*

Agricultural wastes from the farming sector have gained huge attention in automotive industries to be used as reinforcement in MMCs. This contributes to less environmental pollution, inhibiting the use of fossil fuels. The most common reinforcements extracted from the agro wastes are coconut shell ash, rice husk ash, sugarcane bagasse, bread fruit seed hull ash, and egg shell (Kandpal et al. 2021). Coconut shell is found in large quantities in tropical regions of the earth and contains hard phase particulates such as silica (SiO_2), alumina (Al_2O_3), magnesium oxide (MgO), and ferric oxide (Fe_2O_3). These are generally ground to their powder form and heat treated in a furnace to be suitable for a reinforcement phase in the metal matrix. The low density of rice husk ash and the high percentage of silica contents besides other ceramic materials make it a valuable reinforcement agent used in MMCs. Sugarcane bagasse is similar to rice husk ash and also contains a significant amount of SiC. It can be used at a high temperature of around 1600 °C. The mechanical properties can be improved with the addition of bread fruit seed hull ash in the Al-Si-Fe alloy matrix (Aigbodion and Nwigbo 2012). Chicken egg shells co-reinforced with SiC get uniformly distributed in the matrix of Al alloy enhancing the specific strength and lowering the porosity of the MMC. The 7.5% egg shell combined with a 2.5% SiC in Al matrix was found to have the lowest thermal expansion when heated at a temperature of 450 °C in an electric furnace (Sharma and Dwivedi 2017). Industrial waste, namely fly ash is considered the most cost-efficient reinforcement agent in an MMC. The characterization of fly ash mixed with rice husk ash resulted in revamped wear resistant properties (Singh and Singh 2011). A lower concentration of 2.5% fly ash mixed with SiC in the Al359 matrix enhanced the hardness of AMC. Its ductility and lubricous nature were found to increase by adding a higher percentage of fly ash (Sathishkumar et al. 2020). In the past few years, snail shell has acquired a huge relevance as a reinforcement agent in MMCs. The tensile strength and hardness of 16–48 wt% snail shell dispersed in aluminium metal matrix nanocomposite are 236 MPa and 48.3 HRF respectively, which is quite high as compared to the base alloy (Asafa et al. 2016). Various categories of MMCs produced from waste materials are depicted in Fig. 2.2.

2.3 Fabrication and Properties

2.3.1 *Fabrication*

The selection of manufacturing techniques of metal matrix nanocomposites for particular applications is extremely important. Some methods of manufacturing of MMCs are illustrated in Table 2.1.

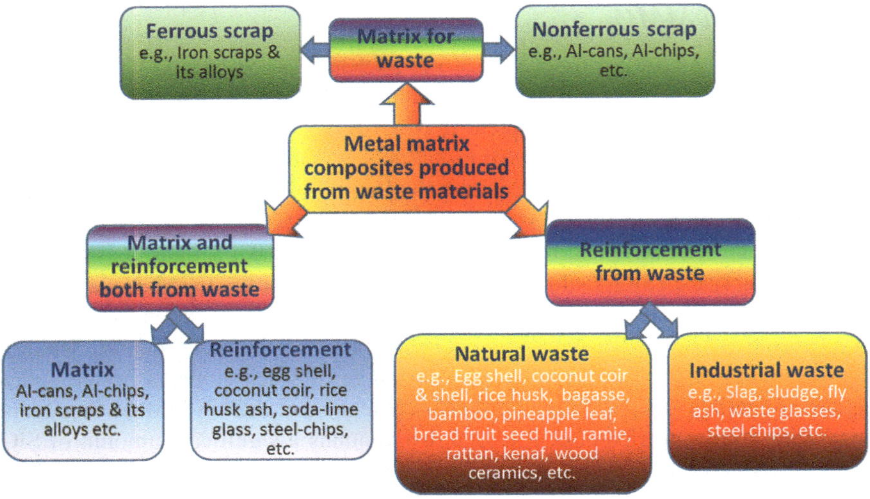

Fig. 2.2 Various categories of MMCs produced from waste materials

2.3.2 Properties

The salient properties of some important reinforced metal matrix nanocomposites based on the type of reinforcements are explained in the following subsections. Nano materials play a crucial role in the improvement of mechanical and multifunctional properties.

2.3.2.1 Properties of Ceramic Particle Reinforced Metal Matrix Nanocomposites

The incorporation of nano-sized reinforcements in metal alloys escalates the strength of the MMCs. Sometimes, the ductility significantly improves with the addition of a small amount of nano reinforcement agents (Thostenson et al. 2005; Dehnavi et al. 2014; Casati and Vedani 2014). On adding TiO_2 nanoparticles with a size of around 30 nm in Al AA2024 alloy matrix, the grain size distribution narrowed down by 66% to around 400 μm. The dendrite arm spacing was reduced by 31% in comparison to the monolithic casting as seen by the optical micrograph study. The presence of α-Al dendrites in the composite castings with an enclosure of intermetallic phases and α-Al eutectic structures was confirmed by scanning electron microscopy (SEM). The porosity of the nano MMC reduced as compared to the monolithic alloy casting. The premixing of TiO_2 with Cu and Al decreased the agglomeration and inhibited the wettability of the particles during the stir casting manufacturing process as confirmed by transmission electron microscopy (TEM). The presence Ti and O atoms in the nanocomposite can be confirmed by the energy dispersive spectroscopy (EDS) study

Table 2.1 Manufacturing techniques of MMCs

Manufacturing process	Specific technique	Technical details	References
Liquid phase processing	Stir casting	In this process, a mechanical stirrer is used to form a vortex for mixing the reinforcement(s) in the molten matrix material	Sahu and Sahu (2018)
	Electromagnetic stir carting	In this method, an electromagnetic field is used to control the flow of liquid metal. Thus, an alternating field induces eddy currents in liquid metal, which interact with the field to provide a Lorentz body force. This force, generally rotational, drives fluid motion This method is most effective for the production of hybrid metal matrix nanocomposites	Kumar (2018)
	Squeeze casting	In this method, a matrix metal melt solidifies under the application of external pressure, mainly to prevent the formation of shrinkage porosity	Casting et al. (2021)
	Compo-casting	It is a kind of liquid state process, in which the reinforcing agents are added to a solidifying melt within its liquid–solid zone at vigorously agitated conditions	Mudasar Pasha and Kaleemulla (2018)
	Spray forming or casting	In this method, the fine molten metal droplets or semi-solid particulates are deposited layer-by-layer onto a substrate to form a bulk composite product or thick coating of the composite after their solidification	Li and Lavernia (2000)
	Reactive processing	In this method, the volume combustion mode is combined with hot consolidation to produce metal matrix composite (MMC) materials with the variation of reinforcements in the bulk metal	Brinkman et al. (1999)

(continued)

Table 2.1 (continued)

Manufacturing process	Specific technique	Technical details	References
	Infiltration	In this method, liquid matrix metal or alloy is infiltrated into the porous forms of nano fibers or whiskers like reinforcing materials. Here, SiO_2 and metal-based mixtures are generally used as a binder to retain the integrity and shape of the porous forms	Hannula et al. (2003), Vasile et al. (2020)
Solid state processing	Mechanical alloying (MA)	In MA, matrix metal powder and particulates, whisker, or short fiber are mixed in dry or wet conditions for long period often using ball milling. The high mechanical energy applied during the milling process results in the formation of the nanocomposites	Hort and Kainer (2006)
	Diffusion bonding	Here under high pressure, the inter diffusion atoms at the metallic surfaces form bonding between the matrix metal and reinforced materials. This fabrication method is often used for Al or Mg MMCs	Hannula et al. (2003)
	Sintering	In sintering, generally, the compact pellets are heated at the desired temperature(s), heating or cooling rate, pressure, atmosphere, and time. Sintered MMCs exhibit higher density values by grain growth and become stronger by forming diffusion boding between the matrix–matrix, matrix-reinforcement, reinforcement-reinforcement grain or particles	Ataollahi Oshkour et al. (2015)
	Spark plasma sintering (SPS)	SPS is an advanced sintering method to produce MMC with the desired quality	Stalin et al. (2020)

(continued)

Table 2.1 (continued)

Manufacturing process	Specific technique	Technical details	References
	Powder metallurgy (PM)	In PM, metal matrix powder is mixed with hard reinforced particulates, whisker, or short fiber followed by cold compaction, canning, degassing, and high temperature consolidation	Hannula et al. (2003)
	Friction stir processing (FSP)	In FSP, a similar principle is used as friction stir welding (FSW). However, instead of joining two materials together, the FSP modifies the local microstructure of monolithic specimens to achieve specific and desired properties by surface modifying the microstructure using reinforced nanoparticles. Here, torque, forces, and heat generation are important factors	Węglowski (2018)
	Physical vapour deposition	In this process, the reinforcements are continuously provided at a region where metal to be deposited at a certain rate of 3–10 μm/min. At the same time, a high partial pressure vapour of metal is produced and injected to the surface. Then the vapour is condensed at the region to produce a coating on the fiber and form a composite layer	Hannula et al. (2003), Vasile et al. (2020)
Two phase processing	Semi-solid phase	In two phase process, the MMC with one metallic and another material, such as a ceramic or metallic material powder mixture is heated up to a semi-solid state, and pressure is applied to form the composites. Here, a two-phase microstructure is obtained upon controlled unidirectional solidification of the eutectic alloy	Wu and Kim (2011), Wu et al. (2010)

(Shayan et al. 2020). The optical microscopy images of 1.5 and 7.5% weight fraction of SiC dispersed in Al7075 alloy matrix showed no cracks or porosity in the MMC formed. The SiC was homogeneously spread throughout the matrix due to the proper setup of process parameters during the stir casting production. Slight clusters of SiC can be seen in Al metal matrix with a bit of porosity from the SEM micrographs. A 1 wt.% of Mg was added to SiC dispersed with the metal matrix to aid the wettability of the MMC mixture. The EDS spectrum of the composite showed a fraction of the Mg constituent due to its dissolution with the SiC in the course of the stir casting process (Gosavi and Jaybhaye 2020).

The reinforcement SiC particle with an average size of 50 μm was merged with 60 μm WC particle in the Al6061 base metal alloy. AMC with varied concentrations of the reinforcement and homogeneity of the SiC/WC particles in Al matrix can be obtained by SEM study. Tribological properties are very important for energy conservation applications. More wear means more loss of materials and waste of extra energy. A Pin-on-disc tribometer can be used to investigate the wear properties of the MMCs. Since the reinforcements possess greater stiffness, the wear rate of the AMCs was found to increase with a greater content of the hard SiC/WC particles (Jayakanth et al. 2021). Titanium powder was mixed with boron carbide (B_4C) to form TiC and TiB_2, respectively, and used together in the Al matrix melt to form AMCs. On performing XRD analysis of the AMC samples, distinct peaks of TiC and TiB_2 were noticed. The even distribution of TiC and TiB_2 clusters inside the metal matrix was observed in FESEM images (Rai et al. 2020). The SEM images taken from magnesium alloy reinforced with SiC nano powders revealed the α-grain structure of silicon. A 4–8 wt.% proportion of alumina was mixed with SiC nano sized particles in an electric furnace before the stir casting manufacturing process. A uniform distribution of the reinforcement as well as porosity was observed at a high magnification range. The porosity might have created due to the air bubbles formation during their stir casting process (Sathishkumar et al. 2021). The surface morphology observed through SEM microscopy disclosed that the uniform distribution of TiC ceramic particles in Al6061 alloy matrix can improve the mechanical properties (Siddappa et al. 2021).

SEM study was done on a metal matrix nanocomposite which consisted of 160 nm Cu nanocrystalline particles reinforced by 30 nm nano-alumina particles. Two different concentrations such as, 1 and 5 vol% of nano-alumina reinforced with nanocrystalline copper were prepared and analyzed. The morphology revealed the Cu grains as grey particulates and alumina as black. Bright and dark field TEM images showed the presence of alumina in both the Cu matrix and its grain boundaries. X-ray spectroscopy analysis confirmed the presence of base Cu nano matrix as well as Al and O due to alumina as a reinforcement agent. AFM study of the topography of the nanocomposite suggested the equiaxed structure of Cu nano particle and validated its diameter as 160 nm (Nachum et al. 2010).

2.3.2.2 Properties of Fiber Reinforced Metal Matrix Nanocomposites

The fiber reinforced MMCs are generally used to improve the strength of the composites. The advanced fiber reinforced metal matrix nanocomposites are making into hybrid composites to improve their mechanical as well as functional properties for specific applications. A hybrid reinforcement of MWCNT mixed with iron (Fe) and nickel (Ni) was dispersed in Al matrix. The optical micrography study shows the presence of Ni nano sized particles agglomerated in the boundary of the parent Al particle matrix. MWCNT was observed in the form of clusters alongside near the layer of Fe particles. Layers of Fe particles were found to be coated over the Al phase which contributed to its magnetic properties. High resolution SEM images suggested that the MWCNTs offered a quite less mechanical enhancement to the AMC (Tugirumubano et al. 2020). Homogeneous distribution of a combined reinforcement of graphene with ZrO_2 in Al6061 alloy matrix was showcased by an optical metallurgical microscope. Porosity observed was null in the MMC. There is an enhancement of interfacial bonding in the metal matrix in the presence of reinforcements (Boppana et al. 2021). Nitinol (NiTi) nanoparticles act as a great functional reinforcement agent used with Al alloy matrix as it possesses shape memory behavior in the presence of varying temperatures. On the increasing percentage of nitinol in the metal matrix, porosity was reduced to a great extent and no cracks were found as captured by the SEM micrography study. There was a reasonable distribution of the reinforcement in the matrix (Chauhan et al. 2021). A new technique known as the molecular level mixing process was adopted replacing the traditional powder metallurgy procedure to form a nanocomposite consisting of Cu and CuO respectively as the base matrix and CNTs as the nano reinforcement. The SEM study portrayed the homogeneous distribution of CNTs inside the matrix, which was not found in the case of powder metallurgy manufacturing wherein the CNTs were only found on the surface of the metals. In some cases, it was mixed with the hard ceramic powder due to the agglomerating nature of CNTs. The TEM micrography showed a clear spherical form of the CNTs uniformly disseminated in the Cu powders. The dislocation density is investigated to be low for the Cu grains after the reinforcement (Cha et al. 2005).

2.3.2.3 Properties of Whisker Reinforced Metal Matrix Nanocomposites

Whisker reinforced MMCs have very high strength and toughness. Two different concentrations 5 and 20% β-Si_3N_{4w} whiskers were used as reinforcements in Al matrix. The SEM images showcased the bulged out rod-like structures of β-Si_3N_{4w} dispersed evenly in the matrix specifying the isotropic nature of the MMC. The smaller size of the whiskers resulted in a harder movement of the dislocation and formed a refined grain boundary structure. This was captured in the TEM images. The peaks of Al progressively widened with the lowering intensities due to the whisker strengthening in the Al matrix as collected from the XRD analysis (Zhang et al. 2019a). SEM micrography was conducted on the worn out surface of as-received $Mg_2B_2O_{5w}$ whiskers coated with a mixture of CuO and ZnO, respectively, in Al matrix alloy after a wear test. The wear test setup consisted of a ball-on-disc wear testing equipment used against a GCr45 steel ball on a load range of 5–25 N. The sliding velocities were in the range of 120–480 m/min. The morphology of the composite showed few cracks and slight grooves in comparison to the wear surface of the parent Al alloy matrix. The inclusion of the whiskers in the form of reinforcement developed convex peaks during the wear test process which inhibited the contact between the alloy matrix and the steel balls. This showed the reduction of adhesion between the surface of the matrix and the steel balls. A change from adhesive to abrasion wear was observed. The chip size tends to be like sheets rather than being granular in case of the worn out surface of the Al matrix. Hence, the wear rate was reduced for the whisker reinforced composites as compared to the Al alloy (Jin et al. 2014). The high magnification SEM micrography study of $MgAlB_4w$ whiskers reinforced in Al metal matrix revealed the three-dimensional hexagonal shape of the dispersed whiskers. Homogeneous distribution of the $MgAlB_4w$ matrix was spotted. The real needle shaped straight structure of the whiskers was investigated and confirmed under a TEM study. Besides, the interface between the whiskers and the matrix was clean. The Mg, Al, and B were detected by the peaks recorded in EDS spectrum. At a higher temperature of approximately 500 °C during the sintering process, the $MgAlB_4w$ whiskers were produced and detected in an XRD analysis. There is a tendency of the coarsening of the whiskers at an increased temperature beyond 500 °C. After the hot extrusion of the MMC, the density improved with a better uniformity in the dissemination of the whiskers inside the matrix. Hence, the ductility was raised to 9.6% and the tensile strength of AMC increased to about 382 MPa making it suitable for advanced structural applications (Li et al. 2021; Huang et al. 2017).

2.3.2.4 Properties of Green Materials Reinforced Metal Matrix Nanocomposites

One of the largest industrial solid waste materials that create huge problems for the power industries and their surroundings is fly ash. Therefore, researchers are desperately trying to recycle this green material as reinforcements in nanocomposites. Various concentrations of SiC and fly ash reinforcements into Al2014 alloy matrix were prepared by a stir casting technique and the morphology of the composite samples were analyzed under SEM microscopy. The images showcased a fairly good distribution of the reinforcement particles inside the matrix. There was the presence of smaller size particles in the grain boundaries. Moreover, there were clusters of reinforcements due to the agglomeration of the fly ash particles with SiC inside the matrix. A greater percentage of fly ash contributed to more agglomeration and porosity in the composite samples (Lal et al. 2020).

2.4 Applications

Metal matrix composites have shown many important applications. Recently, metal matrix nanocomposites are being used as the most fascinating material in energy related applications. The recent advances of metal matrix nanocomposite materials and their applications are depicted in Table 2.2. Some salient features of the most

Table 2.2 Recent advances of metal matrix nanocomposite materials and their applications

Application	Materials	References
Battery- Cathode	Carbon metal fluoride nanocomposite (CMFNC); C-FeF_2; C-NiF_2; V_2O_5/Au	Badway et al. (2003), Wood et al. (1969), Helen et al. (2020), Hongliang et al. (2021)
Battery- Anode	$ZnWO_4$/SnO_2@r-GO; Sb_2Se_3/C; Sn/CNT	Liu et al. (2009), Wang et al. (2019a, (2021, 2012), Brijesh et al. (2020), Kim et al. (2020b), Seo et al. (2012)
Clean energy	TM nitrides (TMNs); TiN/ Si_3N_4 nanocomposite; Ag (NP)/ TiO_2 (T); MoS_2-Go@ Ni-Co- MnO_2; $CoSnO_3$-rGO;	Bang and Suslick (2009), Chi et al. (2019), Deng et al. (2016), Dong et al. (2010), Fechler et al. (2012), Jun et al. (2009), Kaskel et al. (2004), Kim et al. (2017), Pan et al. (2014), Qi et al. (2018), Ramasamy et al. (2012), Lalliansanga et al. (2020), Abdullah and Ling (2010), Saleh and Taufik (2018), Chen et al. (2014)
Fuel cell	Ti/TiO_2NTs/PAIn; Ti/TiO_2NTs/PAIn/Pd; PAni-Co-PPy@TiO_2	Zheng et al. (2015), Kormányos et al. (2020), Xu et al. (2021), Pattanayak et al. (2019)
Supercapacitor	MnO_2@$MnNiCo_4$/ CC; GMC + C nanocomposite	Tamaddoni Saray and Hosseini (2016), Hung et al. (2013)
Thermal energy storage	PCM; MNH-PCM; Carbonaceous PCM; Metal Oxide PCM	Liu et al. (2015), Madarász et al. (2007), Zhang et al. (2019b), Chen et al. (2012), Narayanan et al. (2017)
Wind energy generation	Steel spars, with aluminum shells supported by wooden ribs	Mishnaevsky et al. (2017)
Space crafts	Aluminium Matrix Composites (AMCs); C/SiC ceramic matrix composite	Rawal (2001), Badiey and Abedian (2010), Hocine et al. (2013), Muley et al. (2015), Toor (2017), May et al. (2020)
Automotive parts	Al-7075 alloy reinforced with fly ash and SiC; copper reinforced with graphite; Aluminium-fly ash; Al-SiC; Al-Boron carbide (B_4C)	Bisane et al. (2015), Premnandh et al. (2020), Rohatgi et al. (1992, 1998), Kestursatya et al. (2001), Maddever and Guinehut (2005), Singh et al. (2020)

Fig. 2.3 Important applications of metal matrix nanocomposites

advanced metal matrix nanocomposites in some specific applications are discussed in this section. The main applications of metal matrix nanocomposites are depicted in Fig. 2.3.

2.4.1 *Battery*

2.4.1.1 Cathode Materials

Carbon metal fluoride nanocomposite (CMFNC) is considered a better cathode material to be used in lithium batteries as opposed to pure metal fluorides (Badway et al. 2003). Various chemical methods are used to create CMFNC (Li et al. 2010; Lu et al. 2014a, b; Rui et al. 2015; Yang et al. 2012). Most of these methods produce carbon monoxide as a by-product during the chemical reaction, which is obnoxious (Breitung et al. 2013; Reddy et al. 2012, 2018). A new process referred to as the electrochemical intercalation was used to synthesize C-FeF_2 and C-NiF_2 nanocomposites showing enhanced electrochemical performance as the substrate material for cathodes in rechargeable lithium batteries. The most common cathode material for lithium batteries is CF_x. Due to their high reduction potential of 4.21 V versus Li^+/ Li (Wood et al. 1969) they became thermodynamically most suitable for the intercalation of Fe^{2+} and Ni^{2+}. The platelet-like shape with a uniformly distributed nanocrystalline size of 5–10 nm FeF_2 particles into the carbon matrix leads to a better cycling performance of the C-FeF_2. The reversible capacity of C-FeF_2 as fabricated by the electrochemical intercalation was found to improve with a value of 349 mAh g^{-1}. The reduction potential of NiF_2 (2.94 V) was greater than that of FeF_2 (2.66 V vs. Li^+/ Li). This results in low discharge potential for NiF_2 amounting to −0.3 V whereas for FeF_2 it was −0.1 V (Helen et al. 2020).

In another recent study, low cost orthorhombic phase vanadium pentoxide (V_2O_5) was fabricated as a layered structure with a specific capacity for promising cathode material used in lithium ion micro batteries (Wood et al. 1969; Beke 2011). Furthermore, the addition of Au nanoparticles in V_2O_5 powder, a V_2O_5/Au nanocomposite film (165.3 nm) was developed using pulsed laser deposition technique. Here, the incorporation of Au nanoparticles in V_2O_5 increased the discharge capacity to 140.2 mAh g^{-1}, which is greater than the pure V_2O_5 powder. The cycle stability and specific capacity were also found to increase for this material (Hongliang et al. 2021).

2.4.1.2 Anode Materials

In a recent study, a facile solvothermal process was used to synthesize a nanocomposite comprising of zinc tungstate ($ZnWO_4$) and tin oxide (SnO_2) with reduced graphene oxide (r-GO) as buffers (Wang et al. 2019a). This composite was specifically designed as an anode material for lithium-ion batteries. Their x-ray diffraction results disclosed the monoclinic Wolframite shape of $ZnWO_4$ (JCPDS Card No. 00-015-0774) (Brijesh et al. 2019). Several peaks were also observed for SnO_2 planes at 34.35°, 42.61°, 67.81°, and 71.34°. The presence of reduced graphene oxide (r-GO) was confirmed with an appearance of a wide peak at between 20° and 28°. The D-band at 1349 cm^{-1} and G-band at 1599 cm^{-1} as observed in the Raman spectra validated the existence of r-GO in $ZnWO_4/SnO_2$@r-GO nanocomposite. The mesoporous powder sized ZWSN-10 nanocomposite had shown better electrochemical reaction alongside the faster rate of reaction and rapid diffusion of both ions and electrons compared to the ZWSN-5 and ZWSN-10/GO nanocomposites. The current (mA) and potential (V vs. Li/Li^+) relation indicated a larger surface area in ZWSN-10/GO nanocomposite than $ZnWO_4/SnO_2$ nanocomposites with a higher capacity of 477 $mAhg^{-1}$. It also had good reversibility with a capacity of 530 $mAhg^{-1}$ at a lower current density of 100 mAg^{-1} and 436 $mAhg^{-1}$ at 500 mAg^{-1} higher current density. The discharge capacity of ZWSN-10/GO nanocomposite was 1486 $mAhg^{-1}$ greater than that of base $ZnWO_4/SnO_2$ nanocomposites (Brijesh et al. 2020).

In another recent study by Kim et al., 2020, a simplified high energy ball milling technique was used to fabricate Sb_2Se_3/C nanocomposite, which was used as an anode material, and $LiFePO_4$ (LFP), which was used as cathode material. The size of Sb_2Se_3 particles was found under SEM in a range of 200 nm–1.2 μm. The nano-sized particles within the nanocomposite matrix and the inner particle spacing were found to be 0.325 nm. The homogeneous encapsulation of the Sb_2Se_3 within the carbon matrix was visible under high-resolution TEM (HRTEM). Their EDS mappings also portrayed the individual Sb, Se, and C elements distributed evenly inside the matrix. They had found that with the addition of 20% acetylene black (AB) to the Sb_2Se_3/C nanocomposite, the cyclic performance increased to 410 $mAhg^{-1}$ at 1 Ag^{-1} after 1000 cycles with a capacity retention of 84.2%. At a lower current of 250 mAg^{-1}, the stable charge capacity of the nanocomposite was as much as 545 $mAhg^{-1}$ succeedings 1000 cycles than its theoretical counterpart (Kim et al. 2020b). Further, the incorporation of carbon enhanced the capacity utilization of Sb_2Se_3@20% AB nanocomposite to 63.5% than the base Sb_2Se_3 alloy. Compositing carbon on the Sb_2Se_3@20% AB electrode shows no changes in the Raman peaks in comparison to the Sb_2Se_3 electrodes, which showed a variation in peak sizes decreasing after 100 cycles from 252 and 150 cm^{-1} ascribed as metallic Sb and Li_3Sb. Moreover, the inclusion of 20% AB reduced the change in thickness due to lithiation.

The segregation between Sb and Se phases was inhibited resulting in a smooth reversible reaction and increased cyclic performance. The full cell consisting of LFP and Sb_2Se_3/C nanocomposite possesses a stable 340 $mAhg^{-1}$ capacity and capacity retention of 83.5% at 1.1 Ag^{-1} following 300 cycles. The Coulombic efficiency was high equivalent to 99.7% (Liu et al. 2009; Wang et al. 2019a, 2021).

A combination of tin (Sn) particle size of 2–5 nm with GO nanosheets rendered a capacity of 508 $mAhg^{-1}$ after the 100th cycle (Wang et al. 2009). Its capacity was higher in comparison to pure Sn nanoparticles. A higher reversible capacity of 700 $mAhg^{-1}$ was obtained for Sn/CNT based nanocomposite (Wang et al. 2012; Seo et al. 2012).

The capacity of SnO_2 was calculated to be 781 $mAhg^{-1}$ (Derrien et al. 2007; Winter and Besenhard 1999). The Sn nanoclusters combined with SnO_2 nanowires produced a reversible 845 $mAhg^{-1}$ capacity followed by 400 cycles. In contrast, the initial capacity of pure SnO_2 nanowires was estimated to be 2400 $mAhg^{-1}$ which degraded rapidly within the upcoming 15 cycles. Some Sn/SnO_2 composites presented an initial capacity of around 2800 $mAhg^{-1}$ which after 40 cycles reduced to 490 $mAhg^{-1}$ (Sivashanmugam et al. 2005). The reversible capacity was estimated to be 814 $mAhg^{-1}$ and a Coulombic efficiency greater than 98%. This was due to the high surface area to volume ratio of the SnO_2 nanowires leading to effective alloying and dealloying of Li. The diameter of Sn nanocluster enveloped SnO_2 was around 30–100 nm under SEM but using TEM, the size was found to be 15 nm Sn nanocluster covering SnO_2 nanowires (Meduri et al. 2009).

2.4.2 Clean Energy

Electrocatalysts are used for different clean energy applications. The optimal design and development of nano-sized catalysts escalate the effectiveness in the performance of clean energy systems (Suib 2013). Many researchers are trying to develop highly efficient oxygen-evolution electrocatalysts using transition metal (TM) ceramics. The TM oxides, carbides, and nitrides possess the characteristics of both ceramics and TM, so these are considered promising materials in catalytic usage (Balogun et al. 2017; Choi and Kim 2017; Hargreaves 2013; Klemenz et al. 2018; Mera et al. 2015; Védrine 2017; Zhong et al. 2015). TM nitrides (TMNs) acquire exceptional properties such as extreme hardness, high melting point, outstanding electrical conductivity, and good resistance to corrosion (Oyama 1996). An integral part of the TMNs is titanium nitride (TiN), which interacts rapidly with the electronic structure of platinum (Pt) which leads to improvised durability in performance (Nan et al. 2018). The formation of nano sized TiN with a large surface area along with its proper access to its active sites makes it suitable in the robust application as a catalyst (Bang and Suslick 2009; Chi et al. 2019; Deng et al. 2016; Dong et al. 2010; Fechler et al. 2012; Jun et al. 2009; Kaskel et al. 2004; Kim et al. 2017; Pan et al. 2014; Qi et al. 2018; Ramasamy et al. 2012). Hence, uniform dispersion of TiN nanoscale particles in a covalently bonded Si_3N_4 matrix led to the formation of a TiN/Si_3N_4 nanocomposite. To overcome the appalling mechanical ability due to excessive mesoporosity (Lale et al. 2016; Salameh et al. 2018), activated carbon (AC) was used as a support to develop a preceramic polymer known as PHTiPS2.5 with a composition of $Si_{1.0}Ti_{0.4}C_{2.0}N_{1.0}H_{5.6}$ (Bechelany et al. 2014, 2016; Lale et al. 2017). Two steps of pyrolysis with nitrogen at 400 °C facilitated the cross-linking of polymer around AC producing mSiTiN10 nanocomposite consisting of $Si_{1.0}Ti_{0.5}N_{2.0}O_{0.2}$ which on further annealing at a temperature higher than 1000 °C, formed mSiTiN14 with a composition of $Si_{1.0}Ti_{0.5}N_{1.8}$. The Raman spectrum of the mSiTiN14 nanocomposite powder displayed an absorption peak at 480 cm^{-1} indicating the face-centered cubic (FCC) structure of TiN without any peak for carbon which validated the appearance of carbon as a support matrix in TiN/Si_3N_4 nanocomposite, which is an amorphous shaped matrix of Si_3N_4. Deposition of Pt nanoparticles was introduced into mSiTiN10 and mSiTiN14 nanocomposite samples. The nano scale Pt particles of 6–8 nm were homogeneously distributed in the amorphous TiN matrix.

It was also found that after 90 min of complete dehydrogenation of $NaBH_4$, four equivalent H_2 molecules were produced with the inclusivity of Pt/mSiTiN14 nanocomposite. This proves that the deposition of Pt nanoparticles enhances the catalytic activity of the AC samples. Pyridine adsorption followed by water treatment was performed on mSiTiN14 and mSi_3N_4. The enhanced stability of Si_3N_4 by TiN nanoclusters was confirmed by the Fourier transform infrared (FTIR) spectroscopy. This result in turn supported the hydrolysis reaction meanwhile inhibiting the catalytic capability of the nanocomposite to form protonic hydrogen. The specific surface area (SSA) of Pt/ mSiTiN14 was same as 1248 m^2 g^{-1} with an unchanged BET isotherm even after 5 cycles of recyclability tests. There was no degradation found in the nanocomposite samples indicating the prevention of TiN corrosion by Si_3N_4 matrix (Lale et al. 2020).

In another recent work, a thin film Ag/TiO_2 nanocomposite catalyst was fabricated by a facile template process. The 5–10 nm size of the Ag nanoparticle (NP) was present in the amorphous nanocomposite. The incorporation of Ag NP lessens the band gap energies with a value of 2.88 eV of the Ag/TiO_2 nanocomposite catalysts. With an increase in pH value from 4.0 to 6.0, the photocatalytic removal of Alizarin yellow increased. With a further increase in pH from 6.0 to 10.0, a dip in the removal of Alizarin yellow was noticed. Moreover, the inclusion of scavengers namely 2-propanol, and sodium azide, and bicarbonate in Ag/TiO_2 nanocomposite reduced the percentage removal of Alizarin yellow from aqueous solutions which indicates that the ·OH radical is reactive species in the oxidation process (Lalliansanga et al. 2020).

Another study reported that the metal ions like, Ni^{2+} and Co^{3+} were in-situ doped in MnO_2 to produce a mesoporous Ni-Co-MnO_2 matrix with the addition of MoS_2

resulting in the formation of a nanocomposite. The researchers adopted an ultra-sonication technique to disperse the nanoscale MoS_2 in the octahedral layer of MnO_2 to fabricate a strongly bonded nanocomposite. The major weight loss obtained for 10% additive of MoS_2 in Ni-Co-MnO_2 matrix was 5.05% as deciphered from thermo-gravimetric analysis (TGA). This was greater in comparison to pristine Ni-Co-MnO_2. Their 86% residual mass loss for 10%MoS_2-Go@Ni-Co-MnO_2 nanocomposite indicated its improvised thermal stability (Kanaujiya et al. 2018; Xu et al. 2020). Their degradation of Congo red dye was highest in the case of 10% MoS_2-Go@Ni-Co-MnO_2 nanocomposite catalyst in comparison with 5% nanocomposite and pristine Ni-Co-MnO_2. Besides, the ultraviolet–visible light (UV–Vis) spectrum showed the rapid degradation of congo dye via ultrasonication procedure as compared to conventional oxidation and photocatalytic process (Abdullah and Ling 2010; Saleh and Taufik 2018). Hence, the catalytic performance is enhanced upon the addition of MoS_2 in Ni-Co-MnO_2 matrix via the ultra-sonication method. This in turn enables quick degradation of carcinogenic congo red dye and can be used in textile and paint industries (Radoor et al. 2020). On the other hand, this may negatively affect marine life. (JothiRamalingam et al. 2021; Borthakur and Das 2018; Zhang and Yan 2019).

Another interesting study had utilized semiconductor nanomaterials namely TiO_2 Fe_2O_3, zinc oxide (ZnO), and cadmium sulfide (CdS) for photelectrocatalytic splitting of water in the presence of visible light (Kanaujiya et al. 2018; Han et al. 2016; Hernández et al. 2015; Hwang et al. 2002; Bohn et al. 2012). However, their poor stability and ineffective light harvesting inhibit its usage (Fountaine et al. 2016; Valenti et al. 2016).

In contrast, a novel integrated $CoSnO_3$-rGO nanocomposite using 2.45 nm $CoSnO_3$ particle was synthesized for its application in water splitting in presence of white LED light irradiation. Their I_D/I_G intensity ratio for D-band (at Raman shift 1349 cm^{-1}) to G-band (at Raman shift 1591 cm^{-1}) of $CoSnO_3$-rGO was estimated to be 0.97, was greater than that of pristine r-GO equivalent to 0.94. This described the existence of more defects due to the aggregation of $CoSnO_3$ nanoparticles in the r-GO matrix (Chen et al. 2014). Photoelectrocatalytic water oxidation behaviour was studied by linear sweep voltammetry (LSV) and compared among SnO_2, $CoSnO_3$, and $CoSnO_3$-rGO. The $CoSnO_3$-rGO nanocomposite electrode showed oxidation of water at 1.7 V vs RHE (reference hydrogen electrode) with anodic photocurrent (J) equivalent to 0.416 $mAcm^{-2}$. A low overpotential of 340 mV was obtained at 1.0 $mAcm^{-2}$ at a neutral pH. Besides, they obtained an excellent catalytic activity of $CoSnO_3$-rGO, which was validated by a slope of 68 mV per decade as seen from a Tafel plot. The incident photo to current efficiency (ICPE) was the highest (40.52) in the case of $CoSnO_3$-rGO which explained the effect of utilization of light in improved photoelectrocatalytic performance by band gap engineering. The Faradic efficiency was calculated to be 86% for $CoSnO_3$-rGO which showed a greater amount of oxygen generation in electrolysis under the LED light. As per the Kubelka–Munk theorem, the band gap $CoSnO_3$-rGO was computed to be lowest at 2.85 eV which demonstrated its enhanced light emitting diode (LED) light harvesting capacity(Mohanta et al. 2021).

2.4.3 Fuel Cells

One of the best important applications of nano MMCs is fuel cells (FCs). Among the FCs, direct methanol fuel cell (DMFC) is a very promising candidate replacing conventional fossil fuel-based energy gadgets (Luo et al. 2019; Pei et al. 2017) due to its high volumetric energy density, less emission generation, and ease of storage in comparison to hydrogen fuel (Soleimani-Lashkenari et al. 2017; Zhang et al. 2019c). Conductive polymers have high electrochemical stability and high surface area (Antolini and Gonzalez 2009; Shatla et al. 2018), thus poly(5-aminoindole) (PAIn) was introduced as a support material to the matrix of TiO_2NTs. This leads to the formation of Ti/TiO_2NTs/PAIn nanocomposites by electropolymerization and electrochemical anodization method with PAIn acting as a binder between low electrical conductivity and high anti-corrosion metal oxide TiO_2 (Zheng et al. 2015; Kormányos et al. 2020) and metal composites. Then, the pulsed electrodeposition method was adopted to coat palladium (Pd) on the PAIn. The cyclic voltammetry (CV) curves of the electropolymerization reaction of PAIn on Ti/TiO_2NTs indicated outrageous redox activity on the surface of Ti/TiO_2NTs (Nie et al. 2008; Zhou et al. 2009, 2010) with a perfect synthesis and moderate increment in the thickness of PAIn films. Then, the CV of PAIn was conducted in 0.5 M H_2SO_4, where the peak current density starts to increase as the scan rate escalates. During the cycle, no decomposition of the PAIn can be seen which portrays its good structural stability. The average diameter of Ti/TiO_2NTs was estimated to be 100 nm and showcased the fluffy surface of Ti/TiO_2NTs/PAIn which explains its possession of a large surface area providing exceptional electrochemical ability (Zhang et al. 2014). Besides, the surface area of Ti/TiO_2NTs/PAIn was calculated to be 36.5 $m^2\ g^{-1}$, which was greater than that of Ti/TiO_2NTs (10.3 m^2g^{-1}). The nano-architecture of multibranched Pd NFs with a diameter of 60 nm was observed under SEM microscopy. The electrochemically active surface area (ECSA) of Ti/TiO_2NTs/PAIn/Pd nanocomposite electrode was 119.9 m^2g^{-1}. The high ECSA value justified the synergistic effect among Pd, TiO_2, and PAIn. A high methanol oxidation reaction (MOR) electrochemical activity value of 1040 mAm^2g^{-1} was obtained for the Ti/TiO_2NTs/PAIn/Pd nanocomposite. In addition, the CO poisoning tolerance was found to be 2.60 which stated the exceptional quality of the nanocomposite for its application as an anode catalyst in fuel-cell activities (Xu et al. 2021).

In another recent work, the PAni-Co-PPy@TiO_2 nanocomposite was prepared by the in-situ polymerization technique (Bhaumik et al. 2012; Dutta et al. 2012). The specific surface area (SSA) in correspondence to BET theory was estimated to be 270.895 m^2g^{-1}, which is greater than that of PPy@ TiO_2 and PAni@ TiO_2 nanocomposites. This indicated the greater catalytic activity of the fabricated nanocomposite structure demonstrating a perfect covering of TiO_2 NPs within the mesoporous matrices of (PAni-Co-PPy) copolymer. This also revealed the improved specific capacitance of the nanocomposite. Here, the oxygen reduction rate (ORR) as observed from the CV peaks was 0.435 V at a reduction current value of −0.303 mA for PAni-Co-PPy@TiO_2 nanocomposite. Whereas, at a reduction current of -0.204 mA, the ORR rate for Pt/C catalyst was 0.521 V. The chronoamperometry of PAni-Co-PPy@TiO_2 displayed an initial current of 0.0508 mA and a limiting current of 0.0323 mA showed an enhanced synergistic action of the nanocomposite as compared to Pt/C and other considered catalysts. Furthermore, the charge transfer resistance (R_{ct}) value of PAni-Co-PPy@TiO_2 was the lowest equivalent to 43.53 Ω as derived from the Nyquist plot of the electrochemical impedance spectroscopic (EIS) analysis. This result suggests a large number of active sites in the nanocomposite samples, with a rapid exchange of electrons facilitating a higher oxygen reduction reaction (ORR) catalytic ability. Hence, PAni-Co-PPy@TiO_2 was used as a catalyst for cathode material with the generation of maximum bioelectricity in a microbial fuel cell (MFC) (Pattanayak et al. 2019).

2.4.4 Supercapacitors (Energy Storage Devices)

Another important application of the metal matrix nanocomposites is in supercapacitors as an excellent energy storage device. The supercapacitor (SC) is also known as pseudocapacitor (PC) or ultracapacitor (UC) The SC can broadly be classified into three categories: such as (i) electric double layer capacitors (EDLCs) (generally, carbon based material), (ii) pseudocapacitor (generally, metal oxide and conductive polymer), and (iii) hybrid capacitor (generally, composite, asymmetric and battery type). The SCs are having higher specific capacitance and greater energy density resulting in a rapid and reversible redox reaction as compared to the traditional EDLCs. In general, metal oxides preferably MnO_2, RuO_2, Co_3O_4, and Mo_3O_4 have been considered suitable candidates possessing pseudocapacitive features (Conway 1999; Brousse et al. 2015). Several factors such as the morphology, thickness of the electro-active material, current densities, and output energy affect the overall performance of the PCs (Conway 1999; Yan et al. 2014; Augustyn et al. 2014) and thus related research on developing the SC material and its effectiveness were performed (Huang et al. 2013; Yang et al. 2014). Binary metal oxides namely $NiCo_2O_4$ and $MnCo_2O_4$ showed better cyclic stability and capacity than those of monolithic metal oxides (Dubal et al. 2015; Gomez and Kalu 2013; Li et al. 2012; Tan et al. 2016; Yuan et al. 2012; Zhang et al. 2015a, b). Hence, a multi-step hydrothermal method and then a low temperature annealing process was employed to fabricate a novel three-dimensional (3D) hierarchical nanocomposite structure for its applications as electrodes in supercapacitors. The mesoporous hybrid nanocomposite comprised of ternary metal oxide $MnNiCo_4/MnO_2$ nanosheets and an array of nanowires with a robust adhesion on carbon cloth (CC) (Du et al. 2013; Gao et al. 2015; Guo et al. 2014; Liu et al. 2012; Nagaraju et al. 2015; Rakhi et al. 2012; Shen et al. 2014; Wang and Wang 2013; Wang et al. 2015; Zhi et al. 2013). The surface morphology of the $MnNiCo_4$/carbon cloth (CC) hierarchical structure revealed well-ordered uniform growth of $MnNiCo_4$ nanowires (60–150 nm diameter) vertical to the carbon cloth. The 3D $MnNiCo_4$ was strongly adhered to the carbon cloth shortening the electron path thus facilitating rapid diffusion of ions.

This results in enhanced conductivity of the electrode. The MnO_2 nanosheet shells of thickness 8 nm were observed to fully cover the surface of $MnNiCo_4$ arrays of nanowires. The inclusion of Ni and Co encouraged the availability of surface area in the $MnNiCo_4$ for efficient charge transfer. The area under the CV curve was the highest in the case of MnO_2@$MnNiCo_4$/CC electrode alongside higher redox reaction and superior capacitive performance in comparison to $MnNiCo_4$/CC, $MnCo_2O_4$/CC, and $NiCo_2O_4$/CC. At a current density of 0.8 Ag^{-1}, they computed the specific capacitance of MnO_2@$MnNiCoO_4$/CC as 1931 Fg^{-1}. After 6000 cycles, their capacitance retention was 91.2% at a current density of 6 A g^{-1}. Moreover, at a power density of 200 W kg^{-1}, the energy density of the electrode was evaluated to be 67.1 Wh kg^{-1}. This portrays the potential application of the MnO_2@$MnNiCoO_4$ nanowire array as a high-performance and stable supercapacitor (Tamaddoni Saray and Hosseini 2016).

In another recent study, with the help of the electrophoretic deposition (EPD) method, a flake like MnO_2 (M) was uniformly dispersed in a combination of CNTs (C) and graphene (G) nanoporous framework to form CNT/ MnO_2 (CM), graphene/MnO_2 and a hybrid GMC + C nanocomposite to be used as electrodes in pseudocapacitors. The specific capacitance of 398 Fg^{-1} at 100 mV s^{-1} was deciphered from the CV curves for the GMC + C electrode which was the highest in comparison to CM, GM, GMC, and GM + C. The charge/discharge curves displayed a slight dip in the current-resistance (IR) slope of 6.5 Ω. In addition, at a current density of 3 Ag^{-1}, the specific capacitance of GMC + C was estimated to be 416 Fg^{-1}. Both the bulk resistance (R_s) and charge-transfer resistance (R_{ct}) were the lowest values for GMC + C of 2.1 Ωcm^2 and 2.63 Ωcm^2, respectively which verified the excellent interfacial contact between GMC and CNTs resulting in effective and rapid migration of ions. Cottrell equation Eq. 1 can be used to calculate the diffusion coefficient D of the nanocomposite electrodes via the Chronoamperometry (CA) experiment.

$$i_t = \frac{nFAC_0\sqrt{D_0}}{\sqrt{\pi t}} \tag{2.1}$$

where, n = stoichiometric number of electrons involved in the reaction; F = Faraday's constant (96,485 C/mol), A = area of the electrode surface (cm^2), C_0 = concentration of reactive species (mol/cm), and D_0 = diffusion coefficient (cm^2/s).

The researchers showed that the inclusion of CNTs provided a large surface area to the GMC + C nanocomposites alongside reducing the resistance of the electrode. This exhibits an exceptional capacitive behaviour of the GMC + C electrode corroborated by the high intercalation value of 3.647 × 10^{-8} cm^2/s and low deintercalation value of 2.899 × 10^{-8} cm^2/s. Furthermore, 83.3% retention of specific capacitance with respect to its initial value of 415 Fg^{-1} was observed after 15,000 charge/discharge cycles implying the outstanding cyclic durability of the GMC + C nanocomposite electrode as a supercapacitor (Hung et al. 2013).

2.4.5 Thermal Energy Storage

Phase change materials (PCMs) are gaining a huge attraction in the application of thermal energy storage and its management. To attain the highest effectivity of PCMs, they must possess huge enthalpy of fusion and high thermal conductivity. Organic PCMs (Lu et al. 2015) and inorganic salt hydrates (Graham et al. 2017) were commonly used in thermal management usage in buildings (Lu et al. 2015; Xu et al. 2015) and thermal regulation (Wen et al. 2016) but not feasible to be used in industrial heat processes and solar thermal power plants. The reason was their low thermal conductivity, prone to corrosion, and slow thermal charging/ discharging capability. Previous works reported the thermal conductivity value in the range 0.1–1 Wm^{-1} K^{-1} for organic PCMs and salt hydrates and between 0.5 and 5 Wm^{-1} K^{-1} for salts. Carbonaceous fillers such as graphene (Li 2013) and CNTs (Chen et al. 2012) are thermally conductive in nature. The addition of nanographene (NG) and CNT sponge (CNTS) to the paraffin wax (PW) formed NG/PW and CNTS/PW composites, which are known as PCMs. The thermal conductivity of these PCMs was found to increase with the inclusion of graphene and CNTs. Mostly, the usage of thermal storage is stationary, and thus metallic PCMs as efficient high energy thermal storage materials found their applicability in high scale industrial heat processes. A Bi-Ag nanocomposite comprising of Bi nanoparticles dispersed in an Ag matrix was formed. The energy dispersive X-ray spectroscopy revealed the uniform distribution of Bi nanoparticles in the bulk Ag matrix. The Differential Scanning Calorimetry (DSC) test showed that the Ag matrix acted as an isolation barrier to prevent the coalescence of the nanoparticles during melt-freeze cycles. Moreover, the melting temperature can be modified by changing the size of the nanoparticle diameter. The enthalpy of fusion was found to be size dependent as it varied from 20.1–37.6 J/g_{Bi} with respect to the nanoparticle diameter between 8.1 and 14.9 nm. The thermal conductivity (k, Wm^{-1} K^{-1}) was measured in accordance with the Wiedemann–Franz law as per the following Eq. 2.2 (Chen 2005):

$$k = L\sigma T \tag{2.2}$$

where, σ = electrical conductivity (Sm^{-1}), T = absolute temperature and L_0 = Sommerfeld value for the Lorenz number (L). For most of the metals, L_0 is a reasonable approximation as 2.44×10^{-8} W Ω K^{-2} and expressed in Eq. 2.3.

$$L_o = \frac{\pi^2 k_B^2}{3e^2} \tag{2.3}$$

where, k_B = Boltzmann constant and e = elementary charge. Generally, both phonons, as well as electrons, can conduct heat in solid materials. Thus, the thermal conductivity measured by the Wiedemann–Franz law approach is only via electron contribution.

With an increase in the volume fraction of Bi nanoparticles, the thermal conductivity was observed to decrease. The thermal conductivity was also calculated to be in the order of 10^1 to 10^2 Wm^{-1} K^{-1} which is greater than that of conventional organic and salt hydrate PCMs (10^{-1}-10^0 Wm^{-1} K^{-1}) (Liu et al. 2015).

Magnesium nitrate hexahydrate (MNH) based nanocomposites (Madarász et al. 2007; Zhang et al. 2019b) prepared from combining nano silver, nano copper, and graphene as filler agents also improved the thermal conductivity of pristine PCMs (Wang et al. 2019c).

High conducting metal oxide nanoparticles (NPs) dispersed in a PCM matrix were found to enhance the rate of heat release thus improving the efficiency of the thermal system. The low thermal conductivity of myristic acid (MA) limits their performance as a PCM. The incorporation of 2 wt.% metal oxide NPs such as ZnO, Al_2O_3, TiO_2 and so on in MA formed nano-enhanced PCM (NEPCM). Thus, the thermal conductivity of the NEPCM after the addition of ZnO, Al_2O_3, and TiO_2 raised to 0.24 Wm^{-1} K^{-1}, 0.23 Wm^{-1} K^{-1}, and 0.22 Wm^{-1} K^{-1}, respectively from 0.16 Wm^{-1} K^{-1} of the pristine MA and this raised the thermal conductivity of the NEPCM by 1.50, 1.44, and 1.38, respectively in comparison to the base MA (Ouikhalfan et al. 2019). A melt mixing technique (Narayanan et al. 2017) was adopted to fabricate the PCM metal-oxide nanocomposites. The average size of TiO_2 was around 50 nm as observed under SEM images. The charging and discharging rates incremented to 38 and 75% for 1 wt.% TiO_2 NPs in comparison to the pristine PCMs. Specifically, the TiO_2 based PCM nanocomposite can be considered a suitable candidate for its application in solar water heating due to its capability to maintain high latent heat and enhanced thermal conductivity due to its high surface area and reduced agglomeration with the PCM matrix (Gupta et al. 2020).

2.4.6 *Wind Energy Generation*

Minimizing the utilization of fossil fuels is the primary goal of all the renewable energy sectors. Particularly, the European Union (EU) has been trying to cover at least 20% of its energy needs from renewables by expanding the wind energy capacity at two orders of magnitude. The offshore wind energy capacity of the EU is growing around 21% annually (Mishnaevsky et al. 2012). In 1888, electric power was first generated using wind turbines by Charles F. Brush in Cleveland Ohio, USA, and then, by pioneer Poul La Cour in 1889, in Askov, Denmark. The production of electricity using wind turbines was done by the company S. Morgan-Smith at Grandpa's Knob in Vermont in the USA, 1941, employing steel blades. However, one of the blades crashed after running only a few hundred hours. This is one of the demonstrations of the inherent limitations of metals as a wind blade material. Therefore, the selection of proper materials in wind energy generation applications is still challenging. Although the first turbine blade built with steel failed, in the next attempt, the three blades made of composite materials built from steel spars, with aluminum shells supported by wooden ribs, installed at Gedser coast in Denmark in 1956–1957 worked for several years. Thereafter, in the 1970s, the wind turbines had been mainly produced with many composite blades (Brøndsted et al. 2005). The composite materials were used not only for increasing the strength but also for reducing the mass and corrosion attack by environments. Although the application of metal matrix nanocomposites in wind energy applications is very limited, recent work has reported metal nanoparticle coated on CFs as multifunctional nanocomposites for offshore wind energy applications (Gou et al. 2010). Therefore, MMCs nanomaterials have great potential in wind energy applications.

2.4.7 Space Crafts

In today's world, satellite plays a crucial role in effective and rapid transfer of information across the world for its usage in the weather forecast, agriculture, and various sectors of defense and mass communication. The material selection and structural integrity of the satellite are important factors during the launching of the satellite to space as it must sustain high temperatures in its overall mission and operation. Several other characteristics such as light in weight, high specific strength, and high thermal resistance are vital for the satellite owing to the extremely challenging environment of the space. Al-MMCs (AMCs) are considered an ideal material for robust space applications since they possess high specific strength, high wear resistance, high thermal conductivity, and low density. The SiC_w and germanium (Ge) in the form of whisker particulates are incorporated in discontinuously reinforced aluminium (DRA) to form lightweight AMCs having reduced density and adaptive coefficient of thermal expansion (Rawal 2001). These are generally used to fabricate the satellite structures namely bushings, truss nodes, and thermal planes. Hence, these satellites become feasible and qualified for their applicability in Communication and Global Positioning Systems (GPS) satellites. The weight of the satellite truss structure was reduced by 70% when 6061-T6 Al was reinforced with SiC_p particulates (Badiey and Abedian 2010). The Al reinforced with Ge is an efficient heat spreader that helps in lessening the heat increment in the satellite structures in space (Hocine et al. 2013). The horizontal cross arm of Intelsat-IV was produced from Al reinforced with graphite-epoxy tape. Anik, the Canadian Communication Satellite was also found to be made from Al-honeycombs with graphite layers as the reinforcement. The SiC nanoparticles reinforced Al-matrix composite has shown an exceptional nanocomposite with advanced thermal properties and dimensional stability (Muley et al. 2015; Toor 2017). The body flaps of the NASA/ ESA X-38 crew return vehicle as part of the International Space Station were manufactured from C/SiC ceramic matrix composite (May et al. 2020).

2.4.8 *Automobile*

An eco-friendly, isotropic metal matrix composite can be prepared by different processing techniques, homogeneous particle reinforcement, and further testing and analysis. Mostly, AMCs are preferred by the automotive industries to produce parts such as brake rotors, connecting rods, pistons, and so on (Bisane et al. 2015). The design and material of the piston play an important role in the overall performance of an engine. Certain forces such as inertia and frictional forces along with cyclic gas pressure act on a prolonged basis on the piston thus, damaging it in the long run. Hence, on reinforcing fly ash and SiC with Al-7075 alloy, properties namely hardness, tensile strength, and modulus of elasticity increase. The wear rate characteristics of the MMC also improve. These experimental characteristics were validated by Finite Element Analysis in ANSYS software (Premnandh et al. 2020). MMC formed from copper reinforced with graphite is a good replacement for copper lead bearings used as crankshafts. The wear characteristic of the MMC enhanced is due to the formation of continuous film of graphite used as reinforcement alongside its nontoxic nature. This provides for better self-lubrication properties thus increasing the working life of the bearings (Rohatgi et al. 1992; Kestursatya et al. 2001). Aluminium-fly ash composite syntactic foams developed at the University of Wisconsin-Milwaukee are used as reinforcements in tube shaped frames to improve both the dynamics of vehicles and their energy absorption capability upon impact (Rohatgi et al. 1998; Maddever and Guinehut 2005). The Al-SiC composite is used in brake drums. Moreover, brake pads are being manufactured from aluminium reinforced with Boron carbide (B_4C) (Singh et al. 2020).

2.5 Concluding Remarks

This chapter has described the most advanced metal matrix nanocomposite materials emerging in energy applications. The MMCs nanomaterials are advantageous over conventional MMCs materials due to their excellent bonding and multifunctional capability. The different manufacturing, processing, and characterization techniques of various metal matrix nanocomposites in energy-related applications have been evidently elucidated in this chapter. Different phenomena related to the improvement of strength of nanocomposites have been discussed. Various salient features related to several energy applications and their testing procedures for the nano MMCs have been highlighted. The main advantages of the metal matrix nanocomposites and their future potential in various energy applications have been described thoroughly in this chapter.

References

Abdullah AZ, Ling PY (2010) Heat treatment effects on the characteristics and sonocatalytic performance of TiO2 in the degradation of organic dyes in aqueous solution. J Hazard Mater 173(1–3):159–167. https://doi.org/10.1016/j.jhazmat.2009.08.060

Abudayyeh OK, Nelson C, Bradshaw GK, Whipple S, Wilt DM, Han SM (2019) Crack-tolerant metal composites as photovoltaic gridlines. IEEE J Photovoltaics 9(6):1754–1758. https://doi.org/10.1109/JPHOTOV.2019.2939096

Abudayyeh OK, Nelson C, Han SM, Gapp N, Wilt DM (2015) Silver-carbon-nanotube metal matrix composites for metal contacts on space photovoltaic cells. In: 2015 IEEE 42nd Photovoltaic Specialist Confeence (PVSC 2015), pp 8–10. https://doi.org/10.1109/PVSC.2015.7355662

2. Aigbodion and 3. Nwigbo (2012) Characterization of breadfruit seed hull ash for potential utilization in metal matrix composites for automotive application

Amatucci GG, Pereira N (2007) Fluoride based electrode materials for advanced energy storage devices. J Fluor Chem 128(4):243–262. https://doi.org/10.1016/j.jfluchem.2006.11.016

Andre D, Kim S-J, Lamp P, Lux SF, Maglia F, Paschos O, Stiaszny B (2015) Future generations of cathode materials: an automotive industry perspective. J Mater Chem A 3(13):6709–6732. https://doi.org/10.1039/c5ta00361j

Antolini E, Gonzalez ER (2009) Polymer supports for low-temperature fuel cell catalysts. Appl Catal A Gen 365(1):1–19. https://doi.org/10.1016/j.apcata.2009.05.045

Abbas A, Veeramanikandan R, Song-Jeng H (2021) “Magnesium Metal Matrix Composites and Their Applications.” In Magnesium Alloys. IntechOpen, 2021. https://doi.org/10.5772/intechopen.96241

Asafa TB, Durowoju MO, Oyewole AA, Solomon SO, Adegoke RM, Aremu OJ (2015) Potentials of snailshell as a reinforcement for discarded aluminum based materials. Int J Adv Sci Technol 84:1–8. https://doi.org/10.14257/ijast.2015.84.01

Ataollahi Oshkour A, Pramanik S, Mehrali M, Yau YH, Tarlochan F, Abu Osman NA (2015) Mechanical and physical behavior of newly developed functionally graded materials and composites of stainless steel 316L with calcium silicate and hydroxyapatite. J Mech Behav Biomed Mater 49:321–331. https://doi.org/10.1016/j.jmbbm.2015.05.020

Augustyn V, Simon P, Dunn B (2014) Pseudocapacitive oxide materials for high-rate electrochemical energy storage. Energy Environ Sci 7(5):1597–1614. https://doi.org/10.1039/c3ee44164d

Badiey M, Abedian A (2010) Application of metal matrix composites (MMCs) in a satellite boom to reduce weight and vibrations as a multidisciplinary optimization. In: 27th Congress on Intenational Council of Aeronautical Science (ICAS 2010), vol 3, pp 2512–2523

Badway F, Pereira N, Cosandey F, Amatucci GG (2003) Carbon-metal fluoride nanocomposites. J Electrochem Soc 150(9):A1209. https://doi.org/10.1149/1.1596162

Bakshi SR, Singh V, Seal S, Agarwal A (2009) Aluminum composite reinforced with multi-walled carbon nanotubes from plasma spraying of spray dried powders. Surf Coatings Technol 203(10):1544–1554. https://doi.org/10.1016/j.surfcoat.2008.12.004

Balogun MS, Huang Y, Qiu W, Yang H, Ji H, Tong Y (2017) Updates on the development of nanostructured transition metal nitrides for electrochemical energy storage and water splitting. Mater Today 20(8):425–451. https://doi.org/10.1016/j.mattod.2017.03.019

Bang JH, Suslick KS (2009) Dual templating synthesis of mesoporous titanium nitride microspheres. Adv Mater 21(31):3186–3190. https://doi.org/10.1002/adma.200802309

Bechelany MC, Proust V, Gervais C, Ghisleni R, Bernard S, Miele P (2014) In situ controlled growth of titanium nitride in amorphous silicon nitride: a general route toward bulk nitride nanocomposites with very high hardness. Adv Mater 26(38):6548–6553. https://doi.org/10.1002/adma.201402356

Bechelany MC, Proust V, Lale A, Miele P, Malo S, Gervais C, Bernard S (2016) Nanocomposites through the chemistry of single-source precursors: understanding the role of chemistry behind the design of monolith-type nanostructured titanium nitride/silicon nitride. Chem Eur J 23(4):832–845. https://doi.org/10.1002/chem.201603661

Beke S (2011) A review of the growth of V_2O_5 films from 1885 to 2010. Thin Solid Films 519(6):1761–1771. https://doi.org/10.1016/j.tsf.2010.11.001

Bhaumik M, Maity A, Srinivasu VV, Onyango MS (2012) Removal of hexavalent chromium from aqueous solution using polypyrrole-polyaniline nanofibers. Chem Eng J 181–182:323–333. https://doi.org/10.1016/j.cej.2011.11.088

Bisane V, Sable Y, Dhobe M, Sonawane P (2015) Recent development and challenges in processing of ceramics reinforced Al matrix composite through stir casting process: a review. Int J Eng Appl Sci (IJEAS) 2(10):11–16

Boehlert CJ, Tamirisakandala S, Curtin WA, Miracle DB (2009) Assessment of in situ TiB whisker tensile strength and optimization of TiB-reinforced titanium alloy design. Scr Mater 61(3):245–248. https://doi.org/10.1016/j.scriptamat.2009.03.054

Bohn C, Agarwal A, Walter E, Vaudin M, Herzing A, Haney P, Talin A, Szalai V (2012) Effect of tin doping on alpha-Fe_2O_3 photoanodes for water splitting. J Phys Chem C 116(29):15290–15296

Boppana SB, Aravinda T, Dayanand S, Reddy V, Vijay Kumar M, Gautham R, Nagaraj M (2021) Synthesis and characterization of ZrO_2 and graphene particle reinforced Al6061 metal matrix composites. Mater Today Proc 46:6064–6067. https://doi.org/10.1016/j.matpr.2021.02.570

Borthakur P, Das MR (2018) Hydrothermal assisted decoration of NiS_2 and CoS nanoparticles on the reduced graphene oxide nanosheets for sunlight driven photocatalytic degradation of azo dye: Effect of background electrolyte and surface charge. J Colloid Interface Sci 516:342–354. https://doi.org/10.1016/j.jcis.2018.01.050

Bose S, Pandey A, Mondal A, Mondal P (2019) A novel approach in developing aluminum hybrid green metal matrix composite material using waste eggshells, cow dung ash, snail shell ash and boron carbide as reinforcements. In: Advances in industrial and production engineering. Springer, Singapore, pp 551–562

Breitung B, Reddy MA, Chakravadhanula VSK, Engel M, Kübel C, Powell AK, Hahn H, Fichtner M (2013) Influence of particle size and fluorination ratio of CFx precursor compounds on the electrochemical performance of C-FeF_2 nanocomposites for reversible lithium storage. Beilstein J Nanotechnol 4:705–713. https://doi.org/10.3762/bjnano.4.80

Brijesh K, Bindu K, Shanbhag D, Nagaraja HS (2019) Chemically prepared polypyrrole/$ZnWO_4$ nanocomposite electrodes for electrocatalytic water splitting. Int J Hydrogen Energy 44(2):757–767. https://doi.org/10.1016/j.ijhydene.2018.11.022

Brijesh K, Vinayraj S, Dhanush PC, Bindu K, Nagaraja HS (2020) $ZnWO_4$/SnO_2@r-GO nanocomposite as an anode material for high capacity lithium ion battery. Electrochim Acta 354. https://doi.org/10.1016/j.electacta.2020.136676

Brinkman HJ, Duszczyk J, Katgerman L (1999) Reactive hot pressing of functional graded metal matrix composites. Mater Sci Forum 308–311(May):140–144. https://doi.org/10.4028/www.scientific.net/msf.308-311.140

Brøndsted P, Lilholt H, Lystrup A (2005) Composite materials for wind power turbine blades. Annu Rev Mater Res 35(1):505–538. https://doi.org/10.1146/annurev.matsci.35.100303.110641

Brousse T, Bélanger D, Long JW (2015) To be or not to be pseudocapacitive? J Electrochem Soc 162(5):A5185–A5189. https://doi.org/10.1149/2.0201505jes

Cabana J, Monconduit L, Larcher D, Palacín MR (2010) Beyond intercalation-based Li-ion batteries: the state of the art and challenges of electrode materials reacting through conversion reactions. Adv Mater 22(35):170–192. https://doi.org/10.1002/adma.201000717

Casati R, Vedani M (2014) Metal matrix composites reinforced by nano-particles—a review. Metals (basel) 4(1):65–83. https://doi.org/10.3390/met4010065

Casting D, Composite M, Composite MM, Oxide A, Material M (2021) Fundamentals of metal matrix com- posites casting routes for production of metallic based composite parts corrosion of metal matrix composites applications: magnesium-based metal matrix composites (MMCs) squeeze casting for the production of metallic, pp 2016–2018

Cha SI, Kim KT, Arshad SN, Mo CB, Hong SH (2005) Extraordinary strengthening effect of carbon nanotubes in metal-matrix nanocomposites processed by molecular-level mixing. Adv Mater 17(11):1377–1381. https://doi.org/10.1002/adma.200401933

Chauhan A, Vates UK, Kanu NJ, Gupta E, Singh GK, Sharma BP, Gorrepati SR (2021) Fabrication and characterization of novel nitinol particulate reinforced aluminium alloy metal matrix composites (NiTip/AA6061 MMCs). Mater Today Proc 38:3027–3034. https://doi.org/10.1016/j.matpr.2020.09.326

Chen G (2005) Nanoscale energy transport and conversion: a parallel treatment of electrons, molecules, phonons, and photons. Oxford University Press, Oxford (2005)

Chen L, Zou R, Xia W, Liu Z, Shang Y, Zhu J, Wang Y, Lin J, Xia D, Cao A (2012) Electro- and photodriven phase change composites based on wax-infiltrated carbon nanotube sponges. ACS Nano 6(12):10884–10892. https://doi.org/10.1021/nn304310n

Chen M, Wang H, Li L, Zhang Z, Wang C, Liu Y, Wang W, Gao J (2014) Novel and facile method, dynamic self-assemble, to prepare SnO_2/rGO droplet aerogel with complex morphologies and their application in supercapacitors. ACS Appl Mater Interfaces 6(16):14327–14337. https://doi.org/10.1021/am5036169

Chen W, Yang T, Dong L, Elmasry A, Song J, Deng N, Elmarakbi A, Liu T, Lv HB, Fu YQ (2020) Advances in graphene reinforced metal matrix nanocomposites: mechanisms, processing, modelling, properties and applications. Nanotechnol Precis Eng 3(4):189–210. https://doi.org/10.1016/j.npe.2020.12.003

Chi YM, Mishra M, Chin TK, Liu WS, Perng TP (2019) Fabrication of macroporous/mesoporous titanium nitride structure and its application as catalyst support for proton exchange membrane fuel cell. ACS Appl Energy Mater 2(1):398–405. https://doi.org/10.1021/acsaem.8b01426

Choi EY, Kim CK (2017) Fabrication of nitrogen-doped nano-onions and their electrocatalytic activity toward the oxygen reduction reaction. Sci Rep 7(1):1–9. https://doi.org/10.1038/s41598-017-04597-6

Conway BE (1999) Electrochemical supercapacitors: scientific fundamentals and technological applications. Springer, New York, p 698

Cox ND, Pham M, Gapp N, Rape A, Rossi J, Wilt D, Landi BJ (2016) Electrical characterization of carbon nanotube metal matrix composite solar cell electrodes under mechanical stress. In: 2016 IEEE 43rd photovoltaic specialists conference (PVSC). https://doi.org/10.1109/PVSC.2017.8366276

Dehnavi MR, Niroumand B, Ashrafizadeh F, Rohatgi PK (2014) Effects of continuous and discontinuous ultrasonic treatments on mechanical properties and microstructural characteristics of cast Al413-SiCnp nanocomposite. Mater Sci Eng A 617:73–83. https://doi.org/10.1016/j.msea.2014.08.042

Deng DR, An TH, Li YJ, Wu QH, Sen Zheng M, Dong QF (2016) Hollow porous titanium nitride tubes as a cathode electrode for extremely stable Li-S batteries. J Mater Chem A 4(41):16184–16190. https://doi.org/10.1039/c6ta07221f

Derrien G, Hassoun J, Panero S, Scrosati B (2007) Nanostructured Sn-C composite as an advanced anode material in high-performance lithium-ion batteries. Adv Mater 19(17):2336–2340. https://doi.org/10.1002/adma.200700748

Divakaran AM, Minakshi M, Bahri PA, Paul S, Kumari P, Divakaran AM, Manjunatha KN (2021) Rational design on materials for developing next generation lithium-ion secondary battery. Progr Solid State Chem 62:100298. https://doi.org/10.1016/j.progsolidstchem.2020.100298

Dong S, Chen X, Gu L, Zhou X, Xu H, Wang H, Liu Z, Han P, Yao J, Wang L, Cui G, Chen L (2010) Facile preparation of mesoporous titanium nitride microspheres for electrochemical energy storage. ACS Appl Mater Interfaces 3(1):93–98. https://doi.org/10.1021/am100951h

Du J, Zhou G, Zhang H, Cheng C, Ma J, Wei W, Chen L, Wang T (2013) Ultrathin porous $NiCo_2O_4$ nanosheet arrays on flexible carbon fabric for high-performance supercapacitors. ACS Appl Mater Interfaces 5(15):7405–7409. https://doi.org/10.1021/am4017335

Dubal DP, Gomez-Romero P, Sankapal BR, Holze R (2015) Nickel cobaltite as an emerging material for supercapacitors: an overview. Nano Energy 11:377–399. https://doi.org/10.1016/j.nanoen.2014.11.013

Dutta K, Mahale RY, Arulkashmir A, Krishnamoorthy K (2012) Reversible assembly and disassembly of micelles by a polymer that switches between hydrophilic and hydrophobic wettings. Langmuir 28(26):10097–10104. https://doi.org/10.1021/la301760a

Elshalakany AB, Osman TA, Khattab A, Azzam B, Zaki M (2014) Microstructure and mechanical properties of MWCNTs reinforced A356 aluminum alloys cast nanocomposites fabricated by using a combination of rheocasting and squeeze casting techniques. J. Nanomater. 2014. https://doi.org/10.1155/2014/386370.

Fan G, Geng L, Wu H, Miao K, Cui X, Kang H, Wang T, Xie H, Xiao T (2017) Improving the tensile ductility of metal matrix composites by laminated structure: a coupled X-ray tomography and digital image correlation study. Scripta Mater 135:63–67. https://doi.org/10.1016/j.scriptamat.2017.03.030

Fechler N, Fellinger T, Antonietti M, Fechler N, Fellinger T-P, Antonietti M (2012) Chem Mater 24:713–719

Fountaine KT, Lewerenz HJ, Atwater HA (2016) Efficiency limits for photoelectrochemical water-splitting. Nat Commun 7:1–9. https://doi.org/10.1038/ncomms13706

Fowler JD, Allen MJ, Tung VC, Yang Y, Kaner RB, Weiller BH (2009) Practical chemical sensors from chemically derived graphene. ACS Nano 3(2):301–306. https://doi.org/10.1021/nn800593m

Gao S, Liao F, Ma S, Zhu L, Shao M (2015) Network-like mesoporous $NiCo_2O_4$ grown on carbon cloth for high-performance pseudocapacitors. J Mater Chem A 3(32):16520–16527. https://doi.org/10.1039/c5ta02876k

Gomez J, Kalu EE (2013) High-performance binder-free Co-Mn composite oxide supercapacitor electrode. J Power Sources 230:218–224. https://doi.org/10.1016/j.jpowsour.2012.12.069

S. Goriparti, E. Miele, F. De Angelis, E. Di Fabrizio, R. Proietti Zaccaria, and C. Capiglia, "Review on recent progress of nanostructured anode materials for Li-ion batteries," *J. Power Sources*, vol. 257, pp. 421–443, 2014, doi: https://doi.org/10.1016/j.jpowsour.2013.11.103.

Gosavi SV, Jaybhaye MD (2020) Microstructural studies on aluminium metal matrix composite (Al7075-SiC) fabricated through stir casting process. Mater Today Proc 39:1412–1416. https://doi.org/10.1016/j.matpr.2020.05.094

Gou J, Liang F, Tang Y, Kapat J (2010) Multifunctional nanocomposites for offshore wind energy. In: The twentieth international offshore and polar engineering conference. OnePetro, 20 June 2010

Graham M, Coca-Clemente JA, Shchukina E, Shchukin D (2017) Nanoencapsulated crystallohydrate mixtures for advanced thermal energy storage. J Mater Chem A 5(26):13683–13691. https://doi.org/10.1039/c7ta02494k

Guo D, Yu X, Shi W, Luo Y, Li Q, Wang T (2014) Facile synthesis of well-ordered manganese oxide nanosheet arrays on carbon cloth for high-performance supercapacitors. J. Mater. Chem. A 2(23):8833–8838. https://doi.org/10.1039/c4ta01238k

Gupta N, Kumar A, Dhasmana H, Kumar V, Kumar A, Shukla P, Verma A, Nutan GV, Dhawan SK, Jain VK (2020) Enhanced thermophysical properties of metal oxide nanoparticles embedded magnesium nitrate hexahydrate based nanocomposite for thermal energy storage applications. J Energy Storage 32:101773. https://doi.org/10.1016/j.est.2020.101773

Han S, Pu YC, Zheng L, Hu L, Zhang JZ, Fang X (2016) Uniform carbon-coated CdS core-shell nanostructures: synthesis, ultrafast charge carrier dynamics, and photoelectrochemical water splitting. J Mater Chem A 4(3):1078–1086. https://doi.org/10.1039/c5ta09024e

Hannula SP, Lintula P, Lintunen P, Lindroos T (2003) Processing and properties of metal matrix composites synthesized by SHS. Mater Sci Forum 426–432(3):1971–1978. https://doi.org/10.4028/www.scientific.net/msf.426-432.1971

Hargreaves JSJ (2013) Heterogeneous catalysis with metal nitrides. Coord Chem Rev 257(13–14):2015–2031. https://doi.org/10.1016/j.ccr.2012.10.005

He P, Zhang Q, Huang Q, Huang B, Chen T (2018) Vertically-oriented graphene nanosheet as nano-bridge for pseudocapacitive electrode with ultrahigh electrochemical stability. RSC Adv 8(25):13891–13897. https://doi.org/10.1039/c8ra01537f

Heidarzadeh A, Mohammadzadeh R, Brabazon D (2021) Green materials and production of metallic composite materials. Elsevier Ltd.

Helen M, Fichtner M, Anji Reddy M (2020) Electrochemical synthesis of carbon-metal fluoride nanocomposites as cathode materials for lithium batteries. Electrochem Commun 120:106846. https://doi.org/10.1016/j.elecom.2020.106846

Hernández S, Hidalgo D, Sacco A, Chiodoni A, Lamberti A, Cauda V, Tresso E, Saracco G (2015) Comparison of photocatalytic and transport properties of TiO_2 and ZnO nanostructures for solar-driven water splitting. Phys Chem Chem Phys 17(12):7775–7786. https://doi.org/10.1039/c4cp05857g

Hocine R, Boudjemai A, Boukortt A, Belkacemi K (2013) 3D TLM formulation for thermal modelling of metal matrix composite materials for space electronics systems. In: 2013 6th international conference on recent advances in space technologies (RAST), pp 47–52. https://doi.org/10.1109/RAST.2013.6581254

Hongliang L, Kaiyuan W, Zi Y, Quanchao Z, Yanhua C (2021) V_2O_5-Au nanocomposite film cathode with enhanced electrochemical performance for lithium-ion micro batteries. Chem Phys 544:111111. https://doi.org/10.1016/j.chemphys.2021.111111

Hort N, Kainer KU (2006) Powder Metallurgically Manufactured Metal Matrix Composites

Huang L, Chen D, Ding Y, Feng S, Wang ZL, Liu M (2013) Nickel-cobalt hydroxide nanosheets coated on $NiCo_2O_4$ nanowires grown on carbon fiber paper for high-performance pseudocapacitors. Nano Lett 13(7):3135–3139. https://doi.org/10.1021/nl401086t

Huang L, Wang L, Qian M, Zou J (2017) High tensile-strength and ductile titanium matrix composites strengthened by TiB nanowires. Scr Mater 141:133–137. https://doi.org/10.1016/j.scriptamat.2017.08.007

Hung CJ, Lin P, Tseng TY (2013) Electrophoretic fabrication and pseudocapacitive properties of graphene/manganese oxide/carbon nanotube nanocomposites. J Power Sources 243:594–602. https://doi.org/10.1016/j.jpowsour.2013.06.055

Hunt Jr. WH, Miracle DB (2001) Automotive applications of metal-matrix composites. Composites 21. https://doi.org/10.31399/asm.hb.v21.a0003484

Hwang DW, Kim J, Park TJ, Lee JS (2002) Mg-doped WO3 as a novel photocatalyst for visible light-induced water splitting. Catal Lett 80(1–2):53–57. https://doi.org/10.1023/A:1015322625989

Jamil M, Wei S, Taylor MP, Chen JJJ, Kennedy JV (2021) Hybrid anode materials for rechargeable batteries—a review of Sn/TiO_2 based nanocomposites. Energy Rep 7:2836–2848. https://doi.org/10.1016/j.egyr.2021.05.004

Jayakanth JJ, Mamidi VK, Pugazhenthi R, Anbuchezhiyan G, Ponshanmugakumar A (2021) An investigation on wear properties of SiC/WC strengthened aluminium alloy hybrid composites. Mater Today Proc. https://doi.org/10.1016/j.matpr.2020.12.416

Jayaraj A, Moorthy CVKNSN, Venkataramana VSN, Jaikumar S, Srinivas V (2020) Corrosion, mechanical and thermal properties of aluminium alloy metal matrix nano composites (AA-MMNCs) with multi-walled carbon nanotubes. SN Appl Sci 2(7):1–8. https://doi.org/10.1007/s42452-020-3081-9

Jin P, Chen G, Han L, Wang J (2014) Dry sliding friction and wear behaviors of $Mg_2B_2O_5$ whisker reinforced 6061Al matrix composites. Trans Nonferrous Met Soc China 24(1):49–57. https://doi.org/10.1016/S1003-6326(14)63027-2

JothiRamalingam R, Periyasami G, Ouladsmane M, ALOthman ZA, Arunachalam A, Altalhi T, Radhika T, Alanazi AG (2021) Ultra-sonication assisted metal chalcogenide modified mesoporous Nickel-cobalt doped manganese oxide nanocomposite fabrication for sono-catalytic dye degradation and mechanism insights. J Alloys Compounds 875:160072. https://doi.org/10.1016/j.jallcom.2021.160072

Jun YS, Hong WH, Antonietti M, Thomas A (2009) Mesoporous, 2D hexagonal carbon nitride and titanium nitride/carbon composites. Adv Mater 21(42):4270–4274. https://doi.org/10.1002/adma.200803500

Kala H, Mer KKS, Kumar S (2014) A review on mechanical and tribological behaviors of stir cast aluminum matrix composites. Procedia Mater Sci 6:1951–1960. https://doi.org/10.1016/j.mspro.2014.07.229

Kanaujiya N, Kumar N, Srivastava AK, Sharma Y, Varma GD (2018) One-step synthesized mesoporous MnO_2@MoS_2 nanocomposite for high performance energy storage devices. J Electroanal Chem 824:226–237. https://doi.org/10.1016/j.jelechem.2018.07.046

Kandpal BC, Johri N, Kumar N, Srivastava A (2021) Effect of industrial/agricultural waste materials as reinforcement on properties of metal matrix composites. Mater Today Proc. https://doi.org/10.1016/j.matpr.2021.01.575

Karvanis K, Fasnakis D, Maropoulos A, Papanikolaou (2016) Production and mechanical properties of Al-SiC metal matrix composites. IOP Conf Ser Mater Sci Eng 161(1). https://doi.org/10.1088/1757-899X/161/1/012070.

Kaskel S, Schlichte K, Kratzke T (2004) Catalytic properties of high surface area titanium nitride materials. J Mol Catal A Chem 208(1–2):291–298. https://doi.org/10.1016/S1381-1169(03)00545-4

Kestursatya M, Kim JK, Rohatgi PK (2001) Friction and wear behavior of a centrifugally cast lead-free copper alloy containing graphite particles. Metall Mater Trans A Phys Metall Mater Sci 32(8):2115–2125. https://doi.org/10.1007/s11661-001-0023-z

Kim BG, Jo C, Shin J, Mun Y, Lee J, Choi JW (2017) Ordered Mesoporous titanium nitride as a promising carbon-free cathode for aprotic lithium-oxygen batteries. ACS Nano 11(2):1736–1746. https://doi.org/10.1021/acsnano.6b07635

Kim H-J, Krishna T, Zeb K, Rajangam V, Gopi CVVM, Sambasivam S, Raghavendra KVG, Obaidat IM (2020a) A comprehensive review of Li-ion battery materials and their recycling techniques. Electronics 9(7):1161. https://doi.org/10.3390/electronics9071161

Kim H, Choi W, Yoon J, Um JH, Lee W, Kim J, Cabana J, Yoon W-S (2020b) Exploring anomalous charge storage in anode materials for next-generation Li rechargeable batteries. Chem Rev 120(14):6934–6976. https://doi.org/10.1021/acs.chemrev.9b00618

Klemenz S, Schuch J, Hawel S, Zieschang A-M, Kaiser B, Jaegermann W, Albert B (2018) Synthesis of a highly efficient oxygen-evolution electrocatalyst by incorporation of iron into nanoscale cobalt borides. Chemsuschem 11(18):3150–3156. https://doi.org/10.1002/cssc.201801547

Kntges M, Kunze I, Kajari-Schrder S, Breitenmoser X, Bjørneklett B (2011) The risk of power loss in crystalline silicon based photovoltaic modules due to micro-cracks. Sol Energy Mater Sol Cells 95(4):1131–1137. https://doi.org/10.1016/j.solmat.2010.10.034

Kormányos A, Speck FD, Mayrhofer KJJ, Cherevko S (2020) Influence of fuels and pH on the dissolution stability of bifunctional PtRu/C alloy electrocatalysts. ACS Catal 10(19):10858–10870. https://doi.org/10.1021/acscatal.0c02094

Krishnan D, Kim F, Luo J, Cruz-Silva R, Cote LJ, Jang HD, Huang J (2012) Energetic graphene oxide: challenges and opportunities. Nano Today 7(2):137–152. https://doi.org/10.1016/j.nantod.2012.02.003

Kulova TL (2013) New electrode materials for lithium-ion batteries (review). Russ J Electrochem 49(1):1–25. https://doi.org/10.1134/S1023193513010102

Kumar A (2018) Electromagnetic stir casting: an approach to produce hybrid metal matrix composite (MMC). Automob Eng Mech Eng Res Innov [Online]. Available: https://iilmcet.ac.in/blog/electromagnetic-stir-casting/

Lal S, Kumar S, Kumar A, Patel L, Aniruddha (2020) Fabrication and characterization of hybrid metal matrix composite Al-2014/SiC/fly ash fabricated using stir casting process. Mater Today Proc. https://doi.org/10.1016/j.matpr.2020.11.168

Lale A, Wasan A, Kumar R, Miele P, Demirci UB, Bernard S (2016) Organosilicon polymer-derived mesoporous 3D silicon carbide, carbonitride and nitride structures as platinum supports for hydrogen generation by hydrolysis of sodium borohydride. Int J Hydrogen Energy 41(34):15477–15488. https://doi.org/10.1016/j.ijhydene.2016.06.186

Lale A, Proust V, Bechelany MC, Viard A, Malo S, Bernard S (2017) A comprehensive study on the influence of the polyorganosilazane chemistry and material shape on the high temperature behavior of titanium nitride/silicon nitride nanocomposites. J Eur Ceram Soc 37(16):5167–5175. https://doi.org/10.1016/j.jeurceramsoc.2017.04.001

Lale A, Mallmann MD, Tada S, Bruma A, Özkar S, Kumar R, Haneda M, Francisco Machado RA, Iwamoto Y, Demirci UB, Bernard S (2020) Highly active, robust and reusable micro-/mesoporous TiN/Si3N4 nanocomposite-based catalysts for clean energy: understanding the key role of TiN nanoclusters and amorphous Si3N4 matrix in the performance of the catalyst system. Appl Catal B Environ 272:118975. https://doi.org/10.1016/j.apcatb.2020.118975

Lalliansanga, Tiwari D, Tiwari A, Shukla A, Shim MJ, Lee SM (2020) Facile synthesis and characterization of Ag(NP)/TiO_2 nanocomposite: Photocatalytic efficiency of catalyst for oxidative removal of alizarin yellow. Catal Today. https://doi.org/10.1016/j.cattod.2020.09.010

Li M (2013) A nano-graphite/paraffin phase change material with high thermal conductivity. Appl Energy 106:25–30. https://doi.org/10.1016/j.apenergy.2013.01.031

Li C, Gu L, Tsukimoto S, Van Aken PA, Maier J (2010) Low-temperature ionic-liquid-based synthesis of nanostructured iron-based fluoride cathodes for lithium batteries. Adv Mater 22(33):3650–3654. https://doi.org/10.1002/adma.201000535

Li J, Xiong S, Li X, Qian Y (2012) Spinel $Mn_{1.5}Co_{1.5}O_4$ core-shell microspheres as Li-ion battery anode materials with a long cycle life and high capacity. J Mater Chem 22(43):23254–23259. https://doi.org/10.1039/c2jm35607d

Li Z, Bai Z, Mi H, Ji C, Gao S, Pang H (2019) Biowaste-derived porous carbon with tuned microstructure for high-energy quasi-solid-state supercapacitors. ACS Sustain Chem Eng 7(15):13127–13135. https://doi.org/10.1021/acssuschemeng.9b02303

Li B, Lavernia EJ (2000) Spray forming of MMCs. In: Kelly A, Zweben CM (eds) Pergamon, Oxford, pp 617–653

Li J, Wang F, Shi C, Liu E, He C, Zhao N (2021) High strength-ductility synergy of MgAlB4 whisker reinforced aluminum matrix composites achieved by in situ synthesis. Mater Sci Eng A, 799:140127. https://doi.org/10.1016/j.msea.2020.140127

Li J, Wang F, Shi C, Liu E, He C, Zhao N (2021) High strength-ductility synergy of MgAlB4 whisker reinforced aluminum matrix composites achieved by in situ synthesis. Mater Sci Eng A 799:140127. https://doi.org/10.1016/j.msea.2020.140127

Liang M, Zhi L (2009) Graphene-based electrode materials for rechargeable lithium batteries. J Mater Chem 19(33):5871–5878. https://doi.org/10.1039/b901551e

Liang ZX, Ye B, Zhang L, Wang QG, Yang WY, Wang QD (2013) A new high-strength and corrosion-resistant Al-Si based casting alloy. Mater Lett 97:104–107. https://doi.org/10.1016/j.matlet.2013.01.112

Liu J, Liu F, Gao K, Wu J, Xue D (2009) Recent developments in the chemical synthesis of inorganic porous capsules. J Mater Chem 19(34):6073–6084. https://doi.org/10.1039/b900116f

Liu B, Zhang J, Wang X, Chen G, Chen D, Zhou C, Shen G (2012) Hierarchical three-dimensional $ZnCo_2O_4$ nanowire arrays/carbon cloth anodes for a novel class of high-performance flexible lithium-ion batteries. Nano Lett 12(6):3005–3011. https://doi.org/10.1021/nl300794f

Liu M, Ma Y, Wu H, Wang RY (2015) Metal matrix-metal nanoparticle composites with tunable melting temperature and high thermal conductivity for phase-change thermal storage. ACS Nano 9(2):1341–1351. https://doi.org/10.1021/nn505328j

Liu Z, Yuan X, Zhang S, Wang J, Huang Q, Yu N, Zhu Y, Fu L, Wang F, Chen Y, Wu Y (2019) Three-dimensional ordered porous electrode materials for electrochemical energy storage. NPG Asia Mater 11(1). https://doi.org/10.1038/s41427-019-0112-3

Lu Y, Wen ZY, Jin J, Wu XW, Rui K (2014a) Size-controlled synthesis of hierarchical nanoporous iron based fluorides and their high performances in rechargeable lithium ion batteries. Chem Commun 50(49):6487–6490. https://doi.org/10.1039/c4cc01762e

Lu Y, Wen Z, Jin J, Rui K, Wu X (2014b) Hierarchical mesoporous iron-based fluoride with partially hollow structure: facile preparation and high performance as cathode material for rechargeable lithium ion batteries. Phys Chem Chem Phys 16(18):8556–8562. https://doi.org/10.1039/c4cp00568f

Lu Z, Zhang J, Sun G, Xu B, Li Z, Gong C (2015) Effects of the form-stable expanded perlite/paraffin composite on cement manufactured by extrusion technique. Energy 82:43–53. https://doi.org/10.1016/j.energy.2014.12.043

Luo B, Zhao F, Xie Z, Yuan Q, Yang F, Yang X, Li C, Zhou Z (2019) Polyhedron-assembled ternary PtCuCo nanochains: integrated functions enhance the electrocatalytic performance of methanol oxidation at elevated temperature. ACS Appl Mater Interfaces 11(35):32282–32290. https://doi.org/10.1021/acsami.9b10192

Ma X, Huang C, Moering J, Ruppert M, Höppel HW, Göken M, Narayan J, Zhu Y (2016) Mechanical properties of copper/bronze laminates: role of interfaces. Acta Mater 116:43–52. https://doi.org/10.1016/j.actamat.2016.06.023

Madarász J, Varga PP, Pokol G (2007) Evolved gas analyses (TG/DTA–MS and TG–FTIR) on dehydration and pyrolysis of magnesium nitrate hexahydrate in air and nitrogen. J Anal Appl Pyrolysis 79(1):475–478. https://doi.org/10.1016/j.jaap.2006.12.024

Maddever W, Guinehut S (2005) Use of aluminum foam to increase crash box efficiency. SAE Tech. Pap., no. 724. https://doi.org/10.4271/2005-01-0704

May M, Rupakula GD, Matura P (2020) Non-polymer-matrix composite materials for space applications. Compos Part C Open Access 3. https://doi.org/10.1016/j.jcomc.2020.100057

Meduri P, Pendyala C, Kumar V, Sumanasekera GU, Sunkara MK (2009) Hybrid tin oxide nanowires as stable and high capacity anodes for Li-ion batteries. Nano Lett 9(2):612–616. https://doi.org/10.1021/nl802864a

Mera G, Gallei M, Bernard S, Ionescu E (2015) Ceramic nanocomposites from tailor-made preceramic polymers. Nanomaterials 5(2):468–540. https://doi.org/10.3390/nano5020468

Mhaske MS, Shirsat UM (2021) An investigation of mechanical properties of aluminium based silicon carbide (AlSiC) metal matrix composite by different manufacturing methods. Mater Today Proc 44:376–382. https://doi.org/10.1016/j.matpr.2020.09.746

Miracle DB (2001) Aeronautical applications of metal-matrix composites. Composites 21. https://doi.org/10.31399/asm.hb.v21.a0003485

Mishnaevsky L Jr, Brøndsted P, Nijssen R, Lekou DJ, Philippidis TP (2012) Materials of large wind turbine blades: recent results in testing and modeling. Wind Energy 15(1):83–97. https://doi.org/10.1002/we.470

Mishnaevsky L, Branner K, Petersen HN, Beauson J., McGugan M, Sørensen BF (2017) Materials for wind turbine blades: an overview. Materials 10(11). https://doi.org/10.3390/ma10111285

. Mohanta D, Barman K, Jasimuddin S, Ahmaruzzaman M (2021) Encapsulating band gap engineered $CoSnO_3$ mixed metal oxide nanocomposite in rGO matrix: a novel catalyst towards LED light induced photoelectrocatalytic water oxidation at neutral pH. J Electroanal Chem 880:114830. https://doi.org/10.1016/j.jelechem.2020.114830

Mudasar Pasha BA, Kaleemulla M (2018) Processing and characterization of aluminum metal matrix composites: an overview. Rev Adv Mater Sci 56(1):79–90. https://doi.org/10.1515/rams-2018-0039

Muley AV, Aravindan S, Singh IP (2015) Nano and hybrid aluminum based metal matrix composites: an overview. Manuf Rev 2. https://doi.org/10.1051/mfreview/2015018

Nachum S, Fleck NA, Ashby MF, Colella A, Matteazzi P (2010) The microstructural basis for the mechanical properties and electrical resistivity of nanocrystalline $Cu\text{-}Al_2O_3$. Mater Sci Eng A 527(20):5065–5071. https://doi.org/10.1016/j.msea.2010.04.070

Nagaraju G, Ko YH, Yu JS (2015) Tricobalt tetroxide nanoplate arrays on flexible conductive fabric substrate: facile synthesis and application for electrochemical supercapacitors. J Power Sources 283:251–259. https://doi.org/10.1016/j.jpowsour.2015.02.104

Nan H, Dang D, Tian XL (2018) Structural engineering of robust titanium nitride as effective platinum support for the oxygen reduction reaction. J Mater Chem A 6(14):6065–6073. https://doi.org/10.1039/c8ta00326b

Narayanan SS, Kardam A, Kumar V, Bhardwaj N, Madhwal D, Shukla P, Kumar A, Verma A, Jain VK (2017) Development of sunlight-driven eutectic phase change material nanocomposite for applications in solar water heating. Resour Efficient Technol 3(3):272–279. https://doi.org/10.1016/j.reffit.2016.12.004

Nie G, Zhang Y, Xu J, Zhang S (2008) Low-potential facile electrosyntheses of free-standing poly(5-methoxyindole) film with good fluorescence properties. J Electroanal Chem 622(1):121–127. https://doi.org/10.1016/j.jelechem.2008.05.008

Ouikhalfan M, Sari A, Chehouani H, Benhamou B, Biçer A (2019) Preparation and characterization of nano-enhanced myristic acid using metal oxide nanoparticles for thermal energy storage. Int J Energy Res. https://doi.org/10.1002/er.4856

Oyama ST (1996) The chemistry of transition metal carbides and nitrides. Springer, Netherlands

Samuel Ratna Kumar PS, Robinson Smart DS, John Alexis S (2017) Corrosion behaviour of aluminium metal matrix reinforced with multi-wall carbon nanotube. J Asian Ceram Soc 5(1):71–75. https://doi.org/10.1016/j.jascer.2017.01.004

Pan Z, Xiao Y, Fu Z, Zhan G, Wu S, Xiao C, Hu G, Wei Z (2014) Hollow and porous titanium nitride nanotubes as high-performance catalyst supports for oxygen reduction reaction. J Mater Chem A 2(34):13966. https://doi.org/10.1039/c4ta02402h

Pattanayak P, Papiya F, Kumar V, Singh A, Kundu PP (2021) Performance evaluation of poly(aniline-co-pyrrole) wrapped titanium dioxide nanocomposite as an air-cathode catalyst material for microbial fuel cell. Mater Sci Eng C 118:111492. https://doi.org/10.1016/j.msec.2020.111492

Pei J, Mao J, Liang X, Zhuang Z, Chen C, Peng Q, Wang D, Li Y (2017) Ultrathin Pt–Zn nanowires: high-performance catalysts for electrooxidation of methanol and formic acid. ACS Sustain Chem Eng 6(1):77–81. https://doi.org/10.1021/acssuschemeng.7b03234

Pramod R, Veeresh Kumar GB, Gouda PSS, Mathew AT (2018) A study on the Al_2O_3 reinforced Al7075 metal matrix composites wear behavior using artificial neural networks. Mater Today Proc 5(5):11376–11385. https://doi.org/10.1016/j.matpr.2018.02.105

Prasad SV, Asthana R (2004) Aluminum metal-matrix composites for automotive applications: tribological considerations. Tribol Lett 17(3):445–453. https://doi.org/10.1023/B:TRIL.0000044492.91991.f3

Prashantha Kumar HG, Anthony Xavior M (2014) Graphene reinforced metal matrix composite (GRMMC): a review. Procedia Eng 97:1033–1040. https://doi.org/10.1016/j.proeng.2014.12.381

Premnandh TP, Krishna V, Akshay KR, Pranav S, Kumar MM, Shankar VK (2020) Analysis of hybrid metal matrix composite piston for automobile application. AIP Conf Proc 2283(1):20003. https://doi.org/10.1063/5.0025420

Purant PAN, Joshi PRR, Mhetre PSB, Patil PAA (2020) Applications of metal matrix composites in modern engineering. Int Res J Eng Technol 7(6):4555–4558

Qi B, Zhao X, Wang S, Chen K, Wei Y, Chen G, Gao Y, Zhang D, Sun Z, Li F (2018) Mesoporous TiN microspheres as an efficient polysulfide barrier for lithium–sulfur batteries. J Mater Chem A 6(29):14359–14366. https://doi.org/10.1039/c8ta04920c

Radoor S, Karayil J, Parameswaranpillai J, Siengchin S (2020) Removal of anionic dye Congo red from aqueous environment using polyvinyl alcohol/sodium alginate/ZSM-5 zeolite membrane. Sci Rep 10(1):1–15. https://doi.org/10.1038/s41598-020-72398-5

Rai RS, Palanivel R, David Raja Selvam J (2020) In-situ synthesis and microstructural characterization of AA6061/(TiB2+ TiC) particles in AA6061 aluminium composite. Mater Today Proc 43:2255–2258. https://doi.org/10.1016/j.matpr.2020.12.533

Rakhi RB, Chen W, Cha D, Alshareef HN (2012) Substrate dependent self-organization of mesoporous cobalt oxide nanowires with remarkable pseudocapacitance. Nano Lett 12(5):2559–2567. https://doi.org/10.1021/nl300779a

Ramasamy E, Jo C, Anthonysamy A, Jeong I, Kim JK, Lee J (2012) Soft-template simple synthesis of ordered mesoporous titanium nitride-carbon nanocomposite for high performance dye-sensitized solar cell counter electrodes. Chem Mater 24(9):1575–1582. https://doi.org/10.1021/cm203672g
Ravi Kumar K, Kiran K, Sreebalaji VS (2017) Micro structural characteristics and mechanical behaviour of aluminium matrix composites reinforced with titanium carbide. J Alloys Compd 723:795–801. https://doi.org/10.1016/j.jallcom.2017.06.309
Rawal SP (2001) Metal-matrix composites for space applications. JOM 53(4):14–17. https://doi.org/10.1007/s11837-001-0139-z
Reddy MA, Breitung B, Chakravadhanula VSK, Wall C, Engel M, Kübel C, Powell AK, Hahn H, Fichtner M (2012) CFx derived carbon-FeF_2 nanocomposites for reversible lithium storage. Adv Energy Mater 3(3):308–313. https://doi.org/10.1002/aenm.201200788
Reddy MA, Breitung B, Kiran Chakravadhanula VS, Helen M, Witte R, Rongeat C, Kübel C, Hahn H, Fichtner M (2018) Facile synthesis of C–FeF_2 nanocomposites from CFx: influence of carbon precursor on reversible lithium storage. RSC Adv 8(64):36802–36811. https://doi.org/10.1039/c8ra07378c
Robinson JT, Perkins FK, Snow ES, Wei Z, Sheehan PE (2008) Reduced graphene oxide molecular sensors. Nano Lett 8(10):3137–3140. https://doi.org/10.1021/nl8013007
Rohatgi PK, Ray S, Liu Y (1992) Tribological properties of metal matrix-graphite particle composites. Int Mater Rev 37(1):129–152. https://doi.org/10.1179/imr.1992.37.1.129
Rohatgi PK, Guo RQ, Iksan H, Borchelt EJ, Asthana R (1998) Pressure infiltration technique for synthesis of aluminum—fly ash particulate composite. Mater Sci Eng A 244(1):22–30. https://doi.org/10.1016/S0921-5093(97)00822-8
Rui K, Wen Z, Lu Y, Jin J, Shen C (2015) One-Step solvothermal synthesis of nanostructured manganese fluoride as an anode for rechargeable lithium-ion batteries and insights into the conversion mechanism. Adv Energy Mater 5(7):1–11. https://doi.org/10.1002/aenm.201401716
Sahu MK, Sahu RK (2018) Fabrication of aluminum matrix composites by stir casting technique and stirring process parameters optimization. Adv Cast Technol. https://doi.org/10.5772/intechopen.73485
Salameh C, Moussa G, Bruma A, Fantozzi G, Malo S, Miele P, Demirci UB, Bernard S (2018) Robust 3D boron nitride nanoscaffolds for remarkable hydrogen storage capacity from ammonia borane. Energ Technol 6(3):570–577. https://doi.org/10.1002/ente.201700618
Saleh R, Taufik A (2018) Degradation of methylene blue and congo-red dyes using fenton, photo-fenton, sono-Fenton, and sonophoto-Fenton methods in the presence of iron(II,III) oxide/zinc oxide/graphene (Fe_3O_4/ZnO/graphene) composites. Sep Purif Technol 210:563–573. https://doi.org/10.1016/j.seppur.2018.08.030
Sathishkumar N, Pravinkumar K, Ashwin Prabhu G, Siva M, Sudharsan PL (2020) Investigation of mechanical properties in Al359 hybrid metal matrix composite reinforced with SiC and fly ash. Mater Today Proc. https://doi.org/10.1016/j.matpr.2020.04.024
Sathishkumar P, Deepakaravind V, Gopal P, Azhagiri P (2021) Analysis the mechanical properties and material characterization on magnesium metal matrix nano composites through stir casting process. Mater Today Proc 3–8. https://doi.org/10.1016/j.matpr.2021.01.041
Seo S-D, Lee G-H, Lim A-H, Min K-M, Kim J-C, Shim H-W, Park K-S, Kim D-W (2012) Direct assembly of tin–MWCNT 3D-networked anode for rechargeable lithium ion batteries. RSC Adv 2(8):3315. https://doi.org/10.1039/c2ra00943a
Sharma S, Dwivedi SP (2017) Effects of waste eggshells and SiC addition on specific strength and thermal expansion of hybrid green metal matrix composite. J Hazard Mater 333:1–9. https://doi.org/10.1016/j.jhazmat.2017.01.002
Shatla AS, Hassan KM, Abd-El-Latif AA, Hathoot AA, Baltruschat H, Abdel-Azzem M (2018) Poly 1,5 diaminonaphthalene supported Pt, Pd, Pt/Pd and Pd/Pt nanoparticles for direct formic acid oxidation. J Electroanal Chem 833:231–241. https://doi.org/10.1016/j.jelechem.2018.11.005
Shayan M, Eghbali B, Niroumand B (2020) Fabrication of AA2024–TiO_2 nanocomposites through stir casting process. Trans Nonferrous Met Soc China (English Ed.) 30(11):2891–2903. https://doi.org/10.1016/S1003-6326(20)65429-2

Shen L, Che Q, Li H, Zhang X (2014) Mesoporous NiCo2O4 nanowire arrays grown on carbon textiles as binder-free flexible electrodes for energy storage. Adv Funct Mater 24(18):2630–2637. https://doi.org/10.1002/adfm.201303138

Shirvanimoghaddam K, Hamim SU, Karbalaei Akbari M, Fakhrhoseini SM, Khayyam H, Pakseresht AH, Ghasali E, Zabet M, Munir KS, Jia S, Davim JP, Naebe M (2017) Carbon fiber reinforced metal matrix composites: fabrication processes and properties. Compos Part A Appl Sci Manuf 92:70–96. https://doi.org/10.1016/j.compositesa.2016.10.032

Siddappa PN, Shivakumar BP, Yogesha KB (2021) Synthesis and characterization of cold extruded Al/TiC metal matrix composite produced by stir casting technique. Mater Today Proc. https://doi.org/10.1016/j.matpr.2021.01.757

Singh RP, Singh H (2011) Characterization and comparison of treated and untreated rich hush ash & fly ash for metal matrix composites. J Metall Mater Sci 53:189–196

Singh V, Joung D, Zhai L, Das S, Khondaker SI, Seal S (2011) Graphene based materials: past, present and future. Prog Mater Sci 56(8):1178–1271. https://doi.org/10.1016/j.pmatsci.2011.03.003

Singh H, Singh Brar G, Kumar H, Aggarwal V (2020) A review on metal matrix composite for automobile applications. Mater Today Proc 43:320–325. https://doi.org/10.1016/j.matpr.2020.11.670

Sivashanmugam A, Kumar TP, Renganathan NG, Gopukumar S, Wohlfahrt-Mehrens M, Garche J (2005) Electrochemical behavior of Sn/SnO_2 mixtures for use as anode in lithium rechargeable batteries. J Power Sources 144(1):197–203. https://doi.org/10.1016/j.jpowsour.2004.12.047

Soleimani-Lashkenari M, Rezaei S, Fallah J, Rostami H (2018) Electrocatalytic performance of Pd/PANI/TiO_2 nanocomposites for methanol electrooxidation in alkaline media. Synth Met 235:71–79. https://doi.org/10.1016/j.synthmet.2017.12.001

Stalin B, Ravichandran M, Balasubramanian M, Anand Chairman C, Pritima D, Dhinakaran V (2020) Processing of MMC through conventional sintering and spark plasma sintering process: a review. IOP Conf Ser Mater Sci Eng 988:2020. https://doi.org/10.1088/1757-899X/988/1/012092

Suib SL (2013) New and future developments in catalysis. Elsevier B.V.

Tamaddoni Saray M, Hosseini H (2016) Mesoporous $MnNiCoO_4$@MnO_2 core-shell nanowire/nanosheet arrays on flexible carbon cloth for high-performance supercapacitors. Electrochim Acta 222:505–517. https://doi.org/10.1016/j.electacta.2016.11.003

Tan HT, Sun W, Wang L, Yan Q (2016) 2D transition metal oxides/hydroxides for energy-storage applications. ChemNanoMat 2(7):562–577. https://doi.org/10.1002/cnma.201500177

Thostenson ET, Li C, Chou TW (2005) Nanocomposites in context. Compos Sci Technol 65(3–4):491–516. https://doi.org/10.1016/j.compscitech.2004.11.003

Toor ZS (2017) Applications of aluminum-matrix composites in satellite: a review. J Sp Technol 7(1):6

Tugirumubano A, Go SH, Shin HJ, Santhiyagu Joseph V, Kwac LK, Kim HG (2020) Magnetic and mechanical characterization of Al-MWCNT-Fe-Ni hybrid metal matrix composites. Mater Sci Eng B Solid-State Mater Adv Technol 266:115045. https://doi.org/10.1016/j.mseb.2021.115045

Valenti M, Jonsson MP, Biskos G, Schmidt-Ott A, Smith WA (2016) Plasmonic nanoparticle-semiconductor composites for efficient solar water splitting. J Mater Chem A 4(46):17891–17912. https://doi.org/10.1039/c6ta06405a

Vallés C, Drummond C, Saadaoui H, Furtado CA, He M, Roubeau O, Ortolani L, Monthioux M, Pénicaud A (2008) Solutions of negatively charged graphene sheets and ribbons. J Am Chem Soc 130(47):15802–15804. https://doi.org/10.1021/ja808001a

Vasile BS, Birca AC, Surdu VA, Neacsu IA, Nicoară AI (2020) Ceramic composite materials obtained by electron-beam physical vapor deposition used as thermal barriers in the aerospace industry. Nanomaterials 10(2):370. https://doi.org/10.3390/nano10020370

Védrine JC (2017) Heterogeneous catalysis on metal oxides. Catalysts 7(11). https://doi.org/10.3390/catal7110341

Wang H, Wang X (2013) Growing nickel cobaltite nanowires and nanosheets on carbon cloth with different pseudocapacitive performance. ACS Appl Mater Interfaces 5(13):6255–6260. https://doi.org/10.1021/am4012484

Wang G, Wang B, Wang X, Park J, Dou S, Ahn H, Kim K (2009) Sn/graphene nanocomposite with 3D architecture for enhanced reversible lithium storage in lithium ion batteries. J Mater Chem 19(44):8378. https://doi.org/10.1039/b914650d

Wang B, Luo B, Li X, Zhi L (2012) The dimensionality of Sn anodes in Li-ion batteries. Mater Today 15(12):544–552. https://doi.org/10.1016/S1369-7021(13)70012-9

Wang X, Jiang K, Shen G (2015) Flexible fiber energy storage and integrated devices: recent progress and perspectives. Mater Today 18(5):265–272. https://doi.org/10.1016/j.mattod.2015.01.002

Wang J, Zhou Q, Shao S, Misra A (2016) Strength and plasticity of nanolaminated materials. Mater Res Lett 5(1):1–19. https://doi.org/10.1080/21663831.2016.1225321

Wang F, Li J, Shi C, Zhao N, Liu E, He C, He F (2018) Preparation and mechanical properties of in-situ synthesized nano-$MgAl_2O_4$ particles and $Mg_xAl_{(1-x)}B_2$ whiskers co-reinforced Al matrix composites. Mater Sci Eng A 735:236–242. https://doi.org/10.1016/j.msea.2018.08.054

Wang L, Han J, Kong D, Tao Y, Yang QH (2019a) Enhanced roles of carbon architectures in high-performance lithium-ion batteries. Nano-Micro Lett. 11(1):1–23. https://doi.org/10.1007/s40820-018-0233-1

Wang F, Li J, Shi C, Liu E, He C, Zhao N (2019b) In-situ synthesis of MgAlB4 whiskers as a promising reinforcement for aluminum matrix composites. Mater Sci Eng A 764:138229. https://doi.org/10.1016/j.msea.2019.138229

Wang H, Guo L, Liu K, Song Z, Wu L, Fang M, Li J (2019c) Investigation of magnesium nitrate hexahydrate based phase change materials containing nanoparticles for thermal energy storage. Mater Res Express 6(10):105512. https://doi.org/10.1088/2053-1591/ab386f

Wang S, Yang X, Lee PK, Yu DYW (2021) Mechanically and structurally stable Sb2Se3/carbon nanocomposite as anode for the lithium-ion batteries. J Alloys Compd 874:159859. https://doi.org/10.1016/j.jallcom.2021.159859

Węglowski MS (2018) Friction stir processing—state of the art. Arch Civ Mech Eng 18(1):114–129. https://doi.org/10.1016/j.acme.2017.06.002

Wen R, Huang Z, Huang Y, Zhang X, Min X, Fang M, Liu Y, Wu X (2016) Synthesis and characterization of lauric acid/expanded vermiculite as form-stabilized thermal energy storage materials. Energy Build 116:677–683. https://doi.org/10.1016/j.enbuild.2016.01.023

Wilt DM, Bradshaw G, Gap LN, Abudayyeh OK, Nelson C, Han S, Cox N, Rape A, Landi B, Whipple S (2017) Metal matrix composite solar cell metallization. In: E3S web of conferences, vol 16, p 03001. https://doi.org/10.1051/e3sconf/20171603001

Winter M, Besenhard JO (1999) Electrochemical lithiation of tin and tin-based intermetallics and composites. Electrochim Acta 45(1):31–50. https://doi.org/10.1016/S0013-4686(99)00191-7

Wong EW, Sheehan PE, Lieber CM (1997) Nanobeam mechanics: elasticity, strength, and toughness of nanorods and nanotubes. Science 277(5334):1971–1975. https://doi.org/10.1126/science.277.5334.1971

Wood JL, Badachhape RB, Lagow RJ, Margrave JL (1969) The heat of formation of Poly(carbon monofluoride). J Phys Chem 73(9):3139–3142. https://doi.org/10.1021/j100843a059

Wu Y, Kim GY (2011) Carbon nanotube reinforced aluminum composite fabricated by semi-solid powder processing. J Mater Process Technol 211(8):1341–1347. https://doi.org/10.1016/j.jmatprotec.2011.03.007

Wu Y, Kim GY, Anderson IE, Lograsso TA (2010) Fabrication of A16061 composite with high SiC particle loading by semi-solid powder processing. Acta Mater 58(13):4398–4405. https://doi.org/10.1016/j.actamat.2010.04.036

Xu B, Ma H, Lu Z, Li Z (2015) Paraffin/expanded vermiculite composite phase change material as aggregate for developing lightweight thermal energy storage cement-based composites. Appl Energy 160:358–367. https://doi.org/10.1016/j.apenergy.2015.09.069

Xu Y, Dhainaut J, Rochard G, Dacquin J-P, Mamede A-S, Giraudon J-M, Lamonier J-F, Zhang H, Royer S (2020) Hierarchical porous ε-MnO_2 from perovskite precursor: application to the formaldehyde total oxidation. Chem Eng J 388:124146. https://doi.org/10.1016/j.cej.2020.124146

Xu M, Wang F, Liang X, Shehzad MA, Wu L, Xu T (2021) Poly (5-aminoindole)–modified TiO_2NTs nanocomposites supported palladium as an anode catalyst for enhanced electrocatalytic oxidation of methanol. Electrochim Acta 388:138562. https://doi.org/10.1016/j.electacta.2021.138562

Xue Y, Zheng S, Xue H, Pang H (2019) Metal-organic framework composites and their electrochemical applications. J Mater Chem A 7(13):7301–7327. https://doi.org/10.1039/C8TA12178H

Yan J, Wang Q, Wei T, Fan Z (2014) Recent advances in design and fabrication of electrochemical supercapacitors with high energy densities. Adv Energy Mater 4(4). https://doi.org/10.1002/aenm.201300816

Yang ZH, Wang XY, Liu L, Su XP (2012) Structural, magnetic and electronic properties of FeF2 by first-principle calculation. Trans Nonferrous Met Soc China (English Ed.) 22(2):386–390. https://doi.org/10.1016/S1003-6326(11)61188-6

Yang W, Gao Z, Ma J, Zhang X, Wang J, Liu J (2014) Hierarchical $NiCo_2O_4$@NiO core-shell hetero-structured nanowire arrays on carbon cloth for a high-performance flexible all-solid-state electrochemical capacitor. J Mater Chem A 2(5):1448–1457. https://doi.org/10.1039/c3ta14488g

Yu MF, Files BS, Arepalli S, Ruoff RS (2000) Tensile loading of ropes of single wall carbon nanotubes and their mechanical properties. Phys Rev Lett 84(24):5552–5555. https://doi.org/10.1103/PhysRevLett.84.5552

Yuan C, Li J, Hou L, Zhang X, Shen L, Lou XW (2012) Ultrathin mesoporous $NiCo_2O_4$ nanosheets supported on Ni foam as advanced electrodes for supercapacitors. Adv Funct Mater 22(21):4592–4597. https://doi.org/10.1002/adfm.201200994

Zhang R, Pang H (2021) Application of graphene-metal/conductive polymer based composites in supercapacitors. J Energy Storage 33:102037. https://doi.org/10.1016/j.est.2020.102037

Zhang JZ, Yan Y (2019) Synthesis of mesoporous sphere-like Copper(I) oxide and its enhancement of Congo red photodegradation performance and CO sensing properties. J Taiwan Inst Chem Eng 95:405–415. https://doi.org/10.1016/j.jtice.2018.08.006

Zhang L, Shao J-J, Zhang W, Zhang C, Zheng X, Du H, Yang Q-H (2014) Graphene-based porous catalyst with high stability and activity for the methanol oxidation reaction. J Phys Chem C 118(45):25918–25923. https://doi.org/10.1021/jp508541b

Zhang J, Liu F, Cheng JP, Zhang XB (2015a) Binary Nickel-cobalt oxides electrode materials for high-performance supercapacitors: influence of its composition and porous nature. ACS Appl Mater Interfaces 7(32):17630–17640. https://doi.org/10.1021/acsami.5b04463

Zhang Y, Li L, Su H, Huang W, Dong X (2015b) Binary metal oxide: Advanced energy storage materials in supercapacitors. J Mater Chem A 3(1):43–59. https://doi.org/10.1039/c4ta04996a

Zhang C, Yao D, Yin J, Zuo K, Xia Y, Liang H, Zeng Y-P (2019a) Effects of β-Si3N4 whiskers addition on mechanical properties and tribological behaviors of Al matrix composites. Wear 430–431:145–156. https://doi.org/10.1016/j.wear.2019.05.003

Zhang Y, Sun J, Ma G, Wang Z, Xie S, Jing Y, Jia Y (2019b) Hydrophilic expanded graphite-magnesium nitrate hexahydrate composite phase change materials: understanding the effect of hydrophilic modification on thermophysical properties. Int J Energy Res 43(3):1121–1132. https://doi.org/10.1002/er.4336

Zhang Y, Shi R, Ren J, Dai Y, Yuan Y, Wang Z (2019c) PtFeCu concave octahedron nanocrystals as electrocatalysts for the methanol oxidation reaction. Langmuir 35(51):16752–16760. https://doi.org/10.1021/acs.langmuir.9b03238

Zheng Y, Chen H, Dai Y, Zhang N, Zhao W, Wang S, Lou Y, Li Y, Sun Y (2015) Preparation and characterization of Pt/TiO_2 nanofibers catalysts for methanol electro-oxidation. Electrochim Acta 178:74–79. https://doi.org/10.1016/j.electacta.2015.07.177

Zheng S et al (2017) Transition-Metal (Fe Co, Ni) based metal-organic frameworks for electrochemical energy storage. Adv Energy Mater 7(18):1–27. https://doi.org/10.1002/aenm.201602733

Zheng Y, Zheng S, Xue H, Pang H (2018) Metal-organic Frameworks/graphene-based materials: preparations and applications. Adv Funct Mater 28(47):1–28. https://doi.org/10.1002/adfm.201804950

Zhi M, Xiang C, Li J, Li M, Wu N (2013) Nanostructured carbon-metal oxide composite electrodes for supercapacitors: a review. Nanoscale 5(1):72–88. https://doi.org/10.1039/c2nr32040a

Zhong Y, Xia XH, Shi F, Zhan JY, Tu JP, Fan HJ (2015) Transition metal carbides and nitrides in energy storage and conversion. Adv Sci 3(5). https://doi.org/10.1002/advs.201500286

Zhou W, Zhai C, Du Y, Xu J, Yang P (2009) Electrochemical fabrication of novel platinum-poly(5-nitroindole) composite catalyst and its application for methanol oxidation in alkaline medium. Int J Hydrogen Energy 34(23):9316–9323. https://doi.org/10.1016/j.ijhydene.2009.09.059

Zhou W, Du Y, Ren F, Wang C, Xu J, Yang P (2010) High efficient electrocatalytic oxidation of methanol on Pt/polyindoles composite catalysts. Int J Hydrogen Energy 35(8):3270–3279. https://doi.org/10.1016/j.ijhydene.2010.01.083

Chapter 3
Metal Oxide–Carbon Nanocomposites for Electrochemical Storage

Jyoti Roy Choudhuri

Abstract Metal oxides are one of the interesting candidates for electrochemical energy devices owing to their significant theoretical capacity and promising material as electrode. But the major drawback of lower electrical conductivity and their substantial structural instability during the charge/discharge process are the rapid capacity fading factors. The combination of metal oxides with carbon-based nanostructures give rise to new nanocomposites with the improvement in structural and functional properties to serve the purpose of electrochemical storage application. This book chapter briefly highlights the combination of metal oxides with one-, two-, or three-dimensional (1D, 2D, or 3D) carbon materials. It also discusses the various characteristic (physical, electrical, chemical, and structural properties) features of such unique composites and their application in the field of electrochemical storage. Considerable attention has also been given for the development of easy synthesis methods of these nanocomposites and wide range of possibilities related to their fabrication to achieve a range of applications. Finally, the major challenges and future perspectives related to the synergized nanostructures are summarized. In general, this book chapter will open up new vistas in the field of metal oxide-based nanocomposites with unique and novel futuristic applications in energy storage devices.

Keywords Metal oxides · Electrochemical storage · Carbon · Composite

3.1 Introduction

The shortage in the supply of fossil fuels along with the rise of environmental hazards, are the two alarming concerns of the twenty-first century. The studies on the renewable sources of energy have emerged to a larger extent. These renewable sources have the capability to satisfy the world's energy requirement, but the

J. R. Choudhuri (✉)
BMS Institute of Technology and Management, Bengaluru, Karnataka 560064, India

S. P. Singh et al. (eds.), *Metal Nanocomposites for Energy and Environmental Applications*, Energy, Environment, and Sustainability,
https://doi.org/10.1007/978-981-16-8599-6_3

inevitable problem is the non-spontaneous nature of these energy sources. This motivates the scientific community to shift their attention toward electrochemical storage devices. Electrochemical storage like lithium (or sodium) ion batteries and supercapacitors have gained significant attention and have vast applications from portable electronic devices to electrically driven vehicles. The performance of these devices is profoundly dependent on the characteristics of the electrodes. In this context, it is highly essential to develop novel electrode materials for electrochemical applications.

The high theoretical capacity, good rate performance and long cycle life of lithium-ion batteries and sodium ion batteries are reasons behind its extensive research during the last two decades. But, despite the tremendous effort on the storage systems, they are not sufficient enough for electric vehicles application due to the requirement of higher battery capacity and prolonged stability (Balogun et al. 2016; Bruce et al. 2012; Chen and Shaw 2014). In recent years, there have been tremendous advancements in designing and synthesis of nanostructured electrode materials for energy storage applications. Currently, the commercially available lithium-ion battery involves graphite as anode material, but suffers a drawback of lower theoretical capacity of 372 mAh/g (Gu et al. 2015; Hwang et al. 2018; Sohn et al. 2016a, b; Sohn 2017). On the other hand, due to large size of sodium ion it is difficult to use graphite as anode material (Ge and Fouletier 1988). The carbon based materials used as electrodes in commercialized supercapacitors also suffers from low energy density problems (Frackowiak and Beguin 2001; Stoller et al. 2011). In the context of increased theoretical capacity (Kim et al. 2012; Laruelle et al. 2002; Poizot et al. 2000), metal oxide acquired a significant attention, but suffers a serious drawback of volume expansion during the charging-discharging reaction (Zhang et al. 2008; Jiang et al. 2012). In addition, on account of the lower electrical conductivity of metal oxides, these electrode materials exhibit higher charge transfer resistance, which is unfavourable for output density and rate performance of the battery (Ma et al. 2014; Zhu et al. 2011a).

To resolve the issues related to metal oxides, the development of composites with carbonaceous material is one of the probable solutions. The carbon-based material will allow high electrical conductivity and will be capable of accommodating a good volume change. Thus, the combination of carbonaceous material with metal oxide can lead to enhancement of electrochemical performance with crucial synergistic effects (illustrated in Fig. 3.1) (Seok et al. 2019). This book chapter covers the combination of metal oxides with one-, two-, or three-dimensional (1D, 2D, or 3D) carbon materials. It highlights the various characteristic (chemical, physical, electrical, and structural properties) features of such unique composites and their suitable applications in the field of electrochemical energy devices. The synthesis techniques of these nanocomposites and methodologies related to fabrication to achieve successful applications in different types of energy storage devices are discussed. Lastly, the major challenges and future perspectives related to the synergized nanostructures are briefly concluded.

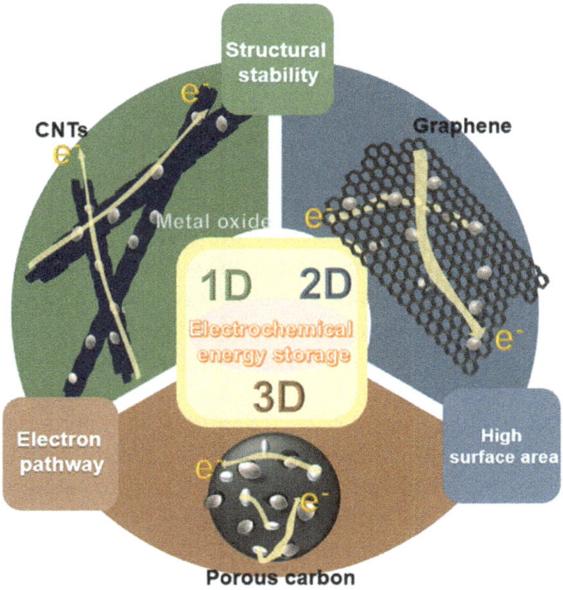

Fig. 3.1 Schematic representation of carbon-based metal-oxide composites of varying dimensions for electrochemical storage device application. Seok et al. 2019 Adapted from Ref. (), copyright permission not required from MPDI

3.2 1D Carbon-Based Metal Oxide Nanocomposite Materials

One-dimensional (1D) carbon-based nanocomposite basically combines the benefits of 1D carbon nanomaterials, which are highly recognized for their superior properties as electrode materials in energy storage applications. The typical formation of nanocomposite is done via decorating the metal oxides homogeneously around the 1D materials like carbon nanotubes (CNTs) or carbon nanofibers (CNFs). The formed nanocomposite displays superior properties due to their unique structure. CNTs have unique anisotropic properties, which facilitates faster transport of axial electron and favours the diffusion of ions with improved rate capability (De las Casas et al. 2012; Goriparti et al. 2014). They also furnish larger surface area for absorption of ions and also influences smooth strain relaxation during the de-absorption of the ions (Landi et al. 2009; Xiong et al. 2013). Thus the compensation of the lower electrical conductivity of metal oxides is achieved by the introduction of CNTs, which promotes a continuous network for the transport of electron (Zhang et al. 2010). A higher electrical conductivity can also be accomplished by relatively less mass loading than the round-shaped nanoparticles leading to a percolation network. The electrical conductivity of Co@Co_3O_4/CNTs at 18 MPa is reported to be improved by 104 times in comparison to pure Co_3O_4 (Park et al. 2013). Thus, metal-oxide/CNT composite empowers higher electrical conductivity with rise in electrochemical performance for energy storage devices (Zhao et al. 2018a). The area of contact between the nanocomposite electrode and the electrolyte is also increased with higher specific surface area of the CNTs. Lastly, the chemical and physical bonding between the CNTs and metal oxides leads to improved mechanical toughness of the nanocomposite in comparison to the pure metal oxide (Zhi et al. 2013). The Co_3O_4/super-aligned CNT (SACNT) composites exhibit 6.4-fold more mechanical stability than the Co_3O_4/Super-P (He et al. 2013). Thus, the superior mechanical strength of Co_3O_4/SACNT allows its suitability as an electrode material in resisting the volumetric changes that take place during the charging/discharging process of the energy storage device. The appropriate fraction of CNT as a filler with metal oxide in the composite material also plays a vital role in faster electronic transport via formation of a percolation network and results in the improvement of performance of the electrochemical energy device (Zhi et al. 2013; He et al. 2010; Higgins et al. 2014). Similarly, the appropriate volume fraction of CNTs in CNT/MnO_2 composite enhance the electrochemical performance of a supercapacitor to a good extent on the basis of percolation theory (Higgins et al. 2014). Thus, the 1D structured Carbon/Metal-Oxide Composite serves as an efficient composite electrode in improving the performance of the energy storage device, with superior mechanical properties and enhanced transfer characteristics of a CNT network.

The synthesis of high-quality CNT/MOs composite materials is one of the challenging tasks for the scientific community. Various approaches have been developed to synthesize CNT/MOs composites. The two major synthetic methods are: (i) in-situ approach, where the growth of CNTs is accomplished within the same process and (ii) ex-situ, or post-growth approach, where the decoration is executed in a different step after the generation of CNTs. But, the major obstacles in the synthesis methods are related to the CNTs. CNTs have a strong tendency towards agglomeration owing to the minimization of the total surface free energy and the lack of suitable functional surface groups leads to poor compatibility with the transition metal oxide (TMO) components (Liu et al. 1998; Datsyuk et al. 2008; Chen et al. 1998; Srivastava et al. 2015). Thus, the activation or functionalization of CNTs is the key step in these synthetic approaches. The conventional functionalization of the CNTs involves the introduction of hydrophilic groups like -COOH and –OH groups by modifying the structure of pristine graphitic CNT. The other methods involve functionalization (via carboxylate ions) of the surface of CNT to facilitate the growth of metal oxide layers (Du et al. 2007) and the use of gel of polymer inducers in the composite material (Yang et al. 2005, 2003; Li et al. 2009). Earlier acid treatment was extensively used to obtain functional groups on the surfaces of CNT (Cao et al. 2011; Liang et al. 2011; Huang et al. 2013), but it leads to poor growth of metal oxide nanostructures on the surface of CNT. The improved methodology formulation includes the treatment of –OH substituted with maleic anhydride, followed by in situ polymerization of styrene. It results in the co-axial growth of polystyrene (PS) layer on CNT, further followed by sulphonation. The obtained material with negative sulfonic groups (–SO_3^-) absorption of positively-charged metal ions (e.g.: Ni^{2+}, Co^{2+}, Zn^{2+}) via electrostatic interactions (Zhou et al. 2016). The ex-situ or building block approach involves the production of two hybrid components separately, depending on the required dimension and morphology. Then, the relevant functional group or linker molecule is added to either the MO nano-particulates (NPs) or the CNTs or both the components. Finally, the linkering agents are combined via various interactive forces (Shirai et al. 2013; Hu et al. 2011; Lu 2007).

In the In-situ approach, the synthesis of one hybrid component is performed in presence of the other, in some cases both the components are generated simultaneously. The major advantage is that the CNTs can provide support and an establishment of continuous amorphous or single-crystalline layers of MO takes place with controlled thickness, or in the form of separate units as NPs, nanorods, or nanobeads. Thus, one component monitors the production of other with suitable size, crystal structure and specific morphology (Shirai et al. 2013; Hu et al. 2011; Yi et al. 2015; Estili et al. 2011) However, the introduction of more binding sites, such as defects and functional groups can be done via acid treatment, ultrasonication, microwave irradiation or g-ray irradiation in the framework of CNTs. The most commonly used approaches to synthesize CNT/MOs composite materials of higher application purpose, are polymer-assist fabrication methods (Deng et al. 2018), hydrothermal methods (Chen et al. 2011a), electrospinning method (Liu et al. 2016), chemical vapor deposition (CVD) method, (Zhu et al. 2011b; Wang et al. 2017) and physical vapor deposition (PVD) method (Wang et al. 2018).

3.3 2D Carbon-Based Metal Oxide Nanocomposite Materials

2D carbon-based metal nanocomposite materials involve the utilization of 2D carbon nanosheet like graphene, having a honeycomb structure with sp^2-hybridized carbon atoms. The unique 2D structure enable several advantages, which plays a key role in enhancing the performance of energy storage devices (Candelaria et al. 2012). The high electrical conductivity and its inherent charge carrier mobility facilitates the kinetics of the electrochemical reaction (Geim and Novoselov 2010; Chen and Tao 2009). Graphene has higher surface area with considerably larger surface-to-volume ratio (Pumera 2011). Thus, it has several active sites for the reaction and favours diffusion of ions with more effectivity. The higher Young's modulus and tensile strength (Lee et al. 2008) provides better cyclic performance and higher stability of the device. However, they suffer from poor volumetric energy density and considerably lowers the rate performance (Wu et al. 2012). The process of agglomeration and re-stacking of graphene nanosheets in the course of electrochemical chemical reaction is the reason behind its declined performance.

As mentioned in the earlier section, electrochemical devices with metal oxide as electrode materials exhibit poor performance due to lower electrical and ionic conductivity. The volume change in the course of the charging/discharging process leads to severe mechanical deformation of the metal oxide layer. Thus, the concept of graphene-based metal oxide nanocomposite is reported to be a superior electrode material for energy storage application (Lian et al. 2010). The nanocomposite material will have superior conductivity and will overcome the induced strain due to volume change during the process of charging/discharging. The modified surface area of graphene in the material reinforces the structural stability of the material and also enhances the electrochemical performance of the energy storage devices.

In the case of graphene/Fe_3O_4 composite material, the growth of Fe_3O_4 nanoparticles on the graphene sheets shows uniformity without any aggregation (Su et al. 2011). This reflects a strong interaction between the sheet and the metal oxide nanoparticles. The graphene-wrapped Fe_3O_4 composite material is reported to exhibit a porous texture of the graphene, which allows it to function as a flexible electrode in lithium-ion battery (Zhou et al. 2010). The porous graphene matrix and the metal oxide confined between the layers of graphene, controls the presence of the metal oxide between the graphene layers. Consequently, the re-stacking of graphene nanosheets is restricted. Furthermore, the composite material showed good structural stability after 30 cycles with very limited change in the average size (Zhou et al. 2010). 2D porous graphene and metal oxide composite material exhibits reduced path length for the ion diffusion, extended surface area for smooth electrolyte penetration and minimizes the strain generated due to volume change during the charging/discharging process (Peng et al. 2018; Zhao et al. 2011). It is found to be comparatively more effective than the non-porous metal oxide composite.

The manufacture of graphene-based nanocomposites usually requires bulk quantities of graphene sheets with a preference towards modified surface structure (Huang et al. 2011; Liang et al. 2014; Tan et al. 2013; Schwab et al. 2012). Thus, the top-down approaches such as, chemical or thermal reduction of the derivatives of graphite, i.e., graphite oxide (GO) (Hamwi and Marchand, 1996) and graphite fluoride (Okotrub et al. 2009) are strategically more efficient in this context. These techniques are capable of yielding low-cost large amounts of graphene-like sheets (Kim et al. 2010), but with defects. They can be easily processed and fabricated to obtain a variety of materials. The primary focus in the synthesis methods is to produce highly reduced graphene oxide (HRG) (Moon et al. 2011; Chua and Pumera 2014) or chemically modified graphene (CMG) (Chen et al. 2012; Barg et al. 2014).

The composite material consisting of reduced graphene oxide and Fe_2O_3 nanocomposite (RGO/Fe_2O_3) is synthesized by undergoing solvothermal reaction in presence of iron precursor and GO. This leads to synchronous deposition of iron oxide nanoparticles and in situ chemical reduction of GO in absence of any reducing agent (Modafferi et al. 2018). The low-cost synthesis of graphene/Fe_3O_4 nanocomposite is reported by Su et. al from graphite oxide (GO) and iron chloride ($FeCl_3.6H_2O$) in the presence of hydrazine hydrate (Su et al. 2011). In this in situ approach, the reduction of GO and the deposition of Fe_3O_4 nanoparticles on GN sheets takes place simultaneously. The nanostructured iron oxides are directly dispersed on the graphene nanosheet without any additional molecular linkers. The SnO_2 and graphene nanocomposite is synthesized by in situ chemical method, involving the treatment of graphene oxide nanosheets suspension with $SnCl_2.H_2O$

solution (Du et al. 2010). The transformation of the attached Sn^{2+} ions with the functional groups to SnO_2 takes place and simultaneously the graphene nanosheets are also converted. The anchored SnO_2 prevents re-stacking of the graphene sheets. Similarly, RGO/Co_3O_4 was synthesized via in situ reduction of cobalt salt with graphene oxide (Shen et al. 2010). But the recently reported microwave-assisted technique with low RGO content is reported to produce nanocomposite material with better electrochemical performance and improved thermal stability (Mussa et al. 2019).

3.4 3D Carbon-Based Metal Oxide Nanocomposite Materials

For an electrochemical storage device, the performance is analysed on the basis of specific capacity, rate capability, energy density, and power density. These properties can be enhanced by maximising the quantity and quality of the embedded active material (Candelaria et al. 2012). But the increased active material quantity leads to the deterioration of the electrochemical performance (Chen et al. 2014). In this context, 3D carbon-based metal nanocomposites have drawn attention owing to their unique properties. The 3D porous carbon with tuneable pore size, suitable wall thickness, large pore volume as well as surface area and unique inner connective network among the pore channel made them ideal candidate to function as a matrix of an active material (Chen et al. 2014). The microporous, mesoporous and microporous characteristic properties of hierarchical porous carbon, is utilized to accomplish better electrochemical performance. Their function includes ion-buffering, channelizing ions on shorter pathways, ion confinement and resisting volume change during the electrochemical reaction. The metal-oxide nanomaterials can be well inserted in the porous structure and establish an interconnection among the pores. This effectively influence the penetration of electrolyte and diffusion of ions, results in an improved structural stability (Guo et al. 2014). On the other hand, the larger surface area through porous channels resulting in a better contact of electrolyte solution with broadened active sites. This also enables higher loading of metal oxide, which enhances the energy and power density (Chang et al. 2012).

The suitability of the hierarchical porous graphitic carbon to serve as an electrode material in case of electrochemical capacitors is reported by Wang's group (Wang et al. 2008). It's larger surface area with appropriate pore volume made it relevant for electrochemical energy devices. The three different porous architecture (macro-, meso-, micropores) with variation in the diameter enable it to function as an ion-buffering reservoir, transporting electrolyte in shorter pathways through the wall and supporting easy confinement of the ions. The MnO/C composite, where the MnO_2 nanoparticles are embedded inside the porous carbon reported to serve as a better anode material in lithium-ion battery than the pure MnO (Cao et al. 2017). The composite structure influences the easy penetration of electrolyte through the internally connected pores and the layers aligned with effective shortening of ion diffusion path. The volume change occurring due to lithiation/delithiation is restricted by the presence of MnO_2 in the carbon matrix. Another composite (Fe_3O_4@C) material where Fe_3O_4 confined in a framework of nanoporous carbon reported by Zhang et. al to function as an anode in lithium-ion battery application by means of self-assembled unit of $Fe(NO_3)_3$/resol/F127 (Zhang et al. 2015a). The morphology of the composite can be varied with variation in the amount of Fe_3O_4 (Fe_3O_4@C-1, Fe_3O_4@C-2, and Fe_3O_4@C-3) contented during the self-assembly process. The variation in the insertion of Fe_3O_4 in the carbon matrix leads to alteration in the pore volume and surface area of the composite material. The composite material also lowers the possibility of Fe_3O_4 to aggregate or separate from the collector. This facilitates larger the loading of Fe_3O_4 into the carbon framework, leading towards extended surface area and pore volume (Yoon et al. 2011). Similarly, SnO_2@CMK-3 composites with ultrafine SnO_2 particles uniformly embedded inside the tubular mesoporous carbon structure shows its suitability to function as an electrode material in lithium-ion batteries (Han et al. 2012).

Various techniques are reported regarding the synthesis of 3D carbon-based metal oxide nanocomposite material with suitable morphology. Petkovich et. al reported the synthesis of nanostructured TiO_2/C composites via combining of colloidal crystal templating, which consist of titanium isopropoxide (source of TiO_2), phenol–formaldehyde sol (source of amorphous carbon) with a block copolymeric surfactant template. Then, followed by thermal process, which leads to the formation of porous structure with TiO_2 nanocrystals inside the carbon matrix (Petkovich et al. 2014). The 3D porous MnO/C mesoscale hybrids are reported to be synthesized by a facile and scalable route involving polyvinyl alcohol-assisted aqueous precipitation and followed by thermal decomposition of precursors (MnO_2 and $H_2C_2O_4.2H_2O$ solution mixture) (Cao et al. 2017). Kang et. al has reported the synthesis of Fe_3O_4@C composite by impregnating the iron oxide precursors in mesocellular carbon foam, followed by heat treatment (Zhang et al. 2015a). This results in the fabrication of Fe_3O_4 nanocrystals with suitable size inside the mesoporous structure. Qiao et al. reported the synthesis of ordered SnO_2/CMK-3 nanocomposites using sonochemical technique (Qiao et al. 2011), which involves the surface modification of CMK-3, treating with $SnCl_2$ solution and followed by ultrasonic irradiation at room temperature.

3.5 Applications

3.5.1 Lithium-Ion Batteries

The commercially available lithium-ion batteries consist of $LiCoO_2$, $LiMn_2O_3$ as cathode and graphite as anode (Yun et al. 2018). However, owing to the low theoretical capacity of graphite, there are many studies where the graphite electrodes are replaced by metal oxides (Fe_2O_3, SnO_2, Fe_3O_4, Co_2O_3, etc.) (Poizot et al.

2000). Despite the advantage of high theoretical capacity, the metal-oxide-based LIB endures from lower electric conductivity and structural instability (Poizot et al. 2000). Thus, carbon-based composites are introduced as anodes, where the carbon material provides higher electrical conductivity, enabling a synergistic effect involving higher theoretical capacity of metal oxide.

As a promising anode material 1D structured carbon nanocomposite like CNT/metal-oxide-based (Zhou et al. 2012) or carbon-fiber/metal-oxide-based (Lv et al. 2015; Lyu et al. 2016) are used for LIBs. The CNTs and networks function as a buffer later to resist the mechanical stress due to volume expansion of metal oxides. They also facilitate conductive pathways, leading to improvement of the electrical conductivity of the composite (Qiu et al. 2015) and also inhibit the aggregation of nano-sized metal oxides which are dispersed on the composite material. Luo et al. reported the use of composite material, composed of Mn_3O_4 with well-aligned carbon nanotubes as cathode in LIB (Luo et al. 2014). The SACNT film plays a crucial role in serving a conduction path for migration of electron and serves as a matrix for the suitable dispersion of Mn_3O_4 nanoparticles. The main advantage of the 1D structured carbon composite is that there is no requirement of a conductive agent (e.g., super-P), a binding agent or a substrate required for cathode. The composite material is reported to have a high capacity (~342 mAh/g) with high current density. The composite reported by Zhou et. al, consisting of single-walled carbon nanotubes (SWCNTs) and Fe_2O_3, shows high capacity (1243 mAh/g) with good current density (Zhou et al. 2012). Meanwhile, the consequence of volume expansion during dilithiation process leads to retention in the capacity at an elevated energy density for multiple cycles. Li et al. reported the betterment of electrical conductivity and structural stability with multi-walled CNTs (MWCNTs) as a matrix with Fe_2O_3 nanoparticles (Li et al. 2015).

2D structure graphene or graphene oxide-based composites are extensively studied as electrode material for energy storage application. Liu et al. reported a composite, consisting of Co_3O_4 and graphene having a theoretical capacity twice that of graphite (Li et al. 2014). It exhibits a multifunctional structure consisting of Co_3O_4 nanowires developed on a 2D graphene membrane. This material can eliminate the requirement of copper current collector and the necessity of binder to enhance the energy density. The material also possesses a good stability for multiple cycles with high capacity. Hu et al. addressed a composite material with reduced graphene oxide (rGO), surrounded by Co_3O_4 encapsulated by porous nanofibers (Hu et al. 2017). The composite can suppress large volume change during cyclic processes and the interconnective network between the constituents results in the acceleration of the mobility of electrons and lithium ions. There are also other reported studies where metal-oxide nanoparticles are implanted in between reduced graphene oxide sheets (Park et al. 2016) or on a graphene surface (Lin et al. 2014) or nanorods (Tao et al. 2012) with improved battery performance. However, Fu et al. demonstrated the formation of composite material by the utilization of Chromium (II) oxide and carbon sheets with glycine as a carbon precursor (Fu et al. 2015), with good electrochemical performance.

The metal oxide and 3D carbon-based composite along with porous structure is reported to exhibit extended specific surface area with good structural stability during the cyclic performance. The composite of MnO dispersed in a 3D porous carbon network, is demonstrated to have good ionic and electrical conductivities via the 3D carbon network and can buffer the large change in volume during the charging/discharging process of the battery (Wang et al. 2014). The electrode material can obtain a capacity of 560.2 mA/g at a high current density of 4 A/g. As mentioned in the earlier section, a composite material (CMK-3 or CMK-5) of embedded SnO_2 nanoparticles in a carbon matrix made of ordered mesoporous carbon can enhance electrochemical performance. The thin-walled CMK-5 is capable of effectively trapping more SnO_2 than CMK-3 and demonstrated to offer a capacity of 1039 mAh/g after a completion of 100 cycles at a current density of 200 mA/g (Han et al. 2012).

3.5.2 Sodium-Ion Batteries

In recent times, sodium (Na) ion Battery (SIB) has gained considerable attention, owing to its higher abundance and lesser cost of sodium in comparison to lithium. Meanwhile, they suffer from various disadvantages like lower standard electrochemical potential, comparatively larger ionic radius with slow dynamic response and lower energy density (Wang et al. 2014, 2019). The large size of sodium also directs to a large change in the volume of the electrode material. Many studies have reported that the composite material with CNTs as matrix, where the metal oxide nanoparticles like SnO_2, Co_3O_4 are dispersed can be used in SIBs for improved performance (Ma et al. 2018; Wang et al. 2013). The CNTs in the composite material are capable of retarding the battery capacity by buffering the change in volume of metal oxide during the cyclic process. Co_3O_4/CNT composite reported to maintain a capacity at 403 mAh/g even after the completion of 100 cycles (Rahman et al. 2015).

The composite consisting of 2D based graphene as a matrix with uniform dispersion and bonding of the metal oxide like Fe_2O_3 nanoparticles can function as an electrode material in SIBs (Jian et al. 2014). The homogeneously distributed Fe_2O_3 nanoparticles on the reduced graphene oxide surface is also reported to exhibit good electrochemical performance with a capacity of 289 mAh/g at 50 mA/g after 50 cycles. The reduced graphene with large specific area enables better contact between sodium ion and the composite material, resulting to a smooth charge transfer reaction with improved electrical conductivity and also enables alleviation of the volume change of the nanoparticles (Liu et al. 2015). The SnO_2/rGO composite is reported to maintain a stable solid electrolyte interface (SEI) during the alloying/dealloying reaction with Na-ion for a charge/discharge process, which leads to improved battery performance (Zhang et al. 2015b). Su et al. showed that SnO_2@graphene nanocomposite exhibits elevated reversible specific capacity of beyond 700 mA h/g with excellent cyclability, and significantly raises the rate performance (Su et al. 2013).

For 1D or 2D structured carbon-based metal oxide nanocomposite, there is a possibility of high contact resistance due to small size of CNTs or graphene. To resolve this issue, Zhao et al. synthesized SnO_2@ carbonized eggshell membrane (CEM) composites, involving the anchoring of SnO_2 nanosheets on a carbonized eggshell membrane (CEM) with a porous architecture as a matrix (Zhao et al. 2018b). The porous 3D architecture improves electronic conductivity and also lead to substitution of copper substrate required for conventional battery electrode (cathode). The porous framework also promotes faster diffusion of electrolyte, with reduced volume change in the metal oxide. Thus, the battery demonstrates improved performance with a good reversible capacity of 656 mA h/g at 0.1 A/g and exhibits excellent cycling stability with better rate capability.

3.5.3 Supercapacitors

Supercapacitors are basically used as energy storage devices that can store and release energy through reversible adsorption and desorption of ions at the electrode–electrolyte interface. They have high power density, with rapid charging and discharging facility (Borenstein et al. 2017). The commercially available carbonaceous electrodes exhibit good cyclic stability, but due to low-capacitance achieve a low energy density (Zhi et al. 2013). On the other hand, low electrically conductive metal oxide electrodes show better charge transfer and sheet resistance (Bélanger et al. 2008). Furthermore, a large reduction in capacitance is observed at a high current density. The volume change during the charging/discharging process led to loss of contact between the electrode and electrolyte, which lowers cyclic stability. Thus, the formation of carbon/metal oxide nanocomposite, enabling a synergistic effect to achieve high energy density is identified to be one of the solutions. 1D structured ZnO/MWNT (Sankapal et al. 2016), NiO/CNTs (Yu et al. 2016), and MnO_2@MWCNT composite fibres (Choi et al. 2016; Shi et al. 2017) are reported to demonstrate better electrical conductivity, good cyclic stability, and shorter pathways for ionic owing to the 1D structure of carbon nanotubes.

In the earlier section, it is seen that graphene oxide and reduced graphene oxide with number of functional groups on the surface, leads to easy synthesis of metal oxide composite. The existing functional groups can function as a redox centre and may get involved in redox reaction. This also contributes to pseudo-capacitance. Xiang et al. showed that the composite involving Co_3O_4 nanoparticles dispersed on reduced graphene can function as an electrode material, where the major contribution of pseudo-capacitance comes from metal oxide (Xiang et al. 2013). The electron transfer between the metal oxide nanoparticles becomes facile due to the presence of reduced graphene matrix, with improved energy density and power density. Similarly, better performance of supercapacitor is observed by the utilization of a composite material consisting of Co_3O_4 nanoparticles and reduced graphene oxide in the form of rolled structure (Zhou et al. 2011) The ionic bonding between the functional groups of graphene oxide and Co^{2+} ions facilitate better electrical conductivity of the electrodes and favours 93% specific capacitance retention after many cycles at a scan rate of 20 mV/s.

The improvement of supercapacitor performance requires the storing of more capacitance with more loading of active materials. Li et al. demonstrated composite with Fe_3O_4 nanoparticles distributed on hollow porous graphitized carbon with a double shell can exhibit excellent electrochemical performance (Li et al. 2016). The hollow porous carbon with larger specific surface area and enhanced electrical conductivity of the metal oxide, can function as a matrix with more loading of uniformly dispersed Fe_3O_4 nanoparticles on the porous architecture. 3D porous architecture in the composite material facilitates easy penetration of electrolyte through the Fe_3O_4 nanoparticles and enhances the cyclic stability with the facile redox reaction at the active sites. Other 3D composite materials like Co_3O_4 nanowires developed on a 3D matrix containing graphene foam or a composite of MnO_2 and CNTs in a 3D sponge architecture can also be utilized for supercapacitor applications (Dong et al. 2012; Chen et al. 2011b).

3.6 Conclusion and Future Prospects

Carbon-based metal oxide nanocomposites have attracted immense attention for energy storage applications owing to their unique structure and excellent properties in the scientific community. In the composite material-based electrodes, the metal oxide component enhances the specific capacitance and energy density. The carbon nanostructure increases the electronic conductivity with good rate capability and provides mechanical stability. In addition, the carbon nanostructures with different dimensions provide unique features in the microstructure and properties of the composite material. The individual nanostructures exhibit particular advantages as follows:

- The 1-D carbon nanostructures provide higher specific area and can support the charge transport properties, which is highly beneficial for the redox reactions taking place in the metal oxide layers. It also effectively influences the homogeneous dispersion of active nanoparticles in the matrix with accelerated electronic/ionic transportation and restricts the aggregation of active nanoparticles for the cyclic process.
- The 2-D carbon nanostructures incorporate superior features such as high surface area, greater electronic conductivity and better mechanical strength. These properties develop attractive features in the composite material with profound application in energy-storage devices.
- The 3D porous carbon favours tuneable pore size with larger pore volume and surface area. It also provides an inner connective network via inserting the metal oxide in the porous architecture for the composite material with higher mechanical stability during the cyclic process.

In the composite material, the carbon nanostructure provides physical support to the metal oxides and its structural features manipulate the architecture of the whole material. The high specific surface area of the carbon empowers the loading of metal oxides to a maximum extent and also helps to accomplish a large area of the carbon-metal interface. Smaller dimension with optimum porosity leads to shortening of the ion diffusion distance towards the electrode surface. The presence of the carbon structure revives the electronic conductivity in the composite material to increase the rate capability of the electrochemical devices. It also enhances the intimate contact between the carbon nanostructure and the current collector to minimize the interfacial resistance to a greater extent. The enhanced interfacial carbon-oxide area and controlled thickness of the metal oxide layer are the key factors to govern the charge transfer capability of the composite electrode.

Despite the substantial advantages, there are still multiple challenges in the synthetic techniques and application of carbon-based nanocomposites. The synthetic methods are complex and usually of high cost which can be a barrier for the large-scale and commercial application of the composite materials. The overall specific capacity of the electrode materials is constrained due to the poor electrochemical activity of the carbon structure. The lower density of the carbon component also limits the volumetric energy density of the carbon-based nanocomposites. So, there lies adequate scope for the development of carbon-based nanocomposites in designing facile synthesis, increasing electrical conductivity and improving structural stability as electrode material. In many cases, design strategies seem to be complicated and are quite time consuming, which can soar as barriers for scaling up production. Thus, simplified strategies are required to develop for large-scale production with proper cost effectiveness and in an eco-friendly way. In addition to that, the electrochemical performance of the carbon-based composite electrodes can be upgraded by involving defect engineering strategies and heteroatom doping. In case of LIBs or NIBs with carbon-based MO composite electrodes, large irreversible capacity is usually noticed during the first cycle due to the formation of solid electrolyte interface (SEI) layer for the first discharge reaction. This issue requires more focus and efforts like execution of artificial SEI layer protection, prelithiation treatment etc. Another issue is the high and sloping voltage plateau of MOs, which affects the overall energy output of an electrochemical device. The structural design of the carbon-based composites gives utmost attention towards increasing the overall specific capacity with better cycling stability, and rate capability. However, more attention should be given on practical aspects like volumetric energy density, initial coulombic efficiency and power density. Thus, the rational structure of the material can provide good electrochemical performance and can solve various practical application needs.

It is well understood that the scientific community would focus on comprehensive research regarding the structural design and the establishment of fundamental relationship between composite structure and electrochemical performance, which may tune the next generation electrode design and promote improved carbon-MO composite for energy storage devices. However, most of the anode materials employed in the commercialized battery in the present time doesn't comprise such fanciful structures. It is feasible that those aspects could be employed for niche applications in the field where there is extreme need of high-performance batteries and high gravimetric energy density is of utmost priority. The reality of such developing field is not so far away, considering the amount of advancement the scientific community have achieved in simplifying the fabrication protocols with the betterment of properties and performances.

References

Balogun M-S, Luo Y, Qiu W, Liu P, Tong YJC (2016) A review of carbon materials and their composites with alloy metals for sodium ion battery anodes. Carbon 98:162–178

Barg S, Perez FM, Ni N, do Vale Pereira P, Maher RC, Garcia-Tunon E, Eslava S, Agnoli S, Mattevi C, Saiz E (2014) Mesoscale assembly of chemically modified graphene into complex cellular networks. Nat Commun 5(1):1–10

Bélanger D, Brousse T, Long J (2008) Manganese oxides: battery materials make the leap to electrochemical capacitors. J Electrochem Soc 17(1):49

Borenstein A, Hanna O, Attias R, Luski S, Brousse T, Aurbach D (2017)Carbon-based composite materials for supercapacitor electrodes: a review. J Mater Chem A 5(25):12653–12672

Bruce PG, Freunberger SA, Hardwick LJ, Tarascon J-M (2012) Li–O 2 and Li–S batteries with high energy storage. Nat Mater 11(1):19–29

Candelaria SL, Shao Y, Zhou W, Li X, Xiao J, Zhang J-G, Wang Y, Liu J, Li J, Cao GJ (2012) Nanostructured carbon for energy storage and conversion. Nano Enegy 1(2):195–220

Cao X, Shi Y, Shi W, Lu G, Huang X, Yan Q, Zhang Q, Zhang H (2011) Preparation of novel 3D graphene networks for supercapacitor applications. Small 7(22):3163–3168

Cao Z, Shi M, Ding Y, Zhang J, Wang Z, Dong H, Yin Y, Yang S (2017) Lotus-root-like MnO/C hybrids as anode materials for high-performance lithium-ion batteries. J Chem Phys C 121(5):2546–2555

Chang P-y, Huang C-h, Doong R (2012) Ordered mesoporous carbon–TiO_2 materials for improved electrochemical performance of lithium ion battery. Carbon 50(11):4259–4268

Chen L, Shaw LL (2014) Recent advances in lithium–sulfur batteries. J Power Sources 267:770–783

Chen F, Tao N (2009) Electron transport in single molecules: from benzene to graphene. Acc Chem Res 42(3):429–438

Chen J, Hamon MA, Hu H, Chen Y, Rao AM, Eklund PC, Haddon R (1998) Solution properties of single-walled carbon nanotubes. Science 282(5386):95–98

Chen Y-J, Xiao G, Wang T-S, Ouyang Q-Y, Qi L-H, Ma Y, Gao P, Zhu C-L, Cao M-S, Jin H-B (2011a) Porous Fe_3O_4/carbon core/shell nanorods: synthesis and electromagnetic properties. J Phys Chem C 115(28):13603–13608

Chen W, Rakhi R, Hu L, Xie X, Cui Y, Alshareef HN (2011b) High-performance nanostructured supercapacitors on a sponge. Nano Lett 11(12):5165–5172

Chen L, Hernandez Y, Feng X, Müllen K (2012) From nanographene and graphene nanoribbons to graphene sheets: chemical synthesis. Angewandte Chemie 51(31):7640–7654

Chen Y, Song B, Li M, Lu L, Xue J (2014) Fe_3O_4 nanoparticles embedded in uniform mesoporous carbon spheres for superior high-rate battery applications. Adv Functional Mater 24(3):319–326

Choi C, Sim HJ, Spinks GM, Lepró X, Baughman RH, Kim SJ (2016) Elastomeric and dynamic MnO_2/CNT core–shell structure coiled yarn supercapacitor. Adv Energy Mater 6(5):1502119

Chua CK, Pumera M (2014) Chemical reduction of graphene oxide: a synthetic chemistry viewpoint. Chem Soc Rev 43(1):291–312

Datsyuk V, Kalyva M, Papagelis K, Parthenios J, Tasis D, Siokou A, Kallitsis I, Galiotis C (2008) Chemical oxidation of multiwalled carbon nanotubes. Carbon 46(6):833–840

De las Casas C, Li W (2012) A review of application of carbon nanotubes for lithium ion battery anode material. J Power Sources 208:74–85

Deng M, Qi J, Li X, Xiao Y, Yang L, Yu X, Wang H, Yuan B, Gao Q (2018) MoC/C nanowires as high-rate and long cyclic life anode for lithium ion batteries. Electrochimica Acta 277:205–210

Dong X-C, Xu H, Wang X-W, Huang Y-X, Chan-Park MB, Zhang H, Wang L-H, Huang W, Chen PJ (2012) 3D graphene–cobalt oxide electrode for high-performance supercapacitor and enzymeless glucose detection. ACS Nano 6(4):3206–3213

Du N, Zhang H, Chen B, Ma X, Liu Z, Wu J, Yang D (2007) Porous indium oxide nanotubes: layer-by-layer assembly on carbon-nanotube templates and application for room-temperature NH_3 gas sensors. Adv Mater 19(12):1641–1645

Du Z, Yin X, Zhang M, Hao Q, Wang Y, Wang T (2010) In situ synthesis of SnO_2/graphene nanocomposite and their application as anode material for lithium ion battery. Electochem Commun 64(19):2076–2079

Estili M, Kawasaki A, Pittini-Yamada Y, Utke I, Michler J (2011) In situ characterization of tensile-bending load bearing ability of multi-walled carbon nanotubes in alumina-based nanocomposites. J Mater Chem 21(12):4272–4278

Frackowiak E, Beguin FJC (2001) Carbon materials for the electrochemical storage of energy in capacitors. Cabon 39(6):937–950

Fu Y, Gu H, Yan X, Liu J, Wang Y, Huang J, Li X, Lv H, Wang X, Guo J (2015) Chromium (III) oxide carbon nanocomposites lithium-ion battery anodes with enhanced energy conversion performance. Chem Eng J 277:186–193

Ge P, Fouletier MJ () Electrochemical intercalation of sodium in graphite. Solid State Ionics 28:1172–1175

Geim AK, Novoselov KS (2010) The rise of graphene. In: Nanoscience and technology: a collection of reviews from nature journals World Scientific, pp 11–19

Goriparti S, Miele E, De Angelis F, Di Fabrizio E, Zaccaria RP, Capiglia C (2014) Review on recent progress of nanostructured anode materials for Li-ion batteries. J Power Sources 257:421–443

Gu Y, Wu A, Sohn H, Nicoletti C, Iqbal Z, Federici JF (2015) Fabrication of rechargeable lithium ion batteries using water-based inkjet printed cathodes. J Manuf Process 20:198–205

Guo S, Lu G, Qiu S, Liu J, Wang X, He C, Wei H, Yan X, Guo Z (2014) Carbon-coated MnO microparticulate porous nanocomposites serving as anode materials with enhanced electrochemical performances. Nano Enegy 9:41-49

Hamwi A, Marchand V (1996) Some chemical and electrochemical properties of graphite oxide. J Phys Chem Solids 57(6–8):867–872

Han F, Li W-C, Li M-R, Lu A-H (2012) Fabrication of superior-performance SnO_2@ C composites for lithium-ion anodes using tubular mesoporous carbon with thin carbon walls and high pore volume. J Mater Chem 22(19):9645–9651

He Y, Huang L, Cai J-S, Zheng X-M, Sun S-G (2010) Structure and electrochemical performance of nanostructured Fe_3O_4/carbon nanotube composites as anodes for lithium ion batteries. Electrochimica Acta 55(3):1140–1144

He X, Wu Y, Zhao F, Wang J, Jiang K, Fan S (2013) Enhanced rate capabilities of Co 3 O 4/carbon nanotube anodes for lithium ion battery applications. J Mater Chem A 1(37):11121–11125

Higgins TM, McAteer D, Coelho JCM, Sanchez BM, Gholamvand Z, Moriarty G, McEvoy N, Berner NC, Duesberg GS, Nicolosi V (2014) Effect of percolation on the capacitance of supercapacitor electrodes prepared from composites of manganese dioxide nanoplatelets and carbon nanotubes. ACS Nano 8(9):9567–9579

Hu Y, Guo C (2011) Applications: carbon nanotubes and carbon nanotubes/metal oxide heterostructures: synthesis, characterization and electrochemical property. Carbon Nanotubes 3–34

Hu R, Zhang H, Bu Y, Zhang H, Zhao B, Yang C (2017) Porous Co_3O_4 nanofibers surface-modified by reduced graphene oxide as a durable, high-rate anode for lithium ion battery. Electrochimica Acta 228:241–250

Huang X, Yin Z, Wu S, Qi X, He Q, Zhang Q, Yan Q, Boey F, Zhang H (2011) Graphene-based materials: synthesis, characterization, properties, and applications. Small 7(14):1876–1902

Huang L, Chen D, Ding Y, Feng S, Wang ZL, Liu M (2013) Nickel–cobalt hydroxide nanosheets coated on $NiCo_2O_4$ nanowires grown on carbon fiber paper for high-performance pseudocapacitors. Nano Lett 13(7):3135–3139

Hwang K, Sohn H, Yoon SJ (2018) Mesostructured niobium-doped titanium oxide-carbon (Nb-TiO_2-C) composite as an anode for high-performance lithium-ion batteries. J Power Sources 378:225–234 (2018).
Jian Z, Zhao B, Liu P, Li F, Zheng M, Chen M, Shi Y, Zhou H (2014) Fe_2O_3 nanocrystals anchored onto graphene nanosheets as the anode material for low-cost sodium-ion batteries. Chem Commun 50(10):1215–1217
Jiang J, Li Y, Liu J, Huang X, Yuan C, Lou XW (2012) Recent advances in metal oxide-based electrode architecture design for electrochemical energy storage. Adv Mater 24(38):5166–5180
Kim H, Abdala AA, Macosko C (2010) Graphene/polymer nanocomposites. Macromolecules 43(16):6515–6530
Kim H, Seo D-H, Kim H, Park I, Hong J, Park K-Y, Kang KJ (2012) Multicomponent effects on the crystal structures and electrochemical properties of spinel-structured M_3O_4 (M= Fe, Mn, Co) anodes in lithium rechargeable batteries. Chem Mater 24(4):720–725
Landi BJ, Ganter MJ, Cress CD, DiLeo RA, Raffaelle RP (2009) Carbon nanotubes for lithium ion batteries. Energy Envion Sci 2(6):638–654
Laruelle S, Grugeon S, Poizot P, Dolle M, Dupont L, Tarascon JJ (2002) On the origin of the extra electrochemical capacity displayed by MO/Li cells at low potential. J Electro Chem Soc 149(5):A627
Lee C, Wei X, Kysar JW, Hone J (2008) Measurement of the elastic properties and intrinsic strength of monolayer graphene. Science 321(5887):385–388
Li, J., Ding, S., Zhang, C., Yang, Z.J.P.: Synthesis of composite eccentric double-shelled hollow spheres. 50(16), 3943–3949 (2009).
Li L, Zhou G, Shan X-Y, Pei S, Li F, Cheng H-M (2014) Co_3O_4 mesoporous nanostructures@ graphene membrane as an integrated anode for long-life lithium-ion batteries. J Power Sources 255:52–58
Li X, Gu H, Liu J, Wei H, Qiu S, Fu Y, Lv H, Lu G, Wang Y, Guo Z (2015) Multi-walled carbon nanotubes composited with nanomagnetite for anodes in lithium ion batteries. RSC Adv 5(10):7237–7244
Li X, Zhang L, He GJC (2016) Fe_3O_4 doped double-shelled hollow carbon spheres with hierarchical pore network for durable high-performance supercapacitor. Carbon 99:514–522
Lian P, Zhu X, Xiang H, Li Z, Yang W, Wang H (2010) Enhanced cycling performance of Fe_3O_4–graphene nanocomposite as an anode material for lithium-ion batteries. Electrochimica Acta 56(2):834–840
Liang Y, Li Y, Wang H, Zhou J, Wang J, Regier T, Dai H (2011) Co_3O_4 nanocrystals on graphene as a synergistic catalyst for oxygen reduction reaction. Nat Mater 10(10):780–786
Liang H-W, Zhuang X, Brüller S, Feng X, Müllen K (2014) Hierarchically porous carbons with optimized nitrogen doping as highly active electrocatalysts for oxygen reduction. Nat Commun 5(1):1–7
Lin J, Raji ARO, Nan K, Peng Z, Yan Z, Samuel EL, Natelson D, Tour J (2014) Iron oxide nanoparticle and graphene nanoribbon composite as an anode material for high-performance Li-ion batteries. Adv Functional Mater 24(14):2044–2048
Liu J, Rinzler AG, Dai H, Hafner JH, Bradley RK, Boul PJ, Lu A, Iverson T, Shelimov K, Huffman C (1998) Fullerene pipes. Science 280(5367):1253–1256
Liu X, Chen T, Chu H, Niu L, Sun Z, Pan L, Sun CQ (2015) Fe_2O_3-reduced graphene oxide composites synthesized via microwave-assisted method for sodium ion batteries. Electrochimica Acta 166:12–16
Liu Y, Zhang N, Yu C, Jiao L, Chen J (2016) $MnFe_2O_4$@ C nanofibers as high-performance anode for sodium-ion batteries. Nano Lett 16(5):3321–3328
Lu K (2007) Rheological behavior of carbon nanotube-alumina nanoparticle dispersion systems. Powder Technol 177(3):154–161
Luo S, Wu H, Wu Y, Jiang K, Wang J, Fan S (2014) Mn_3O_4 nanoparticles anchored on continuous carbon nanotube network as superior anodes for lithium ion batteries. J Power Sources 249:463–469

Lv H, Qiu S, Lu G, Fu Y, Li X, Hu C, Liu J (2015) Nanostructured antimony/carbon composite fibers as anode material for lithium-ion battery. Electrochimica Acta 151:214–221

Lyu H, Liu J, Qiu S, Cao Y, Hu C, Guo S, Guo Z (2016) Carbon composite spun fibers with in situ formed multicomponent nanoparticles for a lithium-ion battery anode with enhanced performance. J Mater Sci 4(25):9881–9889

Ma J, Wang J, He Y-S, Liao X-Z, Chen J, Wang J-Z, Yuan T, Ma Z-F (2014) A solvothermal strategy: one-step in situ synthesis of self-assembled 3D graphene-based composites with enhanced lithium storage capacity. J Mater Chem A 2(24):9200–9207

Ma C, Li X, Deng C, Hu Y-Y, Lee S, Liao X-Z, He Y-S, Ma Z-F, Xiong H (2018) Coaxial carbon nanotube supported TiO_2@ MoO_2@ carbon core–shell anode for ultrafast and high-capacity sodium ion storage. ACS Nano 13(1):671–680

Modafferi V, Fiore M, Fazio E, Patanè S, Triolo C, Santangelo S, Ruffo R, Neri F, Musolino M (2018) Synthesis and characterization of Fe_2O_3/reduced graphene oxide nanocomposite as a high-performance anode material for sodium-ion batteries. Model Measure Control 87:129–134

Moon IK, Lee J, Lee H (2011) Highly qualified reduced graphene oxides: the best chemical reduction. Chem Commun 47(34):9681–9683

Mussa Y, Ahmed F, Abuhimd H, Arsalan M, Alsharaeh E (2019) Enhanced Electrochemical performance at high temperature of cobalt oxide/reduced graphene oxide nanocomposites and its application in lithium-ion batteries. Sci Rep 9(1):1–10

Okotrub A, Asanov I, Yudanov N, Babin K, Gusel'Nikov A, Nedoseikina T, Gevko P, Bulusheva L, Osvath Z, Biro L (2009) Development of graphene layers by reduction of graphite fluoride C2F surface. physica status solidi (b) 246(11–12):2545–2548

Park J, Moon WG, Kim G-P, Nam I, Park S, Kim Y, Yi J (2013) Three-dimensional aligned mesoporous carbon nanotubes filled with Co_3O_4 nanoparticles for Li-ion battery anode applications. Electrochimica Acta 105:110–114

Park S-K, Seong C-Y, Yoo S, Piao Y (2016) Porous Mn_3O_4 nanorod/reduced graphene oxide hybrid paper as a flexible and binder-free anode material for lithium ion battery. Proc ICE Energy 99:266–273

Peng L, Fang Z, Zhu Y, Yan C, Yu G (2018) Holey 2D nanomaterials for electrochemical energy storage. Adv Energy Nanomater 8(9):1702179

Petkovich ND, Rudisill SG, Wilson BE, Mukherjee A, Stein A (2014) Control of TiO_2 grain size and positioning in three-dimensionally ordered macroporous TiO_2/C composite anodes for lithium ion batteries. Inorg Chem 53(2):1100–1112

Poizot P, Laruelle S, Grugeon S, Dupont L, Tarascon JJN (2000) Nano-sized transition-metal oxides as negative-electrode materials for lithium-ion batteries. Nature 407(6803):496–499

Pumera M (2011) Graphene-based nanomaterials for energy storage. Energy Environ Sci 4(3):668–674

Qiao H, Li J, Fu J, Kumar D, Wei Q, Cai Y, Huang F (2011) Sonochemical synthesis of ordered SnO_2/CMK-3 nanocomposites and their lithium storage properties. ACS Appl Mater Interface 3(9):3704–3708

Qiu S, Lu G, Liu J, Lyu H, Hu C, Li B, Yan X, Guo J, Guo Z (2015) Enhanced electrochemical performances of MoO_2 nanoparticles composited with carbon nanotubes for lithium-ion battery anodes. RSC Adv 5(106):87286–87294

Rahman MM, Sultana I, Chen Z, Srikanth M, Li LH, Dai XJ, Chen Y (2015) Ex situ electrochemical sodiation/desodiation observation of Co_3O_4 anchored carbon nanotubes: a high performance sodium-ion battery anode produced by pulsed plasma in a liquid. Nanoscale 7(30):13088–13095

Sankapal BR, Gajare HB, Karade SS, Salunkhe RR, Dubal DP (2016) Zinc oxide encapsulated carbon nanotube thin films for energy storage applications. Electrochimica Acta 192:377–384

Schwab MG, Narita A, Hernandez Y, Balandina T, Mali KS, De Feyter S, Feng X, Müllen K (2012) Structurally defined graphene nanoribbons with high lateral extension. J Am Chem Soc 134(44):18169–18172

Seok D, Jeong Y, Han K, Yoon DY, Sohn H (2019) Recent progress of electrochemical energy devices: metal oxide–carbon nanocomposites as materials for next-generation chemical storage for renewable energy. Sustainability 11(13):3694

Shen J, Hu Y, Shi M, Li N, Ma H, Ye M (2010) One step synthesis of graphene oxide-magnetic nanoparticle composite. J Phys Chem C114(3):1498–1503

Shi P, Li L, Hua L, Qian Q, Wang P, Zhou J, Sun G, Huang W (2017) Design of amorphous manganese oxide@ multiwalled carbon nanotube fiber for robust solid-state supercapacitor. ACS Nano 11(1):444–452

Shirai T, Kato T, Fuji M (2013) Effects of the alumina matrix on the carbonization process of polymer in the gel-casted green body. J Eur Ceramic Soc 33(2):201–206

Sohn HJ (2017) Deposition of functional organic and inorganic layer on the cathode for the improved electrochemical performance of Li-S battery. Kor Chem Eng Res 55(4):483–489

Sohn H, Gordin ML, Regula M, Kim DH, Jung YS, Song J, Wang DJ (2016a) Porous spherical polyacrylonitrile-carbon nanocomposite with high loading of sulfur for lithium–sulfur batteries. J Power Sources 302:70–78

Sohn H, Kim D, Lee J, Yoon SJR (2016b) Facile synthesis of a mesostructured TiO_2-graphitized carbon (TiO_2–gC) composite through the hydrothermal process and its application as the anode of lithium ion batteries. RSC Adv 6(45):39484–39491

Srivastava M, Singh J, Kuila T, Layek RK, Kim NH, Lee J (2015) Recent advances in graphene and its metal-oxide hybrid nanostructures for lithium-ion batteries. Nanoscale 7(11):4820–4868

Stoller MD, Magnuson CW, Zhu Y, Murali S, Suk JW, Piner R, Ruoff RS (2011) Interfacial capacitance of single layer graphene. Energy Envion Sci 4(11):4685–4689

Su J, Cao M, Ren L, Hu C (2011) Fe_3O_4–graphene nanocomposites with improved lithium storage and magnetism properties. J Phys Chem C 115(30):14469–14477

Su D, Ahn H-J, Wang G (2013) SnO_2@ graphene nanocomposites as anode materials for Na-ion batteries with superior electrochemical performance. Chem Commun 49(30):3131–3133

Tan Y-Z, Yang B, Parvez K, Narita A, Osella S, Beljonne D, Feng X, Müllen K (2013) Atomically precise edge chlorination of nanographenes and its application in graphene nanoribbons. Nat Commun 4(1):1–7

Tao L, Zai J, Wang K, Zhang H, Xu M, Shen J, Su Y, Qian X (2012) Co_3O_4 nanorods/graphene nanosheets nanocomposites for lithium ion batteries with improved reversible capacity and cycle stability. J Power Sources 202:230–235

Wang DW, Li F, Liu M, Lu GQ, Cheng H (2008) 3D aperiodic hierarchical porous graphitic carbon material for high-rate electrochemical capacitive energy storage. Communication 47(2):373–376

Wang Y, Su D, Wang C, Wang G (2013) SnO_2@MWCNT nanocomposite as a high capacity anode material for sodium-ion batteries. Electochem Commun 29:8–11

Wang S, Xing Y, Xu H, Zhang S (2014) MnO nanoparticles interdispersed in 3D porous carbon framework for high performance lithium-ion batteries. ACS Appl Mater Interface 6(15):12713–12718

Wang X, Li G, Seo MH, Lui G, Hassan FM, Feng K, Xiao X, Chen Z (2017) Carbon-coated silicon nanowires on carbon fabric as self-supported electrodes for flexible lithium-ion batteries. ACS Appl Mate Interface 9(11):9551–9558

Wang M, Huang Y, Wang K, Zhu Y, Zhang N, Zhang H, Li S, Feng Z (2018) PVD synthesis of binder-free silicon and carbon coated 3D α-Fe_2O_3 nanorods hybrid films as high-capacity and long-life anode for flexible lithium-ion batteries. Energy 164:1021–1029

Wang S, Sun C, Wang N, Zhang Q (2019) Ni-and/or Mn-based layered transition metal oxides as cathode materials for sodium ion batteries: status, challenges and countermeasures. J Mater Chem 7(17):10138–10158

Wu Z-S, Zhou G, Yin L-C, Ren W, Li F, Cheng H-M (2012) Graphene/metal oxide composite electrode materials for energy storage. Nano Enegy 1(1):107–131

Xiang C, Li M, Zhi M, Manivannan A, Wu N (2013) A reduced graphene oxide/Co_3O_4 composite for supercapacitor electrode. J Power Sources 226:65–70

Xiong Z, Yun YS, Jin H-J (2013) Applications of carbon nanotubes for lithium ion battery anodes. Materials 6(3):1138–1158

Yang Z, Niu Z, Lu Y, Hu Z, Han C (2003) Templated synthesis of inorganic hollow spheres with a tunable cavity size onto core–shell gel particles. Angewandte Chemie 115(17):1987–1989

Yang M, Ma J, Zhang C, Yang Z, Lu Y (2005) General synthetic route toward functional hollow spheres with double-shelled structures. Angewandte Chemie 44(41):6727–6730

Yi J, Xue W, Xie Z, Chen J, Zhu L (2015) A novel processing route to develop alumina matrix nanocompositesreinforced with multi-walled carbon nanotubes. Mater Res Bull 64:323–326

Yoon T, Chae C, Sun Y-K, Zhao X, Kung HH, Lee J (2011) Bottom-up in situ formation of Fe_3O_4 nanocrystals in a porous carbon foam for lithium-ion battery anodes. J Mater Chem 21(43):17325–17330

Yu W, Li B, Ding SJ (2016) Electroless fabrication and supercapacitor performance of CNT@ NiO-nanosheet composite nanotubes. Nanotechnology 27(7):075605

Yun Q, Lu Q, Zhang X, Tan C, Zhang H (2018) Three-dimensional architectures constructed from transition-metal dichalcogenide nanomaterials for electrochemical energy storage and conversion. Angewandte Chemie Int Ed 57(3):626–646

Zhang WM, Wu XL, Hu JS, Guo YG, Wan L.J (2008) Carbon coated Fe_3O_4 nanospindles as a superior anode material for lithiumion batteries. Adv Funct Mater 18(24):3941–3946

Zhang W-D, Xu B, Jiang L-C (2010) Functional hybrid materials based on carbon nanotubes and metal oxides. J Mater Chem 20(31):6383–6391

Zhang X, Hu Z, Xiao X, Sun L, Han S, Chen D, Liu X (2015a) Fe_3O_4@ porous carbon hybrid as the anode material for a lithium-ion battery: performance optimization by composition and microstructure tailoring. New J Chem 39(5):3435–3443

Zhang Y, Xie J, Zhang S, Zhu P, Cao G, Zhao X (2015b) Ultrafine tin oxide on reduced graphene oxide as high-performance anode for sodium-ion batteries. Electrochimica Acta 151:8–15

Zhao X, Hayner CM, Kung MC, Kung H (2011) Flexible holey graphene paper electrodes with enhanced rate capability for energy storage applications. ACS Nano 5(11):8739–8749

Zhao Y, Dong W, Riaz MS, Ge H, Wang X, Liu Z, Huang F (2018a) "Electron-sharing" mechanism promotes Co@Co_3O_4/CNTs composite as the high-capacity anode material of lithium-ion battery. ACS Appl Mater Appl 10(50):43641–43649

Zhao X, Luo M, Zhao W, Xu R, Liu Y, Shen H (2018b) SnO_2 nanosheets anchored on a 3D, bicontinuous electron and ion transport carbon network for high-performance sodium-ion batteries. ACS Appl Mater Interface 10(44):38006–38014

Zhi M, Xiang C, Li J, Li M, Wu N (2013) Nanostructured carbon–metal oxide composite electrodes for supercapacitors: a review. Nanoscale 5(1):72–88

Zhou G, Wang D-W, Li F, Zhang L, Li N, Wu Z-S, Wen L, Lu GQ, Cheng H (2010) Graphene-wrapped Fe_3O_4 anode material with improved reversible capacity and cyclic stability for lithium ion batteries. Chem Mater 22(18):5306–5313

Zhou W, Liu J, Chen T, Tan KS, Jia X, Luo Z, Cong C, Yang H, Li CM, Yu TJ (2011) Fabrication of Co_3O_4-reduced graphene oxide scrolls for high-performance supercapacitor electrodes. Phys Chem Chem Phys 13(32):14462–14465

Zhou G, Wang D-W, Hou P-X, Li W, Li N, Liu C, Li F, Cheng H-M (2012) A nanosized Fe_2O_3 decorated single-walled carbon nanotube membrane as a high-performance flexible anode for lithium ion batteries. J Mater Chem 22(34):17942–17946

Zhou H, Zhang L, Zhang D, Chen S, Coxon PR, He X, Coto M, Kim H-K, Xi, K, Ding S (2016) A universal synthetic route to carbon nanotube/transition metal oxide nano-composites for lithium ion batteries and electrochemical capacitors. Sci Rep 6(1):1–11

Zhu J, Zhu T, Zhou X, Zhang Y, Lou XW, Chen X, Zhang H, Hng HH, Yan Q (2011a) Facile synthesis of metal oxide/reduced graphene oxide hybrids with high lithium storage capacity and stable cyclability. Nanoscales 3(3):1084–1089

Zhu C, Yu Y, Gu L, Weichert K, Maier J (2011b) Electrospinning of highly electroactive carbon-coated single-crystalline $LiFePO_4$ nanowires. Angewandte Chemie 123(28):6402–6406

Chapter 4
Metal Nanocomposite Synthesis and Its Application in Electrochemical CO_2 Reduction

Rishabh Sharma, Pradip Kalbar, Simant Kumar Srivastav, Kamlesh Kumar, and Swatantra P. Singh

Abstract To address the immoderate emission of carbon dioxide (CO_2), the development of earth-excessive metals and non-metals based electrocatalyst is of great importance. The development of an active electrocatalyst with cost-effective, efficient, and easy accessibility for electrochemical-based CO_2 reduction (ECR), is a growing field of research. Growing industries and urban populations lead to increased pollution generation, especially air pollution causing serious environmental and health problems and causing deterioration of air quality. Non-conventional energy-based ECR to alternative substitutes with heavy energy densities proves to be an effective route for production and storage of energy and manage the carbon and energy balance. For an efficient yield of chosen products, the modification, outline, technology, and finding of new catalysts are important steps. In the present chapter, we have discussed heterostructures of catalysts, the tunable component for an effective CO_2 conversion, and metal/metal oxide hybrids for enhancing CO_2 reduction capability, and an alliance between different nanocomposites for efficient reduction processes. The accurate and precise tuning of interlinkage between dissimilar metal and metal

R. Sharma · P. Kalbar · S. P. Singh (✉)
Interdisciplinary Program in Climate Studies, Indian Institute of Technology Bombay, Mumbai 400076, India
e-mail: swatantra@iitb.ac.in

P. Kalbar
Centre for Urban Science and Engineering, Indian Institute of Technology Bombay, Mumbai 400076, India

S. K. Srivastav
University Department of Chemistry, L. N. Mithila University, Darbhanga, Bihar 846006, India

K. Kumar
Department of Chemistry, Institute of Science, Banaras Hindu University, Varanasi 221005, India

S. P. Singh
Environmental Science and Engineering Department (ESED), Indian Institute of Technology Bombay, Mumbai 400076, India

Centre for Research in Nanotechnology & Science (CRNTS), Indian Institute of Technology Bombay, Mumbai 400076, India

S. P. Singh et al. (eds.), *Metal Nanocomposites for Energy and Environmental Applications*, Energy, Environment, and Sustainability,
https://doi.org/10.1007/978-981-16-8599-6_4

oxides will enhance the reaction kinetics, maximize binding between intermediates, and accomplish effective ECR.

Keywords Electrocatalysis · Electrochemical CO_2 reduction (ECR) · Composite materials · Nanocomposites · Metal oxide

4.1 Introduction

The exceeding concerns about the warming of the earth are caused by the regular escalation in CO_2 present in the atmosphere. These increased CO_2 concentrations have propelled substantial research interest in carbon conversion and utilization. In fact, in the last few decades, the concentration of CO_2 is going to increase tremendously due to increased fossil fuel combustion, increased population, and thereby increasing demand; this induces climate change and rising sea levels (Lateef and Nazir 2017). Different researches have been carried out on photosynthesis that too artificially for CO_2 reduction using sunlight as the major substrate. Till now, several studies have been carried out on some effective methods such as photocatalytic, thermal treatment, chemical methods, and other electrochemical methods (Woyessa et al. 2021). So, the up-gradation of the energy economy by accessible, inexpensive, and environment-friendly means is a must to ensure a long-lasting supply to satisfy exponentially increasing global energy demand (Zhu and Diao 2011). In this context, catalyst plays a very crucial part inefficient energy conversion of plentiful natural resources such as CO_2, nitrogen to clean fuels, and chemical raw material precursors such as hydrocarbons (HCs), hydrogen, and urea (ammonia). And for the long term, energy storage, long-lasting and nimble electrocatalysts are also required (Lim et al. 2020).

The nanocomposite is a type of composite material because of its superior qualities to single metal nanoparticles (NPs); nanocomposite (NC) materials have attracted a lot of attention and interest from scientists in recent years (Lateef and Nazir 2017). In a mixture or matrix, several materials are mixed to create new qualities, ensuring that at least any of the materials have a size range of 1–100 nm. NC being a composite material, is divided into two parts: (i) continuous reinforcing phase and (ii) discontinuous reinforcing phase. NC can be used in a variety and can be manufactured from any combination of the components that fall into one of these three categories-metals, ceramics, and polymers (Lim et al. 2020). These three materials serve as the building blocks for NC. As a result, the nanocomposite can have a mix of electrical, thermal, mechanical, electrochemical, photocatalytic, and optical characteristics of the constituent materials (Zhao et al. 2015). The different phases of NCs are 0-D called core–shell, 1-D in the form of nanowires and nanotubes, 2-D as lamellar, and 3-D in the form of metal matrix composites (Woyessa et al. 2021). And, based on their qualities of structure, layered composites, filamentary composites, and particulate composites are the three types of NCs. Furthermore, these NCs have evolved as 21st-century materials in every aspect of life; the twentieth century has brought a slew of technological and business advancements (Ray and Pal 2017). Based on the nature of production, the metal nanocomposites are divided into two major subclasses, Natural polymer metallic nanocomposites and synthetic polymer metallic nanocomposites (He et al. 2017).

4.1.1 Natural Polymer-Metal Nanocomposites (NP-MNCs)

Natural polymers have long been material scientists' primary choice for the generation of metal and metal-based blends because of their non-toxic and bio-compatible features (Peng et al. 2019). Chitosan, for example, is a biogenic polymer having many cationic characteristics in non-concentrated acidic solutions, allowing electrostatic adhesion to negatively charged Au-NPs surfaces. It is produced from chitin by partial deacetylation (Zhu and Diao 2011). Transmission electron microscopy (TEM) research confirms that adding a portion of total chitosan solution to Au-colloidal solution results in chitosan surmounted Au-NPs complex system. Another example includes cotton fabric that has been treated with a suspension of silver oxide chitosan nanocomposites has outstanding antibacterial properties (Lim et al. 2020). For wastewater treatment and medicinal/biotechnological uses, ferromagnetic nanoparticles (e.g., maghemite, magnetite) have been implanted in biogenic polymers like chitosan (e.g., separation of microorganisms, extracorporeal blood detoxification, and targeted drug delivery) (Lateef and Nazir 2017).

4.1.2 Synthetic Polymer-Metal Nanocomposites (SP-MNCs)

Poly (vinyl alcohol), a biocompatible polymer with a wide range of applications, has been frequently employed to create nanocomposites with metal nanoparticles (Chen et al. 2020). For example, after solvent evaporation, adding polyvinyl alcohol solution to Ag-NPs suspension generated with citrate ($C_6H_8O_7$) reduction will conveniently result in the creation of a yellow translucent polyvinyl alcohol silver (PVAAg) nano-complex film (Lateef and Nazir 2017). The thermal stability of polymer-metal nanocomposites films has been observed to improve. Gao et al. used a star polymer with a crown ether core, i.e., [poly(styrene)]-dibenzo-18-crown-6-[poly(styrene)], to make silver (Ag) NPs (He et al. 2017).

Metal NPs have been utilized as catalysts in various reactions that are critical in the manufacturing and processing units. On the other hand, bio-metallic catalysts are more important because mixing two different metals can manifest control over selectivity, stability, and activity of the catalyst, and some merger can have combinational effects (Woyessa et al. 2021). This chapter focuses on metal/metal oxide nanocomposites (MNC/MONC) synthesis, mode of formation, and promising properties of these materials, making them suitable for a huge number of utilization in the field of structural and functional applications (Yu et al. 2021). Metal nanoparticles can be incorporated into polymer nanofibers by either electrospinning metal nanoparticle-containing polymer solutions or reducing metal salts or complexes in electrospun polymer nanofibers (Woyessa et al. 2021). Further discussion on the synthesis of metal nanocomposites is in the next section.

The conversion of CO_2 into chemical raw materials using non-conventional energy resources (such as hydro, wind, solar etc.) is a propitious solution to lessen our reliance on conventional resources and minimize environmental problems. Copper, Gold, Silver, Platinum, and Nickel electrocatalysts have been employed as CO_2 reduction reaction routes to create numerous single carbon products, including carbon mono-oxide, formic acid, formaldehyde, methanol, and methane (Diaz et al. 2018). They all have poor selectivity, sluggish kinetics, and minimal efficiency, though dispersed metal nanocatalysts have recently been discovered to exhibit significant activity for CO_2 reduction reactions due to high atomic efficiency, identical electronic properties, and brilliant selectivity. Significant development has been made in recent decades, primarily in the morphology of highly functional electrocatalysts, as well as their CO_2RR mechanism (Birdja et al. 2019; Dimeglio and Rosenthal 2013; Ding et al. 2020). Other factors, essentially electrode design, inner-surface morphology, and process and reaction conditions, must be recognized as important not just from a mechanistic but also from a systemic point of view. In addition to all this, the electrode/electrolyte interface be investigated, and also transfer processes, stability of the catalyst, and electrolyte/solvent interactions (Garg et al. 2020). In the present chapter, we are going to discuss basic methods for the synthesis of metal nanocomposites (MNCs), precursors used, and applications. In the later section, electrocatalysis basic and applications and use of metal nanocomposites as a potential electrocatalyst, recent advancements and sub-surface structure, reaction and process conditions are discussed.

4.2 Methods Used for Synthesizing Metal Nanocomposites (MONCs/MNCs)

Much effort has gone into developing efficient synthetic methods to form extremely stable and well-defined metal nanocomposites over the last decade. The amalgamation of uniformly sized NCs is critical because of their characteristics, which are dependent on their size and dimensions, include optical, magnetic, electrical, and biological properties (Chen et al. 2020; Lateef and Nazir 2017). Synthetic procedures are typically divided into three categories: solution-based synthesis, chemical synthesis, and biological synthesis (Wang et al. 2018). Unlike the first approach, the second one is to divide these manufacturing approaches into two main groups, (a) top to down approach; and (b) bottom to up approach (Lateef and Nazir 2017). Wet methods are included in both the top to down and bottom to up approaches. The manufacturing technique is determined by the NPs desired features, such as size, shape, crystal structure, and other characteristics of composites (Woyessa et al.

2021). Physical approaches, on the other hand, have a distinct advantage, i.e., the manufacture of a huge number of nanocomposites, but the synthesis of nanocomposites of identical particle size is really difficult to get in (Lateef and Nazir 2017). The synthesis of high-quality nanocomposites can be accomplished using a variety of techniques, including the hydrothermal/solvothermal method, filling process, self-assembly method, template-based synthesis, detonation-induced reaction, chemical vapour deposition, sol–gel process, pyrolysis procedure and so on (Yu et al. 2021). For different shapes, NCs synthesis such as nanorods, nanowires, nanotubes, and so on can be achieved by altering reaction conditions (Li et al. 2020).

4.2.1 Sol–Gel Method

Due to its gentle reaction conditions and ability to construct materials from molecular precursors, this method has acquired traction as a feasible method for the formation of nanomaterials (Lateef and Nazir 2017). Material and property variations, the sol–gel technique produces either films or gels or colloidal powder. The sol–gel process may create micro and nanoscale structures. The reaction significantly impacts the ultimate product's size, shape, and structure (Lu and Jiao 2016). The sol–gel approach is also appealing for synthesizing multicomponent nanostructures because the weak reaction kinetics allow for practical morphological engineering of the finished product (Reddy 2017). The other benefit of the technology is that the reactions can be carried out at low temperatures or room temperature (He et al. 2017). In the sol–gel process, inorganic precursors go through a series of chemical reactions before forming a three-dimensional (3D) molecular network (Yuan et al. 2017). One of the most popular routes is through metal alkoxides which are hydrolyzed and condensed to generate bigger metal oxides, which is one of the frequent approaches (Nasrollahzadeh et al. 2015). The coating is made up of molecules that polymerize to produce a coating around these metal oxides (Woyessa et al. 2021). On the nanoscale to micrometer scale, the sol–gel technique allows the coating of surfaces with complicated forms, which other regularly used coatings do not allow (Yu et al. 2021).

4.2.2 Hydrothermal Method

The hydro-thermal technique involves a heterogeneous chemical reaction in a sol mixture (aqueous/non-aqueous) that takes place in a closed system above room temperature and at a pressure greater than 1 atm. (Lateef and Nazir 2017; Yu et al. 2021). Surfactants, capping agents, and mineralizers are used to change the size and properties of the system; this is a common practice. The emerging tendency is to combine this approach with other techniques. Microwave, sol–gel are two techniques that can change the physicochemical and structural properties of a substance (Zhu and Diao 2011). However, the materials can also result in the production of single-phase materials having improved stability (Woyessa et al. 2021). Further, just by small alteration in temperature, pressure, and contact time of reaction, particle size, properties, and phase change morphology as presented in table 4.1.

4.2.3 Co-precipitation Method

The co-precipitation process creates a precipitate that is removed from the solution and is used to synthesize metal oxide nanoparticles, mixed metal nanoparticles, and metal-ceramic nanocomposites (Lateef and Nazir 2017). The salts (inorganic) are used as the forerunner, and to achieve a homogeneous solution, salts are dissolved in water or other solvents. When the critical concentration of ions is reached, the salts begin to precipitate as oxides of hydrogen and/or oxalates. The nucleation and growth phases are followed when the critical concentration of species is attained (Zhu and Diao 2011). The pH, temperature, and salt concentration of a solution are certain factors that have a big impact on particle form, i.e., size and shape of particles. After precipitation, filtration, separation, and washing are performed along with heating at a high temperature to convert hydroxide into oxides with a particular crystal morphology (Lateef and Nazir 2017). The table below lists various forms of metal NCs (metal–metal-based oxides) that have been formed using this method (Lim et al. 2020). Typically, the precipitating media is NaOH, NH_3 or NH_4OH, Na_2CO_3, and other compounds have been used. The usage of surfactants is also a consideration (Woyessa et al. 2021). Avoiding collection, which also affects the particle morphology of composites, is a standard strategy that can be achieved using this method. The approach has the following advantages: low cost, simplicity, water-based reaction, adaptability, and mildness (Lateef and Nazir 2017).

4.2.4 Biogenic Method

In recent years, biogenic production of metal and metal oxide-based NPs have obtained a lot of significance (Ingale and Chaudhari 2013). This fascination stems from the unique features of nanoparticles, which make them ideal for a variety of pharmaceutical, electrochemical and biological applications (Akhtar et al. 2013). The biogenic formation of NPs by seaweeds (a valuable source of bio-active material like polysaccharides), marine algae, plants etc. has the potential to be a simple, green, and environmentally acceptable method. For effective production of metal/metal oxides by using algae, plants, and other organisms or living cells, a thorough understanding of different biological, physical, biochemical and adaptational mechanisms is a prerequisite, and it's also necessary to make it economically competitive and environmentally feasible (Akhtar et al. 2013; Fawcett et al. 2017). These techniques, which are based on green chemistry, have the potential to provide a substitute to

Table 4.1 Showing different metal/metal oxide NCs their precursors, reaction conditions and characterization techniques for the different synthesis methods (adapted from (Lateef and Nazir 2017))

Method used	Nanocomposites	Precursor	Particle size (nm)	Reaction conditions	References
Sol–gel method	TiO_2-O_3	Titanyl sulfate	5–9		Yu et al. (2021)
	TiO_2– Fe_2O_3	TiO_2	4–10	100–1200 °C	Lateef and Nazir (2017)
	Ag-TiO_2	$AgNO_3$, Titanium (IV) isopropoxide	13–20	Calcinated at 550 °C	Zhao et al. (2015)
	TiO_2-lanathal	Titanyl sulphate	7	600 °C, pH 7	Yuan et al. (2017)
	Al_2O_3– SiC	Aluminium nitrate and hexamethylenetetra mine (HMTA)	-		Lateef and Nazir (2017)
	Mullite-SiC	Mullite	70	Varying concentration of mullite precursor	Nasrollahzadeh et al. (2015)
	WO_3- TiO_2	Ti (OBu_4), ammonium tungstate	60	pH 10, calcinated at 400 °C	(Chen et al. 2020)
	Ag-MMT	$AgNO_3$, $NaBH_4$	4–8.5	Varying ratios of silver and centrifugation at *15,000 rpm*	(Kaviyarasu et al. 2015)
	TiO_2-Clay	Ti $(OC_3H_7)_4$, NaOH	1.7–4.5	pH 4	Lateef and Nazir (2017)
Hydrothermal method	Al_2O_3- TiO_2	Al_2Cl_3, NaOH	3.25	pH 8,500 °C	Zhu and Diao (2011)
	ZnO-kaolinite	$ZnCl_2$, NaOH	14.6–48.2	Calcinated at 600 °C,	Nasrollahzadeh et al. (2015)

(continued)

Table 4.1 (continued)

Method used	Nanocomposites	Precursor	Particle size (nm)	Reaction conditions	References
	ZnS-MMT	Hexadecyl trimethyl ammonium bromide (HTAB), Zn $(OOCH CH_3)_2 \cdot 2H_2O$	3.77	170 °C	Lu and Jiao (2016)
	CdS, TiO_2 Montmorillonite	$TiCl_4$, $CdAc_2 \cdot 2H_2O$ and CS $(NH_2)_2$	7	160 °C	Ray and Pal (2017)
	Co-MgO	Co $(NO_3)_2 \cdot 7H_2O$, Mg $(NO_3)_2 \cdot 6H_2O$	–	150 °C, centrifugation at 4100 rpm	Lateef and Nazir (2017)
	$Al_2(OH)_3$ (VO_4)	V_2O_5, Al $(NO_3)_3 \cdot 9H_2O$, and TMA-OH	–		Yu et al. (2021)
	Mixed-phase TiO_2	Titanium chloride, anatase powder	–		Yuan et al. (2017)
Co-precipitation	$MgO.Al_2O_3$	$MgCl_2$, NaOH	4–28	Centrifuged (100 rpm)	Lateef and Nazir (2017)
	$ZnO.SnO_2$	$ZnCl_2$, NaOH	30	pH 6, 105 °C	Lu and Jiao (2016)
	ZnO.CuO	$CuCl_2$, $NaBH_4$	-		Hamrouni et al. (2013)
	MgO.CuO	Mg $(NO_3)_2$, Cu $(NO_3)_2$ and glycine $(NH_2CH_2C OOH)$	20	120 °C	Lateef and Nazir (2017)
	$ZnO.Fe_3O_4$	Fe_3O_4, NaOH	40	pH 7, centrifuge	Nazari and Halladj (2014)
	$NiO \cdot CeO_2 \cdot ZnO$	$ZnCl_2$, NaOH	14–25	120 °C	Lim et al. (2020)
	Ag-Talc	$AgNO_3$, $NaBH_4$	7–12	Varying molar ratios of silver (Ag)	Ray and Pal (2017)

(continued)

Table 4.1 (continued)

Method used	Nanocomposites	Precursor	Particle size (nm)	Reaction conditions	References
	Ag-Activated Carbon	Soluble starch, NaOH	55	pH 8, in N_2 atmosphere	Lateef and Nazir (2017)
	ZeoliteY–Fe_3O_4	Zeolite-Y	< 100	Different temperatures	Kaviyarasu et al. (2015)
	ZnO activated carbon	Hexamethylenetetramine ($(CH_2)_6N_4$; HMTA)	50–200	70, pH 8	Reddy (2017)

traditional manufacturing procedures, which typically involve harmful chemicals and solvents.

4.2.5 Other Methods

Many alternative procedures for producing metal nanocomposites have been explored in addition to the approaches mentioned above (Zhao et al. 2015). Activity-based on capillary effect, electrolysis using condensed-metal arc, sonochemical deposition, micro emulation technology, electrochemical method, electrospinning, and pulsed laser irradiation are some of the other technologies developed over the years (Woyessa et al. 2021). Though several methods have been mentioned in the above discussion, the three methods discussed broadly are relatively fast, convenient, and used on a larger scale as compared to the other method or technologies.

4.3 Electrochemical CO_2 Reduction (ECR)

The electrocatalytic CO_2 reduction-based reaction (CO_2RR) is a model for a clean and efficient long-term CO_2 reduction solution. The process starts with CO_2 activation using electric energy, which results in CO_2 being formed on the catalyst surface. The production of numerous intermediates and products results from the transfer of electrons (e^-) and/or protons (H^+) (Birdja et al. 2019; Bui et al. 2018). Equations below show CO_2 activation and various intermediary products formed. COOH is thought to be the more likely initial intermediate in the creation of CO, while OCHO is thought to be the more likely intermediate in the formation of formic acid (Dimeglio and Rosenthal 2013). Poor selectivity, small current density, less faradaic efficiency (FE), reactive intermediates, huge overpotentials, and poor CO_2 solubility in the electrolyte (solvent) have all hampered the development of electrocatalytic CO_2 reduction processes (Lim et al. 2020). Therefore, the activity is majorly based on the characteristics such as selectivity, durability, stability and activity of catalysts. This can be accomplished by selecting the right catalyst, modifying its structure, and adjusting the electrolyte content (Huang et al. 2018). Electrocatalysts for CO_2 conversion, for example, can stabilize CO_2 or an intermediate by creating a chemical connection between it and CO_2, lowering the negative redox potential (Diaz et al. 2018).

CO_2 Electroreduction Pathway:

$$* + CO_2 + H^+ + e^- \rightarrow *COOH \quad \text{(I)}$$

$$* + CO_2 + H^+ + e^- \rightarrow *OCHO \quad \text{(II)}$$

$$* + CO_2 + e^- \rightarrow *CO^{-2} \quad \text{(III)}$$

$$* + H^+ + 2e^- \rightarrow H \quad \text{(IV)}$$

where * represents the electrocatalyst.

The CO_2 reduction using the electrochemical method is a propitious method for managing intermittent non-conventional energy generation by turning it into a chemically valuable complex while recycling CO_2 that contributes to climate change (Liu et al. 2015). Understanding the paths for CO_2ER's electrochemical changes is crucial for its technological advancement. CO^{2-} and $HCOO^-$, represented by reaction (II) and (III) respectively are CO_2ER's two proton-electron transfer products and are particularly appealing due to the fewer overpotentials required to progress their synthesis and the large faradic efficiencies that may be achieved. The synthesis of CO from CO_2 (reaction (II)) on transition metal catalysts with an aqueous electrolyte, is thought to occur through the COOH species (reaction (I)), whilst the formation of $HCOO^-$ is thought to occur through the bi-dentate oxygen bounded OCHO species, the pair of which occur after an individual coupled proton-electron transfer (CPET) step (Bohra et al. 2019). However, up to this point, we have discussed only the advantages of using different methods for CO_2 reduction, but there are certain challenges also which need to be addressed. In CO_2 reduction, the slow electron transfer results in low current exchange densities, the large overpotential causes small energy efficiencies, and the formation of multiple products with less quantity causes huge energy losses, which are some problems associated with CO_2 reduction (Lu and Jiao 2016; Yu et al. 2021; Zhu and Diao 2011) (Fig. 4.1).

But all these issues are addressed by electrochemical-based carbon dioxide (CO_2) reduction, which is discussed further in coming sections. Moreover, if we compare benefits over limitations, the clear winner is the benefits of CO_2 reduction. As, per the reports of the National Oceanic and Atmospheric Administration (NOAA), the USA, in the year 2019, the overall atmospheric CO_2 set a new record high at a value of 409.8 as compared to 230 ppm globally, and in the year 1960 a growth rate of atmospheric CO_2 is about 0.6 ± 0.1 ppm/year. (NOAA 2020). At such a tremendous increase to prevent global warming and associated climate change the carbon capture

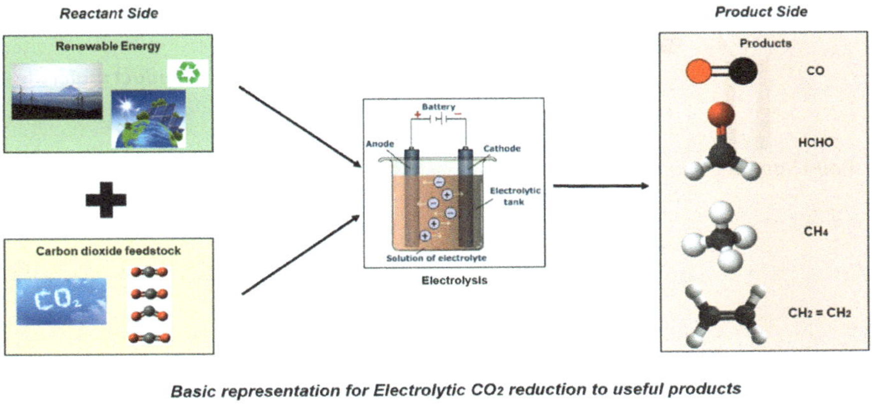

Fig. 4.1 A setup to represent the basic electrochemical reduction of carbon in high-energy derivatives

became the need of the hour, and doing so with the help of the electrochemical method proves to be beneficial and sustainable for the coming future.

4.3.1 Action Mechanism of Electrochemical CO_2 Reduction (ECR)

Heterogeneous chemical reduction of CO_2 takes place electrochemically at the electrode/electrolyte interfaces. The following steps are mainly considered as the backbone for heterogeneous catalytic processes (Garg et al. 2020):

(a) The first step includes the CO_2 adsorption chemically over the surface of the catalyst (cathode) (Garg et al. 2020; Liu et al. 2015).
(b) The second step includes the transfer of electrons and/or migration of protons to break C-O bonds and leads to reduction resulting in C–H bonds (Li et al. 2020).
(c) The third step is initiated with the desorption of product species from the electrode surface and diffusion to the electrolyte as shown in Fig. 4.2.

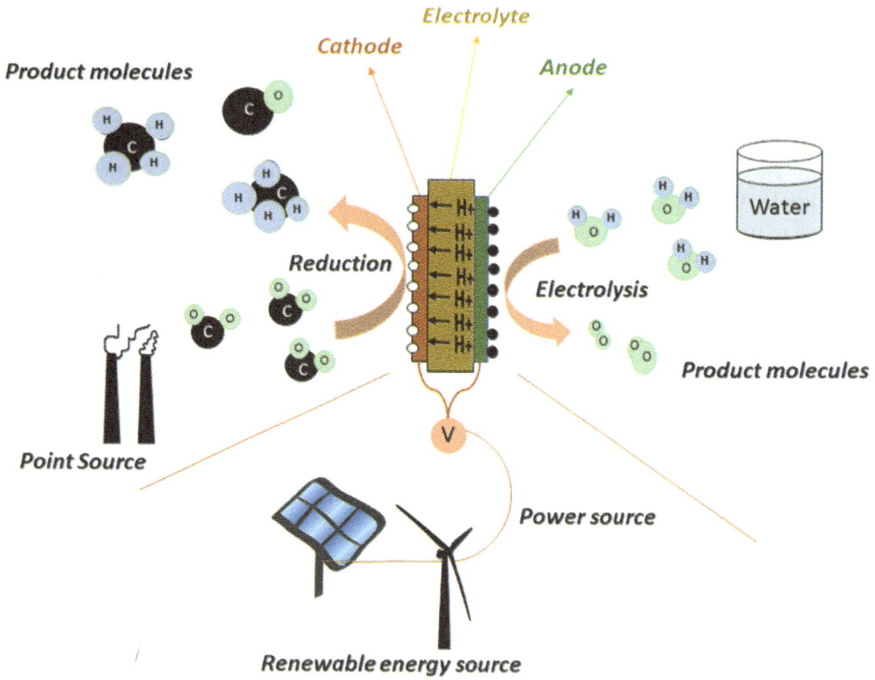

Fig. 4.2 Action mechanism of electrochemical CO_2 reduction (ECR)

And, because of the struggling proton reduction, aqueous electrocatalytic CO_2 reduction is particularly difficult, resulting in low selectivity for carbonaceous products. Other difficulty is CO_2's poor dissolution in water (0.034 M), which affects diffusion-mediated processes negatively. To improve CO_2 dissolution, numerous approaches can be applied, including (1) the use of non-aqueous solutions, (2) lower temperature working conditions, and (3) raised CO_2 partial pressure (Garg et al. 2020; Nasrollahzadeh et al. 2015; Reddy 2017). The ECR's thermodynamics are also influenced by CO_2 pressure. At high pressures, few metal catalysts are inert at atmospheric pressure can convert carbon dioxide to carbon monoxide, hydrocarbons, or formate. Faster reaction speeds, faradic efficiencies, and product suitability result from higher CO_2 flow rates (Chen et al. 2021; Garg et al. 2020). The lack of a basic understanding of carbon reduction is a major hindrance to building active, selective, and stable electrocatalysts. Electrochemical reactions are mediated by the surface, and the absence of a basic understanding of surface-mediated phenomenon causes a key barrier for carbon reduction (Lu and Jiao 2016). The dispersed variety of feasible products and the number of transferred electrons are varied from carbon-monoxide ($2e^-$), and formic acid ($2e^-$) goes up to propane-1-ol ($18e^-$) in the ECR reaction that poses a valuable challenge in interpreting the reaction mechanism (Zhu and Diao 2011). Therefore, a better and clear understanding of how carbon products undergo reduction on catalyst materials is a logical step in determining the proper reaction mechanism of the process.

4.3.2 Electrocatalyst for ECR

To produce efficient electrochemical CO_2 reduction, metallic, molecular, carbon-based and metallic organic frameworks (MOFs) catalysts have been extensively used (Bui et al. 2018). The most researched molecular catalysts, polypyridyl transition metal complexes, have a high CO_2 capacity (Li et al. 2020). In aprotic solvents, the reaction efficiency is reduced and the activity is high. In an aqueous environment, however, their applicability is limited and responsible for low CO_2 response (HER), low stabilities, high concentrations, and recyclability (Birdja et al. 2019; Bui et al. 2018; Sun et al. n.d.). Due to competing hydrogen evolution, the ecosystem is being harmed. The present research models focus on crippling the hybrid and complex system morphology. On multi-walled carbon nanotubes, Wang et al. 2018 (b) added a Co (II) quaterpyridine compound (MW-CNTs) in a liquid solvent environment and got a 100% success rate selectivity, good stability, and faradaic efficiency at reduced catalyst loading, and small overpotential. Similarly, other studies such as Smith et al. (2018) incorporated Fe tetra-(phenyl)-porphyrin into their research.

These findings show that hybrid system designs have immense development potential for CO_2 levels reduction by electrochemical techniques (Lote 2014). A crucial prerequisite for the economic viability of electrocatalytic CO_2RR is the development of capable catalysts with high reactivity and selectivity toward the generation of a useful product (Diaz et al. 2018; Li et al. 2020; Lim et al. 2020). Metal catalysts

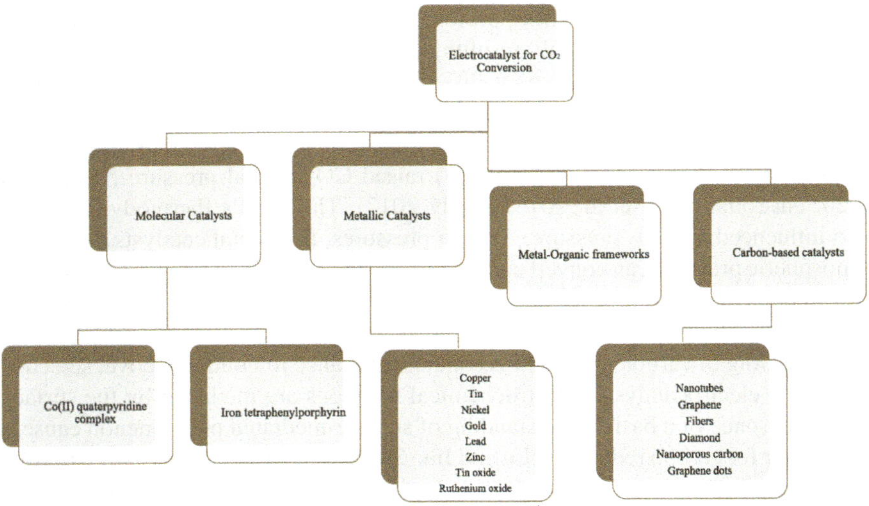

Fig. 4.3 Pathways of CO_2 reduction into different reduced high energy products

are investigated for their selectivity toward their target molecules. (1) Hydrocarbons, HCs (e.g., Cu); (2) Carbon-monoxide, CO (e.g., Au, Ag); (3) formic acid, HCOOH (e.g., Pb, Sn, Zn, Bi) and (4) Hydrogen, H_2 (e.g., Pt, Ni). Whereas, a few of them are restricted due to a variety of factors and others are concerned about their toxicity and environmental impact, while some others are not (Dimeglio and Rosenthal 2013). Figure 4.3 represent the possible pathways for ECR of carbon to achieve a variety of products (Birdja et al. 2019).

In recent studies, Zheng et al. (2019) used Nickle (Ni) atoms on carbon NPs support to obtain nearly 100% selectivity for CO production. Carbon-based materials doped with heteroatoms (e.g., graphene-based fibres, diamond, graphene dots, nanoporous carbon, and nanotubes) have been frequently used as metal-less electrocatalysts. Because of the positive charge on C-atom and negative charge on the heteroatoms, they have huge areas (surface areas), excellent conductivity, great stability, inexpensive, and brilliant activity. Xu et al. (2016) functionalized C-nanotubes (CNTs) with N-atoms and obtained 90% CO generation efficiency and 60-h stabilisation. By stabilising the radical anion, the N-atoms decreased the starting barrier for reduction and enhanced its intricate performance. Carbon-based materials metallic doping have also shown promising activity, but there are still certain drawbacks, such as low selectivity and efficiency (Sun et al., n.d.). The electrode–electrolyte interface can be tuned to overcome these restrictions.

4.3.3 *Kinetic Studies and Particular Limitations of Metal Nanoparticles for Electrocatalysis*

For electrochemical CO_2 reduction, electrokinetic studies play a very crucial role. For example, as per the observations of Chen et al., studies characterizes kinetics on OD-gold (oxide derived) catalysts were shown, including Tafel slope analysis and varying orders of the reaction studies (Woyessa et al. 2021). Below, in Fig. 4.4, the Tafel slope of OD-gold (56 mV/dec) shows a random fall from that of polycrystalline gold particles (114 mV/dec), showing it is intrinsically a better performer than the latter (Yu et al. 2021).

And assuming the intermediates were almost nil, the slope of polycrystalline gold indicates an initiating, rate-determining step for electron transfer, whereas the Tafel slope in the case of OD-gold was representative of a starting electron pre-equilibrium before a rate-determining step (Liu et al. 2015). This result shows the modification of OD-gold was because of its good ability to maintain CO_2 radical, and the reaction intermediate generated from initial transfer of electrons, then polycrystalline gold. Some other searches at the same time by the scientific community include research by Hori et al. and the dependency of first order on both CO_2 and HCO_3^- concentration as indicative from the Fig. 4.4b, c, the author shows a possible mechanism for poly and OD-gold as indicative from the Fig. 4.4d (Lu and Jiao 2016). Pointing out at HCO_3^-,

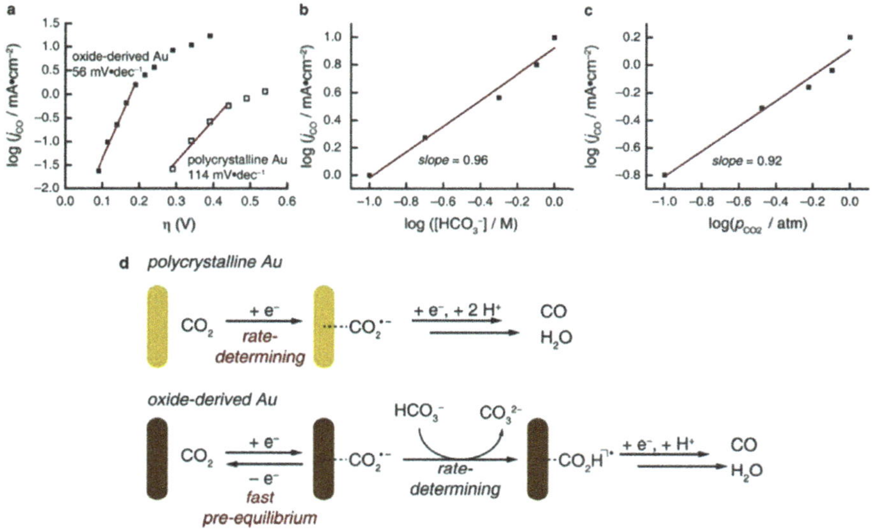

Fig. 4.4 Electro-kinetics of carbon dioxide reduction to carbon-monoxide on oxide-derived gold. Graphs showing carbon-monoxide generation of partial density of current versus **a** potential on OD-gold and polycrystalline gold, **b** $NaHCO_3$ concentration at a static potential, **c** carbon-dioxide partial pressure at a static potential and **d** mechanism for reduction of carbon dioxide (Lu and Jiao 2016)

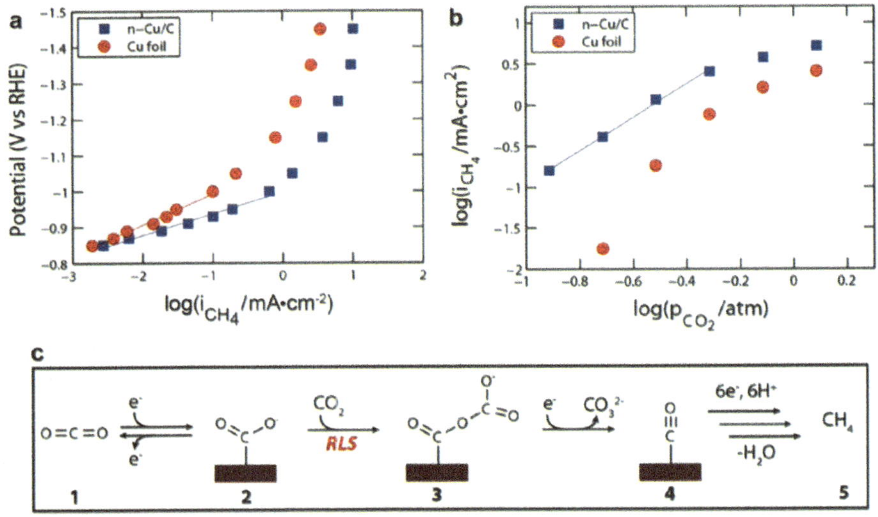

Fig. 4.5 Representation from Tafel analysis, **a** For n-Cu/C and copper foil, **b** methanation current density as a function of the pressure of CO_2 at −1.25 V. **c** Proposed mechanism for ECR of carbon-dioxide to methane, including the rate-limiting step (RLS) (Lu and Jiao 2016)

(Dimeglio and Rosenthal 2013), it acts as a proton donor in the above reaction, i.e., in CO_2RR. The studies of Manthiram et al. show the kinetic behaviors of carbon-braced carbon nanoparticle electrodes in CO_2RR to CH_4 (Lu and Jiao 2016). The n-Cu/C also represents the Tafel slope change to about 65 mV/dec from above 89 mV/dec of polycrystalline Cu foil (Lu and Jiao 2016; Woyessa et al. 2021), representing a fast and convenient single electron transfer process before a rate-limiting and non-electrochemical step.

The methanation reaction in the sequence of the methanation current on carbon-dioxide concentration was established to be an uncommon number of 2 as shown in Fig. 4.5b above, while if we compare with CO_2RR on Cu foil was often considered to follow first-order dependence on CO_2 in aqueous electrolytes (Dimeglio and Rosenthal 2013; Lu and Jiao 2016). When the author joins these particulars, it results in a proposed reaction mechanism (Lu and Jiao 2016) of CO_2RR to CH_4 on n-Cu/C material as indicative from Fig. 4.5c. In this way, the mechanism is proposed for the reduction of CO_2 electrochemically to CH_4 by determining the particular rate-limiting step (Kuhl et al. 2014).

4.4 Future Directions

CO_2 usage, especially when combined with renewable energy, is critical for achieving a sustainable energy future. Recent developments of innovative CO_2 reduction electrocatalysts, new knowledge of CO_2 reduction processes, and CO_2 electrolysis design methodologies were discussed in the present book chapter (Lu and Jiao 2016). These are the most critical issues for moving CO_2 electrolysis technology closer to commercialisation. The following directions, in our opinion, may present new prospects for breaking through CO_2 electrolysis technology in the coming future (Lateef and Nazir 2017).

4.4.1 Hunting for Better CO_2 Catalyst

The search for carbon-dioxide reduction catalysts with good selectivity for better-value compounds is always a worthwhile endeavour. Only CO and $HCOO^-$ can be created particularly using CO_2 electrolysis employing one of the best electrocatalysts at this time. Novel catalysts that can selectively create other best-value compounds from carbon dioxide will undoubtedly offer up new possibilities (Lu and Jiao 2016). Catalysts other than mono-metallics, such as bi-metallic and non-metallic catalysts, could be investigated in the future. Another intriguing field for potential breakthroughs is nanostructuring catalysts. In comparison to their bulk equivalents, many nanostructured catalysts not only enable substantially greater currents but may also exhibit distinctive catalytic characteristics (Lu and Jiao 2016; Woyessa et al. 2021).

4.4.2 In-Depth Understanding of Reaction Mechanism

It's still a puzzle how CO_2 molecules are reduced on the electro-catalytic surface. It's also unclear why HCO_3^- and CO_3^{-2} can't be electrochemically decreased at a significant pace (Kuhl et al. 2014; Lu and Jiao 2016). A thorough understanding of the response mechanisms of ECR may provide solutions to such important problems, allowing for the logical design of future-generation CO_2 reduction electrocatalysts (Lu and Jiao 2016) as well as innovative CO_2-capture-to-CO_2 electrochemical conversion methods (Diaz et al. 2018). Efforts to create a novel in situ spectroscopic methods are greatly appreciated since they allow us to investigate catalytic reactions at the electrode/electrolyte interface. Other characterisation approaches, including time-resolved and surface sensitive techniques, could reveal more about the reaction process (Lu and Jiao 2016).

4.4.3 Maneuvering CO_2-electrode/electrolyte Interface

CO_2 ECR is a multi-electron and multi-proton transfer process (Lu and Jiao 2016). Due to the limited solubility of CO_2 in electrolytes, delivering CO_2 to the cathodic surface effectively turns to simultaneously achieving an excellent current density and a brilliant CO_2 reduction, particularly in an in experiment CO_2 electrolyzer (Woyessa et al. 2021). Unfortunately, there has been relatively little work put into developing CO_2 electrolyzer systems. Future research should look into new mechanisms for introducing CO_2 into cells, as well as effective product separation technologies and membrane/membrane-less designs (Lateef and Nazir 2017).

4.5 Summary and Conclusion

The increasing population serves as an inoculum to positive feedback mechanism for increasing environmental crises, i.e., with the increasing population, the need and requirement for natural resources such as water, air, soil, food, etc., increases and contributes to more environmental problems and pollution. In the present chapter, the sustainable and effective method for capturing carbon is achieved by electrochemical carbon dioxide reduction in high energy density reduced products. Metal and metal oxide nanoparticles are an effective tool for synthesizing and manufacturing strong catalysts that help in the overall CO_2 management process. The increasing Carbon dioxide (CO_2) concentration in the atmosphere serves as an alarming symbol for humankind to take the required steps to minimize the alarming high release of CO2 into the atmosphere. This can only be achieved by sustainable means for carbon capture and storage utilization, and conversion. Recently the Intergovernmental Panel on Climate Change (IPCC) released its sixth assessment report, and the contribution of the first working group, i.e., Climate Change 2021: The Physical Science Basis (IPCC 2021), which indicates the contribution of anthropogenic activities in increasing the global temperature, melting of ice sheets and glaciers, increasing sea level, changing global monsoonal patterns, increased intensity, and frequency of both the floods and heatwaves, changing cropping and agriculture pattern and all this contribute to the climate change. And according to all the studies performed in the sixth assessment report, the carbon-dioxide is found to be one of the most influential greenhouse causing gas, and also its release is much more as compared to many other gases in the atmosphere (IPCC 2021). So, being more concerned for capturing, conserving, and converting the CO_2 into reduced high energy usable products by sustainable and effective methods becomes the need of the hour. The complete CO_2 electrolysis process structure is determined by the required CO_2 reduction product as well as the CO_2 transfer mechanism to the cathode of the cell. This chapter has highlighted a few key challenges linked to the current system, alternatives, synthesis of metal nanocomposites, electrochemical reduction mechanism, electrokinetic of the molecules, future directions for the electrochemical reduction, and utilization of reduced products.

Acknowledgment The corresponding author(SPS) acknowledges the funding received from SERB, Department of Science and Technology, Government of India, and IIT Bombay for carrying out this work.

References

Akhtar MS, Panwar J, Yun YS (2013) Biogenic synthesis of metallic nanoparticles by plant extracts. ACS Sustain Chem Eng 1(6):591–602. https://doi.org/10.1021/sc300118u

Birdja YY, Pérez-Gallent E, Figueiredo MC, Göttle AJ, Calle-Vallejo F, Koper MTM (2019) Advances and challenges in understanding the electrocatalytic conversion of carbon dioxide to fuels. Nat Energy 4(9):732–745. https://doi.org/10.1038/s41560-019-0450-y

Bohra D, Ledezma-Yanez I, Li G, de Jong W, Pidko EA, Smith WA (2019) Lateral adsorbate interactions inhibit HCOO—while promoting CO selectivity for CO_2 electrocatalysis on silver. Angewandte Chemie Int Ed 58(5):1345–1349. https://doi.org/10.1002/anie.201811667

Bui M, Adjiman CS, Bardow A, Anthony EJ, Boston A, Brown S, Fennell PS, Fuss S, Galindo A, Hackett LA., Hallett JP, Herzog HJ, Jackson G, Kemper J, Krevor S, Maitland GC, Matuszewski M, Metcalfe IS, Petit C et al (2018) A ZIF-8 decorated TiO_2 grid-like film with high CO_2 adsorption for CO_2 photoreduction. Environ Chem Lett 18(5):369–375. https://doi.org/10.1016/j.jcou.2018.01.024

Chen C, Khosrowabadi Kotyk JF, Sheehan SW (2018) Progress toward commercial application of electrochemical carbon dioxide reduction. Chem 4(11):2571–2586. https://doi.org/10.1016/j.chempr.2018.08.019

Chen HY, Deng J, Wang Y, Wu CQ, Li X, Dai HW (2020) Hybrid cell membrane-coated nanoparticles: a multifunctional biomimetic platform for cancer diagnosis and therapy. Acta Biomaterialia 112:1–13. https://doi.org/10.1016/j.actbio.2020.05.028

Chen Z, Li R, Zheng X, Liu J (2021) Carbon sequestration of steel slag and carbonation for activating RO phase. Cement Concr Res 139. https://doi.org/10.1016/j.cemconres.2020.106271

Diaz LA, Gao N, Adhikari B, Lister TE, Dufek EJ, Wilson AD (2018) Electrochemical production of syngas from CO2 captured in switchable polarity solvents. Green Chem 20(3):620–626. https://doi.org/10.1039/c7gc03069j

Dimeglio JL, Rosenthal J (2013) Selective conversion of CO_2 to CO with high efficiency using an inexpensive bismuth-based electrocatalyst. J Am Chem Soc 135(24):8798–8801. https://doi.org/10.1021/ja4033549

Ding C, Feng C, Mei Y, Liu F, Wang H, Dupuis M, Li C (2020) Carbon nitride embedded with transition metals for selective electrocatalytic CO2 reduction. Appl Catal B Environ 268:118391. https://doi.org/10.1016/j.apcatb.2019.118391

Fawcett D, Verduin JJ, Shah M, Sharma SB, Poinern GEJ (2017) A review of current research into the biogenic synthesis of metal and metal oxide nanoparticles via marine algae and seagrasses. J Nanosci 2017:1–15. https://doi.org/10.1155/2017/8013850

Garg S, Li M, Weber AZ, Ge L, Li L, Rudolph V, Wang G, Rufford TE, Sun Z., Ma T, Tao H, Fan Q, Han B (2020). Advances and challenges in electrochemical CO2 reduction processes: an engineering and design perspective looking beyond new catalyst materials. J Mater Chem A 3(4):1511–1544. https://doi.org/10.1016/j.chempr.2017.09.009

Hamrouni A, Lachheb H, Houas A (2013) Mater Sci Eng C 178(20):1371–1379

He X, Gan Z, Fisenko S, Wang D, El-Kaderi HM, Wang WN (2017) Rapid formation of metal-organic frameworks (MOFs) based nanocomposites in microdroplets and their applications for

CO_2 photoreduction. ACS Appl Mater Interfaces 9(11):9688–9698. https://doi.org/10.1021/acsami.6b16817
Huang Z, Dong P, Zhang Y, Nie X, Wang X, Zhang X (2018) A ZIF-8 decorated TiO_2 grid-like film with high CO_2 adsorption for CO_2 photoreduction. J CO_2 Utilization 24:369–375. https://doi.org/10.1016/j.jcou.2018.01.024
Ingale AG, Chaudhari AN (2013) Biogenic synthesis of nanoparticles and potential applications: an eco-friendly approach. J Nanomed Nanotechnol 4(2):7. https://doi.org/10.4172/2157-7439.1000165
IPCC (2021) Sixth Assessment Report IPCC. https://www.ipcc.ch/report/ar6/wg1/downloads/report/IPCC_AR6_WGI_Full_Report.pdf
Kaviyarasu K, Maria Magdalane C, Anand K, Manikandan E, Maaza M (2015) Synthesis and characterization studies of MgO:CuO nanocrystals by wet-chemical method. Spectrochimica Acta - Part A Mol Biomol Spectroscopy 142:405–409. https://doi.org/10.1016/j.saa.2015.01.111
Kuhl KP, Hatsukade T, Cave ER, Abram DN, Kibsgaard J, Jaramillo TF (2014) Electrocatalytic conversion of carbon dioxide to methane and methanol on transition metal surfaces. J Am Chem Soc 136(40):14107–14113. https://doi.org/10.1021/ja505791r
Lateef A, Nazir R (2017) Metal nanocomposites : synthesis, characterization and their applications. Sci Appl Tailored Nanostructures 239–256
Li Y, Chu S, Shen H, Xia Q, Robertson AW, Masa J, Siddiqui U, Sun Z (2020) Achieving highly selective electrocatalytic CO_2 reduction by tuning CuO-Sb_2O_3 nanocomposites. ACS Sustain Chem Eng 8(12):4948–4954. https://doi.org/10.1021/acssuschemeng.0c00800
Lim KRG, Handoko AD, Nemani SK, Wyatt B, Jiang HY, Tang J, Anasori B, Seh ZW (2020) Rational design of two-dimensional transition metal carbide/nitride (MXene) hybrids and nanocomposites for catalytic energy storage and conversion. ACS Nano 14(9):10834–10864. https://doi.org/10.1021/acsnano.0c05482
Liu Y, Chen S, Quan X, Yu H (2015) Efficient Electrochemical reduction of carbon dioxide to acetate on nitrogen-doped nanodiamond. J Am Chem Soc 137(36):11631–11636. https://doi.org/10.1021/jacs.5b02975
Lote DA (2014) Literature survey on electrochemical reduction of CO_2. Int J Electron Electr Eng 7(4):341–346. http://www.irphouse.com
Lu Q, Jiao F (2016) Electrochemical CO_2 reduction: electrocatalyst, reaction mechanism, and process engineering. Nano Energy 29:439–456. https://doi.org/10.1016/j.nanoen.2016.04.009
Nasrollahzadeh M, Mohammad Sajadi S, Rostami-Vartooni A (2015) Green synthesis of CuO nanoparticles by aqueous extract of Anthemis nobilis flowers and their catalytic activity for the A3 coupling reaction. J Colloid Interface Sci 459:183–188. https://doi.org/10.1016/j.jcis.2015.08.020
Nazari M, Halladj R (2014) J Taiwan Inst Chem Eng 45(5):2518–2525
NOAA (2020) Climate Gov. https://www.climate.gov/news-features/understanding-climate/climate-change-atmospheric-carbon-dioxide
Peng M, Ci S, Shao P, Cai P, Wen Z (2019) Cu 3 P/C nanocomposites for efficient electrocatalytic CO_2 reduction and Zn–CO_2 battery. J Nanosci Nanotechnol 19(6):3232–3236. https://doi.org/10.1166/jnn.2019.16589
Ray C, Pal T (2017) Recent advances of metal-metal oxide nanocomposites and their tailored nanostructures in numerous catalytic applications. J Mater Chem A 5(20):9465–9487. https://doi.org/10.1039/c7ta02116j
Reddy KR (2017) Green synthesis, morphological and optical studies of CuO nanoparticles. J Mol Struct 1150:553–557. https://doi.org/10.1016/j.molstruc.2017.09.005
Sun Z, Ma T, Tao H, Fan Q, Han B (n.d.) Fundamentals and challenges of electrochemical CO_2 reduction using two-dimensional materials. https://doi.org/10.1016/j.chempr.2017.09.009
Wang Z, Yao J, Li Z, Yang K, Guo J, Zhang S, Sherazi TA, Li S (2018) Bio-inspired fabrication of asymmetric wettability Janus porous membrane for secure F-oil infused F-free-membrane filtration. J Membr Sci 566:161–167. https://doi.org/10.1016/j.memsci.2018.08.071

Woyessa GW, dela Cruz JB, Rameez M, Hung CH (2021) Nanocomposite catalyst of graphitic carbon nitride and Cu/Fe mixed metal oxide for electrochemical CO_2 reduction to CO. Appl Catal B: Environ 291:120052. https://doi.org/10.1016/j.apcatb.2021.120052

Yu J, Wang J, Ma Y, Zhou J, Wang Y, Lu P, Yin J, Ye R, Zhu Z, Fan Z (2021) Recent progresses in electrochemical carbon dioxide reduction on copper-based catalysts toward multicarbon products. Adv Functional Mater 2102151. https://doi.org/10.1002/adfm.202102151

Yuan H, Kong L, Li T, Zhang Q (2017) A review of transition metal chalcogenide/graphene nanocomposites for energy storage and conversion. Chin Chem Lett 28(12):2180–2194. https://doi.org/10.1016/j.cclet.2017.11.038

Zhao Z, Sun Y, Dong F (2015) Graphitic carbon nitride based nanocomposites: a review. Nanoscale 7(1):15–37. https://doi.org/10.1039/c4nr03008g

Zhu M, Diao G (2011) Review on the progress in synthesis and application of magnetic carbon nanocomposites. Nanoscale 3(7):2748–2767. https://doi.org/10.1039/c1nr10165j

Chapter 5
Metal Nanocomposites—Emerging Advanced Materials for Efficient Carbon Capture

Uttama Mukherjee

Abstract The contribution of CO_2 towards greenhouse gas emissions resulting into environmental deterioration, global warming, and as a contaminant impeding the performance of fossil fuels, has led to global efforts to mitigate its emission. Among various technologies employed for carbon capture and sequestration (CCS), the emergence of nanocomposites as state-of-the-art materials has triggered immense scope of research. A nanocomposite is a merger of different materials where at least one of the materials has a size ranging between 1 and 100 nm. The nanomaterials thus incorporated can vary between nanoparticles, nanofibers, carbon nanotubes or activated carbon. The basic building blocks of nanocomposites comprise of metals, ceramics and polymers. Nanocomposites exhibit a combination of properties derived from their constituent components, which renders improved and evolved characteristics to this material. This chapter will focus on the most recent developments, specifically in the domain of metal nanocomposite materials for CO_2 capture. Metal nanocomposites have shown promising adsorbent properties owing to large surface area, plenty of nanopores, enhanced reactivity, porosity and a greater ease of synthesis. Some of the important metal nanocomposites for carbon capture that will be discussed include—iron oxide (Fe_3O_4)-graphene nanocomposite suitable for CO_2 capture at elevated temperatures and pressures; activated carbon infused with Mg, Al, Cu, Ni and mixed metal nanocomposites, and MgO/carbon nanocomposite for improved and faster CO_2 adsorption. In addition, some polymeric nanocomposites such as—polysulfone combined with activated carbon–metal (Ni and Co) nanocomposites and organic polymer membranes infused with amine functionalized SiO_2 or TiO_2 nanoparticles are also included as thermally stable, mechanically strong, energy-efficient and economically viable carbon capture materials.

Keywords CO_2 capture · Metal nanocomposites · Activated carbons · Graphene · Polymeric nanocomposites

---

U. Mukherjee (✉)
Department of Chemistry and Centre for Energy Science, Indian Institute of Science Education and Research, Pune, Dr. Homi Bhabha Road, Pashan, Pune, Maharashtra 411008, India

S. P. Singh et al. (eds.), *Metal Nanocomposites for Energy and Environmental Applications*, Energy, Environment, and Sustainability,
https://doi.org/10.1007/978-981-16-8599-6_5

5.1 Introduction

One of the most challenging environmental issues since decades has been to limit greenhouse gas (GHG) emissions across the globe. The ever-growing population resulting into continuous demand of energy (for power generation, transportation etc.), has given rise to constant enhancement in the CO_2 emissions worldwide (Fig. 5.1) (International Energy Agency (IEA) 2009). Apart from causing environmental pollution, it leads to global warming and abnormal temperature variations. CO_2 is also present as an impurity in energy gas resources such as natural gas, biogas etc. and impairs their performance, requires high energy for transport and corrodes equipment (Sanz-Pérez et al. 2016; George et al. 2016; Huang et al. 2014; Wang et al. 2016; Ben-Mansour et al. 2016). It is anticipated that the atmospheric concentration of CO_2 may reach beyond 550 pm by 2035 (Hileman 2006). Thus, the capture and sequestration of CO_2 is an area of intensive research. Carbon capture and storage or sequestration (CCS) is an integrated process of carbon capture (separating CO_2 from its sources like burning of fossil fuels), its transport to a storage location and storage to appropriate places (underground or in the ocean) or converting it to useful materials (Global CCS Institute 2017). According to the Intergovernmental Panel on Climate Change (IPCC) report 2013–14 (AR5), there is a target to achieve fifty to eighty percent reduction in CO_2 emissions by 2050 and a complete reduction by 2100, which gives CCS a very significant position for clean energy options (IPCC 2013).

Various CCS technologies have been developed so far, such as chemical and physical absorption, membrane separation, chemical looping, physical adsorption and cryogenic distillation etc. (He et al. 2013). These techniques are represented in Fig. 5.2. Traditional chemical absorption is already a developed technology for CO_2 separation but at the same time it is energy intensive, cost-inefficient and its impact

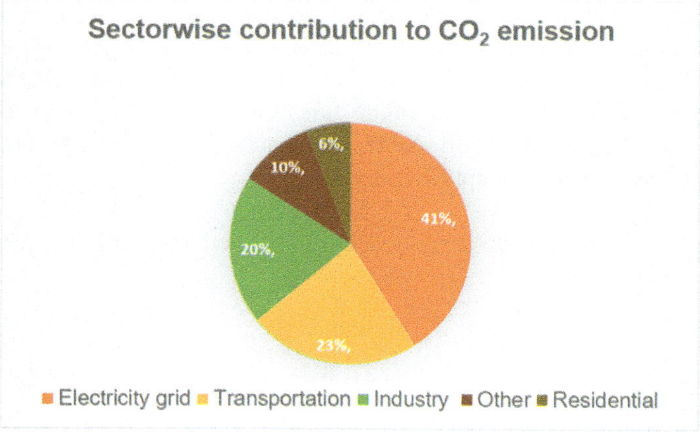

Fig. 5.1 CO_2 emission sources from various sectors (International Energy Agency (IEA) 2009)

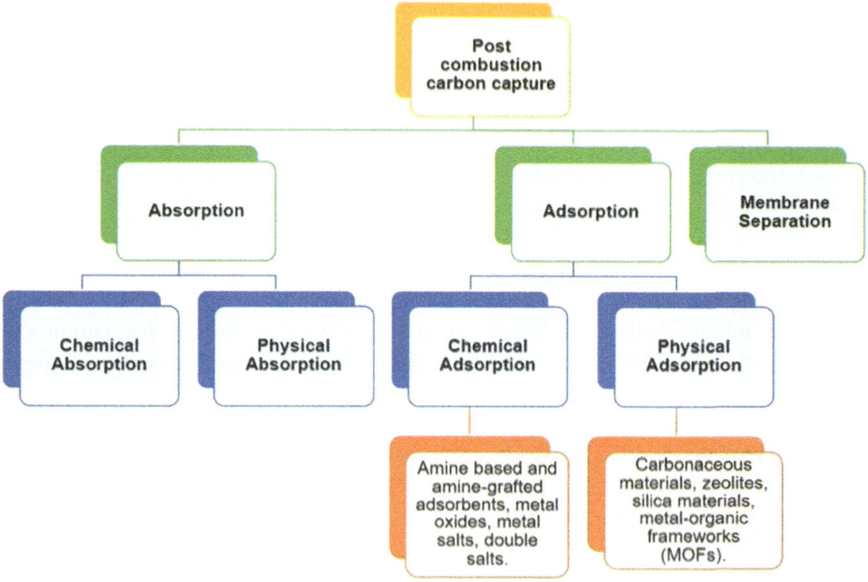

Fig. 5.2 Schematic representation of post-combustion CO_2 capture with major focus on adsorption techniques

on the environment is debatable (He 2018). Among other methods, the technique of adsorption using solid sorbents is one of the most sought-after methods for CO_2 capture on account of its high efficiency, simplicity, low energy requirement and low operational cost (Hao et al. 2011).

Currently, a large number of solid materials including solid amines, organic polymers, metal–organic frameworks (MOFs), carbon-based solids and metal oxides/salts are available as sorbents to capture CO_2 (Jadhav et al. 2007a; Karra and Walton 2010; Mu et al. 2010; Wang and Yang 2012). The prerequisites for a superior adsorbent are high adsorption capacity and fast adsorption kinetics. However, the conventional adsorbents have certain drawbacks such as limited adsorption capacity and low adsorption rates. Thus, it becomes imperative to explore novel advanced materials with improved adsorption characteristics. The advent of nanotechnology with advanced nanomaterials has been one of the most thrilling scientific discoveries in chemical science. Nanotechnology is the manoeuvring of materials at a very miniscule (nanoscopic) scale. Nanomaterials have now emerged as multidisciplinary materials, finding use in chemical engineering, environmental chemistry, material science and other branches of chemical science (Bogue 2011; Zhang et al. 2021; Hassan et al. 2021; Bhat et al. 2021). Nanostructured materials such as nanoparticles, carbon nanotubes, nanosheets and nanofibers etc. have evolved as promising adsorbent materials compared to their bulk counterparts. The nano scale dimensions (1–100 nm) of these materials significantly alter their physical and chemical properties, such as the development of large surface area to volume ratios and vast interfacial reactivity (Twardowski 2007).

In the recent past, the rapid evolution of nanotechnology has given rise to nanocomposites (NC) as state-of-the-art materials and triggered immense scope of research. A nanocomposite is a merger of different materials where at least one of the materials has a size ranging between 1 and 100 nm (Ventra et al. 2004; Camargo et al. 2009a). Thus, there are two parts in nanocomposites

i. Continuous phase (matrix material—metallic/non-metallic/polymer)
ii. Dispersed phase (reinforcing material).

The nanomaterials thus incorporated can vary between nanoparticles, nanofibers, carbon nanotubes (CNTs) or activated carbons (ACs). Incorporating nanoparticles into different bulk materials to form NC has resulted into improved and desired properties compared to single nanoparticles, which in turn, has led to its suitability in environmental applications. For example, carbon nanotubes have shown large length-to-diameter ratio (aspect ratio higher than 1000) together with excellent electrical conductivity and mechanical property (Ounaies et al. 2003; Weisenberger et al. 2003) and nanosheets possess large surface area that is suitable for the fabrication of reinforced polymeric composites (Wang et al. 2012). In addition, they have remarkable catalytic properties such as photo-/thermo-catalysis (Zeng et al. 2015). Thus, there is huge potential to engineer such materials as per requirement. The materials from which NC is prepared can be classified into three major building blocks i.e. metals, ceramics and polymers (Table 5.1). On the basis of their structural properties, the NC can be classified as—Nano-layered composites, nano-filamentary composites and nano-particulate composites.

The superiority of nanocomposites over other composite materials is because of

i. Thermal stability, better electrical conductivity;
ii. Enhanced chemical resistance;
iii. Enhanced mechanical properties—improved ductility without loss of strength, scratch resistance;
iv. Reduced permeability of gases, water and solvents;
v. Inhibits flame and smoke generation;

Table 5.1 Nanocomposite building blocks

Class	Examples	Ref
Metal	Fe–Cr/Al_2O_3 (<10 nm), Ni/Al_2O_3 (50–60 nm), Mg/CNT	Laurent et al. (1994), Chen et al. (2010), Abazari et al. (2020)
Ceramic	Al_2O_3/SiO_2 (~10 nm), Al_2O_3/TiO_2 (~75 nm), Al_2O_3/SiC (5–7.1 μm), Al_2O_3/CNT	Palmero (2015), Choi and Awaji (2005), Klement et al. (2018), Peigney et al. (2000)
Polymer	Thermoplastic/thermoset polymer/layered silicates, polyester/TiO_2, polymer/CNT	Wang et al. (2004), Jordan et al. (2005), Xie et al. (2005)

Table 5.2 Properties of nanomaterials for CO_2 capture (Kumar et al. 2020)

Nanomaterials	Properties				
	Temperature (°C)	Pressure (kPa)	Surface area (m^2/g)	CO_2 adsorption capacity	Ref
CaO–CaO_3	20	–	10.40	99% absorption capacity (1st cycle) and retaining > 50% after 50 cycles	Yang et al. (2009)
TiO_2	30	20	43	(Photocatalytic reduction of CO_2)	Tahir and Amin (2013)
CaO (Si/Ca)	10	150	78	~ 85% molar conversion to $CaCO_3$ (carbonation)	Lu et al. (2009)
Zeolites (CNTs)	20	100	788	1.44 mol/kg	Su et al. (2009)
ACs	25	100	1284	2.23 mol/kg	Lu et al. (2008)

vi. Better optical properties (light transmission and particle size are correlated).

This chapter will focus on the most recent developments, specifically in the domain of metal nanocomposite materials for CO_2 capture. Metal nanocomposites are formed by blending metal nanoparticles with glass, ceramic or polymer. These have shown promising adsorbent properties owing to large surface area, plenty of nanopores, enhanced reactivity, porosity and a greater ease of synthesis (Goharibajestani et al. 2019). Thus, metal NC have attracted state-of-the-art research in various environmental applications including carbon capture (Cortés-Arriagada et al. 2018; Szczęśniak et al. 2017). Table 5.2 shows few properties of nanomaterials that have been used for CCS.

Obtaining nanocomposites with uniform size is essentially important as the various properties of NC materials discussed above are mainly dependent on particle size and dimensions (Rogach et al. 2002). There are many synthetic approaches to obtain NC and the choice depends on the desired properties (Stankic et al. 2016; Khodaei et al. 2011). For example, synthesis of NC can be broadly classified into two parts (Ajayan et al. 2003; Oliveira and Machado 2013)

i. Top-down approach (physical methods); advantage—large amount of NC can be produced but attainment of uniformity in size is difficult.
ii. Bottom-up approach (chemical methods); advantage—particle size can be controlled and different shapes of NC (nano-rods, nanotubes, nano-wires etc.) can be synthesised by varying the reaction conditions.

Types of synthetic methods for metal-based nanocomposite systems with their respective advantages and limitations is presented in Table 5.3.

Table 5.3 Synthetic approaches for metal-based nanocomposites (Camargo et al. 2009b)

Methods	System		Advantages	Limitations	Ref
	Reinforcement	Matrix			
Spray Pyrolysis	W, Fe	Cu, MgO	Effective preparation of ultra-fine, spherical and homogeneous powders in multicomponent systems, reproductive size and quality	High cost associated with producing large quantities of uniform, nanosized particles	Choa et al. (2003)
Liquid Infiltration	Pb, W, Fe, Nb, Al	Cu, C_{60}	Short contact times between matrix and reinforcements; moulding into different and near net shapes of different stiffness and enhanced wear resistance; rapid solidification; both lab scale and industrial scale production	Use of high temperature; segregation of reinforcements; formation of undesired products during processing	Choa et al. (2003), Yoon et al. (2002), Provenzano et al. (1992), Contreras and Lopez (2004), Khalid et al. (2003)
Rapid Solidification Process (RSP)	Pb, X/Zr (X = Si, Cu, Ni)	Al, Fe	Simple; effective	Only metal–metal nanocomposites; induced agglomeration and non-homogeneous distribution of fine particles	Bhattacharya and Chattopadhyay (2004), Bhattacharya and Chattopadhyay (2001), Srinivasan and Chattopadhyay (2004), Branagan (2000), Branagan and Tang (2002)
RSP with ultrasonics	SiC	Al	Good distribution without agglomeration, even with fine particles		Li et al. (2004)

(continued)

Table 5.3 (continued)

Methods	System		Advantages	Limitations	Ref
	Reinforcement	Matrix			
High Energy Ball Milling	$Al_2 O_3$	Cu	Homogeneous mixing and uniform distribution		Ying and Zhang (2000)
CVD/PVD (Chemical/Physical Vapour Deposition)	Mo, W, Pb	Al, Cu	Capability to produce highly dense and pure materials; uniform thick films; adhesion at high deposition rates; good reproducibility	Optimization of many parameters; cost; relative complexity	Choy (2003), Joseph et al. (2005), Chow et al. (1990), Haubold and Gertsman (1992); Holtzt and Provenzano (1994)
Chemical Processes (Sol–gel, Colloidal)	Ag, Au, SiO_2	Fe	Simple; low processing temperature; versatile; high chemical homogeneity; rigorous stoichiometry control; high purity products	Weak bonding, low wear-resistance, high permeability and difficult control of porosity	Cushing et al. (2004), West et al. (2003), Kamat et al. (2002), Roy et al. (1993), Carpenter et al. (2000)

The present chapter comprises of the discussion on metal nanocomposites mainly utilised for CO_2 capture. The following recent developments in the field are included

a. **Iron oxide (Fe_3O_4)-graphene nanocomposite**—Suitable for CO_2 capture at elevated temperatures and pressures (Mishra and Ramaprabhu 2014);
b. **Activated carbon infused with Mg, Al, Cu, Ni and mixed metal nanocomposites**—With improved textural properties and concomitant chemical and physical adsorption (Nowrouzi et al. 2018).
c. **MgO/carbon nanocomposite**—For better and faster CO_2 adsorption (Li et al. 2020).
d. **Polymeric nanocomposites**—As thermally stable, mechanically strong, energy-efficient and economically viable carbon capture materials. The following two are included

    i. Polysulfone combined with activated carbon–metal (Ni and Co) nanocomposites (Nisar et al. 2020a).
    ii. Organic polymer membranes infused with amine functionalized SiO_2 or TiO_2 nanoparticles (Khdary and Abdelsalam 2020).

The discussion on each of the metal nanocomposites included in this chapter is focussed mainly on its synthesis and fabrication, techniques used for its characterization and its carbon capture capacity. All experimental procedures employed for the synthesis and characterization of the metal nanocomposites have not been included in detail in this chapter and can be found in the corresponding articles (Mishra and Ramaprabhu 2014; Nowrouzi et al. 2018; Li et al. 2020; Nisar et al. 2020a; Khdary and Abdelsalam 2020).

5.2 Metal Nanocomposites for Carbon Capture

5.2.1 Iron Oxide (Fe_3O_4)-Graphene Nanocomposite

As already discussed above, among other conventional methods used for CCS, the technique of adsorption using porous sorbents is economically viable as well as effective. Adsorbents such as ACs, zeolites, molecular sieves, MOFs etc. have already been reported for CO_2 capture (Jadhav et al. 2007b; Xu et al. 2003). ACs, with high CO_2 adsorption capacity at higher pressures, have been found to perform better than zeolites (Sircar et al. 1996; Siriwardane et al. 2001; Burchell et al. 1997). Accordingly, the nano-sized materials based on carbon such as carbon nanotubes, graphene etc. also act as CO_2 adsorbents pertaining to their large surface area and porosity (Cinke et al. 2003; Ghosh et al. 2008). CNT-based nanocomposites being more cost-intensive and comparatively low performers, graphene becomes a better choice for CO_2 capture purposes (Mishra and Ramaprabhu 2011a). However, for maintaining CO_2 capture capacity at higher temperatures such as the exhaust of various industries (automobile, steel, cement, power plants etc.), the efficiency of graphene decreases sharply. At this point, the graphene-based nanocomposite materials provide a solution to the problem. It has already been established that CO_2 interacts chemically with metal oxides such as CaO, MgO etc. (Pacchioni et al. 1994; Wang et al. 2011; Baniecki et al. 2009). Mishra and Ramaprabhu (2014) have synthesized a NC by including nanocrystalline iron oxide (Fe_3O_4) particles on the surface of graphene sheets. The resulting NC has shown efficient CO_2 adsorption even at higher temperature and pressures.

5.2.1.1 Synthesis of the Nanocomposite

The synthesis of the Fe_3O_4-graphene nanocomposites included the following two steps

a. Synthesis of graphene (performed via hydrogen induced thermal exfoliation of graphite oxide at 200 °C) (Mishra and Ramaprabhu 2011a; Hummers and Offeman 1958). The obtained graphene sheets were named as hydrogen exfoliated graphene (HEG).
b. Decorating nanocrystalline Fe_3O_4 particles over HEG via a chemical route which utilized $FeCl_3 \cdot 6H_2O$, $FeSO_4 \cdot 7H_2O$ and ammonia solution (Mishra and Ramaprabhu 2010).

5.2.1.2 Characterization of the Nanocomposite

The characterization of the NC was performed using

a. **Microscopic techniques**

   SEM (scanning electron microscopy)—For surface morphology, as depicted in Fig. 5.3a which shows graphene sheets fabricated with Fe_3O_4 nanoparticles.

   TEM (transmission electron microscopy)—For structural morphology, as depicted in Fig. 5.3b which shows homogeneously distributed Fe_3O_4 nanoparticles on the surface of graphene sheets.

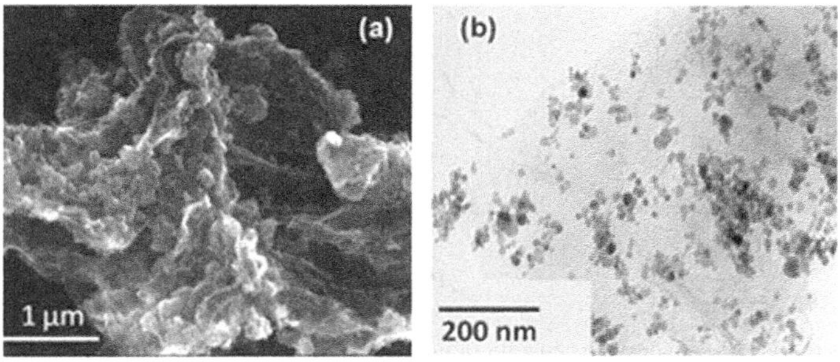

Fig. 5.3 **a** SEM image and **b** TEM image of Fe_3O_4–HEG nanocomposite. Reproduced with permission from Mishra and Ramaprabhu (2014), Copyright 2021, AIP Publishing LLC

b. **FTIR Spectroscopy**

For examining the vibrational characteristics of the synthesized nanocomposite. The authors obtained peaks corresponding to >C=C (1729 cm^{-1}), >C=O (1000–1200 cm^{-1}, broad peak), =CH (1578 cm^{-1}), –OH (3433 cm^{-1}), and Fe–O–Fe (598 cm^{-1}) which indicate the presence of Fe_3O_4 nanoparticles on the surface of graphene (Liu et al. 2009). Further, for the IR spectra of the nanocomposite with adsorbed CO_2, the peaks corresponding to CO_2 stretching (2332 cm^{-1}) and vibrational modes of bicarbonates (O–C–O 1384 cm^{-1}) and carbonates (C–O 1044, 1094 cm^{-1}) were also obtained, which suggest simultaneous physisorption and chemisorption of CO_2 via Fe_3O_4-HEG nanocomposite (Hlaingoo et al. 2005; Yim et al. 2004; Baltrusaitis et al. 2006).

5.2.1.3 CO_2 Capture Capacity

The authors have also depicted the increase in CO_2 capture capacity with pressure at various temperatures (Fig. 5.4). These adsorption capacities were found to be higher than in case of pure HEG (Himeno et al. 2005). They have also emphasized that the entire CO_2 capture process in Fe_3O_4–HEG nanocomposite was a combined physicochemical process with physical adsorption via HEG and chemical interaction of CO_2 with Fe_3O_4 resulting into significantly higher capture capacity. The dual nature of CO_2 adsorption is also supported by FTIR data (previous section). Finally, they presented a comparative analysis of the capture capacity of Fe_3O_4-HEG NC

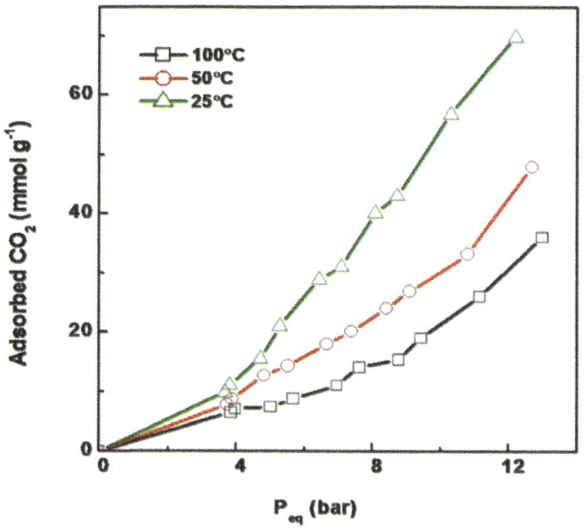

Fig. 5.4 Variation CO_2 adsorption capacity of Fe_3O_4–HEG nanocomposite with pressure (Adsorption isotherm). Reproduced with permission from Mishra and Ramaprabhu (2014), Copyright 2021, AIP Publishing LLC

Table 5.4 CO_2 capture capacity of various solid sorbents compared with Fe_3O_4–HEG nanocomposite (Mishra and Ramaprabhu 2014)

Material	Pressure (bar)/temperature (°C)	Capture capacity (mmol/g)	Ref
Molecular Sieve 13X	12/25	4–4.5	Siriwardane et al. (2001)
Maxsorb AC	35/25	25	Himeno et al. (2005)
ACs	12/25	10–15	Zhang et al. (2010)
13X Zeolite	12/25	3.2	Cavenati et al. (2004)
MCM Silica (Mobil Composition of Matter; mesoporous silica)	12/25	5	Belmabkhout et al. (2009)
MWNTs (Multi-Walled Nanotubes)	11/25	14.3	Mishra and Ramaprabhu (2011b)
Fe_3O_4–MWNTs	11/25	49	Mishra and Ramaprabhu (2011b)
Graphene	11/25	11.7	Mishra and Ramaprabhu (2012)
Fe_3O_4-Graphene	11/25	60	–

material with other commonly used adsorbents, which suggests higher CO_2 capture capacity of Fe_3O_4–HEG (Table 5.4).

5.2.2 *Activated Carbon Infused with Mg, Al, Cu, Ni and Mixed Metal Nanocomposites*

In this work by Nowrouzi et al. (2018), AC as a prominent adsorbent material for CO_2 capture has been utilized and doped with various basic metal oxides (MOs), (Cu, Ni, Mg, Al oxides) forming a nanocomposite material. The significance of the work lies in the following major aspects

i. Use of lignocellulosic waste, i.e. AC derived from Persian ironwood tree found indigenously in parts of Iranian forest.
ii. Surface modification of AC and enhancing its interaction with CO_2 by introducing metal oxides (basic oxides) and forming the NC material.
iii. Incorporating mixed metal oxides (binary metal oxides) with AC (designated as HP5/AlMg8 and HP5/CuNi3; where HP5=AC synthesized at 500 °C and activated with H_3PO_4 for enhanced CO_2 adsorption capacity.

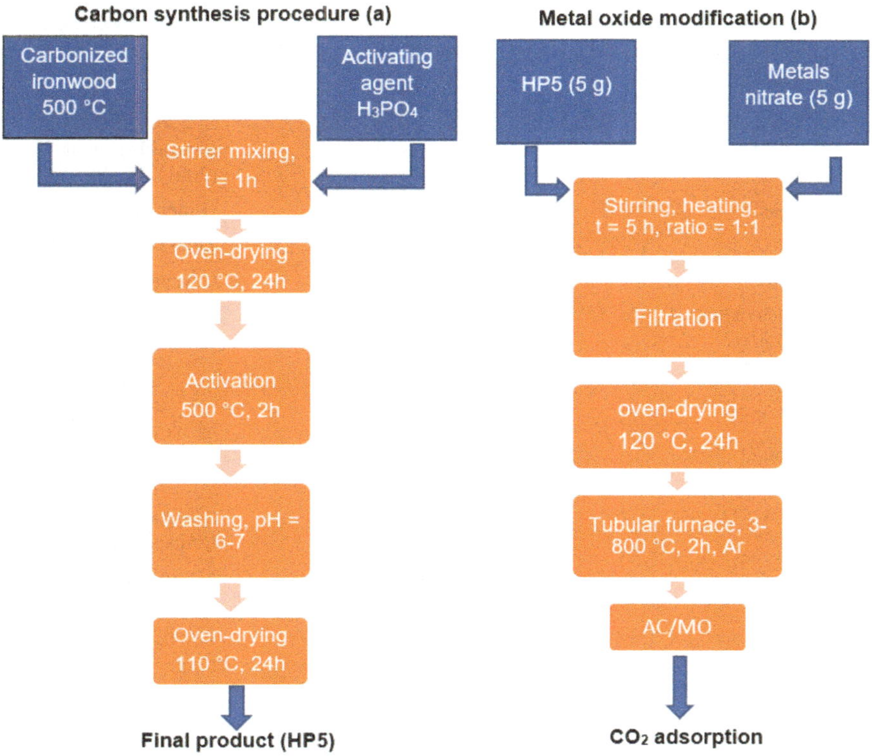

Fig. 5.5 Scheme for the preparation of **a** HP5 and **b** MO modification (Nowrouzi et al. 2018)

5.2.2.1 Synthesis of the Nanocomposite

The synthesis of the AC/MOs comprised of two steps

(a) AC preparation and
(b) MO modification.

The steps involved in the synthesis is schematically presented in Fig. 5.5 (Nowrouzi et al. 2018).

The highlighting features of these AC/MO nanocomposites include

5.2.2.2 Characterization of the AC/MO Nanocomposites

a. **Textural properties**

The properties related to the texture of the NC materials, such as specific surface area, average pore diameter, pore volume etc. is given in Table 5.5. The authors observed that the increase in surface area, micropores and the volume of microporosity can be

Table 5.5 AC/MOs adsorbent materials with their textural properties (Nowrouzi et al. 2018)

AC/MO sorbents	Textural properties						Yield %
	S_{BET}, m^2/g	V_{mic}, cm^3/g	V_{mes}, cm^3/g	V_{total}, cm^3/g	V_{mic}, %	R, nm	
CAC/Al5-1 (CAC = commercial activated carbon)	1043	0.52	0.03	0.56	99.43	0.81	**86**
HP5/Al5-1	1946	1.47	0.02	1.49	98.90	1.51	**87**
CAC/Mg8-1	907	0.43	0.002	0.43	99.53	0.86	**82**
HP5/Mg8-1	1974	1.22	0.02	1.24	98.44	1.70	**83**
CAC/Cu3-1	843	0.40	0.002	0.40	99.56	0.84	**92**
HP5/Cu3-1	1954	1.60	0.02	1.63	98.54	1.56	**91**
CAC/Ni3-1	994	0.46	0.05	0.51	90.20	0.93	**90**
HP5/Ni3-1	1945	1.59	0.04	1.63	97.47	1.79	**92**
CAC/AlMg8-1	1732	0.49	0.07	0.56	88.37	1.16	**91**
HP5/AlMg8-1	1902	1.29	0.015	1.31	99.35	1.42	**93**
CAC/CuNi3-1	1075	0.54	0.045	0.58	92.34	1.14	**90**
HP5/CuNi3-1	1918	1.42	0.15	1.57	90.23	1.37	**93**
CAC	907	0.42	0.065	0.48	86.46	0.80	–
HP5	**1802**	**0.83**	**0.28**	**1.11**	**74.56**	**0.97**	**46**

attributed to the formation of metal oxides on the surface of ACs. This association of MOs on the AC surface leads to a reaction (known as carbothermal reduction) as follows

$$M_nO_m + mC \rightarrow nM + mCO$$

Thus, through this reaction, the metals can be embedded in the pores present in the HP5 structure. Further, it has been established through previous investigations that microporosity is more vital as compared to surface area for physisorption of CO_2 (Heidari et al. 2014). As the physical properties of the AC/MO nanocomposites were found to be better than similar materials (Jang and Park 2012; Barroso-Bogeat et al. 2014), high CO_2 adsorption was observed for these nanocomposites.

b. **Boehm titration**

The authors conducted Boehm titration to analyse the chemical properties of the functional groups on modified ACs and their acidic and basic nature. After MO modification, a significant decrease in acidic surface functional groups together with an increase in the basic sites was observed. In other words, surface acidity was decreased while basicity was enhanced after MO modification. The change from acidic to basic nature of the surface functional groups was attributed to the ion exchange phenomenon between functional groups on AC and MO during the intercalation process (i.e. exchange between two or three H^+ ions from AC surface with a single Cu^{2+}, Ni^{2+}, Mg^{2+} or Al^{3+} ion that leads to low acidity of AC). This change in the surface chemical characteristic of AC from acidic to basic aids in enhanced CO_2 adsorption.

c. **SEM and EDX analysis**

In order to recognize the surface morphology changes and the nature of metals incorporated on the surface of the modified ACs, the authors conducted SEM and EDX (energy dispersive X-ray) analysis. From Fig. 5.6, it can be observed that the surface of HP5 is more irregular (with more cracks, crevices and fissures) as compared to CAC. This irregularity of the HP5 surface was observed to exist even after MO intercalation. It was observed that the MO/CAC surface was smoother and sheet-like with lesser MO particles as compared to MO/HP5 rough surface incorporating a greater number of MO particles. From the EDX data, the presence of significant MO particles along with few other elements (C, O, S, P) were identified. Finally, based on the EDX data, the authors concluded that the high porosity and surface area of HP5 as compared to CAC accounted for enhanced MOs development on the surface of ACs.

5.2.2.3 CO_2 Adsorption Capacity

Figure 5.7 shows the CO_2 adsorption capacity for various MO/ACs at 1 bar and 30 °C compared to unmodified AC. The authors observed that HP5/Cu3-1 has enhanced CO_2 adsorption capacity by 124.5% as compared to unmodified AC. Further, HP5/CuNi3-1 shows higher CO_2 adsorption capacity compared to HP5/AlMg8-1. However, when binary AC/MOs were applied, the CO_2 adsorption capacity was shown to increase only in case of HP5/AlMg8-1 and not for HP5/CuNi3-1 (as compared to single HP5/Cu3-1 or HP5/Ni3-1). This decrease of CO_2 adsorption capacity with binary MOs was ascribed to the blockage of most of the porosity because of the formation of huge volume of the binary MOs. Thus, it was concluded that the application of single Cu or Ni oxide was more favourable for higher CO_2 adsorption capacity as well as an economical practice requiring lesser MO and heat treatment during synthesis.

5.2.2.4 Comparison of the Work with Other Studies

Table 5.6 shows various adsorbents previously developed for CO_2 adsorption using different organic wastes, such as African palm shells, eucalyptus wood, coconut shells and also various MOs loaded on the ACs, majority being MgO and CuO. All of these indicated enhanced CO_2 adsorption capacity by incorporating MOs onto ACs surface. However, among all the previously synthesized sorbents, HP5/Cu3-1 shows highest CO_2 adsorption capacity of 6.78 mmol/g compared to other sorbents enlisted in the table. Thus, the authors have justified that the sorbents developed through their

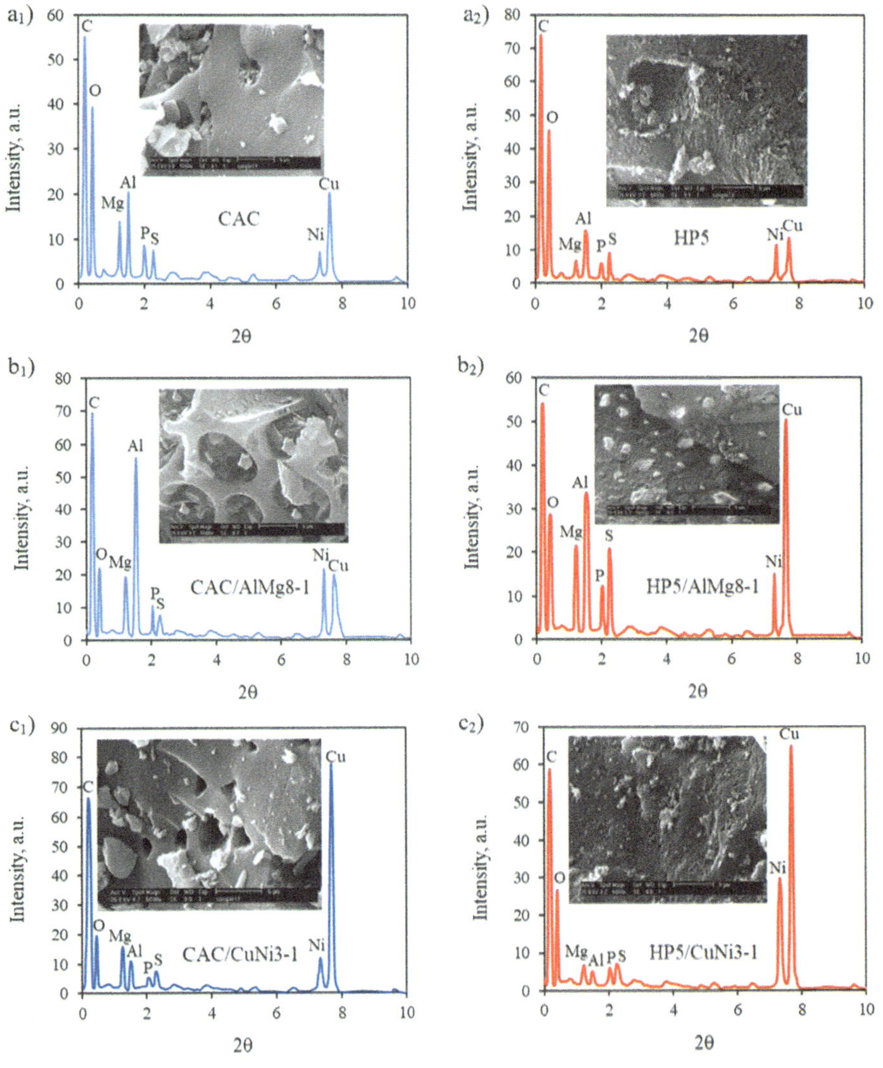

Fig. 5.6 SEM and EDX results of the synthesized CAC/MOs (**a₁**, **b₁**, **c₁**) and HP5/MOs (**a₂**, **b₂**, **c₂**). Reproduced with permission from Nowrouzi et al. (2018), Copyright 2021, Elsevier

work could be a promising new addition to the MO modified AC nanocomposites for CO_2 capture.

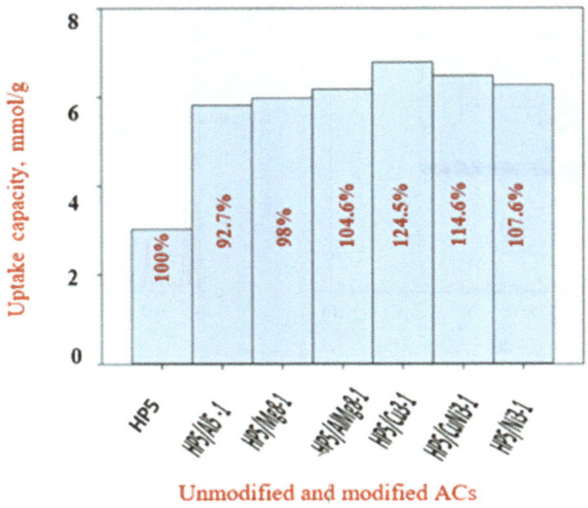

Fig. 5.7 Comparison of CO_2 adsorption capacity on unmodified and modified ACs at 1 bar. Reproduced with permission from Nowrouzi et al. (2018), Copyright 2021, Elsevier

Table 5.6 Adsorption capacity of the sorbents developed in the present work (Nowrouzi et al. 2018) with other investigations at different temperatures (pressure = 1 bar)

Type of adsorbent	Structure	Temp. (°C)	CO_2 adsorption capacity (mmol/g)	Ref
AC	African palm shells	25	4.40	Ello et al. (2013)
	Eucalyptus wood	30	4.01	Heidari et al. (2014)
	Coconut shells	25	2.55	Yang et al. (2011)
MO loaded on AC	CuO	25	0.30	Kim et al. (2010)
	NiO	25	2.23	Jang and Park (2012)
	Cu/Zn	30	2.25	Hosseini et al. (2015)
	Cu/Ce	25	0.77	Li et al. (2010a)
	MgO	0	2.72	Shahkarami et al. (2016)
	MgO	30	5.45	Liu et al. (2013a)
MgO-based adsorbents	Mesoporous MgO	25	4.44	Bhagiyalakshmi et al. (2010)
	MgO + Cs_2CO_3	300	1.90	Liu et al. (2013b)
	MgO-Al_2O_3 aerogel	200	0.50	Han et al. (2014)
	MgO-Al_2O_3	60	1.36	Li et al. (2010b)
Unmodified AC	Persian ironwood biomass	30	3.02	Present study
Modified AC	**HP5/Cu3-1**	**30**	**6.78**	Present study

5.2.3 Hierarchical-Structured MgO/Carbon Nanocomposite

Magnesium oxide (MgO) is preferred as an efficient adsorbent material due to its non-corrosive, non-polluting nature, its cost-efficiency and widespread availability. In the recent past, many MgO-based nanocomposites have been synthesized and utilized for CCS technologies (Kim et al. 2013; Harada et al. 2015; Vitillo 2015). However, the issues regarding complex and tedious fabrication procedures and lower adsorption capacity owing to improper textural properties demand for exploring better strategies to utilize MgO-based nanocomposites. Li et al. (2020) synthesized hierarchical-structured porous MgO/carbon nanocomposite combining solvothermal treatment followed by high temperature pyrolysis. The nanocomposites thus synthesized showed large specific surface area and pore volume, plentiful nanoporous structure and highly distributed ultrafine MgO nanocrystallites together with the implanted carbon material. These characteristics are highly desirable for a CO_2 adsorbent and the MgO/C nanocomposite material was found to have a CO_2 sorption capacity of 28.9 wt% at 27 °C and 1 bar CO_2, faster adsorption, a wide span of working temperature (27–200 °C) and robust character.

This discussion will focus on the following aspects of the MgO/C nanocomposite—Preparation, characterization and CO_2 capture performance.

5.2.3.1 Preparation

The steps involved in the fabrication of MgO/C nanocomposites is illustrated in Fig. 5.8. In Step one, the authors synthesized Mg-PDO complex precursor by reacting Mg acetate with 1,3-propandiol (1,3-PDO) under solvothermal condition. The second step involved thermal annealing of the Mg-PDO complex under inert atmosphere. During thermal annealing, the Mg metal is converted to MgO nanoparticles while the organic ligands (from PDO) are modified to carbon species. In the final product, large number of nanopores were generated resulting into highly porous hierarchical-structured MgO/C nanocomposite.

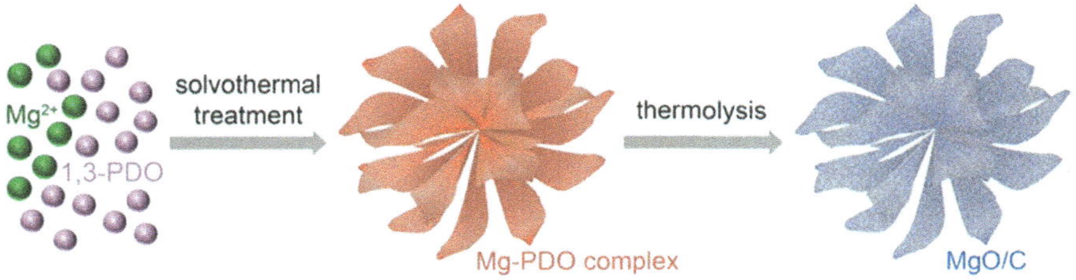

Fig. 5.8 Schematic representation for the synthesis of MgO/C nanocomposite. Reproduced with permission from Li et al. (2020), Copyright 2021, The Royal Society of Chemistry

5.2.3.2 Characterization

a. **Mg/PDO complex precursor**

Figure 5.9 shows the SEM (a) and TEM (b) images of the Mg-PDO complex precursor. The images reveal the hierarchical-structured flower-like particles with nanosheets twisted like petals for the Mg-PDO complex. The XRD, FTIR and TGA analyses of the Mg/PDO complex have also been studied by the authors (not included) which confirmed the presence of Mg, O and C, supporting the chemical composition of the precursor.

b. **MgO/C nanocomposite**

The XRD pattern and the FTIR spectra obtained by the authors for the nanocomposite material are depicted in Fig. 5.10a and b, respectively. The figures reveal full conversion of the Mg-PDO complex into MgO nanoparticles. The FTIR spectra

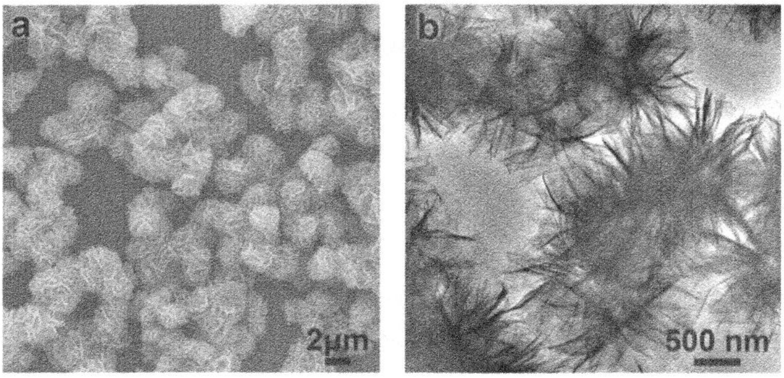

Fig. 5.9 **a** SEM and **b** TEM images of the Mg/PDO complex. Reproduced with permission from Li et al. (2020), Copyright 2021, The Royal Society of Chemistry

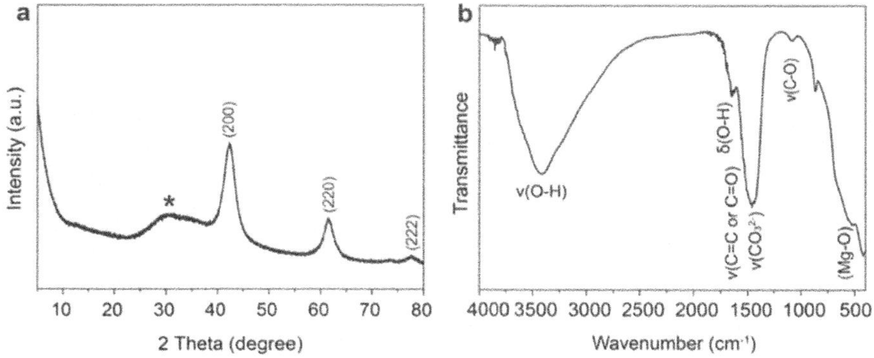

Fig. 5.10 **a** XRD pattern and **b** FTIR spectrum of MgO/C. Reproduced with permission from Li et al. (2020), Copyright 2021, The Royal Society of Chemistry

shows C-H stretching and bending vibrations in the range of 2820–2950 cm^{-1} and 1320–1470 cm^{-1}, respectively. A band around 1040–1130 cm^{-1} corresponds C–O stretching while near 550 cm^{-1} is related to Mg–O stretching vibration. The TEM images of the MgO/C nanocomposite are shown in Fig. 5.11a–c. From the figure, it is clear that the original flower-like hierarchical structure is retained even in the

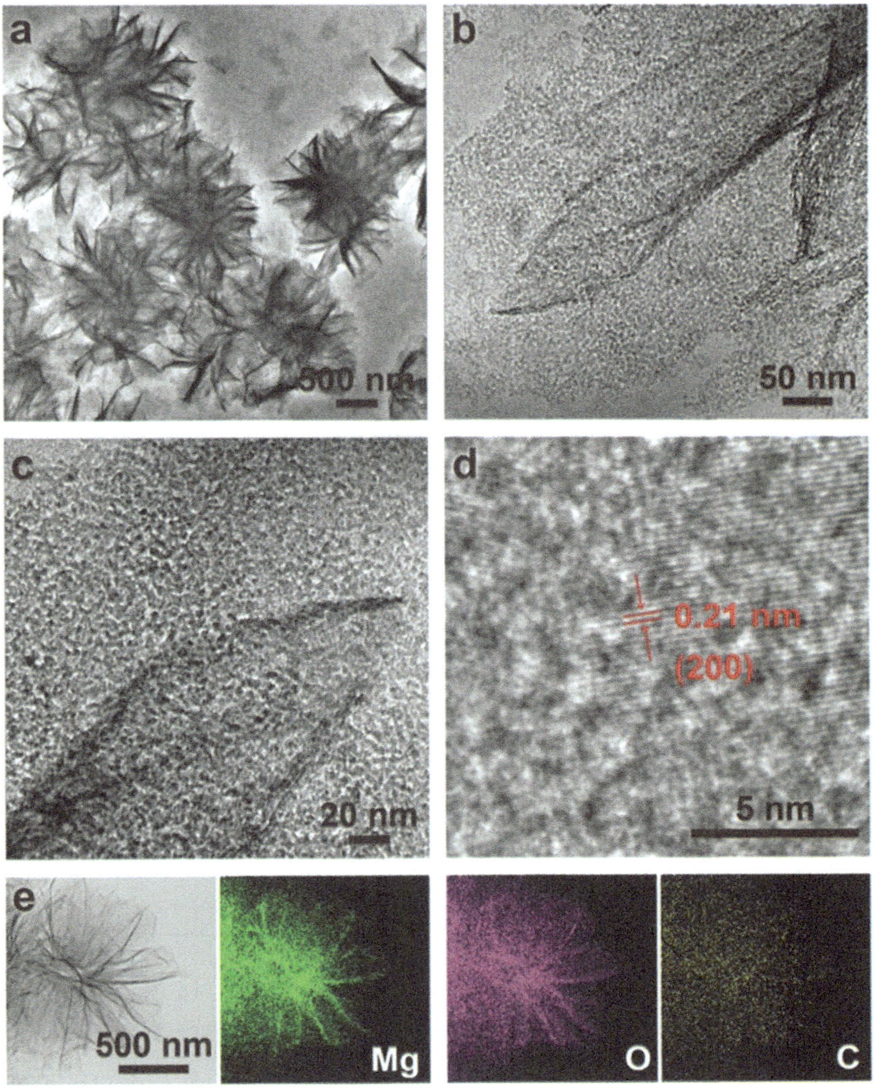

Fig. 5.11 **a**–**c** TEM images and **d** HRTEM image, **e** EDX elemental analysis of the MgO/C nanocomposite. Reproduced with permission from Li et al. (2020), Copyright 2021, The Royal Society of Chemistry

MgO/C nanocomposite. From the TEM image, the particle size of the nanocrystallites is shown to be in the range of 2–3 nm. Besides, the HRTEM image (d) also confirmed the presence of MgO nanocrystallites in the pyrolysis product. The EDX elemental mapping (e) showed the presence of Mg, O and C elements distributed uniformly throughout the whole flower-like particle.

5.2.3.3 CO_2 Capture Performance

As seen from Fig. 5.12a, MgO/C nanocomposite shows CO_2 adsorption capacity of 28.9 wt% which is almost 14 times higher than the commercially available MgO (2 wt%). Further, the authors also evaluated the adsorption capacity of pure MgO (control sample without carbon residue) which shows only 7.30 wt% CO_2 sorption capacity at similar conditions. The excellent CO_2 capture behaviour of MgO/C nanocomposite over pure MgO sample was attributed to its large surface area and pore volume along with nano-sized MgO crystals facilitating gas diffusion via abundant active sorption sites.

Additionally, the authors have also analysed the effect of temperature on the CO_2 adsorption capacity of MgO/C nanocomposites (Fig. 5.12b). From the figure, it is clear that with the rise in temperature, CO_2 capture capacity decreases. Nevertheless, even at 200 °C, MgO/C nanocomposite can still sustain a moderate CO_2 adsorption capacity of 5.80 wt%, exhibiting its wide range of working temperature (27–200 °C) in addition to its fast adsorption rate.

In addition to the above studies, the authors also performed a CO_2 sorption/desorption cycle test for the MgO/C nanocomposite as seen in Fig. 5.13a. It can be observed from the figure that even after 10 successive adsorption–desorption cycles, there is no remarkable decrease in its CO_2 capture capacity, suggesting adequate recyclability of the MgO/C nanocomposite. Further from TEM

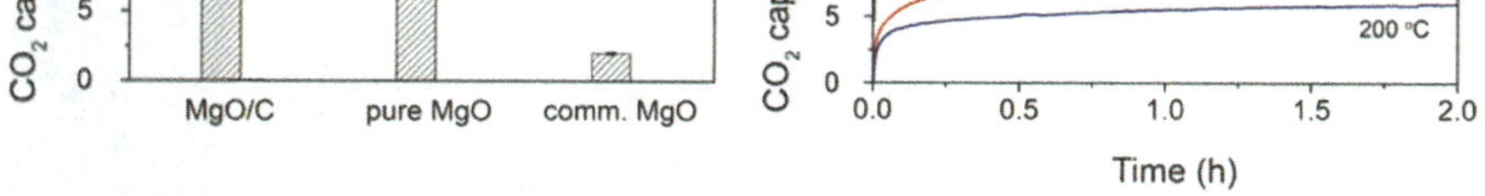

Fig. 5.12 **a** The CO_2 adsorption capacities of Mg sorbents at 27 °C and 1 bar CO_2. **b** The CO_2 capture kinetics of the MgO/C nanocomposite at different temperatures. Reproduced with permission from Li et al. (2020), Copyright 2021, The Royal Society of Chemistry

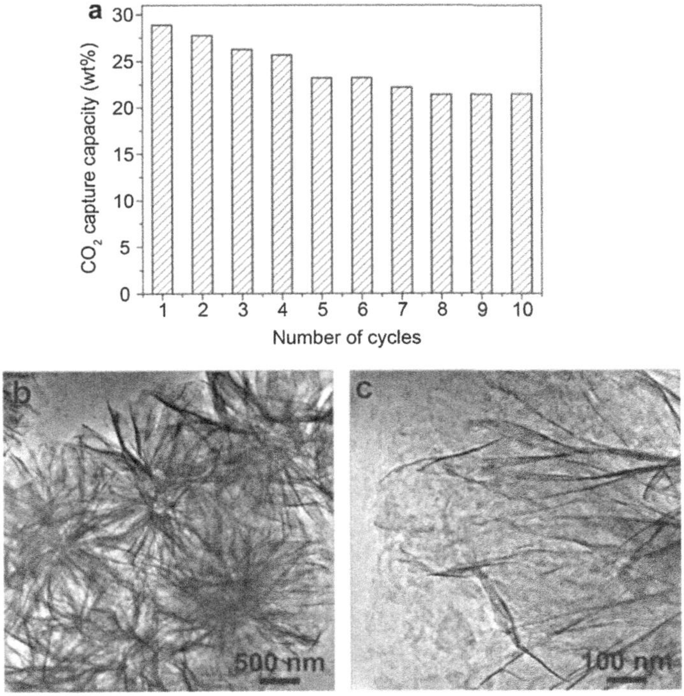

Fig. 5.13 **a** CO_2 sorption/desorption cyclic performance of the MgO/C nanocomposite (at 27 °C and 1 bar CO_2 for 2 h of sorption). **b** TEM images of the spent MgO/C nanocomposite after the CO_2 cycling test. Reproduced with permission from Li et al. (2020), Copyright 2021, The Royal Society of Chemistry

images (Fig. 5.13b), it can be observed that the hierarchical and porous architecture of the spent MgO/C nanocomposite was still maintained even after 10 adsorption/desorption cycles, indicating the strong and durable nature of the synthesized nanocomposites.

5.2.4 Polymeric Nanocomposites

Membrane technology (specifically the polymeric membranes) has gained significant attention in the past few decades as a prominent gas separation technique and utilized successfully for the treatment of natural gas, flue gas, air refining, recovery of hydrogen etc. (Baker 2006). Polymer-based membranes, especially polymeric nanocomposites are one of the advanced materials of choice for CCS technology. These have been found to be superior than the commonly used inorganic membranes such as zeolites, alumina etc. owing to their better permeability, cost-efficiency, facile synthesis and processing, improved mechanical property together with a decent gas separation quality (Ulbricht 2006; Salleh and Ismail 2015).

Polymeric nanocomposites are multicomponent systems where the polymer matrices are combined with nanoparticles. The materials which can efficiently adsorb CO_2, such as activated carbons, modified silica, zeolites and MOFs have been utilized for the development of various types of polymeric nanocomposites (Merkel et al. 2010). Among polymeric nanocomposites, we will first discuss the polysulfone-activated carbon–metal nanocomposites and the second section will include polymeric nanocomposites comprising of modified silica nanoparticles.

5.2.5 Polysulfone Combined with Activated Carbon–Metal (Ni and Co) Nanocomposites

Polysulfones are a group of thermoplastic polymers which are known for their strength and toughness at high temperatures (Muntha et al. 2019). These have widespread applications as membranes and in engineering purposes owing to their mechanical strength and chemical stability (Alosaimi et al. 2017). Thus, the polymeric nanocomposites are an attractive material for membrane-based gas capture technologies.

Herein, we discuss the work of Nisar et al. (2020a) who have synthesized polysulfone-based nanocomposites by utilizing various metal-activated carbons derived from biomass (waste product from wood processing) as filler material. They applied melt-mixing technique (Bao and Tjong 2008) for the synthesis of the polysulfone nanocomposites which showed efficient CO_2 capture performance and superior thermal, mechanical and magnetic properties. The discussion is divided into following sections—Synthesis of the polysulfone nanocomposite, characterisation of the nanocomposite and CO_2 capture performance.

5.2.5.1 Synthesis of the Polysulfone Nanocomposite

The synthesis of the nanocomposite consisted of two parts

a. **Synthesis of activated carbon-based cobalt and nickel from biomass**

The authors used Ayous wood residue (as biomass) as carbon source for the preparation of activated carbons (Sieliechi and Thue 2015). The cobalt (II) and Ni (II) were used as their chloride salts and mixed with water followed by the addition of dried sawdust from biomass. The resulting solution was mixed regularly for up to 2 h at 80 °C and dried overnight for further carbonization at a temperature of 700 °C.

b. **Synthesis of polysulfone nanocomposite**

The nanocomposite was synthesized by mixing appropriate amounts of polymer matrix, carbonized metal and a stabilizer (Irganox1010). The mixing process was performed by gradually varying the quantity of filler material from 0 to 2 wt% and finally the mixture was hot-pressed to form sheets at 325 °C and 200 bar pressure.

5.2.5.2 Characterisation of the Nanocomposite

The major analyses included to characterize the polysulfone nanocomposite are

a. **Elemental analysis of the Ayous biomass and metal-activated carbon**

Table 5.7 shows the elemental constitution of the Ayous biomass and activated carbon with Ni and Co metal. From the table it is clear that as compared to the biomass, the C, O and the ash content of the activated carbon–metal composite was towards a higher end. The ash content is an indication of the metal content in the activated carbon–metal compound. The table also shows that activated carbon-Co composite has the largest surface area, highest pore volume and the smallest pore diameter while the metallic content is highest for CA-Ni composite.

b. **Morphological analysis**

The presence of filler materials and their distribution in the PSF (polysulfone) matrix was established through SEM analysis (Fig. 5.14a–d). A rougher PSF membrane surface was observed after incorporation of metal-activated carbons. Additionally, the nanofillers were not detected in the SEM image, suggesting their uniform distribution in the PSF matrix. The TEM image in Fig. 5.15a shows a uniform distribution of the filler material in the PSF matrix without the formation of any aggregates. The (b), (c) and (d) part of the figure also show that the metal particles (Co and Ni) are circumscribed by amorphous carbons in the polymer matrix. Furthermore, the black spots observed in the figure indicate the presence of nano-sized metal particles in the polymer matrix (Nisar et al. 2020b).

c. **Analysis of thermal stability**

Table 5.7 CNH/O elemental composition and texture properties of Ayous biomass and activated nickel- and cobalt-containing carbons (Nowrouzi et al. 2018)

Samples	% C	% H	% N	% O[a]	% Ash[b]	M-content (%)	S_{BET}[c] ($m^2\ g^{-1}$)	V_{tot}[c] ($cm^3\ g^{-1}$)	D_p[c] (nm)
Biomass	46.05	6.26	0.33	45.78	1.56	0.00	–	–	–
CA–Ni	58.05	2.31	0.15	11.29	28.20	26.64	381	0.321	3.3
CA–Co	53.57	2.49	0.38	19.97	23.60	22.04	619	0.340	2.2

[a] Yield by difference (% O = 100% – % C – % H – % N – % Ash)
[b] Obtained by TGA
[c] S_{BET} = BET surface area; V_{tot} = total pore volume; D_p = average pore diameter, CA = activated carbon

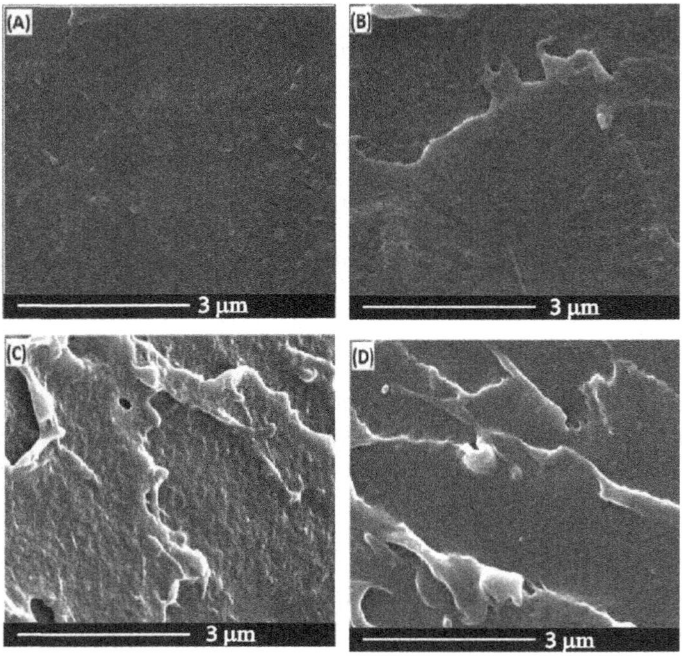

Fig. 5.14 SEM images of **a** Cross-section of PSF-neat, **b** PSF-CA, **c** CA-Ni and **d** PSF-CA-Co. Reproduced with permission from Nisar et al. (2020b), Copyright 2021, The Royal Society of Chemistry

The authors performed TGA analysis (thermal degradation properties) of the synthesized polysulfone nanocomposites and the data is presented in Table 5.8. From the table, it is observed that the maximum degradation temperature (T_{max}) shows a slight enhancement for PSF-CA-Co sample while rest of the samples showed T_{max} value similar to PSF-neat. Further, for the T_{onset} (initial degradation temperature) value, it was observed that incorporation of activated carbons into the polymer matrix (PSF-CA) led to improved (2 °C) T_{onset} value. However, upon addition of metals to PSF-CA, there is a decrease in T_{onset} value. Thus, the metals were considered to be acting as catalysts for an early degradation of the polymer material (Nisar et al. 2016).

5.2.5.3 CO_2 Capture Performance

Figure 5.16 shows the CO_2 capture capacity of the polymer and its nanocomposites (3 bar, 118.15 °C). The authors observed a marginal improvement in CO_2 capture capacity of the nanocomposites upon addition of the filler material. The PSF–CA–Co nanocomposite (activated carbons with cobalt) showed a higher sorption capacity than PSF–CA (activated carbons without metals) which are, in turn, roughly 10% higher than that of PSF. The presence of activated carbons in the polymer structure

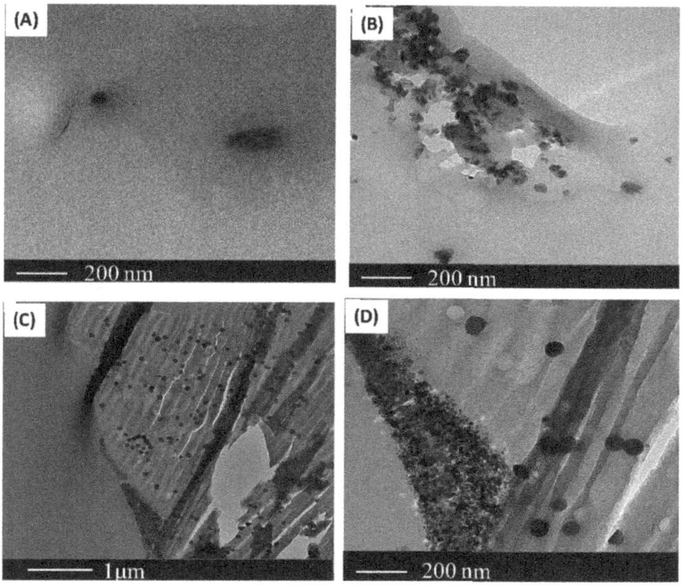

Fig. 5.15 TEM images of **a** PSF-CA-neat, **b** PSF-CA-Co, **c** and **d** PSF-CA-Ni. Reproduced with permission from Nisar et al. (2020b), Copyright 2021, The Royal Society of Chemistry

Table 5.8 Thermal properties of the synthesized polysulfone nanocomposites (Nisar et al. 2020a)

Samples	Filler (%)	Metal[a] (%)	Filler[b] (%)	*T*onset (°C)	*T*max (°C)*T*g[c]	
PSF-neat	0	0	32.5	523	543	190
PSF–CA	2.0	0	32.8	525	544	190
PSF–CA–Ni	2.0	0.5	32.3	521	543	191
PSF–CA–Co	2.0	0.4	34.0	519	546	191

[a] Calculated from the TGA residue
[b] Calculated from the TGA residue
[c] Calculated from the DSC

resulting into a large number of ultra-micropores aids in the CO_2 adsorption process by the nanocomposites (Sethia and Sayari 2015). In addition, the nanocomposite nickel particles (PSF–CA–Ni) showed a slight decrease in the CO_2 capture capacity. This decrease in case of Ni particles was attributed to the formation of aggregates of the fillers in the matrix of PSF. Overall, it was concluded that the presence of the required amount of fillers in the PSF matrix was beneficial for efficient CO_2 capture.

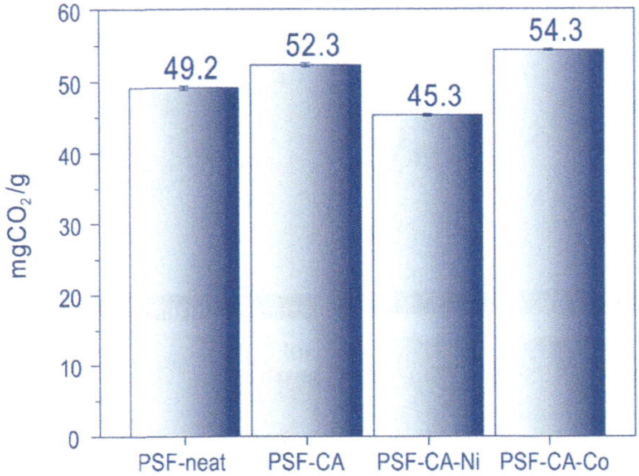

Fig. 5.16 CO_2 capture capacity of PSF-neat and its nanocomposites at 45 °C. Reproduced with permission from Nisar et al. (2020b), Copyright 2021, The Royal Society of Chemistry

5.2.6 Organic Polymer Membranes Infused with Amine Functionalized SiO_2 or TiO_2 Nanoparticles

Polyvinylidene fluoride (PVDF) is a non-reactive thermoplastic polymer produced by the polymerization of vinylidene difluoride. PVDF membrane finds widespread application in scientific research and industrial process due to its excellent characteristics pertaining to chemical resistance, thermal stability and mechanical strength (Khayet et al. 2002). Additionally, its ageing resistance (Wang et al. 2002) and ease of synthesis (Kim et al. 2006) makes it more attractive as a material of choice for membrane separation techniques. It has already been established that the addition of fillers to the polymeric membranes enhances their properties. The materials which show efficient CO_2 capture performance, such as ACs, zeolites, MOFs, amine modified mesoporous silica etc. are suitable as fillers (Assche et al. 2016; Sakpal et al. 2012; Sevilla et al. 2011). Further, other inorganic materials incorporated into PVDF membrane include Al_2O_3, ZrO_2, TiO_2, SiO_2 etc. (Yan et al. 2006; Arthanareeswaran et al. 2008). Among these, silica is the most widely used material owing to its gentle reactivity and its familiarity in chemistry.

In this section, we will discuss the work of Khdary and Abdelsalam (2020) who fabricated nanocomposite membranes (for carbon capture) constituting porous PVDF–HFP (Poly (vinylidene fluoride-co-hexafluoropropylene)) film which was intercalated with amine-functionalized silica particles employing phase separation technique. This discussion highlights the following sections—synthesis of the PVDF-silica nanocomposite membranes, characterisation of the nanocomposite and CO_2 capture performance.

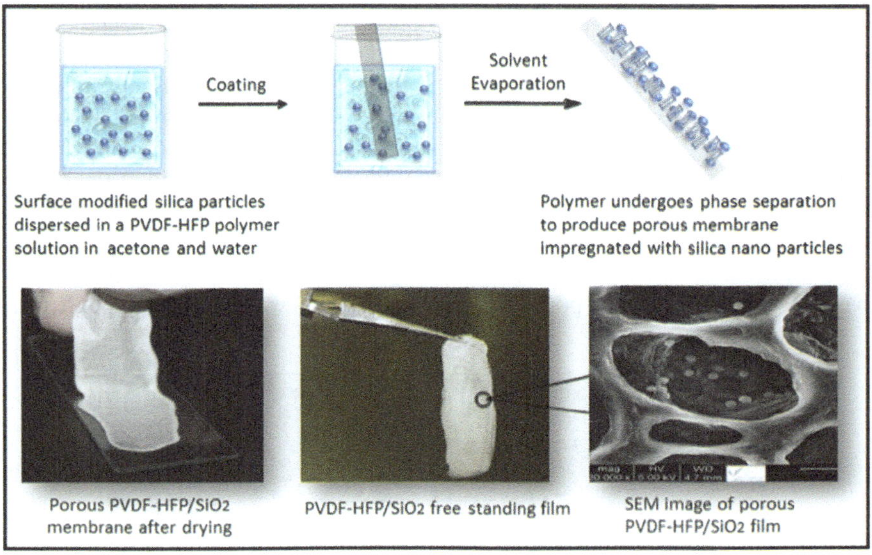

Fig. 5.17 Steps in the synthesis of PVDF-HFP membrane intercalated with silica particles. Reproduced with permission from Khdary and Abdelsalam (2020), Copyright 2021, ScienceDirect (Elsevier)

5.2.6.1 Synthesis of the PVDF-Silica Nanocomposite Membranes

The stepwise preparation process of the nanocomposite membranes is represented in Fig. 5.17. The authors employed phase separation technique for the synthesis using a blend composed of PVDF–HFP/SiO_2, acetone and water. Acetone and water were selected as appropriate solvents because of their specific properties (acetone is a better solvent for PVDF–HFP with a lower flash point while acetone–water are miscible and PVDF–HFP is insoluble in water having a higher flash point). The PVDF–HFP/SiO_2 blend was mixed thoroughly and a glass microscope slide was coated with a thin film of the polymer. After coating, the layer of the thin film was thoroughly dried by completely evaporating acetone and water. This leads to well-developed porous structure and well-embedded SiO_2 nanoparticles in the polymer matrix. Further, the authors observed that the developed nanocomposite membranes after complete drying of the solvents do not stick any further to the glass surface and can be easily lifted to form free standing films. These membranes have been found to be mechanically strong and can be designed to any shape as required for capturing purposes.

5.2.6.2 Characterisation of the Nanocomposite

a. **Morphological analysis**

SEM images of the PVDF-HFP films can be seen in Fig. 5.18, where a. and e. show porous membrane produced from a polymer solution in an acetone/water mixture. It shows irregular-pores with an average pore size of 3.37 ± 0.9 mm. Further, the polymer film was found to be thicker with multiple layers of disordered spherical macro cavities. Upon incorporating SiO_2 particles, a porous film was obtained which contained evenly distributed SiO_2 into the polymer network. In b. and c., the PVDF-HFP membrane morphology can be observed which were loaded with 20% and 40% by weight of SiO_2–AFS (amine modified high surface area nano-silica), respectively. With the enhancement of the weight percentage of SiO_2 from 20 to 40%, a decrease in pore size and depletion in porosity of the polymer film was observed; though it was

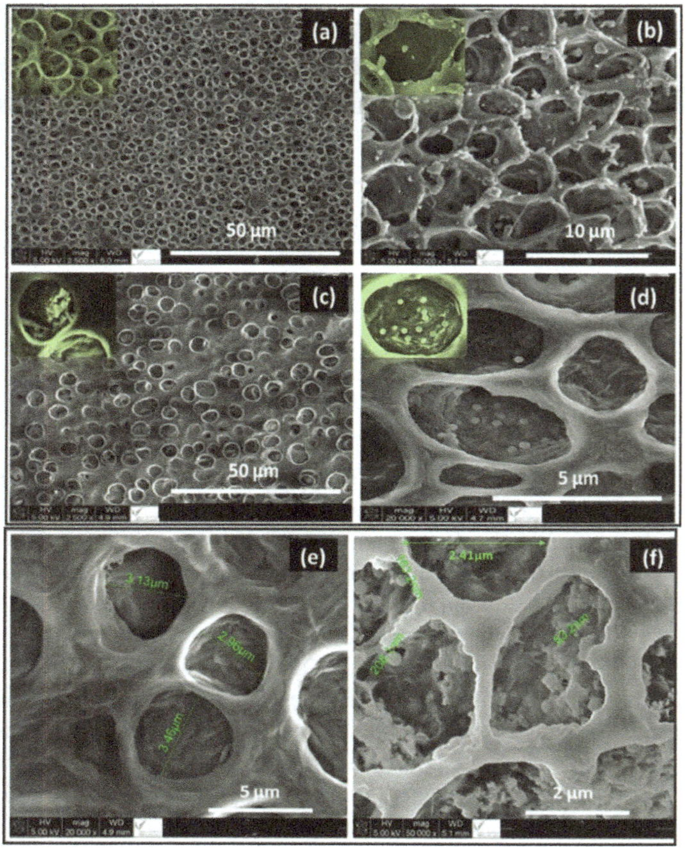

Fig. 5.18 SEM images of porous PVDF-HFP films produced from a solution containing **a** 2 wt% of PVDF–HFP in acetone/water mixture, **b–d** similar to (**a**) after adding SiO_2–AFS particles at **b** 20 wt%, **c** at 40 wt% while (**d**) contains SiO_2–ANS at 20 wt%, **e** 2 wt% of PVDF-HFP in acetone/water mixture at high magnification and **f** contains SiO_2–ANS at 40 wt% at high magnification. Reproduced with permission from Khdary and Abdelsalam (2020), Copyright 2021, ScienceDirect (Elsevier)

concluded that SiO_2 did not block the pores. The morphology of polymer membrane loaded with 20 wt% of SiO_2–ANS (amine modified high surface area nano-silica) is shown is Fig. 5.18d and f.

b. **Elemental Analysis**

EDX was employed to confirm the elemental analysis of the polymer membrane. Two peaks were identified for the PVDF-HFP film without SiO_2 as shown in Fig. 5.19 a. The first peak at Ka = 0.277 keV was attributed to carbon while the second peak at Ka = 0.677 keV was attributed to fluorine. When the PVDFHFP film was impregnated with SiO_2, two more peaks were observed. The peak at Ka = 0.52 keV was found to be that of oxygen and the peak at Ka = 1.74 keV was reported to be that of silicon as shown in Fig. 5.19b–d. The intensity of the Si and oxygen depends on the weight loading of the SiO_2 in the composite. This is clearly indicated by the higher intensity of Si and of oxygen peaks for the film containing 40 wt% of the high surface area SiO_2–AFS than the same film containing 20 wt% of the SiO_2–AFS composite.

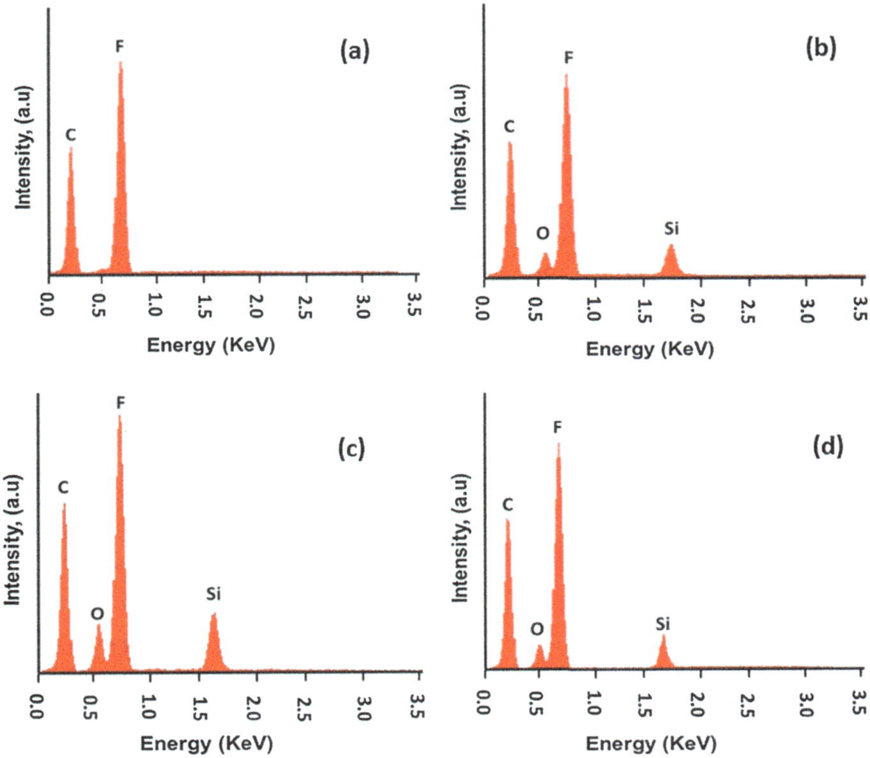

Fig. 5.19 EDX spectroscopy of **a** PVDF–HFP membrane, **b** PVDF–HFP membrane impregnated with 20 wt% of SiO_2–AFS, **c** PVDF–HFP membrane impregnated with 40 wt% of SiO_2–AFS, and **d** PVDF–HFP membrane impregnated with 20 wt% of SiO_2–ANS. Reproduced with permission from Khdary and Abdelsalam (2020), Copyright 2021, ScienceDirect (Elsevier)

Moreover, the intensity of the Si and oxygen peaks is independent of the type of SiO_2 at the same weight loading. This can be seen from the comparable intensities of the Si and oxygen peaks for the membrane comprised 20 wt% of high (SiO_2–AFS) and low surface area SiO_2–ANS as shown in Fig. 5.19b and d, respectively.

5.2.6.3 CO_2 Capture Performance

The authors used TGA technique to evaluate the CO_2 adsorption–desorption behaviour of PVDF–HFP and PVDF–HFP/SiO_2 films. It was observed that the PVDF–HFP film showed poor CO_2 adsorption suggesting its low affinity towards CO_2 uptake. The PVDF–HFP/SiO_2–AFS and PVDF–HFP/SiO_2–ANS films reportedly showed gradual CO_2 adsorption over the first 24 min with further adsorption at a slower rate till the attainment of equilibrium. Table 5.9 shows the maximum range of CO_2 adsorption and the heat of adsorption of the polymer nanocomposites. From the data, it is clear that at the same weight loading, the surface area of the SiO_2 is a dominant factor for CO_2 adsorption capacity. The authors justified the above argument by comparing the CO_2 adsorption capacity of PVDF–HFP/SiO_2–AFS (26.27 mg/g) with that of PVDF–HFP/SiO_2–ANS (12.36 mg/g), where both the films were at 20 wt% SiO_2 loading. Further, the authors also demonstrated that an increase in the weight percent of SiO_2 in the composite also led to a higher the CO_2 adsorption capacity. This could be observed in PVDFHFP/SiO_2–AFS membrane where increment in weight percent from 20 to 40 led to the enhancement of CO_2 uptake from 26.27 to 33.75 mg/g. Additionally, the heat of adsorption data obtained from the DSC heat flow showed similar behaviour where addition of 20 wt% of SiO_2 to the PFDV-HFP (heat of adsorption = 0.126 J/mol) led to an increase in the heat of adsorption to 0.480 J/mole for PVDF–HFP/SiO_2–AFS and 0.342 J/mol for PVDF–HFP/SiO_2–ANS. The authors have reasoned for this increase in the heat of adsorption owing to the strong interaction between amine modified SiO_2 and CO_2. This also explains the surface area of the films being a vital factor directing heat of adsorption values (heat of adsorption of PVDF–HFP/SiO_2–AFS > PVDF–HFP/SiO_2–ANS). Further, as in the case of CO_2 adsorption capacity, the heat of adsorption also increased on enhancing the weight percent of amine-SiO_2, as PVDF–HFP/SiO_2–AFS at 40 wt% showed heat of adsorption value at 0.607 J mol l^{-1}, much higher than at 20 wt% (0.480 J mol l^{-1}).

Table 5.9 Heat of adsorption and CO_2 uptake and for neat PVDF-HFP and PVDF-HFP membrane intercalated with SiO_2 (Khdary and Abdelsalam 2020)

Polymer nanocomposite samples	Heat of adsorption, J/mol	CO_2 adsorption, mg/g
PVDF-HFP Blank	0.126	1.70
PVDF-HFP-20 wt % AFS	0.480	26.27
PVDF-HFP-40 wt % AFS	0.607	33.75
PVDF-HFP-20 wt % ANS	0.342	12.36

5.3 Chapter Summary and Conclusion

This chapter highlights the application of metal nanocomposites as advanced materials for carbon dioxide capture. Various adsorbent materials such as graphene, ACs derived from different sources as biomass (lignocellulosic waste, sawdust etc.) are included to focus on the wide array of adsorbents available for the development of such nanocomposites. The vast range of metals such as Fe, Cu, Co, Al, Ni, Mg etc. that have been utilized to modify the adsorbents are also included in this chapter to emphasize on the significance of metals in altering the properties of the resulting nanocomposite materials. Each work in this chapter has broadly been discussed under three sections—synthesis of the nanocomposites, their characterization using various techniques (SEM, TEM, EDX, FTIR, XRD, TGA etc.) and their CO_2 capture performance. Each nanocomposite described in this chapter has a distinct character for specialized applications, such as Fe_3O_4–graphene nanocomposites, which are shown to be efficient for CO_2 capture at high temperature and pressures. Hence, it is suitable for carbon capture in the exhaust of various industries such as automobile, steel, cement, power plant etc. The inclusion of Fe_3O_4 nanocrystals into the graphene matrix renders higher CO_2 adsorption capacity due to dual (physical via graphene and chemical via Fe_3O_4) mode of adsorption in these nanocomposite materials.

Similarly, metal nanocomposites synthesized utilizing Persian ironwood biomass as a source of ACs and intercalating these with single and binary metals, such as Mg, Al, Cu and Ni has shown improvement in the physical characteristics of the adsorbents with enhanced CO_2 adsorption. The best CO_2 capture performance was depicted by HP5/Cu3-1 and HP5/AlMg8-1 among the various AC/MO nanocomposites. Next, a porous MgO/C nanocomposite with hierarchical architecture has shown efficient carbon capture. A Mg metal–organic complex (Mg–PDO) was used as a precursor to develop this nanocomposite. The prominent features of the MgO/C nanocomposite include high CO_2 adsorption capacity of 28.9 wt% (at 27 °C and 1 bar CO_2), large surface area, plenty of nanopores, large pore volume, hierarchical architecture, fast adsorption kinetics, a broad range of working temperature (27–200 °C), outstanding recyclability and durability.

The last section of this chapter focuses on polymer—metal nanocomposite materials such as polysulfone-based nanocomposites incorporating various metal (Ni and Co)—activated carbons derived from biomass. The inclusion of metal nanoparticles in the polymer matrix has shown to improve thermal, mechanical and magnetic properties in addition to the enhancement of CO_2 capture capacity of the nanocomposite by 10% in contrast to neat polysulfone. Finally, the development of organic polymer membranes infused with amine functionalized SiO_2 nanoparticles (PVDF-HFP/SiO_2) has been discussed. This nanocomposite polymer membrane has shown remarkable CO_2 uptake as amine modified SiO_2 surface provides extremely suitable sorption characteristics.

Thus, the metal nanocomposites comprising of a varied range of matrix materials including ACs (derived from various sources), graphene, carbonaceous materials derived from organic compounds and polymer matrix have been discussed in this chapter. The inclusion of metals remarkably amends the properties of the nanocomposites by increasing microporosity, improved surface characteristics, permeability and specialized metal-CO_2 interactions. Thus, metal nanocomposites, on account of their enhanced CO_2 adsorption characteristics and synthetic flexibility, hold a promising future as state-of-the-art materials for CCS technology. However, some limitations still exist, such as economic viability and complex synthetic techniques. Thus, in subsequent times, focus should be on developing metal nanocomposites which are cost-effective with facile synthesis process and suitable for large scale commercial applications.

References

Abazari S, Shamsipur A, Bakhsheshi-Rad HR et al (2020) Carbon nanotubes (CNTs)-reinforced magnesium-based matrix composites: a comprehensive review. Materials 13(19):4421

Ajayan PM (2003) Bulk metal and ceramics nanocomposites. In: Ajayan PM, Schadler LS, Braun PV (eds) Nanocomposite science and technology. Wiley, pp 1–75

Alosaimi AM, Hussein MA, Abdelaal MY et al (2017) Polysulfone-based modified organoclay nanocomposites as a promising breast anticancer agent. Cogent Chem 3:1–17

Arthanareeswaran G, Sriyamuna Devi TK, Raajenthiren M (2008) Effect of silica particles on cellulose acetate blend ultrafiltration membranes: Part I. Sep Purif Technol 64(1):38–47

Baker RW (2006) Membrane technology and applications, 2nd edn. Wiley, New York, USA

Baltrusaitis J, Jensen JH, Grassian VH (2006) FTIR spectroscopy combined with isotope labeling and quantum chemical calculations to investigate adsorbed bicarbonate formation following reaction of carbon dioxide with surface hydroxyl groups on Fe_2O_3 and Al_2O_3. J Phys Chem B 110:12005–12016

Baniecki JD, Ishii M, Kurihara K et al (2009) Chemisorption of water and carbon dioxide on nanostructured $BaTiO_3$-$SiTiO_3$ (001) surfaces. J Appl Phys 106:054109

Bao SP, Tjong SC (2008) Mechanical behaviors of polypropylene/carbon nanotube nanocomposites: the effects of loading rate and temperature. Mater Sci Eng A 485:508–516

Barroso-Bogeat A, Alexandre-Franco M, Fernández-González C et al (2014) Preparation of activated carbon-metal oxide hybrid catalysts: textural characterization. Fuel Process Technol 126:95–103

Belmabkhout Y, Guerrero RS, Sayari A (2009) Adsorption of CO_2 from dry gases on MCM-41 silica at ambient temperature and high pressure. 1: Pure CO_2 adsorption. Chem Eng Sci 64:3721–3728

Ben-Mansour R, Habib MA, Bamidele OE et al (2016) Carbon capture by physical adsorption: materials, experimental investigations and numerical modelling and simulations. Appl Energy 161:225–255

Bhagiyalakshmi M, Lee JY, Jang HT (2010) Synthesis of mesoporous magnesium oxide: its application to CO_2 chemisorption. Int J Greenhouse Gas Control 4(1):51–56

Bhat A, Budholiya S, Aravind Raj S et al (2021) Review on nanocomposites based on aerospace applications. Nanotechnol Rev 10(1):237–253

Bhattacharya V, Chattopadhyay K (2001) Microstructure and tribological behaviour of nano-embedded Al-alloys. Scripta Mater 44(8–9):1677–1682

Bhattacharya V, Chattopadhyay K (2004) Microstructure and wear behaviour of aluminium alloys containing embedded nanoscaled lead dispersoids. Acta Mater 52(8):2293–2304

Bogue R (2011) Nanocomposites: a review of technology and applications. Assem Autom 31(2):106–112

Branagan DJ (2000) In: Alman DE, Newkirk JW (eds) Powder metallurgy, particulate materials for industrial applications. TMS Publication, St. Louis

Branagan DJ, Tang Y (2002) Developing extreme hardness (>15 GPa) in iron-based nanocomposites. Compos Part A - Appl Sci Manuf 33(6):855–859
Burchell TD, Judkins RR, Rogers MR et al (1997) A novel process and material for the separation of carbon dioxide and hydrogen sulfide gas mixtures. Carbon 35(9):1279–1294
Camargo PHC, Satyanarayana KG, Wypych F (2009a) Nanocomposites: synthesis, structure, properties and new application opportunities. Mater Res 12(1):1–39
Camargo PHC, Satyanarayana KG, Wypych F (2009b) Nanocomposites: synthesis, structure, properties and new application opportunities. Mat Res 12(1):1–39
Carpenter EE, Kumbhar A, Wiemann JA et al (2000) Synthesis and magnetic properties of gold–iron–gold nanocomposites. Mater Sci Eng 286(1):81–86
Cavenati S, Grande CA, Rodrigues AE (2004) Adsorption equilibrium of methane, carbon dioxide, and nitrogen on zeolite 13X at high pressures. J Chem Eng Data 49:1095–1101
Chen J, Huang Y, Wei C et al (2010) A study of Ni-Al_2O_3 nanocomposite coating prepared by jet electrodeposition. Adv Compos Lett 19(5):177–181
Choa YH, Yang JK, Kim BH et al (2003) Preparation and characterization of metal/ceramic nanoporous nanocomposite powders. J Magn Magn Mater 266(1–2):12–19
Choi SM, Awaji H (2005) Nanocomposites: a new material design concept. Sci Technol Adv Mater 6(1):2–10
Chow GM, Holtz RL, Pattnaik A et al (1990) Alternative approach to nanocomposite synthesis by sputtering. Applied Physical Letters 56(19):1853–1855
Choy KL (2003) Chemical vapour deposition of coatings. Prog Mater Sci 48(2):57–170
Cinke M, Li J, Bauschlicher CW Jr et al (2003) CO_2 adsorption in single-walled carbon nanotubes. Chem Phys Lett 376:761–766
Contreras A, Lopez BE (2004) Mg/TiC composites manufactured by pressureless melt infiltration. Scripta Mater 51(3):249–253
Cortés-Arriagada D, Villegas-Escobar N, Ortega DE (2018) Fe-doped graphene nanosheet as an adsorption platform of harmful gas molecules (CO, CO_2, SO_2 and H_2S), and the co-adsorption in O_2 environments. Appl Surf Sci 427:227–236
Cushing BL, Kolesnichenko VL, O'Connor CJ (2004) Recent advances in the liquid-phase syntheses of inorganic nanoparticles. Chem Rev 104(9):3893–3946
Ello AS, de Souza LKC, Trokourey A et al (2013) Development of microporous carbons for CO_2 capture by KOH activation of African palm shells. J CO_2 Utilization 2:35–38
George G, Bhoria N, Alhallaq S et al (2016) Polymer membranes for acid gas removal. Sep Purif Technol 158:333–356
Ghosh A, Subrahmanyam KS, Krishna KS et al (2008) Uptake of H_2 and CO_2 by graphene. J Phys Chem C 112(40):15704–15707
Global CCS institute. The global status of CCS: 2017. https://www.globalccsinstitute.com/wp-content/uploads/2018/12/2017-Global-Status-Report.pdf
Goharibajestani Z, Yürüm A, Yürüm Y (2019) Effect of transition metal oxide nanoparticles on gas adsorption properties of graphene nanocomposites. Appl Surf Sci 475:1070–1076
Han SJ, Bang Y, Kwon HJ et al (2014) Elevated temperature CO_2 capture on nano-structured MgO-Al_2O_3 aerogel: effect of Mg/Al molar ratio. Chem Eng J 242:357–363
Hao GP, Li WC, Lu AH (2011) Novel porous solids for carbon dioxide capture. J Mater Chem 21:6447–6451
Harada T, Simeon F, Hamad EZ et al (2015) Alkali metal nitrate-promoted high-capacity MgO adsorbents for regenerable CO_2 capture at moderate temperatures. Chem Mater 27(6):1943–1949
Hassan T, Salam A, Khan A et al (2021) Functional nanocomposites and their potential applications: a review. J Polym Res 28:36
Haubold T, Gertsman V (1992) On the structure and properties of nanostructured copper-tungsten alloys. Nanostruct Mater 1(4):303–312
He X (2018) A review of material development in the field of carbon capture and the application of membrane-based processes in power plants and energy-intensive industries. Energ Sustain Soc 8:34

He X, Yu Q, Hägg M-B (2013) CO_2 capture. In: Hoek EMV, Tarabara VV (eds) Encyclopedia of membrane science and technology, 3, vol set. Wiley, New York, pp 1–2390

Heidari A, Younesi H, Rashidi A et al (2014) Evaluation of CO_2 adsorption with eucalyptus wood based activated carbon modified by ammonia solution through heat treatment. Chem Eng J 254:503–513

Hileman B (2006) EPA's dioxin review is criticized. Chem Eng News 84:7–8

Himeno S, Komatsu T, Fujita S (2005) High pressure adsorption equilibria of methane and carbon dioxide on several activated carbons. J Chem Eng Data 50:369–376

Hlaing Oo WM, McCluskey MD, Lalonde AD et al (2005) Infrared spectroscopy of ZnO nanoparticles containing CO_2 impurities. Appl Phys Lett 86:073111

Holtzt RL, Provenzano V (1994) Enhanced microhardness of copper-niobium nanocomposites. Nanostruct Mater 4(2):241–256

Hosseini S, Bayesti I, Marahel E et al (2015) Adsorption of carbon dioxide using activated carbon impregnated with Cu promoted by zinc. J Taiwan Inst Chem Eng 52:109–117

Huang K, Zhang XM, Xu Y et al (2014) Protic ionic liquids for the selective absorption of H_2S from CO_2: thermodynamic analysis. AIChE J 60:4232–4240

Hummers WS Jr, Offeman RE (1958) Preparation of graphitic oxide. J Am Chem Soc 80:1339–1339

International Energy Agency (IEA) (2009), in Key World Energy Statistics. Retrieved from: http://www.observatoire-du-nucleaire.org/IMG/pdf/2009-keyworld-aie.pdf

IPCC fourth assessment report: Climate Change 2013 (AR5). https://www.ipcc.ch/site/assets/uploads/2018/03/WG1AR5_SummaryVolume_FINAL.pdf

Jadhav PD, Rayalu SS, Biniwale RB et al (2007a) CO_2 emission and its mitigation by adsorption on zeolites and activated carbon. Curr Sci 92:724–726

Jadhav PD, Chatti RV, Biniwale RB et al (2007b) Monoethanol amine modified zeolite 13X for CO_2 adsorption at different temperatures. Energy Fuels 21:3555–3559

Jang D-I, Park S-J (2012) Influence of nickel oxide on carbon dioxide adsorption behaviors of activated carbons. Fuel 102:439–444

Jordan J, Jacob KI, Tannenbaum R et al (2005) Experimental trends in polymer nanocomposites: a review. Mater Sci Eng, A 393(1–2):1–11

Joseph MC, Tsotsos C, Baker MA et al (2005) Characterisation and tribological evaluation of nitrogen-containing molybdenum-copper PVD metallic nanocomposite films. Surf Coat Technol 190(2–3):345–356

Kamat PV, Flumiani M, Dawson A (2002) Metal-metal and metal-semiconductor composite nanoclusters. Colloid Surf A Physicochem Eng Aspects 202(2–3):269–279

Karra JR, Walton KS (2010) Molecular simulations and experimental studies of CO_2, CO and N_2 adsorption in metal-organic frameworks. J Phys Chem C 114:15735–15740

Khalid FA, Beffort O, Klotz UE (2003) Study of microstructure and interfaces in an aluminium–C60 composite material. Acta Mater 51(15):4575–4582

Khayet M, Feng CY, Khulbe KC et al (2002) Preparation and characterization of polyvinylidene fluoride hollow fiber membranes for ultrafiltration. Polymer 43:3879–3890

Khdary NH, Abdelsalam ME (2020) Polymer-silica nanocomposite membranes for CO_2 capturing. Arab J Chem 13(1):557–567

Khodaei M, Karimzadeh F, Enayati M (2011) Mechanochemically synthesized metallic-ceramic nanocomposite; mechanisms and properties. In: Reddy BSR (ed) Advances in nanocomposites—synthesis, characterization and industrial applications. INTECH Open Access Publisher, pp 157–180

Kim WM, Park N-G, Ryu KS et al (2006) Characteristics of PVdF-HFP/TiO_2 composite membrane electrolytes prepared by phase inversion and conventional casting methods. Electrochim Acta 51(26):5636–5644

Kim B-J, Cho K-S, Park S-J (2010) Copper oxide-decorated porous carbons for carbon dioxide adsorption behaviors. J Colloid Interface Sci 342(2):575–578

Kim TK, Lee KJ, Cheon JY et al (2013) Nanoporous metal oxides with tunable and nanocrystalline frameworks via conversion of metal–organic frameworks. J Am Chem Soc 135(24):8940–8946

Klement R, Svancarek P, Parchoviansky M et al (2018) 4-Al_2O_3-SiC nanocomposites: preparation, microstructure, and properties. In: Low I (ed) Composites science and engineering, advances in ceramic matrix composites, 2nd edn. Woodhead Publishing, pp 49–92

Kumar R, Mangalapuri R, Ahmadi MH et al (2020) The role of nanotechnology on post-combustion CO_2 absorption in process industries. Int J Low-Carbon Technol 15(3):361–367

Laurent C, Demai JJ, Rousset A et al (1994) Fe–Cr/Al_2O_3 metal-ceramic composites: nature and size of the metal particles formed during hydrogen reduction. J Mater Res 9:229–235

Li X, Yang Y, Cheng X (2004) Ultrasonic-assisted fabrication of metal matrix nanocomposites. J Mater Sci 39(9):3211–3212

Li F, Yi H, Tang X et al (2010a) Adsorption of carbon dioxide by coconut activated carbon modified with Cu/Ce. J Rare Earths 28:334–337

Li L, Wen X, Fu X et al (2010b) MgO/Al_2O_3 sorbent for CO_2 capture. Energy Fuels 24(10):5773–5780

Li P, Lin Y, Chen R et al (2020) Construction of a hierarchical-structured MgO-carbon nanocomposite from a metal–organic complex for efficient CO_2 capture and organic pollutant removal. Dalton Trans 49(16):5183–5191

Liu Y, Jiang W, Wang Y et al (2009) Synthesis of Fe_3O_4/CNTs magnetic nanocomposites at the liquid-liquid interface using oleate as surfactant and reactant. J Magn Magn Mater 321:408–412

Liu W-J, Jiang H, Tian K et al (2013a) Mesoporous carbon stabilized MgO nanoparticles synthesized by pyrolysis of $MgCl_2$ preloaded waste biomass for highly efficient CO_2 capture. Environ Sci Technol 47(16):9397–9403

Liu M, Vogt C, Chaffee AL et al (2013b) Nanoscale structural investigation of Cs_2CO_3-doped MgO sorbent for CO_2 capture at moderate temperature. J Phys Chem C 117(34):17514–17520

Lu C, Bai H, Wu B et al (2008) Comparative study of CO_2 capture by carbon nanotubes, activated carbons, and zeolites. Energy Fuels 22:3050–3056

Lu H, Khan A, Pratsinis SE et al (2009) Flame-made durable doped-CaO nanosorbents for CO_2 capture. Energy Fuels 96:1093–1100

Merkel TC, Lin H, Wei X et al (2010) Power plant post-combustion carbon dioxide capture: an opportunity for membranes. J Memb Sci 359:126–139

Mishra AK, Ramaprabhu S (2011) Carbon dioxide adsorption in graphene sheets. AIP Adv 1:032152

Mishra A K, Ramaprabhu S (2014) Enhanced CO_2 capture in Fe_3O_4-graphene nanocomposite by physicochemical adsorption. J Appl Phys 116:064306

Mishra AK, Ramaprabhu S (2010) Magnetite decorated multiwalled carbon nanotube-based supercapacitor for arsenic removal and desalination of seawater. J Phys Chem C 114:2583–2590

Mishra AK, Ramaprabhu S (2011b) Nano magnetite decorated multiwalled carbon nanotubes: a robust nanomaterial for enhanced carbon dioxide adsorption. Energy Environ Sci 4:889–895

Mishra AK, Ramaprabhu S (2012) Polyaniline/multiwalled carbon nanotubes nanocomposite—an excellent reversible CO_2 capture candidate. RSC Adv 2:1746–1750

Mu B, Schoenecker PM, Walton KS (2010) Gas adsorption study on mesoporous metal-organic framework UMCM-1. J Phys Chem C 114:6464–6471

Muntha ST, Ajmal M, Naeem H et al (2019) Synthesis, properties, and applications of polysulfone/polyimide nanocomposite membrane reinforced with silica nanoparticles. Polym Compos 40:1897–1910

Nisar M, Bergmann C, Geshev J et al (2016) An efficient approach to the preparation of polyethylene magnetic nanocomposites. Polymer 97:131–137

Nisar M, Thue PS, Maghous MB et al (2020a) Polysulfone metal-activated carbon magnetic nanocomposites with enhanced CO_2 capture. RSC Adv 10:34595–34604

Nisar M, Thue PS, Heck CA et al (2020b) Polyethylene nanocomposites with Ni Co, and Fe carbon-based magnetic fillers. Polym Eng Sci 60:988–995

Nowrouzi M, Younesi H, Bahramifar N (2018) Superior CO_2 capture performance on biomass-derived carbon/metal oxides nanocomposites from Persian ironwood by H_3PO_4 activation. Fuel 223:99–114

Oliveira M, Machado A (2013) Preparation of polymer-based nanocomposites by different routes. In: Wang X (ed) Nanocomposites: synthesis, characterization and applications. Nova science publishers, pp 1–22
Ounaies Z, Park C, Wise KE et al (2003) Electrical properties of single wall carbon nanotube reinforced polyimide composites. Compos Sci Technol 63(11):1637–1646
Pacchioni G, Ricart JM, Illas F (1994) Ab initio cluster model calculations on the chemisorption of CO_2 and SO_2 probe molecules on MgO and CaO (100) surfaces. A theoretical measure of oxide basicity. J Am Chem Soc 116(22):10152–10158
Palmero P (2015) Structural ceramic nanocomposites: a review of properties and powders' synthesis methods. Nanomaterials 5(2):656–696
Peigney A, Laurent CH, Flahaut E et al (2000) Carbon nanotubes in novel ceramic matrix nanocomposites. Ceramic International 26(6):677–683
Provenzano V, Louat NP, Imam MA et al (1992) Ultrafine superstrength materials. Nanostruct Mater 1(1):89–94
Rogach AL, Talapin DV, Shevchenko EV et al (2002) Organization of matter on different size scales: monodisperse nanocrystals and their superstructures. Adv Func Mater 12(10):653–664
Roy SD, Chakravorty D, Agrawal DC (1993) Magnetic properties of glassmetal nanocomposites prepared by the sol-gel route and hot pressing. J Appl Phys 74(7):4746–4749
Sakpal T, Kumar A, Kamble S et al (2012) Carbon dioxide capture using amine functionalized silica gel. Indian J Chem 51A:1214–1222
Salleh WNW, Ismail AF (2015) Carbon membranes for gas separation processes: recent progress and future perspective. J Membr Sci Res 1:2–15
Sanz-Pérez S, Murdock CR, Didas SA et al (2016) Direct capture of CO_2 from ambient air. Chem Rev 116:11840–11876
Sethia G, Sayari A (2015) Comprehensive study of ultra-microporous nitrogen-doped activated carbon for CO_2 capture. Carbon 93:68–80
Sevilla M, Fuertes AB, Mokaya R (2011) High density hydrogen storage in superactivated carbons from hydrothermally carbonized renewable organic materials. Energy Environ Sci 4(4):1400
Shahkarami S, Dalai AK, Soltan J (2016) Enhanced CO_2 adsorption using MgO-impregnated activated carbon: Impact of preparation techniques. Ind Eng Chem Res 55(20):5955–5964
Sieliechi JM, Thue PS (2015) Removal of paraquat from drinking water by activated carbon prepared from waste wood. Desalin Water Treat 55:986–998
Sircar S, Golden TC, Rao MB (1996) Activated carbon for gas separation and storage. Carbon 34(1):1–12
Siriwardane RV, Shen M, Fisher EP et al (2001) Adsorption of CO_2 on molecular sieves and activated carbon. Energy Fuels 15(2):279–284
Srinivasan D, Chattopadhyay K (2004) Hardness of high strength nanocomposite Al–X–Zr (X= Si, Cu, Ni) alloys. Mater Sci Eng, A 375–377:1228–1234
Stankic S, Suman S, Haque F et al (2016) Pure and multi metal oxide nanoparticles: synthesis, antibacterial and cytotoxic properties. J Nanobiotechnol 14(1):73
Su F, Lu C, Cnen W et al (2009) Capture of CO_2 from flue gas via multiwalled carbon nanotubes. Sci Total Environ 407:3017–3023
Szczęśniak B, Choma J, Jaroniec M (2017) Gas adsorption properties of graphene-based materials. Adv Colloid Interface Sci 243:46–59
Tahir M, Amin NAS (2013) Photocatalytic reduction of carbon dioxide with water vapors over montmorillonite modified TiO_2 nanocomposites. Appl Catal B Environ 142–143:512–522
Twardowski TE (2007) Introduction to nanocomposite materials: properties, processing, characterization. DEStech Publications, Lancaster
Ulbricht M (2006) Advanced functional polymer membranes. Polymer 47:2217–2262
Van Assche TRC, Baron GV, Denayer JFM (2016) Molecular separations with breathing metal-organic frameworks: modelling packed bed adsorbers. Dalton Trans 45(10):4416–4430
Ventra M, Evoy S, Heflin JR (eds) (2004) Introduction to nanoscale science and technology. Springer, US

Vitillo JG (2015) Magnesium-based systems for carbon dioxide capture, storage and recycling: from leaves to synthetic nanostructured materials. RSC Adv 5:36192–36239

Wang L, Yang RT (2012) Significantly increased CO_2 adsorption performance of nanostructured templated carbon by tuning surface area and nitrogen doping. J Phys Chem C 116:1099–1106

Wang P, Tan KL, Kang ET et al (2002) Plasma-Induced immobilization of poly(ethylene glycol) onto poly(vinylidene fluoride) microporous membrane. J Memb Sci 195(1):103–114

Wang C, Guo ZX, Fu S et al (2004) Polymers containing fullerene or carbon nanotube structures. Prog Polym Sci 29(11):1079–1141

Wang Q, Luo J, Zhong Z et al (2011) CO_2 capture by solid adsorbents and their applications: current status and new trends. Energy Environ Sci 4:42–55

Wang X, Pakdel A, Zhang J et al (2012) Large-surface-area BN nanosheets and their utilization in polymeric composites with improved thermal and dielectric properties. Nanoscale Res Lett 7:662

Wang S, Li X, Wu H et al (2016) Advances in high permeability polymer-based membrane materials for CO_2 separations. Energy Environ Sci 9:1863–1890

Weisenberger MC, Grulke EA, Jacques D et al (2003) Enhanced mechanical properties of polyacrylonitrile/multiwall carbon nanotube composite fibres. J Nanosci Nanotechnol 3(6):535–539

West R, Wang Y, Goodson T (2003) Nonlinear absorption properties in novel gold nanostructured topologies. J Phys Chem B 107(15):3419–3426

Xie XL, Mai YW, Zhou XP (2005) Dispersion and alignment of carbon nanotubes in polymer matrix: a review. Mater Sci Eng R 49(4):89–112

Xu X, Song C, Anderson JM et al (2003) Preparation and characterization of novel CO_2 "molecular basket" adsorbents based on polymer-modified mesoporous molecular sieve MCM-41. Micropor Mesopor Mater 62(1–2):29–45

Yan L, Li Y, Xiang C et al (2006) Effect of nano-sized Al_2O_3-particle addition on PVDF ultrafiltration membrane performance. J Memb Sci 276(1–2):162–167

Yang Z, Zhao M, Florin NH et al (2009) Synthesis and characterization of CaO nanopods for high temperature CO_2 capture. Ind Eng Chem Res 48:10765–10770

Yang H, Gong M, Chen Y (2011) Preparation of activated carbons and their adsorption properties for greenhouse gases: CH_4 and CO_2. J Nat Gas Chem 20(5):460–464

Yim WL, Byl O, Yates JT et al (2004) Vibrational behavior of adsorbed CO_2 on single-walled carbon nanotubes. J Chem Phys 120:5377

Ying DY, Zhang DL (2000) Processing of Cu–Al_2O_3 metal matrix nanocomposite materials by using high energy ball milling. Mater Sci Eng 286(1):152–156

Yoon ES, Lee JS, Oh ST et al (2002) Microstructure and sintering behavior of W-Cu nanocomposite powder produced by thermo-chemical process. Int J Refract Met H 20(3):201–206

Zeng M, Li Y, Mao M et al (2015) Synergetic effect between photocatalysis on TiO_2 and thermo-catalysis on CeO_2 for gas-phase oxidation of benzene on TiO_2/CeO_2 nanocomposites. ACS Catal 5(6):3278–3286

Zhang Z, Xu M, Wang H et al (2010) Enhancement of CO_2 adsorption on high surface area activated carbon modified by N_2, H_2 and ammonia. Chem Eng J 160:571–577

Zhang X, Shi X, Gautrot JE et al (2021) Nanoengineered electrospun fibers and their biomedical applications: a review. Nanocomposites 7(1):1–34

Chapter 6
Synthesis of Graphene Based Nanocomposite from Captured Industrial Carbon

A. Geethakarthi, S. G. Dhanushkumar, K. Giftlin Devapriya, B. Mirudhula, L. Monisha, and S. Sanjaikabilan

Abstract Rapid increase in Earth's average temperature has been observed for the past one and half centuries due to industrialization and urbanization. The utilization of fossil fuels in most of the industrial sectors has led to the release of major greenhouse gases (GHG) such as methane (CH_4), carbon dioxide (CO_2), ozone (O_3) and chlorofluorocarbons (CFCs) leading to drastic climate changes and global warming. The varying effects of climate change in all sectors is a notable threat not only to human lives but also to the entire ecosystem. Global initiatives and protocols through Intergovernmental Panel on Climate Change (IPCC) has declared in limiting earth's increased temperature within 1.5 °C to avert the catastrophic effects. CO_2 is considered as the major GHG due to its heat entrapment and longer half-life period. This has kindled and urged many researchers to undertake economic feasible strategies of carbon capturing, sequestration and its conversion. The captured carbon has diverged applications such as building materials, chemicals for commodities, fuel and energy, carbon materials and aquaculture. This chapter focuses on capturing carbon from industrial emissions and converting it into graphene, a carbon allotrope. In recent years, largescale synthesis and exploration on graphene is focussed due its excellent electrical and structural properties. This chapter discusses on the various synthesis process of graphene and its wide range of applications.

Keywords Carbon capture · Climate change · Graphene · Industrial emission · Nanocomposite · Thermal reduction

6.1 Introduction

Ever since industrialisation has occurred by 1760s to till date, fossil fuels have been the main source of energy. Fossil fuels are the most prominent energy source and is dominantly used in electricity generation, industries, transportation and other commercial and municipal sectors (Fig. 6.1). Being an efficient source of energy,

A. Geethakarthi (✉) · S. G. Dhanushkumar · K. Giftlin Devapriya · B. Mirudhula · L. Monisha · S. Sanjaikabilan
Kumaraguru College of Technology, Coimbatore, Tamil Nadu 641049, India

S. P. Singh et al. (eds.), *Metal Nanocomposites for Energy and Environmental Applications*, Energy, Environment, and Sustainability,
https://doi.org/10.1007/978-981-16-8599-6_6

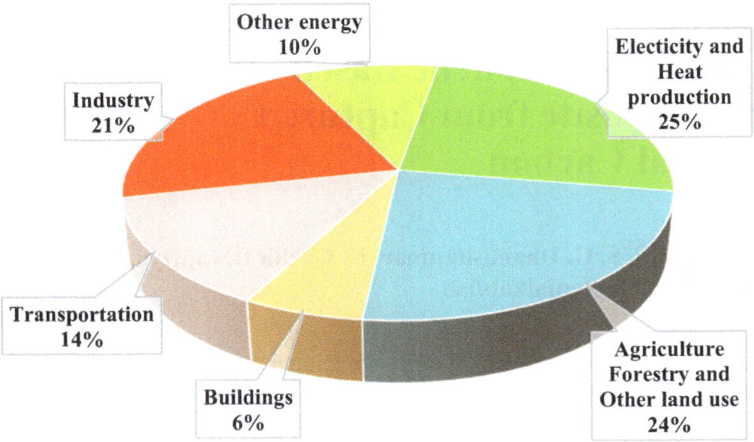

Fig. 6.1 Global greenhouse gases emission from different sectors (2018) (Newsham et al. 2018). *Source* https://www.climatecentral.org/go/cop-21

it is estimated that fossil fuels supply around 80% of the world's energy and would be in high demand for the next half a century (EIA 2015). Though burning of fossil fuels is an unavoidable factor in the society, the huge emission of Greenhouse Gases (GHG) into the atmosphere needs to be focussed. Greenhouse gases behaves like either a cap or a blanket, trapping certain percent of sun's radiation onto the Earth's surface. This increases the earth's mean temperature and ultimately leads to global warming (Al-Mamoori et al. 2017). As per United Nations Framework Convention (UNFC), an increase in 1.5 °C of global average temperature leads to various risks such as rise in sea level, loss of biodiversity and extinction of species including food scarcity and worldwide poverty.

Some of the major GHGs are methane, carbon-dioxide, nitrous-oxide, water vapour and chlorofluorocarbons. Figure 6.2 represents the composition of major GHGs distributed in the atmosphere. Approximately, 50 billion tonnes of GHGs are being emitted from around the world each year [expressed as carbon dioxide equivalents (CO_2eq)]. Out of all greenhouse gases, the most important is carbon dioxide (CO_2), which is formed from combustion of fuels containing carbon.

Carbon dioxide absorbs energy at a wide range of wavelengths between 2,000 and 15,000 nm overlapping with infrared energy. As CO_2 soaks up this infrared energy, it vibrates and re-emits the infrared energy back in all directions. India being a developing country is prone to GHG emissions and ranks third next to China and United States.

From over a decade the carbon-dioxide emissions in the country have risen by an estimated amount of at least 10 million tonnes per year and is expected to rise by the year 2030. Figure 6.3 represents the GHG emission from various sectors in India. Organizational frameworks like IPCC were initiated to offer policymakers periodic scientific assessments on climate change. These frameworks also analysed

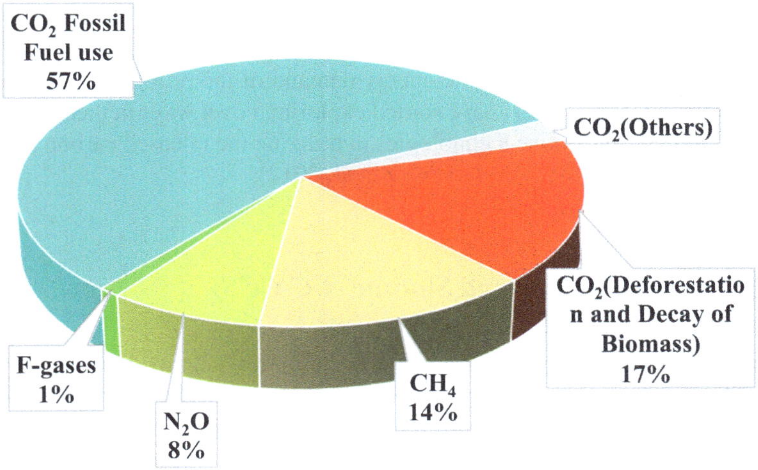

Fig. 6.2 Composition of major greenhouse gases distributed in the atmosphere (2015). *Source* https://www.c2es.org/content/international-emissions/

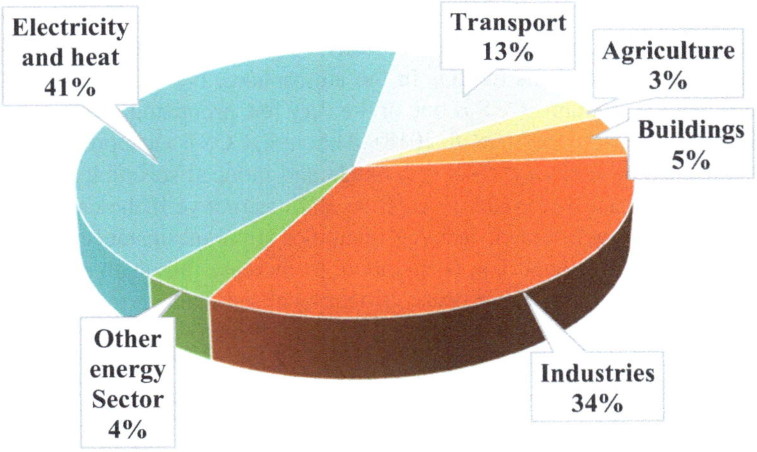

Fig. 6.3 Greenhouse gas emission from various sectors in India (2019). *Source* https://www.climatelinks.org/resources/greenhouse-gas-emissions-factsheet-india

the implications and possible future risks caused by the climate change including the various adaptations and alleviation options.

The Paris Agreement is a legal international treaty on climate change signed by 196 countries in the year 2015 (http://www.c2es.org/international/negotiations/cop21-paris/summary). The main motto of the Paris Agreement is to restrict global warming below 2 °C, ideally to 1.5 °C. To achieve this long-term goal, countries must aim to reduce global peaking of GHG emissions at the earliest to attain a neutral climatic condition by the mid of this century. Apart from such mitigative measures taken to control GHGs, it is also impossible in cutting down the use of fossil fuels in today's industrial world to meet the energy demand of the proliferating population and development. Researchers have started exploring novel ways in the removal and storage of carbon. Innovative technologies in utilizing the released carbon are much focussed beyond the removal (Markewitz et al. 2012).

6.1.1 Carbon Capture and Storage (CC/CCS)

Carbon Capture and Storage (CC/CCS) is a technique which involves selective CO_2 removal from gas streams, compressing it under supercritical conditions and finally sequestrating it in environmental formations such as oil rigs (Boot-Handford et al. 2014). The captured carbon is often stored underground in a process called geological sequestration which involves injecting carbon dioxide into underground rock formation. Considering the past few years, CCS is welcomed as an efficient method to allow the continuous use of fossil fuels without letting the CO_2 emissions into the atmosphere.

CO_2 capture technologies can be used in specific industries like cement manufacturers, iron and steel plants. These industries significantly depend on the fossil fuels, thereby releasing tons of CO_2 in the atmosphere. For these industries, that emit CO_2 in huge amounts, CCS is one of the very few promising methods to lower their emissions (EIA 2015; Bui et al. 2018). Although, CCS is an appreciable control measure, it does pose several risks of leakage and induced seismicity. CCS cause environmental impact that could be massive and destructive if the carbon dioxide leaks out into the environment in large quantities. In some instances, leakage of carbon dioxide underground has been shown to increase the death rate of plants, decrease their growth levels and cause critical localised damage to the ecosystem. Carbon sinking into deep oceans at depths under 3500 m is another storage alternate. Concerns on marine eco-system and non-satisfactory results in CO_2 sinking under high pressure into the environment are not definitely addressed (Metz et al. 2005). Carbon Capture and Utilization (CCU) technology is a more efficient alternative than CCS. The captured carbon can be used in a wide range of domains like building materials, chemical commodities, carbon materials and aquaculture instead of permanently sequestrating it. Table 6.1 represents the characteristics comparison of various carbon materials. This technique is expected to solve the problem of emission-control and has dual advantage in the utilization of carbon resources as reuse materials.

Table 6.1 Characteristics comparison of carbon materials (Kausar 2018)

Carbon based materials	Specific surface area (m^2g^{-1})	Density(gcm^{-3})	Electrical conductivity (Scm^{-1})	Cost	Specific capacitance (Fg^{-1})	
					Aqueous electrolyte	Organic electrolyte
Fullerene	1100–1400	1.72	10^{-8}–10^{-14}	Medium	–	–
CNTs	120–500	0.6	10^4–10^5	High	50–100	<60
Graphene	2630	>1	10^6	High	100–200	80–110
Graphite	10	2.26	10^4	Low	–	–
Activated carbon	1000–3500	0.4–0.7	0.1–1	Low	<200	< 100
Templated porous carbon	500–3000	0.5–1	0.3–10	High	120–350	60–140
Activated carbon fibers	1000–3000	0.3–0.8	5–10	Medium	120–370	80–200
Carbon aerogels	400–1000	0.5–0.7	1–10	Low	100–125	<80

6.1.2 Graphene Nanocomposite

Graphene is a two-dimensional material made up of sp^2 bonded carbon atoms. It is considered as the mother of all the carbon allotropes because of its structural and dimensional arrangement of carbon. For instance, the sweeping of graphene produces carbon nanotubes, the pilling of graphene layers creates graphite and so on. Graphene is a very good conductor of heat and electricity. Compared to steel, graphene is strong and is also harder than diamond. Graphene has the capability to transmit 97.7% of the visible light and graphene also possesses a very high surface to volume ratio. Table 6.2 represents the important properties of graphene (Abergel et al. 2010).

Table 6.2 Properties of graphene (Qureshi and Panesar 2019)

S. No.	Properties	Values
1	Young's modulus	1100 GPa
2	Fracture strength	125 GPa
3	Thermal conductivity	5000 Wm^{-1} k^{-1}
4	Mobility of charge carrier	2×10^5 cm^2Vs^{-1}
5	Specific surface area	2630 m^2g^{-1}
6	Transparency	97.4%
7	Tensile strength	1100 GPa
8	Band gap	0

Because of its excellent properties, graphene finds its applications in various realms like solar cells, fuel cells, chemical sensors, supercapacitors, transparent conducting electrodes, optoelectronics, batteries, photocatalysts, purification of water, biosensors, photonics (Perreault et al. 2015) and the main applications of graphene are summarised in Fig. 6.4. Graphene films are also synthesized using various methods including mechanical exfoliation from highly oriented pyrolytic graphite, chemical exfoliation from bulk graphite, thermal decomposition of carbon terminated silicon carbide and chemical vapour deposition (CVD) (Fig. 6.5).

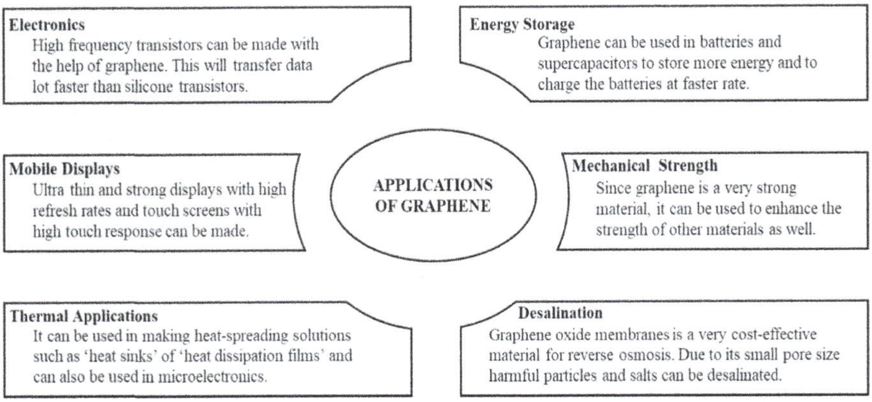

Fig. 6.4 Industrial application of graphene

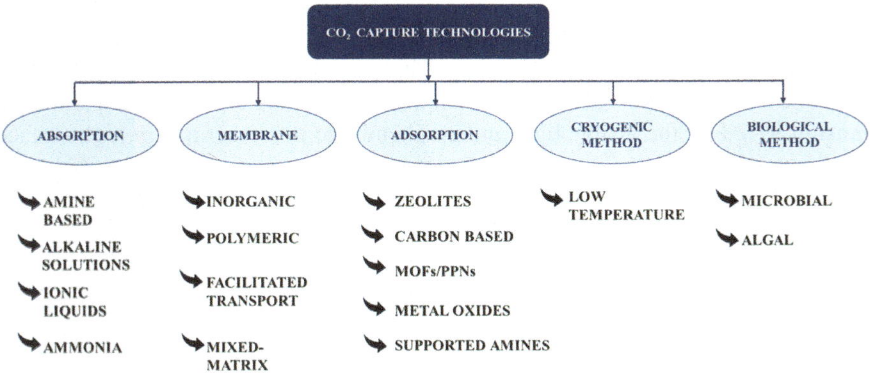

Fig. 6.5 A schematic representation of various CO_2 capture techniques

6.2 CO_2 Capture Technologies and Methodologies

6.2.1 *Absorption*

Absorption is one among the most mature ways of separation of carbon dioxide in the oil industries and some factories that produce chemicals. The precombustion of carbon capture involves physical solvent absorption and the post combustion carbon capture involves chemical solvent absorption. Chemical absorption by aqueous ammonia and different amine-based solvents such as monoethanolamide (MEA), diethanolamine (DEA), N-methyl diethanolamine (MDEA) and alkaline solvents such as $Ca(OH)_2$ and NaOH is the most primitive method for post-combustion capture in industries that include cement, power plants, iron and steel and oil refineries. Physical absorption includes usage of purisol, selexol, rectisol, and fluor. Recently, Ionic Liquids (IL) is considered to be a good alternative for physical solvents (Al-Mamoori et al. 2017). The process of physical absorption involves two steps: Absorption and Stripping process. Figure 6.6 represents the schematic illustration of absorption.

The absorption process usually involves the flue gas being in contact with a specific solvent and the CO_2 is absorbed by the solvent physically. The stripping process involves solvent and captured CO_2, which is exposed to heat and a new solvent is formed with the release of CO_2 from the stripping chamber. Low temperatures and high pressures are the optimum operating conditions for physical absorption. Although absorption is a well-known habitual separation method, the energy required for regenerating the solvent is very high. Other significant barriers are risks of corrosion and large volume of water makeup is also required.

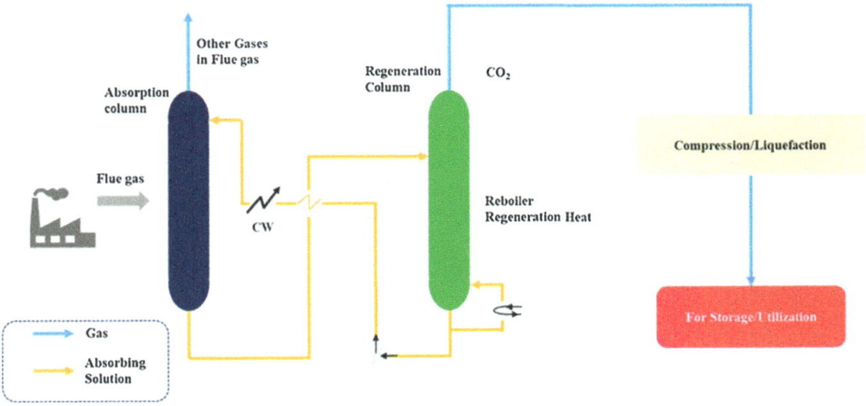

Fig. 6.6 Schematic representation of absorption. *Source* https://www.chiyodacorp.com/en/service/environment/ccs/

6.2.2 Membrane Gas Separation

Usage of membranes to separate gases is an effective technique in areas where the carbon emission is low because it can be easily operated in a periodic system, which is preferred by most industries. Feeding mixture of flue gases and extracting the purified gas can happen simultaneously. This method selectively permeates the desired components and retains the unwanted flue gas and hence resulting in the separation of gas mixtures. The flue gas usually contains CO_2, H_2O (water) and N_2 (Nitrogen). Through the process of condensation, water vapors can be easily removed. The removal of other gases like CO_2 and N_2 requires more efforts, and their removal must be completed before treatments like compression. At ambient atmospheric pressure and moderate temperatures, post-combustion capture separates CO_2/N_2. Figure 6.7 represents the schematic illustration of membrane gas separation. As a result, this method has less hurdles, furthermore, it is more suitable for practical applications. Solution diffusion and surface diffusion are the main parameters on which the segregation of CO_2/N_2 rely. These factors solely depend upon the solubility and adsorb—ability of the gases. The favorable removal of CO_2 in comparison with N_2 using techniques like absorption and adsorption are the advantage in this method. An efficient membrane should be able to separate CO_2 from the industrial emissions and hence should possess high chemical and thermal stability, CO_2 permeability, high levels of CO_2/N_2 selectivity and affordable costs.

Currently, polymer-based membranes such as cellulose acetate, polycarbonates and polyamides are the feasible and cost effective materials for the segregation of CO_2 from the industrial emissions. Polymers that possess high fractional free volume offers remarkable properties like transportation of gas (Mustafa et al. 2016).

The use of silica and zirconia membranes in recent studies have shown high CO_2 permeability with higher gas separation efficiency (Ji and Zhao 2017). The

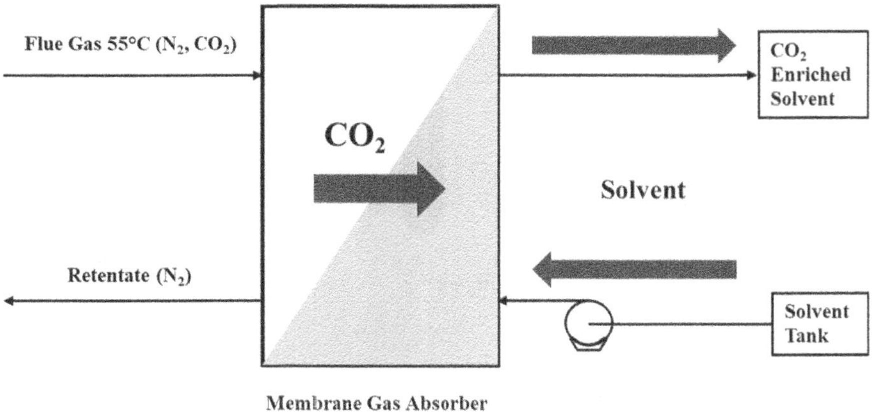

Fig. 6.7 Schematic illustration of membrane gas separation (Scholes et al. 2009). *Source* http://dx.doi.org/10.1016/j.egypro.2009.01.043

characteristics properties of polymeric membranes can be enhanced by using options like mixed-matrix. By adding inorganic materials to a polymer matrix, the thermal and mechanical properties can be enhanced to make it suitable at worst conditions of the environment and the polymer membrane is also stabilized. The low driving force of CO_2 permeation due to the low volume of CO_2 (usually below 15%) is the prominent setback of the membrane separation technique. In addition to this, the brittleness and high-cost of mixed-matrix and membranes makes this method less commercialized. The accumulation of condensable vapors in membrane matrix with changing permeability and selectivity deteriorates the membrane layer and decreases the mechanical strength of the membranes (Mustafa et al. 2016).

6.2.3 *Adsorption*

Carbon capture through adsorption is most dominated and advantageous among post combustion capture techniques (Samanta et al. 2012). The appropriate selection of adsorbent in the separation of CO_2 from the flue gas is atmost significant in considering this technique. An ideal adsorbent is determined by mainly considering the following four factors:

i. Adsorption capacity
ii. Selectivity
iii. Regenerative ability
iv. Adsorption/Desorption pressure.

Adsorption is classified as chemical adsorption and physical adsorption. Chemical adsorption (chemisorption) requires elevated temperature and energy requirements, and covalent molecular bonding plays the major role. The energy requirements of physical adsorption (physisorption) are low compared to chemisorption and is caused due to weal Vander Waals force (Aaron and Tsouris 2005). The process of adsorption has various methods such as:

i. Pressure swing adsorption—adsorption occurs at elevated pressure.
ii. Vacuum swing adsorption—adsorption occurs at elevated pressure in vacuum.
iii. Temperature swing adsorption—adsorption occurs at elevated temperature.

Materials that can adsorb CO_2 are categorized as:

i. Porous
ii. Carbon based materials
iii. MOFs (metal organic framework).

Porous materials such as zeolites can adsorb CO_2 at low temperature and operation cost. Synthetic zeolites can adsorb CO_2 more efficiently. Carbon based materials such as carbon nanotubes have effective regenerative ability and it can adsorb CO_2 efficiently at high temperatures (Kikkinides et al. 1993). MOFs are cost effective and it has high adsorption capacity. The pore size can be tuned for better adsorption and is considered as the most promising material for the CO_2 adsorption (Ben Mansour et al. 2016).

6.2.4 *Cryogenic Separation*

In this technique, the CO_2 laden flue gas is cooled to de-sublimation temperature (-100 to -35 °C) to separate CO_2 from flue gas as solid CO_2. The separated and collected gas is compressed to a final pressure (100 to 200 atm) and delivered as compressed CO_2 stream. The other by-products formed during this separation are used as recuperative heat exchanger to cool the incoming gases. The cryogenic CO_2 capture consumes 30% less energy and cost compared to other effective CO_2 capture processes (Baxter et al. 2009).

6.2.5 *Microbial/Algal Systems*

One of the natural phenomena in the carbon capture is through the absorption of the algae mass. Freshwater algal species like *Oscillatoria*, *Hydrodictyon*, *Oedogonium*, *Chlorella* and *Spirogyra* helps in efficient absorption of CO_2, thereby reducing the carbon footprint. A freshwater pond in Dhanbad, Jharkhand, India is found in a coal mining area, and it contains algal species that can capture CO_2. Having the capability to utilize CO_2 as a carbon source is the major advantage of CO_2 sequestration using algal methods. Higher photosynthesis, higher growth rates, large CO_2 fixation capacities, primitive culture condition and biomass production are some of the highlights for accepting this method. But still, this process is in the research phase and the growth of algae needs more attention as it gets contaminated easily (Anguselvi et al. 2019).

6.3 Graphene Synthesis

The controllable synthesis of elemental carbon has become a significant strategy of developing carbon-based materials with remarkable physicochemical properties. Graphene is one of the most stable carbon allotropes with different structural arrangements of carbon bonds in different patterns. In recent years, graphene has attracted wide range of application due to its electrical and thermal properties. There are many methods for synthesizing graphene. The synthesis of graphene could be performed either through top-down approach or through bottom-up methods. Some of the common top–down approaches are exfoliation and reduction while bottom–up approaches include chemical vapour deposition, epitaxial growth, and pyrolysis (Bhuyan et al. 2016) as represented in Fig. 6.9. A wide range of carbon sources have been successfully used as precursors in various synthesis methods to derive graphene.

Carbon takes position as the sixth element in the periodic table with a ground-state electronic configuration of $1s^2\ 2s^2\ 2P_x{}^1\ 2P_y{}^1\ 2P_z{}^0$. It is for the sake of convenience, that the energy level of $2P_z$ is kept with zero electrons, even though the energy levels is the same as that of $2P_x$ and $2P_y$. Figure 6.8 illustrates the formation of sp^2 hybridization. When the sp^2 electrons of a carbon atom are shared with its neighbouring carbon atoms, a layer of honeycomb network of planar structure called monolayer graphene is formed.

The unit cell of a graphene crystal, indicated by a purple-coloured parallelogram, consists of two carbon atoms, and unit-cell vector cells. The vector cells a1 and a2 have same lattice constant of 2.46 Å. The delocalization and resonance of the electrons help maintain the stability of the planar ring. The valence and conduction band of graphene molecules is formed with zero band gap due to the movement of free electrons that is permitted by the half-filled π bond in monolayer graphene.

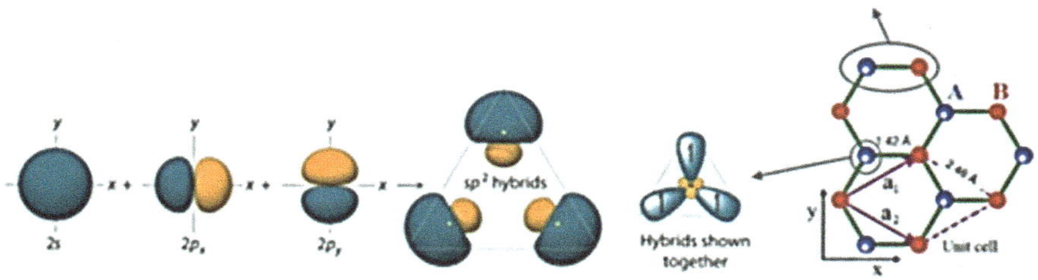

Fig. 6.8 The formation of sp^2 hybrids showing the crystal lattice of graphene (Yang et al. 2018)

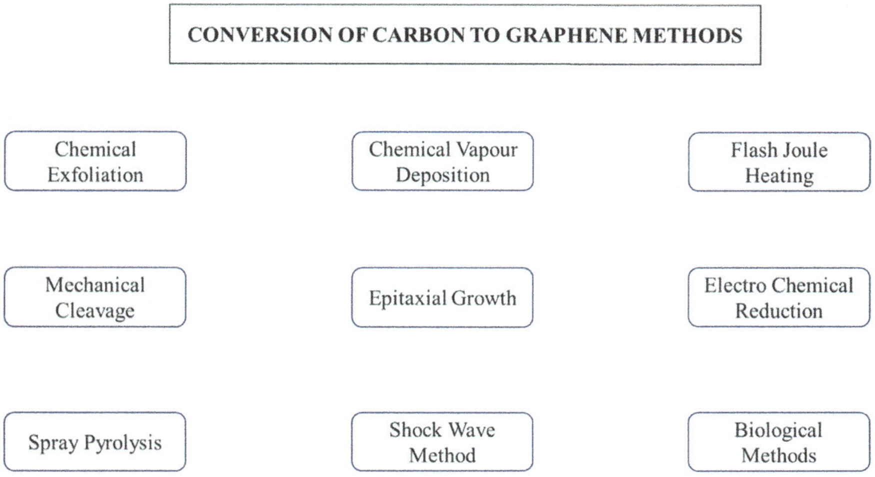

Fig. 6.9 A schematic representation of various methods available for converting carbon to graphene

Further, in graphene materials that are more than a single layer, a weak Van der Waals interaction is observed between the adjacent layers. This interaction is provided by the π-bonds (Zhan et al. 2021). Fig. 6.9 illustrates the various methods available for converting carbon to graphene.

6.3.1 *Chemical Exfoliation*

Chemical exfoliation is one among the most appropriate methods for graphene synthesis. In this method, graphene is manufactured from graphite and graphite intercalation compound by producing colloidal suspensions (Pu et al. 2009). This is found to be an efficient method dealing with the conversion of graphite powder into graphene oxide by oxidation and exfoliation (Parvez et al. 2013). In this process, 0.1 M of H_2SO_4 was used as an electrolyte, graphite flakes and platinum wires were used as anode and cathode respectively. When a positive charge of +10 V is applied to the system for 2 min, the graphite flakes begin to dissolve in the solution. After 2 min, vacuum filtration process is used to exfoliate the graphitic material. To ensure the presence of zero acidity, the collected sample was washed with water multiple times. The resulting powder was then dispersed in the Dimethylformide (DMF) solution that produces exfoliated graphene (EG) sheets. The chemically exfoliated graphene obtained, had a high yield (greater than 80%) and low sheet resistance (4.8 kΩ for a single graphene sheet). Table 6.3 represents the properties of graphene obtained for different graphite electrodes (Pu et al. 2009; Parvez et al. 2013). Another technique was discovered by Lu et al. (2009) to obtain graphene from graphite by chemical exfoliation in ionic liquids (Lu et al. 2009). In this process, the 1-butyl-3-methylimidazolium tetraflouroborate along with water is used as an electrolyte and static potentials of 1.5–15 V were applied to it. To neutralize the obtained exfoliated particles, continuously washing it with water and ethanol is important. This step also helps in eliminating the acidic content (Wang et al. 2020).

6.3.2 *Pyrolysis*

It is a method to synthesize graphene materials using DC high current divergent anode-channel plasma torch.

The most promising method to produce graphene with desired properties is by using plasma. Hydrogen and helium gas are used to produce plasma in this method. In optimum conditions, few layers of graphene with a smaller number of defects were synthesized (Amirov et al. 2006). The source of carbon in this method is natural gas (propane–butane in the ratio 30:70%). The other parameters are listed in Table 6.4. The produced graphene had a wide range of structural and morphological variation depending upon the pressure and flow rate of the gas. Also, the quantity of graphene

Table 6.3 Properties of graphene obtained for different graphite electrodes used in the chemical exfoliation (Li et al. 2016; Jiong et al. 2009)

Electrolytes	Graphite electrodes	Additional conditions	C/O ratios	Properties of graphene obtained			References
				Size	No. of layers	I_D/I_G	
0.1 M K_2SO_4	Graphite foil	Graphite powder HOPG	17.2	0.2–0.6	1–2	0.38	Munuera et al. (2015)
0.2 M $(NH_4)_2SO_4$	Graphite rod	+2 V, DC; sonication	11.7	0.5–1	2	1.76	Sevilla et al. (2016)
10 M H_2SO_4	Expanded graphite	–	7.5	10–30	2–4	0.3	Wu et al. (2014)
0.1 M $(NH_4)_2SO_4$	Graphite flake	Packed in dialysis tubing	4.98	>30	2–7	0.9/1.2	Achee et al. (2018)
0.1 M $(NH_4)_2SO_4$	Graphite + clay	–	NA	NA	3–4	1.2–1.5	Chen et al. (2017)
1.0 M H_2SO_4	Graphite foil	Spark plasma treatment	25.3	NA	1–2	0.16	Gong et al. (2017)
0.1 M Na_2SO_4, 0.1 M NaCl	Graphite foil	–	33.3–5	1	< 5	0.21	Munuera et al. (2018)
0.5 M Na_2SO_4	Graphite foil	0.05 M $CoSO_4$	36	NA	1 and few	0.05	Ejigu et al. (2018)
0.1 M $(NH_4)_2SO_4$	Graphite rod (HOPG)	Thermal reduction up to 800 °C	13.9/13.5	NA	NA	1	Markovic et al. (2016)
0.1 M $(NH_4)_2SO_4$	Graphite rod	Inkjet printing ink	NA	0.5–0.65	2	NA	Miao et al. (2016)

Table 6.4 Technical conditions required to produce graphene (Amirov et al. 2015)

Specification	Details
Power	20–40 kW
Current	200–400 A
Voltage	35–100 V
Plasma gas	Helium, argon
Gas flow rate	0.5–3.75 g/s
Pressure	50–750 Torr
Time of continuous work	20 min

obtained varied from 58 to 95% depending on the optimized experimental conditions (Zhang et al. 2020).

6.3.3 Chemical Vapour Deposition

The general conditions to be considered during the CVD synthesis are:

i. **Carbon source**: Mostly methane is used along with hydrogen as the carbon precursor. Alternative precursors could be ethane, ethanol along with hydrogen.
ii. **Catalyst**: Iron nano particles, nickel foam, gallium vapor or copper.
iii. **Physical conditions**: Low pressure to prevent unwanted side reaction—1–1500 Pa.
iv. **Temperature**: 800–1050 °C.
v. **Carrier gas**: Hydrogen and argon (inert gas) to enhance surface reaction.
vi. **Chamber material**: It should be inert with high melting point. Mostly quartz and glass are used in these cases.

The synthesis of graphene using the CVD method is a two-step process. Firstly, pyrolysis of the carbon precursor to form dissociated carbon atoms. Secondly, the synthesis of graphene using the dissociated carbon atoms from the first step. The first step, i.e., the gas phase is performed on the surface of a substrate to prevent the precipitation of soot. The demand of high temperature levels required for pyrolytic decomposition of the source gas is the major drawback involved in this step. To decrease the reaction temperature, metal catalysts must be used (Pu et al. 2009). Without the usage of a catalyst, the second step would need around 2500 °C of heat. Only with such high temperatures, the dissociated carbon atoms form the structure of graphene. To reduce the temperatures to around 1000 °C, a catalyst must be used. Introducing many components into the chamber is bound to have effects on the reaction within the chamber. This is observed as one of the major drawback in CVD method. Figure 6.10 represents different types of CVD. The graphene film obtained on the catalyst is transferred to SiO_2/Si substrate, which is again transferred to polyethylene terephthalate in order to separate graphene from the catalyst.

Studies have also shown the production of graphene encapsulated Cu nanoparticles with high thermal stability and conductivity using silver as a functional filler. Figure 6.11 represents the schematic illustration of chemical vapour deposition. Inspite of producing high quality graphene, the stumbling block formed during the shedding of graphene from the substrate is one of the major setbacks in this method (Naghdi et al. 2018). The primary reason for this setback is the relationship between graphene and the substrate. Producing uniform layer of graphene is also a difficulty in this method. This is because of convection and diffusion processes affecting the kinetic dynamics of the gas.

The chemical reaction on the substrate is affected as the values keep changing with respect to the space in the reaction chamber (Naghdi et al. 2018). Table 6.5 summarizes the prerequisites and the nature of graphene obtained using CVD.

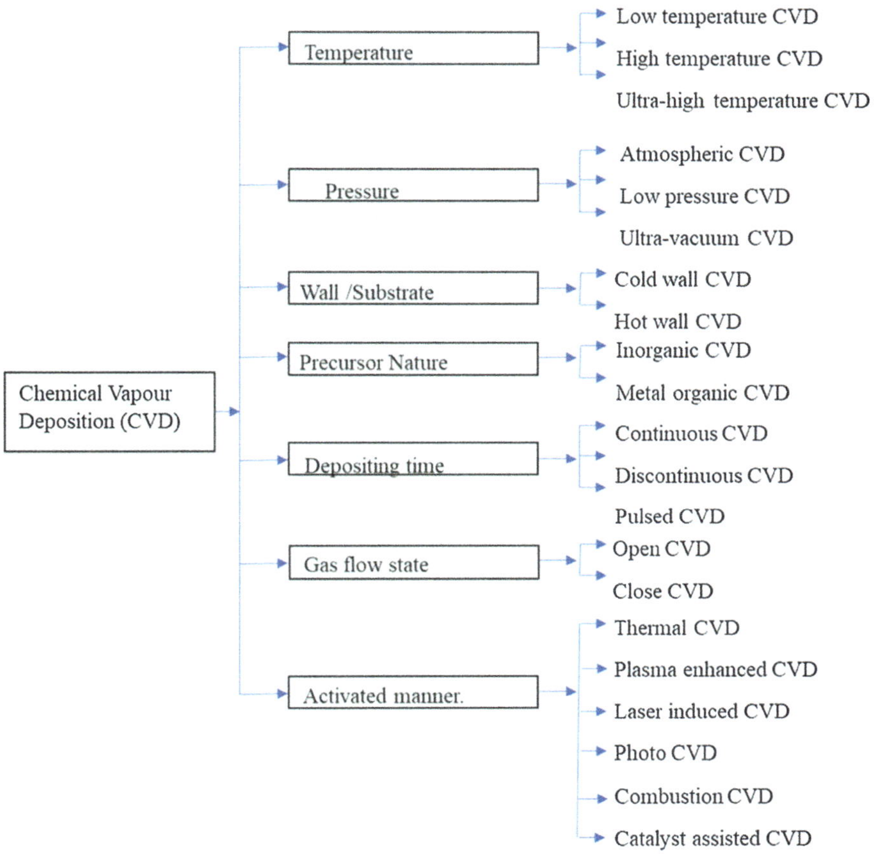

Fig. 6.10 Types of chemical vapour deposition method commonly adopted for graphene synthesis

6.3.4 Electrochemical Reduction

Electrochemical reduction method is the synthesis of considerable amount of graphene from Graphene Oxide (GO). This method is described to be a scalable, fast, economic and an environmentally sustainable pathway. This broad method of electrochemical reduction is further divided into two different types: a one-step route that produces electrochemically reduced graphene oxide (ERGO) directly onto the surface of the electrode by the process of electro chemical reduction of GO (Rummeli et al. 2010). The reduction of GO from an aqueous colloidal suspension produces ERGO. This step is aided with a buffer electrolyte (Liu et al. 2011). Deposition of thin films of GO onto the surface of an electrode is the pre-requisite step for the two-step route. Subsequently, this forms an electrode coated with GO which acts as the substrate (Chen et al. 2011). The substrate electrode coated with GO is then

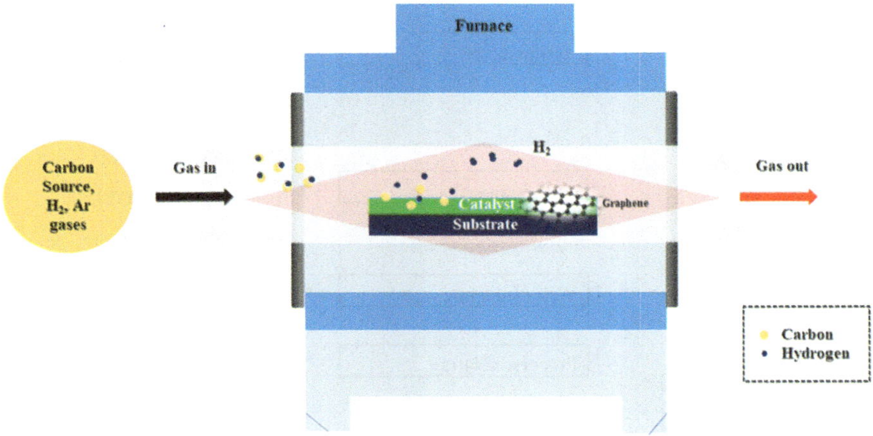

Fig. 6.11 Schematic illustration of chemical vapour deposition (Ago 2015). *Source* http://www.graphenesq.com/whatis/how.asp

exposed to the process of electrochemical reduction. A three-electrode electrochemical system is used for this along with a supporting electrolyte or buffer. Consequently, ERGO films are produced on the electrode substrate (Toh et al. 2014) (see Fig. 6.12).

6.3.5 *Shockwave Synthesis of Graphene*

In this method, graphene is produced transforming a carbonate (carbon source) using a shockwave. Graphene is synthesized by bring calcium carbonate and magnesium (reductant) with the help of a detonation-driven flyer. To obtain an end product with nitrogen doped graphene, ammonium nitrate (nitrogen doping source) was added with the source and reductant. By detonating the main charge of nitromethane (CH_3NO_2) a shock wave compression was produced. The initiation of this was done by a boost charge given through an explosive. This shockwave releases a temperature as high as 2628 K. This increased temperature and pressure are noted to be favourable to produce carbon phases (Yin et al. 2015) and is mentioned in (1).

$$CaCO_3 + Mg + NH_4NO_3 \rightarrow \text{N-doped graphene} \quad (6.1)$$

With remarkable electrocatalytic activity and long-term operation stability in alkaline fuel for oxygen reduction reaction (ORR), the graphene synthesised using the shock wave method seems to act like a metal-free electrode. The parameters like temperature and pressure form the rate limiting factors of graphene synthesis through this method. Shockwave method produces graphene by using explosives, hence this method is not normally recommended. The process stages involved in the conversion of carbon to graphene is shown in Fig. 6.13.

Table 6.5 Prerequisites and the nature of graphene obtained using CVD (Kairi 2017)

S. No.	Process (hot/cold)	Source of the gas	Catalyst used	Temperature	Nature of graphene obtained	References
1	Hot/cold	Methane	Palladium	Defect free, high quality graphene was found at 1000 °C (with a time interval of more than 15 s)	From the shapes of the crystals, it is evident that they are polycrystalline, with single-crystal domains of a few to tens of micrometres in size. Single layered graphene	Lee and Ye (2016)
2	Hot/cold	methane, C_2H_2 and benzene	No catalyst	2500 °C	Multi-layered graphene	Li et al. (2011)
3	Hot/cold	Methane	Ni	600 °C	Multi-layered graphene	Dong et al. 2020)
4	Hot/cold	Benzene	Cu	400 °C	Single layered graphene	Jang et al. (2015)
5	Hot/cold	Ethylene	Ni	550 °C	Monolayer	Ratkovic et al. (2011)
6	Hot/cold	Methane	Cu	1000 °C	Single layered graphene	Kato et al. (2016), Fang et al. (2016)
7	Hot/ Cold	Methane	MgO/Al_2O_3	700 °C	Carbon nanotubes	Zhao et al. (2020)
8	Hot/cold	Methane	Cu with microwave plasma	300 °C	Low sheet resistance graphene	Kim et al. (2013)
10	Hot/cold	Methane, hydrogen	Ni foil 30 micro-meter	900 °C	Single layered	Fujita et al. (2017)
11	Hot/cold	Acetylene	MgO/ZrO_2	Below 500 °C	Mono and multiple layer	Awadallah et al. (2014)

6.3.6 *Epitaxial Growth*

Epitaxial growth is the process of accumulation of either single or multi-layered graphene on SiC substrates under specific conditions like ultrahigh vacuum (UHV) at elevated temperatures. This method of graphene synthesis is generally aided by techniques such as thermal annealing and heating. Figure 6.14a represents the ultra-

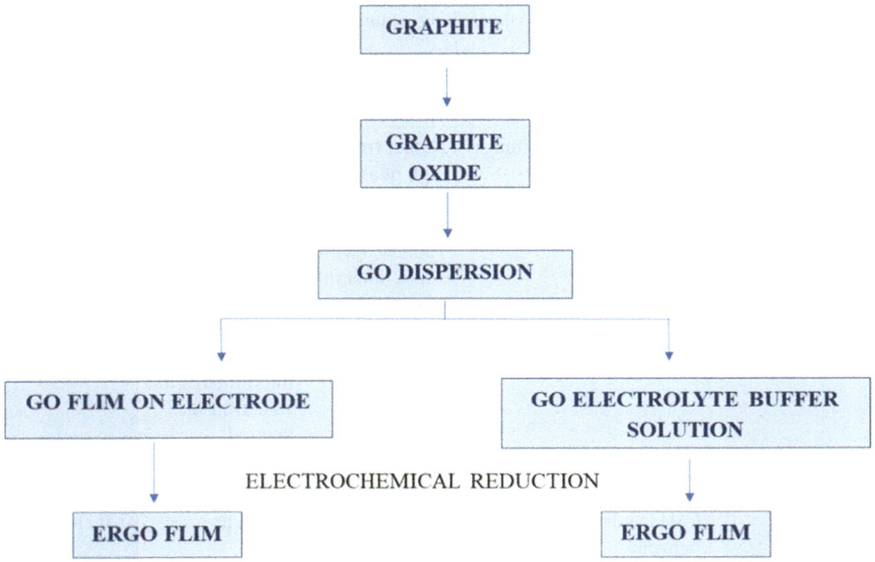

Fig. 6.12 Schematic illustration of electrochemical reduction approach to production of ERGO 2

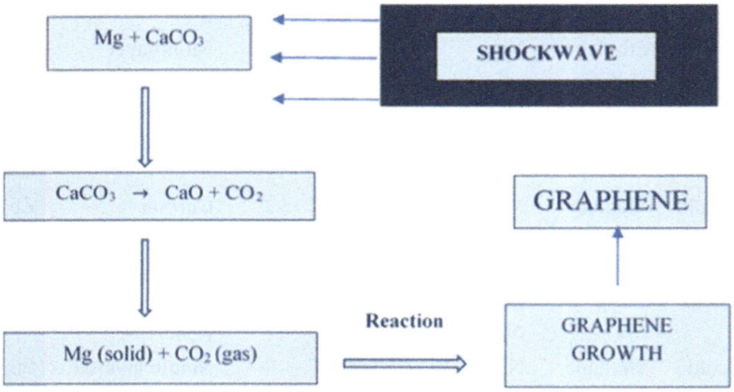

Fig. 6.13 Illustration of synthesizing graphene using shockwave

high vacuum method (b) represent the confinement-controlled sublimation (CCS) process of epitaxial growth.

Depending only upon the annealing temperature and time, single-crystal graphene with controllable number of atomic layers can be epitaxially grown on SiC. Van Bommel (Huang et al. 2014) was the first to discover and observe this method in the year 1975. The silicon or carbon face cause the surface polarity of SiC and is bound to affect the growth of epitaxial graphene (EG). The usage of different SiC polytypes like 3C, 4H and 6H results in variations of the graphene synthesized in this method. Evaporation of Si atoms from the surface starts as the temperature increases. The

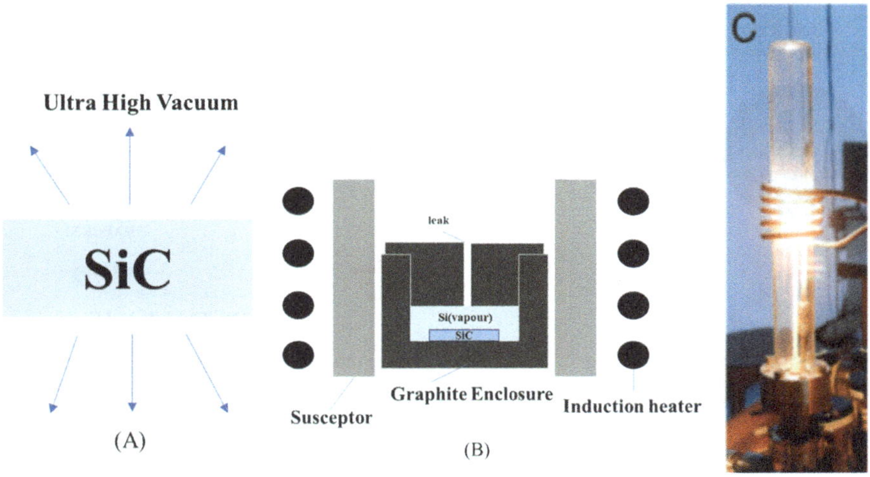

Fig. 6.14 **a** SiC wafer in UHV. **b** Sublimed Si gas is confined in a graphite enclosure so that growth occurs in near thermodynamic equilibrium. **c** Induction furnace. *Source* http://www.graphenesq.com/whatis/how.asp

C-atoms retain and accumulate on the surface forming C-rich layers. These layers range from the interfacial graphene (IG) layer to single-layer EG, bilayer EG and few-layer EG. On increasing the temperatures to more than 800 °C, the carbon atoms from the source gas gain kinetic energy resulting in their deposition over the surface and hence resulting in the formation of graphene. Figure 6.14c represents the induction furnace. The resultant graphene of epitaxial growth method can be used in electronic components and microprocessors. This method is also used for the large-scale production of graphene (Gao et al. 2018).

6.3.7 Mechanical Exfoliation

Mechanical Exfoliation is the method of Peeling off single layer or few layers of carbon atoms from its 3D structures such as graphite, graphene oxide, etc. This method was carried out by placing the carbon source on a sticky material. The constant sticking and peeling through the mechanical force is the driving factor to produce graphene in this method. This method was first discovered in the year 2004 (Gerstner 2010). They used a scotch tape to peel a few layers of carbon atoms from graphite (Van Bommel et al. 1975). Mechanical exfoliation was used for high quality graphene yield. Mechanical exfoliation was not a recommended a commercialized method, because of the longer synthesis period and incapability of large scale production (Pirzado et al. 2019; Luong et al. 2020).

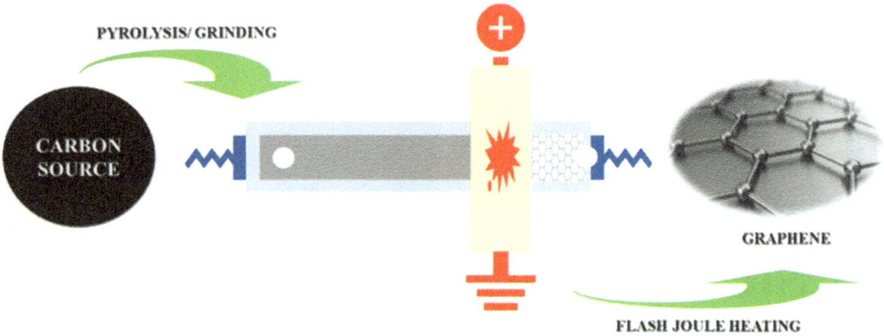

Fig. 6.15 Schematic illustration of synthesis of graphene through flash joule heating method (Chen et al. 2011). *Source* https://doi.org/10.1016/j.carbon.2021.03.020

6.3.8 Flash Joule Heating

Flash joule heating is a bottom-up approach of obtaining graphene from any material that contains carbon. Higher the carbon content, higher is the yield and purity of the synthesized graphene. In this process, the carbon specimen is placed between two electrodes inside a quartz tube under an atmospheric pressure and vacuum of approximately 10 mm Hg. When producing high electric discharge in the electrodes, temperature above 3000 K is created within 100 ms. This exothermic condition separates the carbon in the material, thereby layers of graphene gets deposited in the quartz tube. While using this method graphene can be obtained superfast with low cost (Zhang et al. 2013). Figure 6.15 illustrates the synthesis of graphene through flash joule heating method.

6.3.9 Thermal Reduction

Thermal reduction is an approach for synthesizing graphene using cheap reactants in the presence of carbon dioxide gas. Graphene is produced by reducing CO_2 with the help of reducing agents like magnesium in a horizontal tubular furnace [Fig. 6.16] equipped with a quartz tube or any tube made up of a heat resistant material. The CO_2 gas source from the cylinder is let into the furnace and the setup is heated up to 850 °C. At 850 °C, the reaction takes around 30 min to complete. Then, the power is turned off and the furnace is cooled to room temperature (Zhang et al. 2013). Figure 6.17 represents the processes involved in thermal reduction method. The end product is collected and transferred into a beaker containing hydrochloric acid. The product that is obtained initially consists of graphene, MgO and some quantity of unreacted Mg. MgO and the unreacted Mg are considered as impurities. To remove these impurities, the final product is stirred overnight (at least for10 hours) in an acid solution at room temperatures to eliminate MgO and the unreacted Mg. MgO reacts

Fig. 6.16 Horizontal tubular furnace

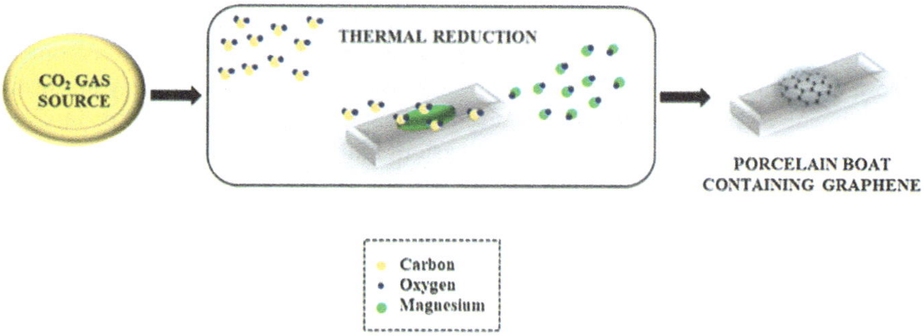

Fig. 6.17 Schematic illustration of thermal reduction

with the HCl acid to form soluble $MgCl_2$. The final product of carbon compound in this step is of high purity. Next the graphene is separated from soluble impurities by serial filtration. Since treating with an acid reduces the pH of the mixture, to neutralize the filtrate it is then filtered and washed using deionized water several times until the filtrate turns out to be neutral or almost neutral (pH 6). To enhance the purity of graphene obtained it is treated with ethanol. Finally, the isolated product is dried under high vacuum overnight at 25 °C (Xing et al. 2015).

This process of thermal reduction involves simple techniques like reduction. The by-products formed are not toxic and can be separated easily from the main product. Since the reacting agents used are cheap, they offer a great affordability and scalability with respect to size and shape. The maintenance of the experimental setup is easy. Moreover, graphene that is got as the final product is more than 90% pure and possess three most needed characteristics such as high conductivity, large specific surface area and relatively high density. The limitation of this process is that huge levels of heat is emitted from the above reaction since the Gibbs free energy of the above reaction is highly negative. Longer synthesis time is also a rate limiting step, though the production process is time efficient, purification stage requires more time to get high purity graphene (Chakrabarti et al. 2011).

6.4 Application of Graphene in Future Technologies

The impact of graphene in diversified research fields has shown a tremendous revolution in today's technology. The development and application of any product could be commercialized based on its production and quality. Graphene has become a substantiate material in most of the industrial sectors including medicine, electronics, water purification, fuel additive, energy, aerospace, defence, transport and infrastructure (Guler and Bag 2020; Stankovich et al. 2006; Kucinskis et al. 2013; Wang et al. 2008; Schedin et al. 1971). This super material is impermeable even to the smallest of atoms (Mittal et al. 2020). It is used in stripping of hydrogen atoms from the electrons and behave as a proton conducting membrane (Novoselov et al. 2005). The silica era's dominance in the electronics and material science are metamorphized by graphene application. Graphene behaves as a dual material under two extremes: as a superconductor and an insulator. Nano-porous graphene is proved to be one of the effective membranes to improve the water permeability. This property has made the graphene to play major role in water purification and desalination technology. Integration of biological and electronics system had led through a breakthrough in medical science. Graphene gadgets are under research towards health monitoring. The antibacterial and antistatic property of graphene has made the cosmetic world to probe into the material.

The super-lightweight graphene sheets absorb two times greater than the Kevlar bullet proof vests. This property of graphene has found application in the parts of drones, racecars, radar signature reduction coatings, ballistic protection materials and in sensors (Fowler et al. 2009; Areshkin et al. 2007). Replacement of steel fibres with carbon nano tubes, carbon fiber reinforced plastics and graphene sheets are under thorough research investigations. Developments are being established to overcome the limitations in the application of graphene supply chain. Graphene has different patterns and structural combination of carbon, hence a standard is necessary to be established for material classification. Production of graphene in large quantity to meet the higher demands is a major economic constraint. A low-cost method of graphene synthesis to resolve the real-life situation in all aspects is essential and is under serious investigation.

6.5 Conclusion

Technologies such as Carbon Capture and Storage (CCS) and Carbon Capture and Utilization (CCU) has become inevitable to achieve negative CO_2 emissions required to attain sustainable climate goals. This can also be a bridge technology that allows the continued use of fossil fuels in industries. CCS technologies could greatly reduce the carbon dioxide emissions. In this chapter, the different available methods for carbon capture—which includes absorption, membrane-based techniques, adsorption techniques, cryogenic method and biological methods that include microbial and algal techniques were presented and discussed. CCS offers a great uncertainty with respect to the leakage of CO_2 from the reservoirs and therefore is a more viable option. The utility of the captured carbon emissions can be utilized and used under various industrial applications like building materials, chemical commodities, carbon materials and aquaculture. Therefore, CCU technology would be considered as more efficient alternative than permanent sequestrating (CCS). This book chapter has also reviewed on the various methods available for the conversion of carbon dioxide to graphene. The significance of converting carbon to graphene is due to its wide range of applications in various realms. The remarkable properties of graphene creates consequential impact in the real field and is extensively used product in many industries. From recent survey, none other nanocomposite material other than graphene has the breadth of superlatives. Graphene has got countless applications in the fields like transport, medicine, electronics, energy, defence and desalination.

References

Abergel DSL, Apalkov V, Berashevich J, Ziegler K, Chakraborty T (2010) Properties of graphene: a theoretical perspective. Adv Phys 59(4):261–482. https://doi.org/10.1080/00018732.2010.487978

Achee TC, Sun W, Hope JT, Quitzau SG, Sweeney CB, Shah SM, Habib T, Green MJ (2018) High-yield scalable graphene nanosheet production from compressed graphite using electrochemical exfoliation. Sci Rep 8(1):14525. https://doi.org/10.1038/s41598-018-32741-3

Ago H (2015) CVD growth of high-quality single-layer graphene. In: Matsumoto K (ed), Frontiers of graphene and carbon nanotubes, pp 3–20. https://doi.org/10.1007/978-4-431-55372-4_1

Al-Mamoori A, Krishnamurthy A, Rownaghi AA, Rezaei F (2017) Carbon capture and utilization update. Energ Technol 5(6):834–849. https://doi.org/10.1002/ente.201600747

Amirov RH, Asinovsky EI, Isakaev EKh, Kiselev VI (2006) Thermal plasma torch for synthesis of carbon nanotubes. High Temp Mater Process 10(2):197–206. https://doi.org/10.1615/HighTempMatProc.v10.i2.30

Amirov RKh, Atamanuk N, Vorobieva NA, Isakaev EH, Shavelkina MB, Shkolnikov EI (2015) Synthesis of graphene-like materials by pyrolysis of hydrocarbons in thermal plasma and their properties. J Phys Conf Ser (Online) 653(1). ISSN 1742-6596. https://doi.org/10.1088/1742-6596/653/1/012161

Anguselvi V, Masto RE, Mukherjee A, Singh PK (2019) CO_2 capture for industries by algae. In: Algae.https://doi.org/10.5772/intechopen.81800

Areshkin DA, Gunlycke D, White CT (2007) Ballistic transport in graphene nanostrips in the presence of disorder: importance of edge effects. Nano Lett 7:204–210. https://doi.org/10.1021/nl062132h

Awadallah AE, Mostafa MS, Aboul-Enein AA, Hanafi SA (2014) Hydrogen production via methane decomposition over Al_2O_3–TiO_2 binary oxides supported Ni catalysts: effect of Ti content on the catalytic efficiency. Fuel 129:68–77. https://doi.org/10.1016/j.fuel.2014.03.047

Baxter L, Baxter A, Burt S (2009) Cryogenic CO_2 capture as a cost-effective CO_2 capture process. Brigham Young University, Provo, UT 846022. Sustainable Energy Solutions Orem, UT 84058

Ben Mansour R, Habib MA, Bamidele OE, Basha M, Qasem NAA, Peedikakkal A, Laoui T, Ali M (2016) Carbon capture by physical adsorption: materials, experimental investigations and numerical modeling and simulations—a review. Appl Energy 161(1):225–255. https://doi.org/10.1016/j.apenergy.2015.10.011

Boot-Handford ME, Abanades JC, Anthony EJ, Blunt MJ, Brandani S, Dowell NM, Fernández JR, Ferrari M-C, Gross R, Hallett JP, Stuart Haszeldine R, Heptonstall P, Lyngfelt A, Makuch Z, Mangano E, Porter RTJ, Pourkashanian M, Rochelle GT, Shah N, Yao JG, Fennell PS (2014) Carbon capture and storage update. Energy Environ Sci 7:130–189. https://doi.org/10.1039/C3EE42350F

Bui M, Adjiman CS, Bardow A, Anthony EJ, Boston A, Brown S, Fennell PS, Fuss S, Galindo A, Hackett LA, Hallett JP, Herzog HJ, Jackson G, Kemper J, Krevor S, Maitland GC, Matuszewski M, Metcalfe IS, Petit C, Puxty G, Reimer J, Reiner DM, Rubin ES, Scott SA, Shah N, Smit B, Martin Trusler JP, Webley P, Wilcox J, Dowell NM (2018) Carbon capture and storage (CCS): the way forward. Energy Environ Sci 11:1062–1176. https://doi.org/10.1039/C7EE02342A

21st session of the conference of the parties to the United Nations framework convention on climate change (COP 21). http://www.c2es.org/international/negotiations/cop21-paris/summary. Accessed 15 Nov 2016

Chakrabarti A, Lu J, Skrabutenas JC, Xu T, Xiao Z, Maguire JA, Hosmane NS (2011) Conversion of carbon dioxide to few-layer graphene. J Mater Chem 21(26):9491–9493. https://doi.org/10.1039/C1JM11227A

Chen L, Tang Y, Wang K, Liu C, Luo S (2011) Electrochemistry communications direct electrodeposition of reduced graphene oxide on glassy carbon electrode and its electrochemical application. Electrochem Commun 13:133–137. https://doi.org/10.1016/j.elecom.2010.11.033

Chen K, Xue D, Komarneni S (2017) Nanoclay assisted electrochemical exfoliation of pencil core to high conductive graphene thin-film electrode. J Colloid Interface Sci 487(1):156–161. https://doi.org/10.1016/j.jcis.2016.10.028

Dong Z, Li B, Cui C, Qian W, Jin Y, Wei F (2020) Catalytic methane technology for carbon nanotubes and graphene. React. Chem. Eng. 5:991–1004. https://doi.org/10.1039/D0RE00060D

Aaron D, Tsouris C (2005) Separation of CO_2 from flue gas: a review. Sep Sci Technol 40(1–3):321–348.https://doi.org/10.1081/SS-200042244

Independent Statistics & Analysis, EIA (2015) U.S. Energy Information Administration—assumptions to the annual energy outlook. U.S. Department of Energy Washington, DC 20585, pp 1–233

Ejigu A, Fujisawa K, Spencer BF, Wang B, Terrones M, Kinloch IA, Dryfe RAW (2018) On the role of transition metal salts during electrochemical exfoliation of graphite: antioxidants or metal oxide decorators for energy storage applications. Adv Funct Mater 28(48):1804357. https://doi.org/10.1002/adfm.201804357

Fang L, Yuan W, Wang B, Xiong Y (2016) Growth of graphene on Cu foils by microwave plasma chemical vapor deposition: The effect of in-situ hydrogen plasma post-treatment. Appl Surf Sci 383:28–32. https://doi.org/10.1016/j.apsusc.2016.04.148

Fowler JD, Allen MJ, Tung VC, Yang Y, Kaner RB, Weiller BH (2009) Practical chemical sensors from chemically derived graphene. ACS Nano 3:301–306. https://doi.org/10.1021/nn800593m

Fujita JI, Hiyama T, Hirukawa A, Kondo T, Nakamura J, Ito SI, Araki R, Ito Y, Takeguchi M, Pai WW (2017) Near room temperature chemical vapor deposition of graphene with diluted methane and molten gallium catalyst. Sci Rep 7(1):1–10. https://doi.org/10.1038/s41598-017-12380-w

Gao E, Lin SZ, Qin Z, Buehler MJ, Feng XQ, Xu Z (2018) Mechanical exfoliation of two-dimensional materials. J Mech Phys Solids 115:248–262. https://doi.org/10.1016/j.jmps.2018.03.014

Gerstner E (2010) Nobel Prize 2010: Andre Geim & Konstantin Novoselov in the year 2004. Nature Phys 6:836. https://doi.org/10.1038/nphys1836

Gong Y, Ping Y, Li D, Luo C, Ruan X, Fu Q, Pana C (2017) Preparation of high-quality graphene via electrochemical exfoliation & spark plasma sintering and its applications. Appl Surf Sci 397:213–219. https://doi.org/10.1016/j.apsusc.2016.11.153

Guler O, Bag N (2020) A short review on mechanical properties of graphene reinforced metal matrix composites. J Mater Res Technol, 1–26. https://doi.org/10.1016/j.jmrt.2020.01.077

Huang H, Chen S, Wee ATS, Chen W (2014) Epitaxial growth of graphene on silicon carbide (SiC). In: Graphene. Woodhead Publishing, pp 3–26. https://doi.org/10.1533/9780857099334.1.3

Jang J, Son M, Chung S, Kim K, Cho C, Lee BH, Ham MH (2015) Low-temperature-grown continuous graphene films from benzene by chemical vapor deposition at ambient pressure. Sci Rep 5:17955. https://doi.org/10.1038/srep17955

Ji G, Zhao M (2017) Membrane separation technology in carbon capture. In: Recent advances in carbon capture and storage, pp 60–89. https://doi.org/10.5772/65723

Lu J, Yang J-X, Wang J, Lim A, Wang S, Loh KP (2009) One-pot synthesis of fluorescent carbon nanoribbons, nanoparticles, and graphene by the exfoliation of graphite in ionic liquids. ACS Nano 3(8):2367–2375. https://doi.org/10.1021/nn900546b

Kato R, Minami S, Koga Y, Hasegawa M (2016) High growth rate chemical vapor deposition of graphene under low pressure by RF plasma assistance. Carbon 96:1008–1013. https://doi.org/10.1016/j.carbon.2015.10.061

Kausar A (2018) Advances in polymer/graphene nanocomposite for biosensor application. Nano World Journal 4(2):23–28. https://doi.org/10.17756/nwj.2018-056

Kikkinides ES, Yang RT, Cho SH (1993) Concentration and recovery of carbon dioxide from flue gas by pressure swing adsorption. Ind Eng Chem Res 32(11):2714–2720. https://doi.org/10.1021/ie00023a038

Kim YS, Lee JH, Kim YD, Jerng SK, Joo K, Kim E, Jung J, Yoon E, Park YD, Seo S, Chun SH (2013) Methane as an effective hydrogen source for single-layer graphene synthesis on Cu foil by plasma enhanced chemical vapor deposition. Nanoscale 5(3):1221–1226. https://doi.org/10.1039/c2nr33034b

Kucinskis G, Bajars G, Kleperis J (2013) Graphene in lithium ion battery cathode materials: a review. J Power Sources 240:66–79. https://doi.org/10.1016/j.jpowsour.2013.03.160

Lee K, Ye J (2016) Significantly improved thickness uniformity of graphene monolayers grown by chemical vapor deposition by texture and morphology control of the copper foil substrate. Carbon 100:441–449. https://doi.org/10.1016/j.carbon.2016.01.032

Li L, Xu J, Li G, Jia X, Li Y, Yang F, Zhang L, Xu C, Gao J, Liuc Y, Fang Z (2016) Preparation of graphene nanosheets by shear-assisted supercritical CO_2 exfoliation. Chem Eng J 284(15):78–84. https://doi.org/10.1016/j.cej.2015.08.077

Li Z, Wu P, Wang C, Fan X, Zhang W, Zhai X, Zeng C, Li Z, Yang J, Hou JG (2011) Chemical vapor deposition growth of graphene using other hydrocarbon sources. Material Science

Wu L, Li W, Li P, Liao S, Qiu S, Chen M, Guo Y, Li Q, Zhu C, Liu L (2014) Powder, paper and foam of few-layer graphene prepared in high yield by electrochemical intercalation exfoliation of expanded graphite. Small 10(7):1421–1429. https://doi.org/10.1002/smll.201302730

Liu C, Wang K, Luo S, Tang Y, Chen L (2011) Direct electrodeposition of graphene enabling the one-step synthesis of graphene–metal nanocomposite films. Small 7(9):1203–1206. https://doi.org/10.1002/smll.201002340

Luong DX, Bets KV, Algozeeb WA, Stanford MG, Kittrell C, Chen W, Salvatierra RV, Ren M, McHugh EA, Advincula PA, Wang Z (2020) Gram-scale bottom-up flash graphene synthesis. Nature 577(7792):647–651. https://doi.org/10.1038/s41586-020-1938-0

Markewitz P, Kuckshinrichs W, Leitner W, Linssen J, Zapp P, Bongartz R, Schreiber A, Mgller TE (2012) Worldwide innovations in the development of carbon capture technologies and the utilization of CO_2. Energy Environ Sci 5(6):7281–7305. https://doi.org/10.1039/C2EE03403D

Markovic ZM, Budimir MD, Kepic DP, Holclajtner-Antunovic ID, Marinovic-Cincovic MT, Dramicanin MD, Spasojevic VD, Perusko DB, Spitalsky Z, Micusik M, Pavlovic VB, Todorovic-Markovic BM (2016) Semi-transparent, conductive thin films of electrochemical exfoliated graphene. RSC Adv 6(45):39275–39283. https://doi.org/10.1039/C6RA04250C

Bhuyan MSA, Uddin MN, Islam MM, Bipasha FA, Hossain SS (2016) Synthesis of graphene. Int Nano Lett 6:65–83. https://doi.org/10.1007/s40089-015-0176-1

Metz B, Davidson O, de Coninck H, Loos M, Meyer L (2005) IPCC Special Report on carbon dioxide capture and storage. Published for the Intergovernmental Panel on Climate Change. Cambridge University press, pp 1–442

Miao F, Majee S, Song M, Zhao J, Zhang S-L, Zhang Z-B (2016) Inkjet printing of electrochemically—exfoliated graphene nano-platelets. Synth Met 220:318–322. https://doi.org/10.1016/j.synthmet.2016.06.029

Mittal SK, Goyal D, Chauhan A, Dang RK (2020) Graphene nanoparticles: the super material of future. Mater Today Proc 28:1290–1294. https://doi.org/10.1016/j.matpr.2020.04.260

Kairi MI, Khavarian M, Bakar SA, Vigolo B, Mohamed AR (2017) Recent trends in graphene materials synthesized by CVD with various carbon precursors. J Mater Sci 53(2):851–879. https://doi.org/10.1007/s10853-017-1694-1

Munuera JM, Paredes JI, Villar-Rodil S, Ayan-Varela M, Pagan A, Aznar-Cervantes SD, Cenis JL, Martinez-Alonso A, Tascon JMD (2015) High quality, low oxygen content and biocompatible graphene nanosheets obtained by anodic exfoliation of different graphite types. Carbon 94:729–739. https://doi.org/10.1016/j.carbon.2015.07.053

Munuera JM, Paredes JI, Villar-Rodil S, Castro-Muniz A, Martinez-Alonso A, Tascon JMD (2018) High quality, low-oxidized graphene via anodic exfoliation with table salt as an efficient oxidation-preventing co-electrolyte for water/oil remediation and capacitive energy storage applications. Appl Mater Today 11:246–254

Mustafa J, Farhan M, Hussain M (2016) CO_2 separation from flue gases using different types of membranes. J Membrane Sci Technol 6(2). https://doi.org/10.4172/2155-9589.1000153

Naghdi S, Rhee KY, Park SJ (2018) A catalytic, catalyst-free, and roll-to-roll production of graphene via chemical vapor deposition: low temperature growth. Carbon 127:1–12. ISSN 0008-6223. https://doi.org/10.1016/j.carbon.2017.10.065

Newsham A, Kohnstamm S, Naess L, Atela J (2018). Agricultural Commercialisation Pathways: Climate Change and Agriculture. https://doi.org/10.13140/RG.2.2.18436.17286

Novoselov KS, Jiang D, Schedin F, Booth TJ, Khotkevich VV, Morozov SV, Geim AK (2005) Two-dimensional atomic crystals. Proc Nat Acad Sci USA 102:10451–10453. https://doi.org/10.1073/pnas.0502848102

Parvez K, Li R, Puniredd SR, Hernandez Y, Hinkel F, Wang S, Feng X, Müllen K (2013) Electrochemically exfoliated graphene as solution-processable, highly conductive electrodes for organic electronics. ACS Nano 7(4):3598–3606. https://doi.org/10.1021/nn400576v

Perreault F, Fonseca de Faria A, Elimelech M (2015) Environmental applications of graphene-based nanomaterials. Chem Soc Rev 44(16):5861–5896. https://doi.org/10.1039/c5cs00021a

Pirzado AA, Le Normand F, Romero T, Paszkiewicz S, Papaefthimiou V, Ihiawakrim D, Janowska I (2019) Few-layer graphene from mechanical exfoliation of graphite-based materials: Structure-dependent characteristics. Chem Eng 3(2):37. https://doi.org/10.3390/chemengineering3020037

Pu NW, Wang CA, Sung Y, Liu YM, Ger MD (2009) Production of few-layer graphene by supercritical CO_2 exfoliation of graphite. Mater Lett 63(23):1987–1989. https://doi.org/10.1016/j.matlet.2009.06.031

Qureshi T, Panesar D (2019) A comparison of graphene oxide, reduced graphene oxide and pure graphene: early age properties of cement composites. In: International conference on sustainable materials, systems and structures (SMSS 2019). New Generation of Construction Materials, 20–22 March 2019, Rovinj, Croatia

Ratkovic S, Vujicic DJ, Kiss E, Boskovic G, Geszti O (2011) Different degrees of weak metal–support interaction in Fe–(Ni)/Al_2O_3 catalyst governing activity and selectivity in carbon nanotubes' production using ethylene. Mater Chem Phys 129(1–2):398–405. https://doi.org/10.1016/j.matchemphys.2011.04.036

Rummeli MH, Bachmatiuk A, Scott A, Bornert F, Warner JH, Volker Hoffman V, Jarrn-Horng Lin J-H, Gianaurelio Cuniberti G, Büchner B (2010) Direct low temperature nano-graphene synthesis over a dielectric insulator. ACS Nano 4(7):4206–4210. https://doi.org/10.1021/nn100971s

Samanta A, Zhao A, Shimizu GK, Sarkar P, Gupta R (2012) Post-combustion CO_2 capture using solid sorbents: a review. Ind Eng Chem Res 51(4):1438–1463. https://doi.org/10.1021/ie200686q

Schedin F, Geim AK, Morozov SV, Hill EW, Blake P, Katsnelson MI, Novoselov KS (1971) Acute toxic effects of drug abuse: diagnosis and treatment. J Florida Med Assoc 58:41–42

Scholes CA, Kentish SE, Stevens GW (2009) The effect of condensable minor components on the gas separation performance of polymeric membranes for carbon dioxide capture. Energy Procedia 1(1):311–317. https://doi.org/10.1016/j.egypro.2009.01.043

Sevilla M, Ferrero GA, Fuertes AB (2016) (2016) Aqueous dispersions of graphene from electrochemically exfoliated graphite. Chem Eur J 22(48):17351–17358. https://doi.org/10.1002/chem.201603321

Stankovich S, Dikin DA, Dommett GHB, Kohlhaas KM, Zimney EJ, Stach EA, Piner RD, Nguyen SBT, Ruoff RS (2006) Graphene-based composite materials. Nature 44:282–286. https://doi.org/10.1038/nature04969

Toh SY, Loh KS, Kamarudin SK, Daud WRW (2014) Graphene production via electrochemical reduction of graphene oxide: synthesis and characterisation. Chem Eng J 251:422–434. https://doi.org/10.1016/j.cej.2014.04.004

Van Bommel AJ, Crombeen JE, Van Tooren A (1975) LEED and Auger electron observations of the SiC (0001) surface. Surf Sci 48(2):463–472. https://doi.org/10.1016/0039-6028(75)90419-7

Wang X, Zhi L, Tsao N, Tomovic Z, Li J, Mullen K (2008) Transparent carbon films as electrodes in organic solar cells. Angew Chem Int Ed 47:2990–2992. https://doi.org/10.1002/anie.200704909

Wang J, Zhao C, Mark LH, Wang X, Li R, Moghimian N, Hamidinejad M, Park CB (2020) Facilitating supercritical CO_2 assisted exfoliation of graphene nanoplatelets with the polymer matrix. Chem Eng J 394:124930. https://doi.org/10.1016/j.cej.2020.124930

Xing Z, Wang B, Gao W, Pan C, Halsted JK, Chong ES, Lu J, Wang X, Luo W, Chang CH, Wen Y (2015) Reducing CO_2 to dense nanoporous graphene by Mg/Zn for high power electrochemical capacitors. Nano Energy 11:600–610. https://doi.org/10.1016/j.nanoen.2014.11.011

Yang G, Li L, Lee WB, Ng MC (2018) Structure of graphene and its disorders: a review. Sci Technol Adv Mater 19(1):631–648. https://doi.org/10.1080/14686996.2018.1494493

Yin H, Chen P, Xu C, Gao X, Zhou Q, Zhao Y, Qu L (2015) Shock-wave synthesis of multilayer graphene and nitrogen-doped graphene materials from carbonate. Carbon 94:928–935. https://doi.org/10.1016/j.carbon.2015.07.078

Zhan D, Yan J, Lai L, Ni Z, Liu L, Shen Z (2021) Engineering the electronic structure of graphene. Adv Mater 24(30):4055–4069 (2021). https://doi.org/10.1002/adma.201200011

Zhang H, Zhang X, Sun X, Ma Y (2013) Shape-controlled synthesis of nanocarbons through direct conversion of carbon dioxide. Sci Rep 3(1):1–8. https://doi.org/10.1038/srep03534

Zhang S, Jiang SF, Huang BC, Shen XC, Chen WJ, Zhou TP, Cheng HY, Cheng BH, Wu CZ, Li WW, Jiang H, Yu H-Q (2020) Sustainable production of value-added carbon nanomaterials from biomass pyrolysis. Nat Sustain 3:753–760. https://doi.org/10.1038/s41893-020-0538-1

Zhao G, Pan X, Zhang Z, Liu Y, Lu Y (2020) A thin-felt Pd–MgO–Al_2O_3/Al-fiber catalyst for catalytic combustion of methane with resistance to water-vapor poisoning. J Catal 384:122–135. https://doi.org/10.1016/j.jcat.2020.01.013

Chapter 7
Metal Organic Frameworks (MOFs) as an Adsorbent Material for CO_2 Capture

Ravi Vaghasia, Miraj Savani, and Bharti Saini

Abstract Due to the uncontrolled release of greenhouse gases into the atmosphere through anthropogenic activities, the planet's temperature and natural ecosystem are being adversely affected globally. Fossil-fueled power plants and transportation are the major sources for the release of CO_2 into the atmosphere. Carbon capture and sequestration (CCS) is the potential alternative for CO_2 mitigation. To capture this CO_2, highly selective and one with high storage capacity adsorbent material is required. The selection of a suitable adsorbent depends on many factors like porosity, chemical stability, surface area, regeneration capacity, heat of adsorption and cost of adsorbent. In recent years, Metal organic frameworks (MOFs) which are highly crystalline materials with large surface areas and good selectivity, have proved to be a class of adsorbent material for carbon capture. The unique structural and chemical characteristics of MOFs all of which have resulted in the largest capture capacity, the most efficient separations and regeneration processes are identified in this review. MOFs' interiors can be precisely engineered to include some beneficial structural characteristics all of which are critical for making higher-performance MOFs. This chapter focuses on the synthesis, performance, and regeneration of MOFs, and the approaches for evaluating MOF material are also summarized.

Keywords CO_2 capture · Metal organic framework · Adsorption · Separation · Selectivity

---

R. Vaghasia · B. Saini (✉)
Department of Chemical Engineering, School of Technology, Pandit Deendayal Energy University, Gandhinagar, Gujarat 382007, India
e-mail: bharti.saini@sot.pdpu.ac.in

M. Savani
Department of Chemical Engineering, The Maharaja Sayajirao University of Baroda, Vadodara, Gujarat 390002, India

S. P. Singh et al. (eds.), *Metal Nanocomposites for Energy and Environmental Applications*, Energy, Environment, and Sustainability,
https://doi.org/10.1007/978-981-16-8599-6_7

Nomenclature

Bbta	1H,5H-benzo(1,2-d:4,5-d′)bistriazolate
BDC	Terephthalate
BDP	1,4-Benzenedipyrazolate
BeBTB	Bromothymol sulfone phthalein
BET	Brunauer-Emmett-Teller
BTtri	1,3,5-Tris(1*H*-1,2,3-triazol-5-yl)benzene
CCS	Carbon capture and sequestration
DAC	Direct air capture
DEF	N,N-diethlyformamide
DH_3PhDC	2′,5′-Dimethyl-3,3″-dihydroxy-[1,1′:4′,1″-terphenyl]-4,4″-dicarboxylate
DMF	*N,N*-Dimethylformamide
DOT	Dioxidoterephthalate
EDSL	Extended dual-site Langmuir
H_3BTC	1,3,5-Benzenetricarboxylic acid
H_4DHBDC	2,5-Dihydroxy-1,4-benzenedicarboxylic acid
HMeIm	2-Methyl imidazole
IAST	Ideal adsorption solution theory
LAG	Liquid-assisted grinding
MA	Melamine
MOFs	Metal organic frameworks
NiNbOFFive	$NiNbOF5(pyrazine)_2 \cdot 2H_2O$
PVSA	Pressure vapor swing adsorption
Qc-5-Cu-α	$Cu(quinoline\text{-}5\text{-}carboxylate)_2$
Q_{st}	Isosteric heat of adsorption
S_{BET}	BET surface area
SBU	Secondary building unit
TADIPA	5,5′-(1*H*-1,2,4-Triazole-3,5-diyl)diisophthalate
tdc	2,5-Thiophenedicarboxylate
TDPAT	2,4,6-Tris (3,5-dicarboxylphenylamino)-1,3,5-triazine
TEA	Triethylamine
TSA	Temperature swing adsorption
V_{pore}	Volume of pore
VSA	Vapor swing adsorption

7.1 Introduction

Given the accelerated world energy demand for economic and civilization improvement, huge reliance on fossil fuels has raised critical environmental problems as huge (CO_2) emissions in the atmosphere, close to 410 ppm in 2019, is the main reason to be the cause for world climate change. Some reason for this kind of situation is development & backlog of modern clean and a pure form of energy sources, like biomass, solar and nuclear power. Hydrogen gas is known as pure, renewable & resilient carrier of energy (Schoedel et al. 2016). Such an option is expected to be a long-time solution to the global supply of energy, however the latest technology for hydrogen generation and storage remains a challenge. Alternatively, the medium option is the use of natural gases that provides lesser emission of CO_2 than fossil fuels (275 vs. 460 g CO_2 per distance travelled in a mile), while trapping and conversion of CO_2 gas should be a quick solution (Schoedel et al. 2016).

The carbon capture process is mainly divided into four modes: Post/after-combustion capture of CO_2, direct air capture (DAC), Capture of CO_2 by pre/before combustion, and oxyfuel heating. Post/after-combustion capture separates CO_2 and other gases after fuel heating. The method that is very effective and widely used is amines scrubs. Another advantage of such a method is that it can apply directly to power plants with regeneration. It has been well built for more than 60 years, & lots of energy stations have already used it for absorbing CO_2 along with high input capacity. This catch system mode can be installed and used at both power plants like natural power plants and coal-fired power plants. But amine dehydration has a problem with amine depletion errors, mechanical distortion, and high energy charges (25–35% of output). The main step in such an approach is developing a highly efficient type of adsorbent materials with high CO_2/N_2 selectivity, good stability, minimal humidity, economic recovery and renewable processes, and low material costs.

Nowadays, traditional attractions like zeolites & carbon (activated) dominate the petrochemical plant with the widespread use for mitigating CO_2 emissions, olefin extraction, natural gas extraction, etc. (Berenstein et al. 2008; Bui et al. 2018). The world commodity market is estimated at approximately $ 11 billion in 2019, 32% and 38% greater than that in 2014 ($ 8.5 billion) and 2013 ($ 8 billion) (BCC Research) respectively for adsorbent materials. So for this condition, the immediate world need for a pure source of energy and sustainability of the environment is urging engineers and scientists for the development of high-quality materials with higher efficiency than conventional ones (Davis 2002). Metal organic frameworks (MOFs), composed of abundant co-ordinating bonds between metals & organic ligands, and are expected to be next-gen marketing materials as an adsorbents (Furukawa et al. 2013). In respect to classical adsorbent type materials, MOFs are subdivided by porosity and their surface area, size of adjusted pore and shape of geometry, and various changes, making them useful in many places, including separation of gas, complex catalysis, purification of water, biological uses, etc. (Trickett et al. 2017). Some good reviews summarize the performance of CO_2 capture of MOFs. Sumida et al. collectively summarized various MOF synthesis techniques and characterization schemes related to capturing CO_2 (Sumida et al. 2012). Seoane et al. comprehensively described MOF membrane systems for the capturing of CO_2 (Seoane et al. 2015). Belmabkhout et al. highlighted less CO_2 capturing adsorbents and also MOFs (Belmabkhout et al. 2016). Lin et al. have reviewed the latest advances in developing MOFs for capturing selected CO_2 emissions from after or post-combustion of flue gas and confining CH_4 for vehicle uses (Lin et al. 2017). In contrast to the last reviews, here we can summarize the application of MOFs for CO_2 capture from an engineering view and assess whether it is possible that these materials are used in industrial CO_2 segregation processes. It is hoped that this review could close the gap between chemical engineering and chemistry in relation to the use of MOFs as adsorbents from one object testing to the active judgment of performance, and ultimately the construction of the CO_2 capture industrial process.

Metal–organic frameworks (MOFs) reticular chemistry have evolved as a strong tool for creating porous materials while finely engineering the interiors to address global energy and environmental concerns (Furukawa et al. 2013). CO_2 capturing and consumption stand out as specific problems that perhaps the MOF industry is ideally positioned to counter (Li et al. 2011; Sumida et al. 2012). The chemistry of MOFs, in particular, has progressed to an extent where the interaction among the MOF structure and the desired characteristics may be modulated sensibly and effectively. MOFs are now viable options for CO_2 applications as a result of four critical breakthroughs during the last 25 years. The first one is reticular chemistry, which involves the functionalization, design, and structural modification of porous, crystalline, and extended solid materials (Furukawa et al. 2013). The development of tailorable solids which perform specified end functions is a direct consequence of reticular chemistry. Second, the combination of molecular and framework chemistry has allowed for molecular-level monitoring and spatially accurate chemical reactions inside extended materials (Jiang et al. 2016). The third is the development of ultra-high porosity. Therefore, enormous surface areas have made it possible to retain and compress enormous volumes of gases in a confined place for lengthy period. The notion of 'heterogeneity within the order,' in which architecturally and chemically diverse constituents are integrated with each other to interact synergistically within the same framework, is the fourth major advancement (Furukawa et al. 2015).

7.2 Capture Systems for CO_2

One of the principal use or application for the capture of CO_2 is mainly to be a larger point of sources available: power plants based on fossil fuel, fuel process plants & some other power plants in industries, mainly to produce or manufacturing of steel, iron materials, cement and some kind of different chemicals and more. To capture CO_2 direct from small-scale and sources that are in transportation means mobile sources and other commercials, as well as residential structure building sectors, might be more costly, expensive, more complex, and difficult when compared to other large kinds of sources. It is very challenging to capture CO_2 from the ambient air of the atmosphere due to the concentration of CO_2 is likely to be more than 100 ppm, and in flue gas, it is about 380 ppm. The most cost saving and effective method to capture CO_2 from the air with the help of increasing rate of biomass growth & which is used in some other industries for the capture of CO_2, are most cost-effective and based on foreseeable techniques and technologies (Kim et al. 2017). In general, three types of basic general systems are used to capture CO_2 from fossil fuels plants and/or biomass plants:

- Pre/before combustion CO_2 separation
- Post/after-combustion CO_2 separation
- Oxygen-fuel combustion for CO_2 mitigation
- Direct air capture.

7.2.1 Pre/Before Combustion CO_2

The Capture of CO_2 using the pre-combustion method is like making a reaction of a fuel with some oxygen gas or atmospheric air and/or vapour to make generally a 'synthesis gas (syngas)' or 'type of fuel gases' combination of gases like hydrogen and carbon monoxide. CO_2 is being separated from the catalytic reactor in which carbon monoxide undergoes reaction when it comes in contact with some steam in the catalytic type of reactor, which is also known as a shift converter, to release CO_2 & to generate more and more hydrogen gas. This is done mainly by the chemical or physical method of absorption process, which results in fuel that is rich in hydrogen, and it is further used in some applications like in furnaces, boiler, gas-based turbines, gas engines, and fuel gas cells. Pre-combustion has methods that have more efficient benefits, like lower consumption of energy means less energy required. However, the range of temperature and effectiveness related to H_2-rich gas turbine fuel becomes a bigger problem that needs a lot of attention. Further, the challenging problems are the enormous cost of capital and resistance of the society to make a new structure or construction for it. The other alternative method is being used sometimes for the generation of syngas which includes investing in chemical cycle loops (Dipl et al. 2019; Kim et al. 2017). The reactions involved in pre combustion capture are:

$$\text{Steam reforming: } C_xH_y + x \cdot H_2O \leftrightarrow x \cdot HCO + (x + y/2) \cdot H_2$$

$$\text{Partial oxidation: } C_xH_y + (x/2) \cdot O_2 \leftrightarrow x \cdot CO + (y/2) \cdot H_2$$

$$\text{Water-gas shift: } CO + H_2O \leftrightarrow CO_2 + H_2$$

7.2.2 Post/After-Combustion Capture of CO_2 or Post-combustion Capturing of CO_2

This process mainly requires the separation and removal of CO_2 gas from a mixture of flue gases, which contains mainly nitrogen & carbon dioxide, before making exhaust or emission into atmospheric air. After combustion capture of CO_2 is likely to be a more effective and more feasible method on a shorter scale of time because there are lots of new proposed techniques and technologies which can be used to retrofit existing fossil fuel which is used by power plants. One of the new such methods involved, uses low temperature cooled and CO_2-rich flue gases for feeding bioreactors to generate microalgae biomass which can be more effectively used as a biofuel. One other advantage of this kind of method to capture CO_2 is that when the unit of capturing CO_2 gas is closed or being shut down due to condition of emergency, you can still produce electricity, which is impossible in all other cases, means in other capturing methods. Post combustion technique has not been used commercially, generally to remove or for separation of tiny little particles or contaminants like Hg and NO_x/SO_x gases, however, materials used for the separation of CO_2 require a lot of changes in it, mainly with those methods which includes larger scale preparation and more costs of regeneration (Li et al. 2011; Kim et al. 2017).

7.2.3 *Capture of CO_2 by Oxygen-Fuel Type Combustion/Oxy-Fuel Combustion Capture*

In such type of method of removing or capturing CO_2, which involves oxy-fuel combustion, the purest form of oxygen is being used to make combustion rather than the use of atmospheric air, which results in the flue gas that contains mainly H_2O and CO_2. When this fuel is being burnt in the presence of the purest form of oxygen, the temperature of flame becomes excessively large, but with this, we can reuse or recycle H_2O and CO_2 rich flue gases to moderate this. And the mixture of carbon dioxide and water is very much easy to separate as the water can easily be condensed out. Oxygen gas is generally generated or produced with the lowest range of temperature (cryogenic), separation of air, and new techniques to supply oxygen gas into the fuel, like as membranes & chemically looped cycles are continuously being developed. The Major disadvantage of such kind of option requires the purest form of oxygen that can generally be obtained with the help of the separation of air or by use of some other novel technologies which are available at that time. But by the use of a basic type or conventional cryogenic system to separate air to generate oxygen becomes a more expensive type of option (Li et al. 2011; Kim et al. 2017) (Fig. 7.1).

7.2.4 *Direct Air Capture (DAC)*

DAC, which assesses MOF adsorbents under atmospheric conditions, is another burgeoning research field for multicomponent CO_2 collection (400 ppm, 298 K, 1 bar). The objective of DAC technique is to collect CO_2 from the atmosphere,

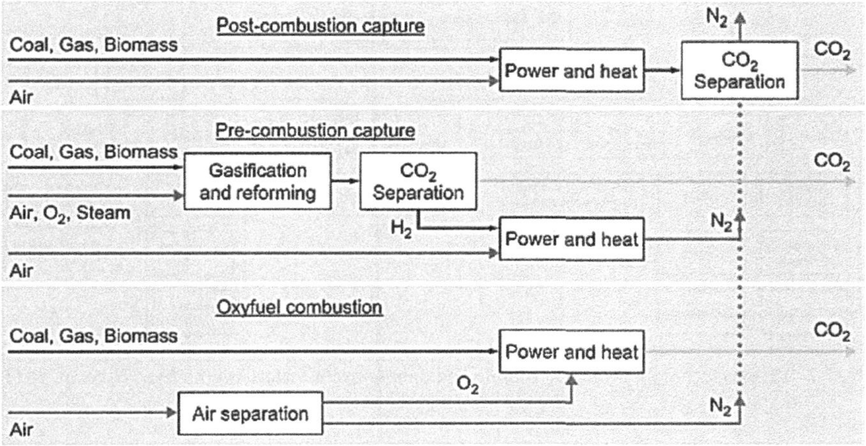

Fig. 7.1 Schematic representing three different processes for carbon capture (Li et al. 2011)

whereas the aim of flue gas capture is to scrub CO_2 off the exhaust gas. One of the most significant advantages of DAC is its ability to reduce CO_2 emissions from both dispersed and point sources. Furthermore, DAC facilities are not constrained by regional constraints and did not have to deal with significant levels of pollutants like SO_x and NO_x in flue gas. At 400 ppm, 298 K, and 1 bar, zeolite 13X has a low CO_2 absorption capacity of 0.035 mmol g^{-1} due to the lack of strong interactions. Under DAC circumstances, even the most potential MOFs for CO_2 capture, such as Mg-MOF-74 and SIFSIX-3-Zn, have CO_2 uptake capabilities of less than 0.2 mmol g^{-1}. SIFSIX-3-Cu has a strong CO_2 absorption capacity of 1.24 mmol g^{-1} at extremely low partial pressures pertinent to air capture by adjusting the pore size to 3.5. Nevertheless, given the significant chemisorption sites within these adsorbents, amine-doped MOFs and silicas have the maximum CO_2 uptake capability for DAC assessment (Shekhah et al. 2014; Kumar et al. 2015; Sanz-Pérez et al. 2016).

7.3 Types of CO_2 Capture Techniques

The system to capture CO_2 uses many known techniques for the separation of gases that are integrated with some general capture systems for capturing of CO_2 just like technologies used for gas separation of CO_2 like the pre-combustion method for capturing of CO_2, after combustion method to capture CO_2 and capturing of CO_2 by oxy-fuel combustion. A Basic summary of such types of methods of separation is given here with more details. Here Fig. 7.2 schematically shows the technique and method which is generally being useful for the separation of the CO_2 in each type of case (Kim et al. 2017).

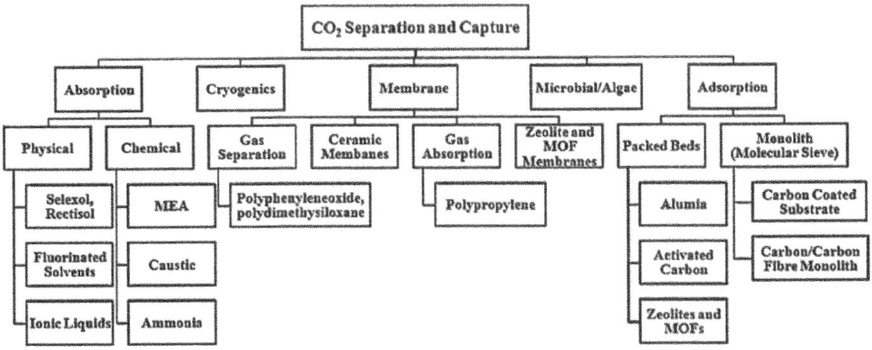

Fig. 7.2 Various techniques for CO_2 capture and their relevant materials employed (Li et al. 2011)

7.3.1 Separation of CO_2 Using Sorbents/Solvent

In such type of method for separation can be achieved by bypassing of the gas which contains CO_2 gas in close contact with the liquid type of absorbent or solid types of absorbent, which is useful for capturing the CO_2 gas. For a particular general scheme of diagram Fig. 7.3a shows, the sorbent loaded along with the captured

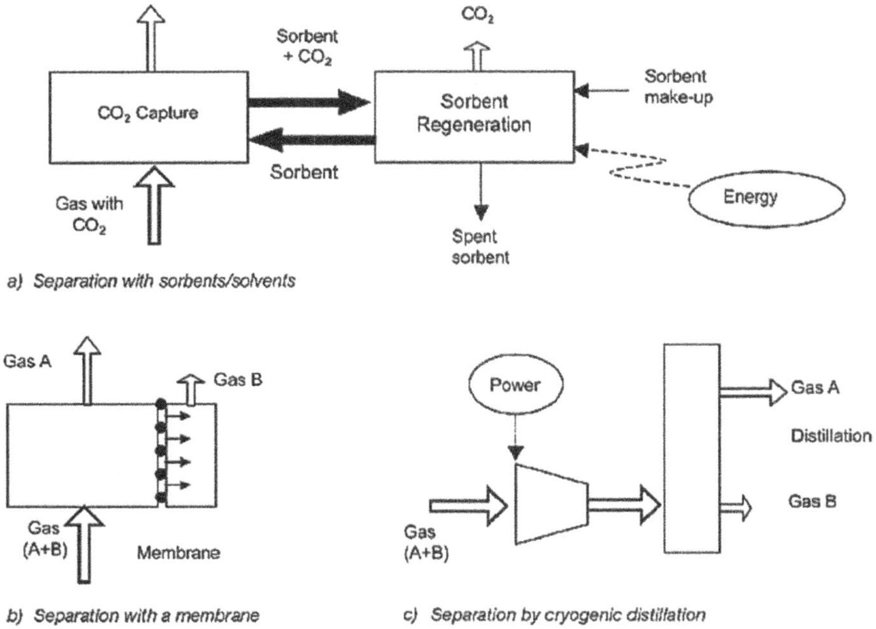

Fig. 7.3 General schematic representing some crucial separation technologies for CO_2 capture. The feed gas stream may be CO_2, H_2 and O_2 (Kelly et al. 2017)

CO_2 is passed to other different vessel, where it makes exhaust of the CO_2 gas (regeneration method) after burning, after a reduction in pressure or after the change of any condition of sorbent. After the regeneration process, the resulting sorbent is sent back for capturing more amount of CO_2. For some other types of variants of such kind of system, where the sorbent is a solid part and does not pass within vessels due to the sorption and regeneration being done with the help of cyclic changes (in terms of temperature or pressure) in the vessel itself where sorbent is kept. A making-up of the flow of pure new sorbent is required the most to compensate for the natural decay or reduction of activity as well as sorbent loss. For some other conditions, the sorbent can be used as a solid oxide that gives reactions in a vessel along with fossil fuel generating heat and CO_2 gas. The used sorbent is further passed to a second vessel in which it is re-oxidized in the air for making up of fresh sorbent from the used sorbent. Figure 7.3 shows governing of such an important type of system for capturing of CO_2, which contains large commercial options such as chemical as well as physical adsorption. Some other processes depend on the fresh liquid type of sorbents, or solid newly generated sorbents are developed with the goal to overcome such types of limitations of systems that are existing. The main big problem of such CO_2 capturing method is that the flow of sorbent within these vessels of Fig. 7.3a is large because it has to match with the large flow of CO_2 processed in the power plant in industries. That is why the size of the equipment and the required energy for the generation of a sorbent is huge and tends to convert into efficiency penalty and added expense. Also, in some systems which use costly sorbent materials, there is always some kind of issues related to increased cost associated with to purchase of sorbent & the threat to exhaust or discard the sorbent particles (Kim et al. 2017).

7.3.2 Membrane Based Separation of CO_2

The type of material that is specially manufactured which permits permeation of gases through them (Fig. 7.3b). To select this kind of membrane for some other different gases, one can closely connect with material type and nature; but gas flowing throughout the membrane is specifically passing with the help of pressure difference along with the membrane material. So for the separation using membrane, highly pressurized vapour or steam is mainly used. Some other different kinds of materials of the membrane (polymeric, ceramic, metallic) which may be used in applications to capture CO_2 to generally separate O_2 and H_2 from the vapour of the flue gas. To make more concentrated CO_2 from the wide range of flow processes and to separate O_2 from the air in the atmosphere. But this separation based on membrane has many commercial applications in a market and in power plants of industries. However, industries did not apply this for the large-scale & lesser-cost needs for the capture of CO_2. Nowadays, all research and development departments are making huge progress towards the completion of the goal to achieve CO_2 gas capturing on huge scales worldwide (Kim et al. 2017).

7.3.3 Separation of CO_2 Based on Cryogenic Distillation

By a series of some processes like compression, expansion, and cooling gas can be converted into a liquid phase. After the gas is converted in a liquid phase, then in the distillation column, some of the components of gas may be separated. For example, take the case of air, such processes are now a day generally used on a huge scale commercially. In the case of O_2 (Oxygen), it can be separated from the atmospheric air by following the schematic diagram of Fig. 7.3c, and that can be applied in a wide range of capture systems of CO_2 (Oxygen-fuel type combustion and pre-combustion capture). The main problem for such kinds of systems is larger streamflow of oxygen is needed. Separation can also be done by a refrigeration system for the separation of CO_2 gas from different types of gases. It is also used for the separation of dirt and impurities from the comparatively high quality of CO_2 stream flows, and such an example is here that from oxygen-fuel type burning and combustion & to remove CO_2 from natural gas and synthesized gases that have undergone a change of shift conversion from CO to CO_2 (Kim et al. 2017).

7.4 Metal Organic Framework Synthesis

A summary of some of the methods of preparation of MOF is shown in the diagram. Some of MOF syntheses are kind-a based on the liquid synthesis, in which different metals solutions and ligands solution are mixed with each other, or the solvents are incorporated into a solid salts and ligands mixture in a reaction vessel. The choice of solvents in this liquid reaction can be depending on various factors like regeneration, melting, strength, etc. It also plays a major role in determining thermodynamics and the ability to activate a specific reaction. In addition to the merging of the water phase, the researchers also experimented with the solid phase of the MOF because it is faster and easier, but the solid compaction is always faced with the difficulty of finding a single crystal, and thus finding product geometry, that is somehow easier for resolving reactions of phase (Dey et al. 2014) (Fig. 7.4).

7.4.1 Solvothermal Synthesis

In this, the solvothermal reaction is performed in vessels that are closed at the pressure above the solvent boiling point. Many initial components can deal with chemical change under conditions belonging to solvothermal, and those are associated with the generation of nano morphologies which are not possible in simple or general ways. In many cases, the solvents which are for the solvothermal reaction have a very high boiling point and are organic in nature. Liquid chemicals largely used as a solvent are ethanol, diethyl formamide, acetonitrile, acetone, dimethylformamide,

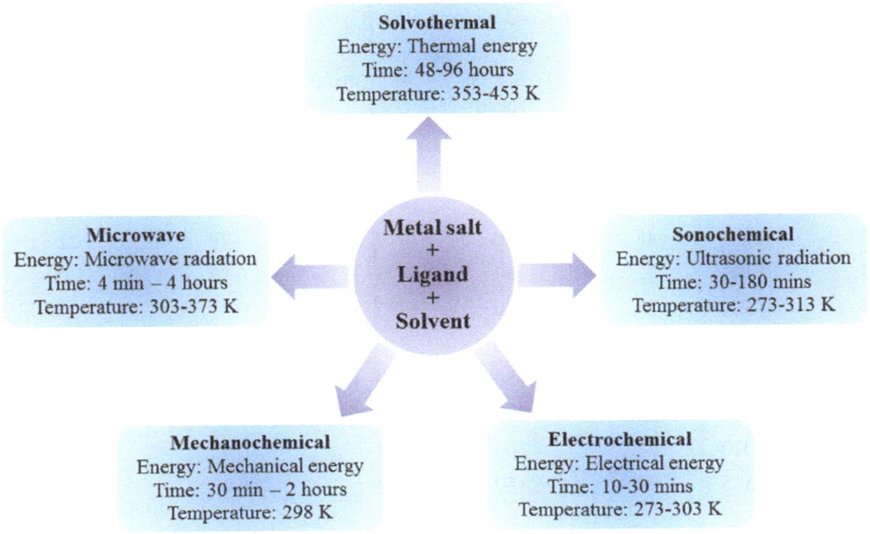

Fig. 7.4 Operating conditions during the synthesis of MOFs for various method

methanol, etc. Solvents mixture has also been used to favour the solubility of a variety of initial components. The solvothermal reaction can perform in a wide range of temperatures, depending on the reaction conditions needed (Wang and Ying 1999). Typically, glass containers are used for low temperatures, and vessels with the Teflon interior are employed for the reactions, which are to be carried out at above 400 K temperature (Fig. 7.5).

Table 7.1 shows the study of HKUST-1 for its synthesis optimization investigation (Kim et al. 2012). In the synthesis, 17 g of copper (II) nitrate hydrate (98%) and 8.5 g of H_3BTC (1,3,5-benzenetricarboxylic acid) (95%) were dissolved in 500 mL of ethanol. The concentration of H_3BTC was 0.101 M, and the ratio (moles) of Cu salt:ligand was 9:5 in an ethanol solution. The mixture was heated for 24 h

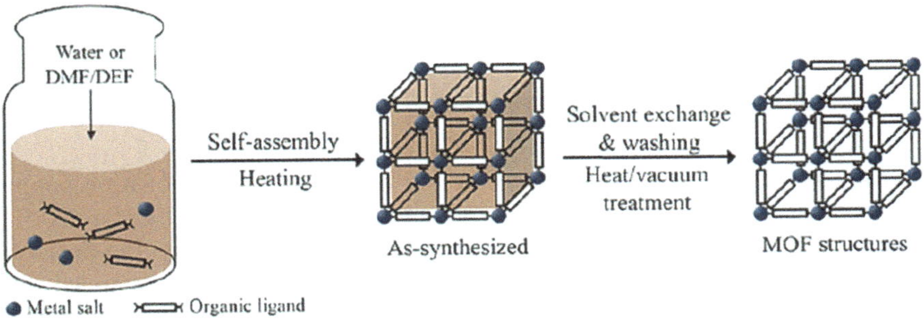

Fig. 7.5 Conventional solvothermal synthesis of MOF structure (Lee et al. 2013)

Table 7.1 Morphological properties and yield of HKUST-1 at varied synthesis parameters (Kim et al. 2012)

S. No.	H_3BTC (M)	Synthesis time (h)	Stirrer speed (RPM)	S_{BET} (m^2/gm)	V_{pore} (cm^3/g)	Yield (%)
1	0.101	12	150	1854	0.78	21
2	0.101	24	150	1893	0.78	26
3	0.101	48	150	1877	0.78	22
4	0.101	72	150	1873	0.78	20
5	0.505	24	150	1395	0.62	34
6	0.808	24	150	1282	0.51	61
7	1.010	24	150	1250	0.47	74
8	1.010	24	200	1481	0.60	61
9	1.010	24	300	1737	0.72	40
10	1.515	24	150	1049	0.43	79

under reflux with continuous stirring at 300 rpm. The product was filtered out after completion of the solvothermal reaction and rinse washed with DI water and with ethanol. The product was dried at 372.5 K for 6 h in a vacuum oven. The BET surface area did not vary of the samples after 12–72 h. Although the higher yield was obtained up to some extent after 24 h reflux and there was a steady growth of crystals over the period of time, and the size of crystals was in the range of 1.5–3.5 μm. BET surface area decreased, and the yield increased after 24 h of synthesis time by increasing the concentration of H_3BTC from 0.101 to 1.515 M (Sr. no. 2, 5, 7, 10), and crystals size also increased. The speed of stirring also played a major role in improving the texture properties and high yield. Small crystals were produced when the speed of stirring was increased from 150 to 300 rpm. It is because of fine dispersion formation (Mateovic et al. 2002). The surface area of HKUST-1 increased from 1250 to 1735 m^2/g by changing the stirrer speed from 300 to 150 rpm, indicating the influence of stirrer speed on the surface is of HKUST-1, which makes the speed of stirred an important parameter to tune in the properties of MOF especially textural.

7.4.2 *Microwave Assisted Synthesis*

Microwave synthesis technique is applied for the faster synthesis of nano-porous materials under the influence of hydrothermal conditions (Lagashetty et al. 2007). Narrow particle size, morphology control, and phase selectivity are the potential advantages of this method. The commercial microwave equipment is equipped with a pressure controller, fibre optic temperature controller, and adjustable power outputs. In this method, a substrate mixture is transferred to a vessel made of Teflon, which is sealed and is to be placed in a microwave unit and heated for a given time at the pre-set temperature. Which results in quick heating of the liquid phase (Ranjan and Tsapatsis 2009; Klinowski et al. 2011; Zornoza et al. 2011; Hindelang et al. 2012 (Fig. 7.6).

Table 7.2 represents the synthesis of Co-MOF-74 by microwave- assisted solvothermal synthesis with varying synthesis parameters to study the change in the morphological properties and the yield of MOF (Cho et al. 2012). In microwave-assisted synthesis of Co-MOF-74, 0.044 mmol of H_4DHBDC (2,5-dihydroxy-1,4-benzenedicarboxylic acid) and 0.148 mmol of cobalt nitrate hexahydrate ($Co(NO_3)_2 \cdot 6H_2O$) is mixed in 18 mL of solvent mixture (DMF:ethanol:water = 1:1:1 (v/v/v%)) (DMF = *N,N*-dimethylformamide) and is kept 35 mL tube, and kept microwave oven sealed (Max. 300 W). The mixture was first heated at 130 °C and at room temperature the mixture was cooled. Separate the red–orange crystals by centrifugation. Decant and replenish the synthesized product for 4 times for 2 days by washing with DMF three times. Vacuum dry the product at 250 °C for 6 h to remove the solvent. Table

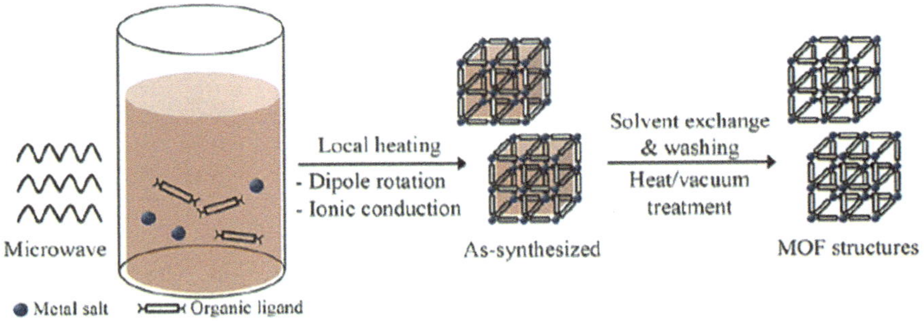

Fig. 7.6 Microwave-assisted solvothermal synthesis of MOF structures (Lee et al. 2013)

Table 7.2 Morphological properties and yield of Co-MOF-74 at varied synthesis parameters (Cho et al. 2012)

S. No.	Synthesis time (h)	Power (W)	S_{BET} (m^2/gm)	V_{pore} (cm^3/g)	Yield (%)
1	60	180	1318	0.49	7
2	60	180	1319	0.51	45
3	60	120	1402	0.71	58
4	60	150	1411	0.59	63
5	30	180	1193	0.49	69
6	40	180	1155	0.47	75
7	60	180	1314	0.51	76
8	70	180	1273	0.51	74
9	30	210	123	0.01	84
10	40	210	1270	0.65	72
11	60	210	1321	0.50	74

Table 7.3 Morphological properties and yield of ZIF-8 at varied synthesis parameters (Cho et al. 2013)

S. No.	Batch size (mL)	Zn^{2+} (mmol)	TEA (mL)	NaOH (10 M, mL)	pH	S_{BET} (m^2/gm)	V_{pore} (cm^3/g)	Yield (%)
1	50	2	–	–	6.4	1370	0.51	60
2	50	2	0.3	–	7.5	911	0.43	57
3	50	2	0.5	–	8.8	1300	0.48	59
4	50	2	0.7	–	9.2	1245	0.42	56
5	50	2	–	3.8	9.3	643	0.39	62
6	50	10	2.5	–	7.1	1198	0.60	22
7	50	10	2.5	1	9.3	1253	0.61	93
8	50	10	0.8	2.4	9.1	1213	0.67	96
9	50	10	0.4	3.5	9	1223	0.82	92
10	50	60	0.4	15	9.5	1249	0.71	93
11	50	30	0.4	7.8	9.1	1454	0.66	60
12	50	60	0.4	15	9.1	1266	0.51	80
13	600	720	4.5	132	9.2	1141	0.56	50
14	600	720	4.5	132	9.2	1174	0.50	85
15	600	720	4.5	132	9.2	915	0.43	84

7.3, summarizes the various textural properties of Co-MOF-74. The best textural properties and yield were found at 130 °C and 180 W microwave power for 1 h (Sr. no. 7) with 1314 m^2/g BET surface area and 76% yield based on ligand. Also, the size of Co-MOF-74 was relatively smaller compared to Co-MOF-74 prepared solvothermally.

7.4.3 Sonochemical Method for Synthesis

It is the process in which molecules start to initiate chemical reactions because of the use of powerful ultrasonic rays (25 kHz–15 MHz). Ultrasound promotes physical or chemical changes as a result of cavitation processes involves the formation, growth as well as the rapid collapse of a bubble in liquids, creating tropical temperatures for a short period of life, having high temperatures and pressures (5000 K and 1000 bar respectively), which results in extreme heating and cooling at fast rates (>10^{10} K/s) resulting in the production of fine crystallites (Son et al. 2008; Aslani and Morsali 2009; Jung et al. 2010; Mohanty et al. 2010; Carson et al. 2011; Kim et al. 2011; Dharmarathna et al. 2012) (Fig. 7.7).

The sonochemical method was employed for the synthesis of ZIF-8 using TEA (Trimethylamine) and NaOH. DMF was used as a solvent (Cho et al. 2013). Using

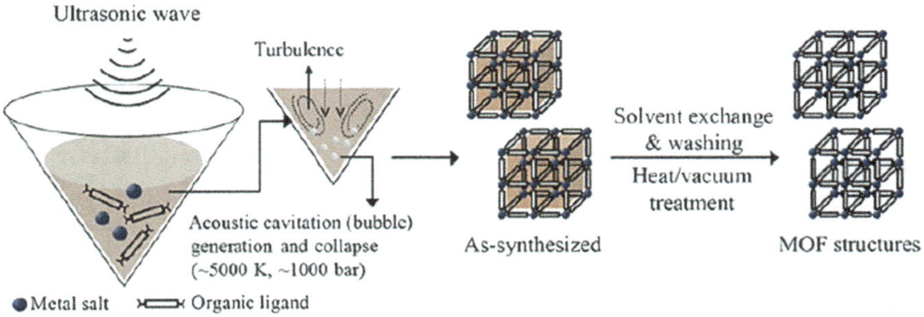

Fig. 7.7 Sonochemical synthesis of MOF structures (Lee et al. 2013)

0.67 g zinc nitrate hexahydrate ($Zn(NO_3)_2 \cdot 6H_2O$, 2 mmol) and 0.167 g HMeIm (2-methyl imidazole) (98%, 2 mmol) and 50 mL of DMF, the substrate was made and stirred to obtain a clear solution. After that, TEA (99%) was added, and the mixture was transferred to a horn-type reactor, which is fitted with a sonicator with adjustable power o/p (max. 500 W). The synthesized ZIF-8 is washed with DMF. Then ZIF-8 was placed in methanol. The samples were filtered and vacuum dried at 80 °C. In Table 7.3, it can be seen that the best textural property of ZIF-8 was observed with 0.5 mL of TEA (Sr. no. 3). The size of ZIF-8 obtained was 400 times smaller than the one produced by the conventional solvothermal method, although textural properties were the same. ca. 9 was the optimal pH observed with 0.5/0.7 mL TEA. In the 5th set, better textural properties were obtained with NaOH (aq, 10 M). When the substrate was used 5 times (sr. no. 6) better textural property was observed, but the yield was very less ~22%. By the use of 10 M NaOH, which showed rapid improvement in the yield of product (93%) (Sr. no. 7). The same yield was obtained, i.e., 93% with just 0.4 mL TEA which is very costly (Sr. no. 10). Recycled mother liquid at Zn^{2+}:MeIM = 2:2 with added make-up zinc salt and ligand to maintain the substrate ratio as fresh showed a very high yield of 80%. For a large batch of 1 L also the yield obtained was 85% (Sr. no. 14) which is ca. 1,200 times greater than the one obtained by solvothermal synthesis.

7.4.4 Electrochemical Synthesis

In the electrochemical synthesis of MOFs, the metal ions required during MOF synthesis are obtained from the dissolution of anode rather than metal salts which are used in other techniques, and these metal ions react with the linker molecules, which are already dissolved and a conducting salt in the reaction mixture. Protic solvents are used to prevent the cathodic deposition of metal ions. H_2 is produced during the synthesis (Martinez Joaristi et al. 2012) (Fig. 7.8).

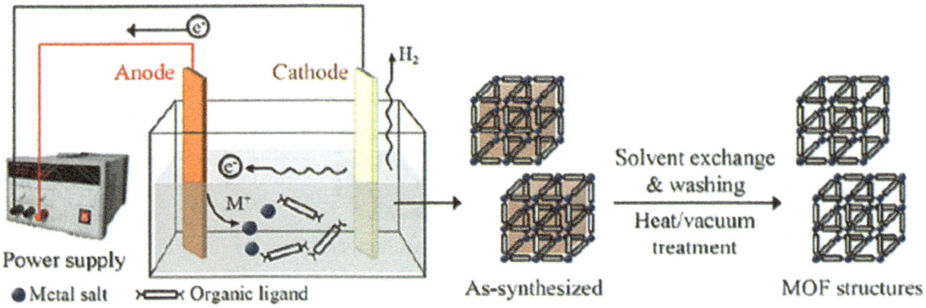

Fig. 7.8 Electrochemical synthesis of MOF structures (Lee et al. 2013)

7.4.5 Mechanochemical Synthesis

No solvent is required for the synthesis of MOFs while following the method of synthesis. The method of initiating chemical reactions with mechanical energy is known as mechanochemistry. Modern synthesis chemistry is fascinated by the production of bonds that is easy, cost-effective, and uses environmentally acceptable chemical pathways. Liquid-assisted milling, in which a tiny quantity of solvent is introduced to a solid reactant, has recently been utilized for the fast synthesis of MOFs (Beldon et al. 2010). Friscic and colleagues demonstrated it just by adjusting the excess solvent in the LAG (Liquid-assisted grinding) process, and one can make 1D, 2D, and 3D composite polymers out of the same reaction mixture. This technique has been used to synthesize other zeolitic imidazolate based MOFs (Friščić et al. 2013) (Fig. 7.9).

7.5 Gas Adsorption Testing

Common methods employed to evaluate the performance of adsorbents, especially those which are porous in nature (MOFs), by the scientists and chemists are based

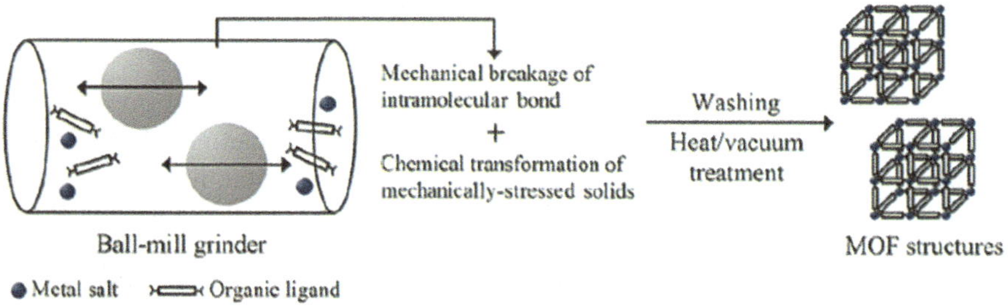

Fig. 7.9 Mechanochemical synthesis of MOF structure. (Lee et al. 2013)

on single-component gas sorption isotherms & X-ray diffraction methods (Chandler et al. 2008; Uribe-romo et al. 2010; Marshall et al. 2015; Liao et al. 2017). Such methods are able to expose small sorption sites and gas molecule structures. But research conducted on the separation of different gas combinations by MOFs remains limited. Here we examine engineering approaches in actual gas sorption performance analysis from the literature in order to establish a connection between diverse materials and engineering ideas.

7.5.1 Isosteric Heat of Adsorption (Q_{st})

For instance, the interactions between adsorbates and adsorbents, as well as their quantitative measurement, are critical since they are linked to the regeneration process (Sircar et al. 1999). The majority of the adsorption process is exothermic, and according to Clausius-Clapeyron's formula, the two assumptions are used in the analysis: the gas phase as in bulk is considered to be ideal, and the volume of adsorbed phase is disregarded (Baekelandt and Baekelandt 1996). As a result, isotherm measurements must be done at a lower pressure, and isotherms at three distinct temperatures must be considered. Adsorbate interactions with adsorbents may be explained using its amplitude and variations as a cover function. The virial equation (Eq. (7.1)) is used to calculate the isosteric heat of adsorption (Q_{st}) by measuring the sorption isotherms of the gas at various temperatures and doing arithmetic operations (Dinca et al. 2010). Equation (7.2) may be used to calculate Qst if the appropriate parameters are used.

$$\ln P = \ln N + \frac{1}{T}\sum_{i=0}^{m} a_i N^i \sum_{i=0}^{m} b_i N^i \tag{7.1}$$

$$Q_{st} = -R\sum_{i=0}^{m} a_i N^i \tag{7.2}$$

where pressure is P (bar), N is the quantity adsorbed (mmol g^{-1}), T denotes temperature (K), R is gas constant (8.314 JK^{-1} mol^{-1}), a_i and b_i are coefficients of the virial equation, n and m represents the adequate number of coefficients required to describe the isotherms.

7.5.2 Calorimetric Method

The calorimetric method is used when high adsorption enthalpy or hysteresis, the indirect evaluation of adsorption values from the isothermal gas detection rates can be problematic (Zielinski et al. 2007; Walton et al. 2008). For the complex and highly exothermic adsorption systems, Navrotsky and his workers came up with the method for MOFs (cyclodextrin based) for determining the enthalpy of adsorption, which was not reliable on any model (Wu et al. 2013). The calculation is very simple, where each injected CO_2 results in a characteristic calorimetric peak and the area under it is measured, and that value is used to evaluate the adsorption enthalpy for each mole of CO_2. For the case of cyclodextrin-based MOFs, three different behaviours were observed pertaining to irreversible and reversible chemical sorption and also physisorption. So, by this method, one can understand the chemical interactions at the molecular level among sorbate and sorbent, which is not possible in the case when we determine enthalpy from isothermal gas uptake.

7.5.3 Ideal Adsorption Solution Theory (IAST)

The CO_2 design capturing system's method is based on a single element as well as an equilibrium sorption mixture (Raksajati et al. 2016). To get isotherms for single components, volumetric gas sorption equipment is used, but the derivation of adsorption isotherms for mixtures by breakthrough experiment is one of the most difficult and time-consuming tasks (Yang et al. 2013; Billemont et al. 2017). IAST is the sole technique to define multicomponent adsorption isotherms from single-component adsorption isotherms, and it was invented in 1965 (Myers and Monson 2014). Three key reasons guided the development of the theory: (Walton et al. 2008) (i) the adsorbent is consistent throughout; (ii) sorbent molecules in a solution have equal reach to the whole surface of adsorbent; and (iii) The adsorption phase is indeed the ideal solution, with especially high molecular interactions. According to Talu (2013) Myers and Monson's (2014) recent study, assumptions (i) and (ii) are unnecessary, and also the 2D pressure distribution is useless. However, in two instances, IAST is unable to properly forecast the isotherm for a gas mixture: (Cessford et al. 2012) (i) sorbents with a broad range of size, cohesion, surface charge, or adsorption interactions; and (ii) sorbents with random sites, like cationic zeolite. In these conditions, new models, like the Toth model isotherm (Terzyk et al. 2003), which is a model that provides an advanced balance in addition to the Langmuir isotherm model, are far more likely to explain diverse systems.

7.5.4 Breakthrough Experiment

The equilibrium data for adsorption is important in many ways: (1) testing the advanced adsorbent type materials, (2) design procedure, and (3) validating the theoretical base of composite adsorption (Bara et al. 2007). In fact, volumetric apparatus combined with a gas chromatograph might be used to measure co-adsorption schemes of CO_2 and other gases (Hamon et al. 2009; Yang et al. 2013; Billemont et al. 2017). Mason et al. we are developing a highly efficient adsorption tool that is able to measure the equilibrium adsorption schemes of gas mixtures in real flue-gas conditions (Mason et al. 2015). However, testing the effectiveness of material for multiple cycle performance is a challenging process, and large-scale exposure tests are required. Dynamic column breakthrough testing is indeed an industry standard for measuring adsorption equilibrium data for the mixture and testing for gas separation potential for adsorbents (Samanta et al. 2012). Predicting ternary N_2, CO_2, and H_2O adsorption isotherms of MOFs is done using the Extended dual-site Langmuir (EDSL) model.

7.6 Post-combustion CO_2 Capture Criteria

The discovery of reticular chemistry was a major step forward in the chemistry for capturing CO_2 (Li et al. 2011; Sumida et al. 2012; Furukawa et al. 2013). Indeed, the versatility with which a given MOF may be tailored allows for the refinement and improvement of CO_2 capture characteristics. The bulk of research into creating MOFs for CO_2 capturing is being centred on reversible adsorption, which reduces the requirement for total energy throughout regeneration and addresses one of the primary drawbacks of conventional sorbents like monoethanolamine solutions. MOF structures have subsequently been routinely created, fine-tuned, and thoroughly researched as a consequence. The structural characteristics that are being explored in the construction of MOFs for the capture of CO_2 under post-combustion circumstances are outlined in this section. The existence of reachable coordinatively unsaturated metal sites within pores; integration of heteroatoms, and also covalently linked functionality; interactions of MOF building units; pore hydrophobicity; and some hybrid approach are among these features. Apart from the typical examples mentioned of each area described below, we should highlight that there is a vast body of research that was not able to be in it.

7.6.1 Metal Sites (Coordinatively Unsaturated Metal Sites)

Lewis acidity, that being provided by partial positive charges on metal sites, is used by MOFs containing coordinatively unsaturated metal site. The isosteric heat of adsorption (Q_{st}) for CO_2 of MOFs with these sites is quite high at low pressures, resulting in significant selectivity and uptake of CO_2. When discussing CO_2 capture off flue gas, this would be essential because CO_2 has a higher quadrupole moment and is more polarizable than N_2, resulting in higher electrostatic interaction with the metal sites exposed. M-MOF-74 (M = Mg(ii), Zn(ii), Mn(ii), Fe(ii), and Ni(ii), among others, is among the most deep-studied series of MOFs (Caskey et al. 2008; Britt et al. 2009; Bao et al. 2011; Märcz et al. 2012; Ye et al. 2013; Wang et al. 2014; Mason et al. 2015). Mg-MOF-74, the series' benchmark material, has the greatest absorption capacity (27.5%wt) yet observed under normal circumstances (298 K and 1 bar) (Bao et al. 2011). M-MOF-74 is made up of 1D secondary building units (SBUs) linked with 2,5-DOT (DOT = Dioxidoterephthalate) linkers that construct hexagonal outlets with a diameter of 12 Å. Upon the eviction of the foreign molecules, every M ion is just five-coordinated, enabling easy CO_2 attachment towards the sixth coordinating site (in an end-on way). This CO_2 binding mechanism is most common in MOFs with coordinatively unsaturated metal sites (Nguyen et al. 2014). The relatively high Q_{st} value with 47 kJ mol^{-1} of Mg-MOF-74 shows the usefulness of this featured characteristic (Caskey et al. 2008). Despite the fact that dynamic adsorption capacity correlated really well along with the thermodynamic uptake observations in a test of breakthrough gas data, moisture loading was observed to decrease CO_2 uptake of the Mg-MOF-74 (Kizzie et al. 2011). This seems to be due to the competitive coordination of water molecules to the coordinately unsaturated metal sites, and it is a difficult problem to solve when utilizing MOFs with such structural characteristics. Despite a key synthetic finding to 'preserve' the coordinatively unsaturated metal sites, there seems to be little remaining to investigate in actual post-combustion CO_2 capturing applications for such a structural characteristic. The coordinatively unsaturated metal sites, on the other hand, can be used to anchor heteroatoms capable of eradication of competitive water adsorption.

7.6.2 Heteroatoms

MOFs have a high polarity and, in certain circumstances, a nucleophilic tendency, the inculcation of heteroatoms as part of a functionality bound covalently bound functionality, has demonstrated a lot of potential for significant CO_2 interactions. Aromatic amines are an example of functionality that has been studied. CAU-1 ($Al_4(OH)_2(OCH_3)_4(BDC\text{-}NH_2)_3{\cdot}xH_2O$, is one such framework comprising of aluminium hydroxide lumps linked by BDC-NH_2, giving a high heat of adsorption and selectivity (CO_2/N_2) value, as well as low CO_2 adsorption of 15 wt% under ambient conditions (Si et al. 2011). The N atom in heteroaromatic amines is investigated because it is a non-coordinating component of the aromatic ring that can bind with any external molecules. One such heteroaromatic amine is Zn(btz), having two non-coordinating nitrogen atoms of tetrazolate, which has a CO_2 uptake of 18 wt% at ambient conditions (Cui et al. 2012), which shows enhancement in CO_2 uptake due to the incorporation of such a heteroatomic site. One similar kind of 3D framework is bio-MOF-11 (Verma et al. 2010). This has various functionalities, like pyrimidine, aromatic amine, and diazole groups, and in this, pyrimidine and diazole are employed for linkages in the framework. Due to presence of free heteroatomic and free aromatic amine both in the pores and the internal pores resulted in the higher attraction of CO_2 with the framework. At ambient temperature and 1.1 bar pressure, it had a 15.3% uptake capacity and a selectivity of 75 (CO_2/N_2) (Bell 1964; Verma et al. 2010).

To increase CO_2 uptake, MOFs grafted with dialkyl amines also showed great potential. It is beneficial because the chemisorptive mechanism involved in CO_2

adsorption leads to better selectivity and absorption at a pressure (1–2 bar) that is similar to flue gas pressure. In the presence of moisture, CO_2 may also be adsorbed competitively. One such example of dialkyl amines grafting is ethylenediamine (en) which was used for grafting on Cu(II) in the framework of CuBTTri ($H_3[(Cu_4Cl)_3(BTTri)_8]$ (Demessence et al. 2009). Cu_4Cl is the SBU in the structure, which is connected by a tritopic triazolate with one non-coordinating N atom. At very low pressures of 0.06 bar and 298 K, the en-functionalized CuBTTri showed 1.6 wt% CO_2 uptakes versus 0.92 wt% for the parent framework. This is owing to the dialkylamine-functionalized CuBTTri having a larger Q_{st} value of 90 kJ mol^{-1} over unfunctionalized CuBTTri, which has a Q_{st} value of 20 kJ mol^{-1}. In the same study, it showed how a stoichiometry of N,N-dimethylethylenediamine (mmen) when attached to Cu(ii) of CuBTTri (SBU), which is a secondary amine, enhances the CO_2 uptake to 15.4 wt% at ambient condition, which has 15% higher uptake compared to the parent material. And also showed very high selectivity. The results confirmed a very high Q_{st} value (96 kJ mol1), indicating a chemisorptive mechanism (McDonald et al. 2011). Mg-IRMOF-74-II (Mg_2(dobpdc), coordinatively unsaturated metal sites, are also used for grafting mmen (McDonald et al. 2012). However, all these dialkyl amines have proved their effectiveness for CO_2 selectivity with no moisture present. It is a logical development for alkylamine-functionalized frameworks to establish covalent connections with CO_2 rather than H_2O. The organic linker Mg-MOF-74, known as Mg-IRMOF-74-III ($Mg_2(DH_3PhDC)$ (DH_3PhDC = 2′,5′-dimethyl-3,3″-dihydroxy-[1,1′:4′,1″-terphenyl]-4,4″-dicarboxylate), was functionalized covalently by varied primary, secondary, tertiary, and aromatic amines to investigate its CO_2 uptake properties from flue gas containing moisture (Fracaroli et al. 2014). Primary alkylamine proved to be most effective, with CO_2 uptake of 12.7 wt% at room temperature and 1.1 bar pressure. Thus, the focus should be on increasing the density of alkylamines in MOF using heteroaromatic amine.

7.6.3 SBU Based Interaction

SBU comprises of non-metallic functional groups which have a strong interaction towards CO_2, which is similar to the unsaturated metallic ions covalently bonded. SIFSIX-1-Cu and SIFSIX-2-Cu-I, at ambient temperature and pressure, it possesses a CO_2 uptake capacity of 19.1 wt% and 19.2 wt%, respectively, which is the greatest for a MOF without any unsaturated metallic ions (Burd et al. 2012; Nugent et al. 2013). In the presence of H_2O, monodentate hydroxide was employed as a capping ligand and demonstrated a high CO_2 uptake capacity (Serhan et al. 2019). Especially, MAF-X25 ($MnII_2Cl_2$(bbta) (bbta = 1H,5H-benzo(1,2-d:4,5-d′)bistriazolate), MAF-X25ox, and MAF-X27ox, the addition of monodentate hydroxide to a series of MOFs increased CO_2 uptake capacity by 50% when compared to the parent MOF (Serhan et al. 2019). The CO_2 adsorption was found constant in the presence of water, too owing to the selective binding of CO_2. Although, the regeneration of this SBU-based MOF is still questionable.

7.6.4 Hydrophobicity

Enhancing the hydrophobicity of MOF is a good alternative for the simple elimination of water from the vicinity of the pores. The hydrophobic MOFs of zeolitic imidazolate frameworks (ZIFs), ZIF-300, ZIF-301, and ZIF-302, showed similar uptake capacity of CO_2 under humid and dry conditions, and also ZIFs showed excellent recyclable performance for 3 consecutive runs with constant performance due to their hydrophobic nature (Nguyen et al. 2014). Hydrophobicity can also be added to MOF by post-synthesis modifications. But, this kind of hydrophobic MOFs which relies on the hydrophobicity of the MOF, has lower CO_2 uptake due to weaker binding or interaction of CO_2 with the framework. Nonetheless, this research is headed in the correct direction in terms of practical applicability. There is a need to design MOFs which has hydrophobicity, unsaturated metal ions, heteroatoms in which these binding sites will be shielded due to the hydrophobicity of the MOFs from water.

7.6.5 Hybrid Approach

The aim of this kind of approach is to prepare a framework that has a maximum number of strong binding sites for strong CO_2 interaction. The MOFs synthesized using Cu(ii) SBUs having a triazole linker with non-coordinating N atoms, ($Cu_6(TADIPA)_3(DABCO)$) and [$Cu_6(TADIPA)_3(H_2O)_6$] (TADIPA = 5,5′-(1*H*-1,2,4-triazole-3,5-diyl)diisophthalate) (TATB = 4,4′,4″-*s*-triazine-2,4,6-triyl-tribenzoate); the triazole- and pyridine-having linker, (Cu(Me-4py-trz-ia) a triazine linker in CuTDPAT ($Cu_3(TDPAT)(H_2O)_3$ (TDPAT = 2,4,6-tris (3,5-dicarboxylphenylamino)-1,3,5-triazine) or CuTATB-60 ($Cu_6(H_2O)_6(TATB)_4$ and by incorporating amide functionalities to add more moieties (Kim et al. 2011; Dasgupta et al. 2012; Li et al. 2012b; Nguyen et al. 2014). This final mixed linker MOF, $Zn_2(tdc)_2(MA)$ (MA = Melamine, tdc = 2,5-thiophenedicarboxylate), exhibited very high CO_2 uptake (27.0 wt%) at ambient conditions. Binding sites are not yet identified for this structure (Lu et al. 2016). This hybrid approach has great potential for high capacity.

7.7 CO_2 Capture from Air and Natural Gas

7.7.1 Capture of CO_2 from Confined Spaces

Particular settings that include the transportation of individuals inside a restricted area provide unique problems in terms of CO_2 accumulation owing to breathing. CO_2 generation in enclosed environments, like spaceships and undersea submarines, could be harmful to inhabitants after a long period. CO_2 levels in such situations were related to headaches and lethargy, prompting the imposition of a 1% CO_2

permissible exposure limit (Law et al. 2010). Lower Carbon collection, on the other hand, is a substantial problem owing to the complexity in developing materials having robust enough interactions at minimal exposure without requiring a big energy input during regeneration. NbOFFIVE-1-Ni [$NiNbOF_5(pyrazine)_2 \cdot 2H_2O$], which has 2-D Ni(II) pyrazine sheets cross-linked containing $(NbOF_5)^{2-}$ ions has lately been demonstrated to be a feasible substance for such an application (Bhatt et al. 2016). NbOFFIVE-1-Ni is found to be able to show 8.2 wt% CO_2 uptake under dynamic breakthrough circumstances with 1% CO_2 in dry N_2 due to its polar pore walls (which induce high interaction) and perfectly sized square channels (that is suitable pockets for CO_2 uptake). At 75% humidity level, the adsorption was decreased to 5.6 wt% (Bhatt et al. 2016). Though some MOFs functionalized by dialkyl amines perform better than NbOFFIVE1Ni with similar circumstances, the reduced input energy required to regenerate NbOFFIVE1Ni seems to be an appealing characteristic for CO_2 absorption at low concentrations (McDonald et al. 2012; Fracaroli et al. 2014). Although bearing in view the fundamental structural characteristics that make the MOF interesting, there's also a wide area to cover for developing novel materials with better performance over NbOFFIVE1Ni.

7.7.2 CO_2 Capture from Natural Gas Upgrading

Since CO_2 is corrosive as in the presence of water and decreases efficiency, removal of CO_2 from CH_4 is an essential step in natural gas sweetening (Cavenati et al. 2006; Férey et al. 2011). Similar structural characteristics of MOFs utilized during post-combustion CO_2 collection hold herein, only with the extra benefit of altering the pore size to utilize the differing kinetic diameters of both CO_2 (3.3 Å) and CH4 (3.8 Å) (Li et al. 2012a). UTSA-49 is one of the strongest MOF, having the claimed selectivity of 33.7 for an equimolar blend of CO_2 and CH_4 at room temperature (Xiong et al. 2014). Its uncoordinated heteroaromatic N atoms and two kinds of tiny pore apertures (2.9 Å $\times$ 3.6 Å and 3.6 Å $\times$ 4.0 Å), which together promote CO_2 absorption above CH_4, are indeed essential characteristics for this tetrazolate framework. The mixed-metal ZIF-204 has demonstrated towards being competent at preferentially collecting CO_2 out of a ternary gas stream of CO_2, CH_4, and H_2O—circumstances which have been suitable at biogas sources (Nguyen et al. 2016). Although the material's dynamical CO_2 uptake seemed low (1.6 wt%) in this study, the presence of moisture had no effect on its performance. In addition, two polymorphic MOFs, Qc-5-Cu-α and –β (Cu(quinoline-5-carboxylate), were prepared and shown to be susceptible to size-selectively adsorbing CO_2 at high humidity. It is indeed a fascinating finding since the pore size was chosen to favour CO_2 over CH_4 molecular sieving (Chen et al. 2016). Several works also looked at how various functionalities and structural characteristics of MOFs, such as hydroxyl groups, coordinatively unsaturated metal sites, and ionic structures, might improve CO_2/CH_4 selectivity (Bae et al. 2008; Caskey, Wong-Foy and Matzger 2008; Chen et al. 2011). Since MOF-based separations have progressed towards the situation where even water barrier for selective CO_2 capturing has been overcome, the next step must undoubtedly to concentrate here on reduction of additional troublesome natural gas contaminants like H_2S. There have been few studies on either the longevity of MOFs also in existence of H_2S, much again the selective CO_2 uptake in the context of H_2S alone (Jin et al. 2010; Hamon et al. 2011; Vellingiri et al. 2016). Certainly, that will be the upcoming major problem to tackle until MOFs can be fully used for a natural gas upgrading option.

7.8 Regeneration of MOFs

The complete CO_2 capture operation necessitates the development of an integrated adsorption and desorption system that allows effective material regeneration as well as CO_2 handling (like injection, storage, and utilization) (Mason et al. 2011). A typical regeneration process would need a minimum amount of energy for eliminating the large percentage of adsorbed CO_2 while causing no capacity deterioration. Towards this aim, the lower the CO_2 bonding energy, the more effective the regeneration will be since it will consume lower energy and create the minimum amount of disruption to the framework. Current adsorbent recycling technology focuses on adsorption pressure (PSA), vacuum swing adsorption (VSA), and temperature swing adsorption (TSA) (Olajire 2010; Yu et al. 2012). The CO_2 is absorbed in a certain pressure in PSA and VSA and then liberated via reducing the pressure. The distinction among these two techniques is that for a VSA approach, CO_2 is adsorbed at low pressures, whereas in a PSA system, high pressures, generally varying between 8 and 28 bar, are used (Sircar and Golden 2011). PSA is better suited for Capturing CO_2 during the pre-combustion step since pre-combustion exhaust gas flows are pressured, while VSA is ideally adapted for ambient pressure post-combustion stream since they are not pressurized. PVSA is often a cross between the two methods, having a higher input pressure and perhaps lower desorption pressure. CuBTTri, Mg-MOF-74, MOF-177, BeBTB $[Be_{12}(OH)_{12}(BTB)_4]$ (BeBTB = Bromothymol sulfone phthalein) and Co(BDP) (BDP = 1,4-benzenedipyrazolate) were used to test the usage of PSA for MOFs in the initial stages, with single-component isotherms for CO_2 and H_2 measured at 40 bar (Herm et al. 2011). CuBTTri (BTtri = 1,3,5-tris(1*H*-1,2,3-triazol-5-yl)benzene) and Mg-MOF-74 are the MOFs containing coordinatively unsaturated metal sites, also had the greatest operating capabilities as well as beat zeolite 13X and even activated carbon. Following that, a pelletized version of MIL-53(Al) [Al(OH)(BDC)] (BDC = Terephthalate) (Ferreira et al. 2015; Serra-Crespo et al. 2015) was used to imitate an industrial PSA process. Around 4 bar at 303 K, having a desorption pressure near 0.1 bar, 92.8% of the methane intake was retrieved, having a purity of 99.4%, with the flow of 87% CH_4 and 13% CO_2.

Heating the sorbent to force out adsorbed molecules is another popular technique for adsorbent regeneration (Berger and Bhown 2011; Mason et al. 2011). Porous solids (2 Jg^{-1} K^{-1}) have lesser thermal capacities than monoethanolamine mixtures (3–4 Jg^{-1} K^{-1}), making the TSA procedure appealing. MOF-177 and MOF-74 (McDonald et al. 2015) were both tested for their capacity to be regenerated using TSA as in with MOFs (Mason et al. 2011).

Only a few other research has been done on the regeneration of MOFs utilizing a VSA technique so far. In comparison to TSA, VSA is inefficient while using it with HKUST-1 ($Cu_3(BTC)_2$). The inability of such coordinatively unsaturated metal sites to regenerate was blamed for this (Ye et al. 2013). HKUST-1 is subjected to a feed of 13–16% CO_2 in dry N_2 at 2 bar and 308 K, and desorption initiating at 0.15 bar in a separate experiment. HKUST-1 had capacity-2.22 mmol g^{-1} and a CO_2 recovery of 63% based on this (Dasgupta et al. 2012). UiO-66 had also been tested as a sorbent for the PVSA operation for both dry and wet modelled flue gas blends in some other work. The intake gas was pressurized to 2 bar pressure at 328 K using 15% CO_2 and 9% water, and desorption was carried out at 0.15 bar (Andersen et al. 2013). As the water was involved, however, only poor quality (60%) CO_2 was recovered, and CO_2 capacity dropped by 25%.

7.9 Conclusion

Metal–organic frameworks could be the material that can make clean coal a reality. Incredible development in the field of MOFs has been observed in the last decade. Since 2000, gas storage has been one of the main aspects of MOF studies. In methane storage, substantial progress has been made. If the research subject becomes even more popular, several different forms of application will arise, the field's outlook is really bright. Virtually limitless MOFs with diverse structures and chemical compositions are feasible, but for industrial application, the synthesis of those with high hydrothermal stability should be treated preferentially. Synthesis of MOF is often limited to small scale production and is only adequate for its structural analysis or to study its crystal structure, and is not optimized. So, in order to prepare a MOF a reliable recipe is highly suitable. For practical CO_2 capture assessment and process studies, MOFs must be pelletized and packed into columns in industrial practice. MOF-based CO_2 capture study has reached a point where engineering assessment and process design are already becoming increasingly crucial in order to advance MOFs as adsorbents closer to real-world applications. As a result, we anticipate additional research from the MOF chemistry and engineering communities in the future, paving the way for interesting process system engineering for MOF adsorbents for industrial-scale CO_2 capture applications. Along with advancements in CO_2 capturing and separation, could close the carbon cycle loop by allowing anthropogenic CO_2 to be collected, processed, and reused as a reusable source of energy. To completely address the CO_2 problem, it is evident that reticular chemistry, as typified by MOFs, would continue to play a vital role in developing novel, strong materials, particularly those with structures constructed from synergistic constituents.

References

Andersen A et al (2013) On the development of Vacuum Swing adsorption (VSA) technology for post-combustion CO_2 capture. Energy Procedia 37(1876):33–39. https://doi.org/10.1016/j.egypro.2013.05.082

Aslani A, Morsali A (2009) Sonochemical synthesis of nano-sized metal-organic lead(II) polymer: a precursor for the preparation of nano-structured lead(II) iodide and lead(II) oxide. Inorg Chim Acta 362(14):5012–5016. https://doi.org/10.1016/j.ica.2009.08.011

Bae Y et al (2008) Separation of CO_2 from CH_4 using mixed-ligand metal–organic frameworks. Langmuir 18:8592–8598

Baekelandt BG, Baekelandt BG (1996) Functional theory functional theory. Nat Commun 4664(1):1159–1166. https://doi.org/10.1038/s41467-020-19024-0

Bao Z et al (2011) Adsorption of CO_2 and CH_4 on a magnesium-based metal organic framework. J Colloid Interface Sci 353(2):549–556. https://doi.org/10.1016/j.jcis.2010.09.065

Bara JE et al (2007) Enhanced CO_2 separation selectivity in oligo(ethylene glycol) functionalized room-temperature ionic liquids. Ind Eng Chem Res 46(16):5380–5386. https://doi.org/10.1021/ie070437g

Beldon PJ et al (2010) Rapid room-temperature synthesis of zeolitic imidazolate frameworks by using mechanochemistry. Angew Chem 122(50):9834–9837. https://doi.org/10.1002/ange.201005547

Bell LN (1964) Thermodynamics of photosynthesis. Biophysics 9(3):316–321

Belmabkhout Y, Guillerm V, Eddaoudi M (2016) Low concentration CO_2 capture using physical adsorbents: are metal-organic frameworks becoming the new benchmark materials? Chem Eng J 296:386–397. https://doi.org/10.1016/j.cej.2016.03.124

Berenstein D, Cotta R, Leonardi R (2008) Numerical tests of AdS/CFT at strong coupling. Gphys Rev D Particles Fields Gravitation Cosmol. https://doi.org/10.1103/PhysRevD.78.025008

Berger AH, Bhown AS (2011) Comparing physisorption and chemisorption solid sorbents for use separating CO_2 from flue gas using temperature swing adsorption. Energy Procedia 4:562–567. https://doi.org/10.1016/j.egypro.2011.01.089

Bhatt PM et al (2016) A fine-tuned fluorinated MOF addresses the needs for trace CO_2 removal and air capture using physisorption. J Am Chem Soc 138(29):9301–9307. https://doi.org/10.1021/jacs.6b05345

Billemont P et al (2017) IAST predictions vs co-adsorption measurements for CO_2 capture and separation on MIL-100 (Fe). Adsorption 23(2–3):225–237. https://doi.org/10.1007/s10450-016-9825-6

Britt D et al (2009) Highly efficient separation of carbon dioxide by a metal-organic framework replete with open metal sites. Proc Natl Acad Sci USA 106(49):20637–20640. https://doi.org/10.1073/pnas.0909718106

Bui M et al (2018) Carbon capture and storage (CCS): the way forward. Energy Environ Sci 11(5):1062–1176. https://doi.org/10.1039/c7ee02342a

Burd SD, Ma S, Perman JA, Sikora BJ, Snurr RQ, Thallapally PK, Tian J, Wojtas L, Zaworotko MJ (2012). Highly Selective Carbon Dioxide Uptake by [Cu(bpy-n)2(SiF6)] (bpy-1 = 4,4′-Bipyridine; bpy-2 = 1,2-Bis(4-pyridyl)ethene). 2, 6–10

Carson CG et al (2011) Sonochemical synthesis and characterization of submicrometer crystals of the metal-organic framework Cu[(hfipbb)(H2hfipbb)0.5]. Cryst Growth Des 11(10):4505–4510. https://doi.org/10.1021/cg200728b

Caskey SR, Wong-Foy AG, Matzger AJ (2008) Dramatic tuning of carbon dioxide uptake via metal substitution in a coordination polymer with cylindrical pores. J Am Chem Soc 130(33):10870–10871. https://doi.org/10.1021/ja8036096

Cavenati S, Grande CA, Rodrigues AE (2006) Removal of carbon dioxide from natural gas by vacuum pressure swing adsorption. Energy Fuels 20(6):2648–2659. https://doi.org/10.1021/ef060119e

Cessford NF, Seaton NA, Düren T (2012) Evaluation of ideal adsorbed solution theory as a tool for the design of metal-organic framework materials. Ind Eng Chem Res 51(13):4911–4921. https://doi.org/10.1021/ie202219w

Chandler BD et al (2008) Mechanical gas capture and release in a network solid via multiple single-crystalline transformations. Nat Mater 7(3):229–235. https://doi.org/10.1038/nmat2101

Chen K et al (2016) Tuning pore size in square-lattice coordination networks for size-selective sieving of CO_2. Angew Chem 128(35):10424–10428. https://doi.org/10.1002/ange.201603934

Chen Z et al (2011) Significantly enhanced CO_2/CH_4 separation selectivity within a 3D prototype metal-organic framework functionalized with OH groups on pore surfaces at room temperature. Eur J Inorg Chem 14:2227–2231. https://doi.org/10.1002/ejic.201100034

Cho HY et al (2012) CO_2 adsorption and catalytic application of Co-MOF-74 synthesized by microwave heating. Catal Today 185(1):35–40. https://doi.org/10.1016/j.cattod.2011.08.019

Cho HY et al (2013) High yield 1-L scale synthesis of ZIF-8 via a sonochemical route. Microporous Mesoporous Mater 169:180–184. https://doi.org/10.1016/j.micromeso.2012.11.012

Cui P et al (2012) Multipoint interactions enhanced CO_2 uptake: azeolite-like zinc-tetrazole framework with 24-nuclear zinc cages. J Am Chem Soc 134(46):18892–18895. https://doi.org/10.1021/ja3063138

Dasgupta S et al (2012) CO_2 recovery from mixtures with nitrogen in a vacuum swing adsorber using metal organic framework adsorbent: a comparative study. Int J Greenhouse Gas Control 7:225–229. https://doi.org/10.1016/j.ijggc.2011.10.007

Davis ME (2002) Ordered porous materials for emerging applications. Nature 417(6891):813–821. https://doi.org/10.1038/nature00785

Demessence A et al (2009) J Am Chem Soc 131:8784–8786

Dey C et al (2014) Crystalline metal-organic frameworks (MOFs): synthesis, structure and function. Acta Crystallogr Sect B Struct Sci Crystal Eng Mater 70(1):3–10. https://doi.org/10.1107/S2052520613029557

Dharmarathna S et al (2012) Direct sonochemical synthesis of manganese octahedral molecular sieve (OMS-2) nanomaterials using cosolvent systems, their characterization, and catalytic applications. Chem Mater 24(4):705–712. https://doi.org/10.1021/cm203366m

Dipl HH, Pr T, Biowissenschaften T (2009). Chemical looping for synthesis gas and power generation with CO_2 capture—pilot plant study and process modeling.

Dinca M, Dailly A, Liu Y, Brown CM, Neumann DA, Long JR (2010). Hydrogen Storage in a Microporous Metal-Organic Framework with Exposed Mn^{2+} Coordination Sites

Férey G et al (2011) Why hybrid porous solids capture greenhouse gases? Chem Soc Rev 40(2):550–562. https://doi.org/10.1039/c0cs00040j

Ferreira AFP et al (2015) Methane purification by adsorptive processes on MIL-53(Al). Chem Eng Sci 124:79–95. https://doi.org/10.1016/j.ces.2014.06.014

Fracaroli AM et al (2014) Metal-organic frameworks with precisely designed interior for carbon dioxide capture in the presence of water. J Am Chem Soc 136(25):8863–8866. https://doi.org/10.1021/ja503296c

Friščić T et al (2013) Real-time and in situ monitoring of mechanochemical milling reactions. Nat Chem 5(1):66–73. https://doi.org/10.1038/nchem.1505

Furukawa H et al (2013) The chemistry and applications of metal-organic frameworks. Science 341(6149). https://doi.org/10.1126/science.1230444

Furukawa H, Müller U, Yaghi OM (2015) "Heterogeneity within order" in metal-organic frameworks. Angewandte Chemie—Int Ed 54(11):3417–3430. https://doi.org/10.1002/anie.201410252

Hamon L et al (2009) Co-adsorption and separation of CO_2-CH_4 mixtures in the highly flexible MIL-53(Cr) MOF. J Am Chem Soc 131(47):17490–17499. https://doi.org/10.1021/ja907556q

Hamon L et al (2011) Molecular insight into the adsorption of H_2S in the flexible MIL-53(Cr) and rigid MIL-47(V) MOFs: infrared spectroscopy combined to molecular simulations. J Phys Chem C 115(5):2047–2056. https://doi.org/10.1021/jp1092724

Herm ZR et al (2011) Metal-organic frameworks as adsorbents for hydrogen purification and precombustion carbon dioxide capture. J Am Chem Soc 133(15):5664–5667. https://doi.org/10.1021/ja111411q

Hindelang K et al (2012) Tandem post-synthetic modification for functionalized metal–organic frameworks via epoxidation and subsequent epoxide ring-opening. Chem Commun 48(23):2888–2890. https://doi.org/10.1039/c2cc16949e

Jiang J, Zhao Y, Yaghi OM (2016) Covalent chemistry beyond molecules. J Am Chem Soc 138(10):3255–3265. https://doi.org/10.1021/jacs.5b10666

Jin L et al (2010) Relationship of tumor necrosis factor gene polymorphism and acute graft-versus-host disease after unrelated allogeneic hematopoietic stem cell transplantation. Zhonghua Nei Ke Za Zhi [chin J Internal Med] 49(4):320–324

Jung DW et al (2010) Facile synthesis of MOF-177 by a sonochemical method using 1-methyl-2-pyrrolidinone as a solvent. Dalton Trans 39(11):2883–2887. https://doi.org/10.1039/b925088c

Kelly M et al (2017) Capture of CO_2. IPCC Special Report on Carbon dioxide Capture and Storage

Kim J et al (2012) Bench-scale preparation of Cu_3 (BTC) 2 by ethanol reflux: synthesis optimization and adsorption/catalytic applications. Microporous Mesoporous Mater 161:48–55. https://doi.org/10.1016/j.micromeso.2012.05.021

Kim J et al (2011) Control of catenation in CuTATB-n metal-organic frameworks by sonochemical synthesis and its effect on CO_2 adsorption. J Mater Chem 21(9):3070–3076. https://doi.org/10.1039/c0jm03318a

Kim Y, Jang H, Kim J, Lee J (2017). Prediction of storage efficiency on CO_2 sequestration in deep saline aquifers using artificial neural network. Applied Energy, 185, 916–928. https://doi.org/10.1016/j.apenergy.2016.10.012

Kizzie AC, Wong-Foy AG, Matzger AJ (2011) Effect of humidity on the performance of microporous coordination polymers as adsorbents for CO_2 capture. Langmuir 27(10):6368–6373. https://doi.org/10.1021/la200547k

Klinowski J et al (2011) Microwave-assisted synthesis of metal-organic frameworks. Dalton Trans 40(2):321–330. https://doi.org/10.1039/c0dt00708k

Kumar A et al (2015) Direct air capture of CO_2 by physisorbent materials. Angewandte Chemie—Int Ed 54(48):14372–14377. https://doi.org/10.1002/anie.201506952

Lagashetty A et al (2007) Microwave-assisted route for synthesis of nanosized metal oxides. Sci Technol Adv Mater 8(6):484–493. https://doi.org/10.1016/j.stam.2007.07.001

Law J, Watkins S, Alexander D (2010) In-flight carbon dioxide exposures and related symptoms: associations, susceptibility and operational implications. NASA Technical Report (June), pp 1–21

Lee YR, Kim J, Ahn WS (2013) Synthesis of metal-organic frameworks: a mini review. Korean J Chem Eng 30(9):1667–1680. https://doi.org/10.1007/s11814-013-0140-6

Li B et al (2012a) Enhanced binding affinity, remarkable selectivity, and high capacity of CO_2 by dual functionalization of a rht-type metal-organic framework. Angewandte Chemie—Int Ed 51(6):1412–1415. https://doi.org/10.1002/anie.201105966

Li JR et al (2011) Carbon dioxide capture-related gas adsorption and separation in metal-organic frameworks. Coord Chem Rev 255(15–16):1791–1823. https://doi.org/10.1016/j.ccr.2011.02.012

Li JR, Sculley J, Zhou HC (2012b) Metal-organic frameworks for separations. Chem Rev 112(2):869–932. https://doi.org/10.1021/cr200190s

Liao P et al (2017) A (19 – 22), 1196(June), pp 1193–1196

Lin Y et al (2017) Metal-organic frameworks for carbon dioxide capture and methane storage. Ad Energy Mater 7(4). https://doi.org/10.1002/aenm.201601296

Lu Y, Dong Y, Qin J (2016) Porous pcu-type Zn(II) framework material with high adsorption selectivity for CO_2 over N_2. J Mol Struct 1107:66–69. https://doi.org/10.1016/j.molstruc.2015.11.045

Märcz M et al (2012) The iron member of the CPO-27 coordination polymer series: synthesis, characterization, and intriguing redox properties. Microporous Mesoporous Mater 157:62–74. https://doi.org/10.1016/j.micromeso.2011.12.035

Marshall RJ et al (2015) Single-crystal to single-crystal mechanical contraction of metal-organic frameworks through stereoselective postsynthetic bromination. J Am Chem Soc 137(30):9527–9530. https://doi.org/10.1021/jacs.5b05434

Martinez Joaristi A et al (2012) Electrochemical synthesis of some archetypical Zn^{2+}, Cu^{2+}, and Al^{3+} metal organic frameworks. Cryst Growth Des 12(7):3489–3498. https://doi.org/10.1021/cg300552w

Mason JA et al (2011) Evaluating metal-organic frameworks for post-combustion carbon dioxide capture via temperature swing adsorption. Energy Environ Sci 4(8):3030–3040. https://doi.org/10.1039/c1ee01720a

Mason JA et al (2015) Application of a high-throughput analyzer in evaluating solid adsorbents for post-combustion carbon capture via multicomponent adsorption of CO_2, N_2, and H_2O. J Am Chem Soc 137(14):4787–4803. https://doi.org/10.1021/jacs.5b00838
Mateovic T et al (2002) The influence of stirring rate on biopharmaceutical properties of Eudragit RS microspheres. J Microencapsul 19(1):29–36. https://doi.org/10.1080/02652040010055289
McDonald TM et al (2011) Enhanced carbon dioxide capture upon incorporation of N, N′-dimethylethylenediamine in the metal-organic framework CuBTTri. Chem Sci 2(10):2022–2028. https://doi.org/10.1039/c1sc00354b
McDonald TM et al (2012) Capture of carbon dioxide from air and flue gas in the alkylamine-appended metal-organic framework mmen-Mg_2(dobpdc). J Am Chem Soc 134(16):7056–7065. https://doi.org/10.1021/ja300034j
McDonald TM et al (2015) Cooperative insertion of CO_2 in diamine-appended metal-organic frameworks. Nature 519(7543):303–308. https://doi.org/10.1038/nature14327
Mn E et al (2010) 'M. Dinca 2006', (25), pp 1–8. Available at: papers://c438f4a5-b4d4–4968-bf1f-63e00e5cec7c/Paper/p4581
Mohanty P, Linn NMK, Landskron K (2010) Ultrafast sonochemical synthesis of methane and ethane bridged periodic mesoporous organosilicas. Langmuir 26(2):1147–1151. https://doi.org/10.1021/la902239m
Myers AL, Monson PA (2014) Adsorption as the basis for thermodynamic analysis. Adsorption 20(4):591–622
Nguyen NTT et al (2014) Selective capture of carbon dioxide under humid conditions by hydrophobic chabazite-type zeolitic imidazolate frameworks. Angew Chem 126(40):10821–10824. https://doi.org/10.1002/ange.201403980
Nguyen NTT et al (2016) Mixed-metal zeolitic imidazolate frameworks and their selective capture of wet carbon dioxide over methane. Inorg Chem 55(12):6201–6207. https://doi.org/10.1021/acs.inorgchem.6b00814
Nugent P et al (2013) Porous materials with optimal adsorption thermodynamics and kinetics for CO_2 separation. Nature 495(7439):80–84. https://doi.org/10.1038/nature11893
Olajire AA (2010) CO_2 capture and separation technologies for end-of-pipe applications—a review. Energy 35(6):2610–2628. https://doi.org/10.1016/j.energy.2010.02.030
Raksajati A, Ho MT, Wiley DE (2016) Understanding the impact of process design on the cost of CO_2 capture for precipitating solvent absorption. Ind Eng Chem Res 55(7):1980–1994. https://doi.org/10.1021/acs.iecr.5b03633
Ranjan R, Tsapatsis M (2009) Microporous metal organic framework membrane on porous support using the seeded growth method. Chem Mater 21(20):4920–4924. https://doi.org/10.1021/cm902032y
Samanta A et al (2012) Post-combustion CO_2 capture using solid sorbents: a review. Ind Eng Chem Res 51(4):1438–1463. https://doi.org/10.1021/ie200686q
Sanz-Pérez ES et al (2016) Direct capture of CO_2 from ambient air. Chem Rev 116(19):11840–11876. https://doi.org/10.1021/acs.chemrev.6b00173
Schoedel A, Ji Z, Yaghi OM (2016) The role of metal–organic frameworks in a carbon-neutral energy cycle. Nat Energy 1(4):1–13. https://doi.org/10.1038/nenergy.2016.34
Seoane B et al (2015) Metal-organic framework based mixed matrix membranes: a solution for highly efficient CO_2 capture? Chem Soc Rev 44(8):2421–2454. https://doi.org/10.1039/c4cs00437j
Serhan M et al (2019) Total iron measurement in human serum with a smartphone. In: AIChE annual meeting, conference proceedings, 2019-Novem. https://doi.org/10.1039/x0xx00000x
Serra-Crespo P et al (2015) Preliminary design of a vacuum pressure swing adsorption process for natural gas upgrading based on amino-functionalized MIL-53. Chem Eng Technol 38(7):1183–1194. https://doi.org/10.1002/ceat.201400741
Shekhah O et al (2014) Made-to-order metal-organic frameworks for trace carbon dioxide removal and air capture. Nat Commun 5(May):1–7. https://doi.org/10.1038/ncomms5228

Si X et al (2011) High and selective CO_2 uptake, H_2 storage and methanol sensing on the amine-decorated 12-connected MOF CAU-1. Energy Environ Sci 4(11):4522–4527. https://doi.org/10.1039/c1ee01380g

Sircar S et al (1999) Isosteric heat of adsorption: theory and experiment. J Phys Chem B 103(31):6539–6546. https://doi.org/10.1021/jp9903817

Sircar S, Golden TC (2011) Purification of hydrogen by pressure swing. Sep Sci Technol (September 2012), 37–41

Son WJ et al (2008) Sonochemical synthesis of MOF-5. Chem Commun 47:6336–6338. https://doi.org/10.1039/b814740j

Sumida K et al (2012) Carbon dioxide capture in metal-organic frameworks. Chem Rev 112(2):724–781. https://doi.org/10.1021/cr2003272

Talu O (2013) Net adsorption of gas/vapor mixtures in microporous solids. J Phys Chem C 117(25):13059–13071. https://doi.org/10.1021/jp4021382

Terzyk AP et al (2003) Developing the solution analogue of the Toth adsorption isotherm equation. J Colloid Interface Sci 266(2):473–476. https://doi.org/10.1016/S0021-9797(03)00569-1

Trickett CA et al (2017) The chemistry of metal-organic frameworks for CO_2 capture, regeneration and conversion. Nat Rev Mater 2(8):1–16. https://doi.org/10.1038/natrevmats.2017.45

Uribe-romo FJ et al (2010) Capture properties of zeolitic imidazolate frameworks. Accounts Chem Res 43(1):58–67. Available at: http://www.ncbi.nlm.nih.gov/pubmed/19877580

Vellingiri K, Deep A, Kim KH (2016) Metal-organic frameworks as a potential platform for selective treatment of gaseous sulfur compounds. ACS Appl Mater Interfaces 8(44):29835–29857. https://doi.org/10.1021/acsami.6b10482

Verma S, Mishra AK, Kumar J (2010) The many facets of adenine: coordination, crystal patterns, and catalysis. Acc Chem Res 43(1):79–91. https://doi.org/10.1021/ar9001334

Walton KS et al (2008) Understanding inflections and steps in carbon dioxide adsorption isotherms in metal-organic frameworks. J Am Chem Soc 130(2):406–407. https://doi.org/10.1021/ja076595g

Wang CC, Ying JY (1999) Sol-gel synthesis and hydrothermal processing of anatase and rutile titania nanocrystals. Chem Mater 11(11):3113–3120. https://doi.org/10.1021/cm990180f

Wang LJ et al (2014) Synthesis and characterization of metal-organic framework-74 containing 2, 4, 6, 8, and 10 different metals. Inorg Chem 53(12):5881–5883. https://doi.org/10.1021/ic500434a

Wu D et al (2013) Direct calorimetric measurement of enthalpy of adsorption of carbon dioxide on CD-MOF-2, a green metal-organic framework. J Am Chem Soc 135(18):6790–6793. https://doi.org/10.1021/ja402315d

Xiong S et al (2014) A new tetrazolate zeolite-like framework for highly selective CO_2/CH_4 and CO_2/N_2 separation. Chem Commun 50(81):12101–12104. https://doi.org/10.1039/c4cc05729e

Yang Q et al (2013) A water stable metal-organic framework with optimal features for CO_2 capture. Angew Chem 125(39):10506–10510. https://doi.org/10.1002/ange.201302682

Ye S et al (2013) Post-combustion CO_2 capture with the HKUST-1 and MIL-101(Cr) metal-organic frameworks: adsorption, separation and regeneration investigations. Microporous Mesoporous Mater 179:191–197. https://doi.org/10.1016/j.micromeso.2013.06.007

Yu CH, Huang CH, Tan CS (2012) A review of CO_2 capture by absorption and adsorption. Aerosol Air Qual Res 12(5):745–769. https://doi.org/10.4209/aaqr.2012.05.0132

Zielinski JM et al (2007) High pressure sorption isotherms via differential pressure measurements. Adsorption 13(1):1–7. https://doi.org/10.1007/s10450-007-9005-9

Zornoza B et al (2011) Functionalized flexible MOFs as fillers in mixed matrix membranes for highly selective separation of CO_2 from CH_4 at elevated pressures. Chem Commun 47(33):9522–9524. https://doi.org/10.1039/c1cc13431k

Chapter 8
Exploration of Amine Based Nanofluids as a Potential Solvent for Post-combustion CO_2 Capture

Fenil Patel, Umang Sutariya, Anirban Dey, Bharti Saini, and Sweta Balchandani

Abstract The substantial increase in CO_2 emission over the last few decades is due to the combustion of fossil fuels for energy generation and it is responsible for the global warming issue across the world. The alternative energy sources are not sufficient to achieve the international targets for CO_2 emission reduction. Carbon capture and sequestration (CCS) technology can be used as a potential tool for the significant reduction of CO_2. CCS is a process which involves the capture of CO_2 from large sources like fossil fuel station and transportation of CO_2 via pipeline to the deep underground storage site. Three main techniques can be implemented for CO_2 capture; Pre combustion capture, Post-combustion capture, and Oxy-fuel combustion. Conventional amines such as monoethanolamine (MEA), diethanolamine (DEA), and N-methyldiethanolamine (MDEA) are generally used for carbon capture. However the usage of amines solvents for absorption process suffers from various drawbacks such as high regeneration energy requirement, toxicity, solvent loss and equipment corrosion. Recent studies showed that Nano-fluids can give more absorption efficiency than conventional solvents. In this chapter, we reviewed and highlighted recent advances in the application of Nano-fluids in carbon capture. Amine based Nano-fluids can be considered as a potential alternative solvent because of its higher absorption capacity as well as enhanced mass transfer performance.

Keywords CO_2 · Absorption · Nano-fluids · Mass transfer

F. Patel · U. Sutariya · A. Dey (✉) · B. Saini · S. Balchandani
Department of Chemical Engineering, Pandit Deendayal Energy University, Gandhinagar, Gujarat 382007, India
e-mail: anirban.dey@sot.pdpu.ac.in

S. P. Singh et al. (eds.), *Metal Nanocomposites for Energy and Environmental Applications*, Energy, Environment, and Sustainability,
https://doi.org/10.1007/978-981-16-8599-6_8

8.1 Introduction

During the previous 10–15 decades, the use of petroleum products has been increased rapidly to fulfill the energy demands around the world. The excessive use of fossil fuels has raised the emission of various greenhouse gases in the atmosphere, especially carbon dioxide (CO_2). The atmospheric concentration of carbon dioxide has been doubled from 200 ppm to around 400 ppm over the last 4–5 decades (Lindsey 2020). According to recent study, emission from various industrial processes and consumption of fossil fuels will increase the estimated CO_2 emission to a historical high value of 40 billion metric tons in the near future (Ritchie and Roser 2020). As a result of increased greenhouse gas emissions, some important global concerns such as global warming, rise in sea level, and ocean acidification have risen. In the recent past, various innovations have been done to reduce the CO_2 emission into the atmosphere. Carbon capture and storage (CCS) is the most potential tool to reduce the direct emission of CO_2 gas into the atmosphere (Lackner 2009). Post-combustion carbon capture in fossil fuel plants is the most practical technology available right now. In this method, CO_2 gas is captured from flue gases using different solvents after the combustion process is done (Wang et al. 2011). Figure 8.1 shows the different technologies available for post combustion carbon capture (Mondal et al. 2012; Xu et al. 2014; Brunetti et al. 2010; Tsouris and Aaron 2005).

The primary choices for post-combustion carbon capture are water and alkaline amine-based solvents because of their chemical reactivity towards CO_2 gas (Gary 2009). Monoethanolamine (MEA), Diethanolamine (DEA), methyldiethanolamine (MDEA), and 2-amino-2-methyl-1-propanol (AMP) are the most commonly used amines in industries for carbon capture (Zhu et al. 2012; Liu et al. 2019; Kenig and Vaidya 2007; Li et al. 2013). But the use of the amine scrubbing method has many drawbacks like corrosion in plant equipments, high regeneration cost, plant

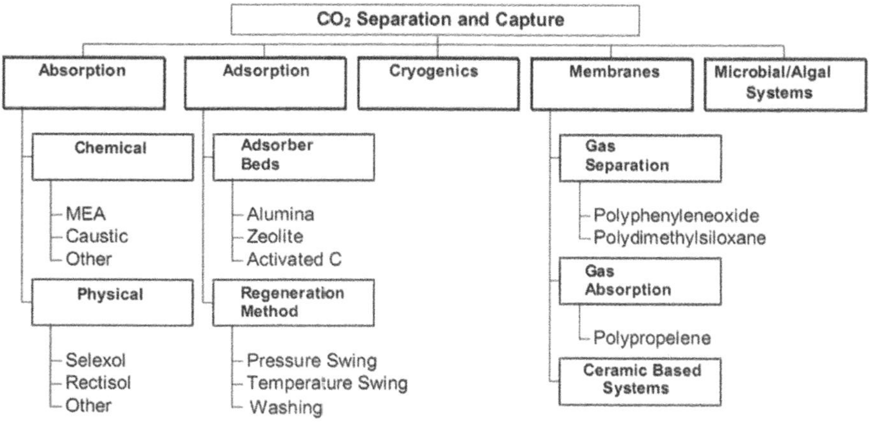

Fig. 8.1 Post combustion carbon capture technologies. Reproduced from Wang et al. (2011)

equipment and loss of amines during regeneration (Arshadi et al. 2019). The use of the amine solvents generates a huge amount of wastewater containing toxic heavy metals which are difficult to treat and raises environmental concerns (Seo et al. 2020). However, recent studies shown that nanofluids can be used as a potential solvent for post combustion carbon capture (Saba Ashrafmansouri and Esfahany 2014). Because nanoparticles have a high specific surface area, a modest amount of nanofluids in the solvent can improve CO_2 absorption, solvent stability, thermal stability, and gas transfer coefficients dramatically (Krishnamurthy et al. 2006; Jiang et al. 2019). It can be used as a catalyst or promoters along with the existing amine solvents. Use of nanofluids could reduce the environment and energy issues. Amine-based Nano-fluids are homogenous mixture of nanoparticles and amine solutions.

Metals, metal oxides, and non-metal oxide elements such as Ni, Fe, SiO_2, CuO, Fe_3O_4, Al_2O_3, and TiO_2 are mixed with amine solutions to create homogeneous suspensions (Suresh and Komati 2008; Rahmatmand et al. 2016; Lu et al. 2013). An external magnetic field can be used to disperse magnetic nanoparticles such as Fe_3O_4 and NiO_2 into solutions in a non-uniform manner (Nabipour et al. 2010). As a general rule, base fluids such as monoethanolamine (MEA), diethanolamine (DEA), methyldiethanolamine (MDEA), and piperazine (PZ) have been used (Jiang et al. 2014; Wang et al. 2016). The single-step and two-step approaches, as indicated in Fig. 8.2, are the most generally utilised methods for preparing nano-fluids (Mukherjee and Paria 2013; Jin Hwang et al. 2008).

In last 2 decades lots of researchers have done various studies on application of nanofluids for carbon capture (Park et al. 2006; Lee et al. 2015; Kim et al. 2008). Wang et al. (2016) studied the various nanofluids such as SiO_2, Al_2O_3, carbon nanotubes (CNT) etc. inside a flat-stirred vessel for CO_2 absorption characteristics (Park et al. 2006). Yu et al. (2016) studied the SiO_2 and Al_2O_3 nanofluids along with MDEA, MEA and PZ solvents in Wetted wall column (Yu et al. 2016). Seo et al. (2020) examined nickel nanoparticles as a catalyst along with MEA solvent for limited mixing and high mixing conditions in microreactor (Seo et al. 2020). This book chapter aims to review and summarize the recent advances in the use of nanofluids for carbon capture. Various aspects such as preparation of nanofluids by different methods, enhancement mechanism for nanofluids, different factors affecting the carbon capture, role of nanofluids on desorption and solvent regeneration are discussed thoroughly in the subsequent section.

8.2 Enhancement Mechanism

The result of the investigation reveals that adding nanoparticles to base fluid enhances CO_2 absorption rates when compared to base fluid alone. The mechanism underlying the use of nanofluids to increase CO_2 absorption rates have yet to be fully understood. Researchers frequently offer the three methods depicted in Fig. 8.3, including bubble breaking effect, shuttle effect, and boundary mixing effect.

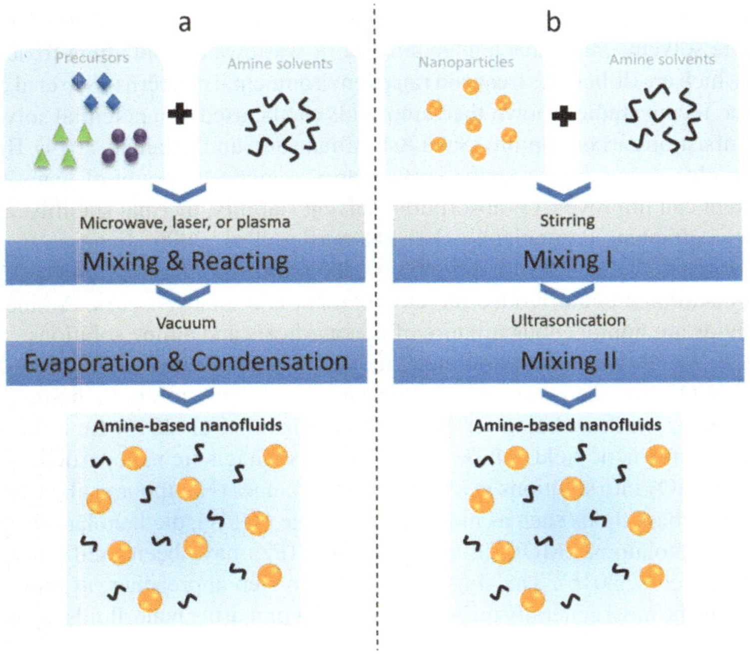

Fig. 8.2 Methods for the preparation of Amine based nanofluids. **a** Single step method **b** two step method. Reproduced from Wei Yu (2019)

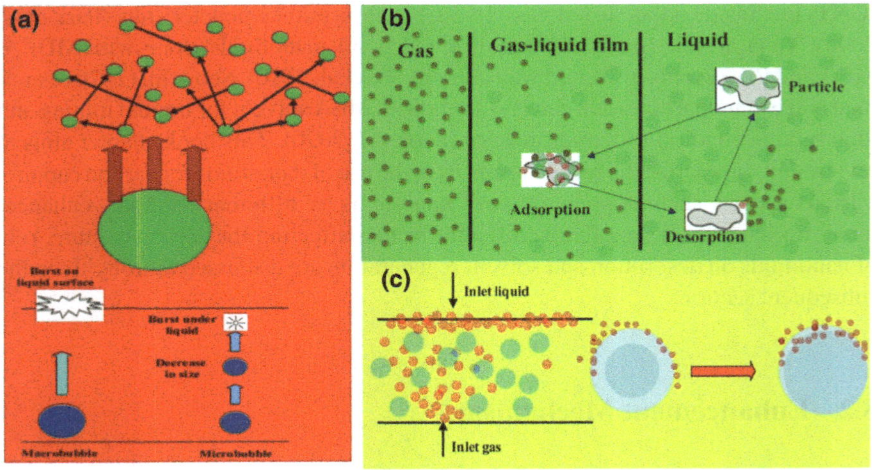

Fig. 8.3 CO_2 absorption rate enhancement mechanism. **a** Bubble breaking effect. **b** Shuttle effect. **c** Boundary mixing effect. Reproduced from Hassan et al. (2020)

8.2.1 Bubble Breaking Effect

Using nozzles, gas is entrained into the liquid phase in a bubbling bed reactor (BR), resulting in small bubbles with a large gas–liquid interfacial area (Craig 2004). The bubbles collide as they go through the liquid. In a liquid medium, when two bubbles collide, the layer of liquid between them thins out to submicron thickness over time. The film may grow thin enough to break, resulting in bubble coalescence, depending on the balance of surface tension, hydrodynamic, and surface forces. With a mixed liquid phase, the formation of bubbles in gas–liquid dispersions will be hampered. According to the suppression of bubble coalescence approach, tiny particles can change the particular interfacial area and have a considerable impact on the overall mass transfer coefficient (MTC). Though, it is difficult to comprehend the impact on the MTC. Kim (2015) investigated this phenomenon in depth and found out the reduced bubble diameters in nanofluids compared to the pure absorbent (Kim et al. 2014). Wang et al. studied the effective gas–liquid contact area to characterize the bubble breaking effect (Wang et al. 2016). For nanoparticle suspensions, the results demonstrated an increase in the gas–liquid contact area. The role of this phenomenon in the overall enhancement of mass transfer in the nanofluids has yet to be defined, despite multiple studies on the subject.

8.2.2 Shuttle or Grazing Effect

The shuttle effect considers the interactions of gas, liquid, and solid at the same time. Kars et al. (1979) initially proposed the grazing or shuttle effect in 1979. The rate of gas absorption in liquids is increased by solid particles in a three-phase system comprising gas, liquid, and solid (Kluytmans et al. 2003). Gas travels from the surface of a gas–liquid mixture to the liquid bulk in this event. As the name implies, nanoparticles operate as shuttles, boosting the transit of CO_2 gas from the surface of a gas–liquid mixture to the liquid bulk (Brilman et al. 1998). During the diffusion flow, CO_2 is absorbed by nanoparticles having a high adsorption ability for the gas component. After that, it desorbs back into the liquid bulk. Particles then travel back and forth between the higher and lower concentration zones. The mass transfer coefficient could be greatly increased as a result of this. Because of the huge surface area of the particles, even at low nanoparticle concentrations, the shuttle effect is substantial. The coefficient of CO_2 diffusion in the liquid phase is measured for exploration of the shuttle effect.

8.2.3 Boundary Mixing Effect

Figure 8.2c depicts the boundary mixing effect, also called the hydrodynamic effect mechanism (Whitman 1923). The gas–liquid reaction's mass transfer resistance can be classified into three types: gas-film resistance, liquid-film resistance and interfacial resistance, and as indicated in the two-film theory. The boundary mixing effect arises from the fact that nanoparticles aid in re-mixing the mass transfer boundary layer, resulting in a thinner effective liquid-film (Kim 2015). The significant Brownian motion and micro convection movement of nanoparticles support this hypothesis. The contact between the particles and the interface causes a decrease in the diffusion layer thickness. Kim et al. investigated migration of particles in Nano-fluids and discovered that particle migration caused additional thermal eddy currents, potentially reducing the conduction sublayer (Kluytmans et al. 2003).

It is important to notice that CO_2 absorption rate enhancement isn't caused by only one mechanism, but rather by a combination of two or more mechanisms. The enhanced CO_2 absorption was attributed by Jung et al. Both the bubble breaking and boundary mixing effects are affected by this (Jung 2012). At a low CO_2 flow rate, the bubble breaking effect, according to Wang et al., contributed more to the CO_2 absorption rate than the other two mechanisms (Jiang et al. 2014). At the same time, the interaction of nanoparticles, gas, and liquid is likely to be dominated by these three mechanisms. The contributions of each phase, however, are still up for further research.

8.3 Enhancement of CO_2 Absorption Rate Using Nano-Fluids

To improve CO_2 absorption, various metallic, nonmetallic, and metal oxide, nanoparticles have been examined, including Al_2O_3, SiO_2, TiO_2, MgO, CuO, Cu, Ni, Fe_3O_4, and CNTs nanoparticles (Rahmatmand et al. 2016). When compared to the result obtained with pure amine solvent, nano-fluids have been demonstrated to enhance CO_2 absorption rates by 2–93% (Suresh and Komati 2008). The enhancement factor (E), which measures the increase in CO_2 absorption rate, has been used to analyse and evaluate the impacts of various nanoparticles. It's the ratio of CO_2 absorption in Nano-fluids to CO_2 absorption in base fluid. An equilibrium cell (EC), a wetted-wall column and a bubbling rector (BR), are commonly used to determine the enhancement factor (E) (WWC). In this section we have discussed the numerous research works that had been done regarding Nano-fluids application for CO_2 capture.

Seo et al. (2020) evaluated the catalytic efficiency of nickel nanoparticles (NiNPs) for CO_2 capture in Monoethanolamine (MEA) solvent as potential activator. The rate of CO_2 absorption in Monoethanolamine (MEA) has been examined in the presence of nickel nanoparticles (NiNPs) in two different mixing conditions (i.e. high mixing and limited mixing) to simulate actual capture processes in the packed

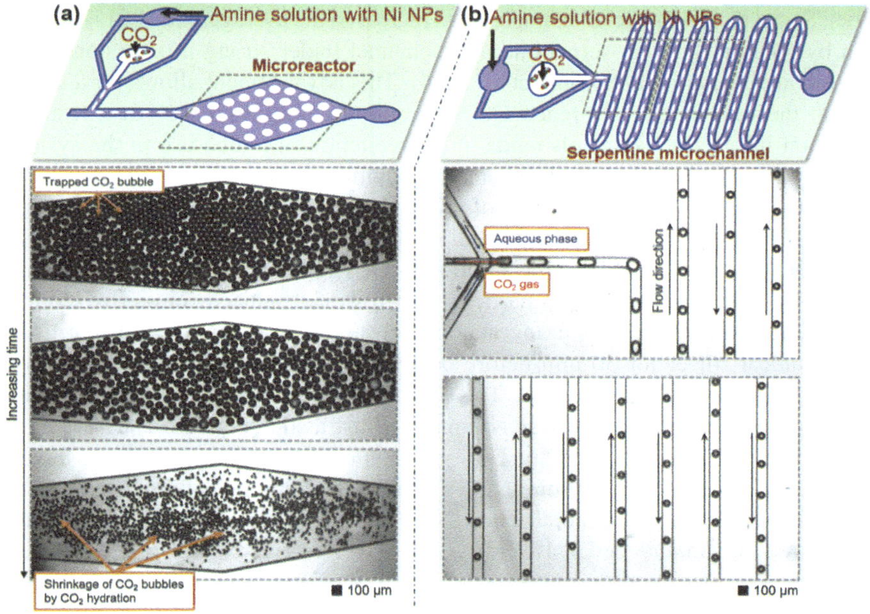

Fig. 8.4 Schematic diagram of evaluation process for CO_2 absorption in the microfluidic approaches. **a** Design of a microreactor. **b** Design of a long serpentine microchannel and images of CO_2 microbubbles at two different locations. Reproduced from Seo et al. (2020)

column reactors for industrial CO_2 absorption process. Because of this study, they have employed a micro-reactor and a long serpentine micro-channel (Zheng et al. 2019). The schematic diagram of the two test reactors and associated methodologies for evaluating CO_2 collection capability are shown in Fig. 8.4. A micro-reactor for limited mixing conditions (Fig. 4a) and for high mixing (Fig. 4b), a long serpentine microchannel is used. In all mixing settings, a flow concentrating shape with two inlets (a continuous phase on both side channels) and one exit was used to generate a sequence of CO_2 microbubbles (a discrete phase in the centre channel). Using CO_2 microbubbles in aqueous MEA, the catalytic absorption performance of NiNPs for CO_2 is assessed, which are monitored for size changes as they absorb CO_2. The concentration of MEA in water solution was varied from 1.25 to 10% in all experiments, keeping the NiNPs concentration constant at 30 mg/l (Seo et al. 2017). To improve the stability of CO_2 bubbles, 625 mg/L sodium dodecyl sulphate (SDS) was used: SDS significantly minimises the coalescence of bubbles caused by collision and bubble explosion (Xu et al. 2009). For all of the experiments, SDS concentration was maintained at a steady level.

In the microreactor's limited condition, the average rate of CO_2 absorption using Nickel nanoparticles is increased by 34%, according to the researchers. This enhancement is primarily due to Brownian motion-driven catalytic CO_2 absorption (Noyes

1996; Seo et al. 2018).Nickel nanoparticles increased the average rate of CO_2 absorption by 54% in the long serpentine microchannel under strong mixing conditions (Suresh and Komati 2008; Wang et al. 2014). This advancement allows for a reduction in the time it takes to reach CO_2 absorption equilibrium and, as a result, a reduction in reactor size. It also aids in the reduction of amine use in the system, which is good in reducing environmental difficulties.

Yu et al. (2016) studied the effects of SiO_2, Al_2O_3, TiO_2, and CuO nanoparticles in three different amine solvents (30% MDEA, 30% MEA, and 1.5 mol/l PZ) (Yu et al. 2016). It has been observed that the liquid-side mass transfer coefficient increases as a result of Brownian movement of particle and Brownian-induced convective motion. Nano-fluids were made by mixing nanoparticles with a base amine solvent using a mechanical stirrer for 30 min before being dispersed separately using a Microfluidizer high shear fluid processor. The CO_2 absorption rates of several nano-fluids were measured using a wetted-wall column, as illustrated in Fig. 8.5. The absorption experiment was done at 40 °C, the CO_2 gas flow rate was 3 L/min, and the solvent flow rate was kept at 128 mL/min in the wetted-wall column, resulting in a laminar, ripple-free flow (Fang et al. 2009; Zhu et al. 2020). In the mixed gas, the CO_2 inflow concentration is maintained at 10%.

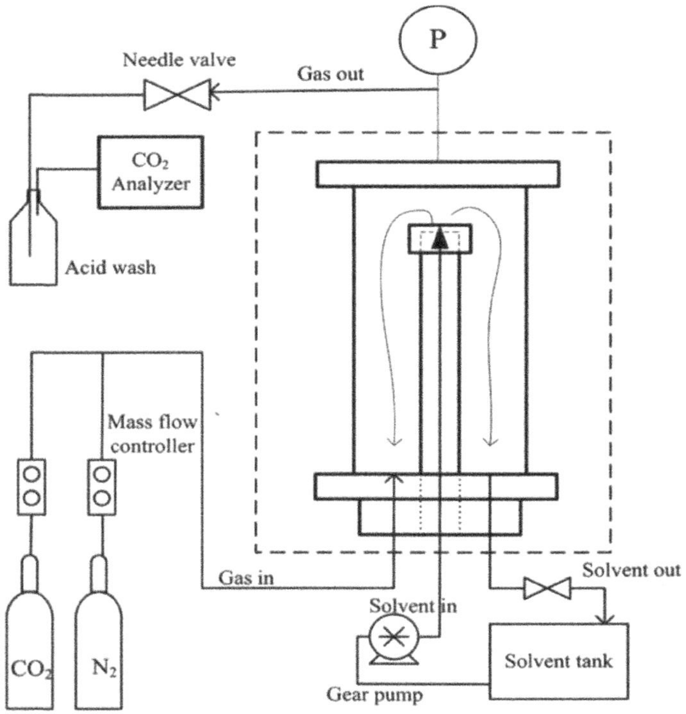

Fig. 8.5 Experimental setup for CO_2 absorption (Wetted-wall column). Reproduced from Yu et al. (2016)

The enhancing effects of different nanoparticles are diverse. The improved effect for these nanoparticles is in the following order: $TiO_2 > SiO_2 > CuO > Al_2O_3$. $TiO_2 > SiO_2 > CuO > Al_2O_3$ (Jiang et al. 2014; Salimi et al. 2015).This enhanced effect is largely due to two methods. On the one hand, the shuttle effect will aid CO_2 absorption for nanoparticles with strong CO_2 adsorption, such as TiO_2 nanoparticles, where the particles carry more CO_2 across the gas and liquid interface through adsorption and desorption. While, TiO_2 and SiO_2, have stronger inherent hydrodynamic properties than CuO and Al_2O_3, at the same mass fraction, this results in less coagulation of nanoparticles. Figures 8.6 and 8.7 shows the effect of SiO_2 and TiO_2 nanoparticles with different amine solvents on Enhancement factor (E) respectively.

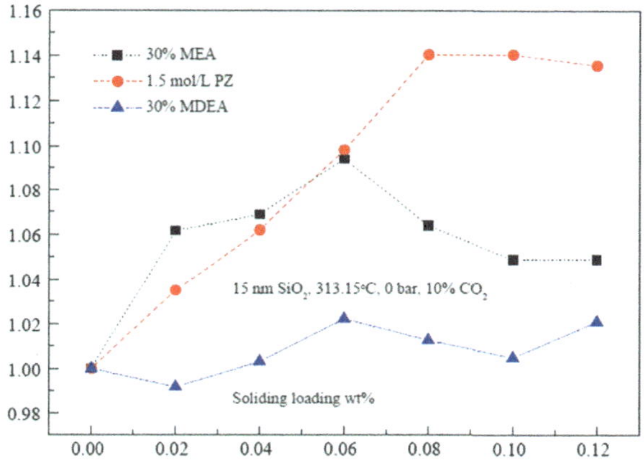

Fig. 8.6 Change in the CO_2 absorption enhancement factor with solid loading of SiO_2 nanoparticles in MEA, MDEA and PZ nanofluids Reproduced from Yu et al. (2016)

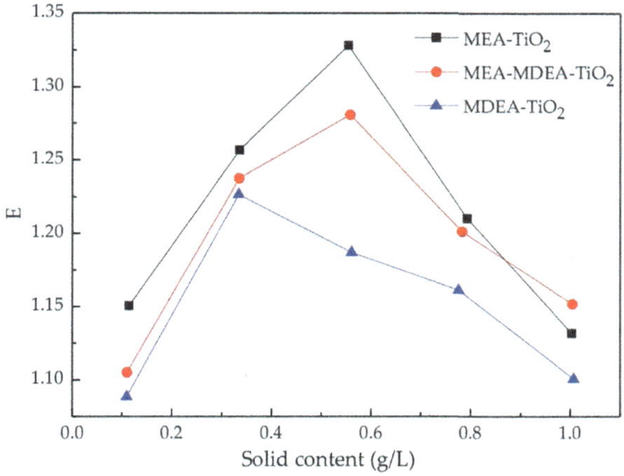

Fig. 8.7 Variations in enhancement factor in TiO_2 nanoparticles along with different amines. Reproduced from Yu et al. (2016)

When compared to simple nano-fluids, nanoferro-fluids can have a more significant enhancing effect on CO_2 absorption rate when exposed to external magnetic fields (Salimi et al. 2015). The effect of Fe_3O_4 nanoparticles in MDEA solvent on CO_2 absorption rate was investigated by Komati et al. (2008) They reported a 92.8% increase in enhancement factor (E) for a fluid volume fraction of around 51 wt% MDEA and a Fe_3O_4 volume fraction of about 0.39% (Suresh and Komati 2008). Although it varied depending on the field source, the enhancement factor increased as the external magnetic field intensity increased. The rate of CO_2 absorption was higher with an alternative current magnetic field that alternated direction 50–60 times per second than with a direct current magnetic field. The increase in CO_2 absorption rate is explained by the increased local velocity surrounding the particles. It is generally created by the quicker movement of magnetic nanoparticles driven by an external magnetic field. Arshadi et al. (2019) investigated the effect of anchoring inorganic and organic compounds with amine-groups onto modified Fe_3O_4 nanoparticles for CO_2 absorption utilising water as a base solvent.

Rahmatmand et al. (2016) compared how different nanofluids improved CO_2 absorption (Al_2O_3, CNT, Fe_3O_4, SiO_2). At a higher concentration of 0.1 wt%, Al_2O_3 and SiO_2 nanoparticles showed higher enhancement in CO_2 absorption rate, whereas CNT and Fe_3O_4 nanofluids showed superior enhancement in CO_2 absorption rate at a low concentration of 0.02%. CNT nanoparticles were also shown to be more effective in MDEA than DEA solutions, with the CO_2 adsorption rate increases by up to 23%. Darabi (2017) studied the CO_2 absorption rate in a hollow fibre membrane using SiO_2/water and CNT/water solutions. CNT nanoparticles absorbed CO_2 at a rate that was 16% higher than SiO_2 nanoparticles. This is attributed to the CNT nanoparticles' superior adsorption capabilities when compared to SiO_2 nanoparticles. According to Lu et al. (2013), both Al_2O_3 and CNT nanoparticles improve CO_2 absorption rates, with CNT outperforming Al_2O_3. Apart from the experiments mentioned above, nanofluids containing metallic, metal oxide, and nonmetallic nanoparticles are also employed to capture CO_2. Table 8.1 shows the results of employing different nanofluids to improve CO_2 absorption.

8.4 Factors Affecting the CO_2 Absorption

8.4.1 Solid Loading of Nanoparticles

At the gas–liquid interface, fine particles absorb gas components, travel through the concentration layer, and desorb gases in the liquid bulk, according to Kars and Alper's "Shuttle Mechanism". Because more particles can take up more gases, they have a bigger impact on CO_2 absorption. At 40 °C and atmospheric pressure, Kim et al. (2008) and Yu et al. (2016) studied the CO_2 absorption using 15 nm SiO_2 and 20 nm Al_2O_3 nanoparticles in 30% MEA solutions. Two nanoparticles, as illustrated in Fig. 8.8, have a similar enhancing effect on CO_2 mass transfer. For 15 nm

Table 8.1 Use of different nanoparticles and base fluids for CO_2 absorption rate enhancement

Nanoparticles	Concentration	Base fluid	Enhancement (%)	References
Ni	30 mg/l	MEA	34–54	Seo et al. (2020)
TiO_2	60 mg/l	MEA	33	Basavaraj Devakki (2020)
TiO_2	35 mg/l	MDEA	22	Basavaraj Devakki (2020)
SiO_2	0.08 wt%	PZ	15	Yu et al. 2016)
SiO_2	0.06 wt%	MEA	10	Yu et al. 2016)
Al_2O_3	0.02 wt%	MEA	8	Yu et al. (2016)
MgO	0.08 wt%	MEA	10	Jiang et al. (2014)
Al_2O_3	0.05 wt%	DEA	40	Taheri (2016)
CNT	0.02 wt%	MDEA	23	Rahmatmand et al. (2016)
CNT	0.05 wt%	Water	32	Rahmatmand et al. (2016)
Fe_3O_4	0.39 vol%	MDEA	92.8	Suresh and Komati (2008)
Fe_3O_4	0.02 wt%	MDEA	15	Nabipour et al. (2010)
Fe_3O_4	0.1 wt%	Ammonia	8	Wu (2013)
UiO-66-NH2	0.1 wt%	MDEA	10	Vahidi (2016)
SiO_2	0.05 wt%	DEA	33	Taheri (2016)

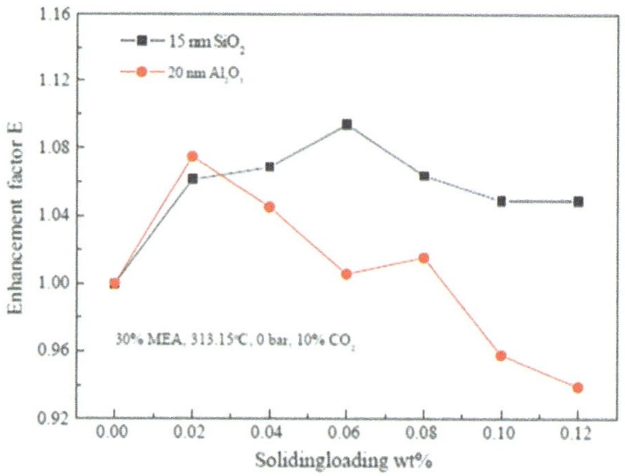

Fig. 8.8 Effects of SiO_2, Al_2O_3 solid loading on the Enhancement factor (E). Reproduced from Yu et al. (2016)

SiO_2 nanoparticles, first increases to the maximum level and then decreases, with the optimum solid loading of 0.06% and the enhancement factor (E) of 1.09% respectively. The best solid loading value corresponding to 20 nm Al_2O_3 nanoparticles has been estimated to be 0.02%, with a 1.08% enhancement factor. The addition of nanoparticles had a detrimental impact on the Enhancement factor when the solid loading was greater than 0.08%. The nanoparticles in the liquid phase cause brownian and convective motion, which improves turbulence between the gas–liquid interface and increases mass diffusivity across the boundary. Brownian motion intensity is determined by a variety of elements such as nanoparticle size, fluid viscosity, temperature, and so on. On the other hand, when nanoparticle concentrations are too high, the probability of particles colliding and accumulating with one another increases, resulting in diluted Brownian and convective motion in the liquid phase.

8.4.2 Size of Nanoparticles

Kim et al. (2008) analyzed the influence of nanoparticle size (30, 70, and 120 nm) on the CO_2 absorption process, however no significant difference has been noticed. It is inferred that even at varied nanoparticle sizes, bubbles fractured to the same size in the same content of nanofluids. Ali et al. (2013). The thermal conductivity of Al_2O_3 nano-fluids was shown to be improved by increasing the nanoparticle size from 11 to 63 nm. It grew in a nonlinear manner as particle size increases. Nagy and Koroknai (2007) explored the impact of brownian particle motion and nanoparticle size on the thermal performance of a micro-channel heat sink (MCHS). Nagy et al. compared and contrasted two mathematical models, one was a pseudo-homogeneous model and the other was a heterogeneous model. With various nanoparticle sizes, the mass transfer rate can be improved, (Nagy and Koroknai 2007). In the case of nanoparticles with small diameters, the increase in gas–liquid mass transfer rate was shown to be larger. However, research on the effect of nanoparticle size on CO_2 absorption is limited. In this particular area, more research is required.

8.4.3 Temperature Conditions

Temperature also plays a pivotal role in the improvement of CO_2 absorption by Nano-fluids. (Olle et al. 2006). The investigation reveals that the enhancement factor first increased and then decreased as the temperature increases. It has been observed that the peak point is achieved at the most suitable temperature and it has the highest enhancement factor. At 10 °C, the CO_2 absorption enhancement of Al_2O_3/NaCl nanofluids was lower than at 20 and 30 °C, according to Lee et al. (2013). The enhancement in the particle motion at higher temperatures, results in the absorption enhancement at 30 °C. This was notably true when the Al_2O_3 concentration was 0.005 vol%.

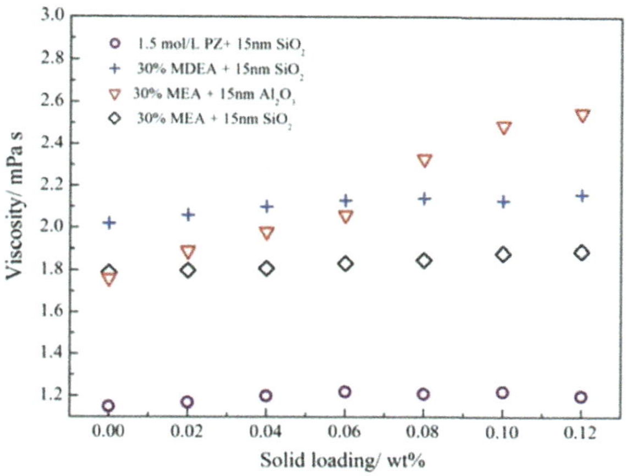

Fig. 8.9 Viscosities of various amine based nanofluids (shear rate = 158 s^{-1}). Reproduced from Yu et al. (2016)

8.4.4 Viscosity of Nanofluids

The solution viscosity is critical in the gas and liquid mass transfer process, particularly in wetted-wall columns. The viscosities of nanoparticle suspensions were determined using a Brookfield viscometer by Yu et al. (2016). Figure 8.9 depicts the viscosities corresponding to different solid loadings of nanoparticles. As demonstrated in Fig. 8.9, the viscosity of all nanoparticles increases as the particle solid loading increases, particularly for Al_2O_3 nanoparticles, obstructing Brownian and convective motion and lowering the Enhancement factor (E). As a result, we can deduce that increasing the viscosity of the solvent has a detrimental impact on CO_2 absorption.

8.4.5 Effects of Other Factors

Other potential factors that influences the enhancement factor (E) for the CO_2 absorption process includes the initial CO_2 concentration, gas and liquid flow rates, and stirrer speed. Increased Brownian motion generated by increased turbulence in the fluid or increased driving force of the temperature or species gradient are all beneficial for boosting the enhancement factor. This is generally caused by increased CO_2 gas concentrations, higher liquid flow rates, and slower CO_2 gas flow rates (Seyf 2012; Samadi 2014; Wei Yu 2019). Hassan and Pashaei (2020) investigated the influence of stirrer speed on the enhancement factor. The results clearly show that the enhancement factor was related to stirrer speed. The enhancement factor increases with increased stirring speed at first, then decreases with increasing stirrer speed, according to the study. The growing enhancement factor phenomena is most likely explained by the numerous mechanisms outlined previously, such as the bubble bursting effect and the hydrodynamic effect, which resulted in an increase in contact area and prevented nanoparticle aggregation and settlement.

8.5 Effect of Nanoparticles on CO_2 Desorption and Solvent Regeneration

Nanoparticles have shown to offer a lot of potential for accelerating CO_2 desorption. CO_2 desorption augmentation by nanoparticles has been the subject of only a few research. Using the experimental setup illustrated in Fig. 8.10, Yu et al. (2016) studied the effects of CuO, SiO_2, TiO_2, and Al_2O_3 nanoparticles on CO_2 desorption of 30 wt% MEA solution. The addition of nanoparticles resulted in a considerable increase in CO_2 desorption rate of over 10%, which they attributed to bubble bursting effects. The desorption ability of MEA solvent with TiO_2 nanoparticles is the best, followed by CuO, SiO_2, and Al_2O_3. The addition of Al_2O_3 nanoparticles, on the other hand, reduces the regeneration extent from 70.8 to 66.6% during 75 min of regeneration, whereas the other three nanoparticles increase the solvent regeneration extent from 70.8 to roughly 75%. In comparison to the pure amine solvent, TiO_2 nanoparticles with MEA solvent reduced desorption time by more than 40% (Fig. 8.11).

By delivering more nucleus cores, nanoparticles will improve the formation of CO_2 bubbles. Furthermore, nanofluids' increased thermal conductivity improves the heat transfer process of solvent regeneration, resulting in great heating efficiency. in the industrial scale, the higher CO_2 desorption rate of these nanofluids will help to reduce solvent regeneration energy, size of the stripping column and heat exchangers,

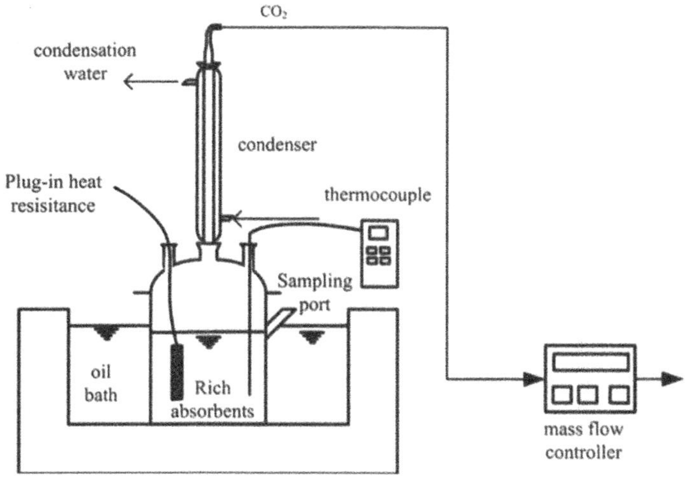

Fig. 8.10 Experimental setup for desorption study. Reproduced from Yu et al. (2016)

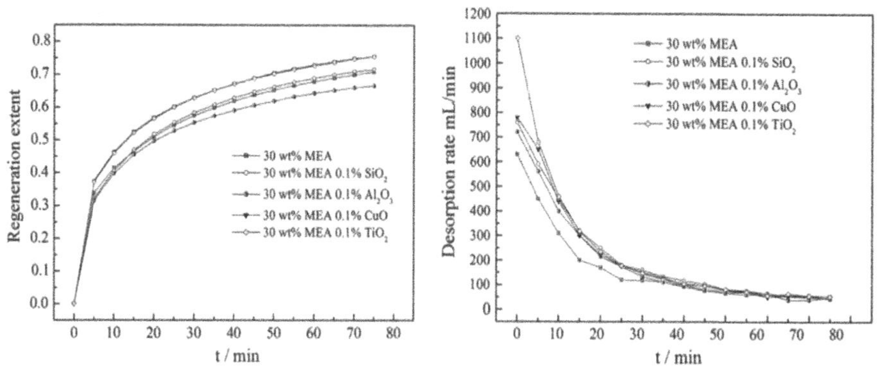

Fig. 8.11 Effects of nanoparticles on regeneration extent and desorption rate of 30% MEA. Reproduced from Yu et al. (2016)

the plant's capital expenditure. The three most essential mechanisms involved in the desorption process by Nano-fluids are the activation energy effect, heat effect, and surface effect. Brownian motion of nanoparticles in the base fluid causes the activation energy effect, which causes more CO_2 bubbles to collide, resulting in more CO_2 being expelled from the rich liquid phase. The thermal effect in increasing CO_2 desorption is linked to an increase in the thermal conductivity of the base fluid. The surface effect, on the other hand, assumes that nanoparticles are deposited on the bubble generation surface during the solvent regeneration process. The nanoparticles also enhance the surface area and roughness, facilitating the desorption of CO_2 bubbles from the surface.

8.6 Conclusion

In recent years, there are growing concerns regarding excessive carbon dioxide (CO_2) emissions that causes the global warming problem. To counter this problem, energy efficient and eco-friendly CO_2 reduction is so important. Use of nanofluids as a potential solvent for carbon capture can significantly enhance the CO_2 absorption rate. This chapter of the book provides an overview of CO_2 absorption using various amine-based Nanofluids. Various methods for preparing amine based nanofluids are discussed thoroughly. Different enhancement mechanisms of grazing effect, The hydrodynamic effects of bubble breaking and boundary mixing are discussed in depth. Future research should concentrate on convective heat transfer and mass transfer experiments, as well as the development of comprehensive models for studying and evaluating results and investigating possible causes. Amine-based nanofluid are determined to be suitable for CO_2 capture from flue gases based on a comprehensive literature study, and It has the ability to improve the conventional amine technologies without affecting the current physical infrastructure of the CO_2

chemical absorption process. Solvents containing catalytic nanoparticles, in particular also shows a greater potential for reducing the solvent regeneration energy. In this chapter, many factors impacting CO_2 absorption rates are thoroughly explored, including nanoparticle concentration and size, gas concentration, nanofluid viscosity, temperature and gas–liquid flow rates.

References

Ali FM (2013) Study of the effect of particles size and volume fraction concentration on the thermal conductivity and thermal diffusivity of Al_2O_3 nanofluids. Int J Phys Sci 9:514–518

Arshadi M, Taghvaei H, Abdolmaleki MK, Lee M, Eskandarloo H, Abbaspourrad A (2019) Carbon dioxide absorption in water/nanofluidby a symmetric amine-based nanodendritic adsorbent. Appl Energy 242:1562–1572

Basavaraj Devakki ST (2020) Experimental investigation on absorption performance of nanofuids for CO_2 capture. Int J Air Conditioning Refrig 28(11)

Brilman DWF, Van Swaaji WPM, Versteeg GH (1998) A one-dimensional instationary heterogeneous mass transfer model for gas absorption in multiphase systems. Chem Eng Process Process Intensif 37:471–488

Brunetti A, Scura F, Barbiere G, Drioli E (2010) Membrane technologies for CO_2. J Memb Sci 359(1–2):115–125

Craig VSJ (2004) Bubble coalescence and specific-ion effects. Curr Opin Colloid Interphase Sci 9:178–184

Darabi M (2017) Gas absorption enhancement in hollow fiber membrane contactors using nanofluids: modeling and simulation. Chem Eng Process Process Intensif 119:7–15

Fang M, Xiang Q, Yu C, Xia Z, Zhou X, Cai D, Wang Z, Yu H (2009) Experimental study on CO_2 absorption by aqueous ammonia solution at elevated pressure to enhance CO_2 absorption and suppress ammonia vaporization. Greenhouse Gas Sci Technol 5(2):7297–7302

Hassan A, Pashaei G (2020) CO_2 absorption into aqueous diethanolamine solution with nano heavy metal oxide particles using stirrer bubble column: hydrodynamics and mass transfer. J Environ Chem Eng 8:104–110

Jiang J, Zhao B, Zhuo Y (2014) Experimental study of CO_2 absorption in aqueous MEA and MDEA solutions enhanced by nanoparticles. Int J Greenh Gas Control 29:135–141

Jiang JZ, Zhang S, Fu XL, Fu L, Liu L, Sun BM (2019) Review of gas–liquid mass transfer enhancement by nanoparticles from macro to microscopic. Heat Mass Transf 55(8):2061–2072

Jin Hwang Y, Lee JK, Lee JK, Jeong YM, Cheong S, Ahn YC, Kim SH (2008) Production and dispersion stability of nanoparticles in Nanofluids. Powder Technol 186:145–153

Jung JY (2012) CO_2 absorption characteristics of nanoparticle suspensions in methanol. J Mech Sci Technol 26:2285–2290

Kars RL, Best RJ, Drinkenburg AAH (1979) The sorption of propane in slurries of active carbon in water. Chem Eng J 17:201–210

Kenig EY, Vaidya PD (2007) CO_2-alkanolamine reaction kinetics: a review of recent studies. Chem Eng Technol 30:1467–1474

Kim H (2015) Experimental investigation on the flow and convective heat transfer coefficient of nanofluids in turbulent region. Korea Aerospace University

Kim WG, Kang HU, Jung KM, Kim SH (2008) Synthesis of silica nanofluid and application to CO_2 absorption. Sep Sci Technol 43:3036–3055

Kim JH, Jung CW, Kang YT (2014) Mass transfer enhancement during CO_2 absorption process in methanol/Al_2O_3 nanofluids. Int J Heat Mass Transf 76:484–491

Kluytmans JH, Van Wachem BG, Kuster BF, Schouten J (2003) Mass transfer in sparged and stirred reactors: influence of carbon particles and electrolyte. Chem Eng Sci 58:4719–4728

Krishnamurthy S, Bhattachary P, Phelan PE (2006) Enhanced mass transport in nanofluids. Nano Lett 6:419–423

Lackner KS (2009) Capture of carbon dioxide from ambient air. Eur Phys J Spec 176:93–106

Lee W (2013) CO_2 absorption enhancement by Al_2O_3 nanoparticles in NaCl aqueous solution. Energy 53:206–211

Lee JS, Lee JW, Kang YT (2015) CO_2 absorption/regeneration enhancement in DI water with suspended nanoparticles for energy conversion application. Appl Energy 143:119–129

Li L, Voice AK, Li H, Namjoshi O, Nguyen T, Du Y, Rochelle GT (2013) Amine blends using concentrated piperazine. Energy Proc 37:353–369

Lindsey R (2020) Climate change: atmospheric carbon dioxide [Online]. Available: https://www.climate.gov/news-features/understanding-climate/climate-change-atmospheric-carbon-dioxide

Liu F, Fang M, Dong W, Wang T, Xia Z, Wang Q, Luo Z (2019) Carbon dioxide absorption in aqueous alkanolamine blends for biphasic solvents screening and evaluation. Appl Energy 233–234:468–477

Lu S, Xing M, Sun Y, Dong X (2013) Experimental and theoretical studies of CO_2 absorption enhancement by nano-Al_2O_3 and carbon nanotube particles. Chin J Chem Eng 21:983–990

Mondal MK, Balsora HK, Varshney P (2012) Progress and trends in CO_2. Energy 46:431–441

Mukherjee S, Paria S (2013) Preparation and stability of nanofluids—a review. IOSR J Mech Civ Eng 9:63–69

Nabipour M, Keshavarz P, Raeissi S (2010) Enquête expérimentale portant sur l'absorption du CO_2 dans les nanofluides Fe_3O_4, MWCNT à base de Sulfinol-M. Int J Refrig 73:1–10

Nagy E, Koroknai B (2007) Enhancement of oxygen mass transfer rate in the presence of nanosized particles. Chem Eng Sci 62:7391–7398

Noyes RM (1996) Transport of carbon dioxide between the Gas Phase and water under well-stirred conditions: rate constants and mass accommodation coefficient. J Phys Chem 100:4167–4172

Olle B, Bucak S, Holmes TC (2006) Enhancement of oxygen mass transfer using functionalized magnetic nanoparticles. Ind Eng Chem Res 45:4355–4363

Park SW, Choi BS, Lee JW (2006) Chemical absorption of carbon dioxide into aqueous colloidal silica solution with diethanolamine. Sep Sci Technol 41:3265–3278

Rahmatmand B, Keshavarz P, Ayatollahi S (2016) Study of absorption enhancement of CO_2 by SiO_2, Al_2O_3, CNT, and Fe_3O_4 nanoparticles in water and amine solutions. J Chem Eng Data 61:1378–1387

Ritchie H, Roser M (2020) CO_2 and greenhouse gas emissions [Online]. Available: https://ourworldindata.org/co2-and-other-greenhouse-gas-emissions

Rochelle GT (2009) Amine scrubbing for CO_2 capture. Science 325:1652–1657

Saba Ashrafmansouri S, Esfahany MN (2014) Mass transfer in nanofluids: a review. Int J Therm Sci 82:84–99

Salimi J, Haghshernasfard M, Etemad SG (2015) CO_2 absorption in nanofluids in a randomly packed column equipped with magnetic field. Heat Mass Transf 51:621–629

Samadi Z (2014) CO_2 absorption using nanofluids in a wetted-wall column with external magnetic field. Chem Eng Technol 37:462–470

Seo S, Nguyen M, Mastiani M, Navarrete G (2017) Microbubbles loaded with nickel nanoparticles: a perspective for carbon sequestration. Anal Chem 89:10827–10833

Seo S, Perez GA, Tewari K, Comas X (2018) Catalytic activity of nickel nanoparticles stabilized by adsorbing polymers for enhanced carbon sequestration. Sci Rep 8

Seo S, Perez GA, Tewari A, Comas X (2020) Catalytic CO_2 absorption in an amine solvent using nickel nanoparticles for post-combustion carbon capture. J CO_2 Utiliz 36:244–252 (2020)

Seyf R (2012) Analysis of Brownian motion and particle size effects on the thermal behavior and cooling performance of microchannel heat sinks. Int J Therm Sci 58:36–44

Suresh AK, Komati S (2008) CO_2 absorption into amine solutions: a novel strategy for intensification based on the addition of ferrofluid. J Chem Technol Biotechnol 83:1094–1100

Taheri M (2016) Simultaneous absorption of carbon dioxide (CO_2) and hydrogen sulfide (H_2S) from CO_2–H_2S–CH_4 gas mixture using amine-based nanofluids in a wetted wall column. J Nat Gas Sci Eng 28:410–417

Tsouris C, Aaron D (2005) Separation of CO_2 from flue gas: a review. Sep Sci 40:321–348

Vahidi M (2016) Preparation of amine functionalized UiO-66, mixing with aqueous N Methyldiethanolamine and application on CO_2 solubility. J Nat Gas Sci Eng 28:651–659

Wang GR, Yang F, Zhao W (2014) There can be turbulence in microfluidics at low Reynolds number. Lab Chip 14:1452–1458

Wang T, Yu W, Liu F, Fang M, Farooq M, Luo Z (2016) Enhanced CO_2 absorption and desorption by monoethanolamine (MEA)-based nanoparticle suspensions. Ind Eng Chem Res 55:7830–7838

Wang M, Lawal A, Stephenson P, Sidders J (2011) Post-combustion CO_2 capture with chemical absorption: a state-of-the-art. Chem Eng Res Des 89:1609–1624

Wei Yu AP (2019) Review of liquid nano-absorbents for enhance CO_2 capture. Nanoscale 11:17137–1156 (2019)

Whitman WG (1923) The two-film theory of gas absorption. Chem Met Eng 29:146–148

Wu WD (2013) Nanoferrofluid addition enhances ammonia/water bubble absorption in an external magnetic field. Energy Build 57:268–277

Xu Q, Nakajima M, Ichikawa S, Nakamura N, Roy P (2009) Effects of surfactant and electrolyte concentrations on bubble formation and stabilization. J Colloid Interface Sci 332:208–214

Xu G, Liang F, Yang Y, Hu Y (2014) An improved CO_2 separation and purification system based on cryogenic separation and distillation theory. Energies 7:3484–3502

Yu W, Wang T, Fang MX, Hei H, Luo ZY (2016) CO_2 absorption/desorption enhanced by nanoparticles in post-combustion CO_2 capture. In: Clean coal technology and sustainable development. Springer, Singapore, pp 591–596

Zheng L, Gaozhe C, Wang S, Liao M, Li Y, Lin J (2019) A microfluidic colorimetric biosensor for rapid detection of Escherichia coli O157:H7 using gold nanoparticle aggregation and smart phone imaging. Biosens Bioelectron 124–125

Zhu D, Fang M, Lv Z, Wang Z, Luo Z (2012) Selection of blended solvents for CO_2 absorption from coal-fired flue gas. Part 1: monoethanolamine (MEA)-based solvents. Energy Fuels 26:147–153

Zhu D, Fang M, Lv Z, Wang Z (2020) Selection of blended solvent for CO_2 absorption from coal fired flue gas. Part 1: monoethanolamine (MEA)—based solvent. Energy Fuels 26(1):147–153

Chapter 9
Rare Earth Element Based Functionalized Electrocatalysts in Overall Water Splitting Reactions

Devidas S. Bhagat, Gurvinder S. Bumbrah, and Wasudeo B. Gurnule

Abstract There are numerous environmental and human health-related challenges with current carbon-based fossil fuel resources and significant in the nation's growing-prosperity. To overcome the potential effects of carbon-based fossil fuel emission scientist community focused on implementing and developing green and clean energy resources like hydrogen energy. Overall, water splitting reactions are future sources of clean hydrogen energy include hydrogen evolution reactions and oxygen evolution reactions. Bifuctionalized nanocomposite employed in overall water reactions such transition metal-based, carbon nanotube, doped graphene. These all-in-one bi-functionalized electrocatalysts have great demand in the industrial market due to their unique bi-functionalized nature, minimizing considerable production costs. The efficacy of electrocatalysts depends upon intrinsic properties include selectivity, durability, renewability, pH, and stability. Herein, we emphases on advances in novel approaches include; synthesis, the electrocatalytic study reported nanocomposite for efficient overall water splitting reactions for generation of the hydrogen and oxygen gasses.

Keywords Hydrogen energy · Water splitting · Rare earth elements · Electrocatalysts · Hydrogen evolution reactions · Oxygen evolution reaction

D. S. Bhagat (✉)
Department of Forensic Chemistry and Toxicology, Government Institute of Forensic Science, Aurangabad, Maharashtra 431 004, India
e-mail: devidas.bhagat@gov.in

G. S. Bumbrah
Department of Chemistry, Biochemistry and Forensic Science, Amity School of Applied Sciences, Amity University, Haryana 122413, India

W. B. Gurnule (✉)
Department of Chemistry, Kamla Nehru Mahavidyalaya, Nagpur, Maharashtra 440 024, India

S. P. Singh et al. (eds.), *Metal Nanocomposites for Energy and Environmental Applications*, Energy, Environment, and Sustainability,
https://doi.org/10.1007/978-981-16-8599-6_9

9.1 Introduction

Carbon-based fossil fuels are valuable sources of energy. They are very commonly used for different applications all over the globe due to their abundance, easy availability, cost-effectiveness, easy transportation, high calorific value, and mass production at low prices (Altawell 2021). However, these fuels are non-renewable sources and contribute to greenhouse gases (especially carbon dioxide) and pollute air and water. These carbon-based fossil fuels significantly contribute to climate change or global warming. In addition to this, serious health-related complications such as chronic asthma, chronic bronchitis, abnormal lung functioning, and cardiovascular diseases have been reported due to the generation of a large number of toxic gases from excessive use of these carbon-based fossil fuels (Inumaru et al. 2021). To overcome the problems raised due to the excessive use of carbon-based fossil fuels, the scientific community is now focused on alternative energy options like renewable energy sources (Haj et al. 2021). Their primary focus is to develop and implement green and clean energy sources for different applications. These green, renewable energy sources include biodiesel, geothermal, wind, solar, and hydropower (Rosen et al. 2021).

Hydrogen energy involves the use of hydrogen and/or hydrogen-containing compounds to generate energy to be supplied to all practical uses required with high energy efficiency, overwhelming environmental and social benefits, as well as economic competitiveness (Acar et al. 2018). Hydrogen energy is widely utilized in sectors including energy production, storage, and distribution; electricity, heat, and cooling for buildings and households; the industry; transportation; and the fabrication of feedstock on a large scale in the world. Its widespread use is due to its renewable sources, low cost, high efficiency, less contaminated emissions, sustainability without affecting the environment and human beings. Hydrogen energy provides clean, reliable, affordable, and efficient energy systems to sustain the future (Miranda and Miranda 2019).

Hydrogen is not a primary energy resource as fossil fuels. It is an energy carrier. Therefore, it first needs to be isolated from hydrogen-rich material sources. Renewable energy, nuclear, biomass, and fossil fuel-based sources can be used to produce hydrogen (Armaroli and Balzani 2007). However, renewable energy sources are considered as ideal clean and replenishing primary energy sources to produce hydrogen due to their renewable nature, cost-effectiveness, and less harm to the environment and humans (Dresselhaus and Thomas 2001).

Water is an ideal, rich source of hydrogen and can be frequently used to produce hydrogen at a large scale due to its abundance in nature. It is generally produced by electrolysis of water by electrical and/or thermal methods. Water electrolysis is the most commonly used method to isolate hydrogen from water. In this method, an electric current is used to generate gaseous hydrogen and oxygen from liquid water. The application of electric current causes lysis of bonds among the oxygen and hydrogen molecules (Morales-Guio et al. 2014). The water on oxidation gives oxygen gas with the generation of the proton and electron in oxidation half-cell whereas generation of the hydrogen gas by gaining of an electron by proton (H^+). Various energy sources (e.g. combustion, fossil fuels, etc.) can be used to supply electric current for the splitting process (Rajakaruna et al. 2020). However, these energy sources damage the environment, and the efficiency of the process is also less. Therefore, clean, renewable, and affordable energy sources should be used to provide electric current for the electrolysis process (Wang et al. 2015).

Water splitting is the process of catalytic conversion of liquid water to gaseous oxygen and hydrogen. It is used as a benchmark to evaluate different electrocatalysts. The overall reaction of water splitting is (Eq. (9.1)) (Das and Ganguli 2018):

$$H_2O\,(l) \rightleftharpoons H_2(g) + 1/2O_2(g)\ (DG > 0) \tag{9.1}$$

Water splitting is a non-spontaneous process. It is a two-electron process. The minimum potential required to split one water molecule is 1.23 V while the minimum energy required to split one water molecule is 2.46 eV. Most electrocatalysts require a potential higher than 1.23 V to overcome the electro-kinetic barrier of the electrode. Decomposition potential decreases with increasing temperature according to the Gibbs–Helmholtz equation. The high voltage showed an adverse effect on electrolysis. Electrochemical water splitting involves two half-reactions: hydrogen evolution reaction (HER) on the cathode and oxygen evolution reaction (OER) on the anode. It is mentioned in Eqs. (9.2) and (9.3) (Deng et al. 2018).

$$2H^+ + 2e^- \rightleftharpoons H_2(g)\ (\text{water reduction}) \tag{9.2}$$

$$2H_2O\,(l) \rightleftharpoons 4H^+ + 4e^- + O_2(g)\ (\text{water oxidation}) \tag{9.3}$$

In reduction, half-cell proton gains two-electron whereas in oxidation half-cell generation of the four electrons during reaction course. The overall water splitting reactions type of the energy-driven process. This energy may be derived from the photocatalytic process or electrochemical process or a combination of both photoelectrochemical PEC processes (Souleymen et al. 2018).

9.2 Need of Electrocatalysts

The overall water splitting reactions includes oxygen evolution reaction and hydrogen evolution reactions. Hydrogen is the source of clean energy which facilitated sustainable development (Takanabe 2021). Over potential is a measure of kinetic energy barriers and plays a crucial role in HER and OER during water splitting. Electrocatalysts play a major role in both HER and OER in water-splitting reactions (Liu et al. 2021).

An electro-catalyst is a catalyst that increases the rate of oxidation and reduction reactions in an electrochemical cell. It can be heterogeneous or homogeneous.

Heterogeneous electrocatalysts include platinum nanoparticles while homogeneous electrocatalysts include a coordination complex of the enzyme (Nemiwal et al. 2021). Electrocatalysts are used to sustain and complete these half-cell reactions in an efficient manner. It is used in overcoming kinetic energy barriers for electrochemical reactions of water, hydrogen, and oxygen in water splitting cells and fuel cells (Nguyen 2021). Highly effective, efficient, and stable electrocatalysts are required to minimize the overpotentials for HER and OER towards efficient production of oxygen and hydrogen (Yang and Moon 2019).

The design of electrocatalysts depends on the operating conditions of water electrolysis. Proton exchange membrane (PEM) electrolysis, alkaline electrolysis, and high-temperature solid oxide water electrolysis are three main types of electrolysis technologies. Due to high temperature, solid oxide water electrolysis requires high energy consumption. In PEM electrolysis, water splitting is performed using PEM under acidic conditions (Ros et al. 2020). It causes a rise in proton conductivity and a decrease in gas permeability. This approach produces hydrogen in a fast and efficient manner. However, the presence of an acidic medium limits the OER electrocatalysts to noble metal and noble metal oxide catalysts. In alkaline electrolysis, water splitting is performed under alkaline conditions and thereby broadens the selection of electrocatalysts to non-noble metals or metal oxides. The activity of HER in an alkaline medium is usually 2–3 orders of magnitude lower than the activity of HER in an acidic medium. OER catalysts (Iridium and Ruthenium) show a high dissolution resistance in the acidic medium (Mu et al. 2019).

The water-splitting reaction by novel-metal-based catalysts is significantly challenging due to their high cost and shortage of noble metals for utilizations in large-scale electrolyzes. The noble metals are mostly showed dissolution, agglomeration, and poor tolerance which causes poisoning during electrocatalysts. However, non-noble metal-based electrocatalysts such as carbides, phosphides cannot survive in an acidic medium. These non-noble metal-based electrocatalysts show high activity for OER in alkaline medium. These electrocatalysts undergo a compositional and structural transformation during the OER condition (Wang et al. 2021).

Electrolytic water splitting is an uphill reaction. Gibbs's free energy value for water splitting is positive. To accomplish water splitting, it must overcome a significant kinetic barrier. Electrocatalysts play an important role in lowering the kinetic barrier. The performance index is used to evaluate the activity, stability, and efficiency of electrocatalysts (Bessarabov et al. 2018). The activity is characterized by overpotential, Tafel slope, and exchange current density which can be isolated from polarization curves. The Tafel is the type of equation that showed the relation between electrochemical kinetics rate relate to electrochemical reaction due to overpotential (Maeda et al. 2013). Stability is characterized by the changes of overpotential or current over time. The efficiency is characterized by the faradaic efficiency and turnover frequency in terms of experimental results versus theoretical predictions (Mu et al. 2019). Optimal electrocatalysts for different mediums with high catalytic activity, cost-effectiveness, durability, and stability for electrolytic water splitting should be developed.

9.3 Rare Earth Element Based Overall All Water-Splitting Reaction

The rare earth elements possess unique structural properties with redox pairs, abundant vacancies due to unoccupied orbital, excellent oxygen storage ability, and effective chemical and mechanical stability to promote the water-splitting reaction, hydrogen evolution reaction, and oxygen evolution reaction. Doping of rare earth elements in catalysts increases the catalytical activity. Lin et al. (2020) has been demonstrated LaOCl-coupled polymer carbon nitride-based overall water splitting via a one-photon excitation pathway. The one-photon excitation pathway includes coupling polymeric carbon nitride (PCN) semiconductor with LaOCl in overall water splitting reaction. LaOCl combined with PCN acts as photocatalysts in artificial photosynthesis which decomposes and split H_2O into H_2 and O_2, with evolution rates of 22.3 and 10.7 μmol/h respectively. LaOCl and PCN performed simultaneously oxidation and reduction reactions. In reaction mechanism of oxidation and reduction induced fast charge separation and ions migration under the influence of the electrical field. The synthesis of the LaOCl based PCN from radial amiable starting material La_2O_3 and melamine is a temperature-dependent reaction which describes systematic and mechanistically in Fig. 9.1. The LaOCl based PCN probe was obtained by molten-salt mediated process with lanthanum oxide (La_2O_3), melamine, lithium chloride, and potassium chloride. LiCl–KCl acts as the medium for the reaction, it decreases melting point and helps to the formation of the homogenous mixture. Solid PCN + LaOCl composite acts as fast inter-particle charge separation properties by a one-photon excitation pathway and showed photocatalytic activities toward overall water splitting to generate hydrogen and oxygen (Lin et al. 2020).

Rodney et al. (2021) have been designed novel nanocomposite dysprosium doped copper oxide ($Cu_{1-x}Dy_xO$) nanoparticles for both oxygen evolution reaction (OER) and hydrogen evolution reaction (HER). $Cu_{0.98}Dy_{0.02}O$ nanoparticles synthesized by solution combustion technique. Dopping of dysprosium into copper oxide improved surface to volume ratio and conductivity biofunctionalized electrocatalysts showed impressive HER and OER 1 M KOH solution delivering a current density of 10 mA cm^{-2} at a potential of −0.18 V versus RHE for HER and 1.53 V versus RHE for OER. Dy doped CuO electrocatalysts act as a bi-functional catalyst for overall water splitting achieved a potential of 1.56 V at a current density of 10 mA cm^{-2} (Rodney et al. 2021).

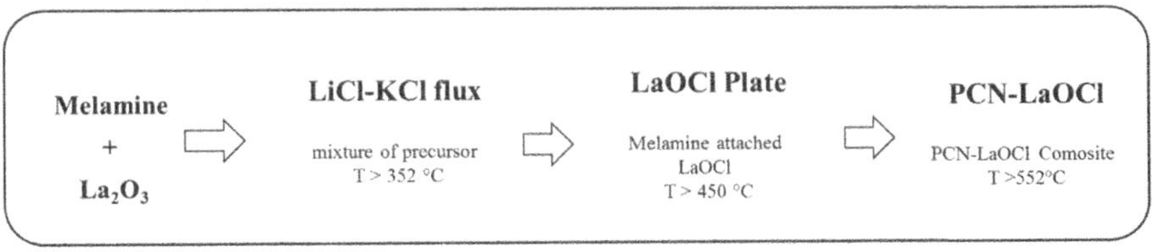

Fig. 9.1 Systematic illustration of the synthesis of LaOCl + PCN composite

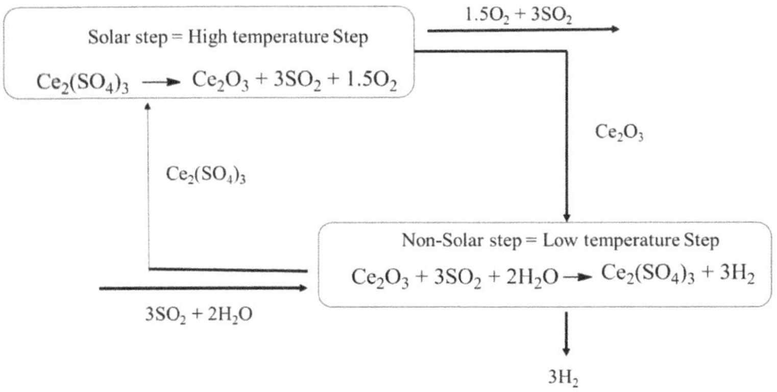

Fig. 9.2 Mechanism of the water-splitting reaction in presence of the cerium oxide - cerium sulfate

$$Ce_2(SO_4)_3 \xrightarrow{\text{Endothermic reaction}} Ce_2O_3 + 3SO_2 + 1.5O_2 \tag{9.4}$$

$$Ce_2O_3 + 3SO_2 + 2H_2O \xrightarrow{\text{Exothermic reaction}} Ce_2(SO_4)_3 + 3H_2 \tag{9.5}$$

Rahul R. Bhosale et al. have been investigated the new composition of nanomaterial for overall water splitting reaction includes cerium oxide–cerium sulfate. They examine the first thermal dissociation of $Ce_2(SO_4)_3$ and re-oxidation of Ce_2O_3 are estimated the thermodynamic equilibrium analysis by HSC software. The starting point of this cycle is the endothermic thermal reduction of $Ce_2(SO_4)_3$ to Ce_2O_3, $3SO_2$, and $1:5O_2$. After the formation of the Ce_2O_3 from $3SO_2$ and $1:5O_2$, the Ce_2O_3 was transferred into the water splitting reactor for the production of the hydrogen, this step of the reaction is highly exothermic. The chemical reaction is illustrated as follows. Figure 9.2 shows the brief diagrammatic mechanism of water splitting and hydrogen production reaction (Bhosale and AlMomani 2020).

P. Ilanchezhiyan et al. have been demonstrated the neodymium (Nd) based oxide perovskite nanostructures showed facial application in biofunctionalized photo-electro-catalyst in overall water splitting reactions. Nd-based perovskite ($Nd_{1\text{-}x}Co_xFeO_3$) nanostructures were synthesized by a hydrothermal protocol which possesses efficient application in photocatalytic and photo-electrochemical (PEC) of overall water splitting reactions into oxygen and hydrogen. The synthesis of the Nd-based oxide perovskite nanostructure $CoFeO_3$ in which cobalt was substituted by the Nd lead to the formation of the ABO_3 class of the perovskite. In synthesized perovskite Co was replaced Nd, it causes boosting in charge carrier property to showed outstanding photocatalytic activity. In dual combination of $Nd_{1\text{-}x}Co_xFeO_3$ nanostructured perovskite also showed up to mark electrocatalytic activity by charge boosting mechanism. Nd–O, CoO, and Co–Fe–O binds boosting the absorption of a light photon and improved electronic conductivity to facilitate the overall water splitting reactions (Ilanchezhiyan et al. 2021).

Rahul R. Bhosale et al. have been investigated the applicability of samarium and erbium oxide in water spitting reactions for generation of the hydrogen and oxygen. The samarium and erbium oxide showed the up mark solar thermochemical water splitting activities. These proposed metal oxides are stable which reused several times. The amount of solar energy needed for the thermal reduction of samarium and erbium oxides was calculated by the following formula. These values were observed for both samarium and erbium cycles is 69.4% respectively.

$$\eta(abs - solar\ energy) = 1 - \left(\frac{\sigma T4}{IC}\right).$$

There two cycles were observed in the proposed mechanism reduction of the Sm_2O_3 to 2Sm and $1.5O_2$ further 2Sm oxidized $3H_2O$ to Sm_2O_3 and $3H_2$. A similar analogous mechanism for the evolution of the hydrogen and oxygen erbium oxide showed as follows;

For Samarium (Sm) Cycle

$$Sm_2O_3 \rightarrow 2Sm + 1.5\,O_2$$
$$2Sm + 3H_2O \rightarrow Sm_2O_3 + 3H_2$$

For Erbium cycle

$$Er_2O_3 \rightarrow 2Er + 1.5\,O_2$$
$$2Er + 3H_2O \rightarrow Er_2O_3 + 3H_2$$

The evolution of oxygen and hydrogen using Sm_2O_3 and Er_2O_3 required steady-state operation conditions and the solar reactor must have perfectly insulated includes emissivity and absorptivity equal to 1. These reaction mechanisms are high exothermic which emits a large amount of heat is evolved, its values were estimated and observed for the erbium cycle (1163.42 kW) whereas for the samarium cycle (1093.20 kW) (Bhosale et al. 2017).

Junying Tang et al. have been proposed the novel europium (Eu) dopped g-C_3N_4 by in-situ methods which acts as electrocatalysts for water splitting reactions. The Eu dopped g-C_3N_4 acts as the efficient photocatalysts the maximum yield of hydrogen was observed near about 2425.7 μmole. Due to dopping of Eu into g-C_3N_4 which increases the efficiency by 4 folds. It is specific towards surface area and also reactive towards visible light radiation. The precursors were used for the synthesis of the Eu dopped g-C_3N_4 includes urea and europium nitrate 4% ($Eu(NO_3)_3 \cdot 6H_2O$). The performance of Eu dopped g-C_3N_4 was examined by Labsolar-6A system with a 300 W Xenon-arc lamp as a light source (Tang et al. 2019). Chaojun Ren et al. has been examined new platinum doped europium oxide photocatalysts for water splitting and selective hydrogen evolution reaction. The synthesis of the photocatalysts has been done by the deposition of the Pt nanocrystals europium oxide in-situ by chemical deposition protocol. Only Eu_2O_3 electrocatalysts showed conduction potential at −

0.82 which is more negative than the H_2O/H_2 conduction electro-potential. The TME analysis of synthesized Pt doped Eu_2O_3 nanocomposite showed the in-situ formation of the 1.81 nm thick layer of Pt on Eu_2O_3 was formed. The Synthesized Pt decorated Eu_2O_3 photocatalysts have been subjected for examination of hydrogen evolution efficacy which was observed under visible light and reached 807.4 μm in two hours (Ren et al. 2019).

Abrar Ahmad et al. have been given new gadolinium (Gd) doped titanium dioxide nanorod as electrocatalysts for overall water splitting reaction. The synthesis of Gd@TiO_2 nanocomposite using readily available starting material urea and Gd $(NO_3)0.6H_2O$ by hydrothermal methods. The FE-SEM and TEM examination showed Gd uniformly distributed in the surface of TiO_2 leads to the formation of the Gd@TiO_2 microsphere. The photoluminescence investigation showed gadolinium-based microsphere acts as an important reducing agent due to surface recombination of electrons and holes. Linear sweep voltammetry analysis showed outstanding results when Gd doped in TiO_2 and increase the efficacy by two-fold in photocurrent density as compared to TiO_2. Titania nanorod array possesses negative potential towards effective charge separation and transportation in the Gd@TiO_2. The schematic diagram explains the mechanism of the Gd@TiO_2 nano rod-based photoelectrocatalysts for efficient water splitting and oxygen evolution reactions and is illustrated in Fig. 9.3 (Ahmad et al. 2020).

Rahul Bhosale et al. investigate the new terbium oxide (TbO_2) electrocatalysts for effective water splitting reactions to the generation of hydrogen and oxygen gasses. The terbium oxide-based water-splitting reaction reduction TbO_2 to Tb and O_2 further Tb on oxidation with water leads to the generation of the TbO_2 and $2H_2$. The redox reactions involved in the Terbium oxide-based water splitting reactions cycle are as follows.

$$TbO_2 \rightarrow Tb + O_2$$

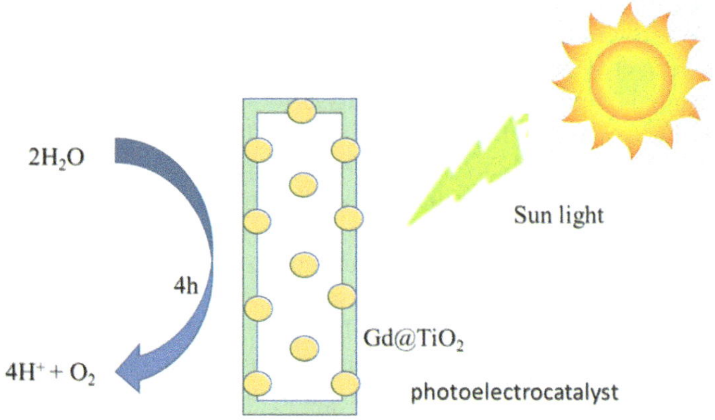

Fig. 9.3 Mechanism of the Gd@TiO_2 nano rod-based photo-electrocatalysts

$$Tb + 2H_2O \rightarrow TbO_2 + 2H_2$$

In the above-mentioned reaction, the first step is endothermic in which the generation of oxygen takes place whereas, the second step observed highly exothermic. In this step, hydrogen gas was evolved after oxidation of the Tb. This investigation was conducted by HSC Chemistry software and databases. The reduction of the TbO_2 to Tb and O_2 decrease significantly its reduction temperature from 2780 to 2280 K. Exergy examination shows that $\eta_{absorption}$ of the terbium oxide in the solar reactor increased notably by a factor of 1.28 due to the decrease in the reduction temperature from 2780 to 2280 K (Bhosale et al. 2016).

Zixia Wan et al. have been fabricated a three-dimensional (3D) iron-dysprosium (Fe-Dy) oxide co-regulated in-situ for selective and effective water splitting reactions. The in-situ catalyst was formed by the by MOF-Ni array on carbon cloths FeDy@MOF-Ni/CC via effective hydrothermal methods. Electrochemical examinations showed that the fabricated Dy@MOF-Ni/CC electrocatalysts need an overpotential of only 251 mV to reach 10 mA cm^{-2} with a small Tafel slope of 52.1 mV dec^{-1}. The observed cell voltage is only 1.57 V in the overall water splitting reaction system which is sufficient to achieve 10 mA cm^{-2}. Synthesized FeDy@MOF-Ni/CC electrocatalyst possess unique 3D rhombus-like geometry. Figure 9.4 showed a systematic pathway that includes reacting species and reaction conditions for the synthesis of the FeDy@MOF-Ni/CC (Wan et al. 2020).

Fengfeng Li et al. demonstrated new rare element-based photocatalysts of Ag_3PO_4@ $HoPO_4$ for effective water splitting reactions to the generation of oxygen and hydrogen gas. The Ag_3PO_4@holmium phosphate photocatalysts synthesis by using readily available starting material $AgNO_3$, ammonia (NH_3) by co-deposition protocol. In this co-deposition, the first formation of the Ag_3PO_4 particles is encapsulated by an amorphous layer of the holmium phosphate which leads to the formation of the heterojunction structure Ag_3PO_4@ $HoPO_4$ composite with the molar ratio of Ho/Ag is 0.05/2.95. The photocatalytic activity was demonstrated in a photo-reactor

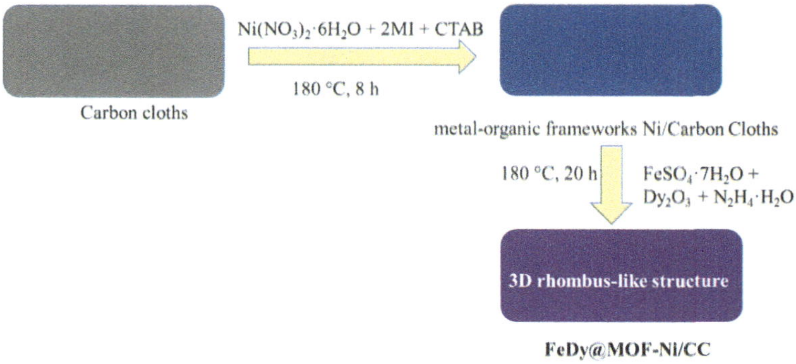

Fig. 9.4 Systematic pathway includes reacting species and reaction conditions for the synthesis of the FeDy@MOF-Ni/CC

at room temperature. BL-GHX-V Xe lamp with a 500 W bulb was used as electric power the light source which resembles sunlight. The reacting mixture included 30 mL of a rhodamine B and methyl thionine chloride solution having a stating concentration was 5 mg/L. Firstly adsorption–desorption equilibrium established half hour. The residual efficacy of catalyst was examined by UV–VIS spectrophotometer (Chen et al. 2017).

Hyekyung Cho et al. have been given new electrocatalysts of TiO_2 nanotubes doped with erbium (Er) for effective hydrogen production from water splitting reactions. Er@TiO_2 possess prominent features includes erbium doping into TiO_2 increase the hydrogen production rate by 2.4 folds, Synthesized Er-doped catalyst have stability and reproducibility and improved the electrocatalytic performance as compared to TiO_2. Er@TiO_2 was synthesized by anodization and electrochemical deposition methods using several quantities of erbium. The sophisticated instrumentation analysis by FE-SEM, TEM, and XRD analysis showed particle size was observed in the range of 24–30 nm. The Er@TiO_2 exhibit photoconversion efficiency at 1.58% and photosensitivity were observed at 115.06 nm. The photocatalytic activity of the Er@TiO_2 NTs has been demonstrated in a hydrogen generation reaction which showed hydrogen production rate is nearly equal to ~17.39 $\mu mol\ h^{-1}\ cm^{-2}$. Synthesized Er@TiO_2 examined for the photocatalytic study includes degradation of methylene blue indicator and it showed outstanding results. The schematic diagram indicates the charge transfer mechanism and generation of H_2 and O_2 gasses by the water-splitting reaction mechanism illustrated in Fig. 9.5 (Cho et al. 2021).

Desiree M.de Los Santos et al. demonstrated and examined the thulium-doped rutile TiO_2 ($Tm_2Ti_2O_7$) nanoparticles for effective photocatalytic activity includes water-splitting reaction to the generation of hydrogen and oxygen gasses. Tm@TiO_2

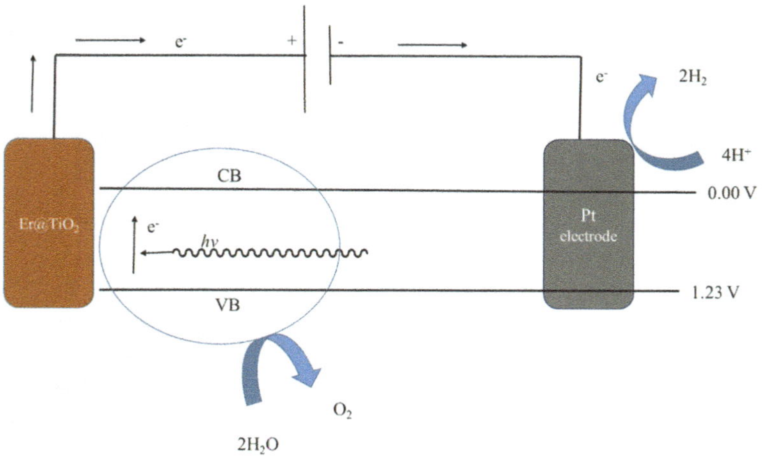

Fig. 9.5 Schematic diagram indicates the charge transfer mechanism and generation of H_2 and O_2 gasses by water splitting reaction mechanism

was obtained by ball mill method in which pure TiO_2 nanocomposite doped with 2.0–4.8% thulium used as a dopant. Synthesized $Tm_2Ti_2O_7$ showed photocatalytic activity in the rutile phase predominantly. Tm-doped TiO_2 showed delay amorphous-anatase–rutile phase conversion at the high annealing temperature. This synthesized nanocomposite also showed applicability in photocatalytic degradation of organic impurities in an aqueous medium (de los Santos et al. 2015).

Umesh Prasad et al. studied the effects of yttrium, ytterbium with tungsten co-dopant in bismuth vanadate photo-anodes to achieve efficient water splitting. The $BiVO_4$ photo-catalyst possesses limited light absorption and charge transport properties. Dual doping of Y^{3+} or Yb^{3+} at the active site of the Bi^{3+} examine to overcome this limitation, it leads to extension in the lambda max up to 550 nm. Dual dopants decrease the bandgap and increase charge transfer spectra. The photo-anode shows complete suppression of recombination of electrons and holes with an improved absorption efficiency up to 85% and efficient charge transfer 69–77%. Y + Yb@ $BiVO_4$ showed significant photo-electrochemical efficacy 5.5 to 6.1 mA cm^{-2} at 1.23 V under sunlight in the K_2HPO_4 solution as electrolyte (Prasad et al. 2020).

Wenrui Zhang et al. have been studied the applicability of ultrathin lutetium oxide (Lu_2O_3) thin film on the crystalline bismuth vanadate ($BiVO_4$) water splitting photoanodes. To facilitate transporter transport through atomic-scale interface control protocol. The Lu_2O_3 1.4 nm thickness interlayer fabricated by pulsed laser deposition protocol which has prominent features includes very few defects at the heterojunction. Lu_2O_3 interlayer modulates improved the electronic conduction between the conduction band and valence band (Zhang et al. 2018).

9.4 Conclusion and Future Perspective

In the proposed study, it has been observed that rare earth elements and doped electrocatalysts possess efficient activity towards overall water splitting reaction, HER, and OER. The doping of f-block elements into the composite of d-block elements leads to improve the efficacy of catalyst. Such as LaOCl-coupled polymer carbon nitride-based, Dysprosium doped copper oxide ($Cu_{1\text{-}x}Dy_xO$) nanoparticles, Nd based perovskite ($Nd_{1-x}CoxFeO_3$) nanostructures, samarium and erbium oxide, europium (Eu) dopped g-C_3N_4, gadolinium (Gd) doped titanium dioxide nanorod, terbium oxide (TbO_2), TbO_2, FeDy@MOF-Ni/CC, of Ag_3PO_4@ $HoPO_4$, Er@TiO_2, thulium-doped rutile TiO_2 ($Tm_2Ti_2O_7$), ytterbium with tungsten co-dopant, ultrathin lutetium oxide, and (Lu_2O_3)Cu1-xDyxO (x ¼ 0.02, 0.03) electrocatalysts showed excellent catalytic activity towards HER and OER reactions in the alkaline medium due to the addition of the Dy^{3+} ions in the CuO.

The presence of Dy^{3+} ions in the CuO increases the vacancy sites and thereby increases the kinetic transfer of oxygen diffusion but an excessive amount of dopant causes lattice distortion, unstable process, and anodic corrosion. Its presence is also useful in minimizing the required overpotential to achieve maximum productivity with greater potential for practical applications. Er-doped CoP is an excellent HER and OER catalyst in an alkaline medium due to its high stability, electron conductivity, and activity. It is observed that the HER and OER performance of CoP electrocatalysts can be improved by incorporating Er in the optimum amount. Er-doped CoP showed the best electrocatalytic performance over Ce, Gd, La, and Nd-doped CoP. In addition to this, Er-doped CoP showed superior HER activity in the acidic medium also (Zhang et al. 2019). Gadolinium-indium-zinc ternary oxide (GIZO) nanostructured electrocatalyst showed excellent HER and OER activities in the alkaline medium due to the presence of Gd^{3+} ions on the surface of the active catalyst. GIZO is a viable material for effective bifunctional electrocatalytic applications with remarkable durability (Ilanchezhiyan et al. 2020).

All lanthanides possess unique structural properties, abundant vacancies due to unoccupied f-orbital, excellent oxygen storage ability, and effective chemical and mechanical stability to promote the water-splitting reaction, hydrogen evolution reaction, and oxygen evolution reaction. Doping of rare earth elements in catalysts increases the catalytical activity. These series of elements possess electrocatalytic activity except promethium. The rare earth element doped electrocatalysts are stable and showed excellent activities for HER and OER in alkaline and acidic media. These rare earth element doped electrocatalysts act as bifunctional electrocatalysts and are a viable candidate in water splitting and can be used as an alternative to current noble metal-based electrodes for large-scale production of hydrogen and oxygen in the water-splitting process. Although the rare earth elements showed outstanding photo-electrocatalysis property it has some limitations includes minimum availability, costly, and sometimes overreaction with hydrogen.

Acknowledgements The authors are especially thankful to Director, Government institute of forensic science, Aurangabad for the constant support, and encouragement.

Consent for Publication Not applicable.

Funding None.

Conflict of Interest The authors declare no conflict of interest, financial or otherwise.

References

Acar C, Dincer I (2018) 1.13 hydrogen energy. In: Dincer I (ed) Comprehensive energy systems. Elsevier, Oxford, pp 568–605

Ahmad A, Yerlikaya G, Zia-ur-Rehman et al (2020) Enhanced photoelectrochemical water splitting using gadolinium doped titanium dioxide nanorod array photoanodes. Int J Hydrogen Energy 45:2709–2719. https://doi.org/10.1016/j.ijhydene.2019.11.117

Altawell N (2021) 1—Energy's general introduction (fossil fuels and renewable energy). In: Rural electrification. Academic Press, pp 1–17

Armaroli N, Balzani V (2007) The future of energy supply: challenges and opportunities. Angew Chemie Int Ed 46:52–66. https://doi.org/10.1002/anie.200602373

Bessarabov D, Millet P (2018) Fundamentals of water electrolysis. In: Bessarabov D, Millet P (eds) PEM water electrolysis. Academic Press, pp 43–73 (Chapter 3)

Bhosale R, Kumar A, AlMomani F (2016) Solar thermochemical hydrogen production via terbium oxide based redox reactions. Int J Photoenergy 2016:9727895. https://doi.org/10.1155/2016/9727895

Bhosale RR, AlMomani F (2020) Hydrogen production via solar driven thermochemical cerium oxide—cerium sulfate water splitting cycle. Int J Hydrogen Energy 45:10381–10390. https://doi.org/10.1016/j.ijhydene.2019.05.159

Bhosale RR, Kumar A, AlMomani F et al (2017) A comparative thermodynamic analysis of samarium and erbium oxide based solar thermochemical water splitting cycles. Int J Hydrogen Energy 42:23416–23426. https://doi.org/10.1016/j.ijhydene.2017.03.172

Chen FL, Li Z, Zhang M, Shen Y, Cai Y, Li Y, He X, Chen C (2017) Ag_3PO_4@holmium phosphate core@shellcomposites with enhanced photocatalytic activity. 34705

Cho H, Joo H, Kim H et al (2021) Improved photoelectrochemical properties of TiO_2 nanotubes doped with Er and effects on hydrogen production from water splitting. Chemosphere 267:129289. https://doi.org/10.1016/j.chemosphere.2020.129289

Das A, Ganguli AK (2018) Design of diverse nanostructures by hydrothermal and microemulsion routes for electrochemical water splitting. RSC Adv 8:25065–25078. https://doi.org/10.1039/C8RA04133D

de los Santos DM, Navas J, Aguilar T et al (2015) Study of thulium doping effect and enhancement of photocatalytic activity of rutile TiO_2 nanoparticles. Mater Chem Phys 161:175–184. https://doi.org/10.1016/j.matchemphys.2015.05.034

de Miranda PEV (2019) Hydrogen energy: sustainable and perennial. In: de Miranda PEV (ed) Science and engineering of hydrogen-based energy technologies. Academic Press, pp 1–38 (Chapter 1)

Deng S, Zhong Y, Zeng Y et al (2018) Hollow TiO_2@Co9S8 core–branch arrays as bifunctional electrocatalysts for efficient oxygen/hydrogen production. Adv Sci 5:1700772. https://doi.org/10.1002/advs.201700772

Dresselhaus MS, Thomas IL (2001) Alternative energy technologies. Nature 414:332–337. https://doi.org/10.1038/35104599

El Haj Assad M, Alhuyi Nazari M, Rosen MA (2021) Applications of renewable energy sources. In: Assad MEH, Rosen MA (eds) Design and performance optimization of renewable energy systems. Academic Press, pp 1–15 (Chapter 1)

Ilanchezhiyan P, Mohan Kumar G, Tamilselvan S et al (2020) Highly efficient overall water splitting performance of gadolinium-indium-zinc ternary oxide nanostructured electrocatalyst. Int J Energy Res 44:6819–6827. https://doi.org/10.1002/er.5423

Ilanchezhiyan P, Mohan Kumar G, Siva C et al (2021) Neodymium (Nd) based oxide perovskite nanostructures for photocatalytic and photoelectrochemical water splitting functions. Environ Res 197:111128. https://doi.org/10.1016/j.envres.2021.111128

Inumaru J, Hasegawa T, Shirai H et al (2021) 1—Fossil fuels combustion and environmental issues. In: Ozawa M, Asano H (eds) Advances in power boilers. Elsevier, pp 1–56

Lin Y, Su W, Wang X et al (2020) LaOCl-coupled polymeric carbon nitride for overall water splitting through a one-photon excitation pathway. Angew Chem Int Ed 59:20919–20923. https://doi.org/10.1002/anie.202008397

Liu G, Dastafkan K, Zhao C (2021) Electrochemical water splitting. In: Heterogeneous catalysts. Wiley, pp 533–555

Maeda K, Domen K (2013) 7.22—Photochemical water splitting using nanostructured metal oxides. In: Reedijk J, Poeppelmeier K (eds) Comprehensive inorganic chemistry II, 2nd edn. Elsevier, Amsterdam, pp 587–614

Morales-Guio CG, Stern L-A, Hu X (2014) Nanostructured hydrotreating catalysts for electrochemical hydrogen evolution. Chem Soc Rev 43:6555–6569. https://doi.org/10.1039/C3CS60468C

Mu B, Wang A (2019) 11—Fabrication and applications of carbon/clay mineral nanocomposites. In: Wang A, Wang W (eds) Nanomaterials from clay minerals. Elsevier, pp 537–587

Nemiwal M, Gosu V, Zhang TC, Kumar D (2021) Metal organic frameworks as electrocatalysts: hydrogen evolution reactions and overall water splitting. Int J Hydrogen Energy 46:10216–10238. https://doi.org/10.1016/j.ijhydene.2020.12.146

Nguyen HL (2021) Metal–organic frameworks for photocatalytic water splitting. Sol RRL 5:2100198. https://doi.org/10.1002/solr.202100198

Prasad U, Prakash J, Kannan AM (2020) Effects of yttrium{,} ytterbium with tungsten co-doping on the light absorption and charge transport properties of bismuth vanadate photoanodes to achieve superior photoelectrochemical water splitting. Sustain Energy Fuels 4:1496–1506. https://doi.org/10.1039/C9SE00978G

Rajakaruna RMPI, Ariyarathna IR (2020) Functionalized metal-based nanoelectrocatalysts for water splitting. In: Mustansar Hussain C (ed) Handbook of functionalized nanomaterials for industrial applications. Elsevier, pp 83–109 (Chapter 4)

Ren C, Li W, Li H et al (2019) Ultrasmall Pt nanoclusters deposited on europium oxide: a newly active photocatalyst for visible-light-driven photocatalytic hydrogen evolution. Appl Surf Sci 480:96–104. https://doi.org/10.1016/j.apsusc.2019.02.192

Rodney JD, Deepapriya S, Robinson MC et al (2021) Dysprosium doped copper oxide ($Cu_{1-x}Dy_xO$) nanoparticles enabled bifunctional electrode for overall water splitting. Int J Hydrogen Energy 46:27585–27596. https://doi.org/10.1016/j.ijhydene.2021.06.014

Ros C, Andreu T, Morante JR (2020) Photoelectrochemical water splitting: a road from stable metal oxides to protected thin film solar cells. J Mater Chem A 8:10625–10669. https://doi.org/10.1039/D0TA02755C

Rosen MA (2021) Renewable energy and energy sustainability. In: Assad MEH, Rosen MA (eds) Design and performance optimization of renewable energy systems. Academic Press, pp 17–31 (Chapter 2)

Souleymen R, Wang Z, Qiao C et al (2018) Microwave-assisted synthesis of graphene-like cobalt sulfide freestanding sheets as an efficient bifunctional electrocatalyst for overall water splitting. J Mater Chem A 6:7592–7607. https://doi.org/10.1039/C8TA01266K

Takanabe K (2021) Photocatalytic water splitting: fundamentals and current concepts. In: Heterogeneous photocatalysis. Wiley, pp 269–286

Tang J, Zhou W, Guo R et al (2019) An exploration on in-situ synthesis of europium doped g-C_3N_4 for photocatalytic water splitting. Energy Proc 158:1553–1558. https://doi.org/10.1016/j.egypro.2019.01.365

Wan Z, He Q, Chen J et al (2020) Dissolution-regrowth of hierarchical Fe-Dy oxide modulates the electronic structure of nickel-organic frameworks as highly active and stable water splitting electrocatalysts. Chinese J Catal 41:1745–1753. https://doi.org/10.1016/S1872-2067(20)63606-3

Wang H, Lee H-W, Deng Y et al (2015) Bifunctional non-noble metal oxide nanoparticle electrocatalysts through lithium-induced conversion for overall water splitting. Nat Commun 6:7261. https://doi.org/10.1038/ncomms8261

Wang S, Lu A, Zhong C-J (2021) Hydrogen production from water electrolysis: role of catalysts. Nano Converg 8:4. https://doi.org/10.1186/s40580-021-00254-x

Yang W, Moon J (2019) Recent advances in earth-abundant photocathodes for photoelectrochemical water splitting. ChemSusChem 12:1889–1899. https://doi.org/10.1002/cssc.201801554

Zhang W, Yan D, Tong X, Liu M (2018) Ultrathin lutetium oxide film as an epitaxial hole-blocking layer for crystalline bismuth vanadate water splitting photoanodes. Adv Funct Mater 28:1705512. https://doi.org/10.1002/adfm.201705512

Zhang G, Wang B, Bi J et al (2019) Constructing ultrathin CoP nanomeshes by Er-doping for highly efficient bifunctional electrocatalysts for overall water splitting. J Mater Chem A 7:5769–5778. https://doi.org/10.1039/C9TA00530G

Part II
Nanocomposites for Energy and Environmental Applications

Chapter 10
Electrospun Nanocomposite Materials for Environmental and Energy Applications

Pooja P. Sarngan, Agasthiyaraj Lakshmanan, and Debabrata Sarkar

Abstract Environmental and energy issues are the two major concerns that are assigned among researchers all over the world. Researchers have introduced many versatile materials of polymers, metals, and metal-oxides matrix nanocomposites in various dimensional structures. Among different nanostructures, one-dimensional nanofibers have captured attention due to their large surface area, directional transmission, interface effect. Electrospinning is a versatile technique for generating continuous nanofibers with tunable chemical composition, diameter, porosity, and its versatility to synthesize a wide variety of materials from polymer to metal oxide. In this chapter, we have focused on the development of semiconductor and polymer-metal composite nanofiber for photocatalysis applications for dyes degradation and water splitting H_2 production. Engineering the junction and interface charge transfer is a significant method to promote charge separation and migration during the photocatalysis process. In the present Coronavirus disease 2019 (COVID-19) pandemic situation, disease transmission prevention is a major task. The demand for a good face mask and the air filter is increased. Afterward, we have discussed the latest development of nanocomposite membranes for the air filter and face mask.

Keywords Electrospinning · 1D nanostructure · Surface plasmon resonance · Photocatalysis · Air filteration · Inorganic-Organic Nano-composites

10.1 Introduction

In the twenty-first century, energy and the environment are the two major global issues that civilization faces. The use of fossil fuels is very much required and has become an integral part of our lives but along with its advantage, it carries a major drawback of non-renewable properties. According to the International Energy Agency's (IGA) World Energy Outlook 2011, worldwide energy consumption would rise by 40% between 2009 and 2035, and fossil fuels—coal, natural gas, and petroleum—will

P. P. Sarngan · A. Lakshmanan · D. Sarkar (✉)
Applied NanoPhysics Laboratory, Department of Physics and Nanotechnology, SRM Institute of Science and Technology, Kattankulathur 603203, India

S. P. Singh et al. (eds.), *Metal Nanocomposites for Energy and Environmental Applications*, Energy, Environment, and Sustainability,
https://doi.org/10.1007/978-981-16-8599-6_10

continue to be the major sources utilized to fulfill humanity's energy demands (The International Energy Agency 2011; Oliveira et al. 2012). However, the present situation indicates a troublesome environmental scenario mainly due to the improper way of utilizing fossil fuels, and overconsumption leads to the emission of hazardous CO_2 gases to the Earth's atmosphere with the contamination of air and water. It is of utmost importance that living beings surviving on this earth need a clean, healthy, and eco-friendly environment (Perera 2018; Child et al. 2018; Nicoletti et al. 2015). This property directed to new challenges among the researchers to invent advanced methods to hitch renewable energy issues and to bring back a clean and toxic-free environment. For these, various researches were carried out from the early 1920s with the photocatalytic experiment, adapted from the natural photosynthesis where a catalyst is enrolled for the decomposition of water into hydrogen and oxygen with lesser harmful by-products emission thereby considered as much efficient for habitat cleaning such as self-cleaning of building materials, antifogging, antibacterial, etc. (Linsebigler et al. 1995; Daneshvar et al. 2004; Kudo et al. 2004). This pioneering work was further developed by Fujishima et al. in 1972 by sensitizing photons on the TiO_2 anode and with Pt cathode electrodes under ultra violet (UV) irradiation, which has led to considerable growth in the field of heterogeneous photocatalysis (Kudo and Miseki 2009; Paul and Giri 2018). Later numerous research works were carried out to study the mechanism and kinetics involved using various spectroscopic techniques, which has also made understood about the properties of TiO_2 catalysts and other factors needed to implement to increase its efficiency and uplift its potential in the environmental remediation. The conversion of photo energy to chemical energy is associated with Gibb's energy reaction. From the viewpoint of the Gibbs free energy change, photocatalytic water splitting is distinguished namely as an uphill and downhill reaction. A downhill reaction occurs when organic compounds are oxidized utilizing oxygen molecules. This reaction is regarded as a photocatalytic water-splitting reaction. The Gibbs free energy, in this case, is -237 kJ/mol (Liu et al. 2014; Ohtani 2014; Chen et al. 2013). Researchers are working increasingly on enhanced functional nanosized materials to meet this condition of photocatalytic water splitting. Among other nanostructures, one-dimensional nanostructures, like nanorods, nanotubes, nanowires, and nanofibers are extensively attracted by their extraordinary physical and chemical properties. Because of their confinement effects, one-dimensional (1D) semiconducting nanowires, nanofibers, and nanotubes, as well as 1D material patterning, can be utilized to improve charge transport and as building blocks in energy and electrical devices (Ge et al. 2016). Till now various methods available to generate 1D nanostructure including electron-beam or focused ion-beam writing, template-assisted synthesis, interface synthesis technique, chemical vapor deposition, lithography, hydrothermal, electrospinning, etc. (Baig et al. 2021; Zhou et al. 2010). With respect to all other methods, electrospinning is a simple and inexpensive technique for the preparation of nanofibers that can be used for a wide range of applications. Electrospinning nanofiber membrane is directly used for water splitting H_2 production, air, and water filtration and oil separation media, etc. (Barhoum et al. 2019).

Particulate particle matter (PM) pollutions are solid or liquid particles in the air is a major toxic form of pollutant which can produce adverse effects on human well-being. Long-term exposure to PM particles in the highly polluted areas was highly linked to death from ischemic heart disease, dysrhythmias, heart failure, and cardiac arrhythmias (Pope et al. 2004). Furthermore, particular biological molecules (bacteria, viruses, and microorganisms) when attached to PM cause direct damage to the atmosphere (Yang 2012). The size of these particles is the reason for human health effects. If the particle size is higher than 10 μm it will be blocked by the nasal cavity. If the particle size is less than 2.5 μm ($PM_{2.5}$) it will enter the lungs and could make different serious health problems. There are many methods available to filter PM_{10} particles. But it is tough to remove fine particles. In the last two years due to the COVID-19 pandemic situation, there are links between disease transmission due to high particulate matter (PM) pollution. Research is ongoing to filter fine particles with the development of new science and technology. Various methods are available to produce nanofiber membranes. Apart from other methods electrospinning is a promising technology to produce nanofiber over the range of 50–2000 nm with high surface area, controlled fiber morphology, and interconnected porous structure (Smoukov et al. 2015). The filter media with these characteristics can achieve high filtration performance and provides a novel filter media at a low cost. This chapter will provide a summary of current developments in electrospun nanofiber membranes for air filtration. After introducing the main technique and materials for air filtration, the structural advantages and filtration mechanisms of fibrous filtration are explained. Also, we underline the types, structural characters, and application performance of the existing electrospun nanofiber filters for air filtration application.

10.1.1 Fundamentals of Photocatalysis

A photocatalyst is a material substance that absorbs photons and accelerates a chemical reaction. An electron from a filled valence band (VB) is excited to the conduction band (CB) upon irradiation of light energy higher than the bandgap (E_g) of a semiconductor and leaves a hole in the VB. These charge carriers, electron and hole pairs (e^- and h^+) take the main part in the catalytic activity. Figure 10.1 (Maeda 2011; Wang and Domen 2020; Yang and Wang 2018) represents the overview of photoelectrochemical (PEC) water splitting. The process involves three succeeding steps during the photosensitization effect: (i) Absorption of light ($h\nu \geq E_g$) to generate the electron–hole pairs when the light is exposed to the surface of semiconductor (referred to as catalyzed photoreaction), (ii) charge separation and migration of the photogenerated charge carriers, (iii) chemical reactions at the surface with photogenerated charge carriers (referred as sensitized photoreaction) (Ibhadon and Fitzpatrick 2013; Pawar et al. 2018).

During photoelectrochemical water splitting, the band edges of the electrode must overlap with the acceptor and the donor states of the water decomposition reaction, thus necessitating that the electrodes should at least have a bandgap of 1.23 V, the

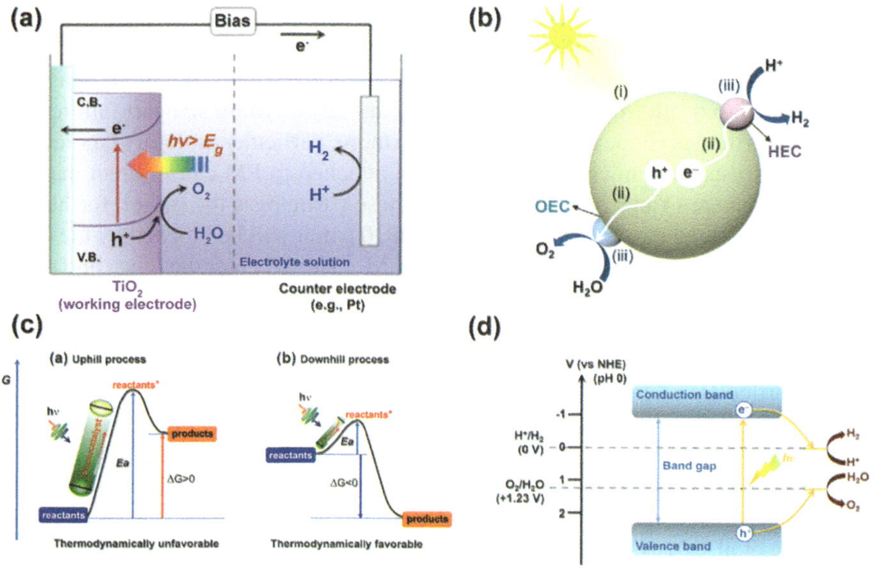

Fig. 10.1 **a** PEC water splitting using a TiO_2 photoanode, and **b** Diagram showing the reactions during water splitting on a semiconductor photocatalyst: (i) light absorption, (ii) charge separation and transport, and (iii) redox reactions. **c** Thermodynamics of uphill and downhill photocatalysis: (i) uphill process, and (ii) downhill process, and **d** Energy diagram for photocatalytic water splitting based on one-step excitation. **a** Reprinted from Maeda (2011), copyright (2011), with permission from Elsevier. **b**, **d** Reprinted from Wang and Domen (2020), copyright (2019), with permission from American Chemical Society. **c** Adapted with permission from Yang and Wang (2018). Copyright (2017) American Chemical Society

reversible thermodynamic decomposition of water (Wenderich and Mul 2016; Al-Mamun et al. 2019; Gao et al. 2020). When the light shines on the semiconductor material, the photogenerated e^- reacts with the protons present and reduces to H_2 and h^+ migrate to the surface, which promotes the oxidization of H_2O to O_2 (Zhu and Wang 2017).

$$2H^+ + 2e^- \rightarrow H_2 \quad 0.00\,V \tag{10.1}$$

$$O_2 + 4e^- + 4H^+ \rightarrow 2H_2O \quad 1.23\,V \tag{10.2}$$

$$2H_2O \rightarrow 2H_2 + O_2 \tag{10.3}$$

The significant factors for a catalyst to empower the activity are its wide band-gap, stability, structural morphology, and active surface charge carriers. The charge transfer from the surface of the semiconductor must be fast enough to prevent photo-corossion. Numerous catalysts of semiconductor metal-oxides have been reported

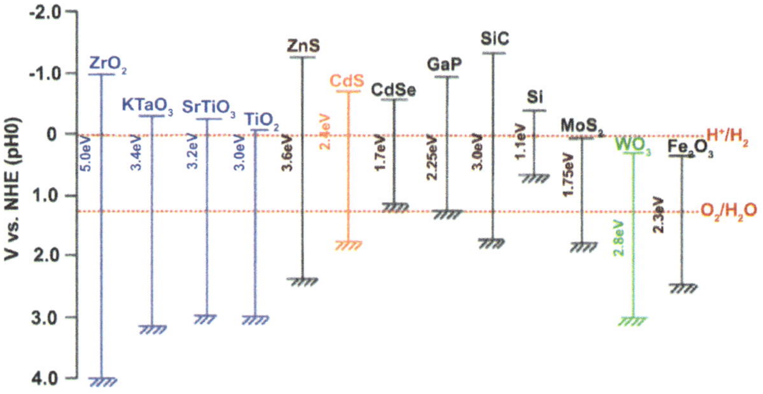

Fig. 10.2 Relationship between band structure of semiconductor and redox potentials of water splitting. Reproduced from Kudo and Miseki (2009), copyright (2008), with permission from Royal Society of Chemistry

in the literature, among which TiO_2, ZnO, V_2O_5, CdS showed promising enhanced results during the reactions with a wide range of applications, represented in Fig. 10.2 (Kudo and Miseki 2009; Zhang et al. 2009). However, the photogenerated charge carriers may recombine with each other, subsequently dissipating energy and hence it reduces the overall reaction efficiency. Wide varieties of modifications have been carried out from past few years to improve the efficiency of catalysts and still it is being continued such as (i) dye sensitization, (ii) surface modification to increase the reaction sites and stability, (iii) multi-layered systems, (iv) doping with wide bandgap semiconductors like TiO_2 by nitrogen, carbon, and sulphur, (v) new semiconductors with metal 3d valence band instead of contributing oxide 2p, etc. (Li and Wu 2015). The electronic structures of the semiconductors are also engineered to derive the functional advantages such as to increase the charge transfer process at the interface, improving the efficiency, developing nanosize material by selecting wide bandgap materials (Li and Chu 2018).

10.1.2 Parameters Affecting the Photocatalysis Process

Effect of dye concentration: An increase in the initial organic dye concentration reduces the degradation percentage. When a greater number of dye molecules are present, more dye molecules were adsorbed on the surface of the catalyst and hence reduces the surface reaction sites. It suppresses the excitation of the number of photons, thereby the electron–hole transfer is minimized along with the production of hydroxyl radicals. Therefore, dye concentration has a strong effect to carry out photocatalytic activity.

Effect of pH: It was studied that pH has a significant impact on the degradation process. The pH parameter shapes the surface condition and the ionization state of the pollutant used. Change in pH affects the isoelectric point pH(i). So if pH < pH(i) is applied, the surface of the catalysts becomes positive, thus attracting the oppositely charged compounds electrostatically. These interactions strengthen the rate of adsorption and conversely.

Effect of light Intensity: Light intensity also plays a vital role in the catalytic process, when the intensity is equal or greater to the material's bandgap energy it will allow the electron–hole transfer. Also, it increases the rate constant value as well as increases the decomposition of the organic pollutant at higher intensity of incident radiation.

Effect of temperature: It has been studied that the degradation process becomes faster as the temperature rises. It allows the formation of radicals and enhances the oxidation states. Hence it allows controlling the recombination mechanism throughout the process (Gnanaprakasam et al. 2015; Topare 2013).

Time Course: With increasing irradiation time, the amount of H_2 and O_2 produced should increase. It is critical to examine not just the value of an activity or the rate of gas evolution, but also the time course. Experiments that are repeated are also instructive.

Turn-over number (TON): Turn-over number is often defined as the number of reacted electrons to the number of atoms in the photocatalysts or the number of atoms at the surface of the photocatalyst. Amounts of H_2 and O_2 present should be sufficient to overcome a photocatalyst. The quantity of evolved H_2 is used to compute the number of reacting electrons.

Quantum yield: Quantum yield is defined as the number of emitted photons to the number of absorbed photons. The definition points out the importance of a photocatalytic process, but it is difficult to figure out the number of photons emitted or absorbed by the catalyst during the reaction. Therefore, the apparent quantum yield calculation is given below

$$Apparent\ Quantum\ Yield = \frac{Number\ of\ reacted\ electrons}{Number\ of\ incident\ photons} \times 100$$

Photoresponse: When a photocatalyst is exposed to light with an energy greater than the bandgap, water splitting occurs. In some instances, even if the material absorbs at the visible spectrum, it may not display photocatalytic activity. In such situations, it is important to determine the material's photoresponse.

10.2 One-Dimensional Semiconductor Catalytic Materials

One-dimensional nanostructure has occupied a major role and is considered a promising candidate with high performance in the field of materials research. The invention of this new structures provoked many new ideas of research along with wide applications owing to its optical, structural, and electronic properties along with its quantum confinement effect (Schneider et al. 2014). This idea and interest in one-dimensional configuration have sparked since the discovery of carbon nanotubes (CNTs) in the early 1990s (Weng et al. 2014). This fascinated many researchers to investigate various other one-dimensional nanostructures such as nanowires, nanotubes, nanobelts, and nanoribbons using different facile synthesis techniques. The transmission of photons, electrons, and protons were found to exhibit along its axial direction made it more specific to control in the making of various devices. Therefore, 1D nanostructures spotlighted as a building block to enhance the next generation research works with broad areas of ample applications along with confined ample phenomena such as in photovoltaics, photocatalysis, electronic devices, sensors, dye-sensitized solar cell (DSSC) etc. (Ge et al. 2016; Zhou et al. 2010; Gupta and Tripathi 2011). Among various types of one-dimensional nanomaterials, electrospun nanofibers represent the most efficient due to flexible structure, large surface area, high mechanical strength, directional transmission, and easily scaled-up. Therefore, in the field of photocatalysis water splitting, pollutant degradation, air filtration, etc. the nanofiber synthesized by the electrospinning process presents broad research prospects.

10.2.1 Electrospinning

Many synthesis steps were carried out with different advanced techniques with well-controlled structural and morphological properties in order to apply in vast applications. Some of which include melt-blown, phase separation, template synthesis, solvothermal, self-assembly, electrospinning, etc. Electrospinning is a novel way of processing tiny threads by utilising electrostatic forces. Some well-known applications that work similarly to the electrospinning mechanism are electrostatic precipitators and pesticide sprayers. Electrostatic fibre production has got a lot of attention for its potential to develop fine fibres. Fibers produced using the electrospinning method have diameter in the range of 30 nm to several micro meters, whereas fibers obtained using other procedures have diameter in the range of 500 nm up to a few microns (Zhang et al. 2005). Because of these extremely appreciable properties as well as its simplicity and affordability, electrospinning has become the most popular technique for the production of nanofibers. With this technique both the randomly arranged nanofibers as well as well- aligned structured nanofibers can be produced. Also, this technique enables the production of various structures and morphology such as beaded, ribbon, porous, and core–shell fibers. The concept of electrospinning started in early 1600 by William Gilbert. Later many researchers came forward with various new techniques to overcome the difficulties faced during the experiments. Afterward, Anton Formhals has developed the equipment towards the commercial level which turned to a great achievement and allowed to further development in the production of fiber mats. Later in the 1960's, Taylor made a detailed study on the formation of a pendant drop on the tip of the needle and focused on the ejection of the fluid. During the electrospinning process, an electric field is applied between a polymer solution droplet and a grounded collector. Taylor found out that the pendant drop gets elongated as the voltage is applied and forms a cone shaped with an angle of 49.3° in order to balance the surface tension of the polymer solution, which later named as Taylor cone (Patil et al. 2017; Xue et al. 2019). The schematic representation of electrospinning set up and the formation of Taylor cone is represented in Fig. 10.3a, b (Lakshmanan et al. 2021; Al-Hazeem 2018). With the gradual increase of electric voltage leads to the elongation of the hemispherical surface of the droplet

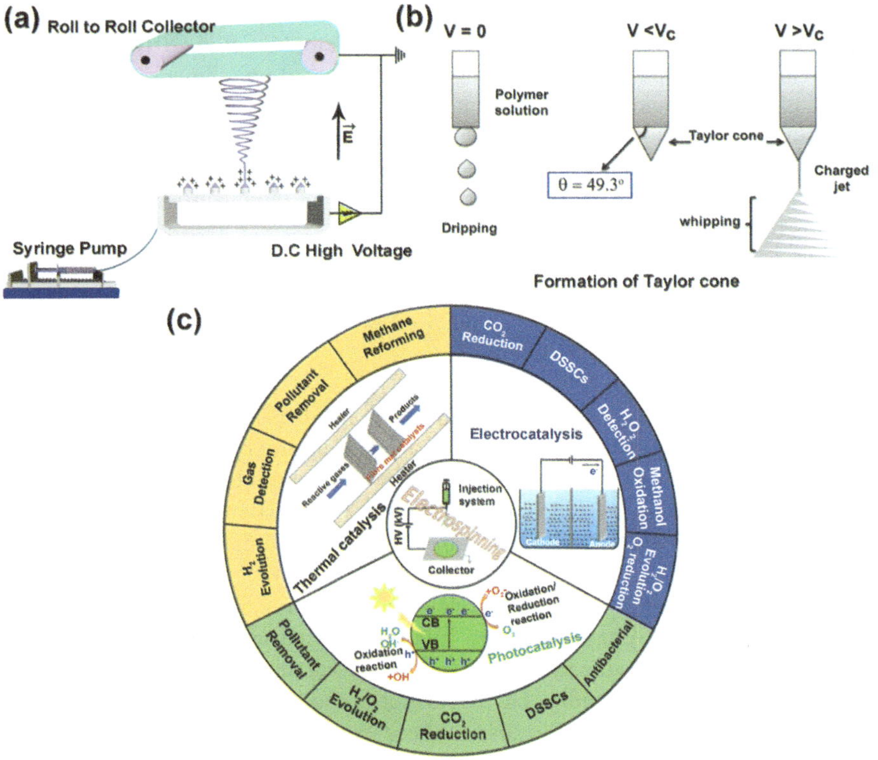

Fig. 10.3 Schematic illustration of electrospinning setup. **b** Illustrates the effects by electric field applied to a solution in a capillary and **c** Represents the applications using electrospun nanofiber. **a** Adapted with permission from Lakshmanan et al. (2021). Copyright (2021) American Chemical Society. **b** Reprinted Al-Hazeem (2018). Copyright (2018), with permission from American Chemical Society. **c** Adapted with permission from Zhu and Nie (2021). Copyright (2020) Elsevier

and when the electrostatic potential overcomes the surface tension of the polymer droplet, a fine charged jet is ejected from the tip of the Taylor cone. Under the electrostatic potential the charged jet undergoes bending instability and accelerates to stretch into long fiber. The technical simplicity and adaptability of the electrospinning process are its advantages. This simply designed apparatus consists primarily of a high voltage electric source with positive or negative polarity, a syringe pump with tubes to carry the solution from the syringe to the nozzle tip, and a conducting collector such as aluminium. The collector can be made of any shape according to the requirements, like a flat plate, rotating drum, etc. (Xue et al. 2019) Fig. 10.3c represents the applications of electrospun nanofibers (Zhu and Nie 2021).

10.2.1.1 Processing Parameter to Control Electrospun Nanofiber

In terms of generating nanofibers, the electrospinning method must maintain a high voltage electric field in order to produce electrically charged jets. In this method the polymer melt is used with the appropriate concentration of solutions. Therefore, after the application of voltage the process undergoes various steps. (i) Formation of the pendant drop at the tip of the nozzle; (ii) charging of the liquid droplet and formation of Taylor cone; (iii) emanation of the charged jet along a straight line; (iv) thinning of the jet undergoes vigorous chaotic motion or bending instability due to the excessive number of charges produced in the presence of an electric field; and (iv) solvent evaporation enables solidification and collection of the jet as solid fiber(s) on a grounded collector. To obtain the required fine fibers the electrospinning techniques should be controlled with the process, solution and ambient parameters. Several factors have a great influence on the nanofiber formation.

Applied Voltage: In electrospinning, if the strength of the applied voltage (kV) is strong leads to a decrease in fiber diameter, it can cause a transformation in the shape of fiber initiation tip. Hence it affects the structural morphology. Also reported that an increase in the current leads to beaded morphology.

Nozzle to Collector Distance: The nozzle to collector distance is an important factor to be maintained which controls the evaporation rate, whipping interval and deposition time hence the fiber morphology and its structure is affected. The lesser nozzle to accumulator distance will lead to wet fibers and could generate beaded morphology.

Polymer Flow Rate: One of the significant process parameters is the polymer solutions flow rate which have an impact on the velocity at which the jet initiates and on the transfer rate of the material. Slower the flow rate provides lesser fiber diameter (in nm) as well as lesser generation of beads. Whereas an increase in the flow rate causes fibers with more beads and wider diameter with pore size 50–190 nm.

Humidity: It has been examined that an increase in humidity will increase the number of pores on the surface of electrospun fibers and also brings a change in fiber diameter by controlling the solidification process during the ejection of the jet.

Solution Viscosity: To generate continuous fine fibers and to prevent the breakage during the spinning or during collection it is important to maintain the solution viscosity and surface tension. The increase in solution concentration and viscosity indicates an increase in fiber diameter. Whereas the low viscosity of the solution causes electrospraying and therby a greater number of particle formation occurs.

Solution Conductivity: For the fabrication of long, aligned, less diameter, and straight fibers, an increase in conductivity of the solutions is required. If the conductivity is less then there generate beads. Hence it has a great influence on the fiber morphology and its structure.

Solvent Volatility: In electrospinning solvent volatility also plays a vital role in the phase separation process. The more volatile solvent has the ability to evaporate faster than the lesser one. It was also found that a highly volatile solution will have pores with 100 nm sizes.

10.3 One-Dimensional TiO_2 Nanofiber

Among the different photocatalytic semiconductors, Titanium dioxide (TiO_2) have wide spread application being a good photocatalyt due to its high photo-activity, high efficiency, chemical stability, mechanical robustness, photo-durability, less toxic, low cost, etc. Titanium dioxide exist in three crystalline forms such as anatase, rutile and brookite. Studies have reported that, among this anatase and rutile exhibits as good photocatalysts. Anatase and rutile have tetragonal structure whereas brookite exists with an orthorhombic structure (Albetran et al. 2014). TiO_2 nanofibers (TNFs) are widely used in various applications and extensive research in the field has led to the development of several fabrication strategies for TiO_2 nanofibers with controllable morphologies (Reghunath et al. 2021). Nanofiber properties can be managed to provide greater flexibility in the end product's surface functionality. The thermal stability of titanate nanofibers is quite high. Calcining titanate nanofibers leads to a change in the crystal structure while maintaining the morphology of the nanofibers upto about 1000 °C. However, the TiO_2 is limited for some reasons such as its wide band gap i.e., it is allowed only to the UV region which is less than 4% of the solar spectrum. As the energy distribution of solar light is 4% UV light, 44% visible light and 52% infrared (IR) light. Hence the energy spectrum is a major limiting factor for practical uses. And also, the recombination time of electron–hole pairs is short. To overcome this, many methods have been introduced by enhancing the light absorptivity and by developing active sites (Ge et al. 2016; Tian et al. 2014).

10.3.1 Enlargement of the Photocatalytic Active Surface

The photocatalytic activity depends on the external surface (or the crystal facet) of the catalyst being exposed to the reaction. In other words, crystal facets of TiO_2 play a significant role in photocatalysis because each crystal facet of TiO_2 is unique in terms of surface and stereochemistry due to the different arrangement of the surface atoms on the facets and due to the preferential flow of photogenerated charge carriers. Each facet has its own contribution in catalysis applications.

Enhancement of the 1D TiO_2nanostructures surface area: On the photocatalyst surface the photocatalyst causes photocatalytic degradation of pollutant and photocatalytic water splits for hydrogen production. A considerable quantity of surface area for organic compound absorption must be provided for highly efficient photocatalysis. For good photocatalytic performance, enlarging the specific surface area of the 1D nanostructure is a significant difficulty. For these two effective approaches were introduced. The first is to design nanoparticles on the surface of 1D TiO_2, while the second is to produce in situ, through an acidic corrosion process, a stuffed surface, with nanoparticles dispersed with the same composition (Tian et al. 2014). Because of the excellent mechanical properties of the nanoparticles on the surface of the 1D nanostructure, the surface area is significantly increased, the absorption mechanism is endowed owing to the high surface energy, and thereby the nanoparticles on the surface of the 1D TiO_2 nanostructure set up surface heterojunctions. These steps will enhance the surface area, surface energy as well as suppress the recombination rate of e^-–h^+ pairs.

Enhancement of light absorption: One of the important challenges is to enhance the absorption of the catalyst, and hence to significantly raise its output to various levels of applications. For this the basic steps employed are:

(i) By making a junction with noble metals- commonly called Surface Plasmon Resonance
(ii) By doping in semiconductors catalyst
(iii) By adding dye.

The surface plasmon resonance is the unified oscillation of conducting electrons in metal environment. This technique has greatly influenced in the research field, which has provoked the efficiency widely as a catalyst. Here noble metal particles of Ag, Au, Pt, Pd and their alloys are mostly used, which are assembled on the surface of the 1D nanostructure. Many findings proved that metal nanoparticles act as an antenna. In this fundamental technique the energy transfer occurs in two different steps; direct electron–hole separation and plasmonic energy-induced electron–hole separation this suppresses the electron–hole pair recombination and allows the absorption from UV to visible spectrum region and in some cases even to the near-infrared (NIR) region.

10.3.2 *Metal (Au, Ag, Pt, Ru, Rh etc.)-TiO_2Composite Nanofiber for Energy and Environment*

10.3.2.1 Localized Surface Plasmon Response

Remarkable research works and developments have been done on surface plasmon resonance both in the matter of instrumentation developments and for various applications such as for environmental remediation and bio-molecular interaction. Surface plasmon resonance (SPR) is the collective oscillation of conduction electrons which is induced at the interface when the frequencies of photons fit the frequency of surface electrons with restoring force. This collective oscillation is initially accelerated by the electromagnetic wave (light) and then the metals interact with electric field produced by the incident light as shown in Fig. 4a (Kriegel et al. 2017; Yesudasu et al. 2021; Mukherjee et al. 2013) This fundamental technique was visible effective in the noble metals, which enhanced the absorption due to the high density of charge carriers (Luther et al. 2011). The wavelength of the resonant photon varies depending on the property of metals. The surface plasmon activity and the resonant photon wavelength depends on the size and form of the metallic nanostructures (Kelly et al. 2003; Linic et al. 2011). Figure 10.5 represents the decoration of metals with semiconductors that enhanced the absorption and enhances interfacial charge transfer. In zero-dimensional nanostructures, the free carriers are confined to a small volume which localizes the surface plasmon. When the resonance condition is met, this phenomenon is called localized surface plasmon resonance (LSPR) and the resonant frequency can be written as

$$\omega_{LSPR} = \sqrt{\frac{\omega_p^2}{1 + 2\varepsilon_m} - \gamma^2} \tag{10.4}$$

whereas, for bulk materials the oscillation of charge carriers is called plasma frequency, ω_p and it is defined as:

$$\omega_p = \sqrt{\frac{ne^2}{\varepsilon_o m^*}} \tag{10.5}$$

where, ε_m is the dielectric constant of medium and γ is the carrier damping, e is the unit of elementary charge, ε_o is the permittivity of free space, n is related to the carrier density and m^* is the effective mass of the charge carriers (Kriegel et al. 2017). The LSPR phenomenon is tuned mainly by controlling the charge density (n) by attuning the dielectric constant (ε_m) of the material. Au, Ag, Pt, Pd, etc. are the prevalent noble metals used for enhancing the activity at the surface level as represented in Fig. 4b (Mukherjee et al. 2013). Hence, the unique ability of noble metals allowed the absorption in wide range of electromagnetic spectrum when integrate with other wide band gap semiconductors which enabled for various applications, more prominently

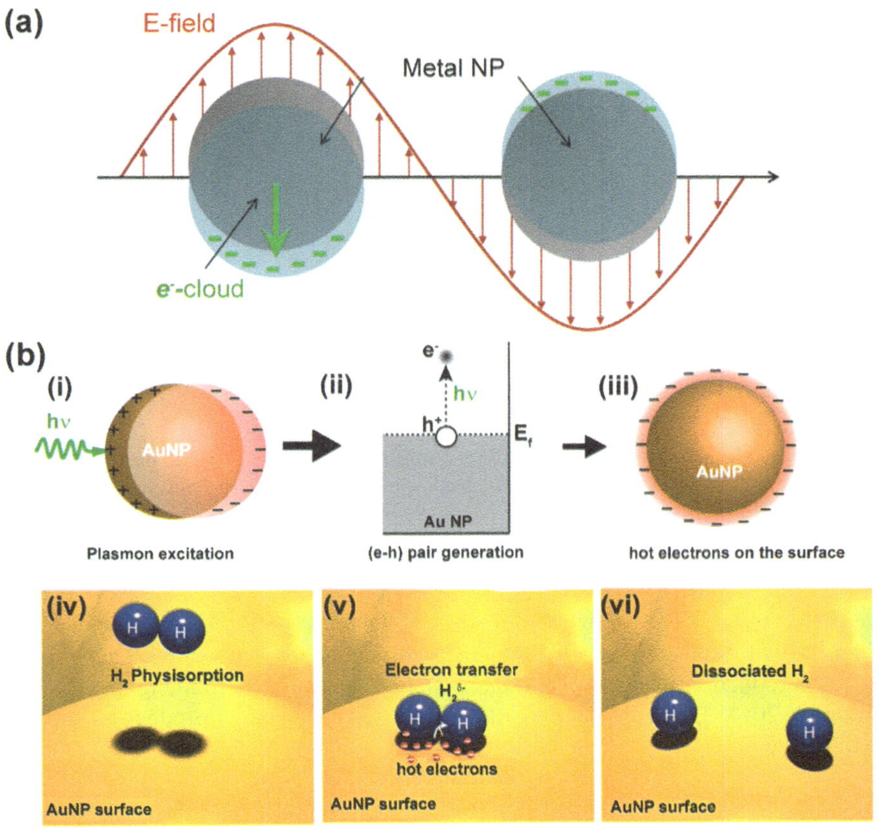

Fig. 10.4 **a** Schematic of plasmon oscillation for a spherical metallic NP showing the displacement of the conduction electron charge cloud relative to the core in an oscillating electric field, and **b** Schematics of plasmon-induced hot electron generation on AuNP and mechanistic representation of H_2 dissociation on the AuNP surface. **a** Reprinted with permission from Kriegel et al. (2017). Copyright 2017 Elsevier; Figure **b** Reprinted with permission from Mukherjee et al. (2013). Copyright 2012 American Chemical Society

for H_2 production and for dye degradation (Paul and Giri 2018; Mukherjee et al. 2013; Furube and Hashimoto 2017; Nalbandian et al. 2015).

10.3.2.2 *Metal TiO_2Nanofiber Composite for Photocatalytic H_2Production*

Hydrogen energy produced from the fossil fuels has become an important energy for environmental and social benefits. Whereas due to the decrease in the production of fossil fuel another effective way was introduced, i.e.; the H_2 production using photocatalytic water splitting. This technique was found environment friendly, gained

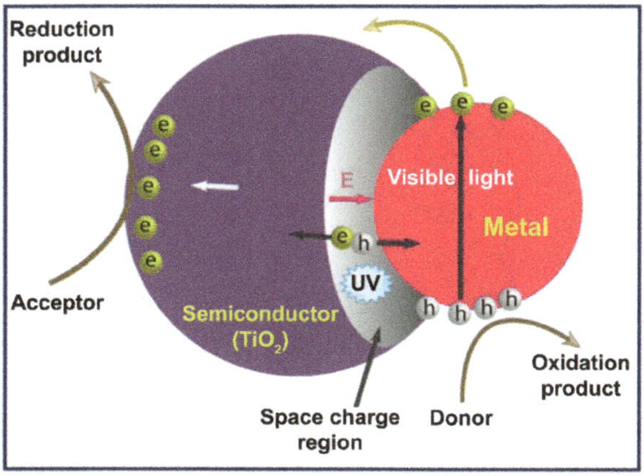

Fig. 10.5 A schematic of interface between a semiconductor and a plasmonic metal NP, charge separation, and photocatalytic activity. Adapted with permission from Paul and Giri (2018). Copyright 2018 Elsevier

much attention with the use of semiconductor material. The technique was later explored extensively, by the deposition of noble metals, addition of sacrificial agent, etc. Here, we discuss some of the promising researches carried out for the hydrogen energy production using a plasmonic system (Yu et al. 2011; Meng et al. 2017; Ge et al. 2017; Bumajdad and Madkour 2014). For hydrogen production through water splitting, semiconductor requires the CB potential at the appropriate energy level such that conduction band electron can reduce the proton into H_2 molecule. The plasmonic metal particles Pt and Pd on to the mixed phase TiO_2 nanofibers enhance the H_2 production through water splitting (Wu et al.). The N doping in the mixed phase nanofiber creates a N–Ti bonds in the lattice site and produces a p state level. This enables the activity to carry out at lower energies. The reactivity of metal-TiO_2 nano composite was analysed in an ethanol–water mixture under UV-A and UV-B illumination. Under UV-B, N-doped TiO_2(B) nanofibers coated with Pt nanoparticles produced the most H_2 at 7000 μmol/gh, owing to the smaller particle size and higher dispersion property than Pd (Wu et al. 2011). By suggesting the catalytic activity of Ag-doped TiO_2 nanofibers for greatly improved water splitting, Barakat et al. proposed a new path via calcination. The resonance impact of surface plasmon localization in the visible spectral range is greatly influenced due to the incorporation of silver nanoparticles, represented in Fig. 6a (Barakat et al. 2020). Many heterojunctions were produced, allowing electrons to be trapped and hence the recombination rate to be reduced (Barakat et al. 2020). For the photoelectrochemical water splitting Nyugen et al. fabricated TiO_2 nanofibers using electrospinng method and to study the plasmonic metal effect, gold nanoparticles were well decorated on the surface of nanofibers. To study the impact of the SPR effect photocurrent density measurements were carried out under green light (540 nm) illumination, which showed an increase in photocurrent for Au nanoparticles. The Fermi level

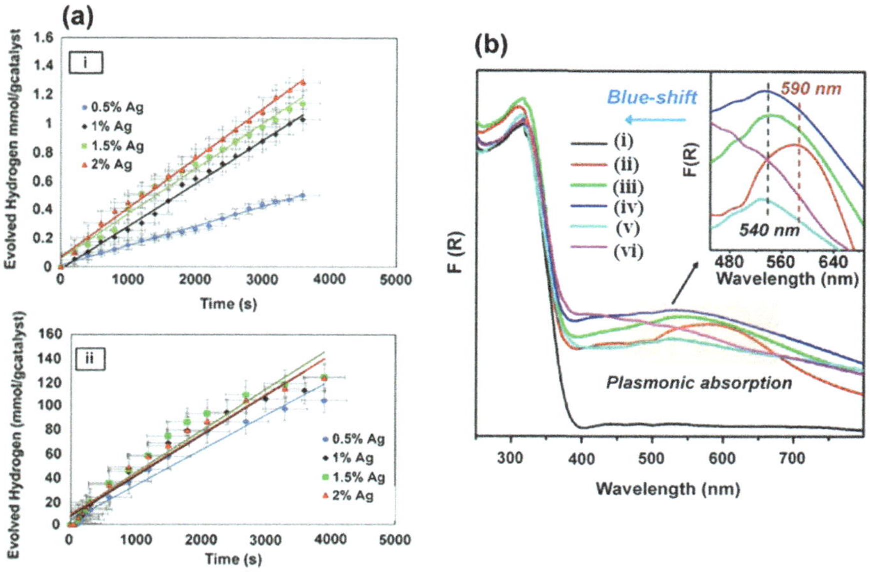

Fig. 10.6 **a** Effect of silver content on the hydrogen evolution rate (mmol/g catalyst), in case of utilizing (i) Ag-doped nanoparticles, and (ii) nanofibers as photocatalyst. **b** UV–vis absorption spectra of the as-electrospun NF: (i) TiO_2, (ii) Au_1/TiO_2, (iii) $Au_{0.75}/Pt_{0.25}/TiO_2$, (iv) $Au_{0.5}/Pt_{0.5}/TiO_2$, and (v) $Au_{0.25}/Pt_{0.75}/TiO_2$, and (vi) Pt_1/TiO_2. **a** Adapted with permission from Barakat et al. (2020). Copyright 2020 Elsevier. **b** Adapted with permission from Zhang et al. (2013). Copyright 2013 American Chemical Society

of Au is lower than the conduction band of TiO_2. Therefore, under UV irradiation the photogenerated electrons from the conduction band of TiO_2 can easily migrate towards Au and hence provide active sites for PEC water splitting (Nguyen et al. 2020). Apart from the single metal plasmonic system, the Au/Pt ratio on the TiO_2 nanofiber plasmonic system has also been reported. The sample containing 25% Au and 75% Pt on the nanofiber had a good H_2 generation rate of 11.658 μmol h^{-1} in presence of L-ascorbic acid. The photocatalytic activity of H_2 production and CO_2 reduction is enhanced due to the surface plasmon resonance effect of Au and the electron sink ability of Pt is represented in Fig. 6b (Zhang et al. 2013).

10.3.2.3 *Metal Nanocomposite TiO_2Nanofiber for the Environment*

Due to the rapid growth of industrial sectors lot many organic dyes released from the waste of textile, paper and various other industries have polluted the available freshwater. Therefore, the day-by-day increase of toxicity in water is harmful to aquatic animals as well as causes serious health issues to human beings. Among the wide variety of techniques, catalytic degradation was found more efficient and environment friendly. Some of the research progress using fibrous membranes are discussed

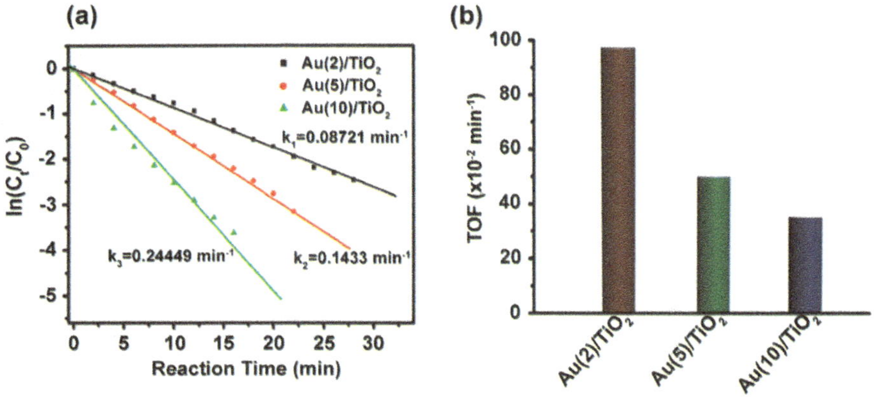

Fig. 10.7 **a** Plots of $\ln(C_t/C_0)$ against the reaction time of the reduction of 4-NP over different Au/TiO_2 hybrid nanofibers, and **b** TOF of the Au/TiO_2 nanofiber catalysts. Reprinted from Hao et al. (2015), copyright (2015), with permission from Elsevier

below (Thavasi et al. 2008; Karthikeyan et al. 2020). The catalytic activity was proposed and studied for 4-nitrophenol (4-NP) to 4-aminophenol (4-AP) in sodium borohydride aqueous solution by Hao et al. Here the efficiency was determined with different loading weight% of Au on mesoporous TiO_2 nanofiber. As the amount of Au loaded into the hybrid fibers increases, the Au particles become larger and denser. The apparent reaction rate constant rises, while the catalyst's turnover frequency (TOF) value falls but it was noticed that the TiO_2 crystallite sizes stay intact, shown in Fig. 10.7 (Hao et al. 2015). The excellent electronic conducting properties and work functions of the metal alloys on the semiconductor enhance the activity. The plasmonic system comprising of PdAg alloy-TiO_2 nanofibers was reported for the electro-oxidation of methanol. The presence of PdAg induced the localization of the photoexcited electrons and hence the amalgamation of these noble metals reduced the recombination rate and could release more positive holes for methanol oxidation (Ju et al. 2013). Similarly, the incorporation of bi-metallic (Au and Ag) nanoparticles on mesoporous TiO_2 nanofibers was demonstrated to degrade MB dye. Figure 10.8 represents the presence of Au nanoparticles built excellent stability and Ag own larger extinction coefficient (Chattopadhyay et al. 2019). Here they exposed the noble nanoparticles under visible light which emits hot electrons to the conduction band of TiO_2 between a metal-Ti interface through heterojunction and act as an electron sink. This combination showed enhancement in the visible region absorption. The AgAu-mTiO_2-H sample showed excellent solar light absorption activity (Chattopadhyay et al. 2019). Misra et al. explored the photocatalytic performance by tuning the shell thickness of Au@Ag bi-metal core–shell nanoparticles which was immersed on electrospun TiO_2 nanofiber. Due to the high work function, the shell thickness of Au was controlled which allowed the transfer of an electron from Ag to Au. This enhanced the surface plasmon properties. The MB dye degradation was

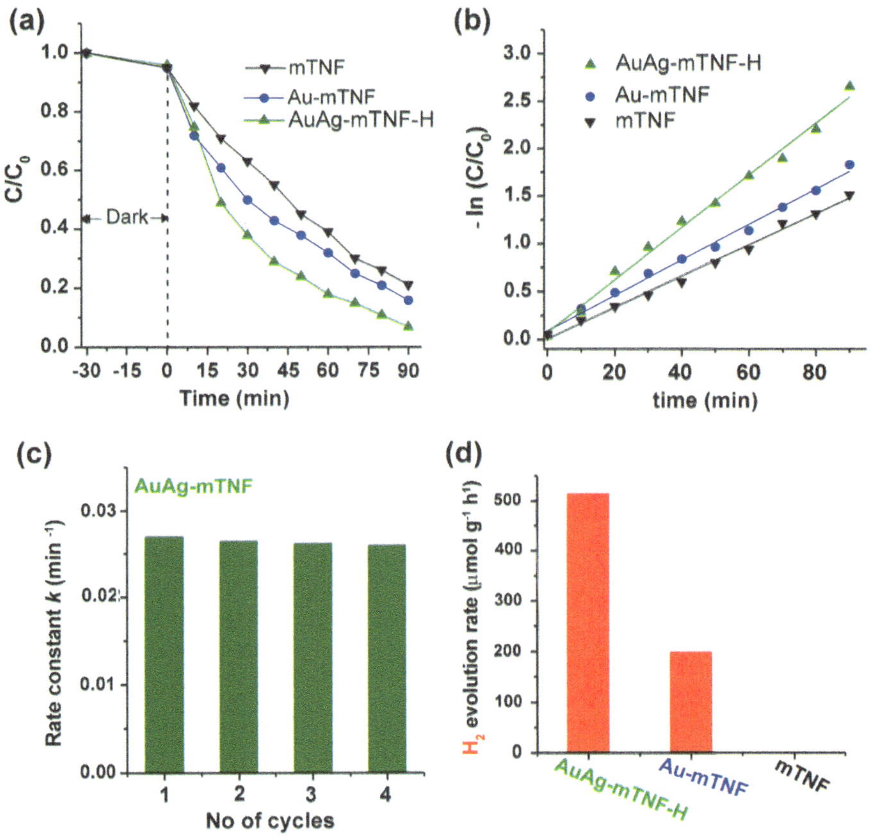

Fig. 10.8 Photoactivity studies: **a** UV–vis spectral analysis for the evaluation of MB photodegradation by AuAg-mTNF-H and controls. **b** Pseudo first order kinetic rate plots. **c** Reusability of AuAg-mTNF-H sample, and **d** H_2 evolution rate of AuAg-mTNF-H and comparative data using Au-mTNF and mTNF as controls. Reprinted from Chattopadhyay et al. (2019), copyright (2019), with permission from American Chemical Society

studied for Au@Ag@TNF sample with lower thickness and showed good catalytic performance under UV light illumination (Misra et al. 2017).

10.4 One Dimensional Polymer Nanofiber to Filter Air Pollutant

Air Pollution: Air pollution becomes serious havoc for human and environmental aspects. Particularly particulate matter (PM) pollution, volatile organic compounds (VOC) pollution resulted in a rising impact on human health and eco systems (Bell et al. 2011; Ren et al. 2017). Air filtration technology is the major technology with different benefits; such as high performance, economy with low consumption of energy, and wide range of application. Fiber based filtration media with different nanocomposites have a various porous structure. This filter media is efficient for high PM capture with sufficient air and light transmission. However, the conventional commercial filter is inefficient for fine particles due to their structural disadvantages like, micron size fiber diameter, large pore size, and low porosity. To achieve high efficiency the thickness and basis weight of the conventional filter is increased but this increment affects the air flow and high energy consumption. For fiber based filtration technology the fiber diameter with the range of nanometers (nanofibers) with open pore structure and high porosity can increase the filtration performance.

10.4.1 Structural Advantages of Nanofiber Air Filter

Woven, non-woven, and membrane filters are the three major filtration media available for air filtration. Among them non-woven fabric filter media has unique characteristics such as high permeability, high specific surface area, controllable pore size distribution as well as small pore size and practically effective for capture PM particles. As a result of these, nonwoven filtering media has been developed; such as spunbonded fibers, glass fibers, melt-blown fibers, and others. Due to their microsized dimensions, spunbonded fibers, glass fibers, melt-blown fiber filters generally suffer from the drawbacks of high pore size and poor porosity. In comparison to micronfibers with electrospun nanofibers, the electrospun nanofibers emerge as a sophisticated nanomaterial with structural characteristics such as tiny pore size, open stacking pores, and highly adjustable porosity. Furthermore, the nanofiber filter has a long life and high dust holding capacity with multifunctional air filtration application, making them for ideal options for air filtration. Although, the electrospun nanofibers has high filtration efficiency for $PM_{2.5}$ particles it is inefficient for ultrafine particles (particle diameter less than 100 nm). To rectify this Ding et al. applied an electrohydrodynamic technique called electrospinnng/netting (Ding et al. 2006, 2011). It is a combination of electrospinning nanofiber and nanonets in the range of 20 nm diameter. The resultant nanofiber/nanonet membranes leads to the outstanding performance to capture ultrafine particles.

10.4.2 Filtration Mechanisms

The filtration process of single fiber is classified into two states one is steady-state and another one is unsteady state. In steady-state the filtration efficiency and pressure drop remain constant over time while other intrinsic properties of filtration materials depends on the nature of PM and air flow. In unsteady state, the filtration efficiency and pressure drop change over time as the particles accumulate on the filters. The

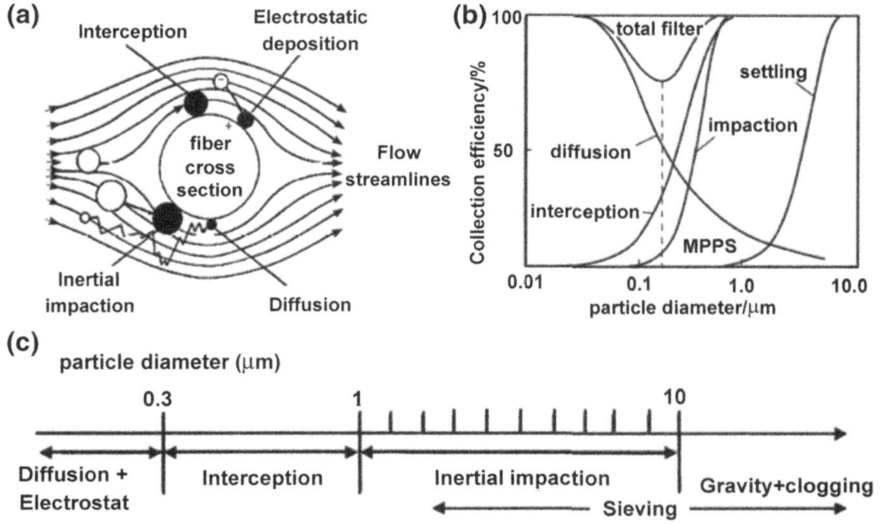

Fig. 10.9 **a** Main five filter mechanisms of single fiber. **b** Collection efficiency of single fiber with different particle diameters. **c** The effect of filtration mechanism with respect to diameter of the particles. MPPS, most penetrating particle size. **a, b** Reprinted from Yang (2012), copyright (2012), with permission from Elsevier. **c** Adapted with permission from Barhate and Ramakrishna (2007). Copyright 2007 Elsevier

unsteady state is too complicated and lacks systematical theories that can provide an exact prediction for the actual filtering process (Qin and Wang 2006). The steady-state of single fiber filtration is classified in five mechanisms such as interception, inertial impaction, diffusion, electrostatic and gravity effect (Yang 2012; Qin and Wang 2006; Zhu et al. 2000, 2017; Bull 2008). It is shown in Fig. 10.9.

Interception: When air streamline of particular particle move close to the fiber; the particle could contact with the fiber and is removed from the air streamline by van der Waals forces. This mechanism is called interception. The probability of direct interception is related to the ratio of particle diameter and fiber diameter. Particle diameter less than 1 μm are captured by interception due to its higher degree of molecular mobility (Yang 2012; Ramskill and Anderson 1951).

Inertial Impaction: The particle inertia is so high with the sudden change of air streamline; the particle is unable to follow the air streamline. That time the particle will deviates from air stream line and deposited to the nanofiber membrane. This mechanism mostly important for the particle size is in the range of 0.3–1 μm (Yang 2012).

Diffusion: The diffusion mechanism of PM capture is occurring due to the Brownian motion of particles. When the size is too low it won't follow air stream line and they will move randomly and colliding with one another; which leads to the particle deposition on the fiber membrane. This mechanism is mostly suitable for particle size less than 0.1 μm (Yang 2012; Ramskill and Anderson 1951).

Electrostatic force: Electrostatic force can change the path of the particle and traps the particles on the fiber surface. In electrostatic force the field will applied to the particle or/and the fibers. This mechanism will traps the particle of size over in the range of submicrometer. The mechanism enhancing the efficiency of filters and also maintains the pressure drop of the filter membrane (Sahay et al. 2012; Wang 2001).

Gravitational force: The effect of gravitational force for PM particle removal is negligible for fine particles. Therefore this mechanism does not have much importance for high arresting air filters (Yang 2012).

10.4.3 Polymer Nanofiber Air Filters

In the last two decades different types of polymer nanofiber membrane is used for high efficiency air filter media produced by electrospinning. Since the electrospun nanofiber has high surface area, uniform pore size, controlled fiber morphology, high surface adhesion, and light weight etc., which allows to capture PM particles easily. Different types of polymers like polyacrylonitrile (PAN), polyvinylpyrrolidone (PVP), polyvinyl alcohol (PVA), polyamide (PA), polyurethane (PU), polysulfone (PSU), and chitosan, etc., are used with unique properties for different filtration devices such as personal respirators (Example: face mask), indoor air cleaners, and vehicle cabin filters. In this section advanced polymer based nanofiber filters have discussed.

10.4.3.1 Single-Component Polymer Nanofiber Air Filter

Different air filter membranes are prepared for PM filtration by using different single component polymers such as PAN, PVP, PA, polystyrene (PS), PSU, etc., Among other polymers PA and PAN polymers have excellent physical and chemical properties to fabricate high arresting air filter membranes for PM filtration. Gibson et al. (2001) proved experimentally and theoretically electropsun nanofiber membranes are highly efficient for trapping aerosol particles (Gibson et al. 2001). Barhate et al. first time investigated the effects of electrospinning parameters on structural and transport properties of of nanofiber membranes (Barhate et al. 2006). Another polymer Nylon 6, has been used to prepare nanofiber air filter membrane for High-Efficiency Particulate Air (HEPA) and Ultra Low Particulate Air (ULPA) filtration which has low fiber diameter in the range of 200 nm, low pore size 0.24 μm and low basis weight 10.75 g m^{-2}. This filter membrane has excellent efficiency (99.993%) than the commercialized HEPA (99.97%) filter membrane for ultra-fine particles (0.3 μm). These results have proven that electrospun nanofiber membranes are extremely attractive for aerosol filtration (Ahn et al. 2006). In the last decade PAN is a commonly used polymer material for air filtration. Yun et al. prepared electrospun nanofiber with an average diameter of 270–400 nm and compared with commercial filters by measuring the penetration of monodisperse NaCl particles (Yun et al. 2007). It is reported 85% optical transmittance PAN nanofiber membrane has 96% high efficiency $PM_{2.5}$ removal. Even after 100 h, the PAN nanofiber membrane remains the same efficiency. The results proved PAN nanofiber membrane is a promising nanofiber membrane for the preparation of high efficiency air filters (Liu et al. 2015). PVP water soluble synthetic polymer were prepared by electrospinning and shown high efficiency due to low fiber diameter. PVP has good mechanical property and high chemical resistance (Morozov and Mikheev 2012). In addition, there are other single component polymers such poly(ethylene oxide) (PEO) (Leung et al. 2010), polyurethane (PU) (Sambaer et al. 2011), cellulose (Grafe and Graham 2003), prepared by electrospinning and shown excellent efficiency for air filtration performance.

10.4.3.2 Composite Polymer Nanofiber Air Filter

Up to date, many studies have been done on the single component nanofiber air filter. However single component nanofiber has low fiber diameter, high surface area, and low pore size, high filtration efficiency, but still lacks on other important properties such as high-pressure drop, flexibility, antifouling properties, ultralight, etc. To rectify these problems two or more polymers are combined and produced composite polymeric nanofiber with additional characteristics. Wang et al. prepared polyvinyl chloride/polyurethane (PVC/PU) blended polymers fiber membrane with different weight ratios to improve filtration performance and mechanical properties as shown in Fig. 10a. With blending PVC and PU they improved tensile strength up to 9.9 MPa. The pure PVC fiber membrane has low tensile strength due to the nonbonding structure of PVC fibers sliding through frictional entanglement at lower stress. After introducing PU into PVC fiber membrane, the mechanical behavior is increased and follows Hooks law (with inelastic limit, stress is directly proportional to strain) and it reaches yield point which confirms the change of plastic to elastic. This behavior occurs coexist of bonding and non-bonding cross-structure in PVC/PU nanofiber membrane (Wang et al. 2013). Superamphiphobic nanofibrous membrane fabricated by combining PAN and PU composites and incorporating new synthesized fluorinated PU (FPU) with PAN/PU composite nanofiber membrane. PAN/PU/FPU nanofiber membrane has excellent antifouling properties. With the addition of FPU, the water contact angle and oil contact angle was regularly increased to >150° (Fig. 10b). The superamphiphobic property of PAN/PU/FPU nanofibrous membrane occurs due to low surface energy and hierarchically nano-scaled rough structure on the surface. Furthermore, as compared to pure PAN/PU membranes, the as-prepared PAN/PU/FPU nanofiber membranes had improved filtration performance against oil and salt aerosol particles, demonstrating the importance of amphiphobicity in air filtration (Wang et al. 2014). Zhang et al. prepared an anti-deformed poly(ethylene oxide)@polyacrylonitrile/polysulfone (PEO@PAN/PSU) composite membranes with binary structures for high filtration performance is given in Fig. 10c.

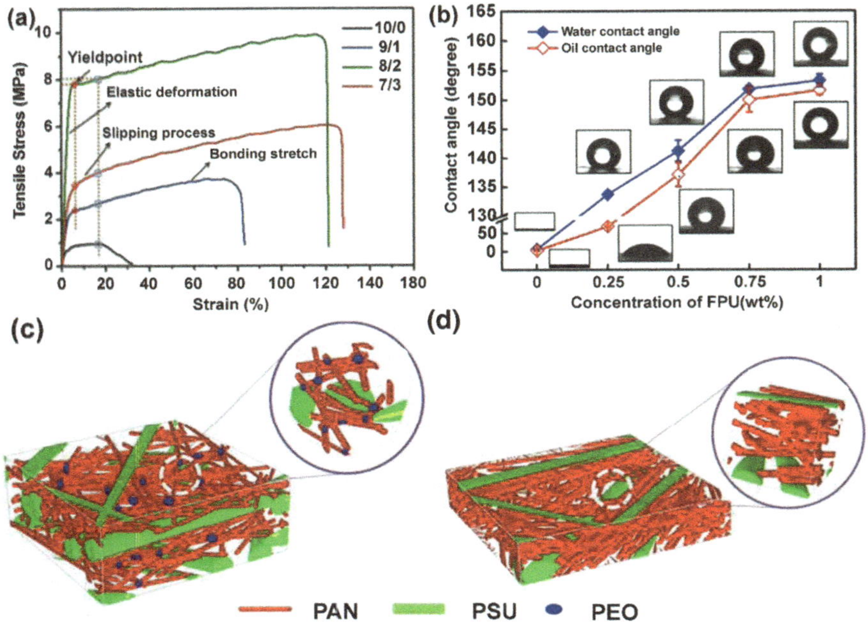

Fig. 10.10 **a** Stress–strain curves of PVC/PU fibrous membranes, **b** Water and oil contact angles of PAN/PU fibrous membranes with different FPU concentrations, **c** and **d** is the representation of PEO@PAN/PSU and PAN/PSU fiber membrane in view of their bulk statical properties. **a** Reprinted with permission from Wang et al. (2013). Copyright 2013 Elsevier; **b** Reproduced from Wang et al. (2014), copyright (2014), with permission from Elsevier; **c**, **d** are modified with permission from Zhang et al. (2016). Copyright 2016, American Chemical Society

To prepare PEO@PAN/PSU composite membrane they used the multi-jet electrospinning method and physical drying process to form bonding. At, first they studied PAN/PSU blended nanofiber membranes with different jet ratios for suitable pore size and packing density, after that PEO was added into the composite membranes, to form the bonding/no bonding that gave the membranes a hollow structure that was stable and anti-deformed (Zhang et al. 2016).

10.4.3.3 Polymer-Inorganic Nanofiber Membrane

Scientists and engineers are motivated to produce specific protective materials with a variety of unique qualities due to the catastrophic dangers of air pollution, particularly PM pollution, to human health. Wang et al. developed a novel air filtration membrane with the combination of poly(lactic acid)/titania (PLA/TiO_2) by using electrospun membrane shown in Fig. 11a, b. The hybrid PLA/TiO_2 nanofiber membrane has high filtration performance (99.99%) and enhancing the antibacterial

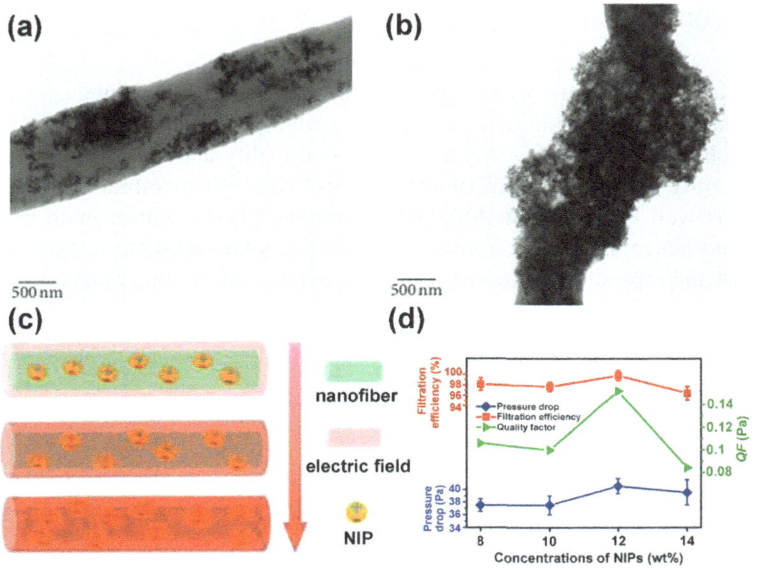

Fig. 10.11 Representation of TEM images of PLA/TiO_2 with **a** 1.5 and **b** 2 wt% concentration of TiO_2 NP, **c** Schematic representation of the mechanism of the diminished shielding effects for electric field around fibers with the increment of NIPs, and **d** Filtration efficiencies, pressure drops and quality factors of PVDF/NIPs-x fibrous media. **a, b** Adapted with permission from Wang et al. (2016). Copyright 2016 Hindawi; **c, d** are reprinted from Zhao et al. (2017), copyright (2017), with permission from American Chemical Society

activity up to 99.5% with a relatively high-pressure drop (128.7 Pa) The antibacterial of PLA/TiO_2 occurs owing to a high surface-to-volume ratio, as the photochemical reaction mainly occurs on its surface (Wang et al. 2016). Vanangamudi et al. combined Al_2O_3 and Ag and prepared hydrophobic, antibacterial PVDF-Ag-Al_2O_3 nanofibrous membrane by electrospinning. Because of the high surface-to-volume ratio, this complex membrane performed well against bacteria. It's worth noting that alumina has antibacterial action as well, but it's not as strong as silver. When Al_2O_3 is added to an electrospun fibrous membrane, the exposure of Ag nanoparticles on the surface is reduced. As a result, the zone of inhibition shrinks slightly as the Al_2O_3 content increases. Increases (Vanangamudi et al. 2015) Zhao et al. developed a low-resistance hybrid nanofiber-based air filter by electrospinning negative ion powder-doped PVDF hybrid nanofiber membranes capable of releasing negative ions and effectively trapping $PM_{2.5}$ pollutants from the air shown in Fig. 10.11. They found that the addition of negative ion powder into PVDF nanofiber membranes has high surface potential because of the high electronegativity of fluorine. So that higher releasing amounts of negative ions (from 789 to 2818 ions cm^{-3}) causes the reducing diameter of the fiber. This method endowed a new way to produce high-efficiency filtration media for different air filtration applications (Zhao et al. 2017).

10.5 Conclusion

In summary, we have provided an overview of the emerging one-dimensional metal based nanocomposites using electrospinning technique, which shows unique properties such as effective surface area, more functionality and higher chemical reactivity to move on for vast areas of applications. The fiber membranes for catalytic activity as well as for the air filtration has shown good attention with high efficiency and hence, is studied from the past several years. Therefore, here we have discussed and represented basic mechanisms and the outstanding outcomes that has been published for the environmental and energy applications using electrospun nanocomposite materials.

Appendix

Abbreviations

COVID-19	Coronavrus Disease 2019
IGA	International Energy Agency's
1D	One-dimensional
PM	Particulate particle matter
VOC	Volatile organic compounds
LSPR	Localized surface plasmon resonance
UV	Ultraviolet
PEC	Photoelectrochemical
HEPA	High-Efficiency Particulate Air
TiO_2	Titanium dioxide
ZnO	Zinc oxide
V_2O_5	Vanadium pentoxide
CdS	Cadmium sulfide
Au	Gold
Ag	Silver
Pt	Platinum
Pd	Palladium
Rh	Rhodium
Ru	Ruthenium
CNT	Carbon nanotubes
PAN	Polyacrylonitrile
PVP	Polyvinylpyrrolidone
PVA	Polyvinyl alcohol
PI	Polyamide
PSU	Polysulfone

References

Ahn YC, Park SK, Kim GT et al (2006) Development of high efficiency nanofilters made of nanofibers. Curr Appl Phys 6:1030–1035. https://doi.org/10.1016/j.cap.2005.07.013

Albetran H, Haroosh H, Dong Y et al (2014) Phase transformations and crystallization kinetics in electrospun TiO_2 nanofibers in air and argon atmospheres. Appl Phys A 116:161–169. https://doi.org/10.1007/s00339-014-8246-1

Al-Hazeem NZA (2018) Nanofibers and electrospinning method. In: Nov Nanomater -Synth Appl, George Kyzas, ed., InTechOpen, pp 191–210. https://doi.org/10.5772/intechopen.72060

Al-Mamun MR, Kader S, Islam MS, Khan MZH (2019) Photocatalytic activity improvement and application of UV-TiO_2 photocatalysis in textile wastewater treatment: a review. J Environ Chem Eng 7:103248. https://doi.org/10.1016/j.jece.2019.103248

Baig N, Kammakakam I, Falath W (2021) Nanomaterials: a review of synthesis methods, properties, recent progress, and challenges. Mater Adv 2:1821–1871. https://doi.org/10.1039/D0MA00807A

Barakat NAM, Erfan NA, Mohammed AA, Mohamed SEI (2020) Ag-decorated TiO_2 nanofibers as Arrhenius equation-incompatible and effective photocatalyst for water splitting under visible light irradiation. Coll Surf A Physicochem Eng Aspects 604:125307. https://doi.org/10.1016/j.colsurfa.2020.125307

Barhate RS, Ramakrishna S (2007) Nanofibrous filtering media: filtration problems and solutions from tiny materials. J Membr Sci 296:1–8. https://doi.org/10.1016/j.memsci.2007.03.038

Barhate RS, Loong CK, Ramakrishna S (2006) Preparation and characterization of nanofibrous filtering media. J Membr Sci 283:209–218. https://doi.org/10.1016/j.memsci.2006.06.030

Barhoum A, Pal K, Rahier H et al (2019) Nanofibers as new-generation materials: from spinning and nano-spinning fabrication techniques to emerging applications. Appl Mater Today 17:1–35. https://doi.org/10.1016/j.apmt.2019.06.015

Bell JNB, Power SA, Jarraud N et al (2011) The effects of air pollution on urban ecosystems and agriculture. Int J Sust Dev World 18:226–235. https://doi.org/10.1080/13504509.2011.570803

Bull K (2008) Cabin air filtration: helping to protect occupants from infectious diseases. Travel Med Infect Dis 6:142–144. https://doi.org/10.1016/j.tmaid.2007.08.004

Bumajdad A, Madkour M (2014) Understanding the superior photocatalytic activity of noble metals modified titania under UV and visible light irradiation. Phys Chem Chem Phys 16:7146–7158. https://doi.org/10.1039/C3CP54411G

Chattopadhyay S, Bysakh S, Mishra PM, De G (2019) In situ synthesis of mesoporous TiO_2 nanofibers surface-decorated with AuAg alloy nanoparticles anchored by heterojunction exhibiting enhanced solar active photocatalysis. Langmuir 35:14364–14375. https://doi.org/10.1021/acs.langmuir.9b02361

Chen Z, Deutsch TG, Dinh HN et al (2013) In: Chen Z, Dinh HN, Miller E (eds) Efficiency definitions in the field of PEC BT—photoelectrochemical water splitting: standards, experimental methods, and protocols. Springer. New York, pp 7–16

Child M, Koskinen O, Linnanen L, Breyer C (2018) Sustainability guardrails for energy scenarios of the global energy transition. Renew Sustain Energy Rev 91:321–334. https://doi.org/10.1016/j.rser.2018.03.079

Daneshvar N, Salari D, Khataee AR (2004) Photocatalytic degradation of azo dye acid red 14 in water on ZnO as an alternative catalyst to TiO_2. J Photochem Photobiol A Chem 162:317–322. https://doi.org/10.1016/S1010-6030(03)00378-2

de Oliveira MCK, Teixeira A, Vieira LC et al (2012) Flow assurance study for waxy crude oils. Energy Fuels 26:2688–2695. https://doi.org/10.1021/ef201407j

Ding B, Li C, Miyauchi Y et al (2006) Formation of novel 2D polymer nanowebs via electrospinning. Nanotechnology 17:3685–3691. https://doi.org/10.1088/0957-4484/17/15/011

Ding B, Wang X, Yu J, Wang M (2011) Polyamide 6 composite nano-fiber/net functionalized by polyethyleneimine on quartz crystal microbalance for highly sensitive formaldehyde sensors. J Mater Chem 21:12784–12792. https://doi.org/10.1039/C1JM11847A

Furube A, Hashimoto S (2017) Insight into plasmonic hot-electron transfer and plasmon molecular drive: new dimensions in energy conversion and nanofabrication. NPG Asia Mater 9:e454–e454. https://doi.org/10.1038/am.2017.191

Gao C, Low J, Long R et al (2020) Heterogeneous single-atom photocatalysts: fundamentals and applications. Chem Rev 120:12175–12216. https://doi.org/10.1021/acs.chemrev.9b00840

Ge M, Cao C, Huang J et al (2016) A review of one-dimensional TiO_2 nanostructured materials for environmental and energy applications. J Mater Chem A 4:6772–6801. https://doi.org/10.1039/C5TA09323F

Ge M, Cai J, Iocozzia J et al (2017) A review of TiO_2 nanostructured catalysts for sustainable H_2 generation. Int J Hydrogen Energy 42:8418–8449. https://doi.org/10.1016/j.ijhydene.2016.12.052

Gibson P, Schreuder-Gibson H, Rivin D (2001) Transport properties of porous membranes based on electrospun nanofibers. Coll Surf A Physicochem Eng Aspects 187–188:469–481. https://doi.org/10.1016/S0927-7757(01)00616-1

Gnanaprakasam A, Sivakumar VM, Thirumarimurugan M (2015) Influencing parameters in the photocatalytic degradation of organic effluent via nanometal oxide catalyst: a review. Indian J Mater Sci 2015:601827. https://doi.org/10.1155/2015/601827

Grafe T, Graham KM (2003) Nanofiber webs from electrospinning. https://doi.org/10.5772/intechopen.7206

Gupta SM, Tripathi M (2011) A review of TiO_2 nanoparticles. Chin Sci Bull 56:1639. https://doi.org/10.1007/s11434-011-4476-1

Hao Y, Shao X, Li B et al (2015) Mesoporous TiO_2 nanofibers with controllable Au loadings for catalytic reduction of 4-nitrophenol. Mater Sci Semicond Process 40:621–630. https://doi.org/10.1016/j.mssp.2015.07.026

Ibhadon AO, Fitzpatrick P (2013) Heterogeneous photocatalysis: recent advances and applications. Catalysts 3

Ju J, Chen X, Shi Y et al (2013) A novel TiO_2 nanofiber supported PdAg catalyst for methanol electro-oxidation. Energy 59:478–483. https://doi.org/10.1016/j.energy.2013.06.058

Karthikeyan C, Arunachalam P, Ramachandran K et al (2020) Recent advances in semiconductor metal oxides with enhanced methods for solar photocatalytic applications. J Alloys Compd 828:154281. https://doi.org/10.1016/j.jallcom.2020.154281

Kelly KL, Coronado E, Zhao LL, Schatz GC (2003) The optical properties of metal nanoparticles: the influence of size, shape, and dielectric environment. J Phys Chem B 107:668–677. https://doi.org/10.1021/jp026731y

Kriegel I, Scotognella F, Manna L (2017) Plasmonic doped semiconductor nanocrystals: properties, fabrication, applications and perspectives. Phys Rep 674:1–52. https://doi.org/10.1016/j.physrep.2017.01.003

Kudo A, Miseki Y (2009) Heterogeneous photocatalyst materials for water splitting. Chem Soc Rev 38:253–278. https://doi.org/10.1039/B800489G

Kudo A, Kato H, Tsuji I (2004) Strategies for the development of visible-light-driven photocatalysts for water splitting. Chem Lett 33:1534–1539. https://doi.org/10.1246/cl.2004.1534

Lakshmanan A, Gavali DS, Thapa R, Sarkar D (2021) Synthesis of CTAB-functionalized large-scale nanofibers air filter media for efficient $PM_{2.5}$ capture capacity with low airflow resistance. ACS Appl Polym Mater 3:937–948. https://doi.org/10.1021/acsapm.0c01203

Leung WW-F, Hung C-H, Yuen P-T (2010) Effect of face velocity, nanofiber packing density and thickness on filtration performance of filters with nanofibers coated on a substrate. Sep Purif Technol 71:30–37. https://doi.org/10.1016/j.seppur.2009.10.017

Li J, Chu D (2018) 4-energy band engineering of metal oxide for enhanced visible light absorption. In: Lin Z, Ye M, Wang MBT-MPM for E (eds) Woodhead publishing in materials. Woodhead Publishing, pp 49–78

Li J, Wu N (2015) Semiconductor-based photocatalysts and photoelectrochemical cells for solar fuel generation: a review. Catal Sci Technol 5:1360–1384. https://doi.org/10.1039/C4CY00974F

Linic S, Christopher P, Ingram DB (2011) Plasmonic-metal nanostructures for efficient conversion of solar to chemical energy. Nat Mater 10:911–921. https://doi.org/10.1038/nmat3151

Linsebigler AL, Lu G, Yates JT (1995) Photocatalysis on TiO_2 surfaces: principles, mechanisms, and selected results. Chem Rev 95:735–758. https://doi.org/10.1021/cr00035a013

Liu B, Zhao X, Terashima C et al (2014) Thermodynamic and kinetic analysis of heterogeneous photocatalysis for semiconductor systems. Phys Chem Chem Phys 16:8751–8760. https://doi.org/10.1039/C3CP55317E

Liu C, Hsu P-C, Lee H-W et al (2015) Transparent air filter for high-efficiency $PM_{2.5}$ capture. Nat Commun 6:6205. https://doi.org/10.1038/ncomms7205

Luther JM, Jain PK, Ewers T, Alivisatos AP (2011) Localized surface plasmon resonances arising from free carriers in doped quantum dots. Nat Mater 10:361–366. https://doi.org/10.1038/nmat3004

Maeda K (2011) Photocatalytic water splitting using semiconductor particles: history and recent developments. J Photochem Photobiol C: Photochem Rev 12:237–268. https://doi.org/10.1016/j.jphotochemrev.2011.07.001

Meng A, Zhu B, Zhong B, et al (2017) Direct Z-scheme TiO_2/CdS hierarchical photocatalyst for enhanced photocatalytic H_2-production activity. Appl Surf Sci 422:518–527. https://doi.org/10.1016/j.apsusc.2017.06.028

Misra M, Singh N, Gupta RK (2017) Enhanced visible-light-driven photocatalytic activity of Au@Ag core–shell bimetallic nanoparticles immobilized on electrospun TiO_2 nanofibers for degradation of organic compounds. Catal Sci Technol 7:570–580. https://doi.org/10.1039/C6CY02085B

Morozov VN, Mikheev AY (2012) Water-soluble polyvinylpyrrolidone nanofilters manufactured by electrospray-neutralization technique. J Membr Sci 403–404:110–120. https://doi.org/10.1016/j.memsci.2012.02.028

Mukherjee S, Libisch F, Large N et al (2013) Hot electrons do the impossible: plasmon-induced dissociation of H_2 on Au. Nano Lett 13:240–247. https://doi.org/10.1021/nl303940z

Nalbandian MJ, Greenstein KE, Shuai D et al (2015) Tailored synthesis of photoactive TiO_2 nanofibers and Au/TiO_2 nanofiber composites: structure and reactivity optimization for water treatment applications. Environ Sci Technol 49:1654–1663. https://doi.org/10.1021/es502963t

Nguyen VN, Nguyen MV, Nguyen TH et al (2020) Surface-modified titanium dioxide nanofibers with gold nanoparticles for enhanced photoelectrochemical water splitting. Catalysts 10

Nicoletti G, Arcuri N, Nicoletti G, Bruno R (2015) A technical and environmental comparison between hydrogen and some fossil fuels. Energy Convers Manag 89:205–213. https://doi.org/10.1016/j.enconman.2014.09.057

Ohtani B (2014) Revisiting the fundamental physical chemistry in heterogeneous photocatalysis: its thermodynamics and kinetics. Phys Chem Chem Phys 16:1788–1797. https://doi.org/10.1039/C3CP53653J

Patil J V, Mali SS, Kamble AS et al (2017) Electrospinning: a versatile technique for making of 1D growth of nanostructured nanofibers and its applications: an experimental approach. Appl Surf Sci 423:641–674. https://doi.org/10.1016/j.apsusc.2017.06.116

Paul KK, Giri PK (2018) In: Wandelt KBT-E of IC (ed) Plasmonic metal and semiconductor nanoparticle decorated TiO_2-based photocatalysts for solar light driven photocatalysis. Elsevier, Oxford, pp 786–794

Pawar M, Topcu Sendoğdular S, Gouma P (2018) A brief overview of TiO_2 photocatalyst for organic dye remediation: case study of reaction mechanisms involved in Ce–TiO_2 photocatalysts system. J Nanomater 2018:5953609. https://doi.org/10.1155/2018/5953609

Perera F (2018) Pollution from fossil-fuel combustion is the leading environmental threat to global pediatric health and equity: solutions exist. Int J Environ Res Publ Health 15

Pope CA, Burnett RT, Thurston GD et al (2004) Cardiovascular mortality and long-term exposure to particulate air pollution. Circulation 109:71–77. https://doi.org/10.1161/01.CIR.0000108927.80044.7F

Qin X-H, Wang S-Y (2006) Filtration properties of electrospinning nanofibers. J Appl Polym Sci 102:1285–1290. https://doi.org/10.1002/app.24361
Ramskill EA, Anderson WL (1951) The inertial mechanism in the mechanical filtration of aerosols. J Coll Sci 6:416–428. https://doi.org/10.1016/0095-8522(51)90013-X
Reghunath S, Pinheiro D, Sunaja Devi KR (2021) A review of hierarchical nanostructures of TiO_2: advances and applications. Appl Surf Sci Adv 3:100063. https://doi.org/10.1016/j.apsadv.2021.100063
Ren Y, Qu Z, Du Y et al (2017) Air quality and health effects of biogenic volatile organic compounds emissions from urban green spaces and the mitigation strategies. Environ Pollut 230:849–861. https://doi.org/10.1016/j.envpol.2017.06.049
Sahay R, Kumar PS, Sridhar R et al (2012) Electrospun composite nanofibers and their multifaceted applications. J Mater Chem 22:12953–12971. https://doi.org/10.1039/C2JM30966A
Sambaer W, Zatloukal M, Kimmer D (2011) 3D modeling of filtration process via polyurethane nanofiber based nonwoven filters prepared by electrospinning process. Chem Eng Sci 66:613–623. https://doi.org/10.1016/j.ces.2010.10.035
Schneider J, Matsuoka M, Takeuchi M et al (2014) Understanding TiO_2 photocatalysis: mechanisms and materials. Chem Rev 114:9919–9986. https://doi.org/10.1021/cr5001892
Smoukov SK, Tian T, Vitchuli N et al (2015) Scalable liquid shear-driven fabrication of polymer nanofibers. Adv Mater 27:2642–2647. https://doi.org/10.1002/adma.201404616
Thavasi V, Singh G, Ramakrishna S (2008) Electrospun nanofibers in energy and environmental applications. Energy Environ Sci 1:205–221. https://doi.org/10.1039/B809074M
The International Energy Agency (2011) World energy outlook 2011. http://www.worldenergyoutlook.org. Accessed 13 May 2011
Tian J, Zhao Z, Kumar A et al (2014) Recent progress in design, synthesis, and applications of one-dimensional TiO_2 nanostructured surface heterostructures: a review. Chem Soc Rev 43:6920–6937. https://doi.org/10.1039/C4CS00180J
Topare N (2013) A review on parameters affecting the photocatalytic degradation of dyes using photocatalysts. 66th Annual Session of Indian Institute of Chemical Engineers Hosted at Institute of Chemical Technology, Mumbai
Vanangamudi A, Hamzah S, Singh G (2015) Synthesis of hybrid hydrophobic composite air filtration membranes for antibacterial activity and chemical detoxification with high particulate filtration efficiency (PFE). Chem Eng J 260:801–808. https://doi.org/10.1016/j.cej.2014.08.062
Wang C-S (2001) Electrostatic forces in fibrous filters—a review. Powder Technol 118:166–170. https://doi.org/10.1016/S0032-5910(01)00307-2
Wang Q, Domen K (2020) Particulate photocatalysts for light-driven water splitting: mechanisms, challenges, and design strategies. Chem Rev 120:919–985. https://doi.org/10.1021/acs.chemrev.9b00201
Wang Z, Pan Z, Wang J, Zhao R (2016) A novel hierarchical structured poly(lactic acid)/titania fibrous membrane with excellent antibacterial activity and air filtration performance. J Nanomater 2016:6272983. https://doi.org/10.1155/2016/6272983
Wang N, Raza A, Si Y et al (2013) Tortuously structured polyvinyl chloride/polyurethane fibrous membranes for high-efficiency fine particulate filtration. J Coll Interface Sci 398:240–246. https://doi.org/10.1016/j.jcis.2013.02.019
Wang N, Zhu Z, Sheng J et al (2014) Superamphiphobic nanofibrous membranes for effective filtration of fine particles. J Coll Interface Sci 428:41–48. https://doi.org/10.1016/j.jcis.2014.04.026
Wenderich K, Mul G (2016) Methods, mechanism, and applications of photodeposition in photocatalysis: a review. Chem Rev 116:14587–14619. https://doi.org/10.1021/acs.chemrev.6b00327
Weng B, Liu S, Tang Z-R, Xu Y-J (2014) One-dimensional nanostructure based materials for versatile photocatalytic applications. RSC Adv 4:12685–12700. https://doi.org/10.1039/C3RA47910B

Wu M-C, Hiltunen J, Sápi A et al (2011) Nitrogen-doped anatase nanofibers decorated with noble metal nanoparticles for photocatalytic production of hydrogen. ACS Nano 5:5025–5030. https://doi.org/10.1021/nn201111j

Xue J, Wu T, Dai Y, Xia Y (2019) Electrospinning and electrospun nanofibers: methods, materials, and applications. Chem Rev 119:5298–5415. https://doi.org/10.1021/acs.chemrev.8b00593

Yang C (2012) Aerosol filtration application using fibrous media—an industrial perspective. Chin J Chem Eng 20:1–9. https://doi.org/10.1016/S1004-9541(12)60356-5

Yang X, Wang D (2018) Photocatalysis: from fundamental principles to materials and applications. ACS Appl Energy Mater 1:6657–6693. https://doi.org/10.1021/acsaem.8b01345

Yesudasu V, Pradhan HS, Pandya RJ (2021) Recent progress in surface plasmon resonance based sensors: a comprehensive review. Heliyon 7:e06321. https://doi.org/10.1016/j.heliyon.2021.e06321

Yu J, Hai Y, Cheng B (2011) Enhanced photocatalytic H_2-production activity of TiO_2 by $Ni(OH)_2$ cluster modification. J Phys Chem C 115:4953–4958. https://doi.org/10.1021/jp111562d

Yun KM, Hogan CJ, Matsubayashi Y et al (2007) Nanoparticle filtration by electrospun polymer fibers. Chem Eng Sci 62:4751–4759. https://doi.org/10.1016/j.ces.2007.06.007

Zhang Y, Lim CT, Ramakrishna S, Huang ZM (2005) Recent development of polymer nanofibers for biomedical and biotechnological applications. J Mater Sci Mater Med 16:933–946. https://doi.org/10.1007/s10856-005-4428-x

Zhang H, Chen G, Bahnemann DW (2009) Photoelectrocatalytic materials for environmental applications. J Mater Chem 19:5089–5121. https://doi.org/10.1039/B821991E

Zhang Z, Wang Z, Cao S-W, Xue C (2013) Au/Pt nanoparticle-decorated TiO_2 nanofibers with plasmon-enhanced photocatalytic activities for solar-to-fuel conversion. J Phys Chem C 117:25939–25947. https://doi.org/10.1021/jp409311x

Zhang S, Liu H, Yin X et al (2016) Anti-deformed polyacrylonitrile/polysulfone composite membrane with binary structures for effective air filtration. ACS Appl Mater Interfaces 8:8086–8095. https://doi.org/10.1021/acsami.6b00359

Zhao X, Li Y, Hua T et al (2017) Low-resistance dual-purpose air filter releasing negative ions and effectively capturing $PM_{2.5}$. ACS Appl Mater Interfaces 9:12054–12063. https://doi.org/10.1021/acsami.7b00351

Zhou W, Liu H, Boughton RI et al (2010) One-dimensional single-crystalline Ti–O based nanostructures: properties, synthesis, modifications and applications. J Mater Chem 20:5993–6008. https://doi.org/10.1039/B927224K

Zhu S, Nie L (2021) Progress in fabrication of one-dimensional catalytic materials by electrospinning technology. J Ind Eng Chem 93:28–56. https://doi.org/10.1016/j.jiec.2020.09.016

Zhu S, Wang D (2017) Photocatalysis: basic principles, diverse forms of implementations and emerging scientific opportunities. Adv Energy Mater 7:1700841. https://doi.org/10.1002/aenm.201700841

Zhu C, Lin C-H, Cheung CS (2000) Inertial impaction-dominated fibrous filtration with rectangular or cylindrical fibers. Powder Technol 112:149–162. https://doi.org/10.1016/S0032-5910(99)00315-0

Zhu M, Han J, Wang F et al (2017) Electrospun nanofibers membranes for effective air filtration. Macromol Mater Eng 302:1600353. https://doi.org/10.1002/mame.201600353

Chapter 11
Popular Synthesis Roots of Metal Nanocomposites

Ritesh Kumar Chourasia, Ankita Srivastava, Nitesh K. Chourasia, and Narendra Bihari

Abstract The development of human society and economy, the rapid growth of the global population have significantly aggravated the energy crisis due to excess energy consumption. Even, this ecological destruction has become a serious threat to a sustainable society. Thus, the exploration of renewable energy along with ecological conservation is becoming the global spotlight. Nowadays, the global society is truly focused on energy-related research. The best alternative for energy is renewable energy. However, the main concern with this type of energy is that it is not available continuously. Hence, it is necessary to store this type of energy when it is available so that it can be utilized whenever needed. Novel materials like metal oxide nanomaterial and nanocomposites, Transition metal chalcogenides nanocomposites, Metalo-Graphene nanocomposites, etc. provide opportunities to address these challenges for various applications like electrochemical performance as electrode material for supercapacitors, for energy storage and conservation and in the ecological direction, especially as sensitive layers of gas sensors, sorbents, and photocatalysts. Further, Demand for ecological protection is gaining more public attention and legislative support. The development in industrial and technological sectors results in severe ecological issues, such as environmental contamination and energy shortage. Therefore, the development of these novel nanocomposites that can effectively act towards ecological remediation is necessary to overcome the detrimental ecological impacts. Thus, these efforts must alleviate the reliance and dependency on the consumption of fossil fuels. Therefore, in this chapter, the various popular and effective synthesis roots as Laser ablation, arc discharge, vapor–liquid–solid (VLS) techniques, scalable electron beam irradiation, liquid exfoliation, etc. method will be discussed in detail.

R. K. Chourasia (✉) · N. Bihari
University Department of Physics, Lalit Narayan Mithila University, Darbhanga 846004, India

A. Srivastava
Department of Physics, Faculty of Science and Humanities, Darbhanga College of Engineering, Darbhanga 846005, India

N. K. Chourasia
School of Physics, Indian Institute of Science Education and Research, Thiruvananthapuram, Kerala 695551, India

S. P. Singh et al. (eds.), *Metal Nanocomposites for Energy and Environmental Applications*, Energy, Environment, and Sustainability,
https://doi.org/10.1007/978-981-16-8599-6_11

Keywords Transition metal chalcogenide nanocomposites · MOCVD · VLS · Laser ablation · Arc discharge · Scalable electron beam irradiation

11.1 Introduction

As human society and economy evolve, the escalating energy crises, as well as the accompanying environmental deterioration and ecological catastrophe, have become severe risks to a sustainable society (Goodenough 2015; Wang et al. 2017; Xiong et al. 2015; Li et al. 2016a). As a result, clean and renewable energy research is becoming a global priority. Solar energy, wind energy, geothermal energy, and other renewable energy sources have all received a lot of attention. This can help to reduce reliance on fossil fuel consumption. However, maximizing the use of intermittent renewable energy sources necessitates the development of improved energy storage and conversion technologies (Wang et al. 2017; Li et al. 2016b; Shan et al. 2016; Cheng et al. 2017). The supercapacitors, solar cells, electrochemical rechargeable batteries, lithium-air batteries, lithium-sulfur batteries (LSBs), lithium-ion batteries (LIBs), and other electrocatalysis processes [e.g., evolution reaction/oxygen reduction (OER/ORR), hydrogen evolution reaction (HER)] have already gained featured attention (Zhang et al. 2016a; Ma et al. 2016). Even though great research in the domains of energy storage and conversion has flourished all over the world, the use of high-efficiency electrode materials, electrocatalysts, and photocatalysts still have to be pushed forward (Sheng et al. 2016). Due to their tunable stoichiometric compositions (Lai et al. 2016), unique crystal structures (Wang and Xu 2016), and rich redox sites, as well as relatively higher electrical conductivity in comparison to their transition metal chalcogenides (TMCs, mainly sulfides and selenides) over the past decades, transition metal chalcogenides (TMCs, mainly sulfides and selenides) have attracted increasing research interest as potential electrode materials for energy storage and conversion TMCs, for example, have a greater theoretical special capacity than regular anode materials (graphite) in LIBs based on the insertion/reinsertion mechanism (Yu et al. 2016; Rui et al. 2014; Huang et al. 2013). Concerning metal-air batteries, TMCs have demonstrated excellent electrocatalytic efficiency (Ganesan et al. 2015). For dye-sensitized solar cells (DSSCs), a key type of power conversion system, the TMCs used as counter electrodes generally give cells with Pt electrodes a superior power conversion performance. Electric and photoelectron-chemical water separation is seen as a promising path for hydrogen production. Water division as in metal-air battery applications, TMCs have also been a type of important catalyst for water dissociation, and due to their seminal characteristics, special band structures, and unique electronic set-up, they show a comparable catalytic reactivity for those of precious catalysts (Zou and Zhang 2015). However, the challenges of limited special surface area, inferior reactivity, low electron/ion transfer rate, and high recombination rate of electrons and holes (Kagkoura et al. 2017) will remain for practical applications of TMCs in energy devices. Due to its unique physicochemical features, graphene as a two-dimensional honeycomb sp^2-hybridized carbon nanosheets with single-atom thickness has attracted a lot of attention in the energy research arena. According to a theoretical study, perfect graphene has a particular surface area of 2600 m^2/g (Tian et al. 2015; Gao et al. 2016), which is substantially larger than that of standard graphite powder (10 m^2/g) and carbon black (900 m^2/g) (Bonaccorso et al. 2015). Furthermore, graphene's excellent electrical conductivity, mechanical strength and flexibility, and charge carrier mobility make it a promising material for next-generation energy storage and conversion (Lai et al. 2016; Wang and Xu 2016; Geim and Novoselov 2007; Lee et al. 2008; Geim 2009; Xiang et al. 2016; El-Kady et al. 2016; Zhao et al. 2017; Ji et al. 2015; Qiu et al. 2016; Peng et al. 2016a). Because of graphene's two-dimensional thin sheet structure, which makes it an ideal support for the formation of inorganic nanomaterials (Lv et al. 2016; Chen et al. 2015; Peng et al. 2016b), graphene-based nanostructured materials have been widely researched with the development of graphene technologies in electrochemical experiments. Synthetic approaches for creating hybrid graphene-based nanomaterials for cutting-edge energy storage and conversion applications have recently been reviewed (Qiu et al. 2016; Peng et al. 2016a, b; Lv et al. 2016; Chen et al. 2015; Raccichini et al. 2015). Graphene can be used as a matrix to prevent inorganic materials from self-aggregating by providing numerous pores for mass transport and interconnecting electrically conductive routes for electron transmission (Peng et al. 2016b). As a result, graphene-based materials have been bestowed with increased capabilities derived from both separate counterparts, allowing them to effectively overcome the deficiencies of independent components. As a result, TMC-graphene composites can be used as effective electrode active materials for enhanced energy storage and conversion. In this chapter, The material structures and synthetics of TMCs/graphene are described in various energy applications and their electrochemical performance is emphasized. The main design concepts for electrode materials are also provided for future energy devices.

Furthermore, in terms of electronic structure, physical, chemical, and electromagnetic properties, nanomaterials based on metal oxides (nanostructured and nano dispersed) represent a diverse family of materials (Fierro 2006). Metal oxide nanoparticles and nanocomposites based on them are becoming more widely employed in applied ecology, particularly as adsorbents and photocatalysts, as well as a material for the manufacturing of environmental monitoring systems. Adsorption materials made of nanosized metal oxides have a large specific surface area, high capacity, quick kinetics, and specific affinity for a variety of pollutants (Kumar and Gopinath 2016; Zhang et al. 2016b; Santhosh et al. 2016; Werkneh and Renein 2019; Cerro-Lopez and Méndez-Rojas 2017). Nanostructured metal oxides can be employed in photocatalytic processes to oxidize organic molecules that are not biochemically degraded, and it is regarded as the most promising prior treatment of aqueous solutions by their usage (Cerro-Lopez and Méndez-Rojas 2017; Pelaez et al. 2012). Metal oxide nanostructures employed as sensitive layers of chemo resistive gas sensors in environmental monitoring have high levels of the sensory signal because of the wide surface area; hence a larger capacity for adsorption (Xu et al. 1991; Miller et al. 2014, 2006). As a result of the major advantages over bulk analogs and of course since they have a high promise of new forms of adsorbents, photocatalysts, and sensitive layers of gas sensors based on them, nanosized metal oxide materials are of considerable interest. Nanostructured and nano-distributing oxides, however, are also far less likely to result in contamination of the environment by nanoparticles. In this scenario, nanocomposites of metal oxide—a highly intriguing form of nanomaterial due to its features may be promising and may, by order, outweigh its phase characteristics. The use of the latter prevents the loss of nanoparticles of metal oxide owing to the stability of the nanoparticles in the composite material matrix which, after the completion of the sorption and photocatalysis process, has a beneficial effect on the separation process (Santhosh et al. 2016). Furthermore, nanocomposites have certain, often unique, physical, and chemical characteristics because of their structure, they may be used in a broad range of disciplines, including the development of novel materials for medicine, energy, and ecology. This study aims at considering the characteristics and possible usage of the nanomaterials and nanocomposites of metal oxides for environmental applications based on TiO_2, ZnO, SnO_2, ZrO_2, and Fe_3O_4. The present book chapter considers their physical, chemical, absorptive, and photocatalytic characteristics to ensure their more effective use is understood. We also evaluated the best synthesizing methods of individual oxides for facilitating their use and the available ways to create nanocomposite metal oxide materials in the literature.

11.2 Popular Synthesis Roots (for Metal Nanostructure/Nanocomposites)

Diverse techniques have been devised in recent years to create nanostructured composites of metal sulfides and selenides, as well as multilayer metal dichalcogenides, using various growth processes. High reaction temperatures or significant activation energies are required to overcome the activation barrier associated with bending, which is a property of most of the nanostructures. Even though different synthetic methods for chalcogenide nanocomposites have been created, their development and mechanism are still under debate. Several approaches were used in the last few decades, including:

- Laser ablation
- Arc discharge
- Microwave induced plasma
- Scalable Electron beam irradiation
- Metal–organic chemical vapor deposition (MOCVD)
- Vapor–liquid–solid (VLS) growth
- Liquid exfoliation

were utilized for the synthesis of metal nanocomposites structures. Further, we shall take a glimpse of these methods in the following subsections.

11.2.1 Laser Ablation

Laser ablation is the process of removing material from a surface using laser irradiation. The phrase "laser ablation" is used to distinguish between "laser evaporation," which is the heating and evaporation of material in thermodynamic equilibrium, and "laser ablation," which is the creation of nonequilibrium vapor/plasma conditions at the surface by a strong laser pulse. Figure 11.1 depicts a common laser ablation schematic diagram.

A pulsed laser (CO_2 laser, Nd-YAG laser, ArF excimer laser, or XeCl excimer laser) and an ablation chamber are the two most important components of a laser ablation system. The tremendous intensity of the laser beam causes significant light absorption on the target's surface, causing the temperature of the absorbing material to rapidly rise. As a result, the material on the target's surface vaporizes, resulting in a laser plume. Without any chemical reaction, evaporated materials can condense into clusters and particles in some situations. In certain situations, the vaporized substance interacts with the reactants supplied to produce new materials. The nanoparticles can then be deposited on a substrate using a drop-coating or screen-printing method. Parilla et al. produced approx. 90 Metal Chalcogenide Nanostructures nano-octahedra of MoS_2 using the same approach, which was later expanded to $MoSe_2$ (Parilla et al. 1999), using same synthesis methodology, $NiCl_2$ and fullerene-like particles by laser ablation (Hacohen et al. 1998, 2002, 2003; Parilla et al. 1999; Sen et al. 2001; Schuffenhauer et al. 2005; Nath et al. 2004). Sen et al. (2001)

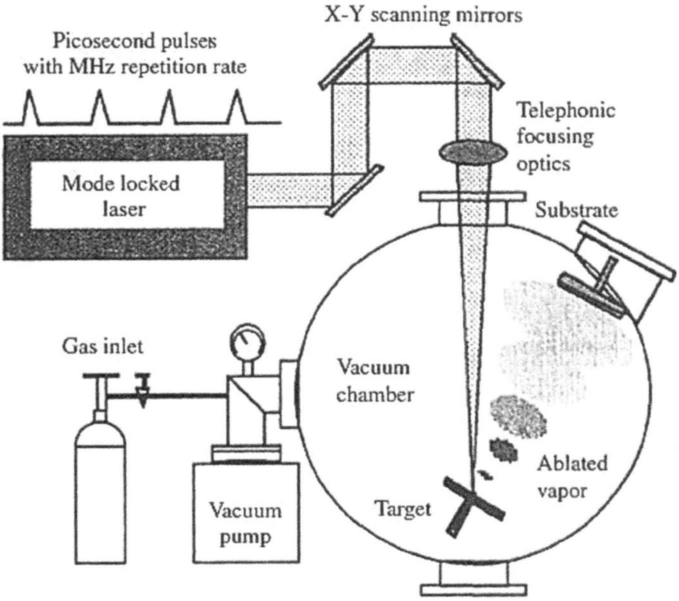

Fig. 11.1 Schematic nanostructure formation through LASER ablation techniques

obtained metal-filled and hollow nanostructures of MoS_2 and WS_2 a few years later. The hexagonal WS_2 nanostructure was created by laser ablation in water (Hu et al. 2006a). Hong et al. (2003) synthesized made up of SnS_2 and SnS.

11.2.2 Arc Discharge

The arc discharge approach has also been used for (carbon) fullerene production. Figure 11.2 depicts the root's schematic configuration. The source material is vaporized through an arc discharge between electrodes, followed by condensation, nucleation, and development of nanoparticles in arc discharge synthesis. Metal or metal oxide nanoparticles, CNTs, fullerenes, or graphene films are all made using the arc discharge process. This approach yields very little amounts of metal nanomaterials (milligrams), yet it is critical for the synthesis of CNTs and fullerenes, as well as core–shell and encapsulated structures. Transition metals (iron, copper, and nickel), iron-carbon core–shell structures, and silica-coated cobalt nano-capsules are just a few examples of metal and ceramic nanomaterials created using this approach. The electrodes are made of source material (anode) and carbon or tungsten (cathode) in the arc discharge method. The background gas for plasma source and chemical reaction (inert or reactive), a quenching gas to cool the vapor for condensation, a source of arc current are also needed in the arc discharge method. Further, if needed, a source of passivation or encapsulation of the nanomaterials produced are among the components. A solid metal electrode or a combination of source materials with or without carbon can be used as the source material. For an inert gas plasma source, the background gas is generally argon or helium; for a reactive gas, the background

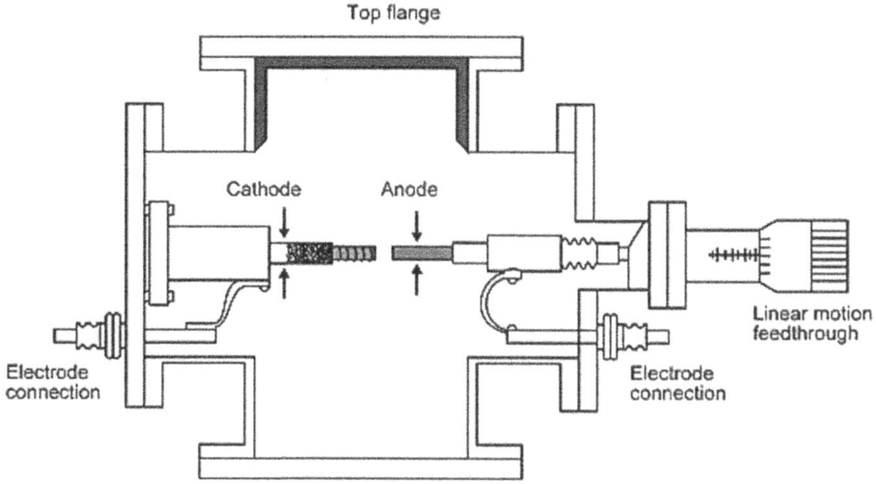

Fig. 11.2 Synthesis of nanocomposites using arc-discharge technique

gas is usually methane, oxygen, hydrogen, ethanol, or diborane, which is ionized to create plasma but may also be a reactant. Methane, for example, may serve as both a supply of carbon for core–shell structures and a source of hydrogen for plasma. The reaction takes place in a high-vacuum environment, and the temperature around the electrode rises quickly before swiftly falling to generate a sharp temperature gradient that allows quenching to take place. Evaporation of the source material causes supersaturation, which is followed by fast quenching, early nucleation, growth by condensation, and eventually collection, as in other vapor phase syntheses. High pressure, high temperature, and submerged arc discharge are some of the arc discharge technologies available. Chhowalla et al. (2003) were the first to use this synthesis technique to effectively synthesize MoS_2 nanostructures in the form of a thin film with good lubricating properties. Following that, other fullerene-like particles, some of which were filled with other materials, such as CoS within WS_2 nanostructures, were produced (Hu et al. 2004, 2006b; Sano et al. 2003; Si et al. 2005; Alexandrou et al. 2003). Filled IFs are particularly intriguing since they may have an advantage over hollow particles in terms of pressure load stability.

11.2.3 Microwave-Induced Plasma

Figure 11.3 depicts the gas phase processes used in the microwave plasma process. The reactive molecules in plasma are partially ionized or dissociated. This is a highly active state and, as a result, thermodynamically conceivable, but kinetically blocked processes can be conducted at lower temperatures under normal conditions. In an oscillating electrical field of frequency f, the energy transfer E to a charged particle with mass m is:

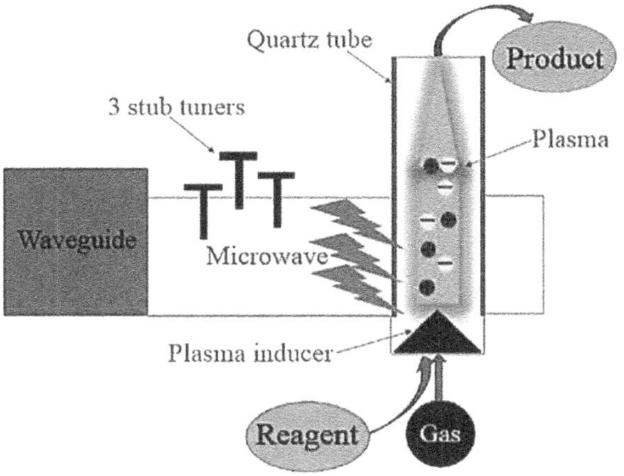

Fig. 11.3 Synthesis of nanocomposites using microwave induced plasma technique

$$E = (1/(m.f^2)) \quad (11.1)$$

Because electrons have a lower mass than ions or radicals, they receive substantially more energy than the latter. As a result, the "temperature" of free electrons is significantly higher than that of ions (Li et al. 2016a). Higher frequencies result in decreased energy transfer, allowing for lower temperatures. Collisions with uncharged molecules and atoms must be considered in a plasma. As a result, the relationship is changed:

$$E = \frac{1}{m}.\frac{z}{f^2 + z^2} \quad (11.2)$$

The energy transfer is a function of the gas pressure as the collision frequency increases with rising gas pressure. For $z < f$, the energy transfer increases with the gas pressure, peaks at $f = z$, and falls for $z > f$. There are two typical industrial frequencies: 2.45 and 0.915 GHz. A non-equilibrium plasma is formed with relatively low overall temperatures in the system due to the drastically varied energy content of the individual species. The temperatures (measured directly behind the plasma zone) of the 2.45 GHz equipment can be adjusted between 100 and 500 °C, while the temperatures of the 0.915 GHz equipment can be set between 500 and 800 °C. In comparison to traditional gas phase reactors, these temperatures are low. The pressure in the system can be varied between 10 and 50 mbar. The particles' residence period in the plasma zone is measured in milliseconds under these experimental conditions. The particles generated in the plasma have electric charges of the same sign, which is a unique property of this process. As a result, they reject one another, preventing particle growth through agglomeration. As a result, the powders' particle size distribution is extremely narrow. Because of the electric charges, a cascaded process can coat the particles individually with a second phase. Broad diversity of chemicals is made available by the microwave-induced plasma technique. MQ_2 (M = Mo, W; Q = S, S), partly also in the form of fullerene-like particles, was achieved by reacting to $M(CO)_6$ with H_2S or with $SeCl_4$ (Vollath and Szabo 2000, 1998). MQ_2 (M = Mo, W). In addition, SnS_2 and ZrS_2 nanoclusters were developed with the same configuration. The plasma produced by microwaves of H_2S and N_2/H_2 (Brooks et al. 2006) were derived from nanoparticles of WO_3, ZrS_3, and HfS_3, and Brooks et al. were obtained in both WS_2 and HfS_2, and ZrS_2, respectively.

11.2.4 Scalable Electron Beam Irradiation

In the electron beam process (Fig. 11.4), the product is bombarded with energy-efficient electrons, which causes the target material to move into a cascade of such electrons. The e-beam technique uses its radiation source from high-energy electrons. The electrons produced by conventional power are speeded up by an accelerator to close to the speed of light. The electrons are concentrated in a scanning horn of a given

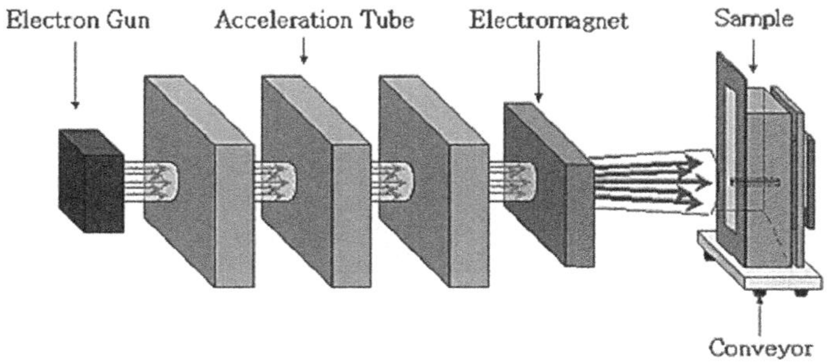

Fig. 11.4 Synthesis of nanostructure using scalable electron beam irradiation lithographic technique

size and scanned in a sweeping motion to create an electron drape. The product is subsequently transmitted at a strictly controlled and measured rate through the scanning drape. Behind the radiation shield itself, often a massive concrete structure is the procedure that stops the cell from exiting the radiation. The accelerated electrons inactivate microorganisms when scanning happens. Samples can produce electron beam irradiation flaws and layered materials subsequently folding. The MoS_2 nanostructure has consequently been synthesized (Jose-Yacaman et al. 1996). Jose-Yacaman has shown that facets are twisted in the inside region of these onion-like structures (small radii). The electron beam irradiation of appropriate precursor materials (Galván et al. 2000, 2001; Remskar et al. 2002) could produce NbSe2, MoSe2, and silver-containing NBS_2 NTs. The NTs containing nanorods were blended with NbSe2. Surface Functionalization and Growth Mechanism (Brooks et al. 2006).

11.2.5 Metal–Organic Chemical Vapor Deposition (MOCVD) Technique

The MOCVD approach, as shown in Fig. 11.5, is a well-known method to produce nano-thin films and nanoparticles. An MOCVD gas-phase reaction is one technique to eliminate solid-state diffusion as a reaction parameter. The chalcogenide nanoparticles, created by thermal decomposition of 92 Metal Chalcogenide Nanostructures, are synthesized in a gaseous period as a reaction of a naked metal, matching to a corresponding metal carbonyl and sulfur or selenium, respectively. One could expect that the lamellar sheet pieces continue to develop and loosen at their ends until they roll up into onion or nanotube structures in a subsequent growth step. The special benefits of this strategy are: The general process may be extended to other MQ_2 members if volatile precursors are available. The procedure with the help of microreactors can be scaled up for synthesizing large quantities. The application of very poisonous H_2S and H_2Se reactants is prevented. Quantitatively, an easy and rapid synthesis of selenides (and even of tellurides) is conceivable without H_2Se (H_2Te) thermally

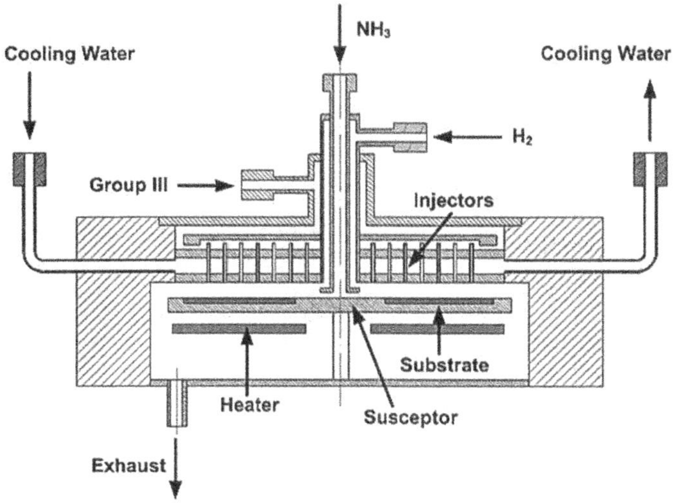

Fig. 11.5 Synthesis of metal nanocomposites using MOCVD technique

solubilized. Following that MOCVD technique the use of metal carbonyls and basic sulfur and selenium as starting materials for the creation of nanoparticles (Etzkorn et al. 2005; Zink et al. 2007) (where M = Mo, W, Re, and Q = S or Se).

11.2.6 Vapor–Liquid–Solid (VLS) Growth

The vapor–liquid–solid (VLS) approach (as depicted in Fig. 11.6) is the mechanism for growth from chemical vapor deposition of one-dimensional structures like

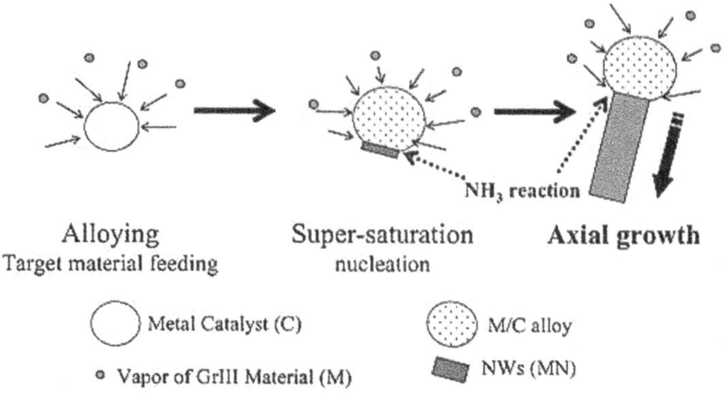

Fig. 11.6 Schematic representation of the vapor–liquid–solid (VLS) growth synthesis technique

nanowires. The development of glass is often very sluggish through direct adsorption of the gas phase on a solid surface. This is bypassed with the introduction of a catalytic liquid alloy phase that can quickly adsorb steam to the levels of supersaturation, which then enables crystal growing from nuclear seeds in the liquid–solid interface. In the VLS method, the physical qualities of nanowires generated in this fashion rely on the size and physical characteristics of the liquid alloy predictably.

A famous technique for the synthesis of semiconductor nanowires composed of chalcogenide metal is the VLS growth process. Catalytic growth with metal seed (typically Au) (Gudiksen and Lieber 2000) has produced nanowire growth. The Au catalyst nanoparticles are supported in these experiments on a SiO_2 substrate surface, and a Laser ablation is used to produce the reactants from a solid objective. Ideally, growth under the metal droplet occurs. The nanowires are therefore cultivated at low temperatures, where non-catalyzed side surfaces are kinetically impaired to grow. The VLS mechanism, in which a liquid droplet acts as a preferential sink for the growth factors that precipitate at the liquid–gas phase boundary, is commonly used to explain unidirectional growth. Many factors influence the VLS mechanism; surface tension and electronegativity are two that have a direct impact on the interaction between catalyst and reactive components. Composition, the pace of development, and decomposition are all factors that have an impact on the droplet. SnS_2 nanotubes were synthesized using the VLS method. Initial research focused on employing tin metal (melting point 232 °C) as a self-catalyst via the vapor transfer method (Toyama 1966). The production of SnS_2 nanoparticles was achieved by using tin metal as a self-catalyst. However, SnS_2 Nanotubes were created since Bi and Sn are known to form a eutectic combination at approximately 150 °C when Bi droplets were utilized as catalysts (Yella et al. 2009). Normally, one catalytic droplet contributes to the formation of one structure, such as a nanowire or nanotube, in a VLS process. As a result, numerous additional metal chalcogenides have been synthesized using this method.

11.2.7 Liquid Exfoliation

Solids with strong in-plane chemical bonding but weak out-of-plane van der Waals bonds are classified as layered materials. Such materials can be sheared parallel to the in-plane direction or extended normally to it. These procedures, at their most extreme, produce nanometre-thin—even atomically thin—sheets that bear no resemblance to the bulk precursor. Exfoliation (Fig. 11.7) or delamination are terms used to describe the process of producing exceedingly thin sheets from layered predecessors, however, we shall use the former term in this chapter. The sheets generated are known as nanosheets, with the term "nano" referring to the thickness magnitude. Although in the ideal case such nanosheets consist of single monolayers, they are often manifested as incompletely exfoliated flakes comprising a small number (< 10) of stacked monolayers.

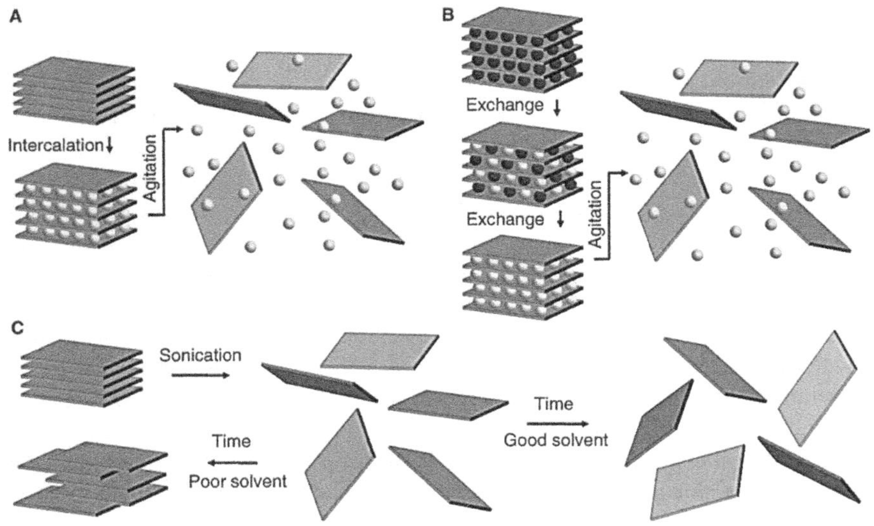

Fig. 11.7 Schematic steps in liquid exfoliation in layered nanostructure synthesis

Like carbon graphene (Geim 2009), layered metal chalcogenides can also be exfoliated to utilize their full potential. For example, Bi_2Te_3 on exfoliation showed enhanced thermoelectric efficiency due to suppression in thermal conductivity (Poudel et al. 2008). Typically, topological insulators on exfoliation reduced the conductance. Also, electronic properties get changed when the number of layers is reduced; for example, the indirect bandgap of bulk MoS_2 becomes direct bandgap materials on exfoliation to a few layers-thick flakes (Splendiani et al. 2010). Like carbon graphene, the exfoliation of layered 100 Metal Chalcogenide Nanostructures can also be mechanically synthesized on a small scale (Novoselov et al. 2005; Ayari et al. 2007) but liquid-phase exfoliation methods are more successful to explore many applications (Ruoff 2008). These liquid exfoliation methods (Fig. 11.7) provide the basis to synthesize novel hybrid and composite materials (Joensen et al. 1986; Liu et al. 1984). A very general solution-based exfoliation method has been reported by Coleman and co-workers for almost all kinds of layered metal chalcogenides. According to this method, layered compounds such as MoS_2, WS_2, $MoSe_2$, $MoTe_2$, $TaSe_2$, $NbSe_2$, $NiTe_2$, BN, and Bi_2Te_3 can be successfully dispersed and exfoliated in most of the common solvents and can be deposited as individual flakes or films (Matte et al. 2010). The main procedure involves the sonication of commercial MoS_2, WS_2, and BN powders in several solvents sorted based on different surface tensions. The sheets were purified using centrifugation and decanting the supernatant. These exfoliated products showed enhanced optical properties in solvents with surface tension close to γ-40 mJ/m^2 (Coleman et al. 2011). The results can be explained using the Hansen solubility parameter theory (Bergin et al. 2008, 2009). According to this theory, successful solvents for the exfoliation are the ones with dispersive, polar, and H-bonding components of the cohesive energy density within certain well-defined ranges. This can be concluded that those solvents which minimize the energy of exfoliation can be successfully used to obtained single sheets.

11.3 Conclusion

In this chapter, we have successfully explained all possible and popular and effective synthesis techniques somehow in detail for metal oxide nanoparticles/nanocomposites, metal chalcogenide nanostructures keeping in mind the attention about their superior mechanical properties, their unique electronic behavior, and their high potential in making technologically advanced nanodevices (such as supercapacitor for energy storage, smart generation solar cell, electrochemical rechargeable batteries, etc.) Among different classes of nanostructures and nanocomposites, layered metal chalcogenides nanostructures are of our major interest throughout the present chapter, for a variety of applications ranging from nanoelectronics or source material for energy applications, nanotribology, and heterogeneous catalysis. These nanoparticles are metastable phases. Therefore, equilibrium methods are necessary to prevent the formation of the thermodynamically stable bulk phase. On the other hand, high energies are needed to "knit" together with the folded layers. Several physical techniques such as Laser ablation and arc discharge are used for the synthesis of these inorganic structures. This present chapter includes the discussion on the growth mechanism of metal chalcogenides synthesized using different methods (MOCVD, VLS, and scalable electron beam irradiation techniques, etc.) and various protocols for their surface functionalization to improve the processability in technological energy applications.

References

Alexandrou I, Sano N, Burrows A, Meyer RR, Wang H, Ikirkland A, Kiely CJ, Amaratunga GAJ (2003) Structural investigation of MoS_2 core-shell nanoparticles formed by an arc discharge in water. Nanotechnology 14:913

Ayari A, Cobas E, Ogundadegbe O, Fuhrer MS (2007) Realization and electrical characterization of ultrathin crystals of layered transition-metal dichalcogenides. J Appl Phys 101:014507

Bergin SD, Nicolosi V, Streich PV, Giordani S, Sun Z, Windle AH, Ryan P, Niraj NP, Wang ZT, Carpenter L, Blau WJ, Boland JJ, Hamilton JP, Coleman JN (2008) Towards solutions of single-walled carbon nanotubes in common solvents. Adv Mater 20:1876

Bergin SD, Sun Z, Rickard D, Streich PV, Hamilton JP, Coleman JN (2009) Multicomponent solubility parameters for single-walled carbon nanotube–solvent mixtures. ACS Nano 3:2340

Bonaccorso F, Colombo L, Yu G et al (2015) Graphene, related two-dimensional crystals, and hybrid systems for energy conversion and storage. Science 347:1246501

Brooks DJ, Douthwaite RE, Brydson R, Calvert C, Measures MG, Watson A (2006) Synthesis of inorganic fullerene (MS_2, $M = Zr$, Hf and W) phases using H_2S and N_2/H_2 microwave-induced plasmas. Nanotechnology 17:1245

Cerro-Lopez M, Méndez-Rojas MA (2017) Application of nanomaterials for treatment of wastewater containing pharmaceuticals. In: The handbook of environmental chemistry, vol 66. Springer, Cham

Chen K, Song S, Liu F et al (2015) Structural design of graphene for use in electrochemical energy storage devices. Chem Soc Rev 44:6230–6257

Cheng XB, Yan C, Chen X et al (2017) Implantable solid electrolyte interphase in lithium-metal batteries. Chem 2:258–270

Coleman JN, Lotya M, O'Neill A, Bergin SD, King PJ, Khan U, Young K, Gaucher A, De S, Smith RJ, Shvets IV, Arora SK, Stanton G, Kim H-Y, Lee K, Kim GT, Duesberg GS, Hallam T, Boland JJ, Wang JJ, Donegan JF, Grunlan JC, Moriarty G, Shmeliov A, Nicholls RJ, Perkins JM, Grieveson EM, Theuwissen K, McComb DW, Nellist PD, Nicolosi V (2011) Two-dimensional nanosheets produced by liquid exfoliation of layered materials. Science 331:568

El-Kady MF, Shao Y, Kaner RB (2016) Graphene for batteries, supercapacitors and beyond. Nat Rev Mater 1:16033

Etzkorn J, Therese HA, Rocker F, Zink N, Kolb U, Tremel W (2005) Metal-organic chemical vapor deposition synthesis of hollow inorganic-fullerene-type MoS_2 and $MoSe_2$ nanoparticles. Adv Mater 17:2372

Fierro JLG (ed) (2006) Metal oxides: chemistry and application. Taylor & Francis, CRC Press, Boca Raton, FL

Galván DH, Kim JH, Maple MB, Avalos-Borja M, Adem E (2000) Formation of $NbSe_2$ nanotubes by electron irradiation. Full Sci Technol 8:143

Galván DH, Kim JH, Maple MB, Adem E (2001) Effect of electronic irradiation in the production of $NbSe_2$ nanotubes. Full Sci Technol 9:225

Ganesan P, Prabu M, Sanetuntikul J et al (2015) Cobalt sulfide nanoparticles were grown on nitrogen and sulfur codoped graphene oxide: an efficient electrocatalyst for oxygen reduction and evolution reactions. ACS Catal 5:3625–3637

Gao Y, Zhang Y, Zhang Y et al (2016) Three-dimensional paper-like graphene framework with highly orientated laminar structure as a binder-free supercapacitor electrode. J Energy Chem 25:49–54

Geim AK (2009) Graphene: status and prospects. Science 324:1530–1534

Geim AK, Novoselov KS (2007) The rise of graphene. Nat Mater 6:183–191

Goodenough JB (2015) A chemistry and material perspective on lithium redox flow batteries towards high-density electrical energy storage. Energy Storage Mater 1:158–161

Gudiksen MS, Lieber CM (2000) Diameter-selective synthesis of semiconductor nanowires. J Am Chem Soc 122:8801

Hacohen YR, Grunbaum E, Tenne R, Sloand J, Hutchison JL (1998) Cage structures and nanotubes of $NiCl_2$. Nature 395:336

Hacohen YR, Popovitz-Biro R, Grunbaum E, Prior Y, Tenne R (2002) Vapor-liquid–solid (VLS) growth of $NiCl_2$ nanotubes via reactive gas laser ablation. Adv Mater 14:1075

Hacohen YR, Popovitz-Biro R, Prior Y, Gemming S, Seifert G, Tenne R (2003) Synthesis of $NiCl_2$ nanotubes and fullerene-like structures by laser ablation: theoretical considerations and comparison with MoS_2 nanotubes. Phys Chem Chem Phys 5:1644

Hong SY, Popovitz-Biro R, Prior Y, Tenne R (2003) Synthesis of SnS_2/SnS fullerene-like nanoparticles: a superlattice with polyhedral shape. J Am Chem Soc 125:10470

Hu JJ, Bultman JE, Zabinsky JS (2004) Inorganic fullerene-like nanoparticles produced by arc discharge in water with potential lubricating ability. Tribol Lett 17:543

Hu JJ, Zabinski JS, Sanders JH, Bultman JE, Voevodin AA (2006) Pulsed laser syntheses of layer-structured WS_2 nanomaterials in water. J Phys Chem B 110:8914

Hu JJ, Sanders JH, Zabinsky JJ (2006b) Synthesis and microstructural characterization of inorganic fullerene-like MoS_2 and graphite-MoS_2 hybrid nanoparticles. J Mater Res 21:1033

Huang X, Zeng Z, Zhang H (2013) Metal dichalcogenide nanosheets: preparation, properties, and applications. Chem Soc Rev 42:1934–1946

Ji J, Li Y, Peng W et al (2015) Advanced graphene-based binder-free electrodes for high-performance energy storage. Adv Mater 27:5264–5279
Joensen P, Frindt RF, Morrison SR (1986) Single-layer MoS_2. Mater Res Bull 21:457
Jose-Yacaman M, Lopez H, Santiago P, Galvan DH, Garzon IL, Reyes A (1996) Studies of MoS_2 structures produced by electron irradiation. Appl Phys Lett 69:1065
Kagkoura A, Skaltsas T, Tagmatarchis N (2017) Transition metal chalcogenides/graphene ensembles for light-induced energy applications. J Chem Eur 23:12967–12979
Kumar SR, Gopinath P (2016) Nano-bioremediation applications of nanotechnology for bioremediation. In: Remediation of heavy metals in the environment. CRC Press, Boca Raton
Lai Z, Chen Y, Tan C et al (2016) Self-assembly of two-dimensional nanosheets into one-dimensional nanostructures. Chem 1:59–77
Lee C, Wei X, Kysar JW et al (2008) Measurement of the elastic properties and intrinsic strength of monolayer graphene. Science 321:385–388
Li X, Chen Y, Huang H et al (2016a) Electro-spun carbon-based nanostructured electrodes for advanced energy storage—a review. Energy Storage Mater 5:58–92
Li W, Liu J, Zhao D (2016b) Mesoporous materials for energy conversion and storage devices. Nat Rev Mater 1:16023
Liu C, Singh O, Joensen P, Curzon AE, Frindt RF (1984) X-ray and electron microscopy studies of single-layer TaS_2 and NbS_2. Thin Solid Films 113:165
Lv W, Li Z, Deng Y et al (2016) Graphene-based materials for electrochemical energy storage devices: opportunities and challenges. Energy Storage Mater 2:107–138
Ma TY, Dai S, Qiao SZ (2016) Self-supported electrocatalysts for advanced energy conversion processes. Mater Today 19:265–273
Matte HSSR, Gomathi A, Manna AK, Late DJ, Datta R, Pati SK, Rao CNR (2010) MoS_2 and WS_2 analogues of graphene. Angew Chem Int Ed 49:4059
Miller TA, Bakrania SD, Perez C, Wooldridge MS (2006) Nanostructured tin dioxide materials for gas sensor applications. Funct Nanomater 30:1–24
Miller DR, Akbar SA, Morris PA (2014) Nanoscale metal oxide-based heterojunctions for gas sensing: a review. Sens Actuators B Chem 204:250–272
Nath M, Rao CNR, Biro RP, Yaron AA, Tenne R (2004) Synthesis of fullerene-like tantalum disulfide nanoparticles by a gas-phase reaction and laser ablation. Chem Mater 16:2238
Novoselov KS, Jiang D, Schedin F, Booth TJ, Khotkevich VV, Morozov SV, Geim AK (2005) Two-dimensional gas of massless Dirac fermions in graphene. Proc Natl Acad Sci USA 102:10451
Parilla PA, Dillon AC, Jones KM, Riker G, Schulz DL, Ginley DS, Heben MJ (1999) The first true inorganic fullerenes? Nature 397:114
Pelaez M, Nolan NT, Pillai SC et al (2012) A review on the visible light active titanium dioxide photocatalysts for environmental applications. Appl Catal B 125(21):331–349
Peng HJ, Wang DW, Huang JQ et al (2016a) Janus separator of polypropylene-supported cellular graphene framework for sulfur cathodes with high utilization in lithium-sulfur batteries. Adv Sci 3:1500268
Peng L, Zhu Y, Li H et al (2016b) Chemically integrated inorganic-graphene two-dimensional hybrid materials for flexible energy storage devices. Small 12:6183–6199
Poudel B, Hao Q, Ma Y, Muto A, Dresselhaus MS, Lan Y, Vashaee D, Chen G, Minnich A, Chen X, Ren Z (2008) High-thermoelectric performance of nanostructured bismuth antimony telluride bulk alloys. Science 320:634
Qiu HJ, Liu L, Wang Y (2016) Template-directed fabrication of 3D graphene-based composite and their electrochemical energy-related applications. Sci Bull 61:443–450
Raccichini R, Varzi A, Passerini S et al (2015) The role of graphene for electrochemical energy storage. Nat Mater 14:271–279
Remskar M, Mrzel A, Jesih A, Levy F (2002) Metal-alloyed NbS_2 nanotubes synthesized by the self-assembly of nanoparticles. Adv Mater 14:680
Rui X, Tan H, Yan Q (2014) Nanostructured metal sulfides for energy storage. Nanoscale 6:9889–9924

Ruoff R (2008) Calling all chemists. Nat Nanotechnol 3:10
Sano N, Wang H, Chhowalla M, Alexandrou I, Amaratunga GAJ, Naito M, Kanki T (2003) Fabrication of inorganic molybdenum disulfide fullerenes by arc in water. Chem Phys Lett 368:331
Santhosh C, Velmurugan V, Jacob G, Jeong SK, Grace AN, Bhatnagar A (2016) Role of nanomaterials in water treatment applications: a review. Chem Eng J 306:1116–1137
Schuffenhauer C, Parkinson BA, Jin-Phillipp NY, Joly-Pottuz L, Martin JM, Popovitz-Biro R, Tenne R (2005) Synthesis of fullerene-like tantalum disulfide nanoparticles by a gas-phase reaction and laser ablation. Small 1:1100
Sen R, Govindaraj A, Suenaga K, Suzuki S, Kataura H, Iijima S, Achiba Y (2001) Encapsulated and hollow closed-cage structures of WS_2 and MoS_2 prepared by laser ablation at 450–1050 °C. Chem Phys Lett 340:242
Shan XY, Li F, Wang DW et al (2016) The smart era of electrochemical energy storage devices. Energy Storage Mater 3:66–68
Sheng T, Xu YF, Jiang YX et al (2016) Structure design and performance tuning of nanomaterials for electrochemical energy conversion and storage. Acc Chem Res 49:2569–2577
Si PZ, Zhang M, Zhang ZD, Zhao XG, Ma XL, Geng DY (2005) Synthesis and structure of multi-layered WS_2 (CoS), MoS_2 (Mo) nanocapsules and single-layered WS_2 (W) nanoparticles. J Mater Sci 40:4287
Splendiani A, Sun L, Zhang Y, Li T, Kim J, Chim C-Y, Galli G, Wang F (2010) Emerging photoluminescence in monolayer MoS_2. Nano Lett 10:1271
Tian GL, Zhang Q, Zhao MQ et al (2015) Fluidized-bed CVD of unstacked double-layer templated graphene and its application in supercapacitors. AIChE J 61:747–755
Toyama M (1966) Kinetics of the vapor growth of II-VI compounds crystals. Jpn J Appl Phys 5:1204
Vollath D, Szabo DV (1998) Synthesis of nanocrystalline MoS_2 and WS_2 in a microwave plasma. Mater Lett 35:236
Vollath D, Szabo DV (2000) Nanoparticles from compounds with layered structures. Acta Mater 48:953
Wang M, Xu YX (2016) Design and construction of three-dimensional graphene/conducting polymer for supercapacitors. Chin Chem Lett 27:1437–1444
Wang H, Zhu QL, Zou R et al (2017) Metal-organic frameworks for energy applications. Chem 2:52–80
Werkneh AA, Renein ER (2019) Applications of nanotechnology and biotechnology for sustainable water and wastewater treatment. In: Water and wastewater treatment technologies, energy, environment, and sustainability. Springer, Singapore
Xiang Z, Dai Q, Chen JF et al (2016) Edge functionalization of graphene and two-dimensional covalent organic polymers for energy conversion and storage. Adv Mater 28:6253–6261
Xiong J, Han C, Li Z et al (2015) Effects of nanostructure on clean energy: big solutions gained from small features. Sci Bull 60:2083–2090
Xu C, Tamaki J, Miura N, Yamazoe N (1991) Grain-size effects on gas sensitivity of porous SnO_2-based elements. Sens Actuators B Chem 3(2):147–155
Yella A, Mugnaioli E, Panthoefer M, Therese HA, Kolb U, Tremel W (2009) Bismut-katalysiertes Wachstum von SnS_2-Nanoröhren und deren Stabilität. Angew Chem 121:6546
Yu XY, Yu L, Lou XWD (2016) Metal sulfide hollow nanostructures for electrochemical energy storage. Adv Energy Mater 6:1501333
Zhang X, Cheng X, Zhang Q (2016a) Nanostructured energy materials for electrochemical energy conversion and storage: a review. J Energy Chem 25:967–984
Zhang Y, Wu B, Xu H et al (2016b) Nanomaterials-enabled water and wastewater treatment. NanoImpact 3–4:22–39
Zhao G, Li X, Huang M et al (2017) The physics and chemistry of graphene-on-surfaces. Chem Soc Rev 46:4417–4449

Zink N, Pansiot J, Kieffer J, Therese HA, Panthöfer M, Rocker F, Kolb U, Tremel W (2007) Selective synthesis of hollow and filled fullerene-like (IF) WS_2 nanoparticles via metal-organic chemical vapor deposition. Chem Mater 19:6391

Zou X, Zhang Y (2015) Noble metal-free hydrogen evolution catalysts for water splitting. Chem Soc Rev 44:5148–5180

Chapter 12
Green Polymers Decorated with Metal Nanocomposites: Application in Energy Storage, Energy Economy and Environmental Safety

Abhay Nanda Srivastva, Nisha Saxena, and Manish Kumar

Abstract Eco-friendly polymers exhibit distinctive physiochemical properties after their transformation in metal nanocomposites. Significant attention by the researchers have been paid in developing polymeric metal composites at nano scale molecular size due to their promising potential for energy storage, environmental remediation, electromagnetic absorption, environment safety etc. and thus green metal nanocomposites have applications in resolving both energy and environmental issues. Easy processing and low cost production of these polymeric materials also attract green chemists to develop novel polymeric metal composites. The significant progress in processing, characterization and modelling of nano sized materials create many opportunities for technical support in the development and designing of biodegradable polymers with metal nancomposite systems. But still there are lots of challenges viz. limited availability of high-quality nanofillers, their high production cost, difficulties to scale up, lack of deep knowledge, understanding and predictive capability in the area of key processes, structures and properties, which are needed to fully utilization of these materials for the commercial benefits which are remained unaddressed. Keeping in view to provide a comprehensive details for researchers to find a better solution of above challenges, in this chapter, centre of attention has been created on the latest and advanced field of research in which designing of green materials of biodegradable polymers with metal nanocomposites and their utilization for different devices such as development of electrochemical cells and batteries for energy storage,

A. N. Srivastva (✉)
Department of Chemistry, Nitishwar Mahavidyalaya, A constituent Unit of B.R.A. Bihar University, Muzaffarpur 842002, India

University Department of Chemistry, B.R.A. Bihar University, Muzaffarpur 842002, India

N. Saxena
Department of Chemistry, M.R.M. College, A Constituent Unit of L.N. Mithila University, Darbhanga 846004, India

M. Kumar
Experimental Research Laboratory, Department of Physics, A.R.S.D. College, University of Delhi, New Delhi 110021, India

S. P. Singh et al. (eds.), *Metal Nanocomposites for Energy and Environmental Applications*, Energy, Environment, and Sustainability,
https://doi.org/10.1007/978-981-16-8599-6_12

electro-chromic devices for conservation of energy, gaseous capturing devices for safety of environment have been discussed.

Keywords Biodegradable polymer · Nano-composites · Energy efficient materials · Environment protection

12.1 Introduction

Since a long time, researcher have been paying efforts towards the development and application of environment friendly and sustainable bio-reinforced composites for use in different area of thrust such as energy storage, automotive, construction, environmental conservation, food packaging, bio-safety and biomedical etc. It has been elucidated by several researches (Kausar 2021; Njuguna et al. 2008) that addition of metal nanocomposites in base green polymers demonstrate perked up properties that make them functional with more suitability in various areas of human wellness with natural sustainability. Nanomaterials and nanocomposites are contentiously in the service of society by providing the solutions to technological and environmental issues in the almost all fields of science, from water treatment to energy conversion, and waste management to medical division (Pandey et al. 2017; Mantione et al. 2017; Kumar and Srivastva 2021; Dahl 2007; Hutchison 2008). Nanocomposites are the materials possessed with nano scaled two or more materials with different properties which may be ceramics, metals or polymers (Haghi 2013; Twardowski 2006). Polymer nanocomposites are extensively studied for their versatile applications ranging from energy storage, electromagnetic (EM) absorption, transportation and safety, defence system, sensing and actuation, information industry, environmental remediation etc. (Ganesan and Jayaraman 2014; Tao et al. 2013; Moon et al. 2011; Wei et al. 2011, 2013; Pérez-Juste et al. 2013; Kumar et al. 2013; Kokate et al. 2013; Kao et al. 2013). The greater demand of nanocomposites must be accompanied by green polymer synthesis system for technological developments with sustainability. For achieving the aim of sustainable growth, chemists, material scientists and material manufacturers are strained to minimizing the use of non-eco friendly practices during the development and production of materials of need related to technological developments. To attain the goal of such a progression, green chemistry and green chemical processes are being coupled with latest developments in chemical sciences and industries to reduce the generation of hazardous waste and also to yield improved quality products as per the need of technical advancements (Kao et al. 2013). Green polymers, which are mainly long-chain molecules, synthesized by large numbers of repeating units of same structures, facilitate as the matrix improvement by various types of nano sized fillers like metal nanofillers on green polymers. Such types of metal/s containing nanocomposites are termed as eco-friendly polymer metal nanocomposites (Zhu et al. 2011; Guo et al. 2006). In continuous phase towards bond with nanofillers, the green polymer matrix show a crucial role in finding the process ability, tensile strength, shear, compressive properties and heat resistance of the green polymer nanocomposites (Wang et al. 2011). In green polymer nanocomposite, green nanofillers have been used to reinforcing functionality of dispersed phase (Haghi and Zaikov 2013). These green nanofillers have been used to improve the mechanical, thermal, electrical, magnetic, gas barrier properties of the polymers (Zhu et al. 2010; Zhang et al. 2012, 2013a, b; Gu et al. 2013). With compared traditional fillers viz. carbon fibres, wood sheets, glass fibres, inorganic mineral particles and metal particles etc., green nanofillers are more advantageous due to their great specific interfacial, higher attainable loads with whole compliance and more manageable interfacial interactions (Hu et al. 2014). This type of nanofillers have been utilized in green polymer nanocomposites together with quantum dots (semiconductors, metals, metal oxides), carbon nanotubes (CNTs), nanowires, dendrimers, layered silicates and metal oxide nanoparticles (Njuguna et al. 2014). The combination of nanofillers in polymers can provide change in properties from synergistic effects of every component in the green polymer nanocomposites. These nanofillers are capable of providing both structural and functional strength to the polymer nanocomposites like worth mentioning improvements in the mechanical strength is witnessed in the polymer nanocomposites by the addition of such types of nanofillers in intrinsic high modulus (Thakur et al. 2012). The continuous dispersion of the nanofillers with great interfacial associations of matrix along with nanofillers is basically accountable in favour of the significantly increased mechanical behaviours (Zhou et al. 2014). The covalent bonding of matrix with filler are considered as the most effective mean to amplify the interfacial shear stress for the improvement of stress relocation. While, accredited to the tremendous thermal and electrical conductivity in the nanofillers, thermally and electrically conductive polymers are prepared by extended applications of sensors, EMI shielding materials, electrodes, and stimuli responsive materials (Mohammadnezhad and Keikavousi 2020). Among the several applications of different type of nanocomposite materials, we are summarising here, in this chapter, the role of green polymer based metal nanocomposites in solving the problem of the very thrust area of scientific research and global need of energy storage, energy economy and safety for environment.

12.1.1 Green Approach for the Synthesis of Metal Nano Particles (MNPs)

Green chemistry is widely applied for the synthesis of several type of new organic, inorganic and nanomaterials as well as playing a role of eco-friendly synthesis as a substitute of traditional non-green synthetic routes (Raveendran et al. 2003). Green synthesis of metal nanoparticles and metal nano oxides are in trend now-a-days for maintaining sustainability of environment. Green synthesis of NP_S are mostly based on phytochemical reactions in the absence of toxic solvents and high range temperature (Fig. 12.1) (Iravani 2011; Iravani and Varma 2020; Ghosh et al. 2012).

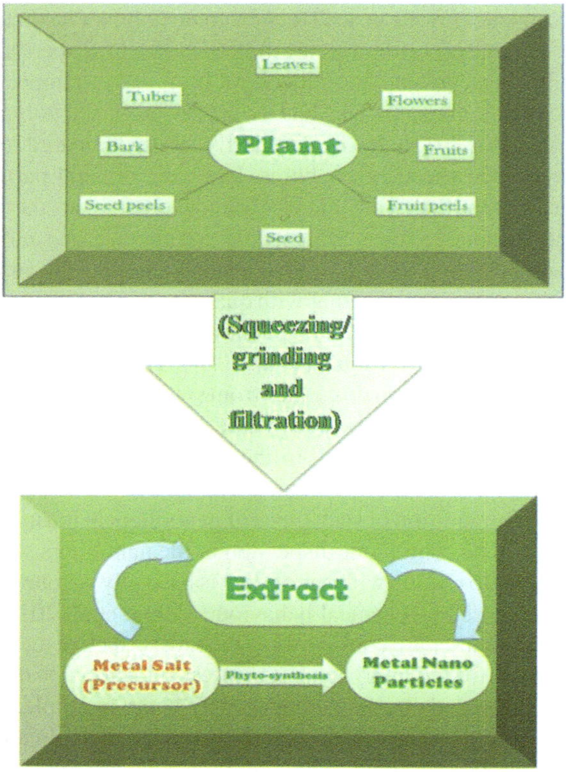

Fig. 12.1 Graphical representation of green synthesis of metal nano-particles

In many of cases, the yield and properties of NP_s are much improved than the particles NPs obtains from traditional non-green methods (Iravani and Varma 2020). The enhanced morphology and low range nano scale size may be the main reason behind improved properties of green synthesized nanoparticles (Ghosh et al. 2012). Since the sustainability is the demand of global community, researchers are giving their best efforts in the development of green nano materials and related eco-friendly techniques concerning technological revolution in this era. Recently synthesized Au nanoparticle show more suitable features to be fit in the frame of nanomaterial with low hazard during green synthesis with *Nyctanthes arbortristis* flower extract (Das et al. 2011). Leaf extract of *Cassia auriculata* was used to facile green gold nanoparticles (Kumar et al. 2011). Edible mushroom extract was utilized as phyto-reducing agent for the synthesis of gold and silver nanoparticles via green approach (Philip 2009). Ag nano particle developed with green approach of using fruit peel extract of *Citrus tangerina*, *Citrus sinensis*, and *Citrus limon* plants (Niluxsshun et al. 2021). Mixed phyto-chemical synthetic approach was followed for the synthesis of copper nanoparticles with enhanced activity potentials (Rajagopal et al. 2021). ZnO nanoparticles were also prepared by phytochemical synthetic route using *Cassia auriculata* flower extract which yielded metal oxide with effective properties (Seshadri 2021). Like the same way, in recent years, numbers of nanoparticles have been synthesized with green synthetic approach and thus motivating the researches to involve their selves in developing the novel route of green synthesis for nanomaterials to fulfil the industrial demand of these highly efficient materials with concern of our environment, globally (Gebreslassie and Gebretnsae 2021; Ahmed et al. 2017; Shafey 2020).

12.1.2 Synthetic Development in Green Polymers

Biodegradable polymers play a crucial role in the field of material science, especially in green nanotechnology (Dmour and Taha 2018). Polymers are greener in behavior depending upon their degradation pattern via biological mean. Green polymers synthesis is basically covenanted with twelve key principle of green approach of chemicals/materials development with minimal side effect on environment and human being (Unterlass 2016; Alberto and Ade 2013). Depending upon the occurrence, green polymers may be obtains directly from nature like polysaccharides (cellulose, chitin, chitosan), proteins (collagen, gelatin, elastin, fibrin, silk, etc.), polyesters (Cutin) etc. and make a class of natural green polymers (Calvert 1975; Spring 1945). Some chemically modified natural polymers are classed in semi-synthetic green polymers (Bhandari et al. 2021). The green polymers synthesized in laboratory are termed as synthetic green polymers, e.g. Polylactic acid, Polyvinyl Alcohol, poly-glycolide, poly(*L*-lactic acid), poly(ε-caprolactone) etc. (Kobayashi 2017). The synthesis practice of green polymers is achieved by direct and indirect approach. During the course of direct synthesis, one or more from melt blending, solution blending, physical mixing processes may be applied, while the indirect approach is followed up by one or more among various available processes viz. chemical polymerization, electro polymerization, photo polymerization, mechano polymerization (Fig. 12.2) (Kobayashi 2017). In recent years, number of green polymers have been synthesized and utilized for their versatile applications sustaining the environment (Table 12.1) (Jothimani et al. 2019). Enzyme catalyzed biodegradable reactive polyesters with renewable itaconic anhydride monomer was reported with improved yield via green approach of synthesis and screened for their potential applications as macro-monomers, tele-chelics or cross-linking agent (Yamaguchi et al. 2014). Valuable facets of enzymatic polyester synthesis were described in terms of conducting green polymer synthesis via ring opening polymerization and poly-condensation (Kobayashi 2015). Direct synthetic approach of melt poly-condensation of lactic acid was reported for the manufacturing of poly(L-lactic acid) polymer (Moon et al. 2000).

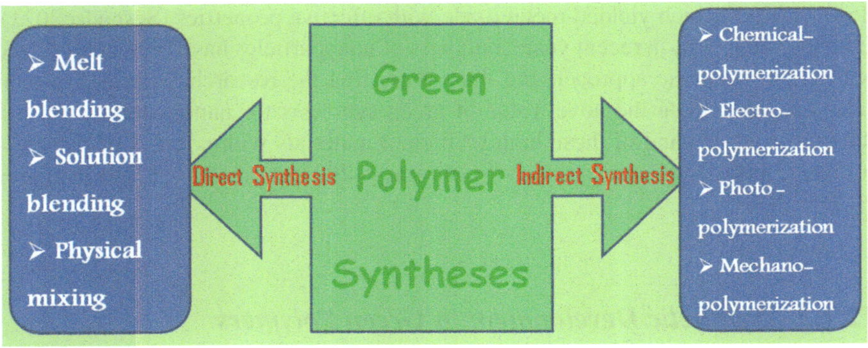

Fig. 12.2 Synthetic approaches of green polymers

Table 12.1 Applications of green polymer-noble metal nanocomposites

Polymer matrix	Metallic nanoparticle/s	Application	References
Polyaniline	Pd	As a sensor for alcohols	(Athawale et al. 2006)
Polyaniline	TiO_2 $NaAlH_4$	As an energy device for hydrogen storage	(Moreira 2013)
Cellulose	ZnO	As an UV sensor	(Mun et al. 2017)
Polyvinyl alcohol	Ag	As a sensor for SPR-based protein	(Ananth et al. 2011)
Polyvinyl alcohol	Ag	In electrical conductivity and dielectric spectroscopic studies	(Mahendia et al. 2010)
Poly(3-hexylthiophene)	Titanium dioxide	In electrochemical property studies	(Almeida et al. 2020)
Cellulose (β 1,4-linked d-glucose rings)	ZnO	As an UV sensor	(Mun et al. 2017)
Polyaniline	CdS	As a photoluminescence material	(Alipour et al. 2021)
Chitosan	ZnO	As a high dielectric material	(Rahman et al. 2018)
Chitosan	ZnO	As optical and photoluminescence material	(Magesh et al. 2018)

12.1.3 Synthetic Development in Green Polymer Metal Nanocomposites (GPMNCs)

The common practice of polymer metal nanocomposite syntheses are applied as Ex-situ process and In-situ process along with adaptation of green approach during the whole process or most of the steps involved in the process (Carotenuto et al. 2004). In ex-situ process, the metal nanoparticles are developed by one of the traditional method like phyto reduction/chemical reduction, precipitation, laser ablation etc. Phyto reduction method is generally applied in green synthesis approach of metal nanoparticles. The stabilizers are applied then to stabilize, capsulate and passivate the metal nanoparticle surface (Carotenuto et al. 2004; Javed et al. 2016). After this, metal nanoparticles are interleaved in the matrix of green polymer. Interleaving of metal nanoparticles in polymer matrix may be processed by intermixing the solutions of both or monomer interactions with metal nanoparticles followed by various casting techniques. In recent years, so many researchers adopted Ex-situ method for developing polymer metal nanocomposites of polyvinyl alcohol, polyvinyl pyrrolidone, polymethyl methacrylate including green polymers and green synthesized nano metal particles (Ananth et al. 2011; Solomon et al. 2007; Mahendia et al. 2010; Ghanbari et al. 2016). In the process of in situ polymer metal nanocomposites synthesis, activated polymer solution or monomer solution is mixed with metal ion/s salt as metal nanoparticle precursors followed by successive chemical reduction steps (Almeida et al. 2020; Aktitiz et al. 2020). Green approach of synthesis like microwave may also be applied with in-situ process (Almeida et al. 2020). In situ process is on consideration by material scientists over the Ex-situ as the method enable researchers to prepare polymeric metal nanocomposites of controlled particle size headed with morphological enhancement and improved properties (Aktitiz et al. 2020; Erabhoina et al. 2021; Magesh et al. 2018). Some polymer matrix structures suitable for metal nano composites preparation are represented in Fig. 12.3 (Magesh et al. 2018; Mun et al. 2017; Alipour et al. 2021; Jianga et al. 2019; Rahman et al. 2018).

Though, the development of polymer metal nano composites via green methods are accepted by the material scientists worldwide for the concern of reducing environmental hazards, but there are some challenges in front of researchers in the field of green material synthesis. The selection of suitable phyto-products having the efficiency of producing lower range metal nano particles is a major issue. For targeted range nano phyto-reduction synthesis, chemical analysis of plant parts or bacterial and fungal cells to be used is required in detail. Interpretation of mechanism related to green synthesis approach of polymeric nano metal composites is still in its formative years. Several green polymers are not compatible with metal particles for composite preparation and vice versa. Some time, the composite product of green polymers and metal nano particles are not so eco-friendly as compared to parent materials. The continuous effort of researchers is in the direction to get rid of with challenges associated to green polymer decorated metal nano composite materials.

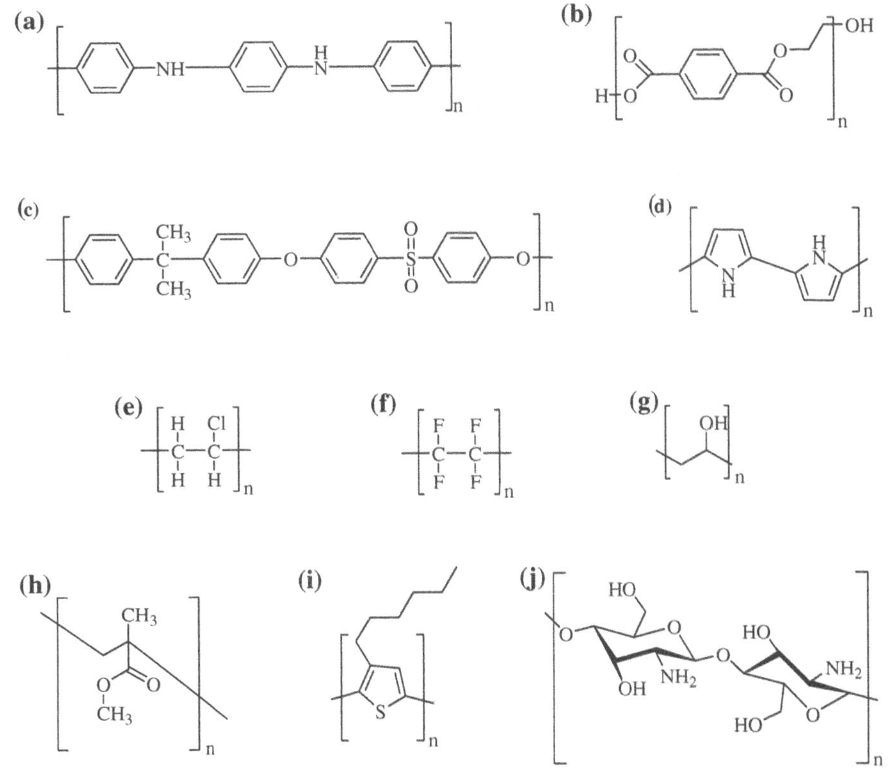

Fig. 12.3 Chemical structure of some polymer matrix used in metal nano polymer composites; **a** Polyaniline, **b** PET, **c** polysulfone, **d** polypyroll, **e** polyvinyl chloride, **f** polytetrafluoroethylene, **g** polyvinyl alcohol **h** polymethylmethacrylate, **i** poly(3-hexylthiophene), **j** chitosan

12.2 Polymer Metal Nanocomposites for Energy Storage Applications

Non-renewable sources of energy like fossil fuels on earth are limited in quantity and also the cause of high level environmental pollution. With the concern of future energy need and protection of earth's environment, researchers are engaged in planning to generate renewable energy like solar energy, wind energy, geothermal energy, hydro energy, tidal/ wave energy and bio-mass energy etc. from their respective natural resources and their storage strategically in eco-friendly manner and thus trying to achieve technological development with maintaining sustainability (Ganesh 2014). The probable dependency on renewable energy resources for fulfilment of high global energy need is convinced by a report published earlier which stated that transformation and storage of total 1 h sun energy coming on earth have the potential to fulfil the whole earth energy demand of energy for ~ 365 days (Ganesh 2014).

Like the same way, other renewable resources may also provide the huge energy of global need, if we able to harness and store it properly with the help of green approaches. For the generation of energy from these eco-friendly resources and the storage of energy, high potency materials are necessitated. Among various types of energy storage materials, nanomorphed materials are playing their role very well. So many nano structured polymeric materials, nanoparticles, nanocomposites, ceramics, metals etc. are in use for energy storage (Buller and Strunk 2016). To keep maintain the environment sustainability, green nanomaterials are in great demand by global industries and among a choice of nanomaterials, green polymer nanocomposites are of quite interest due to their improved energy storage potency, ease of synthesis and eco-friendly behaviour (Kumar et al. 2017). In polymer metal nanocomposites the metal nanoparticle acts as the discontinuous phase and the green polymer base is the continuous phase. A group of researchers designed a well characterized polymer nanocomposite having supercapacitor properties with green synthesized flake like nano zinc oxide (~20 nm) and structurally modified styrene maleic anhydride co-polymer. Structural modification in styrene maleic anhydride was done with thiadiazole moiety. Physico-analytical data are suggestive for the scheming of targeted zinc oxide integrated polymer nanocomposite. The polymer metal nanocomposite possess a specific capacitance of 268.5 F g^{-1} at 0.1 A g^{-1}, comparatively larger that nano zinc oxide and base polymer (Chakraborty et al. 2020). Aqueous medium green root synthetic approach was adopted for the development of nano lithium based poly(o-toluidine) composite by N. Pande and his team. This structurally verified polymer metal nanocomposite has been applied as electrolyte in rechargeable batteries to investigate their charge and discharge performance. The resultant output of tested poly(o-toluidine)-Li nanocomposite was adequate with a power density of 362.88 W/m^3 (Pande et al. 2021). Nanostructured copper metal based polyaniline composite ($CuCr_2O_4$/PANi) possessing high performance energy storage capacity with specific capacity value of a specific capacity of 479.2 C g^{-1} at 2 mV s^{-1} had been utilized as electrode material after their novel synthesis, structural and morphological characterization (Gandla et al. 2017). Eco-friendly modified route of polymer nanocomposites development was processed with synthesis of $BaTiO_3$ nanomaterial in the presence of an aqua soluble epoxy resin and then homogeneously dispersed in poly-(vinylidene fluoride-co-hexafluoropropylene) named polymer matrix with strong adhesion at interfaces. As the $BaTiO_3$ coating load is increased, the dielectric constant of this eco-friendly nanocomposites also increased significantly and the energy storage density was also improved proportionally up to 8.13 J/cm^3 at 20 vol% of $BaTiO_3$. This improved result of metal based nanocomposite may be applied in energy storage with maintaining sustainability (Luo et al. 2015). A group of researchers reported the finding of a comparative study between green-synthesized biogenic *Rhubarb* silver nanoparticles, abbreviated as RS-AgNPs and bio polymer chitosan crosslinked silver nanocomposites, abbreviated as CSHD-AgNCs. The dispersion of silver particles with an average size of 5–50 nm was found throughout the polymer matrix in FCC structure with high stability of approximate one and the half years, even in the absence of any protecting/capping reagent (Palem et al. 2018). In a recent study, green synthesized nanoparticles of ZnS were utilized in fabrication of a nanocomposite film designed with an aqua soluble biodegradable polymer, poly vinyl alcohol. This green nanocomposite film was well characterised for its structural and morphological identification by using various advance methods. To insure the utilization of this polyvinyl alcohol-zinc sulphide metal nanocomposite in energy storage devices, dielectric potential activity was also studied which turned out superior dielectric constant value of 328.93 and dielectric loss value of 6.02 with 3 wt% ZnS nanoparticles homogeneous load on matrix at 50 Hz frequency and 140 °C temperature (Reddy et al. 2019). Dielectric permittivity and breakdown voltage are related to energy storage density (E_d) as per the mathematical expressions (i) and (ii). A simple balance of the heat generated from the induced current and the heat lost from convection over the material yields depends upon material properties, which are of most concern for the designing of large scale storage dielectric capacitor, as well as applied voltage, thickness of polymer film and environmental temperature according to the volumetric energy equation (iii).

$$E_d = \int E \cdot d \cdot P \tag{12.1}$$

$$E_d = \frac{1}{2} \times \varepsilon_0 \times \varepsilon_r \times E_b^2 = \frac{1}{2} \times C \times V_m^2 \times \frac{1}{A.d} \tag{12.2}$$

$$\frac{\kappa(T) \cdot V_{app}^2}{2 \cdot d} = \gamma(T_c - T_0) \tag{12.3}$$

where E = Applied electric field,
P = Polarization, E_b = Breakdown strength,
ε_r = Relative permittivity,
ε_0 = Permittivity of free space (8.854×10^{-14} F cm^{-1}),
$\kappa(T)$ = Electrical conductivity as a function of temperature,
Tc = Critical temperature, γ = Areal thermal conductivity,
V_{app} = Applied voltage,
d = Film thickness,
T_o = Temperature of environment.

To express the impact of each material property/ies, plots of Fig. 12.4 are drawn with changeable one property and taking others as constant which clearly indicates the significance of material properties on capacitor potential.

Polymers have increased breakdown energy field which make them more suitable capacitor applications over the other materials. The foremost drawback of low dielectric constant (5–14) limiting their capacitor potential can be improved by preparing polymer metal nanocomposites (Riggsa et al. 2015). The resulting nanocomposites polymeric materials also allocate greater adaptability than traditional materials used in capacitors due to their lighter weight and flexibility making them appropriate in electronic devices (Riggsa et al. 2015; Liu and Zhai 2014; Yu et al. 2013). In the hope of designing a future material with optimum energy storage efficacy with environment sustainability, the efforts of chemists and material scientists are beyond

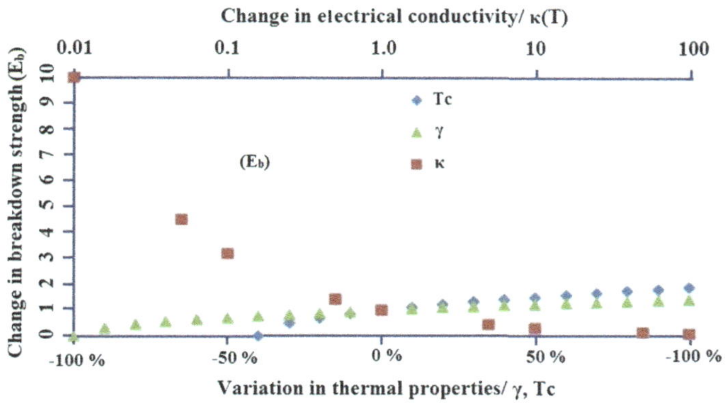

Fig. 12.4 Impact of variation on material properties on breakdown strength (E_b) (Riggsa et al. 2015)

constrains for the development of novel and desired metallic nanocomposites with green polymers (Mallakpour and Ezhieh 2018; Sharma et al. 2019; Sahoo et al. 2018; Ji et al. 2018).

12.3 Polymer Nanocomposites for Energy Economy Applications

For the fulfillment of energy demand per capita in developing countries has been increasing since last two decades due to high demand of energy for industrialization and urbanization activities. The per capita energy demand of India has been reached 36% in 2019 from 25% in 2000 year. This increased demand influenced the consumption of fossil fuels like coal and petroleum oil which reached up to 60 and 30% from 36 and 18% per capita, respectively between the years of 2000 and 2019 (IEA 2021a). As per the data available on the website of Independent Statics and Analysis U.S Energy Information Administration, a huge amount of traditional fuels from non-renewable energy resources are consumed annually for the production of demanded electrical energy in U.S. (https://www.eia.gov/petroleum/weekly/, last accessed 2021/07/20). Though even, other eco-friendly resources like wind energy, solar energy, geothermal, bio mass, hydrogen and fuel cells etc. are also in practice for electrical energy production in U.S, but for fulfilling the high demand of electrical energy in different sectors of country, administration is forced to use non-renewable energy resources. The data mentioned in the report published on website, the average consumption of coal for electricity generation in all sectors is 716582 thousand tons since the year of 2011 to 2020 (https://www.eia.gov/electricity/monthly/epm_table_grapher.php?t=table_2_01_a, last accessed 2021/07/20). Approximate average consumption of 245,360 thousand barrels of liquid petroleum has been reported in the ten year spam of 2011 to 2020 for all the sectors need of electricity generation (https://www.eia.gov/electricity/monthly/epm_table_grapher.php?t=table_2_02_a, last accessed 2021/07/20). Approximate average consumption of 3892 thousand tons of petroleum coke has been reported in the 10 year spam of 2011 to 2020 for all the sectors need of electricity generation (https://www.eia.gov/electricity/monthly/epm_table_grapher.php?t=table_2_03_a, last accessed 2021/07/20). Along with U.S., the dependency on non-renewable resources for electricity production is very high throughout the world (Luceno-Sanchez et al. 2019). To overcome the bad impact of these fossil fuels on environment, concern has been shown by the global community by using eco-friendly alternatives (Manish et al. 2006; Tafarte et al. 2019). The need of energy is now changed with term need of clean energy. In this respect, India has planned well for developing clean energy start-ups in ten year spam (IEA 2021b). The start ups of the year 2010 performed well with providing good solutions in the area of green energy related to solar, hydrogen and bio-energy (IEA 2021b). As per the report published in India Energy Outlook, India is decreasing its dependency on fossil fuels and implementing the eco-friendly methods of energy generation or harvesting like solar energy, wind energy etc. for long run greener sustainability around the earth (IEA 2021c). India is moving towards a great acceptance of solar energy year by year as the date indicates that we are at the position of 62 GW energy production in the year 2020–2025 from solar technology which was just 3 GW in year of 2010–2015 and proposed to achieve the target of 321 GW solar energy production since the year of 2035–2040 (IEA 2021d). Due to technological developments, the levelized cost of solar energy has been reduced 12.8 cents/kWh in the year 2020 from 50 cents/kWh in the year 2010 and targeted to reduce it by 5 cents/kWh up to the year 2030 for residential purpose in U.S. The photovoltaic energy cost reduction is targeted up to 5 cent/kWh, 4 cent/kWh and 2 cent/kWh in residential, commercial and utility sectors, respectively by the year of 2030 (https://www.energy.gov/eere/solar/photovoltaics, last accessed 2021/07/20). Hydrogen energy as a fuel is also promoted by ministry of new and renewable energy of India with the support of eminent Indian scientists developing the better technologies for hydrogen as energy economy (https://mnre.gov.in/new-technologies/hydrogen-energy, last accessed 2021/07/20). Several research projects are supported for minimising the challenges for using hydrogen as a renewable resource of energy and these academic centres, research institutions and industrial R&Ds are doing best utilization with good research outputs (https://mnre.gov.in/new-technologies/hydrogen-energy, last accessed 2021/07/20). United States is also putting his effort for developing technologies in the field of hydrogen as energy economy and has reached up to the limit of 10 million metric tons hydrogen production by using green alternatives like electrolysis, photo-electrochemical cells and biological techniques (https://www.energy.gov/eere/fuelcells/h2scale, last accessed 2021/07/21). Among the four generations (GENs) of materials development with various kinds of moieties, 4GEN is the latest generation popularized among researchers as the era of "inorganics-in-organics". The combinations approach of cost effective, flexible and biodegradable polymers with nanomorph metallic moieties yields nanostructured high potent polymer metal nanocomposite materials of versatile applicability in energy sector (Luceno-Sanchez et al. 2019).

Green polymer metal nanocomposites show their efficacy as a photovoltaic material as well as have the potential of hydrogen storage. In recent year, so many polymeric metal nanocomposite materials of gold, silver, aluminium, lead etc. have been reported with enhanced photovoltaic activity compared to free green polymer matrix (Liu et al. 2017; Torabi et al. 2015; Qian et al. 2015; Kakavelakis et al. 2017). Semiconductor activity of cadmium sulphide possessing a direct band gap of approximate ca. 2.4 electron volt reported much improved after combination with degradable polyaniline showing the efficacy of CdS-PANI nanocomposite in many electrochemical, photo-electro-chemical, sensing and electro-chromic devices (Jindal and Verma 2009; Soudi et al. 2010). Polystyrene-*b*-methylmethacrylate and polystyrene were applied as matrix with nanostructured copper coated alumina dust particles for the development of electrical conductive nanocomposites (Nadeem et al. 2016). Modified polyaniline, poly(aniline-*co*-aminonaphthalenesulfonic acid) degradable polymers were doped with oxide of iron nano metal following up green route of synthesis. Resultant polymer metal nanocomposites were measured for their conductivity and found 5×10^{-1} S/cm which is much greater than conductivity of un-doped polyaniline i.e. 1×10^{-3} S/cm. This enhanced property of developed polymer metal nanocomposite may be useful in photovoltaic activity along with other utilizations (Reddy et al. 2007). The very high enhancement of 377% in conductivity nature of a polyaniline and cerium metal ion (PANI/Ce^{+3}) composite compare to the untainted polymer was reported with a maximum conductance value of 46.76 S/cm at a weight load of twenty percent cerium nitrate on degradable polyaniline matrix (Li et al. 2018). Such type of enhancement in the properties related to conduction creates the interest of researchers in metal nanocomposites designing with probable suitability in photovoltaic, electro-chromic and hydrogen fuel cell devices and thus produces a possibility in energy economy. Research shows a trail that amalgamation of metallic nanoparticles and/or nanoporous constructions with polymers and/or their derivatives, probable materials for H_2 storage with the overcome of in-trained materials. The structural studies of recently synthesized metal nanocomposites revealed the dispersion of nanometals throughout the polyaniline and polypolyetherimide polymer matrices and other studies favored the improved property of metal nanocomposites compare to free polymer matrix. In H_2 sorption test for these polymer metal nanocomposites, considerable increase has been perceived (Beatrice et al. 2018, 1779). Due to the availability of π-conjugate system in green polymer PANI (Polyaniline), it has the high potential of electrical conductivity and thus hydrogen storage. The green metal nanocomposites of PANI have shown much improved potential of hydrogen storage and thus make the suitability of nanometal-PANI composites for hydrogen based energy economy (Moreira 2013). A novel conducting polymeric nanocomposite material for hydrogen storage has been reported. Carbon nanotubes were added with a green polymer Polyaniline as filler and then aluminium powder was additionally added to prepare polymer metal nanocomposite. Carbon nanotubes increased the hydrogen storage capacity as an increase in number of binding sites and the Al-metal powder further enhance the hydrogen sorption tendency of the material (Jurczyk et al. 2007). So, the materials of potent efficiency are needed to resolve the shortcomings of photovoltaic and hydrogen fuel cells and hence to improve their applicability in the area of energy. Many efforts have been paid by the active researchers of different related areas but the approach of green material development suitable for photovoltaic and hydrogen fuel cells are highly appreciable and also the need of global community concerned with sustainable development (Luceno-Sanchez et al. 2019; Lee et al. 2021; Basavaraja et al. 2017).

12.4 Polymer Nanocomposites for Environmental Applications

One of the most emerging topic environmental protection is directly related to the conservancy of our mother planet with survival concern of living being, especially humans, on earth with quite good of their health for enduring period. In the gluttony of luxurious life style, industrial development and urbanization are in progress globally without concerning their side effects on earth's environment (Sapna and Kumar, 2018). By products of various industries or domestic activities governed by human being are destroying the quality of environment and natural resources up to the extent of their unfitness for survival. Natural resources like air, water, soil are being polluted day by day globally. The pollutants which are thrown in these natural resources must be minimized and already contaminated resources should be remediated. These contaminants may be in the form of solid, liquid or gaseous particles suspended or dispersed in air, water and soil. Identification and detection of these pollutants are one of the necessary steps towards their remediation. Among the various remediation techniques, material science and nanotechnology play unique and prominent role for the issues related to environment. Several materials acting as ion exchangers, precipitators, separators, adsorbents etc. are available for detection and remediation of pollutants but, at nano-scale the performance of these materials get enhanced significantly (Blaney et al. 2007). Green polymer nanocomposites are the technically advance product of nano-materials for the resolving environmental issues developed with maintaining sustainability (Singh and Ambika 2018). Metallic nanoparticles embedded with degradable polymers enhance the properties of matrix and vice versa. This augmentation may be due to the availability of extensive pi-conjugation in green polymer (GPs) which permit the required mobility to charges present on polymer backbone (Kumar and Srivastva 2021). The electromagnetic field of metallic nano-particles interrelates with engendered stimulations resulting in the excitation diffusion and charge dissociation. The formation of holes on doping of metallic nano-particles may also enhance the charge mobility within the band gap of polymeric matrix (Ach et al. 2012; Wang et al. 2016). For environmental remediation, generally used nanoparticles are neutral metal/metal oxides integrated with eco-friendly green polymers such as resins, polyaniline, polyacrylamides, cellulose or its derivatives, chitosan, alginate and many more (Sapna and Kumar 2018; Zouboulis and Katsoyiannis 2002; Guo and Chen 2005; Katsoyiannis and Zouboulis 2002). The polymer-based nanocomposite for metal decontamination are available in different forms like candle, mat, membrane, beads, etc. (Chowdhury and Balasubramanian 2014). Water pollution is a major crisis for the availability of drinking water due to its contamination by different types of pollutants which show adverse effects on water quality as well as on human health (Fig. 12.5). Polymer metal nanocomposites show their function in water/waste water treatment with the approach of adsorption, photo-catalysis, monitoring-sensing, disinfection and membrane filtration for contaminants (Fig. 12.6) (Akharame et al. 2018). Harmonized adsorption property of polymer nanocomposites has been investigated due to the presence of nanometals on large surface area of polymers and this optimized behavior is the reason of their suitability for chemical sensing and water decontamination (Khodakarami and Bagheri 2021). Different types of toxic organic, inorganic and microbial pollutants have been removed from water by applying metallic polymer nanocomposites (Khodakarami and Bagheri 2021). An improved monolayer adsorption potential of karaya grafted acrylamide green polymer was observed on addition of nanomorph silicon dioxide. The synthesized green metallic nanocomposite was applied for the removal of methylene blue dye from water (Mittal et al. 2015). Polymer metal nanocomposites with iron oxide nanoparticles, biodegradable chitosan and carbon nanofibers adopting green approach of synthesis, having enhanced adsorption capability were reported. This chitosan/Fe-CNF composite material was applied for efficient removal of Cr(VI) dissolved in water at the rate of 80 mg/g the nanocomposite (Khare et al. 2016). Biodegradable copolymer poly acrylamide with styrene

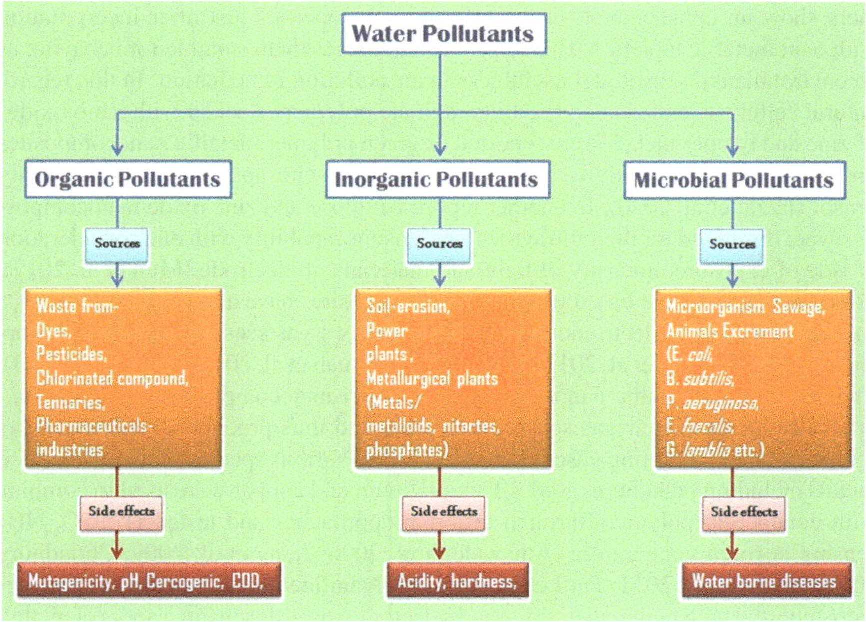

Fig. 12.5 Types of water pollutants, their common sources and side effects

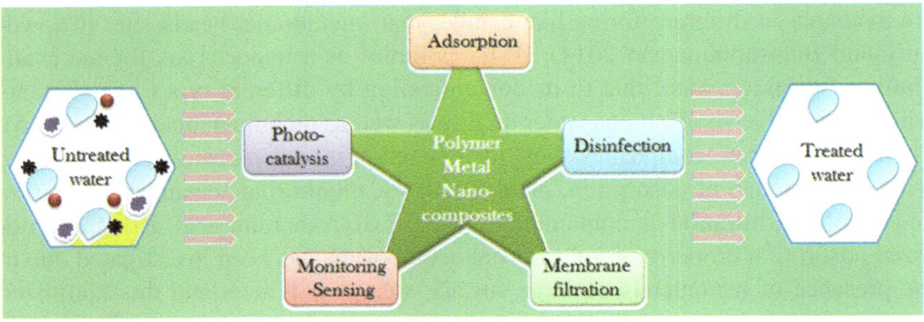

Fig. 12.6 Water treatment process with various potential of polymer metal nanocomposites

sodium sulfonate and nanostructured titanium dioxide were employed to develop nanocomposites and then applied for the adsorption of cesium, cobalt and europium metal ions contaminated in water resources. The rate of adsorption was determined as 120 mg/g, ~ 101 mg/g and ~ 86 mg/g for Cs^{+}, Co^{2+} and Eu^{3+} ions, respectively (Pandey et al. 2017). Green polymer nanocomposite of xylan/poly(acrylic acid) loaded with nano iron oxide particles investigated for its dye adsorbing capacity and found that it can remove 90% methylene blue from polluted water (Sun et al. 2015). Chitosan/Fe_3O_4 and Chitosan grafted Polyaniline /Co_3O_4 nano composites were synthesized with much improved degradation rate than non metallic matrices during their applicability in water treatment (Tang et al. 2014). Biodegradable polymers show an enhancement in their electronic properties just after incorporation with nanometallic moieties. This improvement makes them suitable for behaving as an environmental sensor and usefulness in air pollution remediation. In this regard, natural cellulose, chitosan and sodium alginate polymers stacked with nano oxides of zinc and copper metals were reported as green polymer metallic nanocomposites possessing high conductivity and thermal stability with applicability of humidity sensor (Ezzat et al. 2020). In another report, cellulose and zinc oxide nanocomposites were discussed for their ultraviolet ray sensing capability with due consideration of side of exposure, intensity of light and materials of electrode (Mun et al. 2017). Some other zinc oxide based polymer nanocomposites have also been reported as a suitable sensors for electronic devices designed as a gas sensors, moisture sensors and UV sensors (Yao et al. 2012; Yi et al. 2011; Hsueh et al. 2012; Faure et al. 2013). The insertion of metallic nanoparticles in polymer-nanocomposites (CNT, Gr etc.) gradually increase their gas sensing execution and thus produce a refined sensors for environment polluting gases (Yan et al. 2020). Various species of nanostructured metals including palladium, gold, silver, platinum and copper were used to combine with conducting polymers through different approaches and tested H_2, CO, NH_3 sensing at room temperature (Athawale et al. 2006; Jiang et al. 2009; Choudhury 2009; Su and Shiu 2011; Patil et al. 2015). Polyaniline/Palladium nanocomposites were reported with high sensitivity, good selectivity towards sensing vapors of methyl alcohol, ethyl alcohol and *iso*-propyl alcohol (Athawale et al. 2006). So many other works has been reported with application of green polymer based metallic nanocomposites in sensing of gaseous pollutants (Choudhury 2009; Bai et al. 2015; Zhang et al. 2013c; Hong et al. 2010). An aspect of environmental issue is solid waste due to their huge generation globally. A report has been published regarding the development of waste-derived nanomaterials and their utilization in technologies concerned with monitoring and remediation of environmental pollution. This approach resulted many types of nanomaterials like nanopolymers, nanometallic particles, nanofillers, polymer nanocomposites etc. and thus provides a sustainable solution of solid waste management (Abdelbasir et al. 2020).

12.5 Conclusions

In this era of rapidly changing technology as the demand of development for the fulfilment of basic needs as well as luxurious objects, sustainable development is the only necessitate for protection of our mother earth and our existence on this planet. Green polymer nanocomposites occupied with metallic nanoparticles are the hope for maintaining sustainability with high speed industrial and technological growth. Being the more flexible, more tensile, more conductive, the metal nanocomposites with natural polymers are showing a golden age of their requirement in the highly demanding area of energy production, energy harvesting, and energy storage with green approach. The renewability and degradability characteristics of the polymeric nanocomposites are the main cause of interest of researchers in this material. The application of green polymer metal nanocomposites in environmental remediation also creates the attention of environmental and material scientists towards the development of novel and much potent polymer metal nanocomposites with various kind of fabrication and tailoring of materials which are the key role player in the sustainable development of modern age of contentiously improving technology.

12.6 Future Aspect of Green Polymers Decorated Nanocomposites

As per the discussion in this chapter, significance of green polymer decorated metal nano composites in the field of energy and environment is quite clear. However, the application of these green nano materials are not constrained up to the energy and environment issues only, but also in high demand by various divisions of scientific and industrial organizations due to their high efficiency. In upcoming years, we can assume the wide application of green polymer metal nano composites in drug delivery systems, anti-corrosion blockade glazing, ultra violet shield gels, lubricants, scratch proof painting materials, abrasion resistant materials, fire impeding materials, better-quality fibres and films etc. Industries related to constructions, robotics, pharmaceuticals, food and nutrition, cosmetics, health & hygiene, automobiles, telecommunications, information technology and many others are also hoping towards materials scientists to serve them green metal nano composites as a material of their specific requirements with high grade suitability and durability without disturbing the nature and its components.

Acknowledgements Authors are thankful to the authorities of their respective departments and institutions for providing necessary facilities of computer, internet and libraries to carry out this work.

References

Abdelbasir SM, McCourt KM, Lee CM, Vanegas DC (2020) Waste-derived nanoparticles: synthesis approaches, environmental applications, and sustainability considerations. Front Chem 8:782

Ach BR, Lu J, Beaupré S, Leclerc M, Tao Y (2012) Control of the active layer nanomorphology by using co-additives towards high-performance bulk heterojunction solar cells. Org Electron 13(9):1736–1741

Ahmed S, Chaudhry SA, Ikram S (2017) A review on biogenic synthesis of ZnO nanoparticles using plant extracts and microbes: a prospect towards green chemistry. J Photochem Photobiol B 166:272–284

Akharame MO, Fatoki OS, Opeolu BP, Olorunfemi DI, Oputu OU (2018) Polymeric nanocomposites (PNCs) for wastewater remediation: an overview. Polym Plast Technol Eng 57:1801–1827

Aktitiz İ, Varol R, Akkurt N, Saraç MF (2020) In-situ synthesis of 3D printable mono- and bi-metallic (Cu/Ag) nanoparticles embedded polymeric structures with enhanced electromechanical properties. Polym Test 90:106724

Alberto C, Ade M (2013) Environmental sustainability: implications and limitations to Green Chemistry. Found Chem 16:125–147

Alipour A, Mansour LM, Tashakkorian H (2021) Study of the effect of band gap and photoluminescence on biological properties of polyaniline/CdS QD nanocomposites based on natural polymer. Sci Rep 11:1913

Almeida AIAR, Ferreira LDL, Almeida GC, Calado HDR, Viana MM (2020) Poly(3-hexylthiophene)/titanium dioxide nanocomposites prepared by in-situ polymerization: structure, morphology and electrochemical properties. Synth Met 269:116544

Ananth AN, Umapathy S, Sophia J, Mathavan T, Mangalaraj D (2011) On the optical and thermal properties of in situ/ex situ reduced Ag NP's/PVA composites and its role as a simple SPR-based protein sensor. Appl Nanosci 1:87–96

Anstas PA (1998) Green chemistry: theory and practice. Oxford University Press, Inc., New York

Athawale AA, Bhagwat SV, Katre PP (2006) Nanocomposite of Pd–polyaniline as a selective methanol sensor. Sens Actuator b: Chem 114:263–267

Bai S, Sun C, Wan P et al (2015) Transparent conducting films of hierarchically nanostructured polyaniline networks on flexible substrates for high-performance gas sensors. Small 11:306–310

Basavaraja C, Park JY, Huh DS (2017) Degradable and electrically conductive poly(3,4-ethylenedioxythiophene)/sigma cell cellulose polymer composites. Polym Compos 38:1864–1872

Beatrice CAG, Oliveira AD, Passador FR, Pessan LA (2016) Sulfonated poly(ether imide)/aluminium nanocomposites for hydrogen storage. AIP Conf Proc 1779:040004

Beatrice CAG, de Oliveira AD, Moreira BR, Passador FR, Leiva DR, Pessan LA (2018) Development of polymer nanocomposites for hydrogen storage. In: 22nd world hydrogen energy conference, Rio de Janeiro, RJ, Brazil, June

Bhandari PJ, Reddy MM, Rao KJ, Sandanaraj BS (2021) Rapid chemical synthesis of self-assembling semi-synthetic proteins. J Org Chem 86(13):8576–8589

Blaney LM, Cinar S, Sengupta AK (2007) Hybrid anion exchanger for trace phosphate removal from water and wastewater. Water Res 41:1603–1613

Buller S, Strunk J (2016) Nanostructure in energy conversion. J Energy Chem 25:171–190

Calvert PD (1975) Natural and synthetic polymers. Nature 253:504

Carotenuto G, Nicolais L, Martorana B, Perlo P (2004) Metal–polymer nanocomposite synthesis: novel ex situ and in situ approaches. In: Nicolais L, Carotenuto G (eds) Metal–polymer nanocomposites. Wiley, USA

Chakraborty S, Raj MA, Mary NL (2020) Biocompatible supercapacitor electrodes using green synthesised ZnO/polymer nanocomposites for efficient energy storage applications. J Energy Storage 28:101275

Choudhury A (2009) Polyaniline/silver nanocomposites: dielectric properties and ethanol vapour sensitivity. Sens Actuator b: Chem 138:318–325

Chowdhury S, Balasubramanian R (2014) Recent advances in the use of graphene-family nanoadsorbents for removal of toxic pollutants from wastewater. Adv Colloid Interface Sci 204:35–56

Dahl JA (2007) Toward greener nanosynthesis. Chem Rev 107:2228–2269

Das RK, Gogoi N, Bora U (2011) Green synthesis of gold nanoparticles using *Nyctanthes arbortristis* flower extract. Bioprocess Biosyst Eng 34(5):615–619

Dmour I, Taha MO (2018) Natural and semisynthetic polymers in pharmaceutical nanotechnology. In: Grumezescu AM (ed) Organic materials as smart nanocarriers for drug delivery. William Andrew Publishing, pp 35–100

Erabhoina H, Rosenbach D, Mohanraj J, Thelakkat M (2021) Solid polymer nanocomposite electrolytes with improved interface properties towards lithium metal battery application at room temperature. Electrochim Acta 387:138455

Ezzat HA, Hegazy MA, Nada NA, Osman O, Ibrahim MA (2020) Application of natural polymers enhanced with ZnO and CuO as humidity sensor. NRIAG J Astron Geophys 9:586–597

Faure B, Salazar-Alvarez G, Ahniyaz A, Villaluenga I, Berriozabal G, De Miguel YR, Bergström L (2013) Dispersion and surface functionalization of oxide nanoparticles for transparent photocatalytic and UV-protecting coatings and sunscreens. Sci Technol Adv Mater 14:023001

Gandla D, Sarkar S, Ghanti E, Ghosh S (2017) Copper chromite-polyaniline nanocomposite: an advanced electrode material for high performance energy storage. Electrochim Acta 248:486–495

Ganesan V, Jayaraman A (2014) Theory and simulation studies of effective interactions, phase behavior and morphology in polymer nanocomposites. Soft Matter 10:13–38

Ganesh I (2014) Conversion of carbon dioxide into methanol—a potential liquid fuel: fundamental challenges and opportunities (a review). Renew Sustain Energy Rev 31:221–257

Gebreslassie YT, Gebretnsae HG (2021) Green and cost-effective synthesis of tin oxide nanoparticles: a review on the synthesis methodologies, mechanism of formation, and their potential applications. Nanoscale Res Lett 16:97

Ghanbari D, Salavati-Niasari M, Ghasemi-Kooch M (2016) In situ and ex situ synthesis of poly(vinyl alcohol)–Fe_3O_4 nanocomposite flame retardants. Particuology 26:87–94

Ghosh S, Patil S, Ahire M, Kitture R, Gurav DD, Jabgunde AM, Kale S, Pardesi K, Shinde V, Bellare J, Dhavale DD, Chopade BA (2012) *Gnidia glauca* flower extract mediated synthesis of gold nanoparticles and evaluation of its chemocatalytic potential. J Nanobiotechnol 10:17

Gu H, Guo J, Zhang X, He Q, Huang Y, Colorado HA, Haldolaarachchige N, Xin H, Young DP, Wei S, Guo Z (2013) Giant magnetoresistive phosphoric acid doped polyaniline-silica nanocomposites. J Phys Chem C 117:6426–6436

Guo X, Chen F (2005) Removal of arsenic by bead cellulose loaded with iron oxyhydroxide from groundwater. Environ Sci Technol 39:6808–6818

Guo Z, Henry LL, Palshin V, Podlaha EJ (2006) Synthesis of poly(methyl methacrylate) stabilized colloidal zero-valence metallic nanoparticles. J Mater Chem 16:1772–1777

Haghi K (2013) International conference on composites and nanocomposites. Composites and nanocomposites. Apple Academic Press, Toronto; Waretown, NJ

Haghi AK, Zaikov GE (2013) Update on nanofillers in nanocomposites: from introduction to application. Smithers Rapra Techn., Shawbury

Hong L, Li Y, Yang M (2010) Fabrication and ammonia gas sensing of palladium/polypyrrole nanocomposite. Sens Actuator b: Chem 145:25–31

Hsueh HT, Hsueh TJ, Chang SJ, Hung FY, Hsu CL, Dao BT, Lam KT, Wen KH (2012) A flexible ZnO nanowire-based humidity sensor. IEEE Trans Nanotechnol 11(3):520–525

Hu K, Kulkarni DD, Choi I, Tsukruk VV (2014) Graphene-polymer nanocomposites for structural and functional applications. Prog Polym Sci 39:1934–1972

Hutchison JE (2008) Greener nanoscience: a proactive approach to advancing applications and reducing implications of nanotechnology. ACS Nano 2(3):395–402

IEA (2021a) Key indicators in India as a percentage of global averages, 2000 and 2019. IEA, Paris. https://www.iea.org/data-and-statistics/charts/key-indicators-in-india-as-a-percentage-of-global-averages-2000-and-2019

IEA (2021b) Ten years of clean energy start-ups. IEA, Paris. https://www.iea.org/articles/ten-years-of-clean-energy-start-ups. Accessed 21 July 2021

IEA (2021c) Changes in coal and solar capacity in India in the Stated Policies Scenario, 2010–2040. IEA, Paris. https://www.iea.org/data-and-statistics/charts/changes-in-coal-and-solar-capacity-in-india-in-the-stated-policies-scenario-2010-2040. Accessed 21 July 2021

Iravani S (2011) Green synthesis of metal nanoparticles using plants. Green Chem 13:2638–2650

Iravani S, Varma RS (2020) Sustainable synthesis of cobalt and cobalt oxide nanoparticles and their catalytic and biomedical applications. Green Chem 22:2643–2661

Javed R, Usman M, Tabassum S, Zia M (2016) Effect of capping agents: structural, optical and biological properties of ZnO nanoparticles. Appl Surf Sci 386:319–326

Ji J, Xiong H, Zhu Z, Li L, Huang Y, Yu X (2018) Fabrication of polypyrrole/chitosan nanocomposite aerogel monolith for removal of Cr (VI). J Polym Environ 26:1979–1985

Jiang S, Chen J, Tang J et al (2009) Au nano particles-functionalized two-dimensional patterned conducting PANI nanobowl monolayer for gas sensor. Sens Actuators b: Chem 140:520–524

Jianga X, Zhua X, Chang C, Liuc S, Luoad X (2019) X-ray shielding structural and properties design for the porous transparent $BaSO_4$/cellulose nanocomposite membranes. Int J Biol Macromol 139:793–800

Jindal Z, Verma NK (2009) Electrochemical template-assisted fabrication of CdS micro/nanostructures. Physica E 41:1752–1756

Jothimani B, Venkatachalapathy B, Karthikeyan NS, Ravichandran C (2019) A review on versatile applications of degradable polymers. In: Gnanasekaran D (eds) Green biopolymers and their nanocomposites. Materials horizons: from nature to nanomaterials. Springer, Singapore

Jurczyk MU, Kumar A, Srinivasan S, Stefanakos E (2007) Polyaniline-based nanocomposite materials for hydrogen storage. Int J Hydrogen Energy 32:1010–1015

Kakavelakis G, Alexaki K, Stratakis E, Kymakis E (2017) Efficiency and stability enhancement of inverted perovskite solar cells via the addition of metal nanoparticles in the hole transport layer. RSC Adv 7(21):12998–13002

Kao J, Thorkelsson K, Bai P, Rancatore BJ, Xu T (2013) Toward functional nanocomposites: taking the best of nanoparticles, polymers, and small molecules. Chem Soc Rev 42:2654–2678

Katsoyiannis IA, Zouboulis AI (2002) Removal of arsenic from contaminated water sources by sorption onto iron-oxide-coated polymeric materials. Water Res 36:5141–5155

Kausar A (2021) Progress in green nanocomposites for high-performance applications. Mater Res Innov 25(1):53–65

Khare P, Yadav A, Ramkumar J, Verma N (2016) Microchannel embedded metal–carbon–polymer nanocomposite as a novel support for chitosan for efficient removal of hexavalent chromium from water under dynamic conditions. Chem Eng J 293:44–54

Khodakarami M, Bagheri M (2021) Recent advances in synthesis and application of polymer nanocomposites for water and wastewater treatment. J Clean Prod 296:126404
Kobayashi S (2015) Enzymatic ring-opening polymerization and polycondensation for the green synthesis of polyesters. Polym Adv Technol 26:677–686
Kobayashi S (2017) Green polymer chemistry: new methods of polymer synthesis using renewable starting materials. Struct Chem 28:461–474
Kokate M, Garadkar K, Gole A (2013) One pot synthesis of magnetite–silica nanocomposites: applications as tags, entrapment matrix and in water purification. J Mater Chem A 1:2022–2029
Kumar S, Srivastva AN (2021) Application of carbon nanomaterials decorated electrochemical sensor for analysis of environmental pollutants. In: Srivastva AN (ed) Analytical chemistry: advancement, perspectives and applications. IntechOpen Ltd., London
Kumar VG, Gokavarapu SD, Rajeswari A, Dhas TS, Karthick V, Kapadia Z, Shrestha T, Barathy IA, Roy A, Sinha S (2011) Facile green synthesis of gold nanoparticles using leaf extract of antidiabetic potent *Cassia auriculata*. Colloids Surf B 87(1):159–163
Kumar SK, Jouault N, Benicewicz B, Neely T (2013) Nanocomposites with polymer grafted nanoparticles. Macromolecules 46(9):3199–3214
Kumar P, Kim K-H, Bansal V, Kumar P (2017) Nanostructured materials: a progressive assessment and future direction for energy device applications. Coord Chem Rev 353:113–141
Lee S, Hong Y, Shim BS (2021) Biodegradable PEDOT:PSS/clay composites for multifunctional green-electronic materials. Adv Sustain Syst 2100056
Li Z, Shen Y, Li Y, Zheng F, Liu L (2018) Doping effects of cerium ion on structure and electrochemical properties of polyaniline. Polym Int 67:121–126
Liu SH, Zhai J (2014) A small loading of surface-modified $Ba_{0.6}Sr_{0.4}TiO_3$ nanofiber-filled nanocomposites with enhanced dielectric constant and energy density. RSC Adv 4:40973–40979
Liu Y, Lang F, Dittrich T, Steigert A, Fischer CH, Köhler T, Plate P, Rappich J, Lux-Steiner MC, Schmid M (2017) Enhancement of photocurrent in an ultra-thin perovskite solar cell by Ag nanoparticles deposited at low temperature. RSC Adv 7(3):1206–1214
Luceno-Sanchez JA, Diez-Pascual AM, Capilla RP (2019) Materials for photovoltaics: state of art and recent developments. Int J Mol Sci 20(4):976. https://doi.org/10.3390/ijms20040976
Luo H, Zhang D, Jiang C, Yuan X, Chen C, Zhou K (2015) Improved dielectric properties and energy storage density of poly(vinylidene fluoride-*co*-hexafluoropropylene) nanocomposite with hydantoin epoxy resin coated $BaTiO_3$. ACS Appl Mater Interfaces 7(15):8061–8069
Magesh G, Bhoopathi G, Nithya N, Arun AP, Kumar ER (2018) Tuning effect of polysaccharide chitosan on structural, morphological, optical and photoluminescence properties of ZnO nanoparticles. Superlattices Microstruct 117:36–45
Mahendia S, Tomar AK, Kumar S (2010) Electrical conductivity and dielectric spectroscopic studies of PVA–Ag nanocomposite films. J Alloy Compd 508:406–411
Mallakpour S, Ezhieh AN (2018) Citric acid and vitamin C as coupling agents for the surface coating of ZrO_2 nanoparticles and their behavior on the optical, mechanical, and thermal properties of poly (vinyl alcohol) nanocomposite films. J Polym Environ 26:2813–2824
Manish S, Pillai IR, Banerjee R (2006) Sustainability analysis of renewables for climate change mitigation. Energy Sustain Dev 10:25–36
Mantione D, Agua ID, Sanchez AS, Mecerreyes D (2017) Poly(3,4-ethylenedioxythiophene) (PEDOT) derivatives: innovative conductive polymers for bioelectronics. Polymers 9(8):354
Mittal H, Maity A, Ray SS (2015) Synthesis of co-polymer-grafted gum karaya and silica hybrid organic–inorganic hydrogel nanocomposite for the highly effective removal of methylene blue. Chem Eng J 279:166–179
Mohammadnezhad G, Keikavousi BA (2020) Polymer matrix nanocomposites for heavy metal adsorption: a review. J Iran Chem Soc 17:1259–1281
Moon SI, Lee CW, Miyamoto M, Kimura Y (2000) Melt polycondensation of L-lactic acid with Sn(II) catalysts activated by various proton acids: a direct manufacturing route to high molecular weight poly(L-lactic acid). J Polym Sci a: Polym Chem 38:1673–1679

Moon RJ, Martini A, Nairn J, Simonsen J, Youngblood J (2011) Cellulose nanomaterials review: structure, properties and nanocomposites. Chem Soc Rev 40:3941–3994

Moreira BR (2013) Development of polyaniline/TiO_2 and polyaniline/$NaAlH_4$ nanocomposites for hydrogen storage. M.Sc. thesis, Universidade Federal de São Carlos

Mun S, Kim HC, Ko HU, Zhai L, Kim JW, Kim J (2017) Flexible cellulose and ZnO hybrid nanocomposite and its UV sensing characteristics. Sci Technol Adv Mater 18(1):437–446

Nadeem QA, Rizwan M, Gill R, Rafique M, Shahid M (2016) Fabrication of alumina based electrically conductive polymer composites. J Appl Polym Sci 133(5):42939

Niluxsshun MCD, Masilamani K, Mathiventhan U (2021) Green synthesis of silver nanoparticles from the extracts of fruit peel of *Citrus tangerina*, *Citrus sinensis*, and *Citrus limon* for antibacterial activities. Bioinorg Chem Appl 2:6695734

Njuguna J, Pielichowski K, Desai S (2008) Nanofiller-reinforced polymer nanocomposites. Polym Adv Technol 19:947–959

Njuguna J, Pielichowski K, Zhu H (eds) (2014) Health and environmental safety of nanomaterials: polymer nanocomposites and other materials containing nanoparticles. Woodhead Publishing, Cambridge; Waltham, MA

Palem RR, Ganesh SD, Kronekova Z, Slávíková M, Saha N, Saha P (2018) Green synthesis of silver nanoparticles and biopolymer nanocomposites: a comparative study on physico-chemical, antimicrobial and anticancer activity. Bull Mater Sci 41:55

Pande N, Jambhale A, Purohit V, Jaspal D, Ambekar J, Chauhan NPS (2021) Poly(o-toluidine)-Li nanocomposite: facile preparation and utilization in energy storage devices. Polym Plast Technol Mater 60(2):208–216

Pandey N, Shukla SK, Singh NB (2017) Water purification by polymer nanocomposites: an overview. Nanocomposites 3(2):47–66

Patil UV, Ramgir N, Karmakar N, Bhogale A, Debnath A, Aswal DK, Kothari D (2015) Room temperature ammonia sensor based on copper nanoparticle intercalated polyaniline nanocomposite thin films. Appl Surf Sci 339:69–74

Pérez-Juste J, Pastoriza-Santos I, Liz-Marzán LM (2013) Multifunctionality in metal@microgel colloidal nanocomposites. J Mater Chem A 1:20–26

Philip D (2009) Biosynthesis of Au, Ag and Au-Ag nanoparticles using edible mushroom extract. Spectrochim. Acta A Mol Biomol Spectrosc 73(2):374–381

Qian M, Li M, Shi XB, Ma H, Wang ZK, Liao LS (2015) Planar perovskite solar cells with 15.75% power conversion efficiency by cathode and anode interfacial modification. J Mater Chem A 3(25):13533–13539

Rahman PM, Mujeeb VMA, Muraleedharan K, Thomas SK (2018) Chitosan/nano ZnO composite films: enhanced mechanical, antimicrobial and dielectric properties. Arab J Chem 11:120–127

Rajagopal G, Nivetha A, Sundar M, Panneerselvam T, Murugesan S, Parasuraman P, Kumar S, Ilango S, Kunjiappan S (2021) Mixed phytochemicals mediated synthesis of copper nanoparticles for anticancer and larvicidal applications. Heliyon 7(6):e07360

Raveendran P, Fu J, Wallen SL (2003) Completely "green" synthesis and stabilization of metal nanoparticles. J Am Chem Soc 125(46):13940–13941

Reddy KR, Lee KP, Gopalan AI (2007) Novel electrically conductive and ferromagnetic composites of poly(aniline-*co*-aminonaphthalenesulfonic acid) with iron oxide nanoparticles: synthesis and characterization. J Appl Polym Sci 106:1181–1191

Reddy PL, Deshmukh K, Chidambaram K et al (2019) Dielectric properties of polyvinyl alcohol (PVA) nanocomposites filled with green synthesized zinc sulphide (ZnS) nanoparticles. J Mater Sci Mater Electron 30:4676–4687

Riggsa BC, Adireddy S, Rehm CH, Puli VS, Elupula R, Chrisey DB (2015) Polymer nanocomposites for energy storage applications. Mater Today: Proc 2:3853–3863

Sahoo S, Kalita H, Mohanty S, Nayak SK (2018) Degradation study of biobased polyester-polyurethane and its nanocomposite under natural soil burial, UV radiation and hydrolytic-salt water circumstances. J Polym Environ 26:1528–1539

Sapna, Kumar D (2018) Biodegradable polymer-based nanoadsorbents for environmental remediation. In: Hussain CM, Mishra AK (eds) New polymer nanocomposites for environmental remediation. Elsevier, pp 261–278

Seshadri VD (2021) Zinc oxide nanoparticles from *Cassia auriculata* flowers showed the potent antimicrobial and *in vitro* anticancer activity against the osteosarcoma MG-63 cells. Saudi J Biol Sci 28(7):4046–4054

Shafey AME (2020) Green synthesis of metal and metal oxide nanoparticles from plant leaf extracts and their applications: a review. Green Process Synth 9:304–339

Sharma G, Kumar A, Sharma S, Naushad M, Prakash Dwivedi R, Alothman ZA, Mola GT (2019) Novel development of nanoparticles to bimetallic nanoparticles and their composites: a review. J King Saud Univ Sci 319(2):257–269

Singh PP, Ambika (2018) Environmental remediation by nanoadsorbents-based polymer nanocomposite. Hussain CM, Mishra AK (eds) New polymer nanocomposites for environmental remediation. Elsevier, pp 223–241

Solomon SD, Bahadory M, Jeyarasingam AV, Ruthkowsky SA, Boritz C, Mulfinger L (2007) Synthesis and study of silver nanoparticles. J Chem Educ 84:322–325

Soudi R, Kamal M, Shabaka AA, Abdelrazek EM, Eisa W (2010) Synthesis, characterization and spectroscopic studies of CdS/polyaniline core/shell nanocomposites. Synth Met 160:479–484

Spring FS (1945) Naturally occurring polyesters. Nature 155:272

Su PG, Shiu CC (2011) Flexible H_2 sensor fabricated by layer-by-layer self-assembly of thin films of polypyrrole and modified in situ with Pt nanoparticles. Sens Actuator b: Chem 157:275–281

Sun XF, Liu B, Jing Z (2015) Preparation and adsorption property of xylan/poly(acrylic acid) magnetic nanocomposite hydrogel adsorbent. Carbohydr Polym 118:16–23

Tafarte P, Eichhorn M, Thrän D (2019) Capacity expansion pathways for a wind and solar based power supply and the impact of advanced technology-a case study for Germany. Energies 12:324

Tang X, Tian M, Qu L, Zhu S, Guo X, Han G, Sun K, Hu X, Wang Y, Xu X (2014) A facile fabrication of multifunctional knit polyester fabric based on chitosan and polyaniline polymer nanocomposite. Appl Surf Sci 317:505–510

Tao P, Viswanath A, Li Y, Siegel RW, Benicewicz BC, Schadler LS (2013) Bulk transparent epoxy nanocomposites filled with poly(glycidyl methacrylate) brush-grafted TiO_2 nanoparticles. Polymer 54(6):1639–1646

Thakur VK, Lin MF, Tan EJ, Lee PS (2012) Green aqueous modification of fluoropolymers for energy storage applications. J Mater Chem 22:5951–5959

Torabi NM, Behjal A, Shahpari M, Edalati, S (2015) Development of a silver/polymer nanocomposite interconnection layer for organic tandem solar cells. J Nanophotonics 9(1):093049

Twardowski T (2006) Introduction to nanocomposite materials: properties, processing, characterization. DES Tech Publications Inc., Lancaster, PA

Unterlass MM (2016) Green synthesis of inorganic–organic hybrid materials: state of the art and future perspectives. Eur J Inorg Chem 1135–1156

Wang RM, Zheng SR, Zheng YP (2011) Polymer matrix composites and technology. WP Woodhead Publ, Oxford

Wang L, Wang D, Dong Z, Zhang F, Jin J (2013) Interface chemistry guided long-cycle-life Li–S battery. Nano Lett 13(9):4206–4211

Wang W et al (2016) High efficiency organic bulk-heterojunction solar cells applying a new system of co-additives. Mater Lett 166:227–230

Wei S, Wang Q, Zhu J, Sun L, Lin H, Guo Z (2011) Multifunctional composite core–shell nanoparticles. Nanoscale 3:4474–4502

Yamaguchi S, Tanha M, Hult A, Okuda T, Ohara H, Kobayashi S (2014) Green polymer chemistry: lipase-catalyzed synthesis of bio-based reactive polyesters employing itaconic anhydride as a renewable monomer. Polym J 46:2–13

Yan Y, Yang G, Xu J-L, Zhang M, Kuo CC, Wang SD (2020) Conducting polymer-inorganic nanocomposite-based gas sensors: a review. Sci Technol Adv Mater 21(1):768–786

Yao I, Tseng T, Lin P (2012) ZnO nanorods grown on polymer substrates as UV photodetectors. Sens Actuator a: Phys 178:26–31

Yi J, Lee JM, Park WI (2011) Vertically aligned ZnO nanorods and graphene hybrid architectures for high sensitive flexible gas sensors. Sens Actuator b: Chem 155:264–269

Yu K, Niu Y, Zhou Y, Bai Y, Wang H (2013) Nanocomposites of surface-modified $BaTiO_3$ nanoparticles filled ferroelectric polymer with enhanced energy density. J Am Ceram Soc 96:2519–2524

Zhang X, Wei S, Haldolaarachchige N, Colorado HA, Luo Z, Young DP, Guo Z (2012) Magnetoresistive conductive polyaniline–barium titanate nanocomposites with negative permittivity. J Phys Chem C 116:15731–15740

Zhang X, Alloul O, He Q, Zhu J, Verde MJ, Li Y, Wei S, Guo Z (2013a) Strengthened magnetic epoxy nanocomposites with protruding nanoparticles on the graphene nanosheets. Polymer 54(14):3594–3604

Zhang X, He Q, Gu H, Colorado HA, Wei S, Guo Z (2013b) Flame-retardant electrical conductive nanopolymers based on bisphenol F epoxy resin reinforced with nano polyanilines. ACS Appl Mater Interfaces 5(3):898–910

Zhang J, Liu X, Wu S et al (2013c) One-pot fabrication of uniform polypyrrole/Au nanocomposites and investigation for gas sensing. Sens Actuator b: Chem 186:695–700

Zhou Y, Fu S, Zhang L, Zhan H, Levit MV (2014) Use of carboxylated cellulose nanofibrils-filled magnetic chitosan hydrogel beads as adsorbents for Pb(II). Carbohydr Polym 101:75–82

Zhu J, Wei S, Yadav A, Guo Z (2010) Rheological behaviors and electrical conductivity of epoxy resin nanocomposites suspended with in-situ stabilized carbon nanofibers. Polymer 51(12):2643–2651

Zhu J, Wei S, Ryu J, Guo Z (2011) Strain-sensing elastomer/carbon nanofiber "metacomposites." J Phys Chem C 115(27):13215–13222

Zouboulis AI, Katsoyiannis IA (2002) Arsenic removal using iron oxide loaded alginate beads. Ind Eng Chem Res 41:6149–6155

Chapter 13
Metal Nanocomposites for Energy and Environmental Applications

Hansa, Shalini Sahani, and TaeYoung Kim

Abstract Environmental and Energy issues are the two major issues that our world is confronting nowadays. Utilitarian nanocomposites are rising materials for zillions of Energy and environmental applications. Nanomaterials are a different class of materials composed of metal, metal oxides/sulfides/phosphides (nanostructured and thin films) varying in levels of electronic structure, physical, chemical, and electromagnetic properties. There are several alkaline and alkali earth metal-based nanocomposites that are used in basic heterogeneous catalysis for cleaner fuel production. Whereas transition metal/metal oxide-based nanocomposites (TMMONs) like ZnO, TiO_2, SnO_2, ZrO_2, and Fe_3O_4 are gaining tremendous attention as photocatalysts. Moreover, these transition metal/metal oxide-based nanocomposites consist of lucrative flexible mechanical properties along with extraordinary electrochemical performance to fabricate the natural observing gadgets. Recently, these TMMONs have been majorly employed as active electrodes in electrochemical cells and supercapacitors due to their excellent dielectric properties and high porosity. Even, nanocomposites of metals with carbon-based materials have got a lot of potential in energy storage applications due to their fast power energy delivery, long lifecycle, high power density, besides reasonably high energy density that can fill the gap between the batteries and the conventional capacitors. In this book chapter, we would thoroughly discuss the above-mentioned aspects of various metal/metal oxides-based nanocomposites employed in energy and environmental application.

Keywords Metal nanocomposites · Metal oxides · Nanomaterials · Photocatalysis · Supercapacitors

---

Hansa · S. Sahani (✉) · T. Kim
Department of Materials Science and Engineering, Gachon University, 1342 Seongnam-daero, Sujeong-gu, Seongnam-si, Gyeonggi-do 13120, South Korea

T. Kim
e-mail: taeykim@gachon.ac.kr

S. P. Singh et al. (eds.), *Metal Nanocomposites for Energy and Environmental Applications*, Energy, Environment, and Sustainability,
https://doi.org/10.1007/978-981-16-8599-6_13

13.1 Introduction

Recently, nanocomposite materials have gotten considerable attention from researchers due to their enhanced properties than metal nanoparticles. The nanocomposite is a mixture or matrix, in which unique materials combine to change new properties of the materials confirming that one of the materials has a size in the range of 1–100 nm (Ahmad and Majid 2018; Jeevanandam et al. 2018). There are henceforth two portions of nanocomposite i.e., continuous phase and discontinuous reinforcing phase. Nanocomposites can be organized from any mixture of materials that can be categorized into three elementary structure blocks i.e. metals, ceramics, and polymers (Jeevanandam et al. 2018; Alghamdi et al. 2021). They have an arrangement or have noticeably different mechanical, electrochemical, electrical, catalytic, thermal, and optical properties from the component materials (Gentile et al. 2016; Kamal 2018; Abdel-Karim et al. 2020; Namsheer and Rout 2021). The nanocomposites have altered phases as core–shell, nanowires, nanotubes, lamellar and metal matrix labeled as 0D, 1D, 2D and 3D (Baig et al. 2021). Based on their important characteristics these nanocomposites are categorized as nanolayered composites, nano-filamentary composites, and nano-particulate composites. These nanocomposites have gained the attention of scientists, researchers, and engineers, which had led to the sudden rise in the number of publications related to these materials. Further, these nanocomposites have emerged as smart materials of the twenty-first century that offer many technological and business breakthroughs in all sectors of life (Heiligtag and Niederberger 2013). The focus of this chapter is on possible ways of nanocomposite synthesis and their numerous applications as schematically shown in Fig. 13.1. These materials have encouraging properties, which makes them appropriate for a huge number

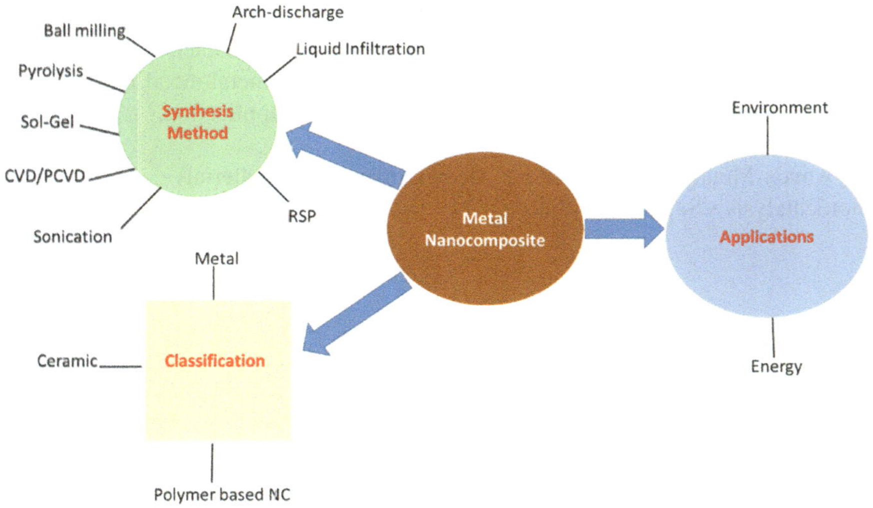

Fig. 13.1 Categories of the metal nanocomposite

of structural and functional applications in energy and the environment. Currently, rising public concern and strict global regulation have challenged the environmental issues related to waste biomass valorization and wastewater treatment (Sharifi et al. 2012). Presently the fast-developing world, waste generation is increasing gradually in significant amounts which affects the ecosystem health and eventually the health of humans. Advancement of processes and technologies for waste management also generating an alternative polluting substance through the development of biodiesel, craft paper, activated carbon along with enhancement of value-added products like pectin extraction, extruded, essential oils, etc. Increased possibility of alternative products increases the overall profits to the manufacturers, maintains environmental stability, sustainable socio-economic development, employment generation (Saifuddin et al. 2013; Jian et al. 2018).

The advanced techniques for the combination of biomass use and photocatalysis represent an innovative approach to reach better sustainability in chemical processes. Publications number related to areas like selective production of fine chemicals from waste and biomass conversion to solar fuels are increasing. Despite the intractable arrangement of many biological waste streams, which impedes technological processing, enormous development has been attained by the utilization of Photocatalytic systems. Presently, the major devices for commercialized energy storage devices are batteries supercapacitors, and fuel cells (Kirubakaran et al. 2009; Wang et al. 2016). Renewable energy storage devices elements such as fuel cells, supercapacitors (SCs), and batteries are the paradigms for the conventional energy recourses and already operating at a large scale. While great attempts have been effected on the development of great-performance fuel cells and Li-ion batteries (Yang et al. 2021), low power capacity and excessive conservation cost have maintained them left from innovative applications. SCs emerged as conventional capacitors and substitutes of batteries and called ultra-capacitors are one of the standing energy storage devices. SCs offer a very high magnitude of energy density and higher order of uptake pulse in comparison to traditional capacitors with better stability and cycle life (Ubaidullah et al. 2020). Furthermore, the distinctive charge storage capacity method provides a huge quantity of charge capacity in a rapid time and consequently delivers better power. Supercapacitors can be applied in several applications, including electric, hybrid vehicles, backup power systems, and industrial energy systems (Lei et al. 2020). Energy storage method SCs are of two types, one is pseudocapacitors, which stores energy based upon surface redox reactions. Another type of SCs belongs to electric double-layer capacitors (EDLCs), where energy stores via the formation of the electric double layer due to reversible adsorption and desorptions of electrolyte ions. Pseudo capacitors even though show a high specific capacitance however cannot maintain performance after long cycling stability including conducting polymers or metal oxides as active material. In contrast, EDLCs capacitors without performance degradation can be charged and discharged with multiple cycles (Kumar et al. 2021). However, in electrochemical double-layer capacitors, ion transportation is quicker as in redox reactions, due to high power density and charge–discharge rate (Saikia et al. 2020). To demonstrate the benefit of supercapacitors structure, performance and mechanism of these methods utilizing various materials have been reviewed due to performance and structural advantage of SCs that are difficult to attain for batteries, SCs has drawn huge attention. Supercapacitors are comprised of two highly porous electrodes submerged into an electrolyte and split by a dielectric membrane which lets ions to pass through, this condition is like the sandwiched structure of capacitors. While employing an external electric field to the device, negative and positive charges accumulate on surfaces of electrodes (Forouzandeh et al. 2020). Ions in the electrolyte solution diffuse across the membrane and get into the pores of the electrodes due to the natural attraction law of opposite charges. The electrodes are arranged in a specific manner, which avoid recombination of ions. Thus, double layer of charges has being established on every single electrode. To achieve good quality capacitance and energy density, higher spongy materials including narrower distance sandwiched between electrodes and higher surface area are required in SCs' electrodes based on their energy storage method. SCs utilize electrodes with thin membrane as a dielectric layer and high surface area based on same basic principles as capacitors (Nandi et al. 2020). As a result, the energy and capacitance would be enhanced significantly. Additionally, they can maintain an extremely high-power density including a small equivalent series resistance (ESR) which is similar to traditional capacitors (Najib and Erdem 2019).

13.2 Metal Nanocomposites

13.2.1 Nanomaterials and Their Composites

Nanomaterials are materials with sizes in nm (less than 1 μm range). But the atomic and molecular building blocks (0.2 nm) of material are accepted as nanomaterials, for instance, crystals with a lattice spacing of nanometers but have macroscopic dimensions overall, are generally ignored (Wang and Mi 2017). The change from microparticles to nanoparticles produces dramatic variations in physical properties. Nanomaterials have a large surface area for a given volume. Because several crucial chemical and physical interactions are controlled by its surface properties. A nanomaterial can have considerably different properties from a larger dimensional material of the same composition. For particles and fibers, the material's diameter is inversely relational to the surface area per unit volume, hence, the lesser the diameter, the bigger the surface area per unit volume (Saleh 2020). Surface area to volume ratio; Nano-sized materials have a comparatively bigger surface area as compared to the same volume or mass of the material created in a larger form Fig. 13.2 (Palchoudhury et al. 2015). While the provided volume is split into tinier pieces the surface area enhances. Therefore, the particle size reduces a greater proportion of atoms observed at the surface compared to those inner sides. Thus, nanoparticles have a considerably greater surface area per provided volume contrasted with larger particles. It makes materials furthermore chemically reactive. Two principal factors such as increased

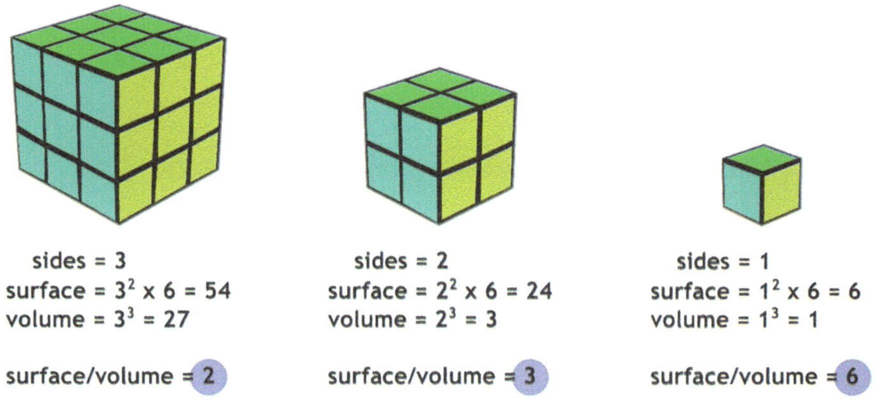

Fig. 13.2 The surface area per unit volume (Palchoudhury et al. 2015)

relative surface area and quantum effects are important for the properties of nanomaterials which make them different as compared to other materials: (Wohlleben et al. 2017).

13.2.2 *Structure and Properties*

The condensed material shows various incredible properties which are considerably distinct as of properties of bulk materials. A few established physical properties of nano-sized materials remain associated with distinct backgrounds: such as, (i) a big portion of surface elements (ii) spatial confinement, (iii) huge surface energy, (iv) decreased defects (Sarma et al. 2019). Nanosized materials need a substantially smaller melting point or phase shift temperature considerably decreased pattern constants, anticipated for a large section of surface atoms across an equal volume of atoms. The mechanical properties of nanosized materials can achieve the hypothetical intensity that is one or two contracts of scale greater than specific crystals appearing in bulk materials. The growth of mechanical strength is basically because of the decreased possibility of imperfections (Yadav et al. 2019). There are some optical properties of nanosized materials that are considerably distinct after bulk crystals. For instance, the optical absorption points of a semiconductor device composed of nanosized particles change to a small wavelength, because of the enhanced bandgap. Due to surface plasmon resonance, the color of metal nanoparticles can vary along with sizes (Li et al. 2020). Electrical conductivity reduces through a decreased dimension because of enhanced surface dispersion. Although electrical conductivity of nanosized materials might be increased considerably, because of the improved microstructure, such as in polymeric strands. Magnetic properties of nanosized materials occur noticeably distinct after bulk materials. Ferromagnetism of bulk materials dissipates as well as removals to superparamagnetic in nanometer range because of large surface energy (Migliore et al. 2018).

13.2.3 *Classification and Synthesis of Nanocomposites*

Nanocomposite material can be categorized corresponding to matrix materials in the various categories as shown in Table 13.1.

Table 13.1 Types of nanocomposites with processing methods

Class	Process	Approach	Methods	References
Ceramic	Powder process	Al_2O_3/Si SiO_2/Ni, Al_2O_3/CNT	• Collection of natural materials mainly fine particles of tiny size, high purity, and uniformity • Blending with ball milling process in natural and aqueous media • Dry using heating, applying lamp, oven, or freeze-drying • Strengthening of strong material using gas pressure sintering or injection molding or hot pressing or slip casting and pressure filtration	(Montanaro and Palmero 2019)
Polymers	Polymer precursor process	Al_2O_3/SiC, SiN/SiC	Combination of a Si-polymeric precursor along with matrix material → pyrolysis process, mixing in a microwave oven, producing the reinforcing particles	(Yu et al. 2002; Jeon and Baek 2010; Parchovianský et al. 2017)
Metal matrix	Sol–gel process	TiO_2/Fe_2O_3, SiO_2/Ni, ZnO/Co, $NdAlO_3/Al_2O_3$, Al_2O_3/SiC	Hydrolysis and polycondensation effects of in-organic molecular precursor dispersed in organic media	(Guo et al. 2018)

13.2.3.1 Ceramic Matrix Nanocomposites (CMNC)

The potential of CMNC, primarily the Al_2O_3/SiC system, was indicted by the groundbreaking research study of Niagara et al. (Din 2019). Many research proved that the strengthening of the Al_2O_3 matrix following the addition of a low fraction of SiC particles. A few investigations described the strengthening method related to the crack associating part of the nano-sized reinforcements. Therefore, integration of elevated intensity nanofibers keen on ceramic matrices needs to be permitted to the formation of enhanced nanocomposite material in conjunction with huge durability and exceptional breakdowns of ceramic materials (Montanaro and Palmero 2019).

13.2.3.2 Metal Matrix Nanocomposites (MMNC)

MMNCs describe materials comprising of alloy matrix or ductile metallic that have several nanosized reinforcement materials. These resources merge into metal and ceramic characteristics, which have toughness and ductility through large intensity and modulus. Therefore, MMNCs remain appropriate used for the fabrication of materials including large power in shear/density activities or elevated maintenance temperature abilities. These materials demonstrate an amazing capacity for use in the field of automotive industries, aerospace, and enhancement for the formation of materials. Both MNCs and CMNCs including CNT nanocomposite materials and carry potential as well as present challenges for actual accomplishment (Guo et al. 2018).

13.2.3.3 Polymer Matrix Nanocomposites (PMNC)

Polymeric materials remain utilized in material manufacturing due to their lightweight nature, simple construction, and frequently flexible nature. These materials have few drawbacks like low strength and low young modules as associated with ceramics and metals. An incredibly useful method to enhance mechanical properties is the application of the whiskers, particles, fibers being as defenses to the polymer matrix (Yu et al. 2002; Jeon and Baek 2010; Parchovianský et al. 2017).

Even Though all nano dimensions, the majority of synthesis methods have three categories of nanocomposite persist nearly equal as in Table 13.1 (Santos Miranda et al. 2006).

There are two approaches for nano-sized materials that is bottom-up and top-down approaches (Hakke et al. 2021). Nanomaterials deal with extremely fine structures which permits us to believe in both the 'top down' or 'bottom up' approaches (Fig. 13.3). In the bottom-up technique, atoms are assembled to form a nanostructure. Whereas in top-down approach the bulk solid material is braked into finer pieces in which it has dimension in nanoscale. The very familiar technique designed for the managing of metal nanocomposite is liquid metal infiltration quick solidifying vapor methods (PCVD and CVD), spray pyrolysis, electro-deposition,

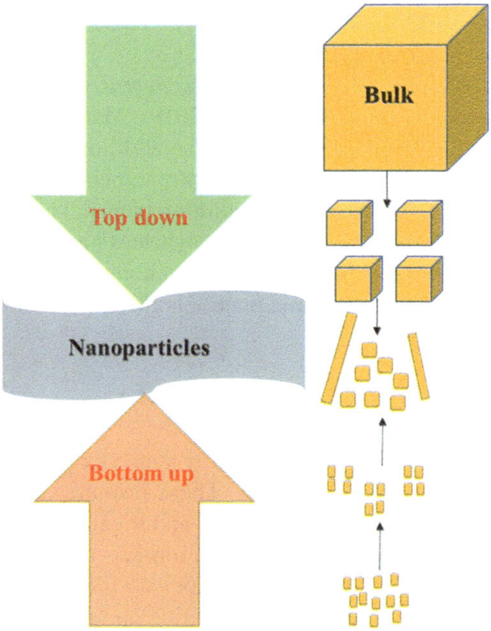

Fig. 13.3 Bottom-up and top-down synthetic routes for nanocomposite (Hakke et al. 2021)

hydrothermal process (Guo et al. 2013), and chemical techniques including sol–gel methods (Somvanshi et al. 2020). Previously, numerous approaches were stated for the synthesis of nanomaterials such as sol–gel, auto-combustion, hydrothermal (Darr et al. 2017), spray pyrolysis, etc. In disparity, hydrothermal processes have been used for the cost-effective production of nanoparticles. In this standard process, the high-pressure pump is used along with room temperature water supply due to parallel heating to increase its temperature > 400 °C or more (Guo et al. 2013). Table 13.2 lists different kinds of systems planned by these methods and Table 13.3 shows the advantages and limitations (Camargo et al. 2009).

13.3 Polymers-Based Nanocomposites

13.3.1 Polymer Matrix and Fillers

A nanocomposite material is a nano aggregate of two or more distinct materials (certainly one of nanoscale or each in nanoscale), having a recognizable interface among them. Current nanocomposite materials are normally optimized to achieve a particular balance of residences for a given variety of applications (Santos

Table 13.2 Methods for metal-based nanocomposites system

Method	Approach	Remarks	References
Spray pyrolysis process	Fe/Cu/MgO	• Addition of inorganic precursors starting materials in a proper solvent to get the liquid source material • Formation of a mist after the liquid source utilizing an ultrasonic atomizer • Utilization of carrier gas to carry the mist into a pre-heated chamber • Vaporization of the droplets in the reaction chamber and confining with a filter, boosting their decomposition to form the respective oxide materials • The selective decline of the metal oxides to generate the respective metallic materials	(Djatoubai and Su 2021; Saha et al. 2020)
Liquid infiltration	W/Cu, Pb/Cu, Nb/Fe, Pb/Fe, Nb/Cu, Al-C_{60}	• Adding of fine reinforcement particles together with the matrix metal material • Thermal treatment will be given, where the matrix melts and surrounds the reinforcements by liquid infiltration • Additional thermal treatment below the matrix melting point, to stimulate, strengthening eliminate internal porosity	(Greenberg et al. 2004)

(continued)

Table 13.2 (continued)

Method	Approach	Remarks	References
Rapid solidification process	Al/X/Zr, Al/Pb (X = Si, Cu, Ni), Fe alloy	• Melting of the metal components together • Maintaining the melt over the critical line of the miscibility gap between the different components for maintaining homogeneity • The rapid solidification of the melt by some process, for example, melt spinning	(Roy Chowdhury et al. 2021)
RSP with ultrasonic		Consumption of ultrasonic for adding or for enhancing wettability among the matrix and reinforcements	(Upadhyay et al. 2014)
Energy ball milling		The milling of the powders collectively till the required nanosized alloy is attained	(Saha et al., 2020)
CVD/PCVD		PCVD • Sputtering of various components to form a vapor-phase • Supersaturating of the vapor phase under inert atmosphere to encourage condensation of metal nanoparticles • Strengthening of the nanocomposite by the thermal treatment with an inert atmosphere CVD Utilization of chemical reactions to form vapors of specific materials, followed by consolidation	(Abuhimd et al. 2020; Alagh et al. 2021)

(continued)

Table 13.2 (continued)

Method	Approach	Remarks	References
Chemical process (sol–gel colloidal)	Au/Fe/Au, Ag/Au, Fe/SiO_2	**Colloidal process** • Chemical reduction of inorganic materials in reaction solution to form metal particles • Strengthening of the dry material • Thermal and drying treatment of the resulting solid in reducing reaction atmosphere, such as hydrogen, to stimulate selective oxide reduction and create the metal component **Sol–gel method** • Generation of two micelle solutions utilizing mesoporous silica comprising 0.1MH$AuCl_4$ (aq.) and 0.6 M $NaBH_4$ (aq.) • Admixing under UV light till full reduction of the gold For Fe/Au-containing nanocomposites • Formation of the iron shell • Synthesis of the second shell and the drying of the powders after second gold coating • Processing of the mixture to get the final product	(Azfar et al. 2020; Navarrete-Magaña et al. 2021)

Table 13.3 Advantages and limitations of metal-based nanocomposites system

Processes	Benefits	Limitations	References
Spray pyrolysis	The effective formation of spherical and ultra-fine and homogeneous powders in multicomponent rea reproductive size and quality	High cost of production-related with producing big quantities of uniform, nanosized particles	(Camargo et al. 2009)
Liquid infiltration	The short contact times among matrix and reinforcements; rapid solidification; both lab-scale and industrial scale production molding into different and near net shapes of different stiffness and enhanced wear resistance	Utilization of high temperature; exclusion of reinforcements; development of undesired products during processing	(Patrianus Khristian Sumule 2012)
Rapid solidification process (RSP)	Very effective and simple	For only metal–metal nanocomposites materials; stimulated agglomeration and non-homogeneous distribution of fine particles	(Lu et al. 2020)
RSP with ultrasonic	Without agglomeration, good distribution even with fine particles		
Energy ball milling	Homogeneous mixing and uniform distribution		(Greenberg et al. 2004)
CVD/PCVD	Capacity to form pure material and highly dense at high deposition rates; uniform thick films; good reproducibility	Improvement of many parameters; relative complexity and cost	(Abuhimd et al. 2020)
Chemical process (sol–gel process)	Very easy, low-temperature process; flexible; effective and simple; accurate stoichiometry control; superior chemical, elevated purity products	Low bonding, weak wear-resistance, elevated permeability and tricky to control porosity	(Camargo et al. 2009)

Miranda et al. 2006).The primary composite classes include organic-matrix composites (OMCs), metallic-matrix composites (MMCs), and ceramic-matrix composites (CMCs). The term natural-matrix composite is commonly assumed to encompass two types of composites: polymer-matrix composites (p.c.) and carbon-matrix composites (Lei et al. 2006). Polymer nanocomposites can be defined as two-segment structures which include polymers and fillers of which at the least one measurement is within the nano-range (1–100 nm). The nano fillers may be one-dimensional nanotubes or nanofibers, two-dimensional clay platelets, or three-dimensional spherical particles (Parida 2018). Researchers are just tried to enhance features of composite materials glance at composites including fillers size, prominent to enhance the micro composites and current challenges. It belongs to the composites wherein a single segment requires nanoscale morphology which includes nanoparticles, nanotubes, and much more structure. Development of these properties can be completed if a proper interface is generated among the nanoparticles and the matrix owing to the diffusion of particles inside the matrix. In nanocomposites, covalent bonds, ionic bonds, Vander Waals forces, hydrogen bonding should occur amongst the matrix and filler additives (Liu et al. 2020). There are some problems with nanoparticles which include the high surface area, the low concentration that is the essential consequence for the macroscopic properties. Throughout the earlier years, polymer nanocomposites have had great significance in academic circles and industry, however, some of the exceptional troubles govern the dispersion of nanoparticles during synthesis. Hence different preparation methods for polymer nanocomposites are explored to avoid this issue (Sadoun et al. 2020). Several nanocomposites might reveal properties outweighed by the interfacial connections and others could display the quantum effects related to nano-dimensional structures. Cost-effective nanoparticle dispersion, mixing with great polymer, and particle interfacial adhesion reduce scattering and accept the exciting possibility of utilizing strong transparent films, coatings, and membranes (Niculescu et al. 2021). These are some benefits like reinforcement of resin resulting in improved tensile strength, textural, impact compression strength, rigidity, and the mixture of these properties. Improved size stability, enhanced fire retardancy, corrosion shield, better-quality electrical properties; decline of dielectric constant, coloring, and enhanced processability; controlled viscosities, excellent mixing, managed orientation of fibers. One of the significant properties of nanocomposite materials is the high strength per density called a specific module and specific strength, individually (Rashidi et al. 2020). Despite numerous benefits of nanotechnology in the field of medicine and electronics has been stated that direct exposure to nanomaterials can affect major health and trouble humans and other organisms (Zhao et al. 2020). The movement of free nanoparticles is not limited, and they can simply be released into the environment causing a significant health risk. In contrast, static nanoparticles would cause no health risk when correctly handled (Marlinda et al. 2020). Because of their extremely small size and relatively massive surface area, nanomaterials may additionally engage with the surroundings in ways that differ from extra conventional materials. Hence a careful inspection should be carried out in the application of nanotechnology for the betterment of human life so that it would not deliver any carcinogenic effect to environmental metabolites (Marlinda et al. 2020).

13.3.2 Structure and Properties

The thermal behavior of polymer is extremely complicated. At elevated temperatures, the behavior of polymers is extra complex because thermally activated rearrangements and movement within and among the chains can occur which might be frequently reversible. Those development are especially responsible for the physical and mechanical properties of polymers. To understand the characterization of the physical and mechanical properties that material changes should be analyzed with proper processing and fabrication process (Jing et al. 2016). If the polymer is amorphous like the glass transition, evaporating temperature or stable-to-liquid transition takes place extremely slowly, moving around a transitional. These transition changes are difficult and fragile and take place across a slender temperature scale called glass transition/phase temperature. The crystalline area continues to be unaffected and behaves as supporting elements according to the pattern and becomes tough and challenging. But heating is maintained, a temperature is extended at which the crystalline zones start to evaporate. Equilibrium crystalline meltdown, factor T_m, instead of polymers resembles the temperature on which remaining crystalline starts melting. Overall modifications in physical condition can be attributed to variations in temperature and molecular weight for amorphous and crystalline polymers (Chen et al. 2016; Vella Durai et al. 2019). Viscous performance and viscoelasticity fluids demonstrate a function resistance i.e. viscosity. Polymeric materials play as both viscous and elastic solids. They are viscoelastic materials. The vital feature of viscoelastic materials is their mechanical properties which depend on time. Viscosity is a portion of the friction and the associated power dissipation between molecules. Polymeric materials because of their macromolecular (long-chain) shape are expected to have high viscosities (Kumar et al. 2017). Mechanical properties of polymers such as the elastic conductivity of polymers are specifically come through the intermolecular bonds among the chain molecules, not through the covalent bonds within. For elastomer and dormers (is a dealer of high-performance plastic compounds), the covalent bonds linking the chains are also relevant. Whenever pressure is exerted on a solid material, the material will deform in reaction to the force. The compressive strength is the capacity of raw material to resist loads of tension to reduce the size. When the forces are affiliated with each other, they are termed compression forces (Durai 2021). A material is inflexible if deformation is instantaneous and constant when the pressure is applied. It can also depend upon the removal of the force, its recuperation is immediate and whole (i.e., the fabric will return to its authentic shape). So, evaluation of mechanical strength takes place and is beneficial to outline a quantity referred to as a strain. The deformation is quantified by way of the stress ε which is defined as the length change (Δl) divided by the initial period and it is dimensionless. In case of a tensile test the pressure is often known as elongation and is normally expressed as a ratio growth in length in comparison to the primary duration (Durai et al. 2020). Stress–strain behavior in the polymer is evaluated if the material is subjected to high-strain deformation, it deforms plastic and, in the end, fails. For sufficiently low stresses, the polymeric materials behave as a linear strong elastic. The point wherein the conductivity starts to be non-linear is known as the proportional limit. The nearby point within the stress–strain arc is called the yield point and indicates the onset of plastic (i.e., everlasting) deformation. The corresponding strain and elongation are called yield electricity and elongation at yield. Further, beyond the yield points the material stretches out significantly and a "neck" is founded; that area is termed as a plastic region. The additional extension starts with an abrupt increase in stress and the greatest upturn of the material. At the breakingpoint, the subsequent stress and strain are known as the ultimate strength and the elongation at break, correspondingly. The stress–strain performance of a polymeric material varies on different factors such as molecular characteristics, micro-structure, strain-rate, and temperature (Li et al. 2018). The observed elastic behavior of solids at low strain-strains is because of the stretching of their chemical bonds that are inherent-short-ranged. Especially in polymers, even though the above mechanism can't be excluded, the elastic behavior is extra complicated because of the chainlike structure of the macromolecules (Ma et al. 2017; He et al. 2018).

13.3.3 Classification of Polymeric Nanocomposites

Polymeric nanocomposites can be generally categorized as:

(i) Nano clay-reinforced Composites: Traditionally, the period clay has been understood to be received of small inorganic particles (soil fraction < 2 mm), not including any significant crystallinity. The clay mineral is also known as phyllosilicate and it is typically of a layered type and a tiny proportion of hydrous, magnesium, or aluminum silicates (Yang et al. 2018).

(ii) Carbon Nanotube-reinforced Composites: Micrometre-size carbon tubes, which are comparable in structure, but not in dimensions to the newly found multi-walled carbon nanotubes, were first discovered in 1960 by Roger Bacon. These termed nanotubes were initially observed and fully characterized in 1991 by Sumio Iijima of NEC Corporation in Japan (Mariano et al. 2014).

(iii) Nano fiber-strengthened compounds: carbon nanofibers (CNF) is a rare type of vapor grown carbon fibers difference in physical properties among traditional carbon (5–10 μm), carbon nanotubes (1–10 nm). Decreased thickness of nanofibers offers a greater surface area along with surface functionalities. Usually, CNF does not acquire concentrical cylinders; the quantity of the fiber can be able to be different after around 100 μm to numerous cm, and the diameter is of the sort of 100–200 nm along with an average aspect ratio greater than 100. The very ordinary form of CNF is shortened cones, although there is a broad limitation on morphologies (Raza et al. 2014).

(iv) Inorganic Particle-reinforced Composites: Nanoparticles are frequently identified as particles of < 100 nm in diameter. Unique nanoparticles have been applied to formulate polymer/inorganic particle nanocomposites, including metals (Al, Fe, Au, Ag, etc.), metal oxides (ZnO, Al, $CaCO_3$, TiO_2, etc.), non-metal oxide (SiO_2), and others (SiC) (Gharbi et al. 2021).

13.4 Nanocomposites for Energy Application

13.4.1 Biodiesel Production

Biodiesel has come up as alluring alternatives to fossil fuels (Sahani and Sharma 2018; Sahani et al. 2020a). Novel nanocatalysts, microbial cultures, and emerging optimization methods are useful in developing the substantial-scale production of biodiesel (Suryanarayana 2018). Biodiesel fabrication through the transesterification method has already been commercialized. The function of nano-catalysts in diverse catalysis has been considered highly efficient in the transesterification reaction. Metal oxides-based nano-catalysts are recognized with unique properties like the huge surface area to volume ratio, superior activity, and product selectivity (Sahani et al. 2018, 2019). Nanomagnetic or mixed metal oxide-based nano-catalyst has performed a significant role in improving the specific surface area, pore size, and average pore diameter (Suryanarayana 2019). Utilizing numerous functionalized groups, the major change of nano-catalysts can be accomplished which eventually improves the general acidity, decreases the discharge of functional metals, and produces further active acidic and basic sets for the catalytic reaction. Certainly, development in nanotechnology considerably restricts the utilization of fossil fuels and as well relieves the pollution intensity. Though, the biodiesel production level acknowledges numerous important tasks like finding feedstock, as a regarding task equivalent to biofuel construction, mostly biodiesel. Biodiesel obtained from edible oils are posing specific limitations reporting to the cost and food-associated problems. In contrast to this, non-edible oils and groups of microalgae reveal improved possibilities because of the esterification method of available fatty acids (Sadoun et al. 2021). As pre-treatment of the feedstock is a needed phase in reducing the development of side-effects and the saponification procedure as it largely hinders the fabrication of biodiesel by influencing its physic-chemical properties consequently decreasing the quality of biodiesel. Examining optimization of response limits and physicochemical properties is imperatively substantial in growing biodiesel production (Stewart and Dingreville 2020). Utilizing optimum volumes of nano-catalyst, reaction factors like methanol to oil ration, catalyst weight, response temperature, and time can be considerably enhanced, thus generally improving the value and amount of biodiesel (Madhu et al. 2017; Sahani et al. 2020b).

13.4.2 Electrode Materials for Lithium-Ion Batteries

Anode and cathode electrode materials are essential for the Li-ion battery. Several latest trials specialize in making the electrode with extended detailed functionality and cycling balanced. For Li-ion batteries, very new materials including Sb_2O_3, TiO_2/MoS_2 need active modifications as anode material. Such materials were observed to be extremely favorable as cathodes along with P14AQ/CNT composite triplite $LiFeSO_4F$. As of these several electrodes, substances get changed via essential doping and coating with other materials. This enables better elevated Li-ions diffusion, improvement in conductivity, ionic mobility, and increased ability at various points of charging/discharging (Liu et al. 2019).

13.4.3 Solid Oxide Fuel Cells

There is a growing interest in the novel and sustainable solid oxide fuel cells (SOFCs) development among the researchers as it offers promising power generation with high energy efficiency, inflated fuel flexibility, and low environmental impact compared to conventional power generation systems. A single cell component of SOFCs is consisting an anode, cathode, and electrolyte which are stacked layer by layer to produce a higher amount of power. The dense ceramic electrolyte transports O_2^- ions by filling the space between the electrode materials. Redox reaction occurs at the side of the electrodes in the presence of fuel operating temperatures of 600–1200 °C. Many ceramic materials have been investigated for SOFCs as electrolytes. Amongst, yttrium-stabilized zirconia (YSZ) material was extensively used as the dense electrolyte in SOFCs technology (Stambouli and Traversa 2002). SOFCs are found to acquire fast electro-catalytic activity while using nonprecious metals.

13.4.4 Supercapacitors

The electroactive material has been recognized as the best candidate for supercapacitor application. Different types of nanocomposite materials are used as electroactive materials including mixed metallic oxides, polymer-based materials, biomass-derived carbon materials, graphene blended with metallic oxide or polymers synthesized through mechanical mixing of metal oxides, and chemical co-precipitation and electrochemical anodic deposition, sol-gel, in situ polymerization and different moist-chemical routes (Majumdar 2021).

It has been seen that great development in terms of surface area, electric and ionic conductivities, unique capacitance, cyclic stability, and energy and electricity density in supercapacitors can be extensively advanced by using nonprecious nanocomposite electroactive materials (Devillers et al. 2014). There are two types of conventional capacitor structures named electrochemical double layer (EDLC) and pseudocapacitors. EDLC capacitor consists of carbon-based materials while pseudo capacitors incorporate metal oxide or conducting polymers as electroactive materials. Compared with the electrochemical double-layer capacitors, pseudocapacitors represent benefits i.e., high energy density and low materials cost, although endure from poor cyclic stability and lower power density. Asymmetric supercapacitors utilizing one electrode of pseudocapacitive materials and the more carbon-based double capacitive materials are of great interest and are of the current research interest (Bigdeloo et al. 2021). Nanocomposite pseudocapacitors have an excellent capacity for asymmetric supercapacitor applications. The important problem is to completely develop nanocomposites' excellent intrinsic properties. Synthesis of nanocomposite electroactive substance and at the same time precisely controlling their chemical composition ratio, micro/nanostructure, phases, surface place, and interfacial characteristics remain difficult (Borenstein et al. 2017; Zhang et al. 2018). Relying on the instruction method and procedure parameters, the assets, and behaviors of the nano-composite electro-active materials can range significantly; consequently, the ability to reproducible synthesize nanocomposite films with consistent properties can be very crucial for electroactive substances. Degradation of the nanocomposite electroactive material restricting from the aggregation of the nano-scale components because of comparatively durable forces between them is major trouble. This can also be attributed to micro/nanostructure variations owing to high charging-discharging cycling and materials impurities because of contamination present throughout fabrication methods. The issue must be solved ahead of their large-scale adoption (Bigdeloo et al. 2021). Most significantly, the costs of materials and their synthesis method must be considerably addressed. However, considering the growth in supercapacitor application, it is estimated that electrode materials derived from nanocomposite make a vast transformation to the energy storage industries (González et al. 2016).

13.5 Nanocomposites for Environmental Application

13.5.1 Waste Biomass Valorization

Bio-economy is derived from productive applications of agricultural products. Bioeconomy by producing sustainable biomass amounts either in conditions of waste or by-product. Biomass is an eco-friendly source but it is restricted due to its production conditions like soil and extra resources (water, nutrients) (Gumisiriza et al. 2017). The needs of the bio-economy involve the ecosystem due to necessary feed supplies to land, ocean to generate useful bioproducts as an alternative to applying fossil fuel-derived feedstocks (Zema et al. 2019). The advancement of industrial biorefineries, bio-based activities, and beginning innovative markets to allocate the bio-based creations produced from these different industrial tools are several major modifications that typical industrial would implement to create a bio-economy. The extremely crucial issue that occurs to start a bio-economy is the source of biomass to sustain the bio-economy. This conversion of plants and grassland to an arable area distributes CO_2 in the environment (Bădescu et al. 2018). Therefore, bio-securing CO_2 as of the atmosphere is an extra useful plan to improve value-additional products. The extreme consumption of land reduces soil potency and soil efficiency (Berbel and Posadillo 2018). Cost-effective and improved manufacture of crops by utilizing composts and pesticides can expand water and soil pollution. These challenges can be exceeded by applying destroyed land, improving productivity, and waste consumption of biomass production. But, applying food waste has its benefits with several key problems like compilation and separation of food waste, minimal expertise among providers of food waste, loading food waste for after use, and efficiently saving plants (Singh and Kumari, 2019). Primarily, biofuels produced after biomass can enhance the area economy by switching to two fossil fuels (gasoline and diesel) hugely used in transportation (Sahani et al. 2018, 2020b). Biodiesel is the substitution of fossil diesel and bioethanol is of gasoline (Sahani and Sharma 2018; Sahani et al. 2019, 2020a). Furthermore, the bio-economy creates innovative service prospects in various fields like biofuels' manufacturing and processing (O'Connor et al. 2021). The bio-economy has numerous negative impacts, but those are not long-lasting and able to easily process. Though, the advantages of the bio-economy for the nature and safety of humankind can be organic for an extended period if we find the solution to the above-mentioned problems and develop those advantages extremely wisely (Luhar et al. 2019).

13.5.2 Wastewater Treatment

Nowadays, wastewater treatment plants have been developed owing to increasing water pollution and its toxicity in the environment. Anthropogenic liquid products produce toxins through the applications of organic fertilizer, inorganic soluble compounds, and the usage of contaminated heavy metals. There are a few usual applications of nanotechnology in this field, such as adsorption, catalytic oxidation, sterilization, and sensing to either monitor and treat the polluted water (Alhazmi and Loy 2021). Nanomaterials are commonly studied as extremely useful adsorbents, photo-catalysts, and purifiers for water treatment. Some nanomaterials can be beneficial for wastewater treatment. To understand the physical, chemical, and biological properties of nanomaterial, molecular interaction between nanomaterials and analytes should be carefully followed (Saleh 2021). So, the appropriate factors such as surface energy, surface charge, and activation are to be considered. Repulsive force of interaction is required to control different nanoparticles. The interface of nanoparticles can affect their behavior and forms aggregates and agglomerates. Some highly porous nanoparticles are meant to absorb water and prevent the melting of salts and other contaminations. Unique categories of nanomaterials are applied in water waste treatment. Figure 13.4 illustrates several approaches to cure wastewater (Yang et al. 2019). Usually, they demonstrate several advantages high efficiency,

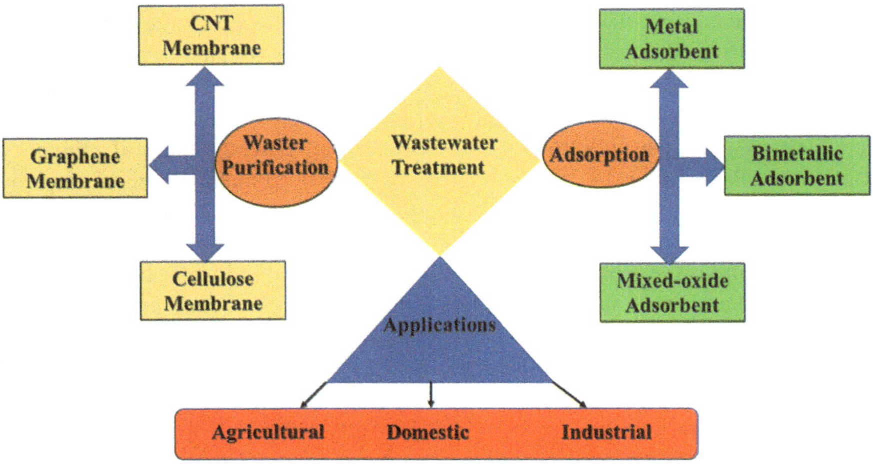

Fig. 13.4 Application of nanomaterials for wastewater treatment (Yang et al. 2019)

fast kinetics, specific attraction towards affected pollutants, improved photocatalytic activity over a wide range, and intense anti-bacterial activity. Initially, nanomaterials are typically volatile, tend to agglomerate because of the existence of van der Waal forces and more interactions (Sudarsanam et al. 2018). Some biopolymers, minerals, activated carbons, or tissues might allow the dispersion and intensity of the filled nanoparticles. They might be encouraging dispersion of pollutants in the hosts, additional than enhancing the interfacial edge. Furthermore, nanocomposites can enhance the compatibility of nanotechnology along with the current structure (Pendergast and Hoek 2011).

13.6 Conclusion

The chapter has emphasized the distinct mechanistic features of nanocomposite materials that provide them rare properties with potential energy and environmental applications. Besides their outstanding physicochemical properties, nanocomposite materials exhibit improved photostability through functionalization. Amongst several synthetic routes, the cost-effective hydrothermal process results in tunable 0D, 1D, 2D, and 3D nanostructures with excellent surface area and porosity. These metal nanocomposite materials consisting of light transition metals and polymers are quite efficient in heterogeneous catalysis for biodiesel production, water splitting, adsorption of pollutants from water, and energy storage i.e., supercapacitor and battery systems.

References

Abdel-Karim R, Reda Y, Abdel-Fattah A (2020) Review—Nanostructured materials-based nanosensors. J Electrochem Soc 167(3):037554. https://doi.org/10.1149/1945-7111/ab67aa

Abuhimd HM, Kiyashko MV, Grinchuk PS (2020) Optimal thickness of the catalytic nickel coating for the CVD synthesis of carbon nanomaterials. Mater Res Express 7(1). https://doi.org/10.1088/2053-1591/ab677d

Ahmad J, Majid K (2018) In-situ synthesis of visible-light responsive Ag_2O/graphene oxide nanocomposites and effect of graphene oxide content on its photocatalytic activity. Adv Compos Hybrid Mater 1(2):374–388. https://doi.org/10.1007/s42114-018-0025-6

Alagh A et al (2021) CVD growth of self-assembled 2D and 1D WS2 nanomaterials for the ultra-sensitive detection of NO_2. Sens Actuators b: Chem 326:128813. https://doi.org/10.1016/j.snb.2020.128813

Alghamdi SS et al (2021) Additive manufacturing of polymer materials: progress, promise and challenges. Polymers 13(5):1–39. https://doi.org/10.3390/polym13050753

Alhazmi H, Loy ACM (2021) A review on environmental assessment of conversion of agriculture waste to bio-energy via different thermochemical routes: current and future trends. Bioresour Technol Rep 14:100682. https://doi.org/10.1016/j.biteb.2021.100682

Azfar AK et al (2020) Comparative study on photocatalytic activity of transition metals (Ag and Ni)doped ZnO nanomaterials synthesized via sol–gel method. R Soc Open Sci 7(2). https://doi.org/10.1098/rsos.191590

Bădescu IS et al (2018) Valorisation possibilities of exhausted biosorbents loaded with metal ions—a review. J Environ Manag 224:288–297. https://doi.org/10.1016/j.jenvman.2018.07.066

Baig N et al (2021) Nanomaterials: a review of synthesis methods, properties, recent progress, and challenges. Mater Adv 2(6):1821–1871. https://doi.org/10.1039/d0ma00807a

Berbel J, Posadillo A (2018) Review and analysis of alternatives for the valorisation of agro-industrial olive oil by-products. Sustainability (switzerland) 10(1):1–9. https://doi.org/10.3390/su10010237

Bigdeloo M et al (2021) Review on innovative sustainable nanomaterials to enhance the performance of supercapacitors. J Energy Storage 37:102474. https://doi.org/10.1016/j.est.2021.102474

Borenstein A et al (2017) Carbon-based composite materials for supercapacitor electrodes: a review. J Mater Chem A 5(25):12653–12672. https://doi.org/10.1039/c7ta00863e

Camargo PHC, Satyanarayana KG, Wypych F (2009) Nanocomposites: synthesis, structure, properties and new application opportunities. Mater Res 12(1):1–39. https://doi.org/10.1590/S1516-14392009000100002

Chen J et al (2016) Enhanced CO_2 capture capacity of nitrogen-doped biomass-derived porous carbons. ACS Sustain Chem Eng 4(3):1439–1445. https://doi.org/10.1021/acssuschemeng.5b01425

Darr JA et al (2017) Continuous hydrothermal synthesis of inorganic nanoparticles: applications and future directions. Chem Rev 117(17):11125–11238. https://doi.org/10.1021/acs.chemrev.6b00417

Devillers N et al (2014) Review of characterization methods for supercapacitor modelling. J Power Sources 246:596–608. https://doi.org/10.1016/j.jpowsour.2013.07.116

Din SH (2019) Nano-composites and their applications: a review. Charact Appl Nanomaterials 2(1):1–9. https://doi.org/10.24294/can.v2i1.875

Djatoubai E, Su J (2021) First spray pyrolysis thin film fabrication of environment-friendly Cu_2BaSnS_4 (CBTS) nanomaterials. Chem Phys Lett 770. https://doi.org/10.1016/j.cplett.2021.138406

Durai SCV et al (2020) Investigation on electrical and structural properties of manganese dioxide nanoparticles. J Nano- and Electron Phys 12(3):4–8. https://doi.org/10.21272/jnep.12(3).03011

Durai V (2021) Investigation of physico-chemical, physical properties and moisture content studies of soil using organic amendment. Int J Agric Technol 17:855–870

Forouzandeh P, Kumaravel V, Pillai SC (2020) Electrode materials for supercapacitors: a review of recent advances. Catalysts 10(9):1–73. https://doi.org/10.3390/catal10090969

Gentile A, Ruffino F, Grimaldi MG (2016) Complex-morphology metal-based nanostructures: fabrication, characterization, and applications. Nanomaterials 6(6). https://doi.org/10.3390/nano6060110

Gharbi B et al (2021) Spray pyrolysis synthesized and ZnO–NiO nanostructured thin films analysis with their nanocomposites for waveguiding applications. Semiconductors 55(1):37–43. https://doi.org/10.1134/S1063782621010085

González A et al (2016) Review on supercapacitors: technologies and materials. Renew Sustain Energy Rev 58:1189–1206. https://doi.org/10.1016/j.rser.2015.12.249

Greenberg R et al (2004) The effect of WS2 nanoparticles on friction reduction in various lubrication regimes. Tribol Lett 17(2):179–186. https://doi.org/10.1023/B:TRIL.0000032443.95697.1d

Gumisiriza R et al (2017) Biomass waste-to-energy valorisation technologies: a review case for banana processing in Uganda. Biotechnol Biofuels 10(1):1–29. https://doi.org/10.1186/s13068-016-0689-5

Guo F et al (2018) A review of the synthesis and applications of polymer-nanoclay composites. Appl Sci (switzerland) 8(9):1–29. https://doi.org/10.3390/app8091696

Guo Y et al (2013) Hydrothermal synthesis of highly fluorescent carbon nanoparticles from sodium citrate and their use for the detection of mercury ions. Carbon 52:583–589. https://doi.org/10.1016/j.carbon.2012.10.028

Hakke V et al (2021) Process intensification approach using microreactors for synthesizing nanomaterials—a critical review. Nanomaterials 11(1):1–21. https://doi.org/10.3390/nano11010098

He J et al (2018) Dynamic deformation behavior of a face-centered cubic FeCoNiCrMn high-entropy alloy. Sci Bull 63(6):362–368. https://doi.org/10.1016/j.scib.2018.01.022

Heiligtag FJ, Niederberger M (2013) The fascinating world of nanoparticle research. Mater Today 16(7–8):262–271. https://doi.org/10.1016/j.mattod.2013.07.004

Jeevanandam J et al (2018) Review on nanoparticles and nanostructured materials: history, sources, toxicity and regulations. Beilstein J Nanotechnol 9(1):1050–1074. https://doi.org/10.3762/bjnano.9.98

Jeon IY, Baek JB (2010) Nanocomposites derived from polymers and inorganic nanoparticles. Materials 3(6):3654–3674. https://doi.org/10.3390/ma3063654

Jian S et al (2018) Nanofibers with diameter below one nanometer from electrospinning†. RSC Adv 8(9):4794–4802. https://doi.org/10.1039/c7ra13444d

Jing D et al (2016) From a novel energetic coordination polymer precursor to diverse Mn_2O_3 nanostructures: control of pyrolysis products morphology achieved by changing the calcination atmosphere. Cryst Growth Des 16(12):6849–6857. https://doi.org/10.1021/acs.cgd.6b00975

Kamal I (2018) Prospects of some applications of engineered nanomaterials: a review. Open Access J Biomed Eng Biosci 2(5). https://doi.org/10.32474/oajbeb.2018.02.000149

Kirubakaran A, Jain S, Nema RK (2009) A review on fuel cell technologies and power electronic interface. Renew Sustain Energy Rev 13(9):2430–2440. https://doi.org/10.1016/j.rser.2009.04.004

Kumar E et al (2017) Preparation and investigation on structural, spectral and electrical properties of polyaniline/manganese dioxide nanocomposites. J Mater Environ Sci 8(10):3490–3495

Kumar S et al (2021) 0D to 3D carbon-based networks combined with pseudocapacitive electrode material for high energy density supercapacitor: a review. Chem Eng J 403:126352. https://doi.org/10.1016/j.cej.2020.126352

Lei SG, Hoa SV, Ton-That MT (2006) Effect of clay types on the processing and properties of polypropylene nanocomposites. Compos Sci Technol 66(10):1274–1279. https://doi.org/10.1016/j.compscitech.2005.09.012

Lei Z et al (2020) Nanoelectrode design from microminiaturized honeycomb monolith with ultrathin and stiff nanoscaffold for high-energy micro-supercapacitors. Nat Commun 11(1). https://doi.org/10.1038/s41467-019-14170-6

Li C, Zheng S, Chen D (2020) Size-dependent isogeometric analysis of bi-directional functionally graded microbeams reinforced by graphene nanoplatelets. Mech Based Des Struct Mach. https://doi.org/10.1080/15397734.2020.1848591

Li P et al (2018) Effects of aggregate types on the stress-strain behavior of fiber reinforced polymer (FRP)-confined lightweight concrete. Sensors (switzerland) 18(10):1–18. https://doi.org/10.3390/s18103525

Liu JK et al (2020) Fabrication of ultrafine W-Cu composite powders and its sintering behavior. J Market Res 9(2):2154–2163. https://doi.org/10.1016/j.jmrt.2019.12.046

Liu T et al (2019) Sustainability-inspired cell design for a fully recyclable sodium ion battery. Nat Commun 10(1):1–7. https://doi.org/10.1038/s41467-019-09933-0

Lu H et al (2020) Dynamic temperature control system for the optimized production of liquid metal nanoparticles. ACS Appl Nano Mater 3(7):6905–6914. https://doi.org/10.1021/acsanm.0c01257

Luhar S et al (2019) Valorisation of glass waste for development of geopolymer composites—mechanical properties and rheological characteristics: a review. Constr Build Mater 220:547–564. https://doi.org/10.1016/j.conbuildmat.2019.06.041

Ma Y et al (2017) Design and application of "J-shaped" stress-strain behavior in stretchable electronics: a review. Lab Chip 17(10):1689–1704. https://doi.org/10.1039/c7lc00289k

Madhu D et al (2017) Synthesis of high-quality biodiesel using feedstock and catalyst derived from fish wastes. J Agric Food Chem 65(10). https://doi.org/10.1021/acs.jafc.6b05608

Majumdar D (2021) Review on current progress of MnO_2-based ternary nanocomposites for supercapacitor applications. ChemElectroChem 8(2):291–336. https://doi.org/10.1002/celc.202001371

Mariano M, El Kissi N, Dufresne A (2014) Cellulose nanocrystals and related nanocomposites: review of some properties and challenges. J Polym Sci Part b: Polym Phys 52(12):791–806. https://doi.org/10.1002/polb.23490

Marlinda AR et al (2020) Recent developments in reduced graphene oxide nanocomposites for photoelectrochemical water-splitting applications. Int J Hydrogen Energy 45(21):11976–11994. https://doi.org/10.1016/j.ijhydene.2020.02.096

Migliore N et al (2018) Effect of the polyketone aromatic pendent groups on the electrical conductivity of the derived MWCNTs-based nanocomposites. Polymers 8(6). https://doi.org/10.3390/polym10060618

Montanaro L, Palmero P (2019) Advances in the field of nanostructured ceramic composites. Ceramics 2(2):296–297. https://doi.org/10.3390/ceramics2020024

Najib S, Erdem E (2019) Current progress achieved in novel materials for supercapacitor electrodes: mini review. Nanoscale Adv 1(8):2817–2827. https://doi.org/10.1039/c9na00345b

Namsheer K, Rout CS (2021) Conducting polymers: a comprehensive review on recent advances in synthesis, properties and applications. RSC Adv 11(10):5659–5697. https://doi.org/10.1039/d0ra07800j

Nandi D et al (2020) Metal/metal oxide decorated graphene synthesis and application as supercapacitor: a review. J Mater Sci 55(15):6375–6400. https://doi.org/10.1007/s10853-020-04475-z

Navarrete-Magaña M et al (2021) Improved photocatalytic oxidation of arsenic (III) with WO_3/TiO_2 nanomaterials synthesized by the sol-gel method. J Environ Manag 282. https://doi.org/10.1016/j.jenvman.2020.111602

Niculescu AG et al (2021) Nanomaterials synthesis through microfluidic methods: an updated overview. Nanomaterials 11(4). https://doi.org/10.3390/nano11040864

O'Connor J et al (2021) A review on the valorisation of food waste as a nutrient source and soil amendment. Environ Pollut 272:115985. https://doi.org/10.1016/j.envpol.2020.115985

Palchoudhury S, Baalousha M, Lead JR (2015) Methods for measuring concentration (mass, surface area and number) of nanomaterials. Front Nanosci. https://doi.org/10.1016/B978-0-08-099948-7.00005-1

Parchovianský M et al (2017) Mechanical properties and sliding wear behaviour of Al_2O_3-SiC nanocomposites with 3–20 vol% SiC. J Eur Ceram Soc 37(14):4297–4306. https://doi.org/10.1016/j.jeurceramsoc.2017.04.051

Parida SK (2018) Polymer nanocomposites and applications a brief review. Int J Sci Res Phys Appl Sci 6(3):75–78. https://doi.org/10.26438/ijsrpas/v6i3.7578

Patrianus Khristian Sumule (2012) No title ال عرب ية ال لغة ت دريس طرق. Экономика Региона 2:32. https://doi.org/10.1351/goldbook.F02547

Pendergast MM, Hoek EMV (2011) A review of water treatment membrane nanotechnologies. Energy Environ Sci 4(6):1946–1971. https://doi.org/10.1039/c0ee00541j

Rashidi K, Moazami-Goudarzi M, Masoudi A (2020) Powder processing, characterization and mechanical properties of Al/GNP composites. Mater Chem Phys 256:123719. https://doi.org/10.1016/j.matchemphys.2020.123719

Raza A et al (2014) Hierarchical porous carbon nanofibers via electrospinning. Carbon Lett 15(1):1–14. https://doi.org/10.5714/CL.2014.15.1.001

Roy Chowdhury S et al (2021) Ternary Al–Mg–Ag alloy promoted palladium nanoparticles as potential catalyst for enhanced electro-oxidation of ethanol. Int J Hydrogen Energy 46(5):4036–4044. https://doi.org/10.1016/j.ijhydene.2020.11.005

Sadoun AM et al (2020) Effect of nano Al_2O_3 coated Ag addition on the corrosion resistance and electrochemical behavior of Cu-Al_2O_3 nanocomposites. J Market Res 9(3):4485–4493. https://doi.org/10.1016/j.jmrt.2020.02.076

Sadoun AM et al (2021) Effect of lattice structure evolution and stacking fault energy on the properties of Cu–ZrO_2/GNP nanocomposites. Ceram Int. https://doi.org/10.1016/j.ceramint.2021.07.129

Saha JK et al (2020) Remarkable stability improvement of zno tft with Al_2O_3 gate insulator by yttrium passivation with spray pyrolysis. Nanomaterials 10(5). https://doi.org/10.3390/nano10050976

Sahani S, Banerjee S, Sharma YC (2018) Study of "co-solvent effect" on production of biodiesel from Schleichera Oleosa oil using a mixed metal oxide as a potential catalyst. J Taiwan Inst Chem Eng 86.https://doi.org/10.1016/j.jtice.2018.01.029

Sahani S, Roy T, Chandra Sharma Y (2019) Clean and efficient production of biodiesel using barium cerate as a heterogeneous catalyst for the biodiesel production; kinetics and thermodynamic study. J Clean Prod 237.https://doi.org/10.1016/j.jclepro.2019.117699

Sahani S, Roy T, Sharma YC (2020a) Smart waste management of waste cooking oil for large scale high quality biodiesel production using Sr-Ti mixed metal oxide as solid catalyst: optimization and E-metrics studies. Waste Manag 108. https://doi.org/10.1016/j.wasman.2020.04.036

Sahani S, Roy T, Sharma YC (2020b) Studies on fast and green biodiesel production from an indigenous nonedible Indian feedstock using single phase strontium titanate catalyst. Energy Convers Manag 203. https://doi.org/10.1016/j.enconman.2019.112180

Sahani S, Sharma YC (2018) 'Economically viable production of biodiesel using a novel heterogeneous catalyst: kinetic and thermodynamic investigations. Energy Convers Manag 171.https://doi.org/10.1016/j.enconman.2018.06.059

Saifuddin N, Raziah AZ, Junizah AR (2013) Carbon nanotubes: a review on structure and their interaction with proteins. J Chem 2013. https://doi.org/10.1155/2013/676815

Saikia BK et al (2020) A brief review on supercapacitor energy storage devices and utilization of natural carbon resources as their electrode materials. Fuel 282:118796. https://doi.org/10.1016/j.fuel.2020.118796

Saleh TA (2020) Trends in the sample preparation and analysis of nanomaterials as environmental contaminants. Trends Environ Anal Chem 28:1–10. https://doi.org/10.1016/j.teac.2020.e00101

Saleh TA (2021) Protocols for synthesis of nanomaterials, polymers, and green materials as adsorbents for water treatment technologies. Environ Technol Innov 24:101821. https://doi.org/10.1016/j.eti.2021.101821

Santos Miranda ME et al (2006) I. The role of N-carboxymethylation of chitosan in the thermal stability and dynamic. Polym Int 55:961–969. https://doi.org/10.1002/pi.2060

Sarma GK, Sen Gupta S, Bhattacharyya KG (2019) Nanomaterials as versatile adsorbents for heavy metal ions in water: a review. Environ Sci Pollut Res 26(7):6245–6278. https://doi.org/10.1007/s11356-018-04093-y
Sharifi S et al (2012) Toxicity of nanomaterials. Chem Soc Rev 41(6):2323–2343. https://doi.org/10.1039/c1cs15188f
Singh A, Kumari K (2019) An inclusive approach for organic waste treatment and valorisation using Black Soldier Fly larvae: a review. J Environ Manag 251:109569. https://doi.org/10.1016/j.jenvman.2019.109569
Somvanshi SB et al (2020) Hyperthermic evaluation of oleic acid coated nano-spinel magnesium ferrite: enhancement via hydrophobic-to-hydrophilic surface transformation. J Alloy Compd 835:155422. https://doi.org/10.1016/j.jallcom.2020.155422
Stambouli AB, Traversa E (2002) Solid oxide fuel cells (SOFCs): A review of an environmentally clean and efficient source of energy. Renew Sustain Energy Rev 6(5):433–455. https://doi.org/10.1016/S1364-0321(02)00014-X
Stewart JA, Dingreville R (2020) Microstructure morphology and concentration modulation of nanocomposite thin-films during simulated physical vapor deposition. Acta Mater 188:181–191. https://doi.org/10.1016/j.actamat.2020.02.011
Sudarsanam P et al (2018) Functionalised heterogeneous catalysts for sustainable biomass valorisation. Chem Soc Rev 47(22):8349–8402. https://doi.org/10.1039/c8cs00410b
Suryanarayana C (2018) Phase formation under non-equilibrium processing conditions: rapid solidification processing and mechanical alloying. J Mater Sci 53(19):13364–13379. https://doi.org/10.1007/s10853-018-2197-4
Suryanarayana C (2019) Mechanical alloying: a novel technique to synthesize advanced materials. Research 2019:1–17. https://doi.org/10.34133/2019/4219812
Ubaidullah M et al (2020) Hydrothermal synthesis of novel nickel oxide@nitrogenous mesoporous carbon nanocomposite using costless smoked cigarette filter for high performance supercapacitor. Mater Lett 266:127492. https://doi.org/10.1016/j.matlet.2020.127492
Upadhyay RK, Soin N, Roy SS (2014) Role of graphene/metal oxide composites as photocatalysts, adsorbents and disinfectants in water treatment: a review. RSC Adv 4(8):3823–3851. https://doi.org/10.1039/c3ra45013a
Vella Durai SC et al (2019) Structural, optical, morphological and thermal properties of CuO nanoparticles prepared by sol-gel technique. J Nano- Electron Phys 11(5):4–7. https://doi.org/10.21272/jnep.11(5).05011
Wang X et al (2016) Facile synthesis ultrathin mesoporous Co_3O_4 nanosheets for high-energy asymmetric supercapacitor. J Alloy Compd 686:969–975. https://doi.org/10.1016/j.jallcom.2016.06.156
Wang Z, Mi B (2017) Environmental applications of 2D molybdenum disulfide (MoS_2) nanosheets. Environ Sci Technol 51(15):8229–8244. https://doi.org/10.1021/acs.est.7b01466
Wohlleben W et al (2017) Reliable nanomaterial classification of powders using the volume-specific surface area method. J Nanoparticle Res 19(2). https://doi.org/10.1007/s11051-017-3741-x
Yadav A et al (2019) Investigating the effects of amine functionalized graphene on the mechanical properties of epoxy nanocomposites. Mater Today: Proc 11:837–842. https://doi.org/10.1016/j.matpr.2019.03.054
Yang J et al (2019) Nanomaterials for the removal of heavy metals from wastewater. Nanomaterials 9(3). https://doi.org/10.3390/nano9030424
Yang X et al (2018) A review on thermally conductive polymeric composites: classification, measurement, model and equations, mechanism and fabrication methods. Adv Compos Hybrid Mater 1(2):207–230. https://doi.org/10.1007/s42114-018-0031-8
Yang Y et al (2021) On the sustainability of lithium ion battery industry—a review and perspective. Energy Storage Mater 36:186–212. https://doi.org/10.1016/j.ensm.2020.12.019
Yu S, Hing P, Hu X (2002) Thermal conductivity of polystyrene-aluminum nitride composite. Compos Part a: Appl Sci Manuf 33(2):289–292. https://doi.org/10.1016/S1359-835X(01)00107-5

Zema DA et al (2019) Wastewater management in citrus processing industries: an overview of advantages and limits. Water (switzerland) 11(12):1–23. https://doi.org/10.3390/w11122481
Zhang L et al (2018) A review of supercapacitor modeling, estimation, and applications: a control/management perspective. Renew Sustain Energy Rev 81:1868–1878. https://doi.org/10.1016/j.rser.2017.05.283
Zhao L, He X, Todoh M (2020) Mechanical behavior of biomimetically mineralized collagen matrix using the polymer-induced liquid precursor process. J Biomech 104:109738. https://doi.org/10.1016/j.jbiomech.2020.109738

Part III
Nanocomposites for Environmental Applications

Chapter 14
Graphene-Polymer Nanocomposites for Environmental Remediation of Organic Pollutants

T. S. Shrirame, J. S. Khan, M. S. Umekar, A. K. Potbhare, P. R. Bhilkar, G. S. Bhusari, D. T. Masram, A. A. Abdala, and R. G. Chaudhary

Abstract The graphene-based polymer nanocomposite is an outstanding material which is attracting researcher's and scientists' attention due to its fascinating physicochemical and thermo-mechanical properties. Even it has been a decade after its discovery, scientists are still exploring the hidden potentials of this material. The present chapter summarizes the most recent studies on the modification of graphene with polymers and the subsequent synthesis and applications of high-quality graphene-polymer nanocomposites. The chapter started with an introduction followed by historical background, conventional and green fabrication strategies for producing high-quality graphene based-polymer nanocomposites including various covalent and non-covalent techniques are summarized and discussed. Finally, the graphene-polymer nanocomposites catalyzed photo-degradation of organic pollutants (OPs) like dyes, pesticides, insecticide for water purification, and future perspective is presented herewith.

Keywords Graphene-polymer nanocomposites · Bioinspired nanocomposites · Green synthesis · Photocatalytic degradation · Water purification · Environmental remediation

T. S. Shrirame · J. S. Khan · M. S. Umekar · A. K. Potbhare · P. R. Bhilkar · R. G. Chaudhary (✉)
Department of Chemistry, Shri Shivaji Science and Arts College, Chikhli 443201, India

G. S. Bhusari
Research and Development Division, Apple Chemie India Private Limited, Nagpur 441108, India

D. T. Masram
Department of Chemistry, Delhi University, Delhi, India

A. A. Abdala
Chemical Engineering Program, Texas A&M University at Qatar, Doha, Qatar
e-mail: ahmed.abdala@qatar.tamu.edu

S. P. Singh et al. (eds.), *Metal Nanocomposites for Energy and Environmental Applications*, Energy, Environment, and Sustainability,
https://doi.org/10.1007/978-981-16-8599-6_14

Abbreviations

NPs	Nanoparticles
NCs	Nanocomposites
PVA	Polyvinyl alcohol
PANI	Polyaniline
PS	Polystyrene
TPU	Thermoplastic polyurethane
PU	Polyurethane
PMMA	Poly (methyl methacrylate)
PE	Polystyrene
PE	Polyethylene

14.1 Introduction

The extensive use of various toxic organic pollutants (OPs) like dyes, pesticides, herbicides, and insecticides by textile, paper printing, and agrochemical industries pollutes water-system and affects an ecosystem. The various routes of water pollution are presented in Fig. 14.1. Elimination of these OPs is a vital practice of wastewater management before being discharged into the environment because of their synthetic nature, complex chemical structure, toxicity, and carcinogenic nature. To make maximum use of the solar spectrum novel visible-light-driven photocatalysts are essential for the complete eradication of noxious OPs from wastewater (Khan and Malik 2014; Singh and Agarwal 2016; Yin et al. 2016; Crock et al. 2013; Pandey et al. 2017).

Presently, graphene-based polymer nanocomposites gaining significant interest, due to their unique properties. Hence, they are widely applicable in a fuel cell, supercapacitors, energy storage, organic transformation, photocatalyst, sensor, and biomedical sectors (Fig. 14.2). Various metal oxide semiconducting photocatalysts have been fabricated and used for the photo-degradation of OPs documented in the literature. The effective photocatalytic degradation under ultra-violet light is often linked with the formation of holes in the valence band via the promotion of electrons to the conduction bands, which is responsible for oxidizing and reducing the OPs. The doping of polymeric materials such as polyaniline (PANI), polypyrrole (PP), polythiophene (PTs), and polyacetylene with different dopants can effectively adsorb many pollutants. Also, offer a better activity than the pristine metal oxide in the photodegradation of OPs (Fu et al. 2019; Armstrong 2015; Tkalya 2012).

On another hand, graphene or graphene-based derivatives have interesting optical properties, fluorescence labels, high dispersibility in various polar solvents, and the ability to attach diverse molecular structures on its surface via hydrogen bonding. These properties facilitate an adsorption of various molecular structures on its surface, leading to better control of the size and the shape of the formed structures. The

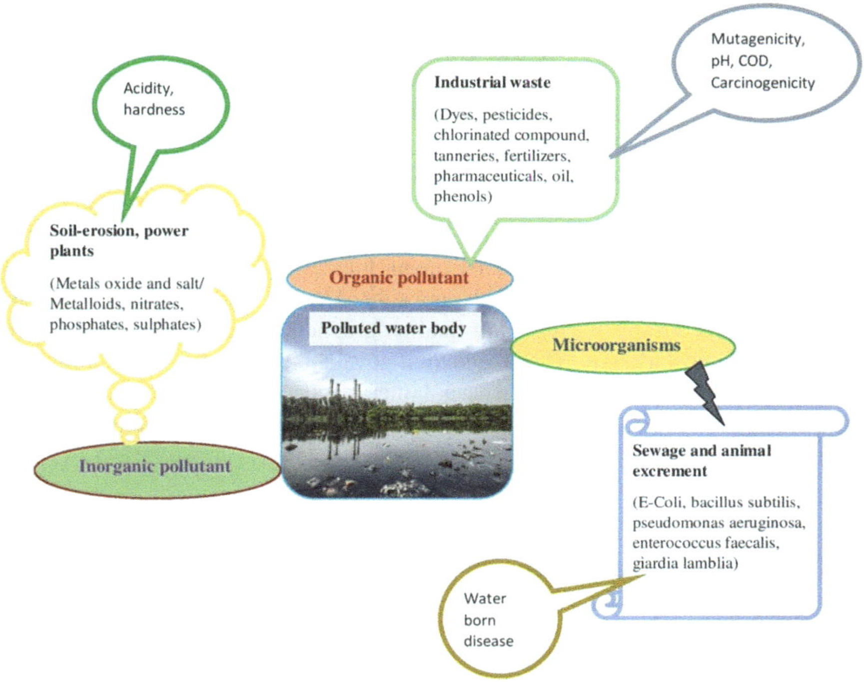

Fig. 14.1 Water pollutants classes, sources, and impact (Pandey et al. 2017)

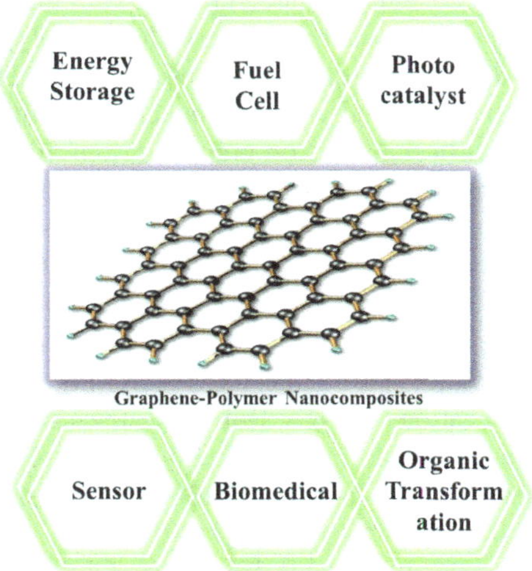

Fig. 14.2 Applications of graphene-based polymer nanocomposites

relative hydrophilicity of graphene due to the highly oxygenated surface facilitates its interaction with aqueous dispersions of metal oxides, leading to the formation of strong chemically bonded graphene-polymer nanocomposites. Graphene-reinforced polymer matrixes encapsulated with metallic nanoparticles (NPs) generally exhibit improved thermal stability, electrocatalysis, and sensing, microelectronic, and excellent photocatalytic properties (Tkalya 2012; Ghorbani et al. 2011; Bin Ahmad et al. 2011).

The present chapter summarizes the most recent studies on the modification of graphene with polymers and the subsequent synthesis and applications of high-quality graphene-based polymer nanocomposites. The conventional and green fabrication strategies for producing high-quality graphene-based-polymer nanocomposites including various covalent and non-covalent techniques are summarized and discussed. Finally, the graphene-polymer nanocomposites-catalyzed photodegradation of organic pollutants (OPs) like dyes, pesticides, insecticide for water purification, and future perspective is presented herewith.

14.2 Historical Background of Graphene-Based Polymer Nanocomposites

The rich history of graphene spans the last forty years of research work. In 2004, Andre Geim and Kostya Novoselov from Manchester University successfully isolated the thinnest, single-layer 2D crystal of graphene. A lot of research has been carried out on expanded and exfoliated graphite NCs including a wide range of polymers like epoxy, polymethyl methacrylate (PMMA), polypropylene (PP), Low-density polyethylene (LDPE), high-density polyethylene (HDPE), polystyrene (PS), nylon, polyaniline (PANI), phenylethynyl-terminated poly-imide, and silicon rubber (Iqbal et al. 2020). The earliest studies on graphene reinforced polymer NCs using graphite as filler follow graphene intercalation chemistry. Primarily GO and graphene intercalation compounds (GICs) are used as the precursor for ascendible approaches to graphene-based composites. In back 1840s, the Go and GICs have been investigated (Potts et al. 2011). In 1958, Podall et al. ascertained that alkali metal-reinforced GIC's could initiate the polymerization of ethylene, styrene, methyl-methacrylate, and isoprene, which shows the same results. Previously, this reaction is the main point of focus for the characterization of PNCs. But, later, it was reported that this method causes exfoliation of host graphite. In 1993, Bunnell proposed incorporating as thin as possible graphite nanoplatelets as filler for PNCs which improve the stiffness of obtained products (Paszkiewicz and Szymczyk 2019).

Most studies have focused on graphene derivatives (GO, rGO, frGO) as an alternative for graphene. Oxidation of graphite produces GO using Hummers and Brodie's method (Botas et al. 2013). Chen et al. (2013) used the modified Hummers method by excluding $NaNO_3$ for eco-friendly GO synthesis. This modification reduced the release of toxic gases such as NO_2/N_2O_2 and increased the yield of the products.

Besides this, wastewater generated in synthesis was disposed of effortlessly because of the inexistence of toxic elements, including Na^+ and NO_3^- ions. Yet several chemical reducing agents have been reported in which very toxic and common one is hydrazine. Owing to toxicity, it limits biobased applications. Green synthesis methods have been explored to eradicate this issue where they used green reducing agents like amino acid, bacteria, vitamin-C, green tea polyphenols, serum albumin, etc. Zhu et al. (Sharma et al. 2018) used an ammonical solution of glucose, fructose, etc., for the reduction of GO. Newly synthesized reduce graphene oxide (rGO) manifest better electrocatalytic activity within a short reduction duration.

GO has an oxygen-containing functional group, which helps in the uniform dispersion of GO in polar solvents. However, reduction of GO tends to increase conductivity. Mittal et al. (2015) reported properties of GO and rGO and primarily focuses on filler. These nanofillers poor dispersion ability poses substantial obstacles to achieving polymer NCs with a promising application at low nanofillers content. Similarly, Supova et al. highlighted the difficulties regarding nanofillers dispersion in a polymer matrix and advances solutions to overcome difficulties such as nanofillers surface modification and additional procedures (grounding, sonication, ultrasonic blending), chemical functionalization, and use of a unique technology such as in situ polymerization. Functionalization of using phenyl isocyanate brings about fine homogeneous dispersion in 1, 2-dichlorobenzene. Graphite reinforced polystyrene nanocomposites synthesized by in situ polymerization observed increased thermal stability and higher glass transition temperature due to substantial interfacial interaction between polymer and graphite nanofillers (Šupová et al. 2011). Direct incorporation of graphene derivatives into polymer has also been reported. The properties of final NCs depend on interfacial bonding between NPs and the polymer matrix. Most of the dispersion method produces NCs that are non-covalent. However, there is growing research for introducing covalent linkage between graphene derivative and polymer matrix to acquire strong-interfacial interaction.

14.3 Synthesis of Graphene-Based Polymer Nanocomposites

The graphene reinforced polymer nanocomposites can be fabricated using both conventional and green methods. The principal aim of adopting a suitable fabrication method is to bring about better homogeneous dispersion of nanofillers into the host polymer matrix. Efficient dispersion of graphene nanofillers is a crucial step that ensures the advancement of polymer chain properties. The diverse synthesis process for the fabrication of graphene reinforced polymer nanocomposites is presented in Fig. 14.3.

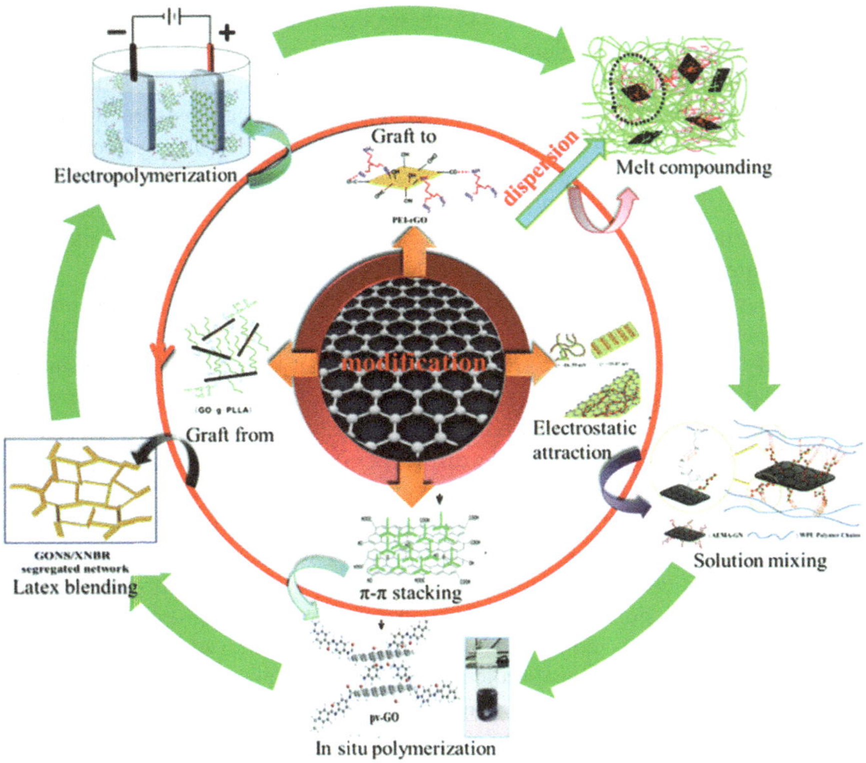

Fig. 14.3 Different fabrication methods of graphene reinforced polymer nanocomposites (Zhang et al. 2015). Reproduced with the permission of the Royal Society of Chemistry publishing in the format Journal/magazine via Copyright Clearance Center

14.3.1 Conventional Techniques

The five different conventional strategies used for fabrication of graphene-based polymer nanocomposites comprising of melt blending, solution compounding, in-situ polymerization, latex mixing, electro-polymerization which is discussed in brief.

14.3.1.1 Solution Blending

In solution blending technique, GO or graphene-based derivatives are dispersed in a suitable organic solvent such as water, acetone, toluene, tetrahydrofuran (THF), chloroform, or dimethylformamide (DMF). The polymer is also dissolved in the same solvent or another miscible solvent. Subsequently, the colloidal suspension and the polymer solution are mixed by sonication, agitation, stirring, or shear mixing until graphene is dispersed homogeneously in the host polymer matrix. Finally, the methods in particular solvent evaporation, filtration, lyophilization, precipitation, etc., are employed to obtain final graphene-based polymer nanocomposites (Zhang et al. 2015). Besides, the use of toxic organic solvents and their total removal from NCs are vital challenges of this method. Extensive use of organic solvents boosts manufacturing costs. Hence, solution compounding is not applicable for wide-reaching industrial production compared to melt blending. Graphene-based polymer nanocomposites synthesized using solution compounding, include graphene/epoxy, graphene/polyvinyl alcohol (PVA) (Sun et al. 2020), graphene/elastomer (Araby et al. 2013), graphene/PANI (Yuan et al. 2012), graphene/PS (Yan et al. 2012), rGO/PS (Kim et al. 2014), rGO/ thermoplastic polyurethane (TPU) (Shen et al. 2017).

14.3.1.2 In-situ Polymerization

In this strategy, graphene or graphene derivatives are dispersed in monomer or prepolymer then polymerization is initiated by adding a suitable initiator under heat or radiation conditions (Chen et al. 2018). In-situ polymerization is advantageous over solution compounding and melts blending because (i) it makes graphene disperse more uniformly in the polymer matrix and provide strong interfacial bonding between filler and matrix, facilitating stress transfer, and (ii) it is environment friendly as it involves limited use of organic solvent. The major constraint of this method is that polymerization is accompanied by an increase in the system's viscosity, leading to a decrease in the polymerization rate. Therefore, this method is yet not applied in industrial production. Numerous graphene-based polymer nanocomposites has been reported using this method *e.g.* graphene/epoxy, graphene/polyurethane (PU), graphene/polystyrene (PS), graphene oxide (GO)/polyvinyl alcohol (PVA), graphene/polymethyl methacrylate (PMMA) (Mao and Wang 2020; Taylor et al. 2013; Verdejo et al. 2011).

14.3.1.3 Melt Blending

It is one of the most commercially attractive method used for the fabrication of graphene reinforced thermoplastic nanocomposites. In this method, graphene or chemically modified graphene in powder form mixed with melted thermoplastic polymer and finally NCs is produced using an extruder under controlled temperature, speed, and time. This method is a simple, versatile, cost-effective, and eco-friendly method most acceptable in industries. Notwithstanding, melt blending is preferred less over solution compounding and in-situ polymerization due to improper nanofillers' dispersion in the polymer matrix (Verdejo et al. 2011). This poor dispersion will result in NCs with less or no enhancement in properties. This method can be applied only when graphene is in dried powder forms e.g. thermally reduced GO. The effect of melts blending on interfacial interaction between GO, and polycarbonate (PC) had been investigated by Shen et al. they observed two different types of interactions. The first one is non-covalent π-π interaction, and the second one is a chemical interaction between carboxyl groups of GO with carbonate group on PC (Shen et al. 2013). A wide range of polymer nanocomposites synthesized by melt mixing has been reported. Blending of exfoliated graphene with polymers like poly (styrene-co-acrylonitrile), polyurethane, polyamide-6, isotactic polypropylene, high-density polyethylene, and polycarbonate are highly thermally stable (Steurer et al. 2009).

14.3.1.4 Latex Mixing

Latex mixing is another successfully develops method to achieve easy dispersion without utilizing any solvent. This method is limited to only conductive materials. In this method, the polymer surface is decorated with graphene then final composites are obtained using corresponding techniques (especially hot pressing). Li et al. (2013) prepared graphene-based PS NCs by latex mixing. They mixed an aqueous solution of GO with PS latex. The later mixture was co-coagulated with sodium chloride to form stabilized suspension, and eventually, reduction of GO was made by adding hydrazine hydrate. Newly synthesized graphene-based PS NCs exhibited improved electrical and mechanical properties. Pang et al. (2010) reported similar studies on the preparation of graphene-based ultra-high molecular weight polyethylene (PE) NCs by latex mixing.

14.3.1.5 Electro-polymerization

Electropolymerization is the easy, short, environment-friendly, and appropriate method to fabricate graphene-based polymer nanocomposites. The NCs synthesized with this method have applications in energy storage devices, such as supercapacitor, batteries, and electrochemical biosensors (Zhang et al. 2015). PANI and carbon nanostructures (CNSs) such as graphene-based polymer nanocomposites were fabricated using in situ electrochemical polymerizations by Petrovski et al. (2017). Electropolymerization was performed using three-electrode systems. Initially, graphene was dispersed in an electrolytic solution by ultra-sonication. The experimental condition was optimized by steady-state approximation and cyclic voltammetry measurement. Finally, electropolymerization of NCs was carried out at the electrode potential of +0.75 V versus saturated calomel electrode (SCE), resulting in good graphene dispersion in PANI. Strong π-π stacking interactions were observed between graphene and the quinoidal structure of PANI. Among all the above discussed conventional methods, solution compounding and in-situ polymerization are preferred over melt-blending because they generate homogeneous NCs with strong graphene-polymer interaction. In solution compounding, solvent removal is difficult. Latex mixing also causes homogeneous graphene dispersion to polymer, but dispersion potential is difficult to confirm in electropolymerization because it depends on the interface between graphene and polymer matrix.

14.3.2 Green Techniques

The conventional fabrication techniques of graphene-based polymer nanocomposites are environmentally hazardous. Consequently, alternative facile, scalable, and environmentally friendly green synthesis approaches have been established. It is categorized into two sections: microwave irradiation synthesis and bioinspired synthesis.

14.3.2.1 Microwave Technique

Over the last few years, microwave (MW) irradiation has been favor above other conventional approaches because of its simplicity, efficiency, and time saving; reduce energy consumption, better product yield, and eco-friendliness. Graphene-based materials exhibit a great magnitude of absorbing MW irradiation efficiently. Li et al. reported MW irradiation for polymeric modification of graphene/GO by poly (allyldiazoacetate-co-acrolein) (Li et al. 2019a). One step in-situ MW assist polymerization (MAP) for Nylon-6/graphene nanocomposites was studied by Pablo et al. (González-morones et al. 2018). The overall method involves high-speed melting of monomer followed by reduction and exfoliation of GO, and finally, the hybridization with Nylon-6 ensues from dielectric heating generated by MW absorption. Pokharel et al. (2014) optimized multi-step MW reduction to obtain rGO to form graphene/epoxy nanocomposites. The rGO at 0.5 wt% loadings in epoxy increased the glass transition temperature of composite by 10 °C. At 0.3 wt% and one wt% loading, the percolation threshold of electrical conductivity was observed. Similarly, Naeem et al. (2021) observed an increment in glass transition temperature from 85.4 to 100 °C at 1.03 vol% of graphene and an electrical percolation threshold at 0.85 wt%. Murugan et al. (2009) presented facile reduction of exfoliated GO into chemically modified graphene nanosheets using MW-assisted solvothermal method and explored graphene/PANI nanocomposites as electrode materials.

14.3.2.2 Bioinspired Technique

The bioinspired fabrication of graphene reinforced polymer nanocomposites can be accomplished using green reducers, including plant extract, cell culture, microbes, enzymes, amino acids, vitamins, and biomolecules. Green reducers also act as capping and stabilizing agents and play an essential role in lessening NPs size (Umekar et al. 2021a). Green reducers are less toxic, non-carcinogenic, non-corrosive, and non-hazardous compared to chemical reducing agents such as hydrazine (Umekar et al. 2021b). In the green method, the first highly oxidized GO is synthesized using the improved Hummers method from graphite flex. Then this GO is reduced into rGO using green reducers. Finally, blending of rGO with polymer matrix was carried out. Figure 14.4 summarizes the reduction of GO with

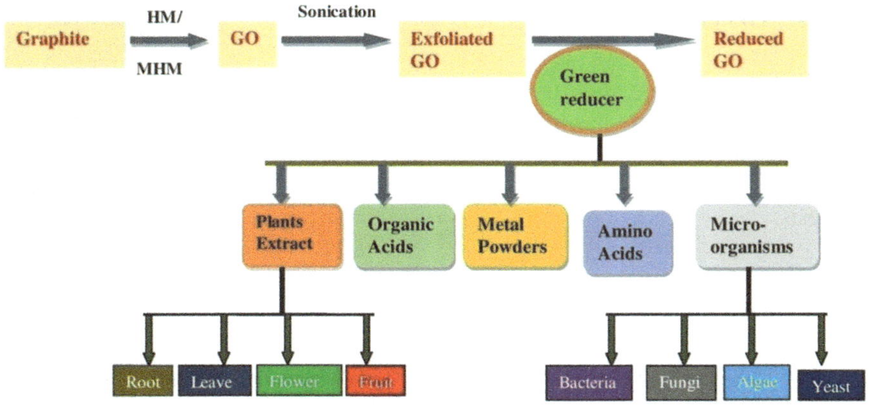

Fig. 14.4 A schematic representation for different routes for green reductions of GO (Chaudhary et al. 2020a; Aunkor et al. 2016)

environmentally friendly agents.

With the evolution in green synthesis, researchers have major focused on plant extract as a reducer. Phytochemicals present in plants converted to quinone in the presence of reactive oxygen. Quinone begins the reduction of an oxygen-containing group on GO. Leaves of *aspidopterys cordata* (Chaudhary et al. 2019a), *Colocasia esculenta, Phyllanthus reticulatus, and Conyza bonariensis* (Potbhare et al. 2019), green tea (Wang et al. 2011), mushroom extract (Muthoosamy et al. 2015), carrot juice (Kuila et al. 2011), rose water (Haghighi and Tabrizi 2013), *Salvadora persica* (Khan et al. 2015) and *Sesbania bispinosa* (Chaudhary et al. 2021) are acted as green reducers to synthesize graphene based nanocomposites. Moreover, organic acids and their salts are some of the best green reducers for the synthesis of rGO. Ascorbic acid, sodium-citrate, caffeic acid, lemon juice etc. have been used as green reductants. Reduction of GO has also been practiced using bacteria and yeast. Reduction of GO by *Escherichia coli* in an anaerobic environment and *Shewanella* in both aerobic and anaerobic conditions have been reported (Akhavan and Ghaderi 2011). The rGO obtained from yeast is of good quality posses' electrical conductivity of 43 S/m. It is believed that NADPH is present in yeast responsible for reducing epoxy ketones in GO (Khanra et al. 2012).

Combining all these, vegetable oil has been used to prepare organic and inorganic filler-based polymer composites from macro to the nanoscale range (Roopan and Madhumitha 2015). Long et al. (2013) were used vitamin C as a green reducing agent in producing PS/rGO composites. GO and mono-disperse PS microsphere dispersed in water, where GO gets absorb on PS microsphere. GO/PS microsphere then reduced using vitamin C. The resultant composites indeed manifest low electrical percolation threshold, high conductivity, and good consent. Yari et al. (2021) modified the PU matrix with rGO nanosheets to strengthen weathering stability. For comparison purposes, they reduced GO *in* two ways; by using *Nettle* leave extract as a green reducer and hydrazine hydrate as a chemical reducing agent. PU/rGO nanocomposites synthesized by the green method observed high weathering stability.

14.4 Properties of Graphene-Based Polymer Nanocomposites

These extraordinary properties of graphene have brought about great engrossment of its reinforcement in the polymer matrix and produce multifaceted polymer nanocomposites. Indeed, graphene tailors the significant improvement in structural (mechanical properties) and functional (photocatalytic, electrical, adsorption/absorption properties) features of resulting polymer nanocomposites. Rheological properties come up with molecular structural information of polymer nanocomposites such as; percolated network formation by nanofillers, dispersion ability, and filler-matrix interaction (Essabir et al. 2019). Reinforcement of nanofillers into polymer matrix causes an elevation in melt viscosity and results in typical rheological properties. Several investigations made by researchers showed that the rheological behavior of graphene-polymer nanocomposites relies on graphene dispersion state, the interaction between graphene-graphene along with graphene and polymer matrix, aspect ratio, and content of graphene (Paszkiewicz and Szymczyk 2019).

The superior electrical conductivity is the most attractive feature of graphene. Owing to the exceptional electrical conductivity and high aspect ratio, the incorporation of a little quantity of graphene in an insulating polymer matrix increases the overall conductivity of NCs. The incorporation of graphene in polymers such as PVC, PP, PMMA, PA12, PVA, PS, PE, and so forth generate electrically conductive graphene polymer nanocomposites. A nonlinear increase in electrical conductivity has been observed at a particular filler loading fraction (Khanam et al. 2015). Xie et al. (2008) successfully compared electrical conductivity between graphene nanosheets and carbon nanotube (CNT). It was foretold that graphene is more efficient in electrical conductivity enhancement than CNTs. Highly conductive graphene polymer composite has a much lower percolation threshold and higher conductivity at small-scale graphene loading.

Furthermore, graphene possesses remarkable mechanical properties, such as high Young's modulus of 1TPa, higher flexibility as well as intrinsic tensile strength. These extraordinary mechanical properties are due to sp^2 hybridized bonding in a hexagonal lattice and active resistance to a variety of in-plane deformations. Hone et al. confirmed graphene as "*the strongest material ever measured*" when they have measured the mechanical properties of single-layered free-standing graphene by using nano indentation in Atomic Force Microscopy (Papageorgiou et al. 2017). Zhao et al. had prepared graphene/PVA composites and observed notable improvement in mechanical properties at lower graphene content. With 1.8 vol% of graphene nanosheets loading, the tensile strength was enhanced by 150%, ten times greater than the pure PVA sample (Zhao et al. 2010).

Graphene probably has the highest in-plane thermal conductivity in just position with any currently studied material. Experimental studies at room temperature show that free-standing monolayer graphene sheets have in-plane thermal conductivity up to 1500–5400 W/m K, which is surprisingly higher than that of diamond, i.e., nearly 2000 W/m K (Paszkiewicz and Szymczyk 2019). Hence, the addition of graphene in polymer gives a solid platform for novel thermal management applications. The elevated thermal conductivity of graphene is attributed to two factors; first, the sp^2 hybridization and lightweight carbon give high-velocity phonon and exceptionally harmonic lattice. Second, the 2D arrangement allows across-plane flexural phonon vibration with little scattering rate on the mirror symmetry of graphene lattice. These two features yield graphene with extremely high thermal conductivity (Huang et al. 2020). In addition, graphene and its derivatives have been considered favorable nanofillers in the gas barrier application of polymer nanocomposites. The incorporation of graphene produces a tortuous path which reduces gas permeability of polymer film and NCs (Luna et al. 2018). The factors which influence gas barrier properties of NCs include aspect ratio, graphene nanosheets orientation, the crystallinity of host polymer matrix, interfacial interaction between graphene and polymer, and dispersion ability. The result of exfoliated graphite nanoplatelets on the gas barrier property of polypropylene was studied by Kalaitzidou et al. They noticed increased barrier properties of polypropylene by at least 20% only three vol% filler loading. Prusty et al. reported 13 times reduction in gas permeability using expanded graphite/PAN nanocomposites with 4 wt% of nanofillers content (Cui et al. 2016).

Besides all the aforementioned magnificent properties of graphene-polymer nanocomposites, the 2D single-layered graphene possess distinctive optical and photonic properties. Many researchers have recently studied the nonlinear optical (NLO) properties of graphene derivatives such as graphene nanoribbons, GO, and graphene nanosheets. Optical limiting (OL) properties of graphene derivative-based NCs have applications in protecting delicate optical sensors and human eyes from powerful lasers. Muralidharan et al. studied OL behavior of rGO/PVA NCs. With the increase in the concentration of rGO in the polymer matrix, the OL properties are found to increases. Accordingly, by tuning the filler concentration, the desired limiting threshold can be attained, and hence the composite brings in applications as potential optical limiters. Gan et al. prepared polyimide (PI) composite film filled with GO nanosheets, GO nanoribbons and GO quantum dots whose NLO and OL properties were investigated at 532 nm. The graphene composite film gives away superior NLO/OL properties compared with their corresponding suspension due to a combination of nonlinear mechanism, charge transfer between PI/graphene material, and the matrices (Muralidharan et al. 2016).

The morphology of GO and rGO covalently grafted with amine-terminated hyperbranched polyamide (HBPA-NH_2) (GO/HBPA) are explored by scanning electron microscopy (SEM) and transmission electron microscopy (TEM). According to Fig. 14.5a, the GO surface was found to be fairly smooth. Whereas, the surface of GO/HBPA is deem to be relatively roughed (Fig. 14.5b). From this it is conclude that, HBPA-NH_2 is successfully incorporated on GO surface. As shown in SEM images Fig. 14.5c, d, GO/HBPA reveals very desultory stacking and isotropic structure in

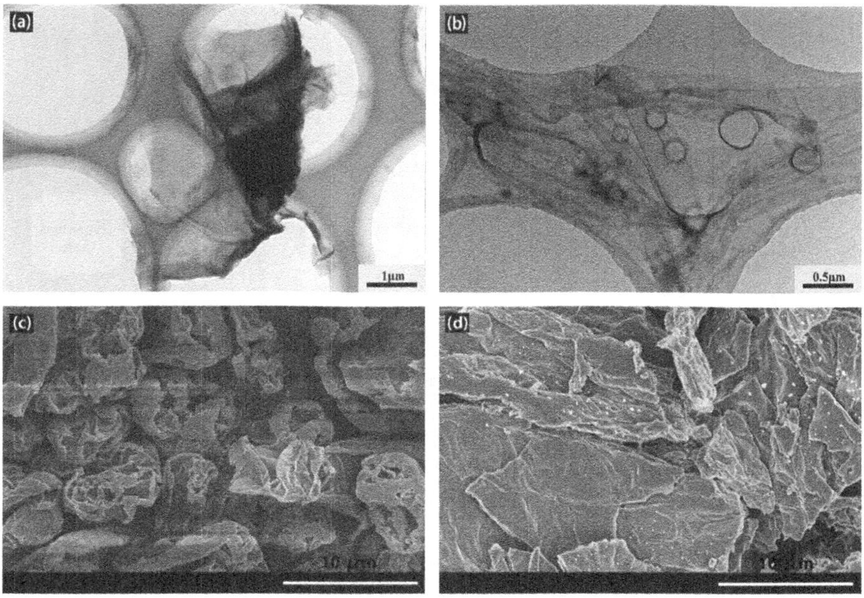

Fig. 14.5 TEM images of **a** GO and **b** GO-HBPA, SEM images of **c** GO and **d** rGO-HBPA (Li et al. 2019b). Reproduced with the permission of the Elsevier publishing in the format Journal/magazine via Copyright Clearance Center

comparison with GO. Besides, GO-HBPA was stacked more loosely and expanded than GO (Li et al. 2019b).

14.5 Photocatalytic Applications of Graphene-Based Polymers Nanocomposites

14.5.1 Photodegradation of Dyes

Polymeric materials such as polyaniline, polypyrrole, polythiophene, and polyacetylene can effectively adsorb many pollutants. The polymers with metal/metal oxide NPs are found better than metal oxide (MO) NPs in the degradation of organic dyes. While, GO has attractive optical properties, fluorescence labels, high dispersibility in various polar solvents, and the ability to attach diverse molecular structures on its surface via hydrogen bonding. These properties facilitate the adsorption of different molecular structures on their surface, leading to better control of the size and the shape of the formed structures. The relative hydrophobicity of GO, due to the highly oxygenated surface facilitates its interaction with aqueous dispersions of TiO_2, leading to the formation of strong chemically bonded MO/GO with PNCs. Anchoring

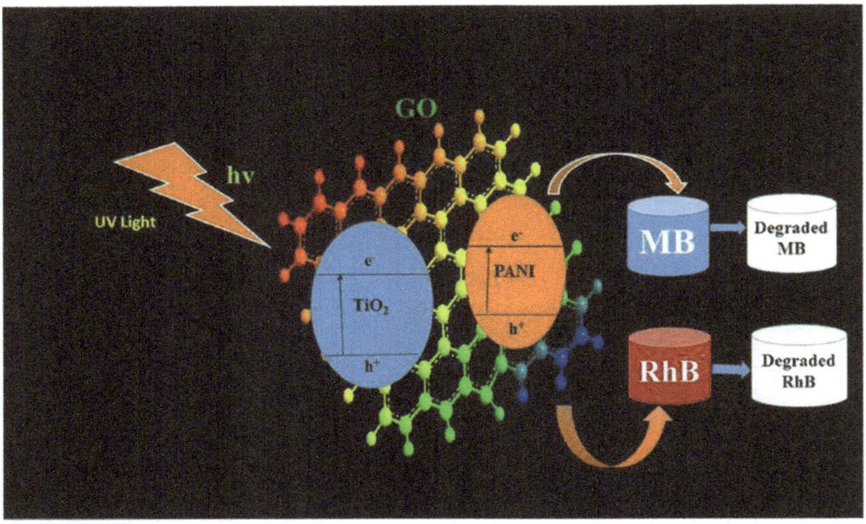

Fig. 14.6 TiO_2/PANI/GO NCs catalyzed degradation of dyes (Reddy et al. 2016)

of polymeric matrixes encapsulated (Fig. 14.6) with metallic oxide NPs on GO generally exhibits enhanced thermal stability, electrocatalysis, sensing, microelectronic, and excellent photocatalytic properties (Reddy et al. 2016).

Kaushal et al. (2020) investigated the variation of pectin concentration on biopolymer pectin-graphene oxide NCs for the photocatalytic degradation of methylene blue (MB) and methyl orange (MO) dyes. They reported nearly 98% of MB and 87% of MO degradation in 25 min and 90 min (Fig. 14.7).

Stacking of conducting polymers (PANI) on GO/rGO and other inorganic semiconductors can deliver matched band structures. Additionally, the recombination rate of photogenerated electron–hole pairs is slightly down. In visible light radiation, PANI shows characteristics of electron donor and hole acceptor (Lee and Chang 2019). Keeping this issue, Wu et al. designed ternary PCs of ZnO/rGO/PANI via in-situ polymerization. The study reveals that the good photodegradation efficiency of MO is found in UV light emission (Wu et al. 2016). The study showed that the enhancement in photodegradation efficacy nearly six times compared to pure ZnO in 60 min of ultraviolet light radiation. This is accredited to the introduction of rGO, which processes the transfer of photogenerated electrons with PANI, which further improves both the absorption of ultraviolet light and the adsorption of dyes. Mitra et al. (2019) described the photocatalytic activity of rGO/PANI NCs with 5 wt% rGO synthesized by in situ polymerization process. Examined the photocatalytic activity of malachite green (MG), Rhodamine-B (RhB), and Congo red (CR) under visible light radiation. They reported the degradation fractions of 99.7, 99.3, and 98.7% for MG, RhB, and CR dyes in 15, 30, and 40 min respectively. PANI/rGO NCs show

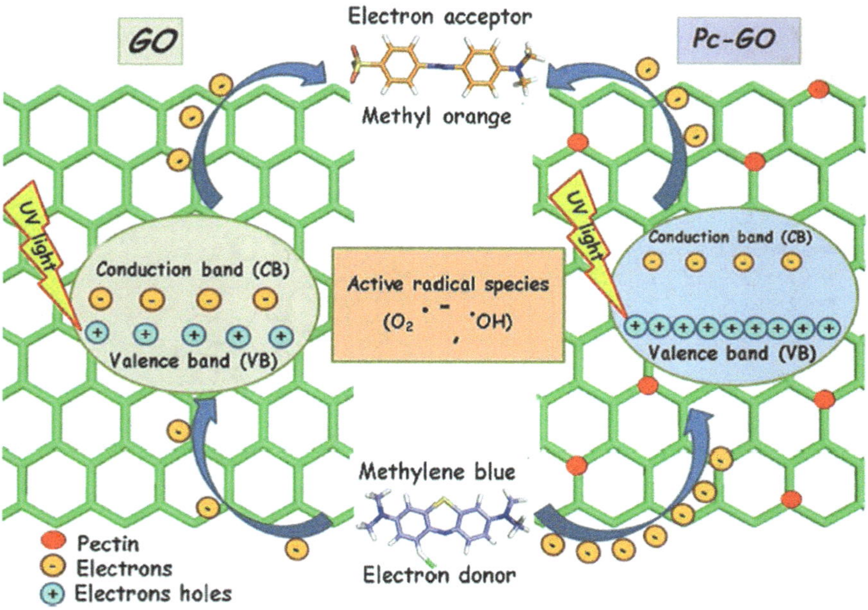

Fig. 14.7 Diagrammatic depiction of the degradation process of MB and MO dyes (Kaushal et al. 2020). Reproduced with the permission of the Elsevier publishing in the format Journal/magazine via Copyright Clearance Center

better photocatalytic activity than pristine PANI and rGO due to its absorption capability regarding the large portion of the solar spectrum and possesses better photo generated electron–hole pair separation.

Chen et al. described the synthesis of TiO_2/Chitosan/rGO NCs as biocompatible photocatalyst via the water-based freeze-casting method. The photodegradation ability of MO was examined in optimum conditions having concentration of initial dye about 250 mg/L, and 100 mL. The degradation profile of MO was observed at nearly 97% within 300 min of UV light irradiation (Chen et al. 2017), basically, chitosan is a natural polysaccharide, which pursued better adsorption ability owing to presence of amino and hydroxyl groups. Chitosan also increases the reaction sites on the surface of TiO_2. Nevertheless, rGO endorses more dispensability due to the synergistic effect and develops better photocatalytic efficiency (Chen et al. 2017; Chaudhary et al. 2020b).

Likewise, Elshahawy et al. synthesized (PVA/PAAc)–rGO–TiO_2 NCs and estimated its photocatalytic efficiency for the decolouration of direct blue 71 dye (DB71) under UV light irradiation. The percent of DB71 degradation was almost 90% up to eight cycles. It shows the complete decolourization of DB71 dye took place at pH six after 40 min in the presence of 2 mL/L H_2O_2 (Elshahawy et al. 2020).

Aminated graphene oxide (NGO) was synthesized by condensation of graphene oxide (GO) and diethylenetriamine (DETA) (Chen et al. 2021). Highly catalytic

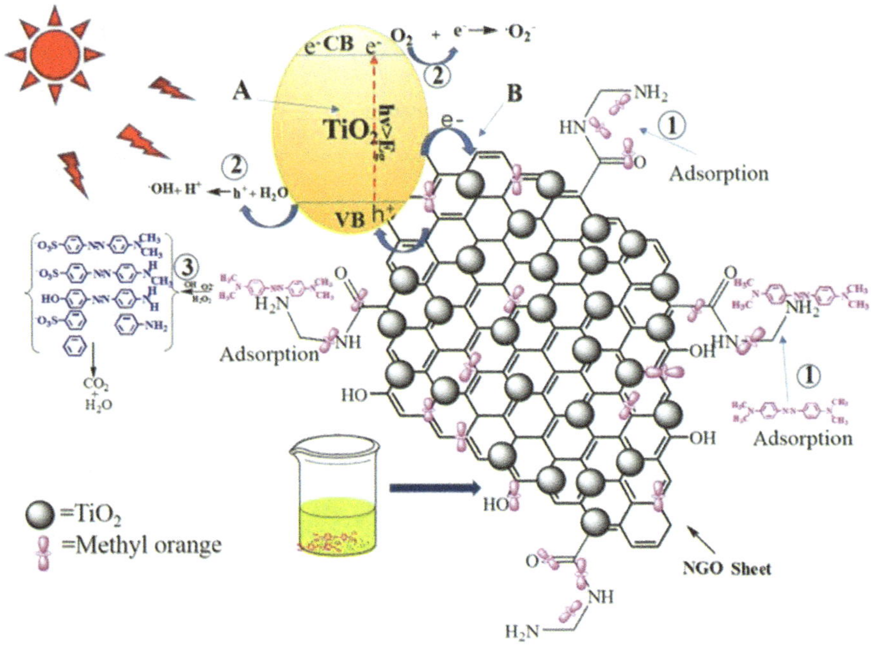

Fig. 14.8 Process and mechanism of NGO-TiO_2 photocatalysis (Chen et al. 2021). Reproduced with the permission of the Elsevier publishing in the format Journal/magazine via Copyright Clearance Center

nanostructured NGO-TiO_2 nanocomposites (Fig. 14.8) were self-assembled by a two-phase system of toluene and water. The NGO-TiO_2 catalyzed degradation of methyl orange dye under simulated solar illumination was examined. The catalytic performance of NGO-TiO_2 improved with increasing NGO loading amount in an appropriate ratio than pristine TiO_2 and GO.

TiO_2/PANI decorated on the graphene oxide nanocomposites hybrid were synthesized with varying TiO_2 percentages via an in-situ chemical oxidative process using ammonium persulphate as an initiator (Baruah et al. 2021). The NCs hybrid exhibited encouraging photocatalytic activity towards degradation of methylene blue (MB) and rhodamine B dye under UV light than pure PANI and TiO_2 NPs. Noteworthy photodegradation was observed about 98.9% by 20% of TiO_2 NPs loading within one hour under short-wavelength UV-light. Polyacrylonitrile (PAN)/β-cyclodextrin (β-CD) NCs membranes immobilized with TiO_2 and GO were prepared by ultrasonic-assisted electrospinning (Zhang et al. 2021). The photocatalytic degradation of methyl orange (MO) and MB under natural sunlight was examined. It was found that PAN/β-CD/TiO_2/GO nanocomposites with an 8:2 mass ratio of TiO_2-to-GO exhibited the best degradation efficiency for the dyes. The degradation efficiency for MB and MO was 93.52% and 90.92%, respectively.

Polythiophene (PTh) and graphene nanoplatelets (GNPs) NCs with different ratios of GNPs (10–50 wt%) were synthesized by oxidative chemical polymerization technique (Noreen et al. 2021a). Examined the photocatalytic activity of bromophenol blue (BPB), PTh-GNPs (50) exhibited three times more (94.1%) photocatalytic degradation than pristine PTh (31.3%) for bromophenol. This is attributed to the improved charge transport and longer recombination time of electron/hole pairs from PTh NPs to GNPs.

14.5.2 Photo-Degradation of Pesticides, Herbicides, Fungicides, and Insecticides

The rapid industrialization, urbanization, economic development, agricultural development, and anthropogenic activities have raised anxieties over the topic such as climate change, wastewater generation, and environmental pollution. From the past several years, growing agricultural activities marked exponential growth in the consumption of pesticides and fertilizers. Pesticides can be derived synthetically or naturally to preserve the agricultural product and eradicate pests and weeds. Using pesticides shields agricultural crops from unwanted plants and externally or internally develops microbes, insects, pathogens, or small organisms. Pesticides are classifieds into different classes, including herbicides, insecticides, bactericides, and fungicides. Pesticides based on chemical structure are again categorized into organochlorines, chlorophenols, carbamates, synthetic pyrethroids, organophosphates, and substituted urea. Pesticides consist of both active as well as inert components. The pesticide component responsible for killing and controlling pests is termed as the active component, and the remaining portion is termed as inert component (Vaya and Surolia 2020). The uncontrolled usage of pesticides has become a dilemma in Asia.

Pesticides originated from the agricultural field may oozes into the soil through wind, rain, and irrigation water. Furthermore, pesticides from soil sources reach the aquatic ecosystem through the wastewater treatment plant, surface runoff, penetration, volatilization, soil erosion, and leaching. The immoderate use of pesticides comes up with various environmental crises and threatens living territories due to their carcinogenic and bio-recalcitrant nature. Similar to OPs, pesticides also belong to categories of polycyclic and halogenated hydrocarbons resistant to chemical and biological degradation and may take years to degrade. The toxic pesticides discharge directly from agricultural fields to nearby water resources, which are dangerous to the marine ecosystem also trouble aquatic life (Sonkusare et al. 2020). The presence of pesticides in water bodies is of worldwide concern. Analogs to marine life, there are various means through which human beings can also get exposed to pesticides. The short-term exposure to pesticides may lead to skin annoyance, allergies, cough and sore throat, respiratory displeasure, vomiting, diarrhea and eye inflammation. Whereas, long term exposure may lead to depression and anxiety, cancer, asthma, Parkinson disease, and cancer in human being (Nazir et al. 2020).

Thus, there is a compelling demand to develop technology that can efficiently degrade these environmentally hazardous pesticides, ecofriendly, and cost-effectively. Several comprehensive studies have been reported an ample number of methods for degradation of pesticides, namely advanced oxidation process (AOPs), electrochemical method, biological method (bioremediation using microbes and phytoremediation operating plants) (Thanaraj et al. 2019), catalytic oxidation, Fenton's reagents (Chen et al. 2007), membrane filtration, adsorption, and nano-based approach. The effectiveness of the electrochemical method is confined due to high treatment cost, insufficient degradation, and overlong process. In addition, the biological process is confined due to several factors, including pH, temperature, culture maintenance, and nutrient availability. The adsorption process is bound with several limitations of generation of hazardous byproducts, expensive sorbent, and capital investment (Yeganeh et al. 2021).

Among all mentioned methods, AOP processes have gained adequate attention as a powerful technique for pesticide remediation. Amid a variety of AOPs, the photocatalytic process has captivated noteworthy consideration for pesticide degradation in water resources by advantages of the process highlighted as; eco-friendly, cost-effective, use of non-toxic chemical catalyst, highly efficient and stable, less amount of sludge production, rapid reaction rate, complete degradation and generation of harmless byproduct.

According to Yeganeh et al., meta-analysis results confirm that the photocatalytic process eliminates pesticides with 93.36% degradation efficiency. The average pooled percentage of photocatalytic degradation of insecticides, fungicides, and herbicides was found to be 93.35, 100, and 90.73 respectively (Yeganeh et al. 2021; Saleh et al. 2020). The remarkable interest in using NPs in photocatalysis is triggered by their larger surface area, the efficient capacity of light absorption, and more reactive sites. Nevertheless, NPs lead to specific trouble in their separation and recycling and their potential release risks to human beings and the environment despite their exceptional properties. To overcome this drawback, the NPs are immobilized on the polymer host surface. The polymers are promising candidates in the photocatalytic field provide the probability of simplistic separation and recycling of photocatalytic material, thus reducing the cost of the procedure and eliminating efforts of the post-treatment separation process. Also, the polymer acts as a support for the immobilization of active semiconductor photocatalytic NPs (such as TiO_2, ZnO, ZrO_2, CdS, etc.) (Melinte et al. 2019).

Graphene reinforced nanocomposites act as potentiality strong photocatalyst and adsorbent for remediation of pesticides from wastewater. The aromatic ring of graphene forms strong π-π interaction with OPs, which is why strong adsorption of different pesticides on GO surface (Frediani and Rosi 2020). Graphitic carbon nitride (g-C_3N_4) is a superior photocatalytic semiconductor that will use approximately 40% of visible light from the total solar spectrum and 2.7 eV of bandgap energy. Vigneshwaran et al. (2019) investigated photocatalytic degradation of chlorpyrifos using metal-free g-C_3N_4 incorporated chitosan (CS) composite. Chlorpyrifos is an organophosphate insecticide used for controlling pests on cotton plants, vegetables, and fruit. The CS/g-C_3N_4 photocatalyst showed pesticide degradation with 85% efficiency. The CS/g-C_3N_4 composite was investigated for the reusability test. Even after five cycles, they reported no loss in photocatalytic activity. Anirudhan et al. (2018) synthesized zinc oxide nanorod incorporated carboxylic graphene/polyaniline composite (ZnR@CGR/PANI) for successful elimination and photocatalytic degradation of pesticides diuron (DU) under visible radiations. They compared photodegradation activities of blank (without catalyst), CGR, and ZnO@CGR/PANI composite. It was observed that blank does not have any degradation activity. Whereas CGR has much higher photodegradation activity compared to blank, also the degradation is achieved within 80 min. But in the case of novel photocatalyst ZnR@CGR/PANI the remarkable enhancement in photocatalytic activity was observed and complete degradation was achieved within 40 min.

In the photocatalytic degradation mechanism of DU pesticide upon irradiation, ZnR absorbs visible radiations from the light source. The Valence band (VB) of ZnR shows the generation of holes and transferred that holes to the highest occupied molecular orbital (HOMO) of PANI. Further, the photogenerated holes again excited to the higher lowest unoccupied molecular orbital (LUMO) of PANI. Finally, the photogenerated electrons from LUMO get transferred to the conduction band (CB) of ZnR. All these processes are responsible for avoiding the recombination of electron–hole pairs. The PL spectrum of the composite argues that the significant decrease in electron–hole recombination rate and enhancement in photocatalytic activity of ZnR is due to the composite formation with PANI and CGR.

These photo-induced electrons of CB are acting as a reducing agent that is transferred to GO surface. Where they react with a dissolved O_2 molecule to gives superoxide anion radical (O_2^-). Superoxide radical would subsequently react with H^+ radical to yield $HO_2^{\cdot}$ radicals. Holes present in VB can readily generate hydroxyl radical ($OH^{\cdot}$) from hydroxide ion (OH^-) adsorbed at the surface or from the water molecule (H_2O). These species are highly reactive and act as a strong oxidizing agent that can degrade pesticides into H_2O, CO_2, and Cl^-. The confirmation regarding CO_2 formation is obtained by bubbling it on decarboxylated water in the presence of phenolphthalein indicators (Maryani and Kustiningsih 2015). Using $AgNO_3$ solution, the presence of Cl^- ion was confirmed. The complete mechanism of photocatalytic degradation of DU on ZnR@CGR/PANI is presented in Fig. 14.9.

14.5.3 Photo-Degradation of Toxic Organic Compounds

Water pollution by toxic organic contaminants is a primary concern of the present time for our ecosystem. Removal of organic pollutant by GO polymer NCs have been considering as promising application because of its great potential to adsorb and degrade contaminants (Tomar et al. 2020; Chaudhary et al. 2019b). An organic conducting polymer PANI has promising photocatalytic applications due to its higher stability and electrical and photoelectrical properties. Hence, many researchers tried to synthesize PANI-based NCs with excellent photocatalytic performance for a wide

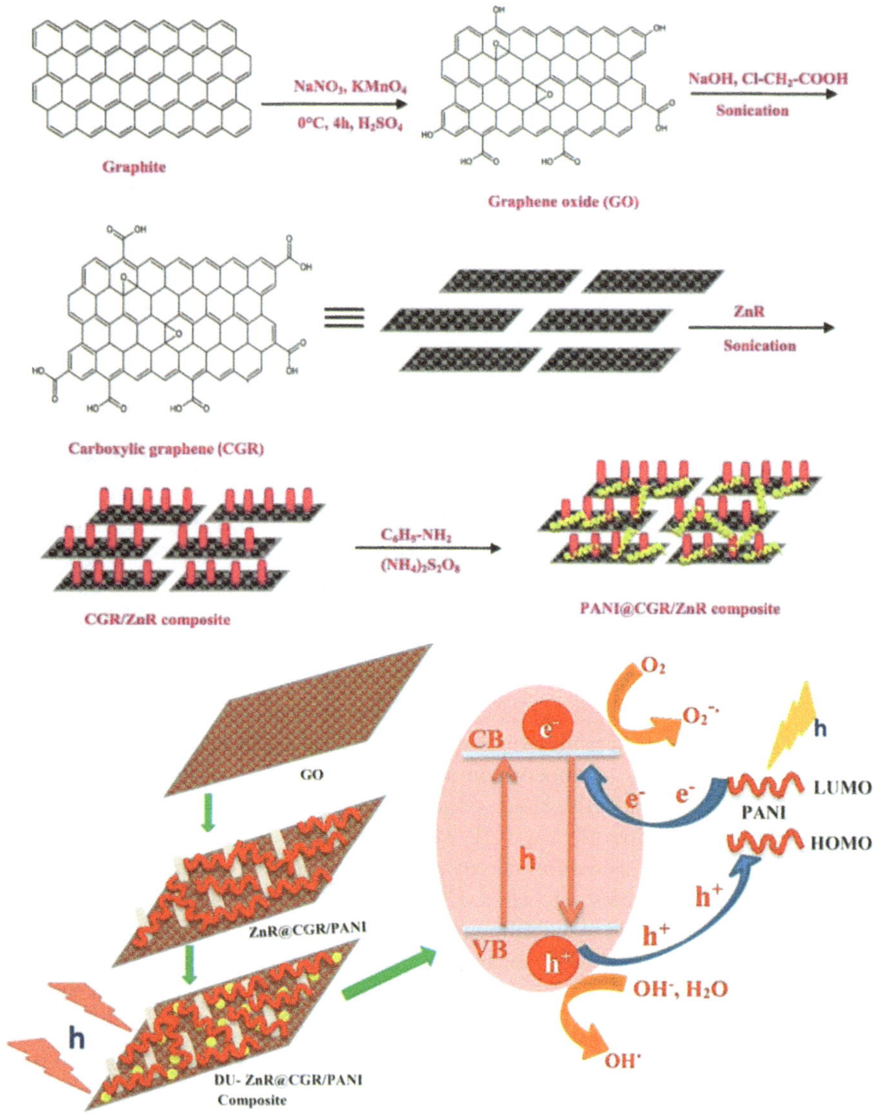

Fig. 14.9 The mechanism of photocatalytic degradation of DU on ZnR@CGR/PANI. Reproduced with the permission of the Elsevier publishing in the format Journal/magazine via Copyright Clearance Center

range of applications. Ameen et al. (2012) synthesized PANI/graphene nanocomposites by in situ polymerizations of aniline with graphene as a successful photocatalyst for photocatalytic degradation of Rose Bengal (RB) dye from water. The newly synthesized PANI/graphene nanocomposites display noteworthy 56% efficient degradation of RB dye within 3 h in the presence of light illumination. The nanocomposites of polymeric GO with polyaniline (PANI), polypyrrole (PPy) and polystyrene (PS) served as promising adsorbents for the removal of actacid orange RL (AO-RL), which is an organic compound and dye. All these three NCs show maximal adsorption of 39.65, 43.99, and 21.10 mg/g within 60 min. Amongst all, the permeability of GO/PPy is highest, whereas the least porosity is observed for GO/PS. Hence, the pollutant adsorption potential of NCs follows the order as GO/PPy > GO/PAN > GO/PS (Noreen et al. 2021b). Ahmad et al. (2020) demonstrated the photocatalytic activity of ternary NCs of rGO supported on polypyrrole (Ppy) and semiconductor CdS for degradation of organic pollutants such as Rhodamine B (Rh B), Reactive Blue-171 (RB-171), and toluene under visible radiations. The conducting polymer Ppy is acting as an active catalyst because of its high polarizability and electrochemical properties. The degradation efficiency of polymer is enhanced by using CdS semiconductors owing to its high adsorption ability and direct bandgap (2.4 eV). Further incorporation of this CdS into rGO increased photodegradation potential due to larger surface area, good electron mobility, and magnificent adsorption properties. Decreased chemical oxygen demand value confirms photocatalytic degradation. Another advantage of this photocatalyst is its stability and reusability. The same photocatalyst can be used frequently for organic pollutant treatment. The pH-sensitive poly (vinyl alcohol)/poly (acrylic acid)/TiO_2/GO nanocomposites were synthesized by radical polymerization for photocatalytic degradation of organic pollutants. The degradation ability of NCs was evaluated by using a UV spectrophotometer. The improvement in photocatalytic degradation of TiO_2 was achieved by combing with higher content of GO. This improvement is due to the acceleration of the electron-transfer process between GO and TiO_2 and the complementary interaction between pollutants and GO (Moon et al. 2013).

14.6 Water Purification Applications of Graphene-Based Polymers Nanocomposites

Pure polymeric membranes have irreversible fouling, low permeability and selectivity, intrinsic hydrophobicity, and short life. To overcome this, polymeric membrane blends with hydrophilic nanomaterials to enhanced water permeation and separation performance. In comparison with other materials, graphene polymer nanocomposites have better efficiency in the area of water purification. To date, several works reported the synthesis of graphene polymer nanocomposites mixed matrix membrane (MMM) such as GO/polyvinylidene fluoride (PVDF), GO/polysulfone (PSF), and GO/polyethersulfone (PES). GO/PVDF MMM comes with enlarged pore size and increased permeability, and superior antifouling performance emerges from increased hydrophilicity. GO/PES MMM was prepared by blending GO first with polyvinylpyrrolidone. Later, PES reported 12° drops in water contact angle, 2.5 times rise in pure water permeability, boost in hydrophilicity and antifouling property, and permeability recovery rate of 90.5%. GO/PSF MMM showed similar results prepared using the wet phase inversion method (Fathizadeh et al. 2017).

Purification ability of poly (vinyl acetate) membrane has been increased ten times with incorporation of 0.07 wt% functionalized graphene. Modified graphene nanofillers reinforced into poly (ethylene glycol) methyl ether methacrylate shows remarkable development in separation, water flux recovery rate, and antifouling performance (Kausar 2018). Intake of heavy metals such as cadmium, mercury, arsenic, lead, chromium, and selenium present in potable water is prone to cause cardio vascular diseases, muscular weakness, cancer, hormone imbalance, neurological disorder, nausea, and loss of appetite (Saleh et al. 2019).The GO/NH_2-β-cyclodextrin nanocomposites were applied for Cr(IV) removal (Fig. 14.10). The mechanism of Cr(IV) removal and synthesis of GO/NH_2-β-cyclodextrin nanocomposites are illustrated in the figure. The NCs allow the rapid adsorption of heavy metal ion Cr(VI) from waste solution with an adsorption efficiency of 120 mg/g. Desorption of Cr(VI) from nanocomposites was done using aqueous NaOH and can be reused for the next adsorption cycle (Gandhi et al. 2016).

Yin et al. (2016) prepared a novel GO containing polyamide thin film nanocomposites (TFN) using an in-situ interfacial polymerization process. The increase in the concentration of GO nanosheets increases the hydrophilicity of the TFN membrane. The increased in water permeate flux under 300 psi were observed from 39.0 ± 1.6 to 59.4 ± 0.4 L/m^2 h, while decreased in rejection of NaCl and Na_2SO_4 were noticed from 95.7 ± 0.6% to 93.8 ± 0.6% and 98.1 ± 0.4% to 97.3 ± 0.3%. Improvement in

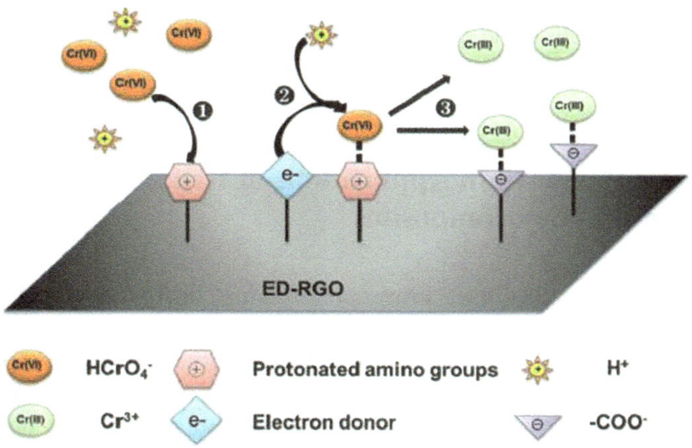

Fig. 14.10 Proposed mechanism of Cr(VI) removal by ED-rGO. Reproduced with the permission of the John Wiley and Sons publishing in the format Journal/magazine via Copyright Clearance Center

GO incorporation conditions increases water permeability and salt rejection capacity of TFN membrane. The polyacrylamide/GO nanocomposites were prepared through chain polymerization used as flocculating agents due to their magnificent cleansing properties. With the increase in GO nanofillers content in NCs, a decrease in η and filtrate and supernatant turbidity was observed. By further increase in GO concentration, there was a noticeable fall in η and turbidity of cleaned water (Manafi et al. 2016).

14.7 Conclusions

In conclusion an article reported the current research status of graphene-polymer nanocomposites synthesis and its application in photodegradation of OPs for environmental remediation. Although this field is still in an early stage of development, the growing interest and huge potential of this field have become apparent. However, we should learn from the unfulfilled expectations of CNT-polymer nanocomposites and keep in perspective the challenges and fundamental issues that need to be resolved. The first challenge relates to graphene and CMG production. The preparation and transfer of high-quality graphene are still not viable in a cost-effective manner. Earlier, numerous efforts have been made to prepare useful graphene-based NCs, and an important improvement achieved. However, despite of the considerable advances, excitement and promise of exfoliated graphene-based polymeric nanocomposites, substantial fundamental research is still necessary to provide a basic understanding of these materials to enable full exploitation of their nanoengineering potential. Despite a large number of combinations of matrices and potential reinforcing nanoelements with different chemistry, size, shape and properties, all graphene-based nanocomposites share common features about fabrication methodologies, processing, morphology characterization and fundamental physics. Thus, the key to preparing advanced graphene-based nanocomposites is the engineering at the polymer-graphene interface. Developing an understanding of the characteristics of this interphase region, its dependence on the graphene surface chemistry, the relative arrangement of constituents and its relationship to the NCs properties is a current research frontier in NCs that unites the interest of scientists in physical chemistry, materials science and engineering. This article highlighted graphene-based polymer nanocomposites and their potentials applications in the coming years for environmental remediation of OPs. Very soon, it is expected that a large number of new graphene-based nanocomposites using different polymer hosts (thermoplastic, thermosetting, and especially commodity polymers) and a wide range of graphene or derivative with different functionalities, size and shape will be reported. For instance the imaginative molecular design of polymeric surfactants has not yet been explored, and can circumvent the problem of solubility, dispersion in polymers and the film-forming ability of graphene. However, we are far from the end of the tunnel in terms of understanding the mechanisms of the enhancement effect in graphene nanocomposites.

References

Ahmad N, Sultana S, Sabir S, Khan MZ (2020) Exploring the visible light driven photocatalysis by reduced graphene oxide supported Ppy/CdS nanocomposites for the degradation of organic pollutants. J Photochem Photobiol A Chem 386:112129. http://doi.org/10.1016/j.jphotochem.2019.112129

Akhavan O, Ghaderi E (2011) *Escherichia coli* bacteria reduce graphene oxide to bactericidal graphene in a self-limiting manner. Carbon 50:1853–1860. https://doi.org/10.1016/j.carbon.2011.12.035

Ameen S, Seo HK, Shaheer Akhtar M, Shin HS (2012) Novel graphene/polyaniline nanocomposites and its photocatalytic activity toward the degradation of rose Bengal dye. Chem Eng J 210:220–228. https://doi.org/10.1016/j.cej.2012.08.035

Anirudhan TS, Shainy F, Mohan AM (2018) Fabrication of zinc oxide nanorod incorporated carboxylic graphene/polyaniline composite and its photocatalytic activity for the effective degradation of diuron from aqueous solutions. Sol Energy 171:534–546. https://doi.org/10.1016/j.solener.2018.06.111

Araby S et al (2013) Electrically and thermally conductive elastomers/graphene nanocomposites by solution mixing. Polymer 55:201–210. https://doi.org/10.1016/j.polymer.2013.11.032

Armstrong G (2015) An introduction to polymer nanocomposites. Eur J Phys 36:63001

Aunkor MTH, Mahbubul IM, Saidur R, Metselaar HSC (2016) The green reduction of graphene oxide. RSC Adv 6:27807–27828. https://doi.org/10.1039/C6RA03189G

Baruah S, Kumar S, Nayak B, Puzari A (2021) Optoelectronically suitable graphene oxide-decorated titanium oxide/polyaniline hybrid nanocomposites and their enhanced photocatalytic activity with methylene blue and rhodamine B dye. Polym Bull 78:1703–1720. https://doi.org/10.1007/s00289-020-03182-8

Bin Ahmad M, Tay MY, Shameli K, Hussein MZ, Lim JJ (2011) Green synthesis and characterization of silver/chitosan/polyethylene glycol nanocomposites without any reducing agent. Int J Mol Sci 12:4872–4884. https://doi.org/10.3390/ijms12084872

Botas C et al (2013) Graphene materials with different structures prepared from the same graphite by the Hummers and Brodie methods. Carbon 65:156–164. https://doi.org/10.1016/j.carbon.2013.08.009

Chaudhary RG, Chouke PB, Potbhare A (2019a) Green fabrication of Zinc oxide nanospheres by Aspidopterys cordata for effective antioxidant and antibacterial activity. Adv Mater Lett 10:355–360. https://doi.org/10.5185/amlett.2019.2235

Chaudhary RG, Bhusari GS, Tiple AD, Rai AR, Somkuvar SR, Potbhare AK, Lambat TL, Ingle PP, Abdala AA (2019b) Metal/metal oxide nanoparticles: toxicity, applications, and future prospects. Curr Pharm Des 25(37):4013–4029. https://doi.org/10.2174/1381612825666191111091326

Chaudhary RG, Potbhare AK, Chouke PB, Rai AR (2020a) Carbon nanomaterials and their nanocomposites with metal oxides: biosynthesis , electrochemical, photocatalytic and antimicrobial applications. In: Magnetic oxides and composites II, vol 83. Material Research Foundation, pp 79–116. http://doi.org/10.21741/9781644900970-4

Chaudhary AU, Abdala A, Lonkar SP, Chudhary RG, Mabrouk A (2020b) Thermal, electrical, and mechanical properties of highly filled HDPE/graphite nanoplatelets composites. Mater Today Proc 29(3):704–708. https://doi.org/10.1016/j.matpr.2020.04.168

Chaudhary RG, Potbhare AK, Tarik Aziz SK, Umekar MS, Bhuyar SS, Mondal A (2021) Phytochemically fabricated reduced graphene oxide-ZnO NCs by *Sesbania bispinosa* for photocatalytic performances. Mater Today Proc 36:756–762. http://doi.org/10.1016/j.matpr.2020.05.821

Chen S, Sun D, Chung J (2007) Treatment of pesticide wastewater by moving-bed biofilm reactor combined with Fenton-coagulation pretreatment. J Hazard Mater 144:577–584. https://doi.org/10.1016/j.jhazmat.2006.10.075

Chen J, Yao B, Li C, Shi G (2013) An improved Hummers method for eco-friendly synthesis of graphene oxide. Carbon 64:225–229. https://doi.org/10.1016/j.carbon.2013.07.055

Chen C, Zhang Y, Zeng J, Zhang F, Zhou K, Bowen CR (2017) Applied surface science aligned macroporous TiO_2/chitosan/reduced graphene oxide (rGO) composites for photocatalytic applications. Appl Surf Sci 424:170–176. https://doi.org/10.1016/j.apsusc.2017.02.137
Chen W, Weimin H, Li D, Chen S, Dai Z (2018) A critical review on the development and performance of polymer/graphene nanocomposites. Sci Eng Compos Mater 25:1059–1073. https://doi.org/10.1515/secm-2017-0199
Chen Y, Wang L, Sun H, Zhang D, Zhao Y, Chen L (2021) Applied surface science self-assembling TiO_2 on aminated graphene based on adsorption and catalysis to treat organic dyes. Appl Surf Sci 539:147889. http://doi.org/10.1016/j.apsusc.2020.147889
Crock CA, Rogensues AR, Shan W, Tarabara VV (2013) Polymer nanocomposites with graphene-based hierarchical fillers as materials for multifunctional water treatment membranes. Water Res 47:3984–3996. https://doi.org/10.1016/j.watres.2012.10.057
Cui Y, Kundalwal SI, Kumar S (2016) Gas barrier performance of graphene/polymer nanocomposites. Carbon 98:313–333. https://doi.org/10.1016/j.carbon.2015.11.018
De Luna MS, Wang Y, Zhai T et al (2018) Nanocomposite polymeric materials with 3D graphene-based architectures: from design strategies to tailored properties and potential applications. Prog Polym Sci 89:213–249. https://doi.org/10.1016/j.progpolymsci.2018.11.002
Elshahawy MF, Mahmoud GA, Raafat AI, Ali AEH (2020) Fabrication of TiO_2 reduced graphene oxide based nanocomposites for effective of photocatalytic decolorization of dye effluent. J Inorg Organomet Polym Mater 1–16. http://doi.org/10.1007/s10904-020-01463-3
Essabir H, Raji M, Bouhfid R, Tiouitchi G et al (2019) Black phosphorous-based polyvinylidene fluoride nanocomposites: synthesis, processing and characterization. Compos Part B Eng 175:107165. http://doi.org/10.1016/j.compositesb.2019.107165
Fathizadeh M, Xu WL, Zhou F, Yoon Y, Yu M (2017) Graphene oxide: a novel 2-dimensional material in membraneseparation for water purification. Adv Mater Interfaces 4:1–16. https://doi.org/10.1002/admi.201600918
Frediani M, Rosi L (2020) Graphene and graphene oxide bounded metal nanocomposite for remediation of organic pollutants. In: Carbon-based material for environmental protection and remediation. http://doi.org/10.5772/intechopen.82334
Fu S, Sun Z, Huang P, Li Y, Hu N (2019) Some basic aspects of polymer nanocomposites: a critical review. Nano Mater Sci 1:2–30. https://doi.org/10.1016/j.nanoms.2019.02.006
Gandhi MR, Vasudevan S, Shibayama A (2016) Graphene and graphene-based composites : a rising star in water purification—a comprehensive overview. ChemistrySelect 4358–4385. http://doi.org/10.1002/slct.201600693
Ghorbani HR, Safekordi AA, Attar H, Sorkhabadi SMR (2011) Biological and non-biological methods for silver nanoparticles synthesis. Chem Biochem Eng Q 25:317–326
González-morones P et al (2018) Exfoliation, reduction, hybridization and polymerization mechanisms in one-step microwave-assist synthesis of nanocomposite nylon-6/graphene. Polymer 146:73–81. https://doi.org/10.1016/j.polymer.2018.05.014
Haghighi B, Tabrizi MA (2013) Green synthesis of reduced graphene oxide nanosheets using rose water and a survey on their characteristics and applications. RSC Adv 3:13365–13371. https://doi.org/10.1039/c3ra40856f
Huang X, Zhi C, Lin Y, Bao H et al (2020) Thermal conductivity of graphene-based polymer nanocomposites. Mater Sci Eng R Rep 142:100577. http://doi.org/10.1016/j.mser.2020.100577
Iqbal AA, Sakib N, Iqbal AKMP, Nuruzzaman DM (2020) Graphene-based nanocomposites and their fabrication, mechanical properties and applications. Materialia 12:100815. http://doi.org/10.1016/j.mtla.2020.100815
Kausar A (2018) Applications of polymer/graphene nanocomposite membranes: a review. Mater Res Innov 8917:1–12. https://doi.org/10.1080/14328917.2018.1456636
Kaushal S, Kaur N, Kaur M, Singh PP (2020) Dual-responsive pectin/graphene oxide (Pc/GO) nanocomposite as an efficient adsorbent for Cr(III) ions and photocatalyst for degradation of organic dyes in waste water. J Photochem Photobiol A Chem 403:112841. http://doi.org/10.1016/j.jphotochem.2020.112841

Khan S, Malik A (2014) Environmental and health effects of textile industry wastewater. Springer, Netherlands, pp 55–71. http://doi.org/10.1007/978-94-007-7890-0

Khan M, Al-Marri AH, Khan M, Shaik MR et al (2015) Green approach for the effective reduction of graphene oxide using Salvadora persica L. root (Miswak) extract. Nanoscale Res Lett 10:1–9. https://doi.org/10.1186/s11671-015-0987-z

Khanam PN, Ponnamma D, AL-Madeed MA (2015) Electrical properties of graphene polymer nanocomposites. In: Graphene-based polymer nanocomposites in electronics. Springer, Cham, pp 25–47. http://doi.org/10.1007/978-3-319-13875-6

Khanra P, Kuila T, Hoon N, Hyeong S, Yu D, Hee J (2012) Simultaneous bio-functionalization and reduction of graphene oxide by baker's yeast. Chem Eng J 183:526–533. https://doi.org/10.1016/j.cej.2011.12.075

Kim JH, Chang WS, Kim D, Yang JR, Han JT (2014) 3D printing of reduced graphene oxide nanowires. Adv Mater 27:157–161. https://doi.org/10.1002/adma.201404380

Kuila T, Bose S, Khanra P, Kumar A (2011) A green approach for the reduction of graphene oxide by wild carrot root. Carbon 50:914–921. https://doi.org/10.1016/j.carbon.2011.09.053

Lee SL, Chang C (2019) Recent developments about conductive polymer based composite photocatalysts. Polymers 11:206. http://doi.org/10.3390/polym11020206

Li H, Wu S, Wu J, Huang G (2013) A facile approach to the fabrication of graphene-based nanocomposites by latex mixing and in situ reduction. Colloid Polym Sci 291:2279–2287. https://doi.org/10.1007/s00396-013-2959-0

Li Y, Tang J, Liu Y, Li T, Ma D, Gao J, Yang J, Zhou Y, Zhang Y (2019a) Microwave assisted polymeric modification of graphite oxide and graphite by poly (allyl diazoacetate-co-acrolein). Mater Des 183:108116. http://doi.org/10.1016/j.matdes.2019.108116

Li J, Zhu W, Zhang S, Gao Q, Li J, Zhangc W (2019b) Amine-terminated hyperbranched polyamide covalent functionalized graphene oxide-reinforced epoxy nanocomposites with enhanced toughness and mechanical properties. Polym Test 76:232–244. https://doi.org/10.1016/j.polymertesting.2019.03.017

Long G, Tang C, Wong K, Man C, Fan M et al (2013) Resolving the dilemma of gaining conductivity but losing environmental friendliness in producing polystyrene/graphene composites via optimizing the matrix-filler structure. Green Chem 15:821–828. http://doi.org/10.1039/c3gc37042a

Manafi M, Manafi P, Agarwal S, Bharti AK, Asif M, Gupta VK (2016) Synthesis of nanocomposites from polyacrylamide and graphene oxide: application as flocculants for water purification. J Colloid Interface Sci 490:505–510. https://doi.org/10.1016/j.jcis.2016.11.096

Mao H, Wang X (2020) Use of in-situ polymerization in the preparation of graphene/polymer nanocomposites. New Carbon Mater 35:336–343. https://doi.org/10.1016/S1872-5805(20)60493-0

Maryani Y, Kustiningsih I (2015) Determination and characterization of photocatalytic products of linear alkyl sulphonate by high performance liquid chromatography and nuclear magnetic resonance. Procedia Chem 17:216–223. https://doi.org/10.1016/j.proche.2015.12.133

Melinte V, Stroea L, Chibac-scutaru AL (2019) Polymer nanocomposites for photocatalytic applications. Catalysts 9:986. https://doi.org/10.3390/catal9120986

Mitra M, Ahamed ST, Ghosh A, Mondal A, Kargupta K, Ganguly S, Banerjee D (2019) Polyaniline/reduced graphene oxide composite-enhanced visible-light-driven photocatalytic activity for the degradation of organic dyes. ACS Omega 4:1623–1635. https://doi.org/10.1021/acsomega.8b02941

Mittal G, Dhand V, Yop K, Park S, Ro W (2015) A review on carbon nanotubes and graphene as fillers in reinforced polymer nanocomposites. J Ind Eng Chem 21:11–25. https://doi.org/10.1016/j.jiec.2014.03.022

Moon Y, Jung G, Yun J, Kim H (2013) Poly(vinyl alcohol)/poly (acrylic acid)/TiO_2/graphene oxide nanocomposite hydrogels for pH-sensitive photocatalytic degradation of organic pollutants. Mater Sci Eng B 178:1097–1103. https://doi.org/10.1016/j.mseb.2013.07.002

Muralidharan MN, Mathew S, Seema A, Radhakrishnan P, Kurian T (2016) Optical limiting properties of in situ reduced graphene oxide/polymer nanocomposites. Mater Chem Phys 171:367–373. https://doi.org/10.1016/j.matchemphys.2016.01.030

Murugan AV, Muraliganth T, Manthiram A (2009) Rapid, facile microwave-solvothermal synthesis of graphene nanosheets and their polyaniline nanocomposites for energy storage. Am Chem Soc 21:5004–5006. https://doi.org/10.1021/cm902413c

Muthoosamy K, Bai RG, Babangida I, Sudheer SM et al (2015) Exceedingly biocompatible and thin-layered reduced graphene oxide nanosheets using an eco-friendly mushroom extract strategy. Int J Nanomed 10:1505–1519. https://doi.org/10.2147/IJN.S75213

Naeem M et al (2021) Epoxy/graphene nanocomposites prepared by in-situ microwaving. Carbon 177:271–281. https://doi.org/10.1016/j.carbon.2021.02.059

Nazir MS, Tahir Z, Hassan S, Ali Z, Akhtar MN, Azam K, Abdullah MA (2020) Remediation of pesticide in water. In: Sustainable agriculture reviews, vol 47. Springer, Cham. pp 271–309. http://doi.org/10.1007/978-3-030-54712-7_8

Noreen H, Iqbal J, Hassan W, Rahman G, Yaseen M, Ur A (2021a) Synthesis of graphene nanoplatelets/polythiophene nanocomposites with enhanced photocatalytic degradation of bromophenol blue and antibacterial properties. Mater Res Bull 142:111435. http://doi.org/10.1016/j.materresbull.2021.111435

Noreen S, Tahira M, Ghamkhar M, Hafiz I et al (2021b) Treatment of textile wastewater containing acid dye using novel polymeric graphene oxide nanocomposites (GO/PAN, GO/PPy, GO/PSty). J Mater Res Technol 14:25–35. https://doi.org/10.1016/j.jmrt.2021.06.007

Pandey N, Shukla SK, Singh NB, Pandey N, Shukla SK, Singh NB (2017) Water purification by polymer nanocomposites: an overview. Nanocomposites 3:47–66. https://doi.org/10.1080/20550324.2017.1329983

Pang H, Chen T, Zhang G, Zeng B, Li Z (2010) An electrically conducting polymer/graphene composite with a very low percolation threshold. Mater Lett 64:2226–2229. https://doi.org/10.1016/j.matlet.2010.07.001

Papageorgiou DG, Kinloch IA, Young RJ (2017) Mechanical properties of graphene and graphene-based nanocomposites. Prog Mater Sci 90:75–127. https://doi.org/10.1016/j.pmatsci.2017.07.004

Paszkiewicz S, Szymczyk A (2019) Graphene-based nanomaterials and their polymer nanocomposites. In: Nanomaterials and polymer nanocomposites. Elsevier, pp 177–216. http://doi.org/10.1016/B978-0-12-814615-6.00006-0

Petrovski A, Paunovic P, Avolio R, Errico ME, Cocca M, Gentile G, Grozdanov A, Avella M, Barton J, Dimitrov A (2017) Synthesis and characterization of nanocomposites based on PANI and carbon nanostructures prepared by electropolymerization. Mater Chem Phys 185:83–90. http://doi.org/10.1016/j.matchemphys.2016.10.008

Pokharel P, Truong Q, Lee DS (2014) Multi-step microwave reduction of graphite oxide and its use in the formation of electrically conductive graphene/epoxy composites. Compos Part B Eng 64:187–193. https://doi.org/10.1016/j.compositesb.2014.04.013

Potbhare AK, Chaudhary RG, Chouke PB et al (2019) Phytosynthesis of nearly monodisperse CuO nanospheres using *Phyllanthus reticulatus*/*Conyza bonariensis* and its antioxidant/antibacterial assays. Mater Sci Eng C 99:783–793 (2019). http://doi.org/10.1016/j.msec.2019.02.010

Potts JR, Dreyer DR, Bielawski CW, Ruoff RS (2011) Graphene-based polymer nanocomposites. Polymer 52:5–25. https://doi.org/10.1016/j.polymer.2010.11.042

Reddy KR, Karthik KV, Prasad SBB, Soni SK, Jeong HM, Raghu AV (2016) Enhanced photocatalytic activity of nanostructured titanium dioxide/polyaniline hybrid photocatalysts. Polyhedron 120:169–174. https://doi.org/10.1016/j.poly.2016.08.029

Roopan SM, Madhumitha G (2015) Green synthesis of polymer composites/nanocomposites using vegetable oil polymer composites. In: Eco-friendly polymer nanocomposites. Springer, New Delhi, pp 495–511. http://doi.org/10.1007/978-81-322-2470-9

Saleh A, Parthasarathy P, Irfan M (2019) Advanced functional polymer nanocomposites and their use in water ultra-purification. Trends Environ Anal Chem 24. http://doi.org/10.1016/j.teac.2019.e00067

Saleh IA, Zouari N, Al-ghouti MA (2020) Removal of pesticides from water and wastewater: chemical, physical and biological treatment approaches. Environ Technol Innov 19:101026. http://doi.org/10.1016/j.eti.2020.101026

Sharma B, Malik P, Jain P (2018) Biopolymer reinforced nanocomposites: a comprehensive review. Mater Today Commun 16:353–363. https://doi.org/10.1016/j.mtcomm.2018.07.004

Shen B, Zhai W, Tao M, Lu D, Zheng W (2013) Enhanced interfacial interaction between polycarbonate and thermally reduced graphene induced by melt blending. Compos Sci Technol 86:109–116. https://doi.org/10.1016/j.compscitech.2013.07.007

Shen B, Li Y, Yi D, Zhai W, Wei X, Zheng W (2017) Strong flexible polymer/graphene composite films with 3D saw-tooth folding for enhanced and tunable electromagnetic shielding. Carbon 113:55–62. https://doi.org/10.1016/j.carbon.2016.11.034

Singh NB, Agarwal S (2016) Nanocomposites: an overview. Emerg Mater Res 5:5–43. https://doi.org/10.1680/jemmr.15.00025

Sonkusare VN, Chaudhary RG, Mondal A, Abdala AA, Bhusari G, Mishra RK, Potbhare AK, Juneja HD (2020) Mesoporous octahedron-shaped tricobalt tetraoxide nanoparticles for photocatalytic degradation of toxic dyes. ACS Omega 5:7823–7835. http://doi.org/10.1021/acsomega.9b03998

Steurer P, Wissert R, Thomann R, Mu R (2009) Functionalized graphenes and thermoplastic nanocomposites based upon expanded graphite oxide. Macromol Rapid Commun 30:316–327. https://doi.org/10.1002/marc.200800754

Sun X et al (2020) Recent progress in graphene/polymer nanocomposites. Adv Mater 33:1–28. https://doi.org/10.1002/adma.202001105

Šupová M, Martynková GS, Barabaszová K (2011) Effect of nanofillers dispersion in polymer matrices: a review. 3:1–25. http://doi.org/10.1166/sam.2011.1136

Taylor P, Das TK, Prusty S (2013) Polymer-plastics technology and engineering graphene-based polymer composites and their applications graphene-based polymer composites and their applications. Polym-Plast Technol Eng 52:37–41. https://doi.org/10.1080/03602559.2012.751410

Thanaraj P, Sivarajasekar N, Sivaramakrishnan M, Prakashmaran J, Sivamani S, Koya A (2019) Graphene family materials for the removal of pesticides from water. In: A new generation material graphene: application in water technology. Springer, Berlin, pp 309–327. http://doi.org/10.1007/978-3-319-75484-0_13

Tkalya E (2012) Graphene-based polymer nanocomposites. Chem Eng Chem. http://doi.org/10.6100/IR724486

Tomar R, Abdala AA, Chaudhary RG, Singh NB (2020) Photocatalytic degradation of dyes by nanomaterials. Mater Today Proc 29(4):967–973. https://doi.org/10.1016/j.matpr.2020.04.144

Umekar MS, Bhusari GS, Potbhare AK, Mondal A, Kapgate BP, Desimone MF, Chaudhary RG (2021a) Bioinspired reduced graphene oxide based nanohybrids for photocatalysis and antibacterial applications. Curr Pharm Biotechnol 22(13):1759–1781. https://doi.org/10.2174/1389201022666201231115826

Umekar MS, Bhusari GS, Potbhare AK, Chaudhary RG (2021b) ZnO decorated reduced graphene oxide nanohybrid by *Clerodendrum infortunatum*. Emerg Mater Res 10:75–84. http://doi.org/10.1680/jemmr.19.00175

Vaya D, Surolia PK (2020) Semiconductor based photocatalytic degradation of pesticides: an overview. Environ Technol Innov 20:101128. http://doi.org/10.1016/j.eti.2020.101128

Verdejo R, Bernal MM, Romasanta LJ, Lopez-manchado MA (2011) Graphene filled polymer nanocomposites. J Mater Chem 21:3301–3310. https://doi.org/10.1039/c0jm02708a

Vigneshwaran S, Preethi J, Meenakshi S (2019) Removal of chlorpyrifos, an insecticide using metal free heterogeneous graphitic carbon nitride (g-C3N4) incorporated chitosan as catalyst: photocatalytic and adsorption studies. Int J Biol Macromol 132:289–299. https://doi.org/10.1016/j.ijbiomac.2019.03.071

Wang Y, Shi Z, Yin J (2011) Facile synthesis of soluble graphene via a green reduction of graphene oxide in tea solution and its biocomposites. ACS Appl Mater Interfaces 3:1127–1133. https://doi.org/10.1021/am1012613

Wu H, Lin S, Chen C, Liang W, Liu X, Yang H (2016) A new ZnO/rGO/polyaniline ternary nanocomposite as photocatalyst with improved photocatalytic activity. Mater Res Bull 83:434–441. https://doi.org/10.1016/j.materresbull.2016.06.036

Xie SH, Liu YY, Li JY (2008) Comparison of the effective conductivity between composites reinforced by graphene nanosheets and carbon nanotubes. Appl Phys Lett 92:8–11. https://doi.org/10.1063/1.2949074

Yan DX, Ren PG, Pang H, Fu Q, Bo Yang M, Li ZM (2012) Efficient electromagnetic interference shielding of lightweight graphene/polystyrene composite. J Mater Chem 22:18772–18774. https://doi.org/10.1039/c2jm32692b

Yari H, Mahdavian M, Ramezanzadeh B, Mahmudzadeh M (2021) Progress in organic coatings enhanced outdoor durability of polyurethane nanocomposite coatings with green reduced graphene oxide nanoplatelets. Prog Org Coat 154:106212. http://doi.org/10.1016/j.porgcoat.2021.106212

Yeganeh MM, Charkhloo E, Sobhi HR, Esrafili A, Gholami M (2021) Photocatalytic processes associated with degradation of pesticides in aqueous solutions: systematic review and meta-analysis. Chem Eng J 130081:2021. https://doi.org/10.1016/j.cej.2021.130081

Yin J, Zhu G, Deng B (2016) Graphene oxide (GO) enhanced polyamide (PA) thin-film nanocomposite (TFN) membrane for water purification. Desalination 379:93–101. https://doi.org/10.1016/j.desal.2015.11.001

Yuan B, Yu L, Sheng L, An K, Zhao X (2012) Comparison of electromagnetic interference shielding properties between single-wall carbon nanotube and graphene sheet/polyaniline composites. J Phys D Appl Phys 45:235108. http://doi.org/10.1088/0022-3727/45/23/235108

Zhang M, Li Y, Su Z, Wei G (2015) Recent advances in the synthesis and applications of graphene-polymer nanocomposites. Polym Chem 6:6107–6124. https://doi.org/10.1039/c5py00777a

Zhang R, Ma Y, Lan W, Sameen DE, Ahmed S, Dai J (2021) Ultrasonics-sonochemistry enhanced photocatalytic degradation of organic dyes by ultrasonic-assisted electrospray TiO_2/graphene oxide on polyacrylonitrile/β-cyclodextrin nanofibrous membranes. Ultrason Sonochem 70:105343. http://doi.org/10.1016/j.ultsonch.2020.105343

Zhao X, Zhang Q, Chen D, Lu P (2010) Enhanced mechanical properties of graphene-based poly (vinyl alcohol) composites. Macromolecules 43:2357–2363. http://doi.org/10.1021/ma902862u

Chapter 15
Graphene and Graphene Oxide-Based Nitrogenous Bases Nanocomposites for the Detection and Removal of Selected Heavy Metals Ions from an Aqueous Medium

Pramanand Kumar and Subrata Das

Abstract Water and soil pollution by heavy metals is significant concerning threats to the environment and human health. Heavy metals alter the functions of different organs; kidneys, liver, etc. The detection/removal of heavy metal ions in an aqueous or non-aqueous medium is a significant challenge for researchers. The Gr/GO-based hybrid materials are beneficial for practical application as adsorbents for removal and chemosensor for detecting heavy metals (Fe, Cu, Cr, Mn, and Pb, etc.). The graphene/graphene oxide (Gr/GO) is recently used widely for functionalization/hybridization and practical applications. Due to very long π-conjugation in structure and high surface-to-volume ratio, the Gr/GO is a significant substrate for functionalization by covalent and noncovalent bonding. The organic compounds (nitrogenous bases) are functionalization molecules for designing the significant nanocomposite. The functionalized materials are generally very selective for the removal/detection of heavy metals. The selective removal of heavy metal ions by any Gr/GO-based nanomaterials does not require utterly demineralized water and maintaining water quality and human health. This chapter highlights the experimental evidence favoring Gr/GO-based hybrid nanomaterial will be considered adsorbent for removing heavy metal ions. These hybrid nanomaterials (Gr/GO-organic/inorganic molecules/particles) have fundamental properties with various potent practical applications. Also, we consider the spectroscopic methods and associated techniques-based removal and quantification of heavy metal ions. It is an attempt to consolidate what are works in progress and completed. This chapter will be helpful for the researcher to find out methods and work done in this area.

Keywords Graphene/graphene oxide · Nitrogenous bases · Heavy metals · Nanomaterials · Spectroscopic determination

P. Kumar · S. Das (✉)
Department of Chemistry, National Institute of Technology Patna, Patna, India
e-mail: subratad@nitp.ac.in

S. P. Singh et al. (eds.), *Metal Nanocomposites for Energy and Environmental Applications*, Energy, Environment, and Sustainability,
https://doi.org/10.1007/978-981-16-8599-6_15

15.1 Introduction

15.1.1 Environmental Backgrounds and Associated Problems

The environment is the nature in which we live and interact with the living and nonliving things surrounding us. Air, water, soil, rocks, and living beings are in our surroundings. Each natural resources are too crucial for survival. Each of them must be pure and healthy for livings and even for nonliving for their existence. For example, acid rain damages the coral reef, historical monuments, and aquatic life. It is due to air pollution caused by So_x, NO_x, Cl^-. It degrades environmental qualities; finely harms living beings (Lefohn and Brocksen 1984; Grennfelt et al. 2020). Water is the most valuable natural resource existing in the universe. The significant quality of drinking water is gaining demand with the development of the human community. Pollution caused by heavy metals is severe concerning problem that persists in the environment (water and soil). Some heavy metals (Cu, Cr, Fe, Mn, etc.) are essential for human beings, fauna, flora, and the environment at a certain concentration. When these metals exceed a certain concentration, they have toxic effects on organisms (Krishna et al. 2009; Wu et al. 2016; Saha and Paul 2016). These metals can go through a different trophic level within the food chain by the process of biomagnification, bioaccumulation. Heavy metals cause severe diseases like Cd; Itai-itai, liver and kidney damage, As; carcinogenic, Hg; minamata, Cr(VI); carcinogenic, Pb; mental retardation, etc. Hence, heavy metals detection in an aquatic resource (groundwater, surface water) is significantly required to monitor environmental pollution levels (Kaushik et al. 2009; Shrivastava et al. 2002; Mohankumar et al. 2016; Pradhan and Kumar 2014; Leita et al. 1995; Sfakianakis et al. 2015). There are various sources for heavy metal ions; natural activities (weathering of rock by air, water, and microbes) and anthropogenic activities (fertilizer industries, textile industry, mining operations, paper manufacturing, metal smelting, pesticides, and electroplating). The discharging of heavy metal-containing sewage into streams leads to the bioabsorption and bioaccumulation of heavy metals into living organisms and causing a dangerous health hazard (Pradhan and Kumar 2014; Leita et al. 1995; Sfakianakis et al. 2015; Dwivedi and Vankar 2014; Kumar et al. 2018). Thus, the current scenario requires the removal of heavy metal ions from the ground and surface water streams at very low levels.

Heavy metals/metals are quantified/determined by advanced instrumentation methods like atomic absorption spectroscopy (AAS), inductively coupled plasma optical emission spectrometer (ICP-OES), and inductively coupled plasma mass spectrometer (ICP-MS), which requires much energy and lots of time and money. Therefore, many researchers are attracted toward the chemical (organic compounds) probes to detect metal ions selectively, sensitivity within a short time duration at very low concentrations.

15.1.2 Nanotechnology and Environment

Nanotechnology is the field of science that deals with materials/particles of nanoscale (10^{-9} M) with various kinds of physiochemical properties; magnetic, optical, antimicrobial, fluorescent, antirusting, biological, and medicinal (Christian et al. 2008; Barcikowski et al. 2007; Rivera Gil et al. 2010; Kolhatkar et al. 2013; Hajipour et al. 2012). The nanotechnological products (nanoparticles/nanocomposites) play various significant roles for protecting the environmental deterioration; curing plants diseases (Hajipour et al. 2012), waste/surface water treatments, as well as groundwater treatment (heavy metal ions removal/detection, organic matters removal/degradation as well as antimicrobial) (Hajipour et al. 2012; Qu et al. 2013; Dimapilis et al. 2018), solid waste management (reducing solid through degradation of solid) (Part et al. 2018; Gitipour et al. 2013; Boldrin et al. 2014), strengthening the metallic product after coating, enhance the durability of raw material (Liu et al. 2015; Shi et al. 2009). Nanotechnology is also used for energy production and conservation (solar shell, (Chang and Wu 2013) supercapacitor, (Chang and Wu 2013; El-Gendy et al. 2017) heat exchanger, (Khedkar et al. 2014; Ravi Kumar et al. 2017), etc.), reduces environmental pollution; air pollution (reduce the emission of hazardous gases from exhaust) (Ozin 2015; Sajeevan and Sajith 2013; Shaafi et al. 2015). Therefore, nanotechnology is a broad field of science and technology that supports various industries and infrastructure and the economy of each country; ultimately, the environment benefits. This report focuses on sensing and removing heavy metals from an aqueous environment using recent nanotechnology progress.

Recent developments in nanotechnology have further increased researcher's interest in developing unique materials, which adsorb and provide innovative systems to improve environmental remediation and metal sensing. Ling et al. reported that the magnesium oxide (MgO) nanoparticles (NPs) stabilized on the surface of N-doped biochar (MgO@N-biochar) for adsorption of Pb^{2+} and tetracycline (Ling et al. 2017). A similar type of work was reported in which the polypropylene $PPy^{+}/TiO_2(O^{-})$ composite was discovered with selective adsorption capacity in ascending order of $Zn^{2+} > Pb^{2+} > Cu^{2+}$ in an aqueous medium (Chen et al. 2018). Another composite $CoFe_2O_4–SiO_2$ (CF–S) was prepared by amino-functionalization; it becomes an efficient adsorbent material for removing the heavy metal ions (Mn^{2+}, Cu^{2+}, Pb^{2+}, and Cd^{2+}) in single or mixed metal ions solution (Ren et al. 2017). Saha et al. reported that the synthesized sulfur-functionalized mesoporous carbons, includes C–S, C=S, –COS, and SO_x functional groups are very efficient for the removal of heavy metals (Hg > Pb > Cd > Ni). The functional groups containing 'S' atoms act as electron-donating groups to metal ions (Saha et al. 2016). The graphene gold-nanoparticles (Gr-Au) hybrid was bio-functionalized with hemoglobin (Hb) by immobilizing on the surface of graphene-AuNP composite to fabricate biosensors for the determination of nitrite (NO_2^{2-}) (Jiang et al. 2014). Moreover, many researchers have developed chemosensors and adsorbents for metal ions; alkali earth metal, post-transition metal, transition metal, novel metals, lanthanide, and actinides ions by and scientists

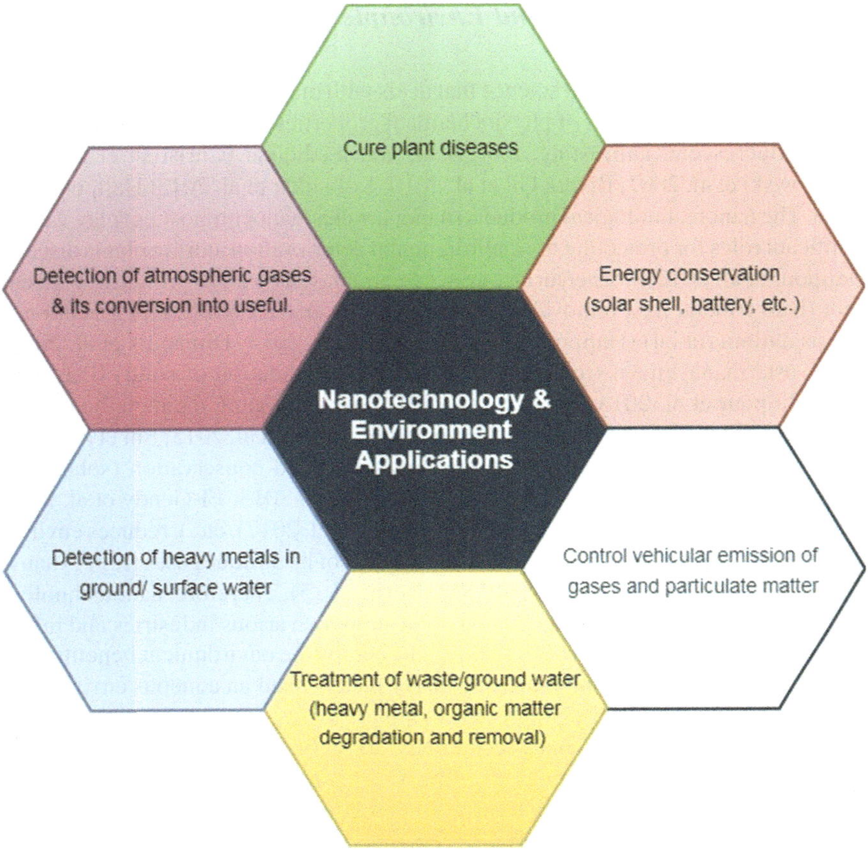

Fig. 15.1 Nanotechnology and various environmental applications

(Jiang et al. 2014; Chang et al. 2014; Jamroz et al. 2019). Various environmental applications of the nanotechnological product are shown in Fig. 15.1.

15.1.3 History and Backgrounds of Graphene/Graphene Oxide and Its Properties

The 'graphene' was first termed in 1986, a flat monolayer carbon sheet, 2D crystal (Geim and Novoselov 2007; Boehm et al. 1986) graphene (Gr) is a 2D, sp^2 hybridized carbon atoms sheet. Its honeycomb-like structure is an elementary building block of other significant allotropes. That may be arranged to form 3D graphite, rolled to produce nanotubes (1D), and wrapped arrangement 0D, known as fullerenes (Allen et al. 2010). In these structures, two or more layers of carbon sheets are arranged

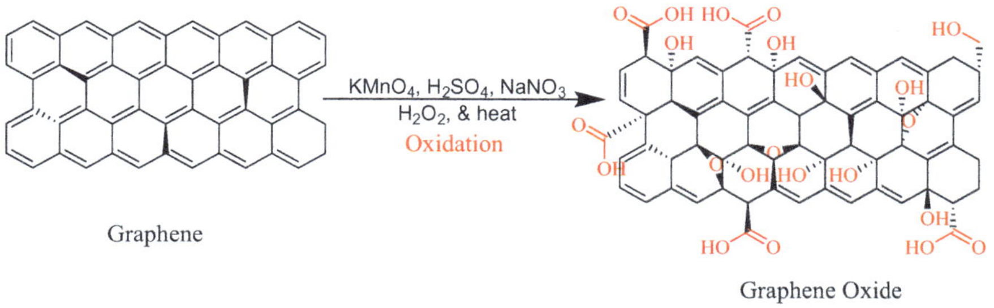

Fig. 15.2 Synthesis of graphene oxide from pristine graphene using modified Hummer's method

in parallel order, similar to graphite. Hence, the carbon sheets that exist as single or independent from each other are described as graphene. The two configurations are generally seen in either armchair or zigzag graphene nanoribbons (GNRs). It shows a variation in their electrical properties; the zigzag configured graphene nanoribbons (GNRs) are metallic, whereas armchairs configured are semiconductor or metallic. The energy bandgap of the armchair is indirectly proportional to GNRs width. The s, p_x, and p_y atomic orbitals of every 'C' atom sp^2 hybridized to form covalent bonds resulted in angles between the C–C–C bond are 120°. The rest p_z orbitals on each carbon atom overlap with its three neighboring electrons to form a band filled with π orbitals, called the valence band, whereas an empty π^* orbitals known as the conduction band. The σ (single) bond formation is carried out with three valence electrons, and the fourth electron forms a π bond (one-third) with its neighboring to produce a carbon–carbon bond in graphene (Boehm et al. 1986; Weiss et al. 2012).

Another hand, graphene oxide is an oxidized form of graphene containing different functional groups C=O (carbonyl), –COOH (carboxylic), OH (hydroxyl), and C–O–C (epoxy), provides sites for different reactions, as shown in Fig. 15.2. The Gr/GO attracted the scientists/researchers due to various extraordinary properties; electrical, mechanical, thermal, optical, drugs carrying into the cell; drug delivery application. These are because of large-scale π-conjugation in the structure. Because of the properties, it has been in the attention of many years for theoretical studies, and recently it has become exciting in experiments. Exceptionally, the Gr/GO has attracted future applications in electronic equipment such as field emitters, transistors, the various component in integrated circuits (IC), sensors, and transparent conducting electrodes. Gr/GO can mobilize the electron (or hole), as well as very low electronic noise (Johnson noise), allowing for practical application in the channels of field-effect transistors (FET). The combined properties, low noise, and excellent electrical make Gr/GO an excellent sensor (Choi 1995; Choi et al. 2010) graphene has vibrant chemistry, can contribute to reactions carried out as a reducer (electron donor) or an oxidizing agent (electron acceptor). The most excellent optical transparency and electrical conductivity encourage the Gr/GO as a material applicable in touch-screens, transparent conducting electrodes, liquid crystal display, organic-based light-emitting diodes (OLEDs), and photovoltaic cells (Sreeprasad et al. 2011;

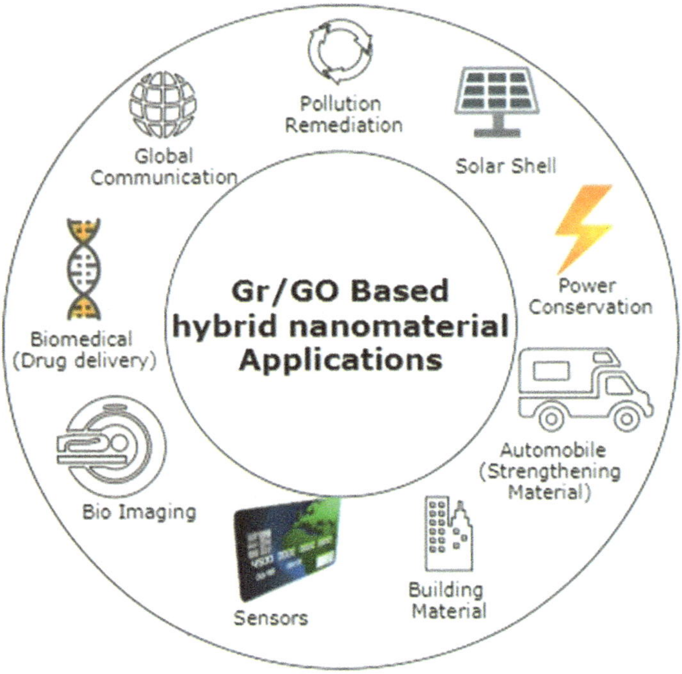

Fig. 15.3 Represents the applications of Gr/GO-based hybrid nanomaterial

Dreyer et al. 2010; Qiao et al. 2019; Atta et al. 2015). Figure 15.3 represents the advanced application of Gr/GO-based hybrid nanomaterials.

15.1.3.1 Synthesis of GO

The graphene oxide is synthesized by oxidizing the natural graphite flakes in the presence of oxidizing agents, known as modified Hummer's method. The natural graphite flakes (1.5 g) and sodium nitrate ($NaNO_3$) (0.75 g) are taken into a beaker/round bottom flask and H_2SO_4 (40 ml) is added to the mixture. The flask is put into an ice bath to maintain the temperature lower than 20 °C, followed by the slow addition of $KMnO_4$ (4.5 g) into the reaction mixture. After that ice bath is removed and maintain the temperature up to 35 °C and continued stirring for 1 h. using a magnetic/mechanical stirrer. Again the temperature is increased to 75 °C and stirred for 3 h. When reaction proceeds the mixture becomes thick. Then distilled water (60 ml) is added and stirred vigorously. The mixture solution is diluted by adding the distilled water. Finally, treated with 30% H_2O_2 till sharp golden-brown color appear. Then filtered the solution, washed with diluted HCl, distilled water, and ethanol to become sulfate and chloride free. The solid content is dried in an oven at about 70–80 °C (Hummers and Offeman 1958).

15.2 Organic Framework Based Functionalized GO/Gr and Its Metal Ions Sensing Application

Functionalization is a process to enhance the properties by combining two or more components in a composite. There are two methods/routes to functionalize the compounds/materials/polymers, i.e., covalent and noncovalent functionalization. The oxidized Gr/CNTs and their equivalents have two active sites, which are significantly favorable for reaction via covalent functionalization. Important sites are the oxygenated groups (–C–O–C and –OH on the basal plane and –COOH at the edge) where new moieties can be an attachment to carry out the esterification, condensation reaction, or reactions directing to oxygen atom containing the lone pair electrons. Another site is the double bond between two carbon atoms on the plane surface of the graphene, which reacts with dienophile or free radical's species (Monteiro et al. 2020). Common reaction methods for covalent functionalization are directing to the all different functional group on graphene oxide as shown in Fig. 15.4: (1) nucleophilic ring-opening reaction (2) acylation reaction (3) cycloaddition reaction (4) diazotization, (5) isocyanate/esterification reaction etc. (Fang et al. 2009).

Noncovalent interaction of molecules is one of the most often applied procedures to prepare graphene nanocomposites. Noncovalent binding between organic compounds and graphene endorses the binding of organic compounds on the surface of graphene. Molecules such as surfactants and polymers can be complexed on graphene surfaces via π-π stacking/electrostatic/hydrophobic interactions, providing

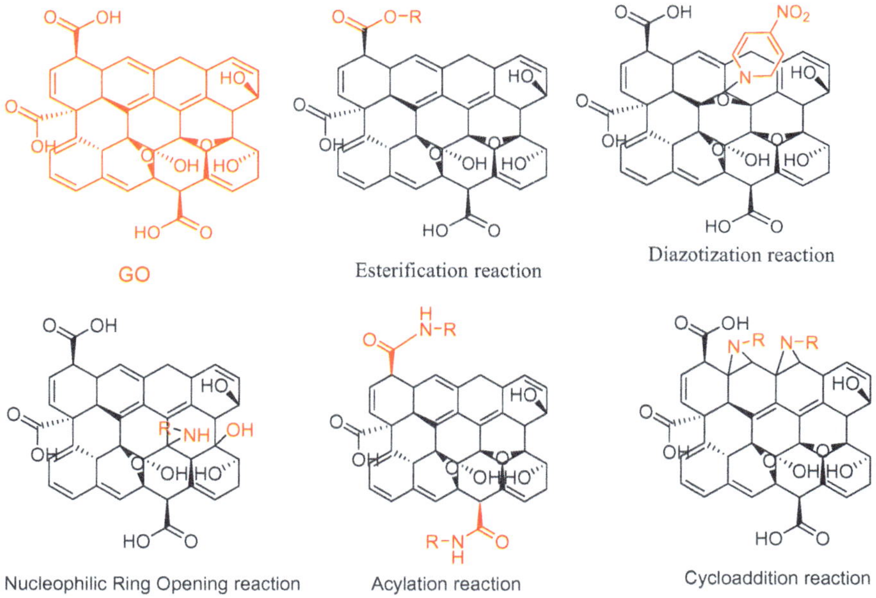

Fig. 15.4 Represents the different mechanisms for covalent functionalization of GO

an effective mechanism to transform graphene surfaces for fabrication of graphene nanocomposites (Fang et al. 2009; Gong et al. 2016). Such methodology is very concerning because it does not interrupt the conjugated sp^2 hybridized carbon network of graphene (Fang et al. 2009; Gong et al. 2016).

Functionalized Gr/GO nanohybrid has various reported applications, as shown in Fig. 15.4. However, in this chapter, we have mainly considered the removal/detection of heavy metal ions. Some recent literature for the detection/removal of heavy metal ions is as follows; Li et al. reported a nafion–graphene nanocomposite as a probe for detecting cadmium (Cd^{2+}) by electrochemical method (Li et al. 2009). Another noble work by Mao et al., the bipyridine functionalized GO as a fluorescent chemosensor for the selective detection of Mn^{2+} in the aqueous medium and a cell (HeLa cell) with very high sensitivity (Mao et al. 2013). This work can also be used for the bio-imaging technique (Gong et al. 2016). Garrido et al. reported the crown [n] ether (aza-9-crown-3 ether) functionalized GO to detect alkali metal ions (Li^{2+}, Na^+ and K^+). Similar work by Wang et al. has been described using crown (Wu et al. 2016) ether to functionalize GO as a sensitive and selective probe to detect Li^{2+}. A similar work reported the derivatives of crown (n) ethers (1-aza-15-crown-5ether, 18 crowns (6) ether) used to functionalize GO, which can detect K^+ (Guo et al. 2014; Olsen et al. 2016; Li et al. 2017).

15.3 Analytical Method for Detection and Removal of Heavy Metals Using Gr/GO Hybrid Nanomaterials

Analytical methods are quantitative and qualitative analysis of any analytes (heavy metal ions, organic compounds) by using different kinds of instruments/methods like spectroscopy [UV–vis spectrophotometry, fluorescence, flame photometry, atomic absorption spectroscopy (AAS), induced coupled plasma spectroscopy (ICP)], chromatography, titrimetric; gravimetric, iodometric, and colorimetry. Spectroscopic methods are the most significantly used techniques and commonly available instruments with high precision and accuracy. In this chapter, the removal and detection of heavy metal ions are being described using spectrophotometric methods.

There are two kinds of procedures to study the removal of heavy metals ions. The first direct measurement of electronic spectra of dyes and heavy metals along with ligand/nanomaterials and another is to remove heavy metal using nanocomposites by filtration then quantify the concentration to find out the C_e (equilibrium concentration). The standard solution of heavy metals is added to the adsorbent materials and shacked. The resulted mixture was filtered at a definite time interval using a syringe filter. The filtrate is finally used to quantify the residual concentration using UV–vis spectrophotometer/fluorescence/AAS/ICP.

15.3.1 UV–Vis Spectroscopy

UV–vis spectrophotometric analysis is simpler to handle, economical, and precise than other electrophoresis, titrimetric, and chromatographic methods. The spectrophotometer is the computer-controlled instrument used to determine the quantitative and qualitative information about a molecule by measuring the light passed through samples. UV–vis spectrophotometer is working on the principle of Lambert–Beer's law, which says that the absorbance of light increases concerning the concentration of compounds. By using the instruments, researchers generally visualize the absorbance band and infer electronic transition level, which occurs during excitation in the presence of UV–vis light. Researchers have recently been attracted to developing a simple method for detecting and removing heavy metal ions and different kinds of organic and inorganic compounds in an aqueous medium by using nanomaterials. Li et al. reported the synthesis of silver nanoflakes and applied them as a colorimetric probe to detect Cr^{3+}. The Cr^{3+} stimulates the changes in the absorbance properties of silver nanoflakes. In the presence of Cr^{3+} silver, nanoflakes get converted into nanoparticles that produce color changes (Li et al. 2019). Nghia et al. described a method for the colorimetric quantification of Cr^{6+} based on graphene oxide (GO) NPs that mimic peroxidase. 3,3,5,5-tetramethylbenzidine (TMB) is oxidized by H_2O_2 in the presence of GO (as a catalyst), gives blue color. In the presence of 8-hydroxyquinoline (8-HQ) color-forming reaction does not occur. Further, upon adding Cr(VI), the blue color appears by engaging TMB with Cr(VI) (Nghia et al. 2019). Kumar et al. has been synthesized the 1,3-dimethyl-5-nitroso-6-aminouracil functionalized rGO (rGO-NO-Ur) hybrid nanomaterial that is useful for the spectrophotometric detection of Cr^{6+} with high precision (Kumar et al. 2020a). Sahraei et al. reported that the beads of the magnetic bio-sorbent hydrogel are prepared by modifying graphene oxide (GO) with polyvinyl alcohol (PVA) and biopolymer gum tragacanth (GT) through the gelation method in the presence of acetone and boric acid. It is an efficient adsorbent for removing positively charged crystal violet (CV) and negatively charged Congo red (CR) dyes and heavy metal ions Pb^{2+} and Cu^{2+} from an aqueous medium. The removal of dyes was studied using UV–vis spectroscopy and heavy metal ions using atomic absorption spectroscopy (AAS) (Sahraei et al. 2017). Le et al. described the synthesis of graphene oxide (GO), polyvinyl alcohol (PVA), and magnetite (Fe_3O_4) (GO/PVA/Fe_3O_4) Nanocomposites based adsorption for removal of cobalt (Co^{2+}) ion. The resulted nanocomposite was applied with a standard solution of Co^{2+} ions, shacked, and filtered at a fixed time interval. The concentration of Co^{2+} is estimated by using NH_4SCN (a ligand) and acetone as a solvent and forms a stable complex (Le et al. 2021).

15.3.2 Fluorescence Spectroscopy

Fluorescence spectroscopy is an essential method for the detection/estimation of heavy metals, organic compounds (dyes, herbicides, antibiotics, biomolecules), and inorganic compounds. This method is based on the fluorescence on/off (quenching/enhancing the fluorescence). There are different mechanisms of fluorescence on/off. The fluorescence quenches in the presence of some specific compound and on in the presence of specific compounds. The fluorescence is the properties of a physiochemical compound that emits/releases energy in light after excitation in the presence of UV–vis light due to π-π* electronic transition (Wang et al. 2017; Nagl et al. 2015). The compound gets excited in the UV–vis light from the ground state, and after that, it releases the excitation energy to come back in the ground state. After adding a specific molecule (target molecule), the site within a molecule gets blocked chromophores due to which the compound is fluorescent, and finally, compounds become fluorescence quenching, the mechanism shown in Fig. 15.5. The fluorescence turns on is based on a mechanism to make accessible the π electron in a compound so that it goes to π-π* electronic transition in the presence of UV–vis light (Romani et al. 2010; Zhu et al. 2012). Many research works showed this mechanism to detect heavy metal ions, organic molecules, etc. Fu et al. reported that the gold nanoparticles (Au-NPs) act as an immune-chromatographic sensor (ICS) for the detection of chromium (Cr^{3+}) based on fluorescence quenching mechanism (Fu et al. 2013). Chauhan et al. described the functionalization of carbon nano-rods (CNR) by the amine (2,2′-(methylenedioxy)-bis(ethylamine)) (EDA) and examined the detection of the Cr^{6+} and Fe^{3+} ions in an aqueous solution based on fluorescence quenching methods (Chauhan et al. 2019). Radhakrishnan et al. developed a new

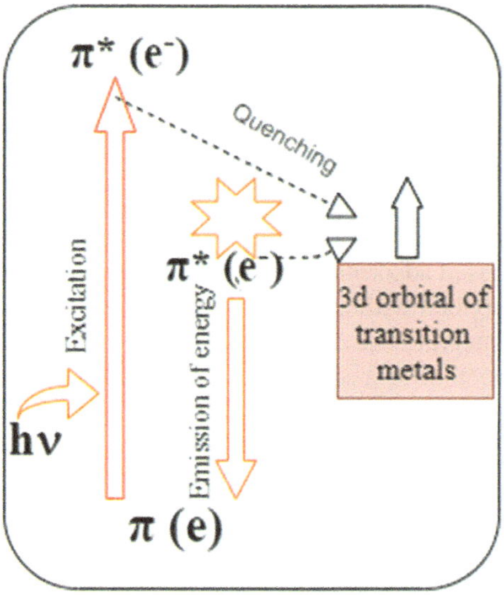

Fig. 15.5 Mechanism of fluorescence quenching (Chauhan et al. 2019)

approach for the green synthesis of carbon dots based on graphitic-carbon nitride ($CDs@g-C_3N_4$) nanohybrid, further potentially applied as a fluorescent "turn-on" sensor for the detection of heavy metal ions (Cu^{2+}, Cr^{6+}, Pb^{2+}). In this work, the CDs are fluorescence material and negatively charged and where $g-C_3N_4$ positively charged. In between these molecules, an electrostatic interaction takes place, which produces a donor–acceptor pair that quench the fluorescent. The 'N' containing compound $g-C_3N_4$ chelates the metal ion at specific pH, and the CDs are become free and get fluorescence on (Radhakrishnan et al. 2020).

15.3.3 Atomic Absorption Spectroscopy

Atomic absorption spectroscopy (AAS) is an analytical instrument/technique for trace/heavy metals estimation (ppb) based on the light absorption of the specific wavelength by the ground state atoms in the flame or graphite furnace. It is famous for the estimation/determination of heavy/trace metals of environmental (soil, ground-water, surface water), biological, geological, mineralogical, and forensic samples. It is a highly sensitive/sophisticated instrument that can measure the trace element per billion (ppb) level with graphite furnace atomization. The removal of heavy metal is generally studied using the AAS technique. Before estimation of heavy metals, it has to be processed to extract/converted into elemental form. For nanomaterial-based heavy metal ions removal, nanomaterials are applied as adsorbent bed material against a standard solution of heavy metal ions and shacked using vertex shaker, filter using 0.45 μm syringe filter at a fixed interval time. Then process the elemental solution for estimation on AAS instruments. Using the result/data, generally find out the reaction order first/the second pseudo-first/pseudo-second-order reaction between the absorbent and metal ions. Some of the work published recently is as follows; a zeolite-NaX nanocomposite was applied for adsorption of Pb(II) (Pandey et al. 2015). Basadi et al. reported the functionalized graphene oxide (GO) with L-cystine (GO@Cystine) as adsorbent for the removal of mercury (Hg) ions from an aqueous medium (Basadi et al. 2021). Perez et al. described a nano-hybrid material consists of graphene oxide (GO), polyethyleneimine (PEI), and chitosan (CS) as an adsorbent for Cr^{6+} and Cu^{2+} removal from aquatic solution. The PEI has numbers of 'N' atoms in their structure. The nitrogen-containing functional group can donate the lone pair electrons. Hence this nanomaterial has very good chelating/adsorbent properties (Nguyen et al. 2006). Musico et al. blended the poly(N-vinyl carbazole) (PVK) with graphene oxide (GO) (PVK–GO), a polymer nanocomposite that can adsorb heavy metal (Pb^{2+}) from an aqueous medium. Graphene, as well as poly(N-vinyl carbazole), have very large double bond conjugation. In this work, the metal ions interact to double bond by π-metal interaction mechanism (Musico et al. 2013). These all works are principally based on the methods mentioned above. However, the AAS is not a common instrument, costly, sophisticated, so we generally focus on UV–vis spectroscopy to study the removal and estimation of heavy metals.

15.3.4 Induced Coupled Plasma Spectroscopy

Induced coupled plasma spectroscopy (ICPS) is a powerful instrument based on optical emission spectrometry. When an analytical sample comes into the exposure of plasma energy as external source energy, the elemental component (atoms) gets excited. The excited atoms come back to the ground state (energy position), which releases the rays are called emissions. The emitted rays (specific photon wavelength) are measured. The position of the photon rays recognizes the types of elements, and the amounts of each element are estimated by measuring the intensity of emitted rays. The process before goes to instrumental analysis, same as atonic absorption spectroscopy, as shown in Fig. 15.6. Janik et al. (2018), described the graphene oxide (GO) modified with amino silanes (3-aminopropyltriethoxysilane (APTES), N-(3-trimethoxysilylpropyl) ethylenediamine (TMSPEDA), and N1-(3-trimethoxysilylpropyl) diethylenetriamine (TMSPDETA)) consist of one, two, or three nitrogen atoms in the molecule for the determination of chromium (VI) (Janik et al. 2018). Baranik et al. (2018) disclosed alumina grafted graphene oxide (Al_2O_3/GO) nanocomposite as a new nano sorbent for As(V) and Cr(III) (Baranik et al. 2018).

15.4 Graphene-Based Nitrogenous Bases Hybrid Nanomaterials and Their Applications

15.4.1 Nitrogenous Bases and Their Coordination Properties

Nitrogenous bases are the major component of nucleic acid [genetic material; deoxyribonucleic acid (DNA), ribonucleic acid (RNA)]. It is classified into two parts, i.e., purines and pyrimidines. The purine is adenine and guanine (double-ring structure), and pyrimidines are thymine, cytosine, and uracil (single-ring structure). The DNA consists of adenine bonded thymine with the help of double bond and guanine with cytosine via a triple bond. However, in the case of RNA, the thymine is being replaced by uracil. The sequence of these nitrogenous bases is important for a living being. Each organism of the universe has the same nitrogenous base, but organisms are different due to their specific sequencing (Ouellette and Rawn 2018a; Sharma et al. 2014).

The nitrogenous bases are the most common aromatic heterocyclic organic compound. Their aromatic structure contains the 'N' and the 'C' atom in the ring, known as a heterocyclic compound. The pyrimidine is the six-membered aromatic ring with a common structure consisting of 4 carbons and two nitrogen atoms in a ring, whereas purine has a pyrimidine ring fused with a five-membered ring (imidazole) at the 5th and 6th position of the pyrimidine ring, shown in Fig. 15.7. The 1,3-pyrimidine has electron (e−) deficiency due to electronegative N atoms in the ring. Due to the presence of these nitrogen atoms, it behaves like an electron donor to

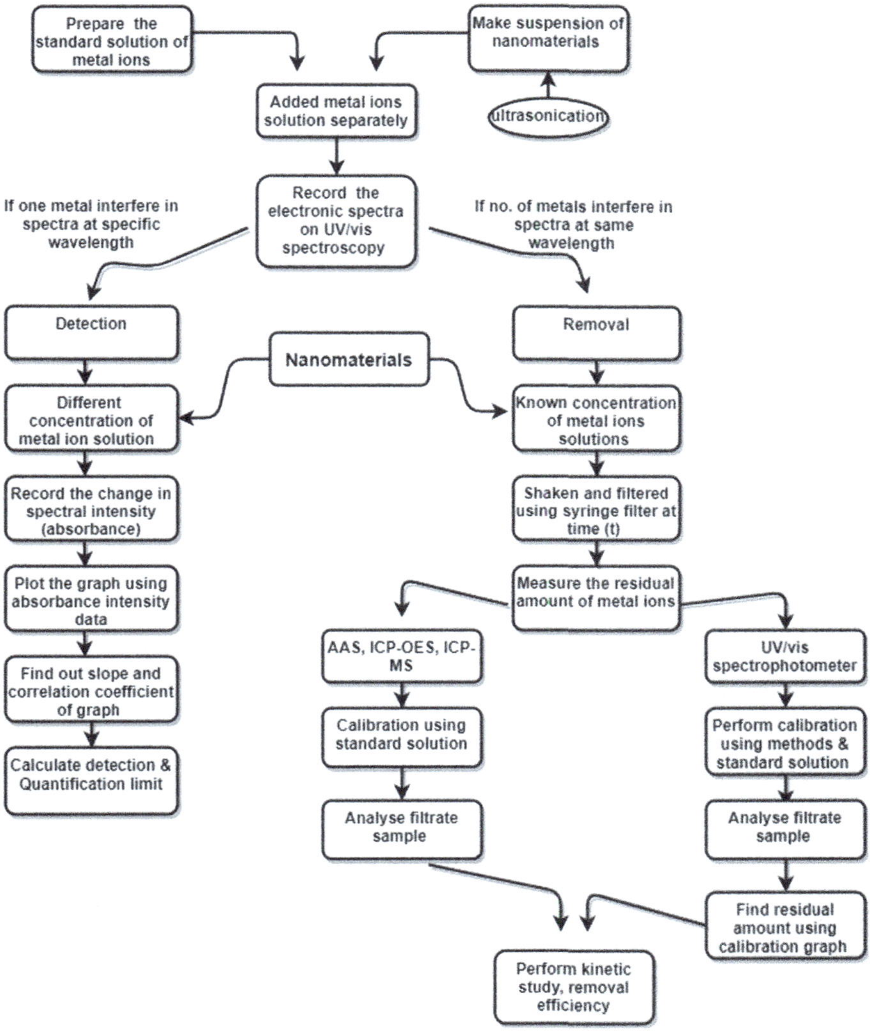

Fig. 15.6 The flowchart diagram of removal and detection of metal ions

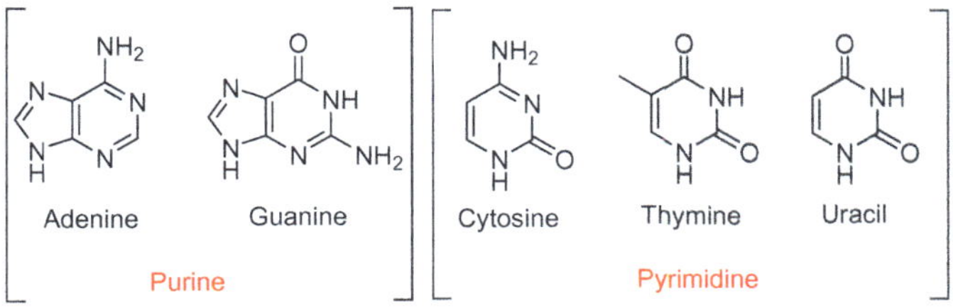

Fig. 15.7 Structures of nitrogenous bases

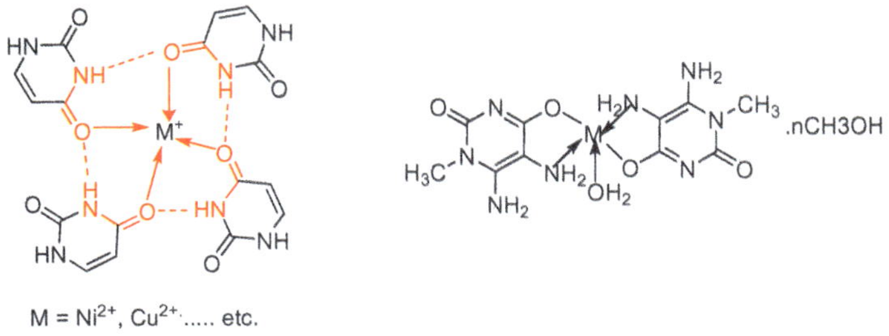

Fig. 15.8 The coordination mechanism of nitrogenous bases (pyrimidine)

electron-deficient molecules. Also, the electron-donating species occur outside the ring structure as C=O and NH_2. So these compound bases behave like ligands to make chelate with the metals ion as coordination compound as shown in Fig. 15.8 (Décout 2020; Richter et al. 1960; Lecca and Ceruti 2008; Taylor et al. 1959; Ouellette et al. 2018b; Mohana et al. 2013).

15.4.2 Nitrogenous Bases Functionalized Gr/GO for the Detection and Removal of Heavy Metal Ions and Their Advanced Applications

The Gr/GO/rGO has different physicochemical properties, and the nitrogenous base has different properties, as mentioned in the above paragraph. After functionalizing the properties of Gr/GO/rGO with nitrogenous base, the electrochemical properties of these compounds have been improved significantly and have an advanced application like; supercapacitor, metal detector, adsorbent materials, etc. Some of the literature is related to advance applications as follows. Liu et al. described the functionalization of GO with adenine via supramolecular interaction. It is based on the noncovalent functionalization of Gr with adenine-bridged aromatic phthalonitrile (AAPN-G). In this work, a new strategy has been developed to prepare functionalized graphene for designing multifunctional materials (Liu et al. 2020). Similar work was reported by Singh et al. in (2012); through the amidation reaction, the oxidized multi-walled carbon nanotubes (MWCNTs) had been functionalized with adenine (Ad-MWCNTs). Uracil substituted ferrocene molecules were complexed with Ad-MWCNTs. The electrochemical behavior/properties of synthesized novel hybrid nanomaterials were investigated using cyclic voltammetry (CV). The auspicious supramolecular interaction occurs between electroactive molecules. Adenine-uracil base pairing in the Ur-Ad-MWCNTs functionalized nanomaterials are very efficient for designing electronic devices (Singh et al. 2012).

Supercapacitors are energy storage devices, and it has attracted researchers due to its very high-power storage capacity, long life cycle compared to batteries. It consists of two electrodes in an electrolyte's medium, in which the ion flows through electrodes. Many conductive materials have been developed based on natural products like cellulose, (Wang et al. 2018) pectin, (Zhou et al. 2020) lignin (Park et al. 2019; Ghosh et al. 2019). The Gr/GO/rGO based hybrid nanomaterial is also used supercapacitors as GO-MOFs derived rGO coating/sandwiching Co_3O_4 composites, (Yin et al. 2016) 'N' doped rGO-poly(p-phenylenediamine), (Wang et al. 2020) an 'N'-doped graphene oxide-pitch oxide, etc. Some nitrogenous base based GO hybrid materials are also used as supercapacitors; El-Gendy et al. have reported adenine-functionalized spongy graphene (SFG) (El-Gendy et al. 2017) and an adenine-functionalized spongy graphene/silver (Ad-FG/Ag) aqueous suspension of nanocomposites, (El-Gendy et al. 2019) an 'N'-doped porous carbon-UiO-66 with different ratios of adenine and 1,4-benzendicarboxylate (H2BDC), act as a supercapacitor (Li et al. 2018). Cao et al. also reported the adenine-modified reduced graphene oxide (AMG) nanocomposite for supercapacitor applications (Cao et al. 2012). The electrochemical reduction of CO_2 to CH_4 was reported by Alinajafi et al. in (2018). In the study, the reduced graphene oxide (rGO) has been modified with adenine through diazonium reaction, and Pt has been deposited to form Pt@Adenine-rGO hybrid nanomaterial by cyclic voltammetry at a specific current range (Alinajafi et al. 2018). The oxidized single-walled carbon nanotubes were functionalized with three nitrogenous bases (uracil, guanine, thymine) and one amino acid (L-alanine). The resulted functionalized nanocomposites have gained excellent adsorption capacity for biomolecules (Silambarasan et al. 2014). Luo et al. have reported microwave-assisted modification of graphene quantum dots with adenine-(A-GQDs), having fluorescence properties, biocompatible for cell imaging, as well as white light-activated antibacterial property. Because of their photoluminescence property and easy biological conjugation, the A-GQDs can be used for multifunctional biomedical applications (Luo et al. 2018).

The nitrogenous bases are naturally abundant biomolecules and having excellent ligating properties due presence of electron donor atoms such as 'N' 'O' and 'S' in the ring or outside the ring. Different auxochrome and chromophore groups are attached in nitrogenous bases and their derivatives provide a large variety of bonding behavior. As a result, several hydrogen bonds, donor and acceptor sites, and other noncovalent bonds increase on the purine/pyrimidine ring. The nitrogen, oxygen, sulfur, halogens atoms containing functional groups help functionalize the Gr/GO/Gr nanotube/Gr quantum dots. After functionalization of Gr/GO and their equivalents, the ligating properties have been changed, as reported in some literature. Nitrogenous bases were used to functionalize GO; resulted material was applied to detect the metal ions.

15.4.2.1 Mercury

Mercury is a silver color metal (liquid), M.P. is 38.7 °C, occurs in the environment comes from both natural and anthropogenic sources. Mercury generally has three forms in terms of valency; Hg^0 (elemental), mercurous (Hg^+) (inorganic), and mercuric cations (Hg^{2+}), as well as organic mercury such as methylmercury, are common in the food chain. The toxicity of mercury depends on its solubility. The Hg^+

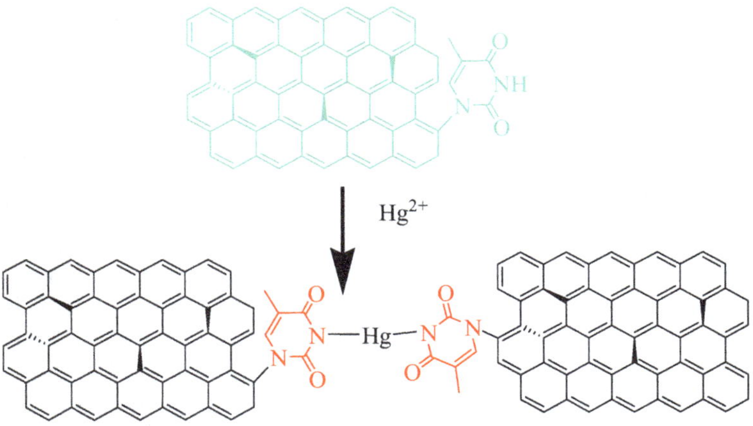

Fig. 15.9 Shows the mechanism to detect the Hg^{2+} ion by thymine functionalized reduced graphene oxide (rGO-T) (Dinda et al. 2015)

is less soluble, so it is less toxic than Hg^{2+} ions. It can be written as toxicity is directly proportional to the solubility of mercury. $HgCl_2$ is very toxic has LD_{50} 0.5 g only. It has severe toxic effects like neurological disorder, mercurous poisoning, redness of the body parts ('Pink disease') with painful extremities (acrodynia). It interacts with the thiol group present in amino acids and proteins, alters the confirmation and functions (Caito and Aschner 2015; Lucchini et al. 2015; Clarkson and Magos 2006; Langford and Ferner 1999). By considering the toxicity of mercury, people want to measure the amount of mercury in drinking water, atmospheric concentration, and finally remove the mercury from natural resources. Some of the researchers worked to detect the amount by using nitrogenous base modified GO Nanocomposites and others nanocomposites, some are mentioned as; Dinda et al. reported the detection of Hg^{2+} and I^- based on thymine functionalized rGO (rGO-T) in an aqueous medium using fluorescence turn off/on the mechanism, as shown in Fig. 15.9. Found interaction rGO-T to Hg^{2+}; form a dimer of rGO-T–Hg-T-rGO and quenched (turned off) the fluorescent activity of rGO-T nanocomposite followed by addition of I^- salt to rGO-T–Hg-T-rGO, the HgI_2 formed, and the fluorescence turned on again (Dinda et al. 2015). Similarly, Wang et al. informed about functionalized GO with AuNPs, cystamine, and thymine (nitrogenous base) as very selective and sensitive sensors against Hg^{2+}, and confirmed the reusability by washing the probes with EDTA (Wang et al. 2016). Thymine was modified with graphene quantum dots (GQDs) and zinc phthalocyanine (ZnPc) separately to obtain nanocomposite of T-GQDs and T-ZnPc, and act as fluorescent probes for detection of Hg^{2+} in an aqueous medium. In this work, T-GQDs are a fluorescent material. When it interacts with an Hg^{2+} ion, it blocks the π electrons; due to this phenomenon, fluorescence turns off. Whereas T-GQDs interacted with T-ZnPc, quenched the fluorescence activity, fluorescence turned on in the presence of Hg^{2+}. The Hg^{2+} ion interacts with T-ZnPC to make a stable complex, and T-GQDs again become pristine, fluorescent turned on (Achadu and Nyokong

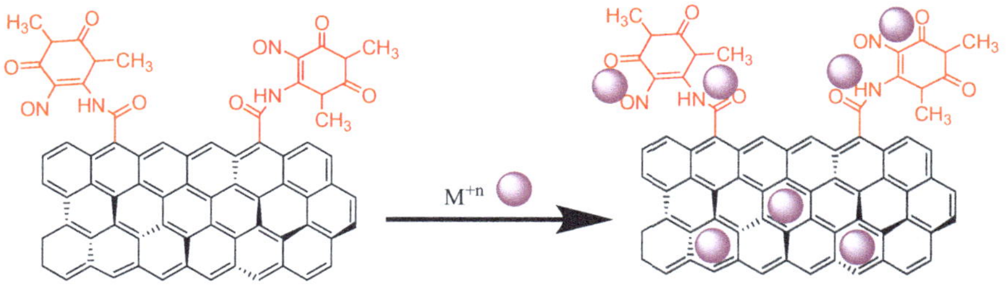

Fig. 15.10 The binding of metal ions on the surface-functionalized rGO/GO nanocomposite

2017). The fluorometric determination of Hg^{2+} was explained via the complexation of thymine and Hg ion $[Hg(T)_2(H_2O)_2]$ on the surface of rGO (Abdelhamid and Wu 2015).

15.4.2.2 Chromium

Chromium (Cr) is an abundant metal present in the earth's crust and persists in the environment. Cr has a wide range of oxidation numbers (I), (III), (IV), (V), (VI). However, Cr(III) and Cr(VI) are stable and, therefore, the most often observed. The hexavalent Cr (Cr(VI)) has a carcinogenic effect declared as group 1 carcinogen to humans by the International Agency for Research on Cancer (IARC) (Barnhart 1997; Costa and Murphy 2019; Zhitkovich 2011).

We recently reported the amino-nitroso-uracil functionalized rGO (rGO-ANU) nanocomposite to detect Ag+ ions in an aqueous medium using UV–vis spectrophotometry (Kumar et al. 2020b). A similar work, 1,3-dimethyl-5-nitroso-6-aminouracil functionalized rGO (rGO-NO-Ur) hybrid nanomaterial, has exquisite sensitivity against Cr^{6+} with high sensitivity precision (Kumar et al. 2020a). The mechanism of metal binding on the surface of functionalized GO nanocomposite via π-metal interaction, described in Fig. 15.10. The procedure to quantify the amount in any aqueous sample has been described in this work. The steps are followed as shown Fig. 15.6. When any metal interacts with nanocomposite, absorption spectra change due to changes in electronic transition under UV–vis light. The various concentration of standard metal ions solution has been added with composite material (concentration constant) and then measure the absorption spectra and get inference about changes occurs. Using the data found after plotting the graphs, calculate the slope of the regression line, as shown in Fig. 15.11. The absorption spectra of ten blank samples have been investigated to determine the standard deviation (SD). Then limit of detection (LOD), and limit of quantification has been calculated using the formula (Patil et al. 2015; Shrivastava and Gupta 2011) as follows Eqs. (15.1) and (15.2);

$$LOD = 3 \times \frac{Slope}{SD} \tag{15.1}$$

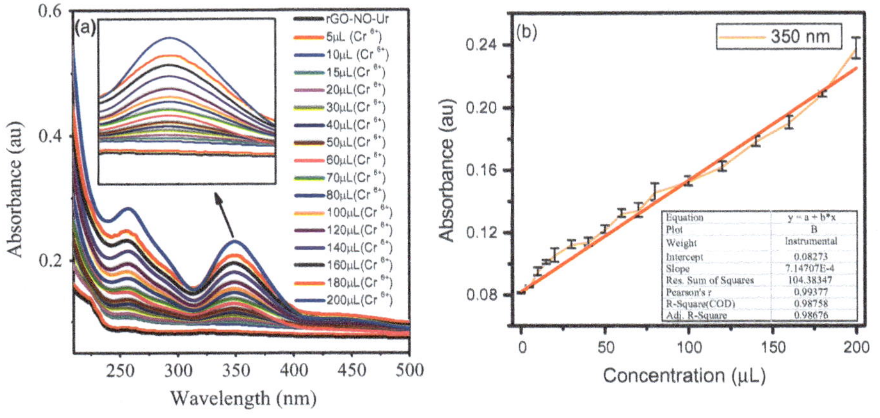

Fig. 15.11 Concentration versus absorbance, **a** spectra of different concentrations of metal ions against composite, **b** regression line plot to find out slope (Kumar et al. 2020a, b)

$$LOQ = 10 \times \frac{slope}{SD} \tag{15.2}$$

15.5 Summary and Future Perspectives

As discussion are made in the chapter, during the recent decades, the noteworthy advancement has been made in the various field of nanotechnology to face the existing environmental issues. The Gr/GO/rGO based nanomaterials are being used to combat environmental problems. The various unique properties of Gr/GO disclosed some new opportunities to improve the number of the process (remediation/estimation/biomedical) for living beings. Even other carbonaceous materials like activated carbon, biochar's, etc., have been used for improvement. However, due to the limited properties of traditional carbon, the researchers attracted the synthesis of Gr/GO-based nanocomposites. Gr/GO is a unique 2D, carbonaceous nanomaterial, which can be extensively used to design different prominent adsorbents, sensors, and catalysts. To enhance the absorptivity, photocatalytic activities, and limitations of graphene as an adsorbent and catalysts for pollutants, a considerable number of Gr/GO-based nanomaterials had been designed/prepared. The adsorption is the process that depends on the properties of synthesized nanomaterials, pollutants (heavy metals, dyes, organic compounds, gases), and the nature of the mediums such as temperature, pH, amount of sorbent material, and coexisting ions via π-π interactions, 'H' bonding, as well as ionic interaction (electrostatic).

The other allotropes of the carbon, such as fullerenes, CNTs, CNRs, are Gr has a similar type of structure. It may show similar activity in specific environmental conditions. The choice to select either Gr/GO or any other carbon nanocomposites is based on availability, cost, and feasibility to prepare. In these aspects, the Gr/GO is the most reliable to use. The pristine graphene is expected theoretically as a highly active material. Still have challenges to reduce GO into pristine graphene to restore the mechanical and electronic properties. The Gr is biocompatible/nontoxic for living beings. So organic compounds functionalized Gr/GO/rGO is extensively used for drug delivery to a cancer cell in vivo. Many works have been reported based on organic frameworks modification of Gr/GO. Their various applications like solar shells, batteries, supercapacitors, biosensors, metal sensors, dye removal/degradation, solid waste reduction/degradation. So, we discussed a specific class of bioactive compounds; nucleic acid (nitrogenous bases) is a class of nitrogen-containing heterocyclic compounds. It is biologically so important as genetic material. These compounds are also chemically crucial because of their active functional group and acts as a ligand. The work is reported as nitrogenous base functionalized GO/Gr hybrid nanomaterials, applied as a supercapacitor, fluorescent probes for detecting heavy metals ions, conversion of gases, and adsorbent material for biomolecules.

This chapter describes environmental issues, especially heavy metals removal/detection as pollutants based on nitrogenous bases modified GO/Gr hybrid nanomaterial from aquatic environment/medium. Gr/GO has unique properties that can significantly lead to development in the field of environmental application. For the last 5–6 years, we have put our best efforts into developing new nitrogenous bases GO/Gr hybrid nanomaterials for the selective detection of heavy metal ions. Although, there are very few reported are available on nitrogenous base functionalized Gr/GO/rGO as an absorbent material for the detection and removal of heavy metals ions. Further involvement is required to work in this direction for the innovations/developing the new materials to face the exiting global environmental challenges and well beings of living beings.

Acknowledgements P. Kumar is expressing sincere gratitude to the NIT Patna, India for financial support as an institutional fellow and providing the infrastructure for research work. Dr. S. Das thanks especially to NIT Patna for providing the research facility and BRNS project grant (54/14/15/2020-BRNS/35054) for financial support.

References

Abdelhamid HN, Wu H-F (2015) Reduced graphene oxide conjugate thymine as a new probe for ultrasensitive and selective fluorometric determination of mercury(II) ions. Microchim Acta 182(9):1609–1617

Achadu OJ, Nyokong T (2017) Application of graphene quantum dots functionalized with thymine and thymine-appended zinc phthalocyanine as novel photoluminescent nanoprobes. New J Chem 41(4):1447–1458

Alinajafi HA, Ensafi AA, Rezaei B (2018) Reduction of carbon dioxide to methanol on the surface of adenine functionalized reduced graphene oxide at a low potential. Int J Hydrogen Energy 43(52):23262–23274

Allen MJ, Tung VC, Kaner RB (2010) Honeycomb carbon: A review of graphene. Chem Rev 110(1):132–145

Atta NF, Galal A, El-Ads EH (2015) Graphene — A Platform for Sensor and Biosensor 13 Applications. In: Biosensors - Micro and Nanoscale Applications. 38–84

Baranik A, Gagor A, Queralt I, Marguí E, Sitko R, Zawisza B (2018) Determination and speciation of ultratrace arsenic and chromium species using aluminium oxide supported on graphene oxide. Talanta 185:264–274

Barcikowski S, Hahn A, Kabashin A V, Chichkov BN (2007) Properties of nanoparticles generated during femtosecond laser machining in air and water. Appl Phys A 87(1):47–55

Barnhart J (1997) Occurrences, uses, and properties of chromium. Regul Toxicol Pharmacol 7(26):3–7

Basadi N, Ghanemi K, Nikpour Y (2021) l-Cystine-functionalized graphene oxide nanosheets for effective extraction and preconcentration of mercury ions from environmental waters. Chem Pap 75(3):1083–1093

Boehm HP, Setton R, Stumpp E (1986) Nomenclature and terminology of graphite intercalation compounds. Carbon 24(2):241–245

Boldrin A, Hansen SF, Baun A, Hartmann NIB, Astrup TF (2014) Environmental exposure assessment framework for nanoparticles in solid waste. J Nanoparticle Res 16(6):2394

Caito S, Aschner M (2015) Neurotoxicity of metals. In: Lotti M, Bleecker MLBT-H of CN (eds) Occupational Neurology. Elsevier, 169–189

Cao H, Wu X, Yin G, Warner JH (2012) Synthesis of Adenine-Modified Reduced Graphene Oxide Nanosheets. Inorg Chem 51(5):2954–2960

Chang H, Wu H (2013a) Graphene-based nanocomposites: preparation, functionalization, and energy and environmental applications Haixin. Energy Environ Sci 6:3483–3507

Chang H, Wu H (2013b) Graphene-based nanocomposites: Preparation, functionalization, and energy and environmental applications. Energy Environ Sci 6(12):3483–3507

Chang J, Zhou G, Christensen ER, Heideman R, Chen J (2014) Graphene-based sensors for detection of heavy metals in water: A review Chemosensors and Chemoreception. Anal Bioanal Chem 406(16):3957–3975

Chauhan N, Anand SR, Aggarwal R, Kaushik J, Shekhawat SS, Sonker AK, Sonkar SK (2019) Soluble non-toxic carbon nano-rods for the selective sensing of iron(iii) and chromium(vi). New J Chem 43(27):10726–10734

Chen J, Yu M, Wang C, Feng J, Yan W (2018) Insight into the Synergistic Effect on Selective Adsorption for Heavy Metal Ions by a Polypyrrole/TiO2 Composite. Langmuir 34(34):10187–10196

Choi SUS (1995) Enhancing thermal conductivity of fluids with nanoparticles. Am Soc Mech Eng Fluids Eng Div FED 231:99–105

Choi W, Lahiri I, Seelaboyina R, Kang YS (2010) Synthesis of graphene and its applications: A review. Crit Rev Solid State Mater Sci 35(1):52–71

Christian P, Von der Kammer F, Baalousha M, Hofmann T (2008) Nanoparticles: structure, properties, preparation and behaviour in environmental media. Ecotoxicology 17(5):326–343

Clarkson TW, Magos L (2006) The Toxicology of Mercury and Its Chemical Compounds. Crit Rev Toxicol 36(8):609–662

Costa M, Murphy A (2019) Overview of chromium(III) toxicology. In: The Nutritional Biochemistry of Chromium (III), Second Edi. Elsevier B.V., pp 341–359

Décout J-L (2020) Molecular sciences and chemical engineering. Elsevier, Amsterdam

Dimapilis EAS, Hsu C-S, Mendoza RMO, Lu M-C (2018) Zinc oxide nanoparticles for water disinfection. Sustain Environ Res 28(2):47–56

Dinda D, Shaw BK, Saha SK (2015) Thymine Functionalized Graphene Oxide for Fluorescence "Turn-off-on" Sensing of Hg2+ and I– in Aqueous Medium. ACS Appl Mater Interfaces 7(27):14743–14749

Dreyer DR, Park S, Bielawski W, Ruoff RS (2010) The chemistry of graphene oxide. Chem Soc Rev 39:228–240

Dwivedi AK, Vankar PS (2014) Source identification study of heavy metal contamination in the industrial hub of Unnao, India. Environ Monit Assess 186(6):3531–3539

El-Gendy DM, Ghany NAA, El Sherbini EEF, Allam NK (2017) Adenine-functionalized Spongy Graphene for Green and High-Performance Supercapacitors. Sci Rep 7(August 2016):1–10

El-Gendy DM, Abdel Ghany NA, Allam NK (2019) Synergistic effect of silver and adenine on boosting the supercapacitance performance of spongy graphene. J Energy Storage 24:100776

Fang M, Wang K, Lu H, Yang Y, Nutt S (2009) Covalent polymer functionalization of graphene nanosheets and mechanical properties of composites. J Mater Chem 19(38):7098–7105

Fu Q, Tang Y, Shi C, Zhang X, Xiang J, Liu X (2013) A novel fluorescence-quenching immunochromatographic sensor for detection of the heavy metal chromium. Biosens Bioelectron 49:399–402

Geim AK, Novoselov KS (2007) The rise of graphene. Nat Mater 6(3):183–191

Ghosh S, Santhosh R, Jeniffer S, Raghavan V, Jacob G, Nanaji K, Kollu P, Jeong SK, Grace AN (2019) Natural biomass derived hard carbon and activated carbons as electrochemical supercapacitor electrodes. Sci Rep 9(1):16315

Gitipour A, El Badawy A, Arambewela M, Miller B, Scheckel K, Elk M, Ryu H, Gomez-Alvarez V, Santo Domingo J, Thiel S, Tolaymat T (2013) The Impact of Silver Nanoparticles on the Composting of Municipal Solid Waste. Environ Sci Technol 47(24):14385–14393

Gong X, Liu G, Li Y, Yu DYW, Teoh WY (2016) Functionalized-Graphene Composites: Fabrication and Applications in Sustainable Energy and Environment. Chem Mater 28(22):8082–8118

Grennfelt P, Engleryd A, Forsius M, Hov Ø, Rodhe H, Cowling E (2020) Acid rain and air pollution: 50 years of progress in environmental science and policy. Ambio 49(4):849–864

Guo J, Lee J, Contescu CI, Gallego NC, Pantelides ST, Pennycook SJ, Moyer BA, Chisholm MF (2014) Crown ethers in graphene. Nat Commun 5(1):5389

Hajipour MJ, Fromm KM, Akbar Ashkarran A, Jimenez de Aberasturi D, Larramendi IR de, Rojo T, Serpooshan V, Parak WJ, Mahmoudi M (2012) Antibacterial properties of nanoparticles. Trends Biotechnol 30(10):499–511

Hummers WS, Offeman RE (1958) Preparation of Graphitic Oxide. J Am Chem Soc 80(6):1339

Jamroz E, Kocot K, Zawisza B, Talik E, Gagor A, Sitko R (2019) A green analytical method for ultratrace determination of hexavalent chromium ions based on micro-solid phaseextraction using amino-silanized cellulose membranes. Microchem J 149:104060

Janik P, Zawisza B, Talik E, Sitko R (2018) Selective adsorption and determination of hexavalent chromium ions using graphene oxide modified with amino silanes. Microchim Acta 185(2):1–8

Jiang J, Fan W, Du X (2014) Nitrite electrochemical biosensing based on coupled grapheme and gold nanoparticles. Biosens Bioelectron 51:343–348

Kaushik A, Kansal A, Kumari, Santosh, Meena S, Kaushik CP (2009) Heavy metal contamination of river Yamuna , Haryana , India : assessment by metal enrichment factor of the sediments. J Hazard Mater 164:265–270

Khedkar RS, Sonawane SS, Wasewar KL (2014) Heat transfer study on concentric tube heat exchanger using TiO2-water based nanofluid. Int Commun Heat Mass Transf 57:163–169

Kolhatkar AG, Jamison AC, Litvinov D, Willson RC, Lee TR (2013) Tuning the Magnetic Properties of Nanoparticles. Int J Mol Sci 14(8):15977–16009

Krishna AK, Satyanarayanan M, Govil PK (2009) Assessment of heavy metal pollution in water using multivariate statistical techniques in an industrial area: A case study from Patancheru, Medak District, Andhra Pradesh, India. J Hazard Mater 167(1–3):366–373

Kumar A, Sharma S, Mehra R, Kanwar P, Mishra R, Kaur I (2018) Assessment of radon concentration and heavy metal contamination in groundwater of Udhampur district, Jammu & Kashmir, India. Environ Geochem Health 40(2):815–831

Kumar P, Bora C, Das S (2020a) Reduced Graphene Oxide/Amino-Nitroso-Uracil Nanocomposite for the Detection of Heavy Metals in Aqueous Environment. In: 2020 International Conference on Emerging Frontiers in Electrical and Electronic Technologies, ICEFEET 2020
Kumar P, Bora C, Kumar B, Sukul PK, Das S (2020b) Synthesis and characterization of 5-nitroso-6-amino-uracil functionalized reduced graphene oxide for the detection of chromium ion in aqueous medium. Synth Met 264:116381
Langford NJ, Ferner RE (1999) Toxicity of mercury. J Hum Hypertens 13(10):651–656
Le TD, Tran LT, Thi H, Dang M, Thu T, Tran H, Tran HV (2021) Graphene Oxide/Polyvinyl Alcohol/Fe3O4 Nanocomposite: An Efficient Adsorbent for Co (II) Ion Removal. J Anal Methods Chem 1(6):6670913
Lecca D, Ceruti S (2008) Uracil nucleotides: From metabolic intermediates to neuroprotection and neuroinflammation. Biochem Pharmacol 75(10):1869–1881
Lefohn AS, Brocksen RW (1984) Acid Rain Effects Research—A Status Report. J Air Pollut Control Assoc 34(10):1005–1013
Leita L, De Nobili M, Muhlbachova G, Mondini C, Marchiol L, Zerbi G (1995) Bioavailability and effects of heavy metals on soil microbial biomass survival during laboratory incubation. Biol Fertil Soils 19(2):103–108
Li J, Guo S, Zhai Y, Wang E (2009) Nafion–graphene nanocomposite film as enhanced sensing platform for ultrasensitive determination of cadmium. Electrochem commun 11(5):1085–1088
Li J, Yim D, Jang W-D, Yoon J (2017) Recent progress in the design and applications of fluorescence probes containing crown ethers. Chem Soc Rev 46(9):2437–2458
Li H, Fu D, Zhang X-M (2018) A novel adenine-based metal organic framework derived nitrogen-doped nanoporous carbon for flexible solid-state supercapacitor. R Soc Open Sci 5(1):171028
Li X, Zhang S, Dang Y, Liu Z, Zhang Z, Shan D, Zhang X, Wang T, Lu X (2019) Ultratrace Naked-Eye Colorimetric Ratio Assay of Chromium(III) Ion in Aqueous Solution via Stimuli-Responsive Morphological Transformation of Silver Nanoflakes. Anal Chem 17 91(6):4031–4038
Ling L-L, Liu W-J, Zhang S, Jiang H (2017) Magnesium Oxide Embedded Nitrogen Self-Doped Biochar Composites: Fast and High-Efficiency Adsorption of Heavy Metals in an Aqueous Solution. Environ Sci Technol 51(17):10081–10089
Liu Y, Cao H, Chen S, Wang D (2015) Ag Nanoparticle-Loaded Hierarchical Superamphiphobic Surface on an Al Substrate with Enhanced Anticorrosion and Antibacterial Properties. J Phys Chem C 119(45):25449–25456
Liu Y, Li R, Liang B, Li C, Hu J, Zeng K, Yang G (2020) Bio-adenine-bridged molecular design approach toward non-covalent functionalized graphene by liquid-phase exfoliation. J Mater Sci 55(1):140–150
Lucchini RG, Aschner M, Bellinger DC, Caito SW (2015) In: Nordberg GF, Fowler BA (eds) Handbook on the toxicology of metals, 4th edn. Academic Press, San Diego, pp 299–311
Luo Z, Yang D, Yang C, Wu X, Hu Y, Zhang Y, Yuwen L, Yeow EKL, Weng L, Huang W, Wang L (2018) Graphene quantum dots modified with adenine for efficient two-photon bioimaging and white light-activated antibacteria. Appl Surf Sci 434:155–162
Mao X, Su H, Tian D, Li H, Yang R (2013) Bipyrene-Functionalized Graphene as a “Turn-On” Fluorescence Sensor for Manganese(II) Ions in Living cells. ACS Appl Mater Interfaces 5(3):592–597
Mohana KN, Prasanna Kumar BN, Mallesha L (2013) Synthesis and biological activity of some pyrimidine derivatives. Drug Invent Today 5(3):216–222
Mohankumar K, Hariharan V, Rao NP (2016) Heavy metal contamination in groundwater around industrial estate vs residential areas in Coimbatore, India. J Clin Diagnostic Res 10(4):BC05–BC07
Monteiro AR, Neves MGPMS, Trindade T (2020) Functionalization of Graphene Oxide with Porphyrins: Synthetic Routes and Biological Applications. Chempluschem 85(8):1857–1880
Musico YLF, Santos CM, Dalida MLP, Rodrigues DF (2013) Improved removal of lead(ii) from water using a polymer-based graphene oxide nanocomposite. J Mater Chem A 1(11):3789–3796

Nagl A, Hemelaar SR, Schirhagl R (2015) Improving surface and defect center chemistry of fluorescent nanodiamonds for imaging purposes—a review. Anal Bioanal Chem 407(25):7521–7536
Nghia NN, Huy BT, Lee YI (2019) Colorimetric detection of chromium(VI) using grapheme oxide nanoparticles acting as a peroxidase mimetic catalyst and 8-hydroxyquinoline as an inhibitor. Microchim Acta 186(1):1–7
Nguyen C, Ruda GF, Schipani A, Kasinathan G, Leal I, Musso-Buendia A, Kaiser M, Brun R, Ruiz-Pérez LM, Sahlberg B-L, Johansson NG, González-Pacanowska D, Gilbert IH (2006) Acyclic Nucleoside Analogues as Inhibitors of Plasmodium falciparum dUTPase. J Med Chem 49(14):4183–4195
Olsen G, Ulstrup J, Chi Q (2016) Crown-Ether Derived Graphene Hybrid Composite for Membrane-Free Potentiometric Sensing of Alkali Metal Ions. ACS Appl Mater Interfaces 8(1):37–41
Ouellette RJ, Rawn JD (2018a) Nucleosides, Nucleotides, and Nucleic Acids. In: Ouellette RJ, Rawn JDBT-OC (Second E (eds). Academic Press, pp 973–1000
Ouellette RJ, Rawn JD (2018b) Arenes and Aromaticity. In: Ouellette RJ, Rawn JDBT-OC (Second E (eds) Organic Chemistry (Second Edition). Academic Press, pp 353–373
Ozin GA (2015) Throwing New Light on the Reduction of CO2. Adv Mater 27(11):1957–1963
Pandey PK, Sharma SK, Sambi SS (2015) Removal of lead(II) from waste water on zeolite-NaX. J Environ Chem Eng 3(4, Part A):2604–2610
Park JH, Rana HH, Lee JY, Park HS (2019) Renewable flexible supercapacitors based on alllignin-based hydrogel electrolytes and nanofiber electrodes. J Mater Chem A 7(28):16962–16968
Part F, Berge N, Baran P, Stringfellow A, Sun W, Bartelt-Hunt S, Mitrano D, Li L, Hennebert P, Quicker P, Bolyard SC, Huber-Humer M (2018) A review of the fate of engineered nanomaterials in municipal solid waste streams. Waste Manag 75:427–449
Patil SR, Nandre JP, Patil PA, Sahoo SK, Devi M, Pradeep CP, Fabiao Y, Chen L, Redshaw C, Patil UD (2015) A uracil nitroso amine based colorimetric sensor for the detection of Cu2+ ions from aqueous environment and its practical applications. RSC Adv 5(28):21464–21470
Pradhan JK, Kumar S (2014) Informal e-waste recycling: Environmental risk assessment of heavy metal contamination in Mandoli industrial area, Delhi, India. Environ Sci Pollut Res 21(13):7913–7928
Qiao Y, Li X, Hirtz T, Deng G, Wei Y, Li M, Ji S, Wu Q, Jian J, Wu F, Shen Y, Tian H, Yang Y, Ren TL (2019) Graphene-based wearable sensors. Nanoscale 11(41):18923–18945
Qu X, Brame J, Li Q, Alvarez PJJ (2013) Nanotechnology for a Safe and Sustainable Water Supply: Enabling Integrated Water Treatment and Reuse. Acc Chem Res 46(3):834–843
Radhakrishnan K, Sivanesan S, Panneerselvam P (2020) Turn-On fluorescence sensor based detection of heavy metal ion using carbon dots@graphitic-carbon nitride nanocomposite probe. J Photochem Photobiol A Chem 389:112204
Ravi Kumar NT, Bhramara P, Addis BM, Sundar LS, Singh MK, Sousa ACM (2017) Heat transfer, friction factor and effectiveness analysis of Fe3O4/water nanofluid flow in a double pipe heat exchanger with return bend. Int. Commun. Heat Mass Transf. 81:155–163
Ren C, Ding X, Li W, Wu H, Yang H (2017) Highly Efficient Adsorption of Heavy Metals onto Novel Magnetic Porous Composites Modified with Amino Groups. J Chem Eng Data 62(6):1865–1875
Richter E, Loeffler JE, Taylor EC (1960) Studies in Purine Chemistry. VIII. A Convenient Synthesis of Hypoxanthines and Adenines1,2. J Am Chem Soc 82(12):3144–3146
Rivera Gil P, Oberdörster G, Elder A, Puntes V, Parak WJ (2010) Correlating Physico-Chemical with Toxicological Properties of Nanoparticles: The Present and the Future. ACS Nano 4(10):5527–5531
Romani A, Clementi C, Miliani C, Favaro G (2010) Fluorescence Spectroscopy: A Powerful Technique for the Noninvasive Characterization of Artwork. Acc Chem Res 43(6):837–846
Saha D, Barakat S, Van Bramer SE, Nelson KA, Hensley DK, Chen J (2016) Noncompetitive and Competitive Adsorption of Heavy Metals in Sulfur-Functionalized Ordered Mesoporous Carbon. ACS Appl Mater Interfaces 8(49):34132–34142
Saha P, Paul B (2016) Assessment of Heavy Metal Pollution in Water Resources and their Impacts : A Review. J Basic Appl Eng Res 3(8):671–675

Sahraei R, Sekhavat Pour Z, Ghaemy M (2017) Novel magnetic bio-sorbent hydrogel beads based on modified gum tragacanth/graphene oxide: Removal of heavy metals and dyes from water. J Clean Prod 142:2973–2984

Sajeevan AC, Sajith V (2013) Diesel Engine Emission Reduction Using Catalytic Nanoparticles: An Experimental Investigation. J Eng 2013:589382

Sfakianakis DG, Renieri E, Kentouri M, Tsatsakis AM (2015) Effect of heavy metals on fish larvae deformities: A review. Environ Res 137:246–255

Shaafi T, Sairam K, Gopinath A, Kumaresan G, Velraj R (2015) Effect of dispersion of various nanoadditives on the performance and emission characteristics of a CI engine fuelled with diesel, biodiesel and blends—A review. Renew Sustain Energy Rev 49:563–573

Sharma V, Chitranshi N, Agarwal AK (2014) Significance and Biological Importance of Pyrimidine in the Microbial World. Int J Med Chem 2014:202784

Shaw G. (2020) Bicyclic 5-6 Systems—Purines. In: Comprehensive Heterocyclic Chemistry II. Elsevier, pp 397–429

Shi X, Nguyen TA, Suo Z, Liu Y, Avci R (2009) Effect of nanoparticles on the anticorrosion and mechanical properties of epoxy coating. Surf Coatings Technol 204(3):237–245

Shrivastava A, Gupta V (2011) Methods for the determination of limit of detection and limit of quantitation of the analytical methods. Chronicles Young Sci 2(1):21–25

Shrivastava R, Upreti RK, Seth PK, Chaturvedi UC (2002) Effects of chromium on the immune system. FEMS Immunol Med Microbiol 34:1–7

Silambarasan D, Iyakutti K, Vasu V (2014) Functionalization of single-walled carbon nanotubes with uracil, guanine, thymine and l-alanine. Chem Phys Lett 604:83–88

Singh P, Ménard-Moyon C, Kumar J, Fabre B, Verma S, Bianco A (2012) Nucleobase-pairing triggers the self-assembly of uracil-ferrocene on adenine functionalized multi-walled carbon nanotubes. Carbon N Y 50(9):3170–3177

Sreeprasad TS, Maliyekkal MS, Deepti K, Chaudhari K, Xavier PL, Pradeep T (2011) Transparent , luminescent , antibacterial and patternable film forming composites of graphene oxide / reduced graphene oxide. ACS Appl Mater Interfaces 3:2643–2654

Taylor EC, Vogl O, Cheng CC (1959) Studies in Purine Chemistry. II. A Facile Synthesis of 2-Substituted Adenines1,2. J Am Chem Soc 81(10):2442–2448

Wang N, Lin M, Dai H, Ma H (2016) Functionalized gold nanoparticles/reduced grapheme oxide nanocomposites for ultrasensitive electrochemical sensing of mercury ions based on thymine–mercury–thymine structure. Biosens Bioelectron 79:320–326

Wang H, Ji X, Li Z, Huang F (2017) Fluorescent Supramolecular Polymeric Materials. Adv Mater 29(14):1606117

Wang C, Wang X, Lu H, Li H, Zhao XS (2018) Cellulose-derived hierarchical porous carbon for high-performance flexible supercapacitors. Carbon N Y 140:139–147

Wang Y, Wang H, Zhang TC, Yuan S, Liang B (2020) N-doped porous carbon derived from rGO-Incorporated polyphenylenediamine composites for CO2 adsorption and supercapacitors. J Power Sources 472:228610

Weiss NO, Zhou H, Liao L, Liu Y, Jiang S, Huang Y, Duan X (2012) Graphene: An emerging electronic material. Adv Mater 24(43):5782–5825

Wu X, Cobbina SJ, Mao G, Xu H, Zhang Z, Yang L (2016) A review of toxicity and mechanisms of individual and mixtures of heavy metals in the environment. Environ Sci Pollut Res 23(9):8244–8259

Yin D, Huang G, Sun Q, Li Q, Wang X, Yuan D, Wang C, Wang L (2016) RGO/Co3O4 Composites Prepared Using GO-MOFs as Precursor for Advanced Lithium-ion Batteries and Supercapacitors Electrodes. Electrochim Acta 215:410–419

Zhitkovich A (2011) Chromium in drinking water: Sources, metabolism, and cancer risks. Chem Res Toxicol 24(10):1617–1629

Zhou Y, Ren X, Du Y, Jiang Y, Wan J, Ma F (2020) In-situ template cooperated with urea to construct pectin-derived hierarchical porous carbon with optimized pore structure for supercapacitor. Electrochim Acta 355:136801
Zhu S, Tang S, Zhang J, Yang B (2012) Control the size and surface chemistry of graphene for the rising fluorescent materials. Chem Commun 48(38):4527–4539

Chapter 16
Metal Organic Frameworks Based Nanomaterial: Synthesis and Applications; Removal of Heavy Metal Ions from Waste Water

Ravindra Singh, Rama Kanwar Khangarot, Ajay Kumar Singh, and Kamlesh Kumar

Abstract Heavy metals in the water are a global environmental issue. Heavy metals in the wastewater are increasing day by day which is mainly caused by various industrial effluents. Heavy metal ions including, Cadmium (II), Arsenic (III and V), Chromium (III and VI), Copper (II), Lead (II), and Mercury (II) are accumulated readily in the environment. It has been created a lot of serious problem to the human health. Thus, removal of heavy metals from the wastewater is one of the major challenges for the scientific community. Various techniques and materials have been developed for removal of heavy metals from the waste water. Recently, Metal organic frameworks (MOFs) based nanomaterial has been synthesized and used for removal of heavy metal ions from waste water. Electrochemical, photochemical energy conversion and storage, biomedical imaging, drug delivery and catalysis, have been investigated. Its unique characteristic properties are accountable for the waste water treatment like easily synthesizable, various size cavities with different-different functional group, surface functional groups, various functionality where host–guest interaction takes place and high surface area which responsible for high absorption capacity. In this chapter, the attention is given to understand the synthesis, chemistry of MOF based nano-composites and its various applications especially, and removal of heavy metals from waste water has been discussed. It is expected that this chapter can be helpful to understand the synthesis of MOF-based nano-materials and its application towards the elimination of heavy metal ions from waste water.

R. Singh (✉)
Department of Chemistry, Maharani Shri Jaya Government Post-Graduate College, Bharatpur, Rajasthan 321001, India

R. K. Khangarot
Department of Chemistry, University College of Science, Mohanlal Sukhadia University, Udaipur, Rajasthan 313001, India

A. K. Singh
Cenral Revenues Control Laboratory, New Delhi 110012, India

K. Kumar
Department of Chemistry, Institute of Science, BHU, Varanasi 221005, India

S. P. Singh et al. (eds.), *Metal Nanocomposites for Energy and Environmental Applications*, Energy, Environment, and Sustainability,
https://doi.org/10.1007/978-981-16-8599-6_16

Keywords MOF-nano material · Synthesis · Applications · Removal of heavy metals

16.1 Introduction

Last few decades, energy and environmental problem, especially heavy metals pollutants have become a serious environmental issue. The Toxic heavy metal ions are increasing in the environment mainly due to industrial processes such as chemical synthesis, metallurgy, mining, electroplating, agriculture and household wastewater. The concentration of heavy metal ions such as cadmium (II), Arsenic (III and V), Chromium (III and VI), Copper (II), Lead (II), and mercury (II) became higher than the permissible limit in the water, shown adverse effect on human health these toxic metals are responsible for lungs, kidney, brain, liver and central nervous system related diseases. As heavy metals are highly toxic and non-biodegradable, are released into the environment by microorganisms. The heavy metals accumulated through the food chain, exerted adverse effects on the environment biological processes and ecological systems.

In recent years, the energy requirement of world has been increased at exponential rates. The fossil fuels are the main source of world energy. However, fossil fuels are being depleted very fast, which have generated environmental issues, such as emission of large amount of CO_2, CO, SO_2, NO_x, CH_4, volatile organic compounds and particulates. In the past several years, electrochemical energy storage/conversion systems based on photocatalysis and electrocatalysis are being used as power supplies for electric vehicles and handy electronics due to produce green and renewable energy. Metal–air batteries, solar cell and H_2–O_2 fuel cells do not release any harmful gases during discharge. Consequently, major efforts have been made to the development renewable energy storage, including rechargeable batteries, supercapacitors, solar cell, fuel cell, wind and hydropower. In this regard, many MOF based nanomaterials have been synthesized and explored for their applications in environmental issues specially elimination of heavy metals from waste water and field of energy. Metal–organic frameworks, a class of crystalline porous materials have synthesized by the self-assembly process of metal ions or metal clusters and organic ligands through covalent coordination linkages. Electrostatic interaction, hydrophobic interaction, acid–base interaction, H-bonding and π–π stacking are responsible for the construction of two/three dimensional (3D) structures of MOFs. Recently, Metal–organic frameworks (MOFs), have received marvelous interest and achieved remarkable developments in the various applications (Long and Yaghi (2009), Ding et al. (2019), Furukawa et al. (2013), Huang et al. (2017a), Li et al. (2018a), Zhou et al. (2012), Jiao et al. (2019), Li et al. (2016a, 2019a), Liang et al. (2019) and Zhao et al. (2019)), in the field of removal of heavy metals from waste water, catalysis (Wu et al. 2016; Corma et al. 2010; Li et al. 2018b; Jiao et al. 2018; Liu et al. 2014; Luo et al. 2019; Huang et al. 2017b; Zhao et al. 2014; Karmakar and Pombeiro 2019; Yang et al. 2017; Liao et al. 2018; Jiao and Jiang 2019; Kaneti et al. 2017), biomedicine

(Lu et al. 2018; Dutta et al. 2019; Wu and Yang 2017; Cai et al. 2015; Lázaro and Forgan 2019; Shieh et al. 2015; Lismont et al. 2017), sensing (He et al. 2019; Qin et al. 2018; Kreno et al. 2011; Rocha et al. 2011; Lustig et al. 2017; Hu et al. 2014; Xu et al. 2017), electrocatalysis, electrochemical and photochemical energy conversion and storage energy applications (Wang et al. 2017, 2020a; Yi et al. 2017; Wu et al. 2019; Li et al. 2020; Downes and Marinescu 2017; Zhang et al. 2019a), gas storage and separation (Zhao et al. 2018; Wang and Zhao 2017; Bachman et al. 2016; Li et al. 2019b; Krause et al. 2016), adsorption of various materials (Esrafili et al. 2018; Shayegan et al. 2019; He et al. 2016; Yamini et al. 2018), as optical sensors, due to their unique properties such as highly ordered, easily functionalizable cavities for specific host–guest interactions, large surface areas, high porosity and tunable metal nodes and organic linkers.

MOFs pores and shape are well arranged, it can be designed by varying the linkers and various metal ions. The properties and structure of the MOFs, basically depends on the choice of metal and linker which are used for synthesis of MOF. Further, MOFs can be amended via post-synthetic modification, to modify for the interaction with guest molecules. The main merit of MOFs, it contains high surface area along with good pore capability over other adsorbent materials, such as carbon-based material and zeolites. Several MOFs display luminescent property, this luminescent feature of MOFs utilized to detect trace amounts of selective metal ions with 'turn on' and 'turn off' mechanism by the advanced spectrophotometric method (Qu et al. 2013; Li et al. 2016b).

The luminescent features of MOFs are based on charge transfer phenomena of metal ions (transition and inner transition metal). The metal ions formed, a bond with organic ligands, to develop an inorganic–organic hybrid material or Metal–organic frameworks for detection of various metal ions (Rath and Vittal 2020; Zhu et al. 2020; Xu et al. 2019). The hybrid material shows the detections ability of very small amounts of metal ions according to the selectivity and sensitivity of the MOFs (Liu et al. 2020) in detection process, the metal ion is acknowledged by the receptor, enhancement or quenching and the luminescence response, can be obtained by electron donor–acceptor or energy transfer. The metal ions may responsible for luminescence property via different interactions or may behave as luminescent centers. Recently, very much focus has been made to synthesis of MOFs based nano-material and explores its applications.

MOFs based nano-material has some unique features due to nano-size particles which exhibit excellent features, such as a quantum effect, small size effect, macro quantum tunnel effect and surface effect. It demonstrates outstanding properties based on the synergetic effect. MOFs based nano-material activity has been tuned with their functionality and surface area which are responsible for suitable sizes for biomedical application, enhancing the adsorption/desorption ability and catalysis, their application in the field of energy and membrane separation-related uses, like, MOFs based nano-material catalysts usually exhibit good catalytic activity than their bulk analogues due to availability of more catalytic sites due to their bigger surface area.

The studies on MOFs based nano-materials have gained great attention to investigate their various applications especially in the field of energy and environmental issues. They show great potential for removal of heavy metal ions from wastewater. This book chapter will outlook for synthesis and potential application of MOFs based nano-material especially in the field of energy and removal of heavy metal ions from waste water.

16.2 Synthesis of MOFs Based Nano-material

16.2.1 Solvothermal Method

Solvothermal method is the most common and valuable method for synthesis of MOFs, based nano-material so far. Recently, Kitagawa and co-workers have synthesized MOFs based nano-material in 2009, using chemical modulators to give desired size and shape of the MOFs based nano-material. It has been observed that chemical modulators are responsible to synthesize desired shape and size MOFs based nano-material via slow down the crystallization process (Fig. 16.1). The whole process can be regulated through coordination bond formation between chemical modulators (such as lauric acid, benzoic acid, and acetic acid), and metal ions/clusters (Tsuruoka et al. 2009). Wang and co-workers have synthesized Hf-based MOFs by controlling hydrolysis process of metal ion precursor in aqueous medium, using acetic acid as a linker to control the coordination site around the metal ion (He et al. 2017a).

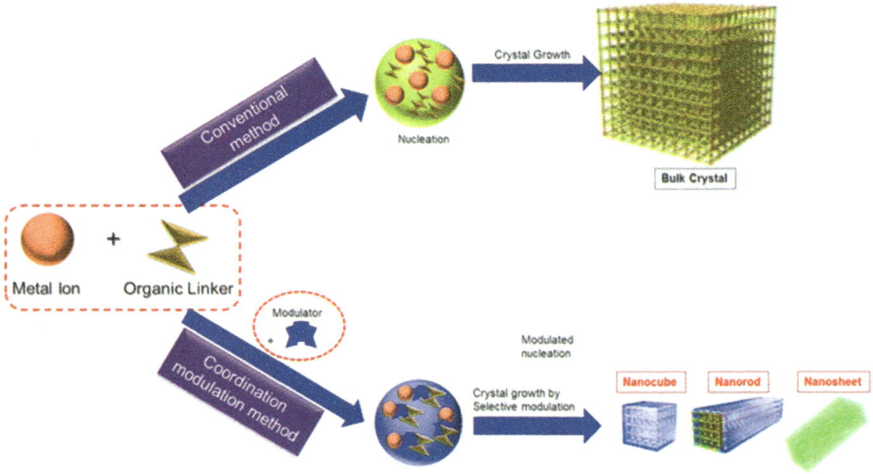

Fig. 16.1 The role of chemical modulators for fabricating MOF nanocrystals. Modified and adopted from Tsuruoka et al. (2009)

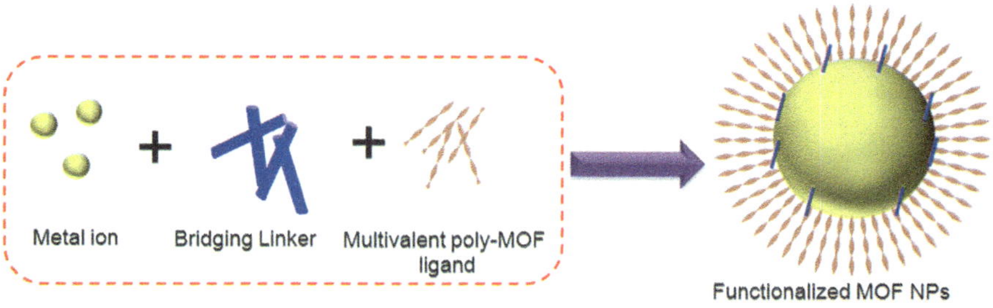

Fig. 16.2 The mechanism of nano metal organic frameworks by using multivalent ligands as chemical modulators

Johnson et al. in 2019 constructed a nano-sized polyMOFs with polyethylene glycol (PEG) methyl ether azide and organic linkers. It has been noticed that linkers were first attached with PEG methyl ether azide. Interestingly, by changing the morphology and size of the connectors and polymer, the length of polymer chains got affected. The polymer length increases with decrease the molecular weight (Gu et al. 2019). By solvothermal method, polyMOF-5 (nanoparticles diameter size ~ 28 nm) were obtained, using PEG5k-L_4, terephthalic acid, and $Zn(NO_3)_2{\cdot}6H_2O$ in N,N-Dimethylformamide. When PEG5k-L_4 polymer replaced by larger molecular weight polymer PEG10k-L_4 the nanoparticles (~20 nm) were formed, has diameter size ~ 28 nm. Further, the polyUiO-66 nanoparticles which has diameter of ~ 36 nm, were prepared by using PEG10k-L_4. It noticed that nanoparticles diameter size decreases with increase in the molecular weight of polymer, used in the synthesis of particular nanoparticle.

Zhang et al. in 2018, has reported some nano metal organic frameworks, were prepared via separating the nucleation and growth process. The framework is constructed with the controlled small portion addition of metal ions as well as ligand solution (Fig. 16.2) (Wang et al. 2018a). It has noticed that synthesis of nano-MOFs controlled by degree of supersaturation of the reaction mixtures.

16.2.2 Microemulsion Method

Nano-MOFs were synthesized by microemulsion method; monodisperse and thermodynamically stable nano-MOFs prepared by solvents with the accomplishment of emulsifiers or surfactants (Ganguli et al. 2010). In this process, size of droplets can be controlled by the concentration of surfactants.

Recently, Mann et al. has prepared nanoparticles by microemulsion method which has Prussian blue color and monodispersed in nature (Vaucher et al. 2000). Lin and co-workers have synthesized $Gd_2(BDC)_{1.5}(H_2O)_2$ nanorods with gadolinium trichloride and bis(methylammonium)benzene-1,4-dic arboxylate by microemulsion method, constructed with hexadecyl trimethyl ammonium bromide in the solvent mixture of

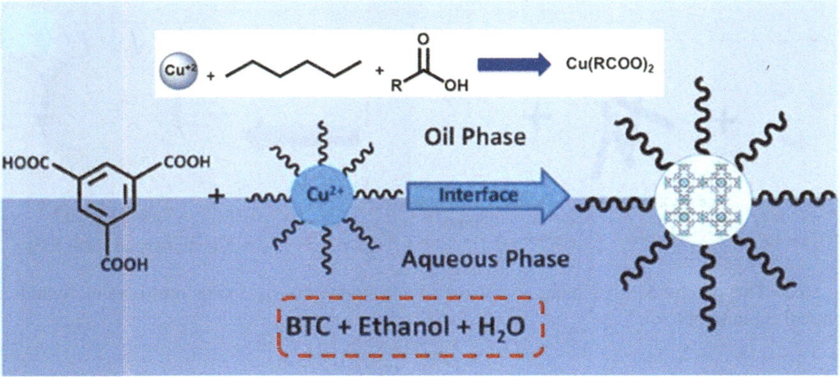

Fig. 16.3 Schematic diagram of synthetic mechanism of HKUST-1 nanocrystals. Modified and adopted from Cai et al. (2019a)

1-hexanol, water and isooctane (Rieter et al. 2006). The nano MOFs size decreased with varying the ratio of water to hexadecyl trimethyl ammonium bromide from 10:1 to 5:1. By this reverse-phase micro-emulsion method, $[Gd(BTC)(H_2O)_3]H_2O$ and $Mn_3(BTC)_2(H_2O)_6$. Nano MOFs have been prepared. In this process, the size of the particles decreases with increase in the concentration of reaction mixture, due to the formation of micelles which lead to the more nucleation sites (Taylor et al. 2008a). Cai and co-workers have reported HKUST-1 nanocrystals, which are uniform in nature, prepared by reverse-phase micro-emulsion technique (Fig. 16.3). The micro-emulsion system is composed of a mixture of oleic acid, sodium hydroxide solution, and n-hexane into ethanol. Further, adding copper ions into the reaction mixture. The copper oleate clusters were produced, then 1,3,5-benzenetricarboxylic acid ligand was added, HKUST-1 nanocrystals were formed. It has noticed that the crystal size can be tuned with the amount of surfactant was added in the reaction mixture (Cai et al. 2019a). Interestingly, HKUST-1 nanospheres (particle size ~ 70 nm) were formed by adding 0.30 mL of oleic acid in the reaction mixture.

16.2.3 Microwave Assisted Preparation

The synthesis of a variety of MOFs has been synthesized by using microwave irradiation as energy source. Microwave-assisted synthesis is based on the interaction of electromagnetic waves with mobile electric charges (e.g. polar solvent molecules or ions in the solution) and therefore, this represents a very energy efficient method of heating. The advantages of this method include fast synthesis, high phase purity of materials, smaller crystal size as well as morphology control. This is due to the direct heating of the solvents and the higher nucleation rate during the reaction.

Hong et al. (2006) have prepared iron oxide-based nanomaterial by microwave assisted method, using $FeCl_3{\cdot}6H_2O$ and $FeCl_2$ metal salt in hydrochloride acid. After the microwave heating Sodium hydroxide was then added for neutralization of acid (Hong et al. 2006). Recently, Sun et al. (2017) have synthesized iron based nanomaterial MIL-88A(Fe) by dispersion of nFe_3O_4 in $FeCl_3{\cdot}6H_2O$ solution and fumaric acid solution, mixture is heated in microwave oven and dried at 100 °C (Sun et al. 2017).

16.3 Applications of Nano Based Metal Organic Frameworks

16.3.1 In Field of Removal of Heavy Metal Ions

So many adsorbents have been reported in literature such as clay, carbon material, polymers with various functionalities for removal of heavy metals. But these materials had shown low binding affinity, weak adsorption capacity towards metals. So, it is necessary to develop new effective adsorbents which can eliminate heavy metals from industrial waste with high selectively and efficiency. To increase the adsorption capability of MOF towards Hg^{2+} ion, Jiang et al. introduced the thioether groups in MOFs, synthesized a Cobalt based MOF which contains thioether arms with free-standing fashion. It is capable to eliminate Hg^{2+} ions from aqueous medium, due to strong interaction of Hg-S (Huang et al. 2017c). Jiang, Hong and co-workers have synthesized sulfur functionalized nanoparticles in the MOF cavity, having particle size around 2.5 nm, which potential used to remove Hg^{2+} from wastewater up to the 99.95% within a minute. It displayed excellent adsorption capacity with high distribution coefficient (2.2×10^7 mL g^{-1}) than other MOF-based mercury adsorbents reported in literature. It also showed very high selectivity towards mercury than other metal ions present in industrial waste effluents due to mercury – sulfur interactions (Liang et al. 2018).

Xu et al. in 2009, has reported MOFs with methylthio group. It was prepared from TMBD ligand with lead nitrate in N, N-dimethylacetamide and acetonitrile solvent mixture by solvothermal method which utilized to adsorb Hg(II) ions, due to the soft–soft interaction between methylthio group and mercury ion (Zhou et al. 2009). Many scientists have explored the potential of MOFs nanosheets for eliminating the heavy metal ions. Recently, Wang et al. prepared MOF nanosheet with [Co(NCS)2(pyz)2]n (pyz = pyrazine) which have 6.5 nm pore size and high surface area of 365 m^2/g. It exhibited high adsorption capability for Hg^{2+}, due to the presence of ions high sulfur content. It shows adsorption of Hg^{2+} ion at both side of nanosheet and formed the 3D structure via mercury-sulfur and cation-π interactions (Wang et al. 2020b). In 2020, Mokhtarian and coworkers have synthesized MOF (ZIF@NiTiO3) by sol–gel method, introduced NiTiO3 into the ZIF MOF. It has particle size 40–50 nm and high Brunauer–Emmett–Teller surface area around

345 m^2/g. It shows excellent adsorption activity for eliminate Pb^{2+} ion (Akbarzadeh et al. 2020). Manos and coworkers in 2020, have prepared Zr-based microporous terephthalate MOFs contains bulky functional group. It showed excellent capture capacity of chromium (VI) from aqueous medium, can be regenerated and reused easily. This MOF has excess chloride ion which replaced by chromium (VI) ions. The crystal structure of MOF reveals that the Cr(VI) oxoanions attach to the Zr_6 moiety by replacing their terminal hydroxide or water ligands. The terephthalate MOFs displayed good chromate and dichromate sorption capabilities; the sorption equilibrium established within very short time about one minute. This excellent adsorption property due to the tendency of the pyridyl-methyl-ammonium moiety and have the pores with labile chloride anions (Rapti et al. 2017). Mohamed and coworkers in 2020, have synthesized Amine-functionalized iron based metal–organic framework nanocomposite (nFe_3O_4@MIL-88A(Fe)/APTMS). It showed excellent activity for selective removal of Cr (VI), Pb (II) and Cd (II) ions from water. It displayed adsorption capacities of Cr (VI), Pb (II) and Cd (II) ions 1092.2, 536.2 and 693.0 mg g^{-1}, respectively. It showed excellent reusability when 0.1 mol L^{-1} hydrochloric acid used (Mahmoud et al. 2020). By solvothermal method, Zhu et al. in 2019, have synthesized nanotubular shapes Tb based metal organic frameworks using tris-benzoic acid ligand. This ligand contains carboxylic acid, imine and amino group which showed good electrostatic interactions and complexation with Pb^{+2} ions. The MOFs exhibited good recyclability and shows 547 mg/g of adsorption capacity for Lead. The XPS studies and Density functional theory calculations explained that nitrogen donor site formed PB-N bond as inner sphere complexes. The amount of binding energy shows the interaction of Tb-MOFs (absorbents) with heavy metal ions which is higher than 1 eV, reveals that lead ion chemically adsorbed on metal organic frameworks due to more favorable transition state (Zhu et al. 2019).

In 2019, Wang et al. prepared a nanomaterial by mixing BUC-21 MOFs with titania nanotube, using mechanical ball-milling. It showed very high photolytic activity as well as adsorption. The Cr(VI) removal efficiency has increased by 100% with 20 min (Wang et al. 2019). Recently, Fan et al. prepared a belt-like shape MOFs with carboxylic groups in 2016, and was used in Fe^{3+} adsorption (Fan et al. 2020). Elimination of As(III) is highly required from the environment. The As(III) has higher toxic effect due to its strong binding to proteins part of various enzymes. For removal of As(III) ion, Deng et al. in 2010, synthesized a MOF framework with bimetallic core which have excellent potential for adsorption process of As(III) (Deng et al. 2010). Recently, amorphous MOFs have been developed for adsorption and oxidation of As^{3+}, simultaneously, because it is having large number of active site (Orellana-Tavra et al. 2015; Horcajada et al. 2006; Bennett et al. 2013; Chapman et al. 2011; Duan et al. 2018). Using temperature-controlled crystallization process, Fe/Mn bimetal-MOFs had prepared, utilizing $FeCl_3$ and $MnCl_2$ compound as the iron and manganese source. It has used to remove As^{+3} ion via oxidation and homogeneous adsorption process.

It displayed adsorption capacity 161.6 mg/g. It contains less coordination site and more voids, which are more favorable for adsorption phenomenon. Initially, Arsenic(III) ion got absorbed on the surface of MOFs through exchange of –OH functional group and formed the hydrogen bond, then As^{+3} get oxidized two electron and converted into As^{+5}. Further, As^{+5} got absorbed on the MOF (Zhang et al. 2019b).

16.3.2 In the Field of Energy and Storage

The nanoMOF potential have been explored in the field of energy. It has used in solar cells, supercapacitors, sodium-ion batteries, lithium-ion batteries (rechargeable) and sodium-ion batteries due to its good electron transferable surface area. The nano-metal organic frameworks are able to facilitate the electron transfer as well as ion diffusion process, which is responsible to enhance the working of nanoMOF as an electrode. Recently, nanoMOFs have been synthesizes by Ho and coworkers with coating. It has been used as flexible counter electrode by carbon cloth with sulfonated polythiophene (Chen et al. 2017). This MOF with porphyrin and zirconium oxo clusters used as active site for I^-/I_3^- redox suttle due to contains of a greater number of pores. The core shell structure constituted with the layer of composite on the carbon cloth. In core shell structure provide the platform for electronics transport phenomenon and shell behaves as a catalyst. The MOF has shown very good activity when 3.0 weight percentage of MOF was used. The battery efficiency increased from 8.91 to 9.75% when dim light was used. Recently, Wang et al. synthesized nanocrystalline MOF, which was utilizes as supercapacitor by layering of nanosized ZIF-67 (size ~ 300 nm) on a carbon cloth material, then further coating of polyaniline. The pores available in MOF, provides a platform to diffusion of electrolyte and polyaniline layer facilitates the redox reaction on metal organic frameworks surface, resultant good surface capacitance was observed (Wang et al. 2015). In 2018, Lan and coworkers had prepared nano-MOF (size ~ 100 nm) based on polyoxymetalate material which was further layered by polypyrrole. It was worked as a highly superconductor. In this material polyoxymetalate moiety showed excellent electron transfer capacity for nanoMOF/polypyrrole part (Wang et al. 2018b). Very recently, MOF based nanomaterial has been prepared and used for sodium-ion batteries. Sodium metal has higher reduction potential and cheaper, abundant on earth than lithium metal; it has capacity of 1160 mAh g^{-1}. So, it can be used as the alternative of lithium-ion batteries, however there is safety issue with sodium metal due to higher reactivity and low melting point. In 2015, Fan et al. prepared a nanocomposite with the ZIF-8 and amorphous carbon nitride by pyrolysis method at 700 °C. It has uniform polyhedron like shape and size, symmetrical N, C, O, and Zn atom distribution in the whole material. It showed 430 mAh g^{-1} capacity at 83 mA g^{-1} and 146 mAh g^{-1} capacity at 8.33 mA g^{-1}. It also retains a capacity after 2000 cycles at a high specific current. This showed high cycling stability and excellent rate, used as anode materials for sodium-ion batteries, due to the large surface area and high nitrogen content (Fan et al. 2015). Basically, nitrogen content in the nanomaterial is the deciding factor for electrochemical efficiency in sodium-ion batteries. Recently, Zheng and coworkers have prepared microporous carbon based MOF at high temperature around 930 °C. It has high specific surface area 1251 m^2 g^{-1} than previous reported compounds;

however it showed 136 mAh g^{-1} capacity at 50 mA g^{-1} which is low because of less nitrogen content in the nanomaterial (Qu et al. 2014).

Chen et al. have prepared from ZIF-67 via an annealing treatment. It showed excellent activity as lithium-ion batteries which displayed reversible capacity with 1080 mAh g^{-1} after 150 cycles at 500 mA g^{-1} (Chen et al. 2016). Recently, Xiao et al. synthesized Pt@UiO-66-NH_2 nanomaterial with platinum nanoparticle inside the cavity. It works as nano catalyst for photocatalytic hydrogen evolution via water splitting reaction due to facile of electron transfer. This nano catalyst has good stability as well as recyclability (Xiao et al. 2016). Wen et al. synthesizes a MOF hybrid nanomaterial from carbon nanotubes and Nickle MOFs. This material shows good electrochemical activity, used as positive electrode with negative electrode of rGO/g-C3N4. This system exhibited high energy density and work as good superconductor (Wen et al. 2015).

16.3.3 In the Field of Bioimaging

Recently, MOF based nano-material has been used for biomedical imaging (Wang 2017). The Iron (III), Manganese (II) and Gadolinium (III) based nano-metal organic frameworks broadly used for magnetic resonance imaging contrast agents because of their unique structure. Lin and coworkers has constructed Gd^{3+} based nanorod ($Gd(BDC)_{1.5}(H_2O)_2$), having size ~ 100 nm in length and 40 nm in diameter, which is able to detect nuclear spin in particular direction in the presence of magnetic field. These materials utilized as the contrast agents for T1 as well as T2-weighted image which exhibits transverse relaxivity (R2) value of 55.6 s^{-1} and longitudinal relaxivity (R1) of 35.8 s^{-1} for per millimolar of gadolinium ion. Previously, many scientists have reported Gd-based nano-metal organic frameworks due to their high magnetic resonance relaxivities which are showing excellent activity as marvelous magnetic resonance imaging agents (Tian et al. 2015; Taylor et al. 2008b; Nishiyabu et al. 2009; Rowe et al. 2009; Qin et al. 2017; Kundu et al. 2016). Recently, researcher has focused on Mn^{2+} based nano-metal organic frameworks over Gd ions for T1-weighted contrast agents, due to lower biological toxicity Mn^{2+} than that of Gd ions (Broome et al. 2007; Thunus and Lejeune 1999). Taylor et al. has synthesized Mn- based nanorod-like MOF which were used as T1-weighted contrast agents for magnetic resonance imaging in liver and spleen in vivo medium (Taylor et al. 2008a).

16.3.4 In the Field of Photodynamic Therapy (Therapeutic Agent for Cancer)

Photodynamic therapy is a binary therapy that involves the combination of visible light and a photosensitizer. Photosensitizer displays cytotoxic effects when it was activated in the presence of light. After irradiation, the photosensitizers got promoted to their excited states and produce singlet oxygen. The biomolecules which are nearby get damaged and this starts a flow of biological response leading to the tumor death. The efficiency of Photodynamic Therapy depends on the nature of the Photosensitizer. The development of efficient photosensitizers have great attention recent time. Very recently, nanomaterial based metal–organic frameworks used for the photodynamic therapy due to their unique properties like high porosity, easy to tune the structure, multiple functionalities and excellent biocompatibility. Lin and coworkers, in 2014, synthesized porphyrinic-based metal–organic frameworks from the reaction between and Hf^{4+} and 5,15-di(pbenzoato)porphyrin via solvothermal method, for the treatment of cancer photodynamic therapy treatment. It has diameter around 100 nm and thickness 10 nm. It is able to generate singlet oxygen, two times more as compared to porphyrin and showed greater photodynamic therapy efficiency for kill the cancer cell located at head and neck of the body (Lu et al. 2014). Kan and et al. in 2018, prepared nanometal organic frameworks based photosensitizer using nanosized UiO-66 and S-ethylthiol ester through post-synthetic method. It showed excellent photodynamic therapy effect on the tumor cell (Kan et al. 2018). Lin and coworkers in 2021, synthesized a core–shell nanostructure metal organic frameworks which were used for photodynamic therapy as well as chemodynamic therapy. It has prepared with NaYF4:20%Yb, 2%Er and layered with copper-based MOF (Cu/ZIF-8) and Cu/ZIF-8 containing rose Bengal, also wrapped by F127. This nano system showed good biocompatibility. In the acidic tumor microenvironment, nano system has been degraded and releases the copper(II) ion and rose Bengal. The eliminated Cu^{2+} ion reacts with glutathione and produce •OH for kill cancer cells (Zou et al. 2021). Lin and coworkers in 2019, prepared a nano-metal organic frameworks (size ~ 100 nm) which was biodegradable and used for photo dynamic therapy via decreasing concentration of glutathione and with supply of oxygen. The •OH and O_2 were produced by MOF in the presence of hydrogen peroxide through Fenton-like reaction (Fig. 16.4). The nanoMOF was responsible to carry the O2 into the tumor cells and showed the photodynamic therapy (Cai et al. 2019b). Park et al. in 2016,

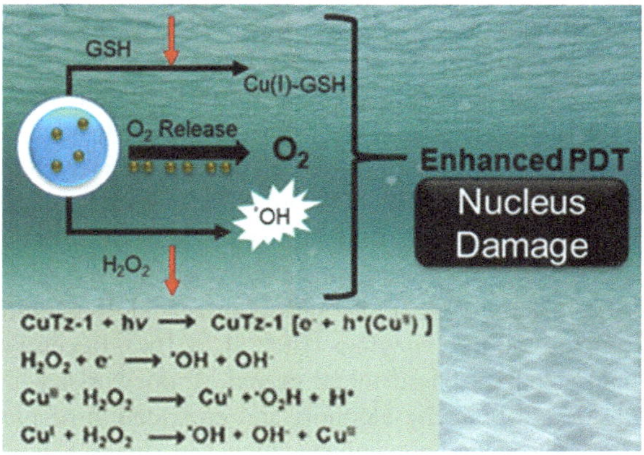

Fig. 16.4 Mechanism of photodynamic therapy. Modified and adopted from Cai et al. (2019b)

prepared UiO-66 nano-metal organic frameworks which was doped with 5,10,15,20-tetrakis(4-carboxylphenyl)-porphyrin as photo sensitizer and the ligand(-bis(5-(4-carbonxyphenyl)-2-methylthien-3-yl)cyclopent-1-ene) works as the photochromic switch, for the control of the production of singlet oxygen. It exhibited only 10% cytotoxicity and showed good photodynamic therapy in vitro medium (Park et al. 2016).

The activity of the photodynamic therapy is restricted by nano-metal organic frameworks towards tumor hypoxia (Lan et al. 2018). Xu and coworkers in 2021, prepared a porphyrin and Fe_3O cluster-based metal organic frameworks which react with intracellular H_2O_2 and generate O_2 via Fenton-like reaction. The produced oxygen was changed into $^1O_2$ in the presence of red light with the help of nano MOFs. When this system treated with cytotoxic T cells, resultant showed decay of tumors cell. Recently, Mao and coworkers developed porphyrinic based nano MOF with the DNA template, exhibited good photodynamic therapy. This MOF contains Zr_6 cluster with tetratopic linker and formed 3D nano porous structure which facilitated the diffusion of $^1O_2$. The core of metal organic frameworks releases the light which was utilized to produce of $^1O_2$ via porphyrinic MOF. The in vitro analysis reveals that these nano assemblies displayed excellent efficiency for cancer cell therapy (He et al. 2017b).

16.3.5 In the Field of Catalysis

Recently, nano-metal organic frameworks have been developed for the catalysis due to availability of various functional active sites in the MOFs, easy separation of the product and high porosity which provides good platform for catalysis (Rogge et al. 2017; Cai et al. 2020). To improve the accessibility of active site and decrease the diffusion barriers, Kitagawa and coworkers prepared various sized Yb-based MOF and check the catalytic activity for isomerization reaction, 1-hexene compound used as substrate. Sub micro-sized Yb-MOF showed better catalytic activity than that of micro-sized Yb-based MOF. This catalytic activity reflected due to the fast diffusion of the substrate to functional catalytic sites (Kiyonaga et al. 2015). In 2015, Li et al. synthesized Zr-based MOF which has mesoporous in nature. It has been analyzed towards the hydrolysis of methyl paraoxon substrate to study the effect of crystallite size for this hydrolysis reaction. It has been observed that the catalytic activity increases with respect to decrease the size of MOF, because the surface area increases with decreasing the nono-particle size (Li et al. 2015). Farha and coworkers also reported a similar type of Zr-based MOF in 2016, which contains large mesopores, its activity investigated as nerve agent for hydrolyzing enzyme and anhydrolase enzyme activity for organophosphorus acid. It exhibited very good catalytic activity compare the free enzymes (Li et al. 2016c). Recently, Qi et al. synthesized various nano sized Cu-based metal organic frameworks which was used for oxidation of allylic and benzylic alcohols and epoxidation of olefins under aerobic condition. The heterogeneous catalyst which has smallest size displayed excellent catalytic reactivity. It is observed that catalytic activity increases with increasing the mono dispersity (Qi et al. 2015). In another classical example also reported by Schmidt and coworkers in 2019. They had synthesized a hybrid of nano-metal organic frameworks and a polymer which have unique stimuli-responsive dispersibility. Notably, it showed very high photocatalytic activity in liquid phase (Lee et al. 2019).

16.4 Challenges and Future Perspectives

Heavy metals in the water are a global environmental issue worldwide. A large amount of toxic materials are present in water ecosystem. The developments of Metal organic frameworks (MOFs) based nonmaterial are extremely challenging for the removal of heavy metals from the waste water as well as for electrochemical, photochemical energy conversion and storage, biomedical imaging, drug delivery and catalysis. However, due to unique features of MOFs based nanomaterial more efforts should be made in the future. In this chapter, the attention is given to understand the synthesis, chemistry of MOF based nano-composites and its various applications especially, and removal of heavy metals from waste water has been discussed. It is expected that this chapter can be helpful to understand the synthesis of MOF-based nano-materials and its application towards the elimination of heavy metal ions from waste water.

References

Akbarzadeh MJ, Hashemian S, Mokhtarian N (2020) J Environ Chem Eng 8:103703
Bachman JE, Smith ZP, Li T, Xu T, Long JR (2016) Nat Mater 15:845
Bennett TD, Saines PJ, Keen DA, Tan JC, Cheetham AK (2013) Chem Eur J 19:7049–7055
Broome DR, Girguis MS, Baron PW, Cottrell AC, Kjellin I, Kirk GA (2007) Am J Roentgenol 188:586–592
Cai W, Chu C-C, Liu G, Wáng Y-XJ (2015) Small 11:4806–4822
Cai X, Xie Z, Pang M, Lin J (2019a) Cryst Growth Des 19:556–561
Cai X, Xie Z, Ding B, Shao S, Liang S, Pang M, Lin J (2019b) Adv Sci 6:1900848
Cai G, Ding M, Wu Q, Jiang H-L (2020) Natl Sci Rev 7:37–45
Chapman KW, Sava DF, Halder GJ, Chupas PJ, Nenoff TM (2011) J Am Chem Soc 133:18583–18585
Chen JF, Ru Q, Mo YD, Hu SJ, Hou XH (2016) Phys Chem Chem Phys 18:18949–18957
Chen T-Y, Huang Y-J, Li C-T, Kung C-W, Vittal R, Ho K-C (2017) Nano Energy 32:19–27
Corma A, García H, Llabrés i Xamena FX (2010) Chem Rev 110:4606–4655
Deng H, Doonan CJ, Furukawa H, Ferreira RB, Towne J, Knobler CB, Wang B, Yaghi OM (2010) Science 327:846–850
Ding M, Cai X, Jiang H-L (2019) Chem Sci 10:10209–10230
Downes CA, Marinescu SC (2017) Chemsuschem 10:4374–4392
Duan YX, Meng FL, Liu KH, Yi SS, Li SJ, Yan JM, Jiang Q (2018) Adv Mater 30:1706194
Dutta S, Kim J, Hsieh PH, Hsu YS, Kaneti YV, Shieh FK, Yamauchi Y, Wu KCW (2019) Small Methods 3:1900213

Esrafili L, Safarifard V, Tahmasebi E, Esrafili M, Morsali A (2018) New J Chem 42:8864–8873
Fan JM, Chen JJ, Zhang Q, Chen BB, Zang J, Zheng MS, Dong QF (2015) Chem Sus Chem 8:1856–1861
Fan W, Liu X, Cheng Y, Qin Y, Yang S, Lu Z, Liu Y, Cao Q, Zheng L (2020) Sep Purif Technol 239:116559
Furukawa H, Cordova KE, O'Keeffe M, Yaghi OM (2013) Science 341:1230444
Ganguli AK, Ganguly A, Vaidya S (2010) Chem Soc Rev 39:474–485
Gu Y, Huang M, Zhang W, Pearson MA, Johnson JA (2019) Angew Chem Int Ed 58:16676–16681
He Y, Wang Y, Yang X, Xie S, Yuan R, Chai Y (2016) ACS Appl Mater Interfaces 8:7683–7690
He T, Xu X, Ni B, Wang H, Long Y, Hu W, Wang X (2017a) Nanoscale 9:19209–19215
He L, Brasino M, Mao C et al (2017b) Small 13(24):1700504
He J, Xu J, Yin J, Li N, Bu X-H (2019) Sci China Mater 62:1655–1678
Hong RY, Pan TT, Li HZJ (2006) Magn Mater 303:60–68
Horcajada P, Serre C, Vallet-Regí M, Sebban M, Taulelle F, Férey G (2006) Angew Chem Int Ed 45:5974–5978
Hu Z, Deibert BJ, Li J (2014) Chem Soc Rev 43:5815–5840
Huang R-W, Wei Y-S, Dong X-Y, Wu X-H, Du C-X, Zang S-Q, Mak TC (2017a) Nat Chem 9:689
Huang Y-B, Liang J, Wang X-S, Cao R (2017b) Chem Soc Rev 46:126–157
Huang N, Zhai L, Xu H, Jiang D (2017c) J Am Chem Soc 139:2428–2434
Jiao L, Jiang H-L (2019) Chem 5:786–804
Jiao L, Wang Y, Jiang HL, Xu Q (2018) Adv Mater 30:1703663
Jiao L, Seow JYR, Skinner WS, Wang ZU, Jiang H-L (2019) Mater Today 27:43–68
Kan J-L, Jiang Y, Xue A, Yu Y-H, Wang Q, Zhou Y, Dong Y-B (2018) Inorg Chem 57:5420–5428
Kaneti YV, Dutta S, Hossain MS, Shiddiky MJ, Tung KL, Shieh FK, Tsung CK, Wu KCW, Yamauchi Y (2017) Adv Mater 29:1700213
Karmakar A, Pombeiro AJ (2019) Coord Chem Rev 395:86–129
Kiyonaga T, Higuchi M, Kajiwara T, Takashima Y, Duan J, Nagashima K, Kitagawa S (2015) Chem Commun 51:2728–2730
Krause S, Bon V, Senkovska I, Stoeck U, Wallacher D, Többens DM, Zander S, Pillai RS, Maurin G, Coudert F-X, Kaskel S (2016) Nature 532:348–352
Kreno LE, Leong K, Farha OK, Allendorf M, Van Duyne RP, Hupp JT (2011) Chem Rev 112:1105–1125
Kundu T, Mitra S, Diaz Diaz D, Banerjee R (2016) ChemPlusChem 81:728–732
Lan G, Ni K, Xu Z et al (2018) J Am Chem Soc 140(17):5670–5673
Lázaro IA, Forgan RS (2019) Coord Chem Rev 380:230–259
Lee H-C, Heil T, Sun J-K, Schmidt BVKJ (2019) Mater Horiz 6:802–809
Li P, Klet RC, Moon S-Y, Wang TC, Deria P, Peters AW, Klahr BM, Park H-J, Al-Juaid SS, Hupp JT, Farha OK (2015) Chem Commun 51:10925–10928
Li B, Wen HM, Cui Y, Zhou W, Qian G, Chen B (2016a) Adv Mater 28:8819–8860
Li J, Wang Q, Guo Z, Ma H, Zhang Y, Wang B, Bin D, Wei Q (2016b) Sci Rep 6:23558
Li P, Moon S-Y, Guelta MA, Lin L, Gómez-Gualdrón DA, Snurr RQ, Harvey SP, Hupp JT, Farha OK (2016c) ACS Nano 10:9174–9182
Li P, Cheng F-F, Xiong W-W, Zhang Q (2018a) Inorg Chem Front 5:2693–2708
Li G, Zhao S, Zhang Y, Tang Z (2018b) Adv Mater 30:1800702
Li D, Xu H-Q, Jiao L, Jiang H-L (2019) EnergyChem 1:100005
Li H, Li L, Lin R-B, Zhou W, Xiang S, Chen B, Zhang Z (2019) EnergyChem 1:100006
Li X, Yang X, Xue H, Pang H, Xu Q (2020) EnergyChem 2:100027
Liang L, Liu L, Jiang F, Liu C, Yuan D, Chen Q, Wu D, Jiang HL, Hong M (2018) Inorg Chem 57(9):4891–4897
Liang Z, Zhao R, Qiu T, Zou R, Xu Q (2019) EnergyChem 1:100001
Liao Y-T, Matsagar BM, Wu KC-W (2018) ACS Sustain Chem Eng 6:13628–13643
Lismont M, Dreesen L, Wuttke S (2017) Adv Funct Mater 27:1606314
Liu J, Chen L, Cui H, Zhang J, Zhang L, Su C-Y (2014) Chem Soc Rev 43:6011–6061

Liu C-H, Guan Q-L, Yang X-D, Bai F-Y, Sun L-X, Xing Y-H (2020) Inorg Chem 59:8081–8098
Long JR, Yaghi OM (2009) Chem Soc Rev 38:1213–1214
Lu K, He C, Lin W (2014) J Am Chem Soc 136:16712–16715
Lu K, Aung T, Guo N, Weichselbaum R, Lin W (2018) Adv Mater 30:1707634
Luo Y-H, Dong L-Z, Liu J, Li S-L, Lan Y-Q (2019) Coord Chem Rev 390:86–126
Lustig WP, Mukherjee S, Rudd ND, Desai AV, Li J, Ghosh SK (2017) Chem Soc Rev 46:3242–3285
Mahmoud ME, Amira MF, Seleim SM, Mohamed AK (2020) J Hazard Mater 381:120979
Nishiyabu R, Hashimoto N, Cho T, Watanabe K, Yasunaga T, Endo A, Kaneko K, Niidome T, Murata M, Adachi C (2009) J Am Chem Soc 131:2151–2158
Orellana-Tavra C, Baxter EF, Tian T, Bennett TD, Slater NKH, Cheetham AK, Fairen-Jimenez D (2015) Chem Commun 51:13878–13881
Park J, Jiang Q, Feng D, Zhou HC (2016) Angew Chem Int Ed 55:7188–7193
Qi Y, Luan Y, Yu J, Peng X, Wang G (2015) Chem Eur J 21:1589–1597
Qin L, Sun Z-Y, Cheng K, Liu S-W, Pang J-X, Xia L-M, Chen W-H, Cheng Z, Chen J-X (2017) ACS Appl Mater Interfaces 9:41378–41386
Qin Z-S, Dong W-W, Zhao J, Wu Y-P, Zhang Q, Li D-S (2018) Inorg Chem Front 5:120–126
Qu K, Wang J, Ren J, Qu X (2013) Chem Eur J 19:7243–7249
Qu QT, Yun JJ, Wan ZM, Zheng HY, Gao T, Shen M, Shao J, Zheng HH (2014) RSC Adv 4:64692–64697
Rapti S, Sarma D, Diamantis SA, Skliri E, Armatas GS, Tsipis AC, Hassan YS, Alkordi M, Malliakas CD, Kanatzidis MG, Lazarides T, Plakatouras JC, Manos MJ (2017) J Mater Chem A 5:14707–14719
Rath BB, Vittal JJ (2020) Inorg Chem 59:8818–8826
Rieter WJ, Taylor KM, An H, Lin W, Lin W (2006) J Am Chem Soc 128:9024–9025
Rocha J, Carlos LD, Paz FAA, Ananias D (2011) Chem Soc Rev 40:926–940
Rogge SMJ, Bavykina A, Hajek J, Garcia H, Olivos-Suarez AI, Sepúlveda-Escribano A, Vimont A, Clet G, Bazin P, Kapteijn F, Daturi M, Ramos-Fernandez EV, Llabrés i Xamena FX, Van Speybroeck V (2017) J Gascon Chem Soc Rev 46:3134–3184
Rowe MD, Thamm DH, Kraft SL, Boyes SG (2009) Biomacromol 10:983–993
Shayegan H, Farahani YD, Safarifard V (2019) J Solid State Chem 279:120968
Shieh F-K, Wang S-C, Yen C-I, Wu C-C, Dutta S, Chou L-Y, Morabito JV, Hu P, Hsu M-H, Wu KC-W (2015) J Am Chem Soc 137:4276–4279
Sun X, Gao G, Yan D, Feng C (2017) Appl Surf Sci 405:52–59
Taylor KM, Rieter WJ, Lin W (2008a) J Am Chem Soc 130:14358–14359
Taylor KM, Jin A, Lin W (2008b) Angew Chem Int Ed 47:7722–7725
Thunus L, Lejeune R (1999) Coord Chem Rev 184:125–155
Tian C, Zhu L, Lin F, Boyes SG, Appl ACS (2015) Mater Interfaces 7:17765–17775
Tsuruoka T, Furukawa S, Takashima Y, Yoshida K, Isoda S, Kitagawa S (2009) Angew Chem Int Ed 48:4739–4743
Vaucher S, Li M, Mann S (2000) Angew Chem Int Ed 39:1793–1796
Wang H-S (2017) Coord Chem Rev 349:139–155
Wang Y, Zhao D (2017) Cryst Growth Des 17:2291–2308
Wang L, Feng X, Ren L, Piao Q, Zhong J, Wang Y, Li H, Chen Y, Wang B (2015) J Am Chem Soc 137:4920–4923
Wang H, Zhu Q-L, Zou R, Xu Q (2017) Chem 2:52–80
Wang XG, Cheng Q, Yu Y, Zhang XZ (2018a) Angew Chem Int Ed 57:7836–7840
Wang H-N, Zhang M, Zhang A-M, Shen F-C, Wang X-K, Sun S-N, Chen Y-J, Lan Y-Q (2018b) ACS Appl Mater Interfaces 10:32265–32270
Wang X, Liu W, Fu H, Yi XH, Wang P, Zhao C, Wang CC, Zheng W (2019) Environ Pollut 249:502–511
Wang K-B, Xun Q, Zhang Q (2020) EnergyChem 2:100025
Wang C, He C, Luo YH, Su S, Wang JY, Hong DL, He XT, Chen C, Sun BW (2020) Chem Eng J 379:122337

Wen P, Gong P, Sun J, Wang J, Yang S (2015) J. Mater. Chem. A 3:13874–13883
Wu MX, Yang YW (2017) Adv Mater 29:1606134
Wu Y-P, Wu X-Q, Wang J-F, Zhao J, Dong W-W, Li D-S, Zhang Q-C (2016) Cryst Growth Des 16:2309–2316
Wu Z, Xie J, Xu ZJ, Zhang S, Zhang Q (2019) J Mater Chem A 7:4259–4290
Xiao JD, Shang Q, Xiong Y, Zhang Q, Luo Y, Yu SH, Jiang HL (2016) Angew Chem Int Ed 55:9389–9393
Xu GW, Wu YP, Dong WW, Zhao J, Wu XQ, Li DS, Zhang Q (2017) Small 13:1602996
Xu C, Bi C, Zhu Z, Luo R, Zhang X, Zhang D, Fan C, Cui L, Fan Y (2019) Cryst Eng Comm 21:2333–2344
Yamini Y, Safari M, Morsali A, Safarifard V (2018) J Chromatogr A 1570:38–46
Yang Q, Xu Q, Jiang H-L (2017) Chem Soc Rev 46:4774–4808
Yi FY, Zhang R, Wang H, Chen LF, Han L, Jiang H-L, Xu Q (2017) Small Methods 1:1700187
Zhang L, Liu H, Shi W, Cheng P (2019a) Coord Chem Rev 388:293–309
Zhang T, Wang J, Zhang W, Yang C, Zhang L, Zhu W, Sun J, Li G, Li T, Wang J (2019b) J Mater Chem A 7:2845–2854
Zhao M, Ou S, Wu C-D (2014) Acc Chem Res 47:1199–1207
Zhao X, Wang Y, Li D-S, Bu X, Feng P (2018) Adv Mater 30:1705189
Zhao J, Liu X, Wu Y, Li D-S, Zhang Q (2019) Coord Chem Rev 391:30–43
Zhou X-P, Xu Z, Zeller M, Hunter AD (2009) Chem Commun 34:5439–5441
Zhou H-C, Long JR, Yaghi OM (2012) Chem Rev 112:673–674
Zhu H, Yuan J, Tan X, Zhang W, Fang M, Wang X (2019) Environ Sci Nano 6:261–272
Zhu Q-Q, Zhou Q-S, Zhang H-W, Zhang W-W, Lu D-Q, Guo M-T, Yuan Y, Sun F, He H (2020) Inorg Chem 59:1323–1331
Zou M, Zhao Y, Ding B, Jiang F, Chen Y, Ma P, Lin J (2021) Inorg Chem Front 8:2624

Chapter 17
Sequestering Groundwater Contaminants via Emerging Nanocomposite Adsorbents

Mitil M. Koli, Kritika Jashrapuria, Anima Johari, and Swatantra P. Singh

Abstract Groundwater resources have seen a potent contamination threat due to a spurt in industrial and anthropogenic activities in recent years. It is the only reliable water resource for a major chunk of rural and semi-urban population and serves as a backbone for drinking water supply. Although chemical quality of groundwater is also influenced by geological conditions of the natural aquifers, the share of industrialization and urbanization is profound. Typical groundwater pollutants such as arsenic, chromium, fluoride, and nitrate are growing causes of concern due to their toxic and lethal nature if consumed in excess to human as well as natural ecosystem. Conventional water treatment technologies are inefficient and uneconomical for sequestering these contaminants. Nanomaterials and nanocomposites are composed of nanometer size particulates. Due to their ample surface area, surface charge and functionalities, they have an extensive research scope for mitigating different contaminants. Carbon-based nanocomposites made up of carbon nanotubes (CNTs) and graphene oxide (GO) have been studied widely in this regard and have resulted in excellent adsorption capacities for different heavy metals and inorganic contaminants. Recently discovered super nano adsorbents—metal–organic frameworks (MOFs) due to its unique cage-like structure also promise better outcomes. In this vein, this chapter provides a holistic approach for investigating the potential of various CNTs, GO, and MOFs nanocomposites/nanohybrids in remediation of four major groundwater pollutants-arsenic, chromium, fluoride, and nitrate. The chapter

M. M. Koli · K. Jashrapuria · S. P. Singh (✉)
Environmental Science and Engineering Department (ESED), Indian Institute of Technology Bombay, Mumbai 400076, India
e-mail: swatantra@iitb.ac.in

S. P. Singh
Centre for Research in Nanotechnology and Science (CRNTS), Indian Institute of Technology Bombay, Mumbai 400076, India

A. Johari · S. P. Singh
Interdisciplinary Program in Climate Studies, Indian Institute of Technology Bombay, Mumbai 400076, India

A. Johari
DST, Science and Engineering Research Board, New Delhi 110070, India

S. P. Singh et al. (eds.), *Metal Nanocomposites for Energy and Environmental Applications*, Energy, Environment, and Sustainability,
https://doi.org/10.1007/978-981-16-8599-6_17

shall help the reader briefly understand the typical synthesis methods, removal mechanism and chemistry, adsorption capabilities and influence of pH on adsorption of these contaminants by different nanocomposites.

Keywords Carbon nanotubes · Graphene oxide · Groundwater · Metal organic framework · Nanocomposite · Nanomaterials

17.1 Introduction

Szent-Győrgi called water the "matrix of life". Due to a finite presence of freshwater sources on the earth's surface, dependance on groundwater for availing potable benefits is inevitable. Groundwater is a precious water resource on earth. However, large-scale industrialization, urbanization, and anthropogenic activities have increased wastewater generation, and its discharge into natural water bodies as well as groundwater further has escalated the freshwater deficit (Koli and Munavalli 2021). Hence, in the twenty-first century an increased global attention on resolving groundwater quality issues is realized (Shaji et al. 2021). Groundwater contaminants can be broadly classified as inorganic and organic contaminants. Typically, metals and metalloids such as barium, cadmium, copper, lead, mercury, uranium, chromium, arsenic, and cations and anions (nitrates and nitrites, sulfates, fluorides, and cyanides) constitute inorganic contaminants. The organic portion includes compounds of hydrocarbons, aromatic organic, chlorinated aliphatic, halogenated aliphatic, pesticides etc. (Sethi and Di Molfetta 2019). Amongst them, arsenic, chromium, fluoride, and nitrate contamination has been widely reported around the globe. World Health Organization (WHO) recommends maximum permissible limits in potable water for the aforementioned contaminants as 0.01, 0.05, 1, and 50 mg/L, respectively.

Arsenic in groundwater forms naturally due to the reductive dissolution of As-bearing minerals (Postma et al. 2016). Anthropogenic activities such as human interventions, mining, coal and petroleum extraction (Shaji et al. 2021) also contribute to arsenic contamination. Arsenic is found in the form of oxyanion of arsenate-As(III) or arsenite-As(V) compounds, both of which are deadly to living creatures as well as the environment (Siddique et al. 2020). Chromium is naturally found in groundwater as double oxides, principally as chromite ($FeO.Cr_2O_3$), crocoite ($PbO.CrO_3$) or chromic oxide (Cr_2O_3) (Sharma et al. 2008). It is also used in the leather tanning, textile dyeing, glass, paints, laundry chemicals industries and chrome plating. Chromium exists in trivalent-Cr(III) and hexavalent-Cr(VI) forms, causing severe health problems such as bronchitis, bronchogenic, carcinoma, liver damage, etc. (Bao et al. 2015). Naturally occurring fluoride-rich igneous and sedimentary rocks and their subsequent weathering are one of the causes for the presence of fluoride in groundwater. In addition, fly ash from the combustion of fossil fuels and phosphate containing fertilizers also elevates fluoride concentration (Sharma et al. 2008). Specific health outcomes viz. effects on the immune system, reproductive and developmental (birth) defects, and effects on the kidney and gastrointestinal tract are caused by excess fluoride intake (Harrison 2005). Natural biogeochemical processes control nitrate leaching into groundwater and thereby govern the presence of nitrate contamination in groundwater. Fertilizers used in agricultural activities are also contributors to the same (Huno et al. 2018; Sacchi et al. 2013). One of the notable health impacts of long-term ingestion of water with elevated nitrate concentrations (i.e., >50 mg/L) is methemoglobinemia, which is popularly called the 'blue baby syndrome' in infants below six months (Huno et al. 2018). Thus, due to perilous impact on human health and the environment, there is an urgent need for developing mechanisms for the efficient removal of these contaminants.

Several methods such as coagulation, adsorption, membrane processes, dialysis, electrocoagulation, and floatation have been used for water treatment polluted with arsenic, chromium, fluoride, and nitrate (Bao et al. 2015; Harrison 2005; Huno et al. 2018; Siddique et al. 2020). Groundwater decontamination via adsorption, though quite popular, has reported limited treatment efficiency by using conventional adsorbents due to their small surface area, the limited number of active sites, deficiency in selectivity, and low adsorption kinetics (Usman et al. 2020). As a solution to the persistent disadvantages of conventional adsorbents, nanomaterials with high surface area coupled with a higher number of active sites, tunable pore size, fast kinetics, and improved surface chemistry have been explored by researchers for environmental pollutants remediation. Moreover, nanotechnology can easily merge with other technologies and modify, endorse or clarify any existing scientific concept, which is why it is a so-called "platform" technology (Kyzas and Matis 2015).

Carbon-based nanomaterials such as carbon nanotubes (CNTs) and graphene oxide (GO) are promising adsorbent materials in water and wastewater treatment, especially in industrial and pharmaceutical-laden wastes (Nasrollahzadeh et al. 2021). The mechanism by which the metal ions are sorbed onto CNTs are attributed to electrostatic attraction, sorption–precipitation, and chemical interaction between metal ions and surface functional groups of CNTs (Anitha et al. 2015). GO contains various oxygen-containing surface functional groups such as hydroxyl, carbonyl, carboxylic, epoxide (Lal et al. 2020). They play a vital role in the removal of various pollutants via mechanisms such as electrostatic attraction ion-exchange, physical adsorption, hydrogen bond and π-π stacking interactions (Wang et al. 2018). Another promising approach to produce tailored adsorbents with tunable pore sizes, numerous active sites, and facile-charge separation are metal–organic frameworks (MOFs) (Gao et al. 2019). MOFs are called super-adsorbents due to their excellent adsorption capacity for sequestering numerous pollutants.

In the following sections, literature based on remediation of major groundwater contaminants; such as, arsenic, chromium, fluoride, and nitrate, using nanocomposites of CNTs, GO, and MOFs. Since, no such review is available dealing with a collation of these nanocomposites targeting specific groundwater contaminants, the following chapter can contribute towards understanding the available nanomaterials, nanocomposites/nanohybrids in this regard. A brief highlight on synthesis methods used for preparing the nanocomposites is discussed. The chapter also describes typical removal mechanisms and chemistry involved in adsorption, influencing parameters and tabulating a summarized form of nanocomposites with respective adsorption capacities.

17.2 Carbon Nanotubes

Carbon nanotubes or CNTs are graphene sheets rolled up into cylinders with diameters as small as one nanometer (Dai 2002). A roll of a single graphene sheet is termed a single-walled carbon nanotube (SWCNT or SWNT) while, multilayered rolls of graphene sheets are known as multiwalled carbon nanotubes (MWCNTs or MWNTs) (Ihsanullah 2019). Walls of pristine CNTs are not reactive, but their fullerene-like tips are considerably more reactive; thus, functionalization of the CNTs ends is often used to generate functional groups (e.g., –COOH, –OH, or –C=O) (Rao et al. 2007). Due to their excellent adsorption capabilities, CNTs have been extensively explored for sequestering various native as well as emerging contaminants.

17.2.1 Arsenic Removal Using CNT Nanocomposites

An iron-oxide coated MWCNT hybrid nanocomposite sorbent was reported for the efficient removal of As(III) and As(V) (Addo Ntim and Mitra 2011). The nanotube was functionalized through a microwave-assisted method. The respective adsorption capacity for As(III) and As(V) was found to be 1.723 and 0.189 mg/g. It was postulated that the anionic arsenic species ($H_2AsO_3^-$, $H_2AsO_4^-$) were adsorbed on the positively charged iron oxide-coated adsorbents through electrostatic attraction. Iron oxide, when exposed to the aqueous medium, the metal ions on the iron-oxide surface complete their coordination shells with hydroxyl groups, which either bind/release H^+ ions depending on the pH, thereby developing a surface charge. This created functional groups of OH^{2+}, OH, and O^- having excellent adsorption properties. The efficiency of arsenic removal was further enhanced by a nanohybrid of MWCNT and zirconia (ZrO_2) (Addo Ntim and Mitra 2012). As(III) was removed to the extent of 2.0 mg/g whereas 5.0 mg/g of As(V) could be mitigated using this nanohybrid. The formation of carboxylic groups and some sulphonation and nitration due to functionalization of MWCNT were impacting factors found for arsenic removal. A study on TiO_2 coated CNT nanocomposite filter was carried out to evaluate As(III) and As(V) removal (Liu et al. 2014). The electro-assisted filter was prepared using the simple filtration steam-hydrolysis method and had a specific surface area of 196 m^2/g, approximately twice greater than that of the CNT network. The filter showed a maximum adsorption capacity of 1.8 and 1.3 mg/g for As(III) and As(V), respectively, at a cell potential of 2 V. The increment in adsorption capacity was established due to internal convection and pore radius range, improved sorption site accessibility due to porosity and TiO_2 dispersion, and reduced TiO_2 negative surface charge due to anodic capacitance. Figure 17.1 describes the typical sorption

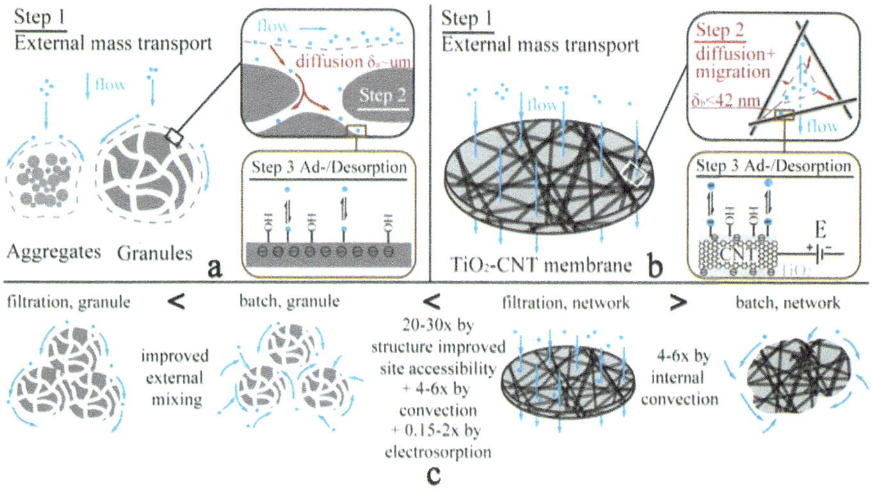

Fig. 17.1 **a** Aggregates of nanoparticles or granular adsorbents. **b** Mechanism of TiO_2–CNT filter. **c** Kinetic comparison of the granular TiO_2 and TiO_2–CNT network (Liu et al. 2014)

mechanism of arsenic on aggregates of nanoparticles (Fig. 17.1a) and TiO_2–CNT filter (Fig. 17.1b). The kinetic comparison of both aggregates and filter membrane is explained via Fig. 17.1c.

Zhang et al. (2017b), investigated a 3D nanostructure of CNTs vertically standing on graphene sheets and iron-oxide nanoparticles decorated on both the graphene and the CNTs (Zhang et al. 2017b). The material performed excellently in removing arsenic from contaminated water. The removal performance was attributed to the high surface-to-volume ratio and open-pore network of the graphene-CNT-iron oxide three-dimensional nanostructure. An in situ doped zero-valent iron (ZVI) onto MWCNTs structure was reported in which the iron source for the synthesis of magnetic MWCNTs was natural α-Fe_2O_3 (Alijani and Shariatinia 2017). The MWCNTs were synthesized using the chemical vapor deposition (CVD) method, and the resultant nanohybrid showed a maximum adsorption capacity of 200 mg/g for As(III) and 250 mg/g for As(V) at 90 and 60 min retention times, respectively. NiO nanoparticles were grown on CNTs by polyol method for preventing the agglomeration problem of nanoparticles (Moradi and Rokni 2017). The NiO/CNT nanocomposite was tested as an adsorbent for evaluating its arsenic removal capacity. The removal efficiency could reach up to 94% for an initial arsenic concentration of 20 ppm. A CNT functionalized magnetron sputtering technique was synthesized by depositing Cu onto the premade CNT membrane (Luan et al. 2019). Four membrane combinations viz. polymeric membrane mixed cellulose ester (MCE), MCE sputtered with Cu, pristine CNT, and CNT sputtered with Cu were compared for their As(V) removal ability. The Cu sputtered CNT had maximum removal above 90% with an adsorption capacity of 45 mg/g, since Cu catalyzed oxidation of As(III) to As(V) followed by adsorptive filtration of As(V) by the membrane. An eco-friendly

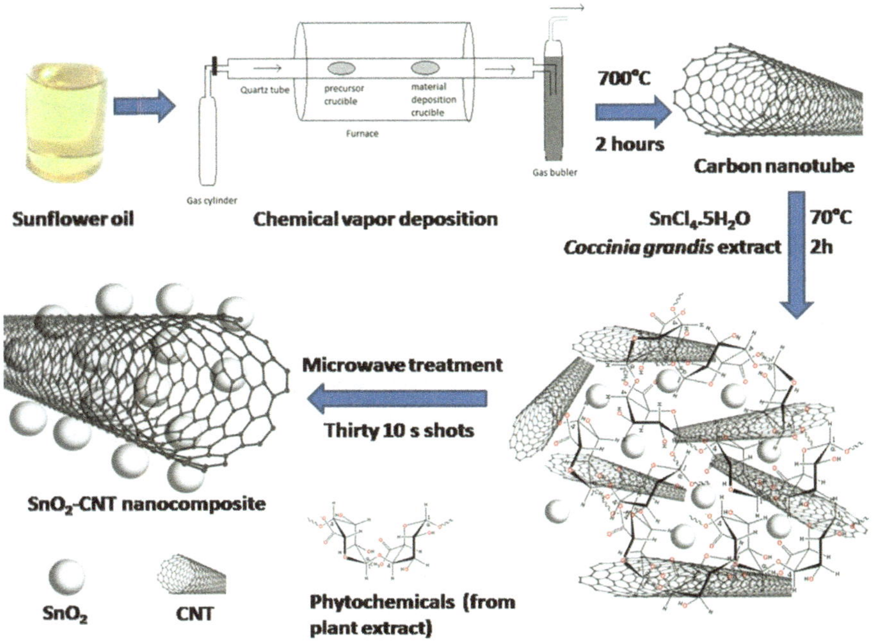

Fig. 17.2 Schematic illustration of SnO_2–CNT nanocomposite synthesis (Ahmaruzzaman et al. 2019)

approach towards the synthesis of nanocomposite adsorbent was tested by Ahmaruzzaman et al. (2019). They synthesized a SnO_2–CNT nano-heterojunction using a biogenic, economically viable, and sustainable strategy using sunflower oil (bio-precursor) to prepare MWCNTs followed by decoration with SnO_2 nanoparticles using *Coccinia grandis* extracts (Ahmaruzzaman et al. 2019). The typical synthesis method is described in Fig. 17.2. The nanocomposite could sequester As(III) to the extent of 106.95 mg/g for an initial arsenic concentration 1 mg/L. The redox environment on the SnO_2–CNT surface was reportedly the active site for oxidation of As(III) to As(V). The surface –OH groups of CNTs facilitated the adsorption process by anion exchange. Furthermore, the nanocomposite displayed potential antimicrobial properties against bacterial and fungal strains, making the nanocomposite an excellent adsorbent in the water treatment process.

17.2.2 Removal of Chromium Using CNT Nanocomposite

The removal of Cr(III) using MWCNTs modified with MnO_2 particles was demonstrated by Mohammadkhani et al. (2016). The modified MWCNTs showed enhanced adsorption capability in the removal of chromium ions from wastewater. Fei et al.

(2017) synthesized reticulated vitreous carbon CNT (RVC-CNT) electrode in the form of a biocathode in a microbial fuel cell (MFC) for improved removal of Cr(VI) ions from wastewater (Fei et al. 2017). The RVC-CNT improved the electron transfer rate and electrical conductivity. Thus, Cr(VI) reduction was enhanced due to the availability of more reaction sites. Efficient removal of Cr(III) from wastewater using nano-modified CNTs with varying conditions of adsorbent dose, pH, agitation speed, and contact time was reported by Atieh et al. (2010). They demonstrated that the physical and chemical properties of modified CNTs had more affinity towards Cr(III) ions to get adsorbed onto M-CNTs and hence enhanced adsorption efficiency of CNTs could be obtained for Cr(III) removal from wastewater Dehghani et al. (2015) determined the role of both SWCNTs and MWCNTs for hexavalent chromium removal experimented under variable conditions of pH, contact time, initial concentration and presence of competing sulfate anions. Sulfate anions tend to compete with Cr(VI) ions for adsorption sites which thereby hindered the removal efficiency of Cr(VI). Santhana et al. (2015) have reported the enhanced adsorption efficiency of MWCNTs by functionalizing them with ionic liquids (ILs). The interaction between ILs and MWCNTs led to formation of IL-oxi-MWCNT adsorbent which acted as a super-adsorbing host for Cr(VI) anions (Santhana et al. 2015). Huang et al. (2015) introduced magnetic properties to MWCNTs to improve their adsorption efficiency for the effective removal of Cr ions. ILs provided electrostatic interactions between quaternary ammonium cations and oxygen-containing carboxyl and hydroxy groups present on the surface of MWCNTs. The cation-π, anion-π, and other electrostatic interactions amongst the MWCNTs-IL complex and chromium ions enhanced the adsorption of chromium ions onto MWCNTs-IL complex. Magnetic CNTs can be easily dispersed in water and separated from water as well due to the presence of a magnet and hence pose like a supermodel for effective removal of heavy metal ions with no post contamination. Ali Atieh (2011) described the use of CNTs supported with activated carbon for efficient removal of chromium (VI) ions from wastewater. It was demonstrated that the adsorption efficiency of carbon nanotubes for Cr removal gets enhanced on the coating with activated carbon. Hence, it can be concluded that CNTs act as robust adsorbent material for the effective removal of heavy metal ions and find great potential in water remediation.

17.2.3 Removal of Fluoride Using CNT Nanocomposite

Wang et al. (2002) demonstrated an experiment for removing fluoride from water using novel CNT-supported alumina (Al_2O_3). A significant finding of this adsorbent was its workability over a broad pH range (6.0–9.0). The adsorbent had a specific surface area of 165 m^2/g, and further, a maximum adsorption capacity of 39.4 mg/g was reached for an alumina loading of 30 wt% at 25 °C. The positively charged aquo group of alumina selectively adsorbed the negatively charged fluoride ions via ligand exchange and ion exchange when the surface is neutral. The potential of MWCNTs was investigated by Ansari et al. (2011). Maximum adsorption (94%) occurred at pH 5 for a detention time of 18 min, and the saturation capacity obtained was 3.5 mg/g. The adsorbent was effective for an initial fluoride concentration less than two mg/L and thereby found to be better than most other adsorbents. Also, the presence of co-anions such as chlorides, sulfates, etc., seem to have a negligible impact on fluoride removal efficiency from water. Dehghani et al. (2016) determined the role of SWCNTs and MWCNTs in the efficient removal of F^- ions from wastewater under different experimental conditions wherein the maximum adsorption capacities of MWCNTs and SWCNTs were found to be 2.83 and 2.4 mg/g, respectively. A possible explanation for the same was the availability of the internal adsorption sites on MWCNTs due to the different pore size diameters (larger surface area) than for SWCNTs.

Recently a nanocomposite of CNT coated with coconut-shell charcoal was studied by Araga et al. (2019). The synthesis was carried out using chemical vapor deposition followed by ball milling of the composite. A high specific surface area of 358 m^2/g was observed, which resulted in almost 77% of fluoride removal efficiency for water contaminated with 4.4 mg/L fluorides. The adsorption mechanism could be explained by Reactions 17.1 and 17.2.

$$-COH + H^+ \leftrightarrow COH_2{}^+ \tag{17.1}$$

$$-COH_2 + F^- \leftrightarrow CF + H_2O \tag{17.2}$$

The overall reaction could be given by Eq. 17.3.

$$-COH + H^+ + F^- \leftrightarrow COH_2{}^+ + H_2O \tag{17.3}$$

where –COH represents the surface hydroxyl group of biomass-derived carbon. The fluoride ions may not only be adsorbed on the outer surface of nanotubes but also entered into the layers of CNTs because CNT caps can be opened up during the ball-milling process, thereby effectively removing fluoride ions even at lower concentrations (<3 mg/L).

17.2.4 Removal of Nitrate Using CNT Nanocomposite

Jodeh et al. (2019) demonstrated the use of magnetic MWCNTs functionalized with chitosan for nitrate removal from wastewater. The removal of $NO_3{}^-$ from wastewater on magnetic multiwalled nanocarbon tubes (MWCNTs) functionalized with chitosan was 88% at the optimum of contact time 50 min. The maximum adsorption capacities were between 119 and 322 mg/g and were highest at 25 °C, which could be attributed to the increased surface area due to the presence of the magnetic species. Tofighy and Mohammadi (2020) studied CNT sheets synthesized via chemical vapor deposition of cyclohexanol and ferrocene in a nitrogen atmosphere at 750 °C and further functionalized using concentrated nitric acid and liquid ammonia. The resultant sheet adsorbent demonstrated maximum monolayer adsorption capacity in the range 90.91–142.80 mg/L for a contact time of 24 h and pH 7. It was concluded that the large nitrate adsorption capacity of the functionalized CNT sheets was mainly due to the functional groups formed on the surface of CNT sheets which can react with nitrate. A copper phthalocyanine supported on functionalized multiwalled carbon nanotubes (CuPc/CNT) was used by Rajmohan and Chetty (2017). The adsorbent was prepared using the impregnation method and then characterized and tested for their electrocatalytic behavior towards nitrate reduction. The researchers conducted electrolysis studies in an H-cell (hybrid cell) which gave a percentage of nitrate removal up to 76 and 45% for CuPc/CNT and CuPc/C, respectively at 24 h contact time. Moradi et al. (2018) modified the conventional CNT with Lanthanum. The La/CNT composite was prepared using CNT loaded by lanthanum nitrate ($La(NO_3)_3$), and Hexamethylenetetramine. The effective removal capacity of nitrate was 6.48 mg/g for an initial concentration of 10 ppm at 210 min retention time. Lanthanum, due to its placement on cavities and carbon nanotubes surfaces, would cause an increase in adsorption capacity of the La/CNT composite (Table 17.1).

17.3 Graphene Oxide

Graphene is a two-dimensional material with carbon atoms in the sp^2 hybridization, which has extremely high thermal conductivity and stiffness (Geim 2009). GO is an oxidized derivative of graphene, which has ample of oxygen-containing functional groups, such as hydroxyl, epoxy, and carboxyl groups, with the 2D structure still intact. Combining GO with nanoparticles or polymer-carriers to prepare GO-based composites is an attractive approach to make the membrane more structurally stable (Han et al. 2021).

17.3.1 Removal of Arsenic Using Graphene Oxide Nanocomposite

A potential application of water-dispersible magnetite modified reduced GO for the removal of arsenic ions from water was reported by Chandra et al. (2010). The researchers synthesized magnetite-reduced graphene oxide (MRGO) composites via a chemical reaction with a magnetite particle size average of 10 nm. MRGO composites were found superparamagnetic at room temperature and had a high binding capacity for arsenic ions. Results described 99.99% removal of arsenic for one ppb initial concentration. The respective maximum adsorption capacity for As(V) and As(III) were 5.83 and 13.10 mg/g. Arsenic ion removal on the MRGO surface was reportedly due to the electrostatic attraction between the positively charged surface

Table 17.1 Removal of arsenic, chromium, nitrate and fluoride using CNT based nanocomposite adsorbents

Metal removed	Nanocomposite adsorbent	Removal capacity	References
As(III) and As(V)	MWCNT coated with iron-oxide	1.723 mg/g and 0.189 mg/g	Addo Ntim and Mitra (2011)
	MWCNT and zirconia (ZrO_2)	2.0 mg/g and 5.0 mg/g	Addo Ntim and Mitra (2012)
	TiO_2/CNT	1.8 mg/g 1.3 mg/g	Liu et al. (2014)
	Zero-valent iron onto MWCNTs	200.0 mg/g 250.0 mg/g	Alijani and Shariatinia (2017)
As(III)	NiO/CNT	94%	Moradi and Rokni (2017)
	SnO_2–CNT	106.95 mg/g	Ahmaruzzaman et al. (2019)
As(V)	Cu onto premade CNT membrane	45 mg/g	Luan et al. (2019)
Cr(III)	MnO_2/MWCNTs	99.01 mg/g	Mohammadkhani et al. (2016)
Cr(III) Cr(VI)	IL-oxidized MWCNTs	85.83 mg/g	Ali Atieh (2011)
F^-	CNT supported alumina	39.4 mg/g	Wang et al. (2002)
	MWCNTs	3.5 mg/g	Ansari et al. (2011)
	MWCNTs and SWCNTs	2.84 mg/g 2.4 mg/g	Dehghani et al. (2016)
	CNT coated with coconut shell charcoal	4.4 mg/g	Araga et al. (2019)
NO_3^-	CNT sheets synthesized via chemical vapor deposition of cyclohexanol and ferrocene in nitrogen atmosphere at 750 °C and further functionalized using concentrated nitric acid and liquid ammonia	90.91–142.80 mg/L	Tofighy and Mohammadi (2012)
	Copper phthalocyanine supported on functionalized MWCNT	76%	Rajmohan and Chetty (2017)
	CNT with lanthanum	6.48 mg/g	Moradi et al. (2018)
	Magnetic multiwalled nanocarbon tubes (MWCNTs) functionalized with chitosan	119–322 mg/g	Jodeh et al. (2019)

of MRGO and the negatively charged arsenic/arsenous acid. Yu et al. (2015) further developed a nanocomposite of magnetic iron oxide/graphene oxide (MGO) with high iron loading (51 wt%) and synthesized using the co-precipitation method. The nanocomposites possessed desirable magnetic as well as adsorptive properties for the removal of As(III) and As(V) ions. The maximum adsorption capacity for both ions was 54.18 and 26.76 mg/g, respectively, which correlated with Freundlich isotherm model. The results suggested that the As(V) became preferentially associated with iron oxides during the adsorption process, and the Fe distribution is directly correlated with the As distribution. Shahrin et al. (2018) studied two types of novel hybrid nanomaterials, i.e., graphene oxide-manganese ferrite (GMF) and titania nanotube-manganese ferrite (TMF), for removing arsenate (As(V)) ions from water via adsorption mechanisms. Results revealed maximum adsorption capacities of GMF and TMF against As(V) are 102.0 and 80.8 mg/g, respectively, when tested using feed water solution containing a concentration of As(V) equal to 400 mg/L. It was observed that under the acidic conditions ($pH < 4$), the presence of the $H_2AsO_4^-$ ions in the aqueous environment and the existence of $-OH_2^+$ ions on the surface of the nanomaterials favoured the As(V) adsorption via electrostatic attraction mechanism. Also, an increase in surface area and the greater number of oxygenated groups in the hybrid nanomaterials improved the adsorption capacity.

17.3.2 Removal of Chromium Using Graphene Oxide Nanocomposite

A facile fabrication and processing of cerium oxide-graphene oxide (CeO_2–GO) hybrid nanocomposites were investigated for chromium removal (Kashinath et al. 2019). The in-situ hexagonal nano cerium oxide particles embedded on the layered surface of GO sheets showed enhanced Cr(VI) removal (better than pure GO and CeO_2). GO/polyamidoamine dendrimers (GO/PAMAMs) composites were used to remove Cr(VI) (Liu et al. 2019). The nanocomposite showed 90.7% removal of Cr(VI) for 30 mg/L initial concentration and pH 2. The adsorption capacity could exceed upto 131.58 mg/g at 293.1 K. It was reported that the Cr(VI) binds to the protonated amine of GO/PAMAMs and then reduces to Cr(III) with the assistance of π-electrons on the carbocyclic six-membered ring of GO. Further, Cr(III) gets released into solution under the electrostatic repulsion between the Cr(III) and the protonated amine groups. Magnetic cyclodextrin–chitosan/graphene oxide (CCGO) was investigated for its chromium removal capacity (Li et al. 2013). Due to high surface area, abundant hydroxyl and amino groups of CCGO, and the magnetic property of Fe_3O_4, the Cr(VI) could be easily and rapidly extracted from water by magnetic attraction. About 67.66 mg/g Cr(VI) was effectively removed using this nanocomposite. The adsorption performance of CCGO strongly depended on the surface charge concentration, specific surface area and pH of the medium. The typical proposed mechanism for adsorption is depicted in Fig. 17.3. A chitosan grafted

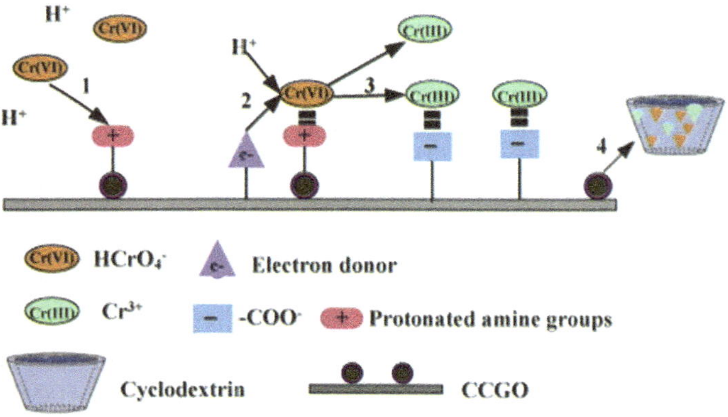

Fig. 17.3 Cr(VI) adsorption mechanism on CCGO nanocomposite (Li et al. 2013)

graphene oxide (CS-GO) nanocomposite was studied in batch mode (Samuel et al. 2019). The CS-GO nanocomposite material was prepared by ultrasonic irradiation technique. The adsorption capacity to the extent of 104.16 mg/g was achieved at pH 2.0, in the contact time of 420 min. The enhanced adsorption capability could be attributed to the presence of abundant active sites in the CS-GO composite material for the adsorption of Cr(VI) metal ions. A batch study using a packed-bed column was explored by using a nanocomposite of polypyrrole graphene oxide (PPy–GO NC) (Setshedi et al. 2015). It was noted that graphite was successfully exfoliated to thin layered graphene sheets and ensured the presence of the PPy in the PPy–GO NC. A maximum adsorption capacity of 625 mg/g at 25 °C was achieved. Moreover, the presence of co-existing ions had no effect on Cr(VI) removal. A reasonable reason for the selectivity of the adsorbent for Cr(VI) is that, at lower pH, the nanocomposite is positively charged and offers a repulsive force to approaching positively charged cations. A highly reductive and magnetic graphene/Fe_3O_4 composite (HR-M-GO/Fe_3O_4) was synthesized via GO oxidizing $FeCl_2$ in situ (Hou et al. 2016). The maximum adsorption capacity of the HR-M-GO/ Fe_3O_4 for Cr(VI) reached 31.8 mg/g, which was greater than the graphene/γ-Fe_2O_3 composite. The possible mechanism of HR-M-GO/Fe_3O_4 for removing Cr(VI) was that Cr(VI) was reduced to Cr(III) by ferrous hydroxide moieties in the graphene structure, and the resulting Cr(III) ions were easily arrested by the negatively charged HR-M-GO/ Fe_3O_4. The nanocomposite was also efficient for sequestering Pb(II), Cu(II), and Zn(II) with about cent percent efficiency.

17.3.3 Removal of Fluoride Using Graphene Oxide Nanocomposite

A novel magnetic adsorbent of iron-aluminium oxide (IAO) nanoparticles anchored on graphene oxide (IAO/GO) was prepared by a one-step method of co-precipitation for fluoride removal from an aqueous solution (Liu et al. 2016). The synthesized IAO/GO revealed a higher adsorption capacity (64.72 mg/g) towards fluoride. Fluoride removal was significant over a wide range of pH from 3 to 9. At pH below 3 and above 10, IAO/GO showed a decline in adsorption capacity since, at pH < 3, fluoride ion is transformed to electrically neutral hydrofluoric acid, which can diminish the electrostatic attraction between the adsorbent surface and fluoride while at pH > 10.0, the adsorbent surface becomes strongly negative, which repels the fluoride ions by electrostatic repulsion, and the high concentration of hydroxyl ions could compete with the fluoride ions to occupy the adsorption sites. The adsorption mechanisms attributed for adsorption of fluoride onto the adsorbent include electrostatic interactions, anion exchange, and inner-sphere complexation. By combining the advantages of GO and IAO, IAO/GO exhibited a high adsorption capacity, good acid-alkali stability, superparamagnetism, and fine selectivity for fluoride, thereby making it a potential candidate for defluoridation studies.

A zirconium-chitosan/graphene oxide (Zr-CTS/GO) based membrane adsorbent was applied for fluoride removal from an aqueous solution by Zhang et al. (2017a). Results exhibited a maximum adsorption capacity of fluoride (29.05 mg/g) within a pH range of 3–11. The adsorbent reached its equilibrium capacity by 45 min. It was suggested that Zr(IV) species reacted strongly with oxygen functional groups from the CTS/GO complex, and Zr-F species formed partly through fluoride ions exchanged with chloride ions and –OH and partly through chemical rearrangement. Sahoo and Hota (2018) proposed a magnetic MgO–$MgFe_2O_4$ nanocomposite anchored onto graphene oxide (GO) substrate (MgO–$MgFe_2O_4$/GO) as well as a MgO–Fe_2O_3 nanocomposite prepared by hydrothermal method as shown in Fig. 17.4. It was revealed that the GO-based nanocomposite achieved a magnetic property by forming new $MgFe_2O_4$ phase. A fast adsorption process was established since the equilibrium was reached within 60 min. It was observed that the presence of GO not only plays a significant role in the structural transformation but also enhances the surface area of the nanocomposites. The possible mechanism of F^- ions adsorbed onto GO-based MgO–$MgFe_2O_4$ adsorbent can be explained by three types of the process such as (i) electrostatic interaction (Eq. 17.4) (ii) complex formation (Eqs. 17.5 and 17.6), and (iii) hydrogen bonding between the hydroxyl group of the GO and fluoride ion (Eq. 17.7).

$$\equiv MgO + H_2O + F^- \rightarrow \equiv MgOH_2{}^+ - F + H_2O \tag{17.4}$$

$$\equiv MgOH_2{}^+ + F^- \rightarrow \equiv MgOH_2{}^+F^- \tag{17.5}$$

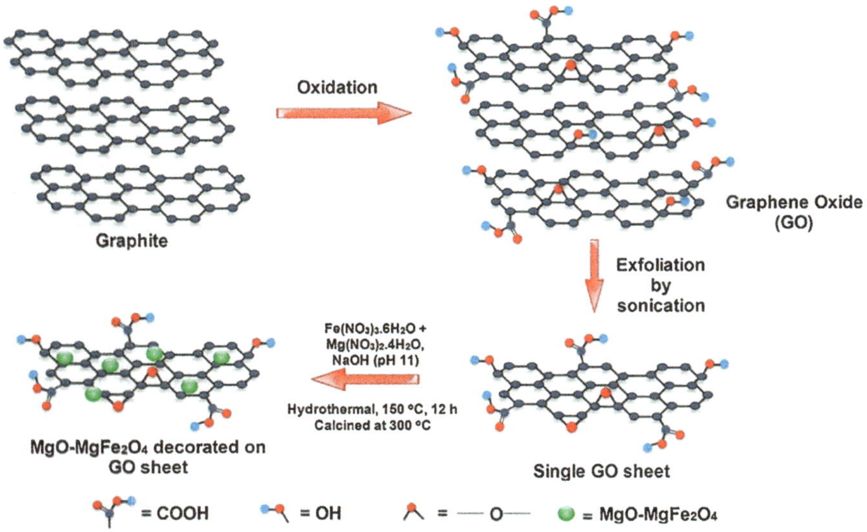

Fig. 17.4 Schematic representation of GO based $MgO–MgFe_2O_4$ nanocomposite (Sahoo and Hota 2018)

$$\equiv MgOH + F^- \rightarrow \equiv Mg{-}F^- + HO^- \quad (17.6)$$

$$\equiv G{-}O{-}H + F^- \rightarrow \equiv G{-}O{-}H\ldots F^- \quad (17.7)$$

Xu et al. synthesized a GO/Al_2O_3 nanocomposite by using the hydrothermal method for defluoridation. The researchers compared the adsorption capacities of GO/Al_2O_3 with γ-Al_2O_3 wherein the former exhibited higher capacity (4.86 mg/g) than the latter (3.04 mg/g) (Xu et al. 2020). The BET surface area of GO/Al_2O_3 is 326.22 m^2/g, and the corresponding pore size centre is 3.7 nm. The presence of GO in the nanocomposite was claimed to have an increment in the adsorbent's specific surface area, which further led to improvement in the adsorption capacity for fluoride from the aqueous solution. The adsorption rate rose during the first 60 min, and the equilibrium state was achieved at 90 min. The removal mechanism of fluoride was attributed to the electrostatic attraction, ligand exchange, ion exchange, and hydrogen-bond interaction between fluoride and OH groups.

17.3.4 Removal of Nitrate Using Graphene Oxide Nanocomposite

Motamedi et al. (2014) studied three nanocomposites viz. Ni@rGO, Co@r-GO, and Fe@r-GO. These magnetically functionalized nanocomposites were quantitatively yielded via $NaBH_4$ reduction of Fe^{3+}, Ni^{2+} and Co^{2+}, in the presence of GO. Basically, Ni and Co nanoparticles (NPs) were coated with graphene oxide (GO) so as to make them comparable to FeNPs in removing nitrate from water. About 99, 95, and 94% of nitrate reduction were possible for the respective three nanocomposites for an initial concentration of 5 ppm at an initial pH of 6.9 and constant ionic strength of 0.01 M KCl. Graphene oxide contains a variety of functional groups including carboxyl, hydroxy, epoxy, and ketone as coordination sites of magnetic nanoparticles to its surface, and if dispersed in water, the negative surface charges adsorb the metal ions through electrostatic attraction. Thus, in situ reduction of metal ions via $NaBH_4$ suggests synergistic coupling between magnetic nanoparticles and GO and as a result, a uniform coating of Fe, Ni, and Co nanoparticles on the surface of r-GO are obtained that is found to be an attribute to the high nitrate removal efficiency of the nanocomposite.

17.4 Metal Organic Framework

Metal–organic frameworks (MOFs) are hybrid organic–inorganic solid materials composed of metal ions or metal-ion clusters and organic ligands (Yin et al. 2020). MOFs have been increasingly applied as functionalizing agents for membranes because of their porosity, high surface area, small particle size, aspect ratio control, tuneability, compatibility with a polymeric network, and exuberance of diverse functional groups (Le et al. 2021). A short insight into the latest available MOFs for arsenic, chromium, fluoride, and nitrate is given in the following sections.

17.4.1 Removal of Arsenic Using MOF Nanocomposite

Arsenic (As(III) and As(V)) removal from aqueous solutions using metal–organic frameworks has been studied extensively. A nano sorbent named zeolitic imidazolate framework-8 (ZIF-8) which is a tetrahedral framework formed by zinc ions and imidazolate ligands with sodalite topology was studied (Jian et al. 2015). The laboratory synthesized ZIF-8 nanoparticles had a high surface area of 1063.5 m^2/g and were 200–400 nm in particle size. A schematic of ZIF-8 synthesis is depicted in Fig. 17.5. The nanoparticles were found to be successful for mitigating both As(III) and As(V) with their adsorption capacities of 49.49 and 60.03 mg/g, respectively, at 25 °C and pH 7.0. ZIF-8 was found to be stable at neutral and alkaline conditions; however, at acidic conditions a large amount of leakage of Zn^{2+} ions was found to be released into water from the sorbent. This hindered the adsorption process. Anions such as SO_4^{2-} and NO_3^- had no significant effect on the arsenic adsorption while the adsorption was significantly inhibited by PO_4^{3-} and CO_3^{2-}. The hydroxyl and amine groups present on ZIF-8 surface contributed to efficient arsenic removal via electrostatic attraction. Further, Liu et al. (2015) in their study demonstrated the

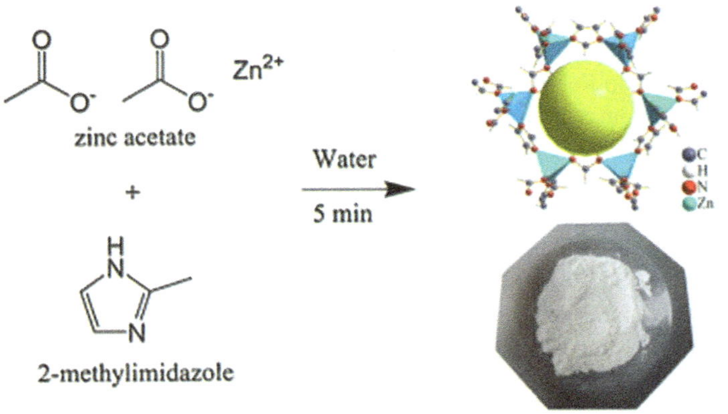

Fig. 17.5 Synthesis of ZIF-8 nanoparticles (Jian et al. 2015)

use of ZIF sorbents with three different morphologies, viz. cubic ZIFs, leaf-shaped ZIFs, and dodecahedral ZIFs. The nanoparticles were prepared in aqueous solution via green methods at room temperature and tested for removal of As(III). Results revealed that the maximum adsorption capacities for cubic ZIFs, leaf-shaped ZIFs, and dodecahedral ZIFs were 122.6, 108.1, 117.5 mg/g, respectively at pH 8.5. At an initial concentration of 200 μg/L, As(III) was rapidly decreased to below 10 μg/L by the three ZIFs in 2 h at an adsorbent dosage of 0.2 g/L. A possible explanation for such high adsorption mechanism. It was evidenced that zinc hydroxyl substitution was the main surface complexation method for As(III) adsorption onto the three ZIFs.

Meanwhile, Fe-based MOF is a good candidate for arsenic removal in water due to the possibility of the formation of "As–O–Fe" bonds (Vu et al. 2015; Zhu et al. 2012). Novel MOFs synthesized at Materials of Institute Lavoisier (MIL) have been found effective for the removal of heavy metals. Accordingly, Fe-based metal–organic framework MIL-88A microrods were recently synthesized by hydrothermal method for studying its applicability for As(V) removal. A maximum of 145 mg/g adsorption of As(V) was possible for an initial arsenic concentration of 100 mg/L at pH 5. Fe-based MIL-88A surfaces generally have abundant OH^- groups on their surface, which can be further exchanged by the ligand $H_2AsO_4^-$ in an aqueous solution (Yang and Yin 2016). Also, when Fe-based MIL-88A is exposed to water, it has an unusually large swelling of ~85% which led to the expansion of its internal channels and cages, which became beneficial for the ion transportation into its interior and resulted in more ligand exchange between $H_2AsO_4^-$ ions and OH^- groups.

A 2D metal–organic framework $[Co_3(tib)_2(H_2O)_{12}](SO_4)_3$ which is hereby called BUC-17 was developed for arsenic removal (Pang et al. 2020). The MOF was prepared by the hydrothermal method. It is revealed that BUC-17 showed higher adsorption capacity toward As(V) with a maximum uptake capacity to the extent of 129.2 mg/g at 298 K. pH values and contamination of foreign ions were emphasized as important influencing factors during the whole adsorption process. The highly efficient adsorption towards arsenate might not be attributed to its porosity but mainly due to electrostatic interaction between the positively charged BUC-17 between pH 4–10 and the As(V) ions.

Abu Tarboush et al. (2018) investigated and compared As(V) removal from aqueous media using the water-stable zinc-metal organic frameworks (Zn-MOF-74) (Abu Tarboush et al. 2018). The MOFs were prepared by two different procedures, first via room-temperature precipitation (RT-Zn-MOF-74) and second by employing solvothermal process (HT-Zn-MOF-74). The nano-sized RT-Zn-MOF-74 exhibited superior performance for As(V) removal to that of HT-Zn-MOF-74, wherein the former showed maximum adsorption uptake 99.0 mg/g while the latter showed 48.7 mg/g uptake of As(V). A higher adsorption uptake took place onto the room temperature synthesized ones due to their small particle size and better dispersion. The endothermic nature of the process confirmed that the mechanism of adsorption occurred via the successive substitution of the different layers of water surrounding the open metal site by the arsenate entity.

A water-stable Zr-metal–organic framework (MOF) (UiO-66) nanoparticles for various defect concentrations was studied by Assaad et al. (2020). The defects in these MOFs were induced by the addition of a monocarboxylic acid modulator [acetic acid or trifluoroacetic acid (TFA)] to the synthesis mixture. The concentrations of the defects were effectively tuned and controlled by changing the type and the amount of the modulator. The most modulated sample (UiO-66-36TFA) which was obtained by using TFA as the modulator, showed an adsorption capacity around 200 mg/g at neutral pH. This result demonstrated that UiO-66-36TFA showed a higher elimination percentage than the non-modulated UiO-66 and reached the equilibrium (100% removal) at lower mass (0.5 mg). The removal mechanism of arsenic was attributed to the free Lewis acid sites formed in the MOF clusters as a result of missing linker defects. In addition, the nanoparticles showed outstanding arsenate selectivity against interfering anions and were efficiently recycled to maintain the same adsorption capacity after five regeneration cycles.

Zhang et al. (2015) developed a robust cationic Zr-cluster-based metal–organic framework removal of $Cr_2O_7^{2-}$ ions from aqueous solution (Zhang et al. 2015). The MOF ZJU-101 was synthesized from MOF-86713 through post-synthetic modification in which methyl groups were added to the pyridyl sites to form the cationic framework. ZJU-101 exhibited a high adsorption capacity of 245 mg/g for $Cr_2O_7^{2-}$ ions within a very short period of time. The high selectivity absorption is attributed to the strong coulombic attraction between framework and $Cr_2O_7^{2-}$, and matching of $Cr_2O_7^{2-}$ with the pores.

17.4.2 Chromium Removal Using MOF Nanocomposite

A dual ultraviolet–visible and electron paramagnetic spectroscopy approach was introduced to determine the Cr speciation within Zr-based metal–organic frameworks

(MOFs) (Saiz et al. 2020). The generation of defect positions at ZrMOFs boost Cr(VI) adsorption, whilst the presence of reductive groups at the organic linkers played a key role to stabilize it as Cr(III) isolated and/or clustered ions. Du et al. (2019) studied the removal of Cr(VI) ions from surface and ground water using photocatalytic reduction of Cr(VI) to Cr(III) (Du et al. 2019). The MOF, UiO-66-NH_2(Zr/Hf) membrane was utilized as a photocatalyst to reduce Cr(VI) ions with high efficiency and easy separation from the treated wastewater. The UiO-66-NH_2(Zr/Hf) MOF membrane photocatalysts were fabricated via a reactive seeding method on a α-Al_2O_3 substrate. The UiO-66-NH_2(Zr) membrane could maintain more than 94% Cr(VI) reduction efficiency after 20 cycles because of its exceptional chemical and water stability. The influences of foreign ions on Cr(VI) reduction were investigated to mimic real lake water, which revealed that no obvious adverse effects can be found with the presence of common foreign ions in surface water. Kaur et al. (2020) reported the fabrication and utilization of a novel 2D zinc-based MOF, $\{[Zn(PA^{2-})(4,4'\text{-bpy})]\text{-}(H_2O)\}_n$ (where PA = pamoic acid and 4,4′-bpy = 4,4′-bipyridine). The bright green fluorescence of Zn-MOF can be quenched upon interaction with a $Cr_2O_7^{2-}$ ion, which implies the MOF's applicability as a Cr(VI) detector through turn-off fluorescence signalling. The limit of detection for fluorometric recognition of Cr(VI) was found to be 4.12 μM, and almost a complete reduction of toxic Cr(VI) ion was achieved.

17.4.3 Fluoride Removal Using MOF Nanocomposite

MOF for fluoride removal has been extensively studied in recent years. The stability and performance of different MOFs in contaminated fluoride solutions were studied (Zhao et al. 2014). These MOFs were. (i) MIL-53(Al, Fe, Cr), (ii) MIL-68(Al), (iii) CAU-1, (iv) CAU-6, (v) UiO-66(Hf, Zr), (vi) ZIFs (7, 8, 9). Amongst them, UiO-66 (Zr) showed maximum adsorption capacity of 41.36 mg/g. The adsorption mechanism followed was Langmuir-based. The adsorbent had a negligible effect due to interference of anions like Cl^-, NO_3^- and SO_4^{2-}, while the adsorption capacity reduced as the concentration of CO_3^{2-} ions increased. The researchers suggested the presence of –OH ions would enhance the defluoridation performance of MOFs significantly. A super adsorbent MOF called aluminium fumarate, AlFu in hydrothermal conditions using aluminium sulfate 18 hydrate and fumaric acid was synthesized (Karmakar et al. 2016). The resultant MOF had a high specific surface area of 1156 m^2/g and was thermally stable up to 700 K. An adsorption capacity of 600 mg/g for fluoride removal was obtained at 293 K for a 30 mg/L F feed solution. Due to the presence of a positive charge on the MOFs at a normal pH of 7, the MOF facilitated significant adsorption of negatively charged fluoride ions. A rice-like granular MOF was synthesized by Wang et al. (2019) via a hydrothermal reaction process. The specific surface area of the MOF was 220 m^2/g which followed the Langmuir isotherm. A maximum fluoride adsorption capacity of 42.19 mg/g was achieved at 298 K. The MOF had an adsorption capacity of 61.8% even after seven regeneration cycles, making it one of the suitable alternatives for fluoride removal.

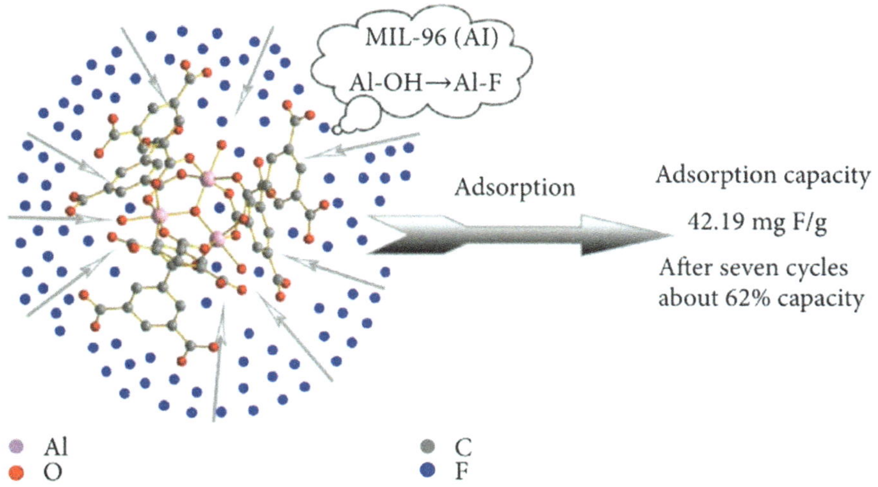

Fig. 17.6 The adsorption process of MIL-96(Al) (Wang et al. 2019)

F^- in water substituted –OH on the adsorbent and combined with the Al atom to achieve the purpose of defluorination, which was a method by means of metal sites in MOFs to remove fluoride. The typical mechanism is illustrated in Fig. 17.6. Another novel adsorbent using Ce(III) ions was studied by Yang et al. (2017). The MOF, Ce-MIL-96, i.e., Ce(III) ion incorporated aluminium-trimeric MOF was synthesized using alcohol-solvent incipient wetness impregnation. The aluminium ions were fixed within the framework of Ce-MIL-96 adsorbent, and the Ce ions entered into the internal space of MIL-96. This disabled the release of aluminium and cerium ions into the treated solution. The MOF gave a maximum adsorption capacity of 38.65 mg/g at 298 K over a wide range of pH (3–10). Moreover, no significant influence due to the presence of Cl^-, NO_3^-, HCO_3^-, PO_4^{3-} and SO_4^{2-} ions was observed. The adsorption capacity was appreciably greater than 70% even after nine regeneration cycles deeming the MOF suitable for fluoride mitigation from groundwater. Two lanthanide-based MOFs were studied by Yang et al. (2017) for fluoride removal from water. $[Ce(L1)_{0.5}(-NO_3)(H_2O)_2]0.2DMF$ had a higher adsorption capacity (103.95 mg/g) compared to $Eu_3(L_2)_2(OH)(DMF)_{0.22}(H_2O)_{5.78}$ (57.01 mg/g) at a pH range of 3–7. The major driving force behind adsorption was due to entropy effect.

17.4.4 Nitrate Removal Using MOF Nanocomposite

Nitrate removal using a metal–organic framework has been explored to a little extent. Mehmandoust et al. (2019) developed MOF-5 using solution and solvothermal methods. The MOF-5 was prepared via the self-assembly of zinc acetate dihydrate, benzene-di-carboxylic acid linker ligand using DMF without and with tri-ethylamine as capping agent at room temperature. High adsorption of nitrate was achieved at acidic pH 4 after 6 h, wherein the specific surface area of MOF-5 was about 988.34 m^2 at room temperature. Not much literature is available for addressing nitrate removal using MOFs.

Metal removed	MOF used	Removal capacity	References
As(III) and As(V)	ZIF-8	49.49 mg/g and 60.03 mg/g	Jian et al. (2015)
As(III)	Dodecahedral shaped ZIF	122.6 mg/g	Liu et al. (2015)
As(V)	MIL-88A	145 mg/g	Yang and Yin (2016)
	$[Co_3(tib)_2(H_2O)_{12}]\ (SO_4)_3$ (BUC 17)	129.2 mg/g	Pang et al. (2020)
	RT-Zn-MOF-74	99 mg/g	Abu Tarboush et al. (2018)
	30 nm UiO66	98.6%	He et al. (2017)
	UiO66-PVDF	267 mg/g	Wan et al. (2020)
	UiO66-PSF-LiCl	87.5% from 100 mg/L As(V)	Tajuddin et al. (2020)
	UiO-66-36TFA	200 mg/g	Assaad et al. (2020)
	Fe_3O_4@MIL-100Fe	18 mg/g	Yang et al. (2016)
	UiO-66-NH_2 MOF on cellulose fibres	84.5%	Hashem et al. (2019)
	UiO-66-NH_2(Zr/Hf)	94%	Du et al. (2019)
	$\{[Zn(PA^{2-})(4,4'\text{-bpy})]\text{-}(H_2O)\}_n$ (where PA = pamoic acid and 4,4′-bpy = 4,4′-bipyridine)	Almost complete rejection	Kaur et al. (2020)
	Chitosan–MOF(UiO-66) composite	94 mg/g	Wang et al. (2016)
	PVDF/chitosan/NMOFs	602.3 mg/g	Pishnamazi et al. (2020)
	Zr-based ZJU-101	245 mg/g	Zhang et al. (2015)
F^-	$[Ce(L1)_{0.5}(-NO_3)(H_2O)_2]0.2DMF/Eu_3(L_2)_2(OH)(DMF)_{0.22}(H_2O)_{5.78}$	103.95/57.01 mg/g	Ke et al. (2017)
	MIL-96(Al)	31.69 mg/g	Wang et al. (2019)
	CaFu	166.11 mg/g	Ke et al. (2018)
	AlFu	600 mg/g	Karmakar et al. (2018)

17.5 Conclusion

Carbon-based nanocomposites and metal–organic frameworks have an excellent potential for groundwater contaminant remediation. Nanocomposites of CNTs have functionalized tips. The formation of tunable functional groups such as carboxylic and hydroxyl enables an enhanced removal of anionic contaminants. Sulphonation and Nitration are typical processes for modulating functional groups onto CNT nanocomposites. CNT composites with nanomaterials like TiO_2 improves the porosity of the material, which creates large number of active sites for the adsorption of contaminants. Further, the anion π–cation π electrostatic interaction too is one of the possible reasons for the improved treatment of water using CNT nanocomposite.

GO nanocomposites show an increment in the number of oxygenated functional groups. The addition of certain nanomaterials in the GO composite increases protonated amine groups, which are responsible for enhanced adsorption capacity. The incorporation of iron-based nanomaterials in GO imparts a negative charge which repels the anionic pollutants. Another mechanism responsible for improved adsorption is the formation of an inner sphere complex with the pollutants with specific nanocomposites.

MOFs such as ZIF-8 contain hydroxyl and amino, functional groups. Thus, the electrostatic attraction becomes a dominant mechanism for adsorption. The nanocomposite of MIL-88-A MOF causes expansion of the cage-like structure, and a ligand exchange between anions is a possible reason for sequestering pollutants. Successive substitution reactions also play an important role in the removal of these anionic contaminants. Therefore, it is possible to excellently remediate groundwater contaminants; such as, arsenic, chromium, fluoride, and nitrate, using these novel nanocomposite adsorbents. The incorporation of these nanocomposites in polymeric membranes can possibly enhance the selective nature of the virgin membrane. GO membrane fabrication needs exploration considering its industrial application. Although these nanomaterials are efficient in their own aspects, research is still required for a large-scale application of the same for the simultaneous removal of mixed contaminants.

Acknowledgement The corresponding author(SPS) acknowledges the funding received from SYST scheme, WTI, Department of Science and Technology, Government of India for carrying out this work.

References

Abu Tarboush BJ, Chouman A, Jonderian A, Ahmad M, Hmadeh M, Al-Ghoul M (2018) Metal-organic framework-74 for ultratrace arsenic removal from water: experimental and density functional theory studies. ACS Appl Nano Mater 1(7):3283–3292. http://doi.org/10.1021/acsanm.8b00501

Addo Ntim S, Mitra S (2011) Removal of trace arsenic to meet drinking water standards using iron oxide coated multiwall carbon nanotubes. J Chem Eng Data 56(5):2077–2083. http://doi.org/10.1021/je1010664

Addo Ntim S, Mitra S (2012) Adsorption of arsenic on multiwall carbon nanotube-zirconia nanohybrid for potential drinking water purification. J Colloid Interface Sci 375(1):154–159. http://doi.org/10.1016/j.jcis.2012.01.063

Ahmaruzzaman M, Mohanta D, Nath A (2019) Environmentally benign fabrication of SnO_2–CNT nanohybrids and their multifunctional efficiency as an adsorbent, catalyst and antimicrobial agent for water decontamination. Sci Rep. https://doi.org/10.1038/s41598-019-49181-2

Ali Atieh M (2011) Removal of Chromium (VI) from polluted water using carbon nanotubes supported with activated carbon. Procedia Environ Sci 4:281–293. http://doi.org/10.1016/J.PROENV.2011.03.033

Alijani H, Shariatinia Z (2017) Effective aqueous arsenic removal using zero valent iron doped MWCNT synthesized by in situ CVD method using natural α-Fe_2O_3 as a precursor. Chemosphere 171:502–511. https://doi.org/10.1016/J.CHEMOSPHERE.2016.12.106

Anitha K, Namsani S, Singh JK (2015) Removal of heavy metal ions using a functionalized single-walled carbon nanotube: a molecular dynamics study. http://doi.org/10.1021/acs.jpca.5b03352

Ansari M, Kazemipour M, Dehghani M, Kazemipour M (2011) The defluoridation of drinking water using multi-walled carbon nanotubes. J Fluorine Chem 132(8):516–520. http://doi.org/10.1016/j.jfluchem.2011.05.008

Araga R, Kali S, Sharma CS (2019) Coconut-shell-derived carbon/carbon nanotube composite for fluoride adsorption from aqueous solution. Clean Soil, Air, Water 47(5):1–9. https://doi.org/10.1002/clen.201800286

Assaad N, Sabeh G, Hmadeh M (2020) Defect control in Zr-based metal-organic framework nanoparticles for arsenic removal from water. ACS Appl Nano Mater 3(9):8997–9008. https://doi.org/10.1021/acsanm.0c01696

Atieh MA, Bakather OY, Tawabini BS, Bukhari AA, Khaled M, Alharthi M, Fettouhi M, Abuilaiwi FA (2010) Removal of chromium (III) from water by using modified and nonmodified carbon nanotubes. J Nanomaterials. http://doi.org/10.1155/2010/232378

Bao Y, Yan X, Du W, Xie X, Pan Z, Zhou J, Li L (2015) Application of amine-functionalized MCM-41 modified ultrafiltration membrane to remove chromium (VI) and copper (II). Chem Eng J 281:460–467. https://doi.org/10.1016/j.cej.2015.06.094

Chandra V, Park J, Chun Y, Lee JW, Hwang I, Kim KS (2010) Water-dispersible magnetite-reduced graphene oxide composites for arsenic removal. ACS Nano 4(7):3979–3986

Dai H (2002) Carbon nanotubes: opportunities and challenges. Surf Sci 500(1–3):218–241. https://doi.org/10.1016/S0039-6028(01)01558-8

Dehghani MH, Taher MM, Bajpai AK, Heibati B, Tyagi I, Asif M, Agarwal S, Gupta VK (2015) Removal of noxious Cr(VI) ions using single-walled carbon nanotubes and multi-walled carbon nanotubes. Chem Eng J 279:344–352. https://doi.org/10.1016/J.CEJ.2015.04.151

Dehghani MH, Haghighat GA, Yetilmezsoy K, Mckay G, Heibati B, Tyagi I, Agarwal S, Kumar Gupta V (2016) Adsorptive removal of fluoride from aqueous solution using single-and multi-walled carbon nanotubes. http://doi.org/10.1016/j.molliq.2016.01.057

Du XD, Yi XH, Wang P, Zheng W, Deng J, Wang CC (2019) Robust photocatalytic reduction of Cr(VI) on UiO-66-NH2(Zr/Hf) metal-organic framework membrane under sunlight irradiation. Chem Eng J 356(September 2018):393–399. http://doi.org/10.1016/j.cej.2018.09.084

Fei K, Song TS, Wang H, Zhang D, Tao R, Xie J (2017) Electrophoretic deposition of carbon nanotube on reticulated vitreous carbon for hexavalent chromium removal in a biocathode microbial fuel cell. Roy Soc Open Sci 4(10). http://doi.org/10.1098/rsos.170798

Gao Q, Xu J, Bu XH (2019) Recent advances about metal–organic frameworks in the removal of pollutants from wastewater. Coord Chem Rev 378:17–31. https://doi.org/10.1016/j.ccr.2018.03.015

Geim AK (2009) Graphene: status and prospects. Science 324(5934). http://doi.org/10.1126/science.1158877

Han Z, Huang L, Qu H, Wang Y, Zhang Z, Rong Q, Sang Z, Wang Y, Kipper MJ, Tang J (2021) A review of performance improvement strategies for graphene oxide-based and graphene-based membranes in water treatment. J Mater Sci 56(16):9545–9574. http://doi.org/10.1007/s10853-021-05873-7

Harrison PTC (2005) Fluoride in water: a UK perspective. J Fluorine Chem 126(11–12):1448–1456. https://doi.org/10.1016/J.JFLUCHEM.2005.09.009

Hashem T, Ibrahim AH, Wöll C, Alkordi MH (2019) Grafting zirconium-based metal-organic framework UiO-66-NH2 nanoparticles on cellulose fibers for the removal of Cr(VI) ions and methyl orange from water. ACS Appl Nano Mater 2(9):5804–5808. https://doi.org/10.1021/acsanm.9b01263

He Y, Tang YP, Ma D, Chung TS (2017) UiO-66 incorporated thin-film nanocomposite membranes for efficient selenium and arsenic removal. J Membr Sci 541(March):262–270. https://doi.org/10.1016/j.memsci.2017.06.061

Hou T, Kong L, Guo X, Wu Y, Wang F, Wen Y, Yang H (2016) Magnetic ferrous-doped graphene for improving Cr(VI) removal. Mater Res Express 3(4). http://doi.org/10.1088/2053-1591/3/4/045006

Huang Z, Wang X, Yang D (2015) Adsorption of Cr(VI) in wastewater using magnetic multi-wall carbon nanotubes. Water Sci Eng 8(3):226–232. http://doi.org/10.1016/J.WSE.2015.01.009

Huno SKM, Rene ER, Van Hullebusch ED, Annachhatre AP (2018) Nitrate removal from groundwater: a review of natural and engineered processes. J Water Supply Res Technol AQUA 67(8):885–902. https://doi.org/10.2166/aqua.2018.194

Ihsanullah (2019) Carbon nanotube membranes for water purification: developments, challenges, and prospects for the future. Sep Purif Technol 209:307–337. http://doi.org/10.1016/J.SEPPUR.2018.07.043

Jian M, Liu B, Zhang G, Liu R, Zhang X (2015) Adsorptive removal of arsenic from aqueous solution by zeolitic imidazolate framework-8 (ZIF-8) nanoparticles. Colloids Surf A 465:67–76. https://doi.org/10.1016/j.colsurfa.2014.10.023

Jodeh S, Ibsharat I, Khalaf B, Hamed O, Jodeh D, Omar D (2019) The use of magnetic multiwalled carbon nanotubes functionalized with chitosan for nitrate removal from wastewater. Chem Afr 2(3):321–333. https://doi.org/10.1007/s42250-019-00056-z

Karmakar S, Dechnik J, Janiak C, De S (2016) Aluminium fumarate metal-organic framework: a super adsorbent for fluoride from water. J Hazard Mater 303:10–20. http://doi.org/10.1016/j.jhazmat.2015.10.030

Karmakar S, Bhattacharjee S, De S (2018) Aluminium fumarate metal organic framework incorporated polyacrylonitrile hollow fiber membranes: spinning, characterization and application in fluoride removal from groundwater. Chem Eng J 334(October 2017):41–53. http://doi.org/10.1016/j.cej.2017.10.021

Kashinath L, Namratha K, Byrappa K (2019) Microwave mediated synthesis and characterization of CeO_2–GO hybrid composite for removal of chromium ions and its antibacterial efficiency. J Environ Sci (China) 76:65–79. https://doi.org/10.1016/j.jes.2018.03.027

Kaur H, Sinha S, Krishnan V, Koner RR (2020) Photocatalytic reduction and recognition of Cr(VI): New Zn(II)-based metal-organic framework as catalytic surface. Ind Eng Chem Res 59(18):8538–8550. https://doi.org/10.1021/acs.iecr.9b06417

Ke F, Liu J, Ma A, Ke F, Jiang J, Yuan Q, Luo Z, Kumar A (2017) As featured in: two lanthanide-based metal-organic frameworks for highly efficient adsorption and removal of fluoride ions from water. 19:2172. http://doi.org/10.1039/c7ce00291b

Ke F, Peng C, Zhang T, Zhang M, Zhou C, Cai H, Zhu J, Wan X (2018) Fumarate-based metal-organic frameworks as a new platform for highly selective removal of fluoride from brick tea. Sci Rep 8:939. http://doi.org/10.1038/s41598-018-19277-2

Koli MM, Munavalli GR (2021) Field-scale baffled and biorack hybrid constructed wetland: effect of fluctuating loading rates and recirculation for domestic wastewater treatment. Int J Phytorem. https://www.tandfonline.com/loi/bijp20. http://doi.org/10.1080/15226514.2021.1895720

Kyzas GZ, Matis KA (2015) Nanoadsorbents for pollutants removal: a review. J Mol Liq 203:159–168. https://doi.org/10.1016/J.MOLLIQ.2015.01.004

Lal S, Singhal A, Kumari P (2020) Exploring carbonaceous nanomaterials for arsenic and chromium removal from wastewater. J Water Process Eng 36(December 2019):101276. http://doi.org/10.1016/j.jwpe.2020.101276

Le T, Chen X, Dong H, Tarpeh W, Perea-Cachero A, Coronas J, Martin SM, Mohammad M, Razmjou A, Esfahani AR, Koutahzadeh N, Cheng P, Kidambi PR, Esfahani MR (2021) An evolving insight into metal organic framework-functionalized membranes for water and wastewater treatment and resource recovery. Ind Eng Chem Res 60:6869–6907. https://doi.org/10.1021/acs.iecr.1c00543

Li L, Fan L, Sun M, Qiu H, Li X, Duan H, Luo C (2013) Adsorbent for chromium removal based on graphene oxide functionalized with magnetic cyclodextrin–chitosan. Colloids Surf B 107:76–83. https://doi.org/10.1016/J.COLSURFB.2013.01.074

Liu H, Zuo K, Vecitis CD (2014) Titanium dioxide-coated carbon nanotube network filter for rapid and effective arsenic sorption. Environ Sci Technol 48(23):13871–13879. http://doi.org/10.1021/es502312t

Liu B, Jian M, Liu R, Yao J, Zhang X (2015) Highly efficient removal of arsenic(III) from aqueous solution by zeolitic imidazolate frameworks with different morphology. Colloids Surf A 481:358–366. https://doi.org/10.1016/j.colsurfa.2015.06.009

Liu L, Cui Z, Ma Q, Cui W, Zhang X (2016) One-step synthesis of magnetic iron-aluminum oxide/graphene oxide nanoparticles as a selective adsorbent for fluoride removal from aqueous solution. RSC Adv 6(13):10783–10791. https://doi.org/10.1039/c5ra23676b

Liu H, Zhang F, Peng Z (2019) Adsorption mechanism of Cr(VI) onto Go/pAMAMs composites. Sci Rep 9. http://doi.org/10.1038/s41598-019-40344-9

Luan H, Teychene B, Huang H (2019) Efficient removal of As(III) by Cu nanoparticles intercalated in carbon nanotube membranes for drinking water treatment. Chem Eng J 355:341–350. https://doi.org/10.1016/j.cej.2018.08.104

Mehmandoust MR, Motakef-Kazemi N, Ashouri F (2019) Nitrate adsorption from aqueous solution by metal-organic framework MOF-5. Iran J Sci Technol Trans A Sci 43:443–449. http://doi.org/10.1007/s40995-017-0423-6

Mohammadkhani S, Gholami MR, Aghaie M (2016) Thermodynamic study of Cr+3 ions removal by 'MnO2/MWCNT' nanocomposite. Orient J Chem 32(1):591–599. https://doi.org/10.13005/ojc/320167

Moradi R, Rokni FF (2017) Synthesis, characterization and performance of NiO/CNT nanocomposite for arsenic removal from aqueous media. Curr Nanosci 13(6):579–585. https://doi.org/10.2174/1573413713666170629152833

Moradi P, Qanavati R, Mirzaei M (2018) Reducing nitrate from aqueous solutions using CNT as a powerful adsorbent: modification by lanthanum, pp 61–70

Motamedi E, Talebi Atouei M, Kassaee MZ (2014) Comparison of nitrate removal from water via graphene oxide coated Fe, Ni and Co nanoparticles. Mater Res Bull 54:34–40. http://doi.org/10.1016/j.materresbull.2014.02.019

Nasrollahzadeh M, Sajjadi M, Iravani S, Varma RS (2021) Carbon-based sustainable nanomaterials for water treatment: state-of-art and future perspectives. Chemosphere 263:128005. http://doi.org/10.1016/J.CHEMOSPHERE.2020.128005

Pang D, Wang P, Fu H, Zhao C, Wang C-C (2020) Highly efficient removal of As(V) using metal–organic framework BUC-17. SN Appl Sci 2(2):1–11. https://doi.org/10.1007/s42452-020-1981-3

Pishnamazi M, Koushkbaghi S, Hosseini SS, Darabi M, Yousefi A, Irani M (2020) Metal organic framework nanoparticles loaded-PVDF/chitosan nanofibrous ultrafiltration membranes for the removal of BSA protein and Cr(VI) ions. J Mol Liq 317:113934. http://doi.org/10.1016/j.molliq.2020.113934

Postma D, Pham TKT, Sø HU, Hoang VH, Vi ML, Nguyen TT, Larsen F, Pham HV, Jakobsen R (2016) A model for the evolution in water chemistry of an arsenic contaminated aquifer over the last 6000 years, Red River floodplain, Vietnam. Geochim Cosmochim Acta 195:277–292. https://doi.org/10.1016/J.GCA.2016.09.014

Rajmohan KS, Chetty R (2017) Enhanced nitrate reduction with copper phthalocyanine-coated carbon nanotubes in a solid polymer electrolyte reactor. J Appl Electrochem 47(1):63–74. https://doi.org/10.1007/s10800-016-1020-7

Rao GP, Lu C, Su F (2007) Sorption of divalent metal ions from aqueous solution by carbon nanotubes: a review. Sep Purif Technol 58(1):224–231. https://doi.org/10.1016/J.SEPPUR.2006.12.006

Sacchi E, Acutis M, Bartoli M, Brenna S, Delconte CA, Laini A, Pennisi M (2013) Origin and fate of nitrates in groundwater from the central Po plain: insights from isotopic investigations. Appl Geochem 34:164–180. https://doi.org/10.1016/J.APGEOCHEM.2013.03.008

Sahoo SK, Hota G (2018) Surface functionalization of GO with $MgO/MgFe_2O_4$ binary oxides: a novel magnetic nanoadsorbent for removal of fluoride ions. J Environ Chem Eng 6(2):2918–2931. https://doi.org/10.1016/j.jece.2018.04.054

Saiz PG, Iglesias N, González Navarrete B, Rosales M, Quintero YM, Reizabal A, Orive J, Fidalgo Marijuan A, Larrea ES, Lopes AC, Lezama L, García A, Lanceros-Mendez S, Arriortua MI, Fernández de Luis R (2020) Chromium speciation in zirconium-based metal–organic frameworks for environmental remediation. Chem A Eur J 26(61):13861–13872. http://doi.org/10.1002/chem.202001435

Samuel MS, Bhattacharya J, Raj S, Santhanam N, Singh H, Pradeep Singh ND (2019) Efficient removal of Chromium(VI) from aqueous solution using chitosan grafted graphene oxide (CS-GO) nanocomposite. Int J Biol Macromol 121:285–292. http://doi.org/10.1016/J.IJBIOMAC.2018.09.170

Santhana A, Kumar K, Jiang S-J, Tseng W-L (2015) Effective adsorption of chromium(VI)/Cr(III) from aqueous solution using ionic liquid functionalized multiwalled carbon nanotubes as a super sorbent. http://doi.org/10.1039/c4ta06948j

Sethi R, Di Molfetta A (2019) Groundwater contaminants. In: Groundwater engineering, pp 169–192. http://doi.org/10.1007/978-3-030-20516-4_9

Setshedi KZ, Bhaumik M, Onyango MS, Maity A (2015) High-performance towards Cr(VI) removal using multi-active sites of polypyrrole–graphene oxide nanocomposites: batch and column studies. Chem Eng J 262:921–931. https://doi.org/10.1016/J.CEJ.2014.10.034

Shahrin S, Lau WJ, Goh PS, Jaafar J, Ismail AF (2018) Adsorptive removal of As(V) ions from water using graphene oxide-manganese ferrite and titania nanotube-manganese ferrite hybrid nanomaterials. Chem Eng Technol 41(11):2250–2258. https://doi.org/10.1002/ceat.201800322

Shaji E, Santosh M, Sarath KV, Prakash P, Deepchand V, Divya BV (2021) Arsenic contamination of groundwater: a global synopsis with focus on the Indian Peninsula. Geosci Frontiers 12(3):101079. http://doi.org/10.1016/J.GSF.2020.08.015

Sharma SK, Petrusevski B, Amy G (2008) Chromium removal from water: a review. J Water Supply Res Technol AQUA 57(8):541–553. https://doi.org/10.2166/aqua.2008.080

Siddique TA, Dutta NK, Choudhury NR (2020) Nanofiltration for arsenic removal: challenges, recent developments, and perspectives. Nanomaterials 10(7):1–37. https://doi.org/10.3390/nano10071323

Tajuddin MHA, Jaafar J, Md Nordin NAH, Ismail AF, Othman MHD, Rahman MA (2020) Metal organic framework mixed-matrix membrane for arsenic removal. Malays J Fundam Appl Sci 16(3):359–362. http://doi.org/10.11113/mjfas.v16n3.1488

Tofighy AM, Mohammadi T (2012) Nitrate removal from water using functionalized carbon nanotube sheets. Chem Eng Res Des 90(11):1815–1822. http://doi.org/10.1016/j.cherd.2012.04.001

Tofighy MA, Mohammadi T (2020) Carbon nanotubes-polymer nanocomposite membranes for pervaporation. In: Polymer nanocomposite membranes for pervaporation. INC. http://doi.org/10.1016/b978-0-12-816785-4.00005-7

Usman M, Zarebanadkouki M, Waseem M, Katsoyiannis IA, Ernst M (2020) Mathematical modeling of arsenic(V) adsorption onto iron oxyhydroxides in an adsorption-submerged membrane hybrid system. J Hazard Mater 400:123221. http://doi.org/10.1016/J.JHAZMAT.2020.123221

Vu TA, Le GH, Dao CD, Dang LQ, Nguyen KT, Nguyen QK, Dang PT, Tran HTK, Duong QT, Nguyen TV, Lee GD (2015) Arsenic removal from aqueous solutions by adsorption using novel MIL-53(Fe) as a highly efficient adsorbent. http://doi.org/10.1039/c4ra12326c

Wan P, Yuan M, Yu X, Zhang Z, Deng B (2020) Arsenate removal by reactive mixed matrix PVDF hollow fiber membranes with UIO-66 metal organic frameworks. Chem Eng J 382:122921. http://doi.org/10.1016/j.cej.2019.122921

Wang S, Li Y, Zhao D, Xu C, Luan Z, Liang J, Wu D (2002) Preparation of alumina supported on carbon nano-tubes and its application in fluoride adsorption from an aqueous solution. Chin Sci Bull 47(9):722–724. https://doi.org/10.1360/02tb9162

Wang C, Lee M, Liu X, Wang B, Paul Chen J, Li K (2016) A metal-organic framework/α-alumina composite with a novel geometry for enhanced adsorptive separation. Chem Commun 52(57):8869–8872. https://doi.org/10.1039/c6cc02317g

Wang S, Li X, Liu Y, Zhang C, Tan X, Zeng G, Song B, Jiang L (2018) Nitrogen-containing amino compounds functionalized graphene oxide: synthesis, characterization and application for the removal of pollutants from wastewater: a review. J Hazard Mater 342:177–191. https://doi.org/10.1016/J.JHAZMAT.2017.06.071

Wang X, Zhu H, Sun T, Liu Y, Han T, Lu J, Dai H, Zhai L (2019) Synthesis and study of an efficient metal-organic framework adsorbent (MIL-96(Al)) for fluoride removal from water. http://doi.org/10.1155/2019/3128179

Xu N, Li S, Li W, Liu Z (2020) Removal of fluoride by graphene oxide/alumina nanocomposite: adsorbent preparation, characterization: adsorption performance and mechanisms. ChemistrySelect 5(6):1818–1828. https://doi.org/10.1002/slct.201904867

Yang J-C, Yin X-B (2016) $CoFe_2O_4$ @MIL-100(Fe) hybrid magnetic nanoparticles exhibit fast and selective adsorption of arsenic with high adsorption capacity. Nature Publishing Group. https://doi.org/10.1038/srep40955

Yang Q, Zhao Q, Ren SS, Lu Q, Guo X, Chen Z (2016) Fabrication of core-shell Fe_3O_4@MIL-100(Fe) magnetic microspheres for the removal of Cr(VI) in aqueous solution. J Solid State Chem 244(Vi):25–30. http://doi.org/10.1016/j.jssc.2016.09.010

Yang X, Deng S, Peng F, Luo T (2017) A new adsorbent of a Ce ion-implanted metal–organic framework (MIL-96) with high-efficiency Ce utilization for removing fluoride from water. Dalton Trans 46. http://doi.org/10.1039/c6dt03934k

Yin B, Sun L, Tang S, Zhou H (2020) Preparation of metal–organic framework/polyvinylidene fluoride mixed matrix membranes for water treatment. Ind Eng Chem Res 59. http://doi.org/10.1021/acs.iecr.0c03648

Yu F, Sun S, Ma J, Han S (2015) Enhanced removal performance of arsenate and arsenite by magnetic graphene oxide with high iron oxide loading. Phys Chem Chem Phys 17(6):4388–4397. https://doi.org/10.1039/c4cp04835k

Zhang Q, Yu J, Cai J, Zhang L, Cui Y, Yang Y, Chen B, Qian G (2015) A porous Zr-cluster-based cationic metal-organic framework for highly efficient Cr2O72-removal from water. Chem Commun 51(79):14732–14734. https://doi.org/10.1039/c5cc05927e

Zhang J, Chen N, Su P, Li M, Feng C (2017a) Fluoride removal from aqueous solution by zirconium-chitosan/graphene oxide membrane. React Funct Polym 114:127–135. https://doi.org/10.1016/j.reactfunctpolym.2017.03.008

Zhang P, Gong JL, Zeng GM, Deng CH, Yang HC, Liu HY, Huan SY (2017b) Cross-linking to prepare composite graphene oxide-framework membranes with high-flux for dyes and heavy metal ions removal. Chem Eng J 322:657–666. https://doi.org/10.1016/j.cej.2017.04.068

Zhao X, Liu D, Huang H, Zhang W, Yang Q, Zhong C (2014) The stability and defluoridation performance of MOFs in fluoride solutions. Microporous Mesoporous Mater 185:72–78. https://doi.org/10.1016/j.micromeso.2013.11.002

Zhu J, Yu X-Y, Jia Y, Peng F-M, Sun B, Zhang M-Y, Luo T, Liu J-H, Huang X-J (2012) Iron and 1,3,5-benzenetricarboxylic metal−organic coordination polymers prepared by solvothermal method and their application in efficient As(V) removal from aqueous solutions. http://doi.org/10.1021/jp212514a

Chapter 18
Metal Nanocomposites Based Sensors for Environmental Pollutions

Nilesh Satpute, Ritika Singh, Kamlesh Shrivas, and Khemchand Dewangan

Abstract The control and compatibility manner combination of nanomaterials that possess distinct features could be a practical approach for improving the properties of resulting nanocomposites. In this concern, metal nanocomposites are emerging multifunctional hybrid nanosystems where nano-sized metal is integrated into a matrix of metals, oxides, nitrides, nanocarbons, etc. These multifunctional metal nanocomposites hold unique and unusual characteristics compared to original nanocomponents, which provide a fantastic outcome in diversified areas of physics, chemistry, biology, environment, and medicine. This chapter discusses the fundamental principle, design strategies, and preparation protocols of metal nanocomposites. Moreover, the sensing methods developed to determine harmful and hazardous metal ions, chemicals, and bio-molecules presence in the environmental, aquatic, and other related samples are summarized.

Keywords Pollutants · Hybrid nanosystems · Nanoparticles

18.1 Introduction

In this modern decade, nanoscience and nanotechnology have become more effective tools for diagnosing and quantitatively evaluating toxic substances present in the air, water, soil, and other environmental samples (Aragay et al. 2011; Dewangan et al. 2020; Lin et al. 2020; Sarma et al. 2019). Recent advancement of nanotechnology allows us to control the size, shape, composition, and porosity of nanomaterials, which offers highly selective and sensitive applications towards the analysis of toxic

N. Satpute · R. Singh · K. Dewangan (✉)
Department of Chemistry, Indira Gandhi National Tribal University, Amarkantak, Madhya Pradesh 484887, India
e-mail: khemchand.dewangan@igntu.ac.in

K. Shrivas (✉)
School of Studies in Chemistry, Pt. Ravishankar Shukla University, Raipur, Chhattisgarh 492010, India

S. P. Singh et al. (eds.), *Metal Nanocomposites for Energy and Environmental Applications*, Energy, Environment, and Sustainability,
https://doi.org/10.1007/978-981-16-8599-6_18

components. Furthermore, hybridizing a minimum of two kinds of pure nanomaterials provides several advantages for evaluating pollutants than using single pristine nanomaterials (Dewangan et al. 2020; Facure et al. 2020; Naseem and Waseem 2021). For this purpose, metals, metal oxides, metal nitrides, and nanocarbons (e.g., graphene oxide (GO), carbon nanotubes (CNT), carbon quantum dots (C-QDs)) based nanostructured materials have been widely hybridized to improve the properties of nanomaterials for sensing harmful contaminates as well as other appications in biomedical, agriculture, environmental, and, energy storage devices.

This book chapter deals with metal-based nanocomposites for sensing various pollutants present in the ecosystem; thus, we will limit our discussion to this aspect only. Pollution is a significant concern of unsafe and unhygienic water, poor quality food, air, and other related problems in the environmental peripheral. In addition, these pollutants cause severe health effects on humans and other living organisms. They are responsible for adverse environmental impacts like global warming, acid rain, ozone depletion, and many others. Among pollutants, heavy metal ions represent one of the deadliest contaminants that can be harmful to human health due to their tendency to amass in the living organism. Some examples of heavy metals (Cr, Ni, Hg, As, Cd, and Pb) ions have been recognized and extensively studied as malignant elements (Crisponi and Nurchi 2015; Nagajyoti et al. 2010; Uchimiya et al. 2020; Wu et al. 2016). At ground level, ozone and sulfur dioxide are the most ubiquitous air pollutants affecting people and agricultural production (Schneidemesser and Monks 2013). Nitrogen oxides, carbon oxides, and many other chemicals (volatile metals and organic compounds) released into the atmosphere through human activity are the major components of air pollutants (Archibald et al. 2017; Liao et al. 2015; Manisalidis et al. 2020). For instance, the burning of wood and fossil fuels like gasoline and coal generate very fine particles. So the breathing with a lungful in dirty air, these ultrafine particles can enter the bloodstream and wreak havoc in organs and cells by promoting swelling and damaging the molecule of life (Pelley 2019; Shy and Finklea 1973). Recently, coronavirus disease of 2019 (COVID-19) is found a major world pandemic and has affected humans since December 2019. The World Health Organization (WHO) considered COVID-19 a pandemic and a global illness caused by a new virus. It is an infectious virus disease, which is spread by air droplets. Some researchers have been reported that air pollution is also a crucial factor for spreading COVID-19 (Dutheil et al. 2020; Schröder 2020).

A vital cradle for survival on earth is fresh and clean water. We depend on freshwater sources like rivers, lakes, wells, and natural springs. At present, these resources are highly contaminated by directly and indirectly human activity. Many contaminants arise from different materials and chemicals from petroleum, pharmaceutical, textile, leather, oil paint, fertilizer, and many other industries are directly drained via a small seaway in these natural water sources. For example, textile industries directly dispose of 10–50% of unused dyes in water. According to the report of WHO, globally, one-third of people do not have access to clean water. WHO also reported in 2010, water pollution is the primary origin of numerous evolving human health issues (Benedict et al. 2017; WHO 2008). Thus, environmental pollutants can affect individual human health over a broad spectrum of biological responses along with severe effects such as death and many incurable diseases.

To analyses these pollutants, there are numerous modern instruments techniques such as atomic absorption spectrometry (AAS), inductively coupled plasma (ICP) spectrometry, voltammetry, surface-enhanced Raman scattering (SERS), mass spectrometer, X-ray fluorescence (XRF), spectrophotometry, and many advanced techniques have been applied for the quantitative and qualitative determination of pollutant species in different types of samples. Most of these techniques are sophisticated and very costly that needs very laborious and time-consuming sample preparation. Also, these techniques are required a well-trained technician to operate these instruments. In addition, these techniques are cumbersome to apply at the sample sources for assessing target pollutants; therefore, they are not appropriate for in-field testing and monitoring of toxic substances present in water and territories.

18.2 General Aspect of Bulk, Nanomaterials, and Nanocomposites

Nanomaterials are engineered materials where the materials have a size in nanometers (nm) scale of at least one dimension (Baig et al. 2021). However, there is the possibility of changes in materials electronic and optical properties at the nanoscale level, which is referred to quantum confinement effect (Connerade 2009). In this phenomenon, electrons and/or holes are trapped in the nano-dimension regime, and nanomaterials exhibit an entirely different density of state, altering the bandgap of materials. The defects are another crucial parameter that arises in nanomaterials. It can be explained as that the nanomaterials are a new category of materials that fall in between solid crystalline and amorphous materials. It is well known that the crystalline bulk materials properties are attributed to their regular periodic arrangement of atoms (or ions or molecules). When the size of crystalline materials is reduced in the nm scale, the traditional systematic arrangement of atoms is reduced in short-range order that introduces grain boundaries, strains, vacancies, dislocations kinds of defects in nanomaterials. Usually, nanomaterials contain a high fraction of defects, about 50%, whereas bulk materials have a fraction of defects in 4–10% (Gleiter 1989). Thus, the whole structure of nanomaterials differs from the corresponding ideal crystal structure. It is documented that a significant deviation from the perfect crystal structure in material means a new kind of electronic and atomic arrangement that results in unique and unusual properties.

In this connection, hybridization of a minimum of two nanomaterials with distinct features is a new concept to introduce massive defects in nanomaterials. The resulting hybrid nanosystem is called a nanocomposite. As a result, nanocomposite formation in one matrix could introduce new sublattices, interfaces, nano-heterojunctions, and novel active centers, as displayed in Fig. 18.1. Consequently, they show the unusual stability, selectivity, and efficiency that might not be available in original hybridizing

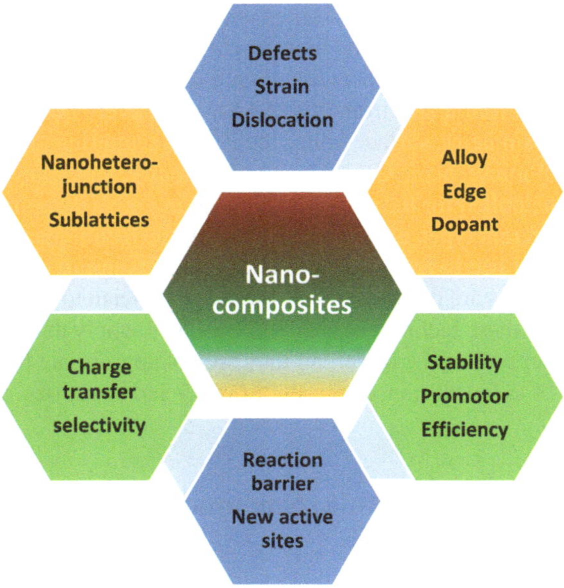

Fig. 18.1 Typical strategies of hybridization of nanomaterials and their roles for enhancing properties

nanomaterials. Therefore, controlled and compatibility ways integrated nanocomposites have been proven for the potential applications in catalysis, sensors, life science, environmental, and many related disciplines.

In this twenty-first century, a variety of nanocomposites have been designed for the detection of pollutants, including harmful and hazardous heavy metals ions, pesticides, organic dyes, and antibiotics (Dewangan et al. 2020; Kao et al. 2013; Khan et al. 2020; Lee and Choi 2019; Lu et al. 2019).

18.3 Scope of the Present Chapter

This chapter discusses the uses of different metal-based nanocomposites as a sensing material to detect environmental pollutants. The synthetic and hybridization approaches to obtain nanocomposites are summarized. The relevant structure and morphology characterization techniques of nanocomposites such as X-ray diffraction (XRD), scanning electron microscopy (SEM), transmission electron microscopy (TEM), energy dispersive X-ray (EDX) spectroscopy, ultraviolet–visible (UV–Vis) spectroscopy, and X-ray photoelectron spectroscopy (XPS) are remarked. It is expected that this chapter comprising with a detailed outline of the preparation of metal nanocomposites and their sensing applications would be helpful to a community of an academician, researchers, and scientists working in the field of different nanosensors, nano biosensor, agrochemical (nano fertilizers and pesticides) including all other that are correlated with the biology and environment applications.

18.4 Methods of Preparation of Pure Metal Nanostructures

The ability to synthesize desired sizes and shapes of nanomaterials is very important in exploring their application. Nanomaterials are fabricated using top-down, bottom-up, and a combination of both strategies (Baig et al. 2021; Lancaster et al. 2020; Merkel et al. 2010). The bottom-up approaches have been extensively studied because they offer more scopes for controlling the structural morphology of the resulting nanomaterials. In this section, some commonly used bottom-up wet-chemical approaches to obtain nanostructured metallic materials are emphasized.

To obtain pure metal (Ag, Au, Cu Pt, Pd) based nanostructured materials, chemical routes are classified into chemical reduction, polyol, microemulsion, sol–gel, hydrothermal, and thermal decomposition. For example, well-established wet-chemical methods for the synthesis of metallic nanoparticles mainly involve reducing the corresponding salts in the presence of reducing and stabilizing agents (Richards and Bönnemann 2005). Generally, reducing agents such as citric acid, glucose, sodium borohydride, hydrazine hydrate, ascorbic acid, 1,2 hexadecanediol, etc., are employed. The citrate ion and borohydride arise from the reduction of citric acid and sodium borohydride, respectively are acting as internal capping agents. Also, external capping agents such as polyvinyl alcohol (PVA), polyvinyl pyrrolidone (PVP), cetyltrimethylammonium bromide (CTAB), polyethylene glycol (PEG), ethylenediaminetetraacetic acid (EDTA), etc. are used to stabilize nanoparticles. Capping agents play a crucial role as stabilizers that inhibit the growth of nanoparticles facets and prevent their aggregation in the synthesis process. Also, they can influence the specific structural development of nanoparticles (Richards and Bönnemann 2005).

Solvothermal/hydrothermal is another extensively used method for the preparation of nanostructured metallic materials (Huang et al. 2009). Herein, the metal salts (or organometallic compounds) aqueous/non-aqueous solution is treated in an autoclave reactor made of an outer-shell with high-quality stainless steel and an inner-cell chamber with Teflon polytetrafluoroethylene (PTFE). On the other hand, the sol–gel technique is used to obtain metal oxides nanomaterials. However, a combination of sol–gel with auto-combustion has been reported to obtain metallic nanoparticles (Jiang et al. 2009). In the first step, they prepared sol and transformed it into a gel by evaporation. The obtained gel was activated in the auto-combustion tube under the Ar/N_2 atmosphere at 300 °C to obtain metal nanoparticles. The polyol method is also a widely employed chemical approach for the shape-controlled synthesis of nanostructured noble metals in the high boiling polyol solvents such as ethylene glycol, diethylene glycol, polyethylene glycol, etc. (Sun and Xia 2002). These polyols not only act as a solvent but also act as a reducing agent and capping agent. The microemulsion technique has also been reported to fabricate specific shapes of metallic nanostructured (Ganguli et al. 2010). In addition, biosynthesis (Njagi et al. 2011), green chemistry (Jadoun et al. 2021), sonochemical synthesis (Hansen et al. 2021), and photochemical preparation (Schmarsow et al. 2020) have been reported to obtain metal nanoparticles.

18.5 Hybridization Protocol of Nanomaterials into Nanocomposites

Based on the inorganic and organic nanomaterials, as shown in Fig. 18.2, nanocomposite can broadly be described in the following two groups:

- Inorganic–inorganic nanocomposites
- Inorganic–organic nanocomposites.

The hybridization of nanomaterials does not simply indicate the physical mixing of nano components makes nanocomposites. The protocol is that the amalgamation should be at the nanoscale range. Nanocomposites can be either homogeneous (metal alloys) or heterogeneous (Au@TiO_2). Thus, the interface regions interconnect two dissimilar nanocomponents with a broad range of unusual combinations with covalent, ionic, metallic, hydrogen, and van der Waal kinds bonds. The choice of interface

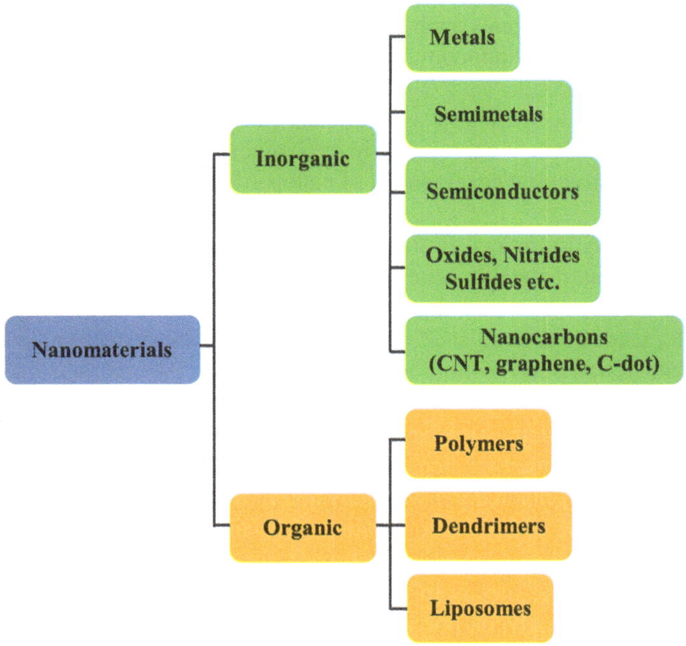

Fig. 18.2 Classification of nanomaterials

and nano components are the crucial factors that decide the final use and application of resultant nanocomposites. Therefore, interplaying between hybridizing nano components could provide unique and unusual potentials to nanocomposites for challenging and required specifications. Amalgamation protocols can be divided into Type I and Type II. Type I is a kind of synchronization where the organic and inorganic components are generally interrelated by weak bonds such as van der Waals and hydrogen bonds. In contrast, Type II amalgamation components are integrated by stronger interactions like ionic, covalent, metallic, or chemical attachment. Due to amalgamation occurs in between a minimum of two distinct materials at nano regime, there is a countless possibility for several kinds of atomic arrangements, defects, and boundaries.

At present, several physical and chemical techniques have been developed to prepare the exact morphology of pure nanomaterials. However, the preparation of multifunctional nanocomposites and their simple and practical applications are still in the laboratory. Mainly, nanocomposites are produced by an in-situ amalgamation process where mixing components are simultaneously grown in one pot via commonly used bottom-up techniques. Also, a combination of top-down and bottom-up approaches has been used to obtain nanocomposites, for example, core–shell-like nanocomposites. First, a top-down process is employed to obtain core particles, and then these seed particles are coated by a bottom-up approach to attain a uniform and precise shell thickness (Gawande et al. 2015). Moreover, sometimes mixing components are separately obtained and mixed together by physical and chemical processes like ball milling, heating, dispersing, sonication, polymerization, and functionalization with appropriate ligand molecules (Rane et al. 2018).

18.6 Characterization Techniques of Nanocomposites

The crystal structure, phase-purity, defects, interface, composition, morphology, and other physiochemical features of nanocomposites are characterized using various commonly used sophisticated instruments and techniques. The elaboration discussion about these instrumentations can be found somewhere; therefore, an outline of routinely used instrumentations is highlighted here. XRD is a multipurpose analytical technique that employs to solve all issues related to the crystal structure of nanocomposites, such as phase formation, lattice parameters, crystal orientation, defects, etc., (Cullity 1978). SEM and TEM are prevalent instrumentation techniques used for the surface topography of nanocomposites. Also, advanced TEM studies like high-resolution TEM images and selected area electron diffraction (SAED) pattern further help to a deep understanding of crystal information of nanocomposites. The EDX spectroscopy generally connected with SEM and TEM instruments provides quantitative/qualitative elemental composition presence in nanocomposites (Williams and Carter 1996). XPS is an essential tool to analyze the surface chemistry of nanocomposites in context to the chemical and electronic state of the elements and their ratio. UV–Vis spectrophotometer is a valuable primary characterization tool for the nanocomposites that exhibit local surface plasmon resonance (LSPR). For example, Ag nanoparticles show characteristics LSPR band appears near 400 nm that could little shift according to the preparation conditions, shape, and size of nanoparticles. Fourier transforms infrared (FTIR) spectroscopy is to study atomic arrangement and their bonds coordination in nanocomposites.

18.7 Metal-Based Nanocomposites Sensors for Environmental Pollutants

According to the International Union of Pure and Applied Chemistry (IUPAC) definition "A chemical sensor is a device that transforms chemical information (composition, presence of a particular element or ion, concentration, chemical activity, partial pressure, etc.) into an analytically useful signal" (Hulanicki et al. 1991). The originated useful singles are generally categorized based on the following properties:

- Physical interactions: absorbance, fluorescence, phosphorescence, conductivity, temperature, mass, and refractive indexed.
- Chemical interactions: chemical reaction, biochemical reactions, electrochemical reactions.

In other words, a sensor is an analytical device that responds toward physiochemical properties change during the interaction between the interest of the subject and the active component of the sensor. Many kinds of sensors have been developed using nanomaterials and nanocomposites for applications in various fields like biomedical, environment remediation, food analysis, public and home security, etc. (Dewangan et al. 2020; Correa et al. 2017; Kaushik et al. 2015). In the following subheading, metal-based nanocomposites have been emphasized as an active component for the fabrication of sensing devices related to environmental remediation. Some of the sensing applications based on nanocomposites are tabulated in Table 18.1.

18.7.1 Metal–Metal Nanocomposite Nanosensors

Generally, metal–metal nanocomposites comprise noble metals (Ag, Au, Cu, Pt) in metallic alloys, intermetallic, and core–shell types of nanostructures. These nanocomposites respond toward target analytes by changing physical properties like color change, fluorescence and phosphorescence emission, quenching, etc. (Min and Wang 2020). For example, fluorescent Au–Ag core–shell nanoparticles were prepared using an eco-friendly procedure. This core–shell nanoparticle ultrasensitively detects Hg^{2+} ions with a limit of detection (LOD) of 9 nM even in the presence of Zn^{2+}, Cd^{2+}, and other divalent metal ions (Guha et al. 2011). A very recent study

Table 18.1 Nanocomposites and their sensing applications

Nanocomposites	Sensing applications	References
Au@Ag core–shell	Visual detection of Hg^{2+}	(Li et al. 2014)
Au–Ag intermetallic	Colorimetric detection of Pb^{2+}	(Sahu et al. 2020)
Mn-Cu-Al	Electrochemical based glucose sensor	(Chandrasekaran and Matheswaran 2020)
CdTe QDs	Fluorescent based Hg^{2+} sensor	(Duan et al. 2009)
Au-Fe_2O_3	Acetone detection	(Yang et al. 2020)
Pd-TiO_2	Butane gas sensing	(Chen et al. 2017)
Ag/polyaniline/GO	Fluorescent based Cr^{4+} sensing	(Ebrahim et al. 2020)
Cu/Polyaniline	Amperometric nitrite sensor	(qian et al. 2021)
Mn doped ZnS	Radical polymerization based insecticide nanosensor	(Wu et al. 2017)
Fe/Ag/CNT	Volatile organic compounds sensor	(Sheikhian and Ghanbarian 2021)
Metal/DNA based nanocomposites	Review article for environment monitoring	(Kumar and Guleria 2020)
Pt/graphene	H_2 gas detection	(Mohammadi et al. 2020)

provides a simple, rapid, and visual recognition and detection of clindamycin antibiotics in the effluent of pharma companies using Au@Ag core–shell nanoparticles (Du et al. 2021). In another report, Au nanorods coated with Ag were demonstrated as effective colorimetric sensor arrays for simultaneous detection and discrimination of five reducing sugars like glucose, fructose, maltose, ribose, and glyceraldehyde (Zhang et al. 2021). Shrivas and groups developed a Cu@Ag core–shell nanoparticles-based smartphone-assisted paper sensor to detect phenthoate pesticide (Fig. 18.3) with a LOD of 15 $\mu g\ L^{-1}$ (Shrivas et al. 2020). In contrast, bimetallic nanoparticles such as FePt, FePd, and FeAu were used as electrochemical sensors for the detection of As^{3+} (Moghimi et al. 2015). Alwan et al. reported that Au nanoparticles on the Si-substrate could be used for biochemical ultra-sensitive sensors to detect tetracycline antibiotic residue in water (Alwan et al. 2020). In addition, intermetallic nanostructured nanocomposites such as AuZr (Endo et al. 2010), TiFe (Dematteis et al. 2021), PtNi (Leonard et al. 2011), PdRh (Furukawa et al. 2014), etc., have been used in many other fields of nanoscience.

18.7.2 Metal-Ceramic and Metal–Semiconductor Nanocomposites

Nowadays, most of the metal nanocomposites being used fall in this category. Herein, metallic nanostructured is integrated into the other nano/bulk matrix of oxides, nitrides, carbides, sulfides, phosphides, etc. These metal-ceramic nanocomposites

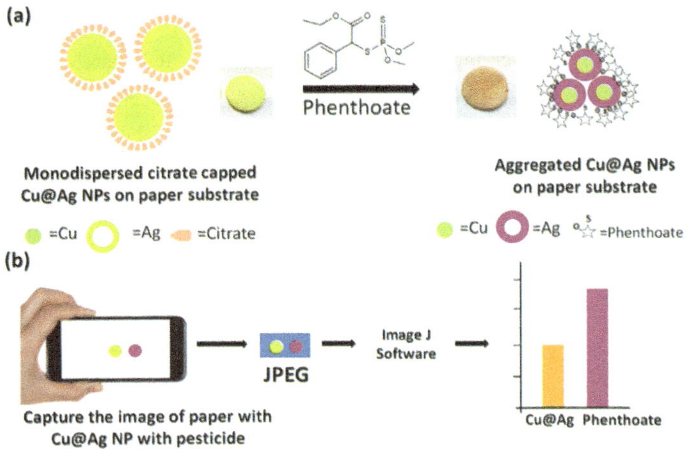

Fig. 18.3 **a** Mechanism for detection of phenthoate on paper-based sensor where the monodispersion and aggregation of Cu@Ag nanoparticles before and after the addition of phenthoate and **b** procedure for the determination of signal intensity of Cu@Ag nanoparticles and with phenthoate by smartphone-assisted paper-based sensor. Reproduced with permission from Shrivas et al. (2020). Copyright (2001) Royal Society of Chemistry

have been widely used in nanosensors due to their stability and ease of fabrication. Some examples of sensing related to the environment are summarized below.

Au nanoparticles embedded in TiO_2 nanofibers using electrospinning procedures were reported for selectively sensing carbon monoxide (CO) at ppb level concentration (Nikfarjam et al. 2017). Similarly, Ag nanoparticles stabilized on the surface of TiO_2 nanorods can be used to clean metal surfaces and fabricate sensing devices (Cozzoli et al. 2004). A critical review of Au-Fe_xO_y hybrid nanocomposites provides information that can potentially be applied for nano-biosensing and various biomedical applications (Leung et al. 2012). In another report, metal–semiconductor (Au-CdTe QDs) nanocomposite is a fascinating class of sensing material that can be used to detect amino acids with LOD 192 nM (Paramanik et al. 2016). Wang et al. demonstrated that C-QDs functionalized with silica aerosol exhibit enhanced performance for NO_2 gas sensing. The nanocomposite of CNT-ZnO thin film shows a highly sensing response towards toluene compared to pure CNT and ZnO (Septiani et al. 2017). Ramalingam et al. reported that a nanocomposite comprising CNT and C_3N_4 exhibits a potential response for simultaneously determining heavy metal ions (Cd^{2+}, Hg^{2+}, Pb^{2+}, and Zn^{2+}) using voltammetry (Ramalingam et al. 2019). The heterostructure of $Cu/CuO/Cu_2O$ microwires nanocomposites was prepared via simple thermal oxidation that shows improved performance for the detection of gases of battery hazardous (Lupan et al. 2020). Mn-doped ZnS QDs were employed as a bifenthrin insecticide nanosensor by an atom transfer radical polymerization reaction mechanism and offered a fast response and low toxicity (Wu et al. 2017). Jin and coworkers used Au-modified MoS_2 monolayer for sensing of DNA molecules through photoluminescence mechanism, allowing rapid, selective, and sensitive detection performance (Jin et al. 2016). A number of review articles are available to design this kind of nanocomposites to detect and eliminate toxic substances from the environment (Ghanbari et al. 2020; Hou et al. 2018; Jeong et al. 2020; Khan et al. 2015; Riazi et al. 2021).

18.7.3 Metal-Polymer Nanocomposites

In this polymer era, metal-polymer nanocomposites have been extensively employed for sensing applications. Figure 18.4 schematically illustrates the metal-polymer nanocomposites-based gas sensing mechanism (Yan et al. 2020). The polymer embedded metallic nanoparticles can sustain more stress/pressure/temperature etc., towards making sensing devices in comparison to other nanocomposites (Esumi et al. 2004; Nandan and Horechyy 2015; Panahi-Sarmad et al. 2020; Shukla and Saxena 2021; Tajik et al. 2021; Wei et al. 2020).

Au@polymer nanoparticles show a highly optical response that can be used for colorimetric sensing of pollutants (Tagliazucchi et al. 2012). Silica nanoparticles grafted on the acrylic acrylamide copolymers displayed ultrafast removal of methylene blue organic dye from wastewater (Saleh et al. 2020). Cu nanoparticles/polymer is an extremely interesting nanocomposite with antifungal and bacteriostatic characteristics that could be applied for the detection of biomolecules (Cioffi et al. 2005). Rastogi et al. prepared an electrochemical sensor using Ag nanoparticles and copolymers for sensitive nitrite (NO_2^-) ion sensing and determination (Rastogi et al. 2014).

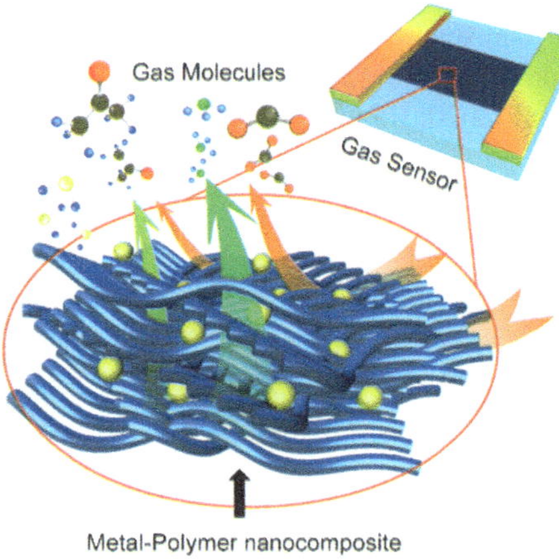

Fig. 18.4 A typical metal-polymer nanocomposite film. Reproduced with an Open Access article from Yan et al. (2020)

Graphene nanoplatelet-polymer nanocomposite offers chemiresistive sensing arrays for the detection and removal of chemical warfare simulants (Wiederoder et al. 2017).

18.7.4 Metal-Nanocarbon Nanocomposites

Metal nanoparticles integrated into the matrix of nanocarbon families such as CNT, C-QDs, and graphene offer a class of promising nanocomposites for various applications. For example, Ag nanoparticles/nitrogen-doped C-QDs nanocomposites received significant attention due to enhanced surface-enhanced Raman scattering (SERS) property (Su et al. 2016), which can be exploited for the potential application of SERS-based detection like uric acid (Wang et al. 2019). In another report, Ag/C-QDs nanocomposite was used as a colorimetric sensor for the detection of methazole in a urine sample (Amjadi et al. 2017). Liu and research groups reported that Ag/Au doped hybrid C-QDs offer effective photothermal therapy of colon cancer treatment (Liu et al. 2020) because this nanocomposite displayed outstanding stability and absorbance at the near-infrared region for the noble phototherapy.

Ag-modified reduced GO-based nanocomposite displays a remarkably fast response for sensing NO_2 gas and food colorant (Li et al. 2020; Xie et al. 2012). Figure 18.5 shows a schematic representation of Ag/reduced GO composite preparation and SERS detection of food colorants. The Fe/Ag/CNT ternary nanocomposite was used to fabricate a sensing device to detect polar organic volatile compounds like methanol (Sheikhian and Ghanbarian 2021). A colorimetric and fluorometric dual-signal nanosensor was developed for the detection of arginine over other amino acids through inhibition of the growth of Au/C-QDs nanocomposite (Liu et al. 2017). Au nanoparticles immobilized in nonreactive and rigid graphitic islands used as electrochemical genosensing for food safety (Brasil de Oliveira Marques et al. 2009). Modified carbon-based nanosensors are broadly used to detect various harmful chemicals and have been studied by multiple research groups (Facure et al. 2020; Jeong et al. 2020; Moosavi et al. 2020; Rahman et al. 2011).

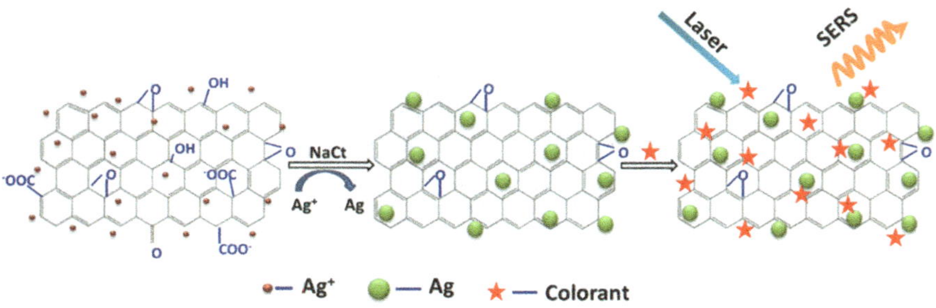

Fig. 18.5 Ag/reduced GO nanocomposite and food colorants sensing by SERS. Reproduced with permission from Xie et al. (2012). Copyright (2012) Elsevier

18.7.5 Metal and Organic/Biomolecules Based Nanocomposites

Organic/bio-molecules are good stabilizing and functionalizing agents of metal nanoparticles that very selectively interact with target analyte molecules. In this concern, lysine functionalized Ag nanoparticles were used for the usual sensing and separation of histidine amino acid and histidine-tagged proteins (Bae et al. 2010). Likewise, Xiong and researchers demonstrated that para-sulfonatocalix[4]arene-modified Ag nanoparticles could be effectively employed as a colorimetric probe for sensing histidine (Xiong et al. 2008). Glutathione capped Au nanocluster is a unique nanosensor for detecting metal ions, anions, and small molecules (Zhang et al. 2015). L-cysteine functionalized C-QDs embedded in a paper matrix exhibited an enantioselective response towards L-lysine among the enantiomers of amino acids (Copur et al. 2019). Miao et al. demonstrated melamine functionalized Ag nanoparticles as the electrochemical highly sensitive sensing probe for easy and fast detection of clenbuterol (Miao et al. 2014). Reduced graphene oxide/chitosan/Ag-nanoparticles nanocomposite was designed for the applications of organic dye molecules decomposition in wastewater (Jiao et al. 2015). Sucralose decorated Ag nanoparticles were obtained using green chemistry and used for naked-eye colorimetric detection of trimethylamine pollutants (Filippo et al. 2013).

Li and group prepared citrate-capped Cu nanoparticles for the colorimetric sensing of Hg^{2+} ion by measuring the change of characteristic absorption peak of Cu nanoparticles observed at 260 nm (Li et al. 2019). Another report used DNA templated Cu nanoparticles as a fluorogenic nanosensor using a smartphone to rapidly detect tuberculosis (Tsai et al. 2019). Double-stranded DNA-Cu nanoparticles were employed as green nano-dye to detect Hg^{2+} and Pb^{2+} ions (Chen et al. 2012; Qing et al. 2014).

18.8 Conclusion

In conclusion, the optimal hybridization of nanomaterials provides unique and unusual characteristics of resulting nanocomposites, which are absent either in the pure nanomaterials or expressively enhanced compared to the parent nanomaterials in the nanocomposites. Thus, hybrid nanosystems are novel materials to improve the merits and demerits of many existing pure nanomaterials. In this chapter, metal-based nanocomposites are discussed with respect to preparation strategies and hybridization protocols. Also, examples and advantages of various metal-based nanocomposites as sensing probes are illustrated for different types of hazardous substances from environmental, biological, food, and agrochemicals and fertilizers.

Thus, metal-based nanocomposites assure a new class of multifunctional materials with high sensitivity and selectivity for sensing pollutants present in the environment. In the future, these materials could hold massive potential for developing lightweight and system-needed sensing devices with remarkable and required properties. In addition, designing recyclable and cost-effective metal-based nanocomposites could be a prodigious achievement toward pollution control and prize war worldwide. It is expected that this chapter will be beneficial to researchers and students for the basic understanding of nanocomposite materials and their applications for the sensing and determination of toxicants present in the environment.

Acknowledgements Dr. Khemchand Dewangan gratefully acknowledges the Department of Science and Technology (DST-FIST) India, for funding and infrastructure.

References

Alwan AM, Wali LA, Hasan KK (2020) A new route for developing highly efficient nano biochemical sensors for detecting ultra-low concentrations of tetracycline antibiotic residue in water. Gold Bull 53:39–46

Amjadi M, Hallaj T, Asadollahi H, Song Z, de Frutos M, Hildebrandt N (2017) Facile synthesis of carbon quantum dot/silver nanocomposite and its application for colorimetric detection of methimazole. Sens Actuator B: Chem 244:425–432

Aragay G, Pons J, Merkoçi A (2011) Recent trends in macro-, micro-, and nanomaterial-based tools and strategies for heavy-metal detection. Chem Rev 111:3433–3458

Archibald AT, Folberth G, Wade DC, Scott D (2017) A world avoided: impacts of changes in anthropogenic emissions on the burden and effects of air pollutants in Europe and North America. Faraday Discuss 200:475–500

Bae DR, Han WS, Lim JM, Kang S, Lee JY, Kang D, Jung JH (2010) Lysine-functionalized silver nanoparticles for visual detection and separation of histidine and histidine-tagged proteins. Langmuir 26:2181–2185

Baig N, Kammakakam I, Falath W (2021) Nanomaterials: a review of synthesis methods, properties, recent progress, and challenges. Mater Adv 2:1821–1871

Benedict KM, Reses H, Vigar M, Roth DM, Roberts VA, Mattioli M, Cooley LA, Hilborn ED, Wade TJ, Fullerton KE, Yoder JS, Hill VR (2017) Surveillance for waterborne disease outbreaks associated with drinking water—United States, 2013–2014. MMWR Morb Mortal Wkly Rep 66:1216–1221

Chandrasekaran NI, Matheswaran M (2020) Unique nonenzymatic glucose sensor using a hollow-shelled triple oxide Mn–Cu–Al nanocomposite. ACS Omega 5:23502–23509

Chen J, Liu J, Fang Z, Zeng L (2012) Random dsDNA-templated formation of copper nanoparticles as novel fluorescence probes for label-free lead ions detection. Chem Commun 48:1057–1059

Chen N, Deng D, Li Y, Liu X, Xing X, Xiao X, Wang Y (2017) TiO_2 nanoparticles functionalized by Pd nanoparticles for gas-sensing application with enhanced butane response performances. Sci Rep 7:7692

Cioffi N, Torsi L, Ditaranto N, Tantillo G, Ghibelli L, Sabbatini L, Bleve-Zacheo T, D'Alessio M, Zambonin PG, Traversa E (2005) Copper nanoparticle/polymer composites with antifungal and bacteriostatic properties. Chem Mater 17:5255–5262

Connerade JP (2009) A review of quantum confinement. AIP Conf Proc 1197:1–33

Copur F, Bekar N, Zor E, Alpaydin S, Bingol H (2019) Nanopaper-based photoluminescent enantioselective sensing of L-Lysine by L-Cysteine modified carbon quantum dots. Sens Actuator B: Chem 279:305–312

Correa DS, Pavinatto A, Mercante LA, Mattoso LHC, Oliveira JE, Riul A (2017) Chemical sensors based on hybrid nanomaterials for food analysis. In: Grumezescu AM (ed) Nanobiosensors. Academic, pp 205–244

Cozzoli PD, Comparelli R, Fanizza E, Curri ML, Agostiano A, Laub D (2004) Photocatalytic synthesis of silver nanoparticles stabilized by TiO_2 nanorods: a semiconductor/metal nanocomposite in homogeneous nonpolar solution. J Am Chem Soc 126:3868–3879
Crisponi G, Nurchi VM (2015) Metal ion toxicity. In: Scott RA (ed) Encyclopedia of inorganic and bioinorganic chemistry. Wiley, pp 1–14
Cullity BD (1978) Elements of X-ray diffraction, 2nd edn. Addison-Wesley Publishing Company Inc., California
Brasil de Oliveira Marques PR, Lermo A, Campoy S, Yamanaka H, Barbé J, Alegret S, Pividori MI (2009) Double-tagging polymerase chain reaction with a thiolated primer and electrochemical genosensing based on gold nanocomposite sensor for food safety. Anal Chem 81:1332–1339
Dematteis EM, Berti N, Cuevas F, Latroche M, Baricco M (2021) Substitutional effects in TiFe for hydrogen storage: a comprehensive review. Mater Adv 2:2524–2560
Dewangan K, Shrivas K, Kurrey R (2020) Chapter 9—Hybrid nanomaterials as chemical sensors. In: Abd-Elsalam KA (ed) Multifunctional hybrid nanomaterials for sustainable agri-food and ecosystems. Elsevier, pp 213–239
Du Q, Jing Z, Qi H, Zuo L, Zhou L, He H, Sun Z (2021) Visual recognition and detection of clindamycin by Au@Ag core–shell nanoparticles. ACS Omega 6:14260–14267
Duan J, Song L, Zhan J (2009) One-pot synthesis of highly luminescent CdTe quantum dots by microwave irradiation reduction and their Hg^{2+}-sensitive properties. Nano Res 2:61–68
Dutheil F, Baker JS, Navel V (2020) COVID-19 as a factor influencing air pollution? Environ Pollut 263:114466
Ebrahim S, Shokry A, Khalil MMA, Ibrahim H, Soliman M (2020) Polyaniline/Ag nanoparticles/graphene oxide nanocomposite fluorescent sensor for recognition of chromium (VI) ions. Sci Rep 10:13617
Endo N, Kameoka S, Tsai AP, Hirata T, Nishimura C (2010) Preparation of nano-composited catalyst from the bulk intermetallic compound $AuZr_3$ with hydrogen absorption. Catal Lett 139:67–71
Esumi K, Isono R, Yoshimura T (2004) Preparation of PAMAM− and PPI−metal (silver, platinum, and palladium) nanocomposites and their catalytic activities for reduction of 4-nitrophenol. Langmuir 20:237–243
Facure MHM, Schneider R, Mercante LA, Correa DS (2020) A review on graphene quantum dots and their nanocomposites: from laboratory synthesis towards agricultural and environmental applications. Environ Sci: Nano 7:3710–3734
Filippo E, Manno D, Buccolieri A, Serra A (2013) Green synthesis of sucralose-capped silver nanoparticles for fast colorimetric triethylamine detection. Sens Actuator B: Chem 178:1–9
Furukawa S, Yoshida Y, Komatsu T (2014) Chemoselective hydrogenation of nitrostyrene to aminostyrene over Pd- and Rh-based intermetallic compounds. ACS Catal 4:1441–1450
Ganguli AK, Ganguly A, Vaidya S (2010) Microemulsion-based synthesis of nanocrystalline materials. Chem Soc Rev 39:474–485
Gawande MB, Goswami A, Asefa T, Guo H, Biradar AV, Peng D-L, Zboril R, Varma RS (2015) Core–shell nanoparticles: synthesis and applications in catalysis and electrocatalysis. Chem Soc Rev 44:7540–7590
Ghanbari MH, Sharafi P, Nayebossadr S, Norouzi Z (2020) Utilizing a nanocomposite consisting of zinc ferrite, copper oxide, and gold nanoparticles in the fabrication of a metformin electrochemical sensor supported on a glassy carbon electrode. Microchim Acta 187:557
Gleiter H (1989) Nanocrystalline materials. Prog Mater Sci 33:223–315
Guha S, Roy S, Banerjee A (2011) Fluorescent Au@Ag core–shell nanoparticles with controlled shell thickness and Hg^{II} sensing. Langmuir 27:13198–13205
Hansen HE, Seland F, Sunde S, Burheim OS, Pollet BG (2021) Two routes for sonochemical synthesis of platinum nanoparticles with narrow size distribution. Mater Adv 2:1962–1971
Hou X, Mu L, Chen F, Hu X (2018) Emerging investigator series: design of hydrogel nanocomposites for the detection and removal of pollutants: from nanosheets, network structures, and biocompatibility to machine-learning-assisted design. Environ Sci: Nano 5:2216–2240

Huang X, Zhang H, Guo C, Zhou Z, Zheng N (2009) Simplifying the creation of hollow metallic nanostructures: one-pot synthesis of hollow palladium/platinum single-crystalline nanocubes. Angew Chem Int Ed 48:4808–4812

Hulanicki A, Glab S, Ingman F (1991) Chemical sensors definitions and classification. Pure Appl Chem 63:1247–1250

Jadoun S, Arif R, Jangid NK, Meena RK (2021) Green synthesis of nanoparticles using plant extracts: a review. Environ Chem Lett 19:355–374

Jeong C, Joung C, Lee S, Feng MQ, Park Y-B (2020) Carbon nanocomposite based mechanical sensing and energy harvesting. Int J Precis Eng Man Greentechnol 7:247–267

Jiang Y, Yang S, Hua Z, Huang H (2009) Sol–gel autocombustion synthesis of metals and metal alloys. Angew Chem Int Ed 48:8529–8531

Jiao T, Zhao H, Zhou J, Zhang Q, Luo X, Hu J, Peng Q, Yan X (2015) Self-assembly reduced graphene oxide nanosheet hydrogel fabrication by anchorage of chitosan/silver and its potential efficient application toward dye degradation for wastewater treatments. ACS Sustain Chem Eng 3:3130–3139

Jin K, Xie L, Tian Y, Liu D (2016) Au-modified monolayer MoS_2 sensor for DNA detection. J Phys Chem C 120:11204–11209

Kao J, Thorkelsson K, Bai P, Rancatore BJ, Xu T (2013) Toward functional nanocomposites: taking the best of nanoparticles, polymers, and small molecules. Chem Soc Rev 42:2654–2678

Kaushik A, Kumar R, Arya SK, Nair M, Malhotra BD, Bhansali S (2015) Organic–inorganic hybrid nanocomposite-based gas sensors for environmental monitoring. Chem Rev 115:4571–4606

Khan M, Tahir MN, Adil SF, Khan HU, Siddiqui MRH, Al-warthan AA, Tremel W (2015) Graphene based metal and metal oxide nanocomposites: synthesis, properties and their applications. J Mater Chem A 3:18753–18808

Khan FSA, Mubarak NM, Tan YH, Karri RR, Khalid M, Walvekar R, Abdullah EC, Mazari SA, Nizamuddin S (2020) Magnetic nanoparticles incorporation into different substrates for dyes and heavy metals removal—a review. Environ Sci Pollut Res 27:43526–43541

Kumar V, Guleria P (2020) Application of DNA-nanosensor for environmental monitoring: recent advances and perspectives. Curr Pollut Rep

Lancaster CA, Scholl WE, Ticknor MA, Shumaker-Parry JS (2020) Uniting top-down and bottom-up strategies using fabricated nanostructures as hosts for synthesis of nanomites. J Phys Chem C 124:6822–6829

Lee WS, Choi J (2019) Hybrid integration of carbon nanotubes and transition metal dichalcogenides on cellulose paper for highly sensitive and extremely deformable chemical sensor. ACS Appl Mater Interfaces 11:19363–19371

Leonard BM, Zhou Q, Wu D, DiSalvo FJ (2011) Facile synthesis of PtNi intermetallic nanoparticles: influence of reducing agent and precursors on electrocatalytic activity. Chem Mater 23:1136–1146

Leung KC-F, Xuan S, Zhu X, Wang D, Chak C-P, Lee S-F, Ho WKW, Chung BCT (2012) Gold and iron oxide hybrid nanocomposite materials. Chem Soc Rev 41:1911–1928

Li L, Feng D, Fang X, Han X, Zhang Y (2014) Visual sensing of Hg^{2+} using unmodified Au@Ag core–shell nanoparticles. J Nanostruct Chem 4:117

Li Q, Wu F, Mao M, Ji X, Wei L, Li J, Ma L (2019) A dual-mode colorimetric sensor based on copper nanoparticles for the detection of mercury-(II) ions. Anal Methods 11:4014–4021

Li Q, Chen D, Miao J, Lin S, Yu Z, Han Y, Yang Z, Zhi X, Cui D, An Z (2020) Ag-modified 3D reduced graphene oxide aerogel-based sensor with an embedded microheater for a fast response and high-sensitive detection of NO_2. ACS Appl Mater Interfaces 12:25243–25252

Liao H, Chang W, Yang Y (2015) Climatic effects of air pollutants over china: a review. Adv Atmos Sci 32:115–139

Lin Y, Cao Y, Yao Q, Chai OJH, Xie J (2020) Engineering noble metal nanomaterials for pollutant decomposition. Ind Eng Chem Res 59:20561–20581

Liu T, Li N, Dong JX, Zhang Y, Fan YZ, Lin SM, Luo HQ, Li NB (2017) A colorimetric and fluorometric dual-signal sensor for arginine detection by inhibiting the growth of gold nanoparticles/carbon quantum dots composite. Biosens Bioelectron 87:772–778

Liu F, Wang X-D, Du S-Y (2020) Production of gold/silver doped carbon nanocomposites for effective photothermal therapy of colon cancer. Sci Rep 10:7618

Lu L, Zhu Z, Hu X (2019) Hybrid nanocomposites modified on sensors and biosensors for the analysis of food functionality and safety. Trends Food Sci Technol 90:100–110

Lupan O, Ababii N, Mishra AK, Gronenberg O, Vahl A, Schürmann U, Duppel V, Krüger H, Chow L, Kienle L, Faupel F, Adelung R, de Leeuw NH, Hansen S (2020) Single CuO/Cu_2O/Cu microwire covered by a nanowire network as a gas sensor for the detection of battery hazards. ACS Appl Mater Interfaces 12:42248–42263

Manisalidis I, Stavropoulou E, Stavropoulos A, Bezirtzoglou E (2020) Environmental and health impacts of air pollution: a review. Front Public Health 8:14–14

Merkel TJ, Herlihy KP, Nunes J, Orgel RM, Rolland JP, DeSimone JM (2010) Scalable, shape-specific, top-down fabrication methods for the synthesis of engineered colloidal particles. Langmuir 26:13086–13096

Miao P, Han K, Sun H, Yin J, Zhao J, Wang B, Tang Y (2014) Melamine functionalized silver nanoparticles as the probe for electrochemical sensing of clenbuterol. ACS Appl Mater Interfaces 6:8667–8672

Min Y, Wang Y (2020) Manipulating bimetallic nanostructures with tunable localized surface plasmon resonance and their applications for sensing. Front Chem 8:411

Moghimi N, Mohapatra M, Leung KT (2015) Bimetallic nanoparticles for arsenic detection. Anal Chem 87:5546–5552

Mohammadi MM, Kumar A, Liu J, Liu Y, Thundat T, Swihart MT (2020) Hydrogen sensing at room temperature using flame-synthesized palladium-decorated crumpled reduced graphene oxide nanocomposites. ACS Sens 5:2344–2350

Moosavi S, Lai CW, Gan S, Zamiri G, Akbarzadeh Pivehzhani O, Johan MR (2020) Application of efficient magnetic particles and activated carbon for dye removal from wastewater. ACS Omega 5:20684–20697

Nagajyoti PC, Lee KD, Sreekanth TVM (2010) Heavy metals, occurrence and toxicity for plants: a review. Environ Chem Lett 8:199–216

Nandan B, Horechyy A (2015) Hairy core–shell polymer nano-objects from self-assembled block copolymer structures. ACS Appl Mater Interfaces 7:12539–12558

Naseem T, Waseem M (2021) A comprehensive review on the role of some important nanocomposites for antimicrobial and wastewater applications. Int J Environ Sci Technol. https://doi.org/10.1007/s13762-13021-03256-13768

Nikfarjam A, Hosseini S, Salehifar N (2017) Fabrication of a highly sensitive single aligned TiO_2 and gold nanoparticle embedded TiO_2 nano-fiber gas sensor. ACS Appl Mater Interfaces 9:15662–15671

Njagi EC, Huang H, Stafford L, Genuino H, Galindo HM, Collins JB, Hoag GE, Suib SL (2011) Biosynthesis of iron and silver nanoparticles at room temperature using aqueous sorghum bran extracts. Langmuir 27:264–271

Panahi-Sarmad M, Noroozi M, Abrisham M, Eghbalinia S, Teimoury F, Bahramian AR, Dehghan P, Sadri M, Goodarzi V (2020) A comprehensive review on carbon-based polymer nanocomposite foams as electromagnetic interference shields and piezoresistive sensors. ACS Appl Electron Mater 2:2318–2350

Paramanik B, Kundu S, De G, Patra A (2016) Structural evolution, photoinduced energy transfer in Au nanocluster–CdTe QD nanocomposites and amino acid sensing. J Mater Chem C 4:486–496

Pelley J (2019) Air pollution may impair children's memory. C&EN 97:4–4

qian W, Guo H, Li X, Zhang G, Zheng J, Li Z (2021) An amperometric nitrite sensing platform with enhanced sensitivity based on copper nanoparticle/nanostructured polyaniline hydrogel nanocomposite. Ionics

Qing Z, Qing T, Mao Z, He X, Wang K, Zou Z, Shi H, He D (2014) dsDNA-specific fluorescent copper nanoparticles as a "green" nano-dye for polymerization-mediated biochemical analysis. Chem Commun 50:12746–12748

Rahman A, Ali I, Zahrani SMA, Eleithy RH (2011) A review of the applications of nanocarbon polymer composites. NANO 06:185–203

Ramalingam M, Ponnusamy VK, Sangilimuthu SN (2019) A nanocomposite consisting of porous graphitic carbon nitride nanosheets and oxidized multiwalled carbon nanotubes for simultaneous stripping voltammetric determination of cadmium(II), mercury(II), lead(II) and zinc(II). Microchim Acta 186:69

Rane AV, Kanny K, Abitha VK, Thomas S (2018) Chapter 5—Methods for synthesis of nanoparticles and fabrication of nanocomposites. In: Mohan Bhagyaraj S, Oluwafemi OS, Kalarikkal N, Thomas S (eds) Synthesis of inorganic nanomaterials. Woodhead Publishing, pp 121–139

Rastogi PK, Ganesan V, Krishnamoorthi S (2014) A promising electrochemical sensing platform based on a silver nanoparticles decorated copolymer for sensitive nitrite determination. J Mater Chem A 2:933–943

Riazi H, Taghizadeh G, Soroush M (2021) MXene-based nanocomposite sensors. ACS Omega 6:11103–11112

Richards R, Bönnemann H (2005) Synthetic approaches to metallic nanomaterials. In: Nanofabrication towards biomedical applications, pp 1–32

Sahu S, Sharma S, Ghosh KK (2020) Novel formation of Au/Ag bimetallic nanoparticles from a mixture of monometallic nanoparticles and their application for the rapid detection of lead in onion samples. New J Chem 44:15010–15017

Saleh TA, Al-Ruwayshid SH, Sarı A, Tuzen M (2020) Synthesis of silica nanoparticles grafted with copolymer of acrylic acrylamide for ultra-removal of methylene blue from aquatic solutions. Eur Polym J 130:109698

Sarma GK, Sen Gupta S, Bhattacharyya KG (2019) Nanomaterials as versatile adsorbents for heavy metal ions in water: a review. Environ Sci Pollut Res 26:6245–6278

Schmarsow RN, dell'Erba IE, Villaola MS, Hoppe CE, Zucchi IA, Schroeder WF (2020) Effect of light intensity on the aggregation behavior of primary particles during in situ photochemical synthesis of gold/polymer nanocomposites. Langmuir 36:13759–13768

Schröder I (2020) COVID-19: a risk assessment perspective. ACS Chem Health Saf 27:160–169

Septiani NLW, Yuliarto B, Nugraha, Dipojono HK (2017) Multiwalled carbon nanotubes–zinc oxide nanocomposites as low temperature toluene gas sensor. Appl Phys A 123:166

Sheikhian L, Ghanbarian M (2021) Constructing a resistive gas sensor based on the highly stable Mil-53(Fe)/Ag/CNT ternary nanocomposite for sensing polar volatile organic compounds such as methanol. J Anal Chem 76:399–412

Shrivas K, Monisha, Patel S, Thakur SS, Shankar R (2020) Food safety monitoring of the pesticide phenthoate using a smartphone-assisted paper-based sensor with bimetallic Cu@Ag core–shell nanoparticles. Lab Chip 20:3996–4006

Shukla P, Saxena P (2021) Polymer nanocomposites in sensor applications: a review on present trends and future scope. Chin J Polym Sci 39:665–691

Shy CM, Finklea JF (1973) Air pollution affects community health. Environ Sci Technol 7:204–208

Su Y, Shi B, Liao S, Zhao J, Chen L, Zhao S (2016) Silver nanoparticles/N-doped carbon-dots nanocomposites derived from siraitia grosvenorii and its logic gate and surface-enhanced raman scattering characteristics. ACS Sustain Chem Eng 4:1728–1735

Sun Y, Xia Y (2002) Shape-controlled synthesis of gold and silver nanoparticles. Science 298:2176–2179

Tagliazucchi M, Blaber MG, Schatz GC, Weiss EA, Szleifer I (2012) Optical properties of responsive hybrid Au@polymer nanoparticles. ACS Nano 6:8397–8406

Tajik S, Beitollahi H, Nejad FG, Dourandish Z, Khalilzadeh MA, Jang HW, Venditti RA, Varma RS, Shokouhimehr M (2021) Recent developments in polymer nanocomposite-based electrochemical sensors for detecting environmental pollutants. Ind Eng Chem Res 60:1112–1136

Tsai T-T, Chen C-A, Yi-Ju Ho N, Yang S, Chen C-F (2019) Fluorescent double-stranded DNA-templated copper nanoprobes for rapid diagnosis of tuberculosis. ACS Sens 4:2885–2892

Uchimiya M, Bannon D, Nakanishi H, McBride MB, Williams MA, Yoshihara T (2020) Chemical speciation, plant uptake, and toxicity of heavy metals in agricultural soils. J Agric Food Chem 68:12856–12869
von Schneidemesser E, Monks PS (2013) Air quality and climate—synergies and trade-offs. Environ Sci: Process Impacts 15:1315–1325
Wang A, Guan C, Shan G, Chen Y, Wang C, Liu Y (2019) A nanocomposite prepared from silver nanoparticles and carbon dots with peroxidase mimicking activity for colorimetric and SERS-based determination of uric acid. Microchim Acta 186:644
Wei Z, Duan H, Weng G, He J (2020) Metals in polymers: hybridization enables new functions. J Mater Chem C 8:15956–15980
WHO (2008) World Health Organization, Geneva
Wiederoder MS, Nallon EC, Weiss M, McGraw SK, Schnee VP, Bright CJ, Polcha MP, Paffenroth R, Uzarski JR (2017) Graphene nanoplatelet-polymer chemiresistive sensor arrays for the detection and discrimination of chemical warfare agent simulants. ACS Sens 2:1669–1678
Williams DB, Carter CB (1996) Transmission electron microscopy: a textbook for materials science. Plenum Press, New York
Wu X, Cobbina SJ, Mao G, Xu H, Zhang Z, Yang L (2016) A review of toxicity and mechanisms of individual and mixtures of heavy metals in the environment. Environ Sci Pollut Res 23:8244–8259
Wu X, Lv X, Wang J, Sun L, Yan Y (2017) Surface molecular imprinted polymers based on Mn-doped ZnS quantum dots by atom transfer radical polymerization for a room-temperature phosphorescence probe of bifenthrin. Anal Methods 9:4609–4615
Xie Y, Li Y, Niu L, Wang H, Qian H, Yao W (2012) A novel surface-enhanced Raman scattering sensor to detect prohibited colorants in food by graphene/silver nanocomposite. Talanta 100:32–37
Xiong D, Chen M, Li H (2008) Synthesis of para-sulfonatocalix[4]arene-modified silver nanoparticles as colorimetric histidine probes. Chem Commun 7:880–882
Yan Y, Yang G, Xu J-L, Zhang M, Kuo C-C, Wang S-D (2020) Conducting polymer-inorganic nanocomposite-based gas sensors: a review. Sci Technol Adv Mater 21:768–786
Yang H, Zhou R, Sun Y, Li P, Zhang W, Zhao Z, Shi J, Hu J, Chen Y (2020) Optimization and gas sensing properties of Au nanoparticle modified α-Fe_2O_3 nanodisk structures for highly sensitive acetone detection. New J Chem 44:16174–16184
Zhang J, Yuan Y, Wang Y, Sun F, Liang G, Jiang Z, Yu S-H (2015) Microwave-assisted synthesis of photoluminescent glutathione-capped Au/Ag nanoclusters: a unique sensor-on-a-nanoparticle for metal ions, anions, and small molecules. Nano Res 8:2329–2339
Zhang X, Wang Z, Liu Z, Liu B, Wu R, Chen Z, Zuo X (2021) New application of a traditional method: colorimetric sensor array for reducing sugars based on the in-situ formation of core-shell gold nanorod-coated silver nanoparticles by the traditional Tollens reaction. Microchim Acta 188:142

Chapter 19
Metal Nanocomposites as Optical Sensor for Ions and Molecules of Environmental Concern

Pranshu Kumar Gupta, Pawan Kumar Sada, Vikas Kumar Sonu, and Abhishek Rai

Abstract Multi phasic metal nanocomposites comprising a zero-dimensional metal nanoparticles phase (0.5–5% w/w) have significantly enhanced the material property and sensing ability of easily recoverable conventional matrices like glass, ceramics, metal, carbon allotropes or polymers. Over the past few years, promising chemical and physical properties of metal nanocomposites, developed via direct in situ chemical, photocatalytic, or thermal reduction of metal salts, over common matrix, or ex situ direct insertion of nanoparticles into the polymer. Colorimetric MNC sensors like metal–metal oxide NPs, quantum dots have depicted a relation between their type, structure, function and sensing performance, via aggregation/decomposition of NPs, fluorescence on/off and ligand-receptor interaction. Functional single/multilayered transition-metal dichalcogenides have been employed both, as a matrix for their decoration exhibiting synergism and quite recently as nanocomposites exhibiting ultra-fast and selective multi-sensor activity over narrow detection limits that are an urgent demand of national, homeland and environment safety. Fabrication of these biosensors endorse their progress in environmental and material aspects. They are employed as bio-analytic sensors against excess of heavy metal ions, nitro-compounds, poly-aromatic hydrocarbons, microbial proteins etc. imparting exclusive photo-chromatic and electronic properties, tuned at grass-root level by adjusting various supramolecular interactions and also the metal-ligands interactions at molecular level. Nanocomposites have several applications mainly in biological arenas, solar cells, communication field, optoelectronic devices etc. In this chapter we would discuss about recently developed nanocomposite-sensors, their synthesis, mechanism for their sensing action and environmental application involving mainly optical sensors.

Keywords Nanocomposites · Sensing · Nanoparticles · Frameworks · Environment · Matrix

---

P. K. Gupta
Department of Chemistry, Institute of Science, Banaras Hindu University, Varanasi 221005, India

P. K. Sada · V. K. Sonu · A. Rai (✉)
Department of Chemistry, Faculty of Science, L.N. Mithila University, Darbhanga 846008, India

S. P. Singh et al. (eds.), *Metal Nanocomposites for Energy and Environmental Applications*, Energy, Environment, and Sustainability,
https://doi.org/10.1007/978-981-16-8599-6_19

19.1 Introduction

Increasing pollution, detiorating health, and energy crises are triads of environmental anthropogeny in the modern world, whereby a special reference to pollution is must owing to its notorious, supportive, direct or indirect involvement in either of these. With the dawn of commercial industrialization, agriculture, and synthetic drug technology, heavy metal ions, anions, dyes, pesticides, gases etc. the condition of natural entities like water, air, and earth have been immensely contaminated paving the way for an irreversible ecological damage (Jin et al. 2017). Reckless pollution of nature particularly, water bodies and consumables by heavy metal ions (HMIs), poly aromatic hydrocarbons (PAHs), industrial effluents, nitro-aromatic compounds (NACs), are a matter of serious global concern owing to their non-biodegradable and toxic nature causing harmful effects on human health even under narrow concentration ranges (Krabbenhoft and Sunderland 2013). HMIs efflux from industries, by fossil fuel combustion, mining, refining, textile printing, production of chemicals, electronics, paints and dyes pose various health hazards (Boening 2000). Therefore, their sensitive, rapid, selective and real-time determination in the natural environment is of much importance.

Quite recently, pollution of water bodies with *Microcystis* sp. affect the quality of water-body making it harmful for consumption owing to its contamination with carcinogenic as well as hepatotoxic group of algal exudates called microcystins, of which microcystin LR is the most common form. Lakes polluted with these needs quick detection and immediate prohibition of its consumption. A huge class of organophosphate, organochlorine pesticides are exhaustively used in agricultural field, not only affects humans through its biomagnification, but pollutes water by agricultural runoff promoting algal bloom and finally its anthropogenic eutrophication. Carcinogenic hydrocarbons as solvents like hexane, toluene, benzene and its derivatives are widely exploited in chemical and dye industries and are released into water bodies which have been reported to effect embryonic development of aquatic fauna. Release of certain drugs like clenbuterol in environment and its exposure to aquatic flora and fauna have shown disastrous effect on stamina, and performance of human beings and animals.

Modern technologies and methods for separating these toxic entities from aqueous phase on the basis of conventional principles such as solvent extraction, ion exchange, chemical precipitation etc. via adsorbents which includes zeolites, porous silica, polymer, clay materials, cellulose, metal organic frameworks (MOF) and others is although convenient and cost effective, but pose limitations of low adsorption rate, efficiency, selectivity, with much lower segregation, regeneration, and large-scale/industrial production ability (Fischer et al. 2017). Conventional bioanalysis techniques, like absorption (UV–vis) spectroscopy, high-performance liquid chromatography (HPLC), liquid chromatography-mass spectrometry (LC–MS), atomic absorption spectroscopy, etc. widely used for the effective determination narrow concentrations of analytes, are expensive, complicated, and time consuming, hence restricting the wider their applications for facile, time-savvy, and accurate bioanalysis and detection of the above-mentioned entities (Pourreza et al. 2014).

Surface-active inorganic NPs and their quantum confinement modulated properties unlike the ones offered by their corresponding bulk counterparts, have attracted modern day scientists for catalysis, sensing, semiconductor technology and medicine (Lu and Yin 2012; Morsi et al. 2021; Hassan et al. 2021). However, problems of polydispersity and aggregation adversely affect their properties and hence many systems such as polymer ligands which are polymeric, microgels and dendrimers have been discovered to prevent their occurrence and to maintain the desired property (Zhang et al. 2004). Scientists have also emphasized over utilization of surface-hydroxylated nanostructured metal oxide (NSMOs) based adsorbents exhibiting huge adsorption capacity particularly owing to high surface-to volume (SA/V) ratio, porosity, defect abundancy, HMI-active site abundancy and reactivity (Henglein 1989). For past few decades, these NSMOs (particularly of magnesium, titanium, cerium, aluminum, iron and manganese) have been prepared and widely utilized for aqueous phase HMI-adsorption, modulated by their particle size and shape (Qiao et al. 2009), but on size reduction from micrometer to nanometer level, the increased surface activity lead to long range interaction mediated particle agglomeration, hampered their adsorption capacity (Jana et al. 2007).

Certain other functional nanomaterials (NMs) like quantum dots (QDs), CNTs/carbon nanofibers (CNFs), nanowires, graphene and more two-dimensional (2D) NMs, have been used to prepare chemo/biosensor platforms employed in environmental and biomedical fields (Kumar et al. 2019). NPs doped glassy carbon electrodes encouraged their use as label-free electrochemical sensors. Fluorescent NPs and QDs facilated high-performance sensing based on fluorescence quenching. Active substrates based on Metal NPs- mediates surface-enhanced Raman scattering determination up to the level of single molecule spectroscopy (Tiwari et al. 2016). However, their applicability turns impractical due to time-consuming analysis and *in-situ* operation. The synthesis of novel colorimetric chemo/biosensors with rapid detection, high sensitivity, and naked-eye sensing capabilities have hence gained much importance (Aldewachi et al. 2018).

Recent advancement in fabrication of 2D graphene and its derivatives led to the exploration of mono- or multi-layered NSs of layered transition metal dichalcogenides (TMDs), known to offer moldable chemical, electrical, and optical properties owing to their large lateral size, small thickness, and versatile physicochemical properties making them an ideal candidate for synthesis of graphene alternative materials (GAMs) (Rao et al. 2013). Among these, molybdenum disulfide (MoS_2) and their NCs find application in fields like energy storage, electronic devices, and biomedical engineering (Chen et al. 2015). Its lamellar mono-layered structure consists of three atomic layers (S–Mo–S), stacked one above the other by the van der Waals forces. Three main types of MoS_2 i.e., ^{2}H-MoS_2(natural, trigonal prismatic), ^{1}T-MoS_2(metastable, semiconducting, coordination flexibility of Mo) and ^{3}R-MoS_2(natural, trigonal prismatic), have been recognized until now on the basis of its crystal structure, whereby the prior two are found to coexist in chemically exfoliated monolayers of MoS_2 (Wypych and Schöllhorn 1992). Point defects, grain boundaries, and edges have noteworthy roles in the field of sensing as they generate opportunity of surface modification and functionalization, profoundly affecting their chemical, electrical, and optical properties (Carbone et al. 2015). Unlike mono-layered graphene, these are three-atom layer NSs with no wrinkle, but smooth morphology. Zhou et al. (2013) systematically studied dislocation and defects in MoS_2 (Zhou et al. 2013). Exfoliating its NSs into one or more than one layer preserves bulk properties and introduce confinement characteristics, hence offering opportunities for sensor applications. Tunable electronic properties of MoS_2 NSs can be comprehended from direct band semiconductive features of ^{2}H type, metallic characteristics of ^{1}T type MoS_2 (Liu et al. 2015). MoS_2 also exhibits excellent optical and quenching photoluminescence (PL) properties, as mono-layered MoS_2 enhances emission instead of quenching it under a suitable concentration in an aggregation-induced emission system, its 2D structure with large specific area makes it an ideal candidate for increasing the effectiveness of sensing devices. With the growing demand for quick, sensitive, and stable sensors, a series of MoS_2-based NCs (MoS_2–NCs) have been prepared in the previous few years, combining the advantages of MoS_2, layered materials and metallic NPs, or by ligand modification of its surface hence to achieve nanohybrids, and these systems have been employed for electrical, catalytic and chemical performances (Tan and Zhang 2015).

Owing to their limited chemical reactivity and high boiling point NAC detection unlike other compounds is highly challenging; but their electron deficient nature is often exploited for their sensing as a donor–acceptor charge transfer (CT) complex (via π-stacking) can be developed using appropriate electron rich sensing molecule, bringing changes in fluorescence emission properties like photo-induced electron transfer process (PET), Turn-on or Turn-off etc., are a potent tool for detection and quantification for NACs (Banerjee et al. 2014). A facile electron transfer can occur when lowest unoccupied molecular orbital (LUMO) of photoexcited NCs is of higher energy than the LUMO of molecules that are electron deficient in nature. Moreover, the CT efficiency would increase when the emission spectrum of the NCs overlaps with the NACs absorption spectrum resulting in decrease of the emission intensity of the prior via luminescence quenching. The degree of luminescence, is also governed by solvent phase reduction potentials of NCs, whereby the redox active species can either be metal ions/metal clusters or organic linkers or both. Solvent molecules tune luminescence properties, structural stability and integrity of NCs dictating their sensing application for an analyte, and it may get affected by framework intervention of solvent molecules to open metal site of NCs. Their emission spectra at times depends on the polarity of solvent, which when high; stabilize the charge separated excited state of molecules, and when low; stabilize charged excited state of molecules which is locally confined. Hence, enhanced solvent polarity facilitates energy transfer from LUMO of NCs to analyte, by lowering the energy of the latter resulting in accelerated luminescent quenching (Cui et al. 2018). Theoretical calculations, via computational methods such as density functional theory, *ab* initio calculations, assists optimization of structures of molecules, spectral analyses, energy spectrum, and activation energy barriers, resonance electron transfer (RET), PET and Dexter electron exchange process (DEEP), etc. Molecular dynamics simulation and Monte Carlo Simulation methods are used to gain insights about different sensing mechanisms. Xiong et al. (2009) reported isoreticular MOFs with a capability to sense RDX explosives and with theoretical evaluation of their capacity via classical molecular dynamics and Grand Canonical Monte Carlo (GCMC) simulation (Xiong et al. 2010). Computational tools can also provide information about non-covalent interactions formation between NACs-co-former molecules during co-crystallization, as in a case reported by Gao et al. (2007) whereby co-crystals of two NACs on simulation, formed hydrogen bonds between them (Zhou et al. 2014).

Coordination polymers (CPs) as a special class, consist of metal atom coordinated to organic, inorganic donor centers or both, via numerous coordination modes, and exo-dentate ligands *i.e.* the ligands having one or more donor sites in the molecule, coordinate to it forming infinite arrays of molecular arrangement. Their dimensionality is totally governed by number of spacer-donor atoms and metal ion coordination capacity (Wang and Cohen 2009). Coordinative unsaturation in metal ions enhance the chances of formation of higher dimensional CPs, and MOF NCs. Ligands to metal ions one, two, three-dimensional coordination forms 1D, 2D, and 3D CPs, array and framework, respectively, (Kitagawa and Uemura 2005). Structural diversity and outstanding tenability results from the self-assembly of poly-dentate bridging ligands having metal-connecting points. can be altered by Modification of shape, size and geometry of the ligands alters pores in these systems making them appropriate candidate for applications in sensing of HMIs, and other organic compounds of environmental concern (Lustig et al. 2017). The non-covalent interactions such as electrostatic interactions, hydrogen bonding $\pi\cdots\pi$ stacking interaction, van der Waals interactions etc. are involved in adsorption of analytes over MOFs, acid–base interaction, or hydrophobic or London force interactions. Usually, adsorption process is collective effect of majority of these supramolecular interactions. Electrostatic interaction is mere a pH dependent interaction, which fall off progressively with distance ($1/r^2$, where r is the distance between the ions). Ligand functionalization on MOFs generate surface charges and influence adsorption ability. The hydrogen bonding also occurs during the adsorption of NACs and other organic compounds over MOFs (Liu et al. 2014). Normally, metal ions with d^9 or d^{10} electronic configurations (for example Cu^{+2} or Zn^{+2}, Cd^{+2}, Cu^{+}, Ag^{+}, etc.) are preferred for construction of Luminescent Molecular Organic Frameworks (LMOF), owing to various reasons like *d–d* transition based luminescence of completely filled *d*-orbitals, structural binding unit based ligand effects or electronic transition, restricted attack by solvent molecules due to no vacant *d*-orbitals and lower reactivity. Due to greater crustal abundance main group metal ions like In^{+3} or Mg^{+2} have been preferred. The presence low lying empty orbitals of main group metal ion, structural integrity and stability based LMOFs is a lively field of research owing to their ease of oxidation or areal hydrolysis. Ln^{+3} based LMOFs have exceptional emission properties that are versatile starting from efficient CT to detection of analyte, but its low natural abundance incites enormous challenge in constructing LMOFs at low cost.

19.2 Types of Nanocomposite Sensors

NCs comprise of two phases, namely the reinforcement and the matrix. Conventionally, there were not much for matrices available so the classification was mere based on the type of reinforced entities. However, most of the senor NCs employed till now are mere metal oxide or metal NPs based NCs, where by these materials have been used as sensors. Hence in this chapter, we preferred classifying these NCs on the basis of their sensing capability or the detection technique involved (Fig. 19.1).

19.2.1 Adsorption Based NC Sensors

Varieties and Strategy. A new class of metal oxide nanocomposites (M_yO_xNCs), as a type of metal nanocomposites (MNCs) system evolved when adsorbing systems like metal oxides were loaded/impregnated over porous supports, or directly grown over solid supports that were comparatively more stable, agglomeration resistant and showed high adsorption and separation efficiency during wastewater treatment, hence improving the overall efficiency and affordability of wastewater treatment (Pan et al. 2009). Hence there has been a pressing need of such stable, environmentally-friendly and adsorption active M_yO_xNCs. Surface active and hydrothermally synthesized MnO_2NCs have gained much scientific and fundamental importance owing to their polymorphic structures, and environmentally benign nature (Jana et al. 2008), for the removal of HMIs and PAHs via adsorption, ion-exchange, surface complexation and co-precipitation from wastewater. Hierarchical Fe_3O_4/MnO_2 NCs by Kim et al. (2013) are capable of aqueous phase Cd^{+2}, Cu^{+2}, Pb^{+2}, and Zn^{+2} detections. Zhang et al. (2009) synthesized core–shell nanoplates (NPIs) of similar oxides via similar route for aqueous phase As^{+3} adsorption (Zhao et al. 2012). Cuia et al. (2016) have synthesized similar magnetic NPIs to detect Pb^{+2}, Cu^{+2}, Cd^{+2}, Zn^{+2} and Ni^{+2} ions from water (Arora 2019). Liang et al. (2017) developed MnO_2 modified biochar via redox reaction of $KMnO_4$ and Mn(II) acetate solution, to detect Pb^{+2} and Cd^{+2} ions (Liang et al. 2017). Zhao et al. (2016) employed commercial Li-ion battery's

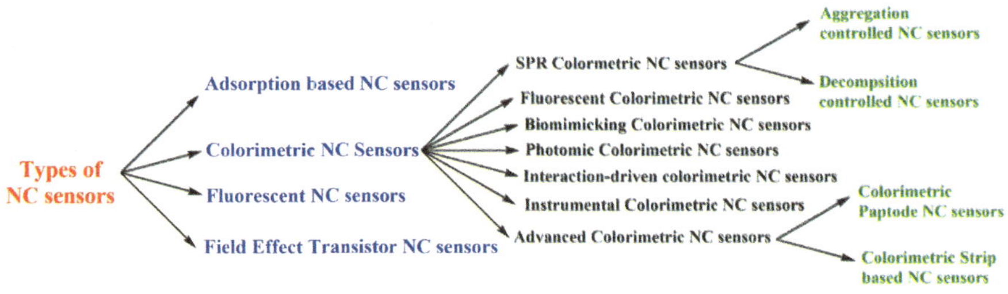

Fig. 19.1 Illustration depicting the classification of metal NCs based upon their sensing capability or principle

anode as carbon support for MnO_2 loading and employed it for Pb^{+2}, Cd^{+2}, and Ag^{+2} ions aqueous phase adsorption (Zhao et al. 2017). MnO_2/polymer adsorbent prepared by Zhu et al. (2019) via phase inversion is capable to remove Ni^{+2} ions from aqueous solutions (Zhu et al. 2019). Minamata disease, mental retardation, renal dysfunction, hepatic injury, etc. is caused by Hg^{+2} bioaccumulation of mercury and also damage the crucial cellular machinery. Its maximum consumable limit (MCT) as recommended by United States Environmental Protection Agency (EPA) is 2.0 ppb in drinking water. Liu et al. (2019) have prepared MnO_2 nanoparticles (NPs), decorated over C_3N_4-nanosheets (NSs) by have been used for detection of Hg^0 (Moghaddam and Pakizeh 2015). MnO_2-coated carbon nanotubes (CNTs) have been used for aqueous phase Hg^{+2} adsorption (Liu et al. 2019). Comparatively much higher Hg^0 adsorption efficiencies were obtained for flowery MnO_2 NCs prepared by light directed hydrolysis at room temperature within a wide range of pH values.

Characterization. After HMI-M_xO_y NCs complexation, a chemical heterogeneity develops and the spin–orbit O 1s as well as vibrational frequency peaks for transition metal oxide in M_xO_y shifts toward the lower binding energy, owing to the association of HMI with surface hydroxyl of M_xO_y NCs. The participation of hydroxyl groups in the adsorption process is indicated by the decrease in intensity of the M-OH group. Doublets owing to the spin–orbit coupling in high resolution HMI (n − 1) *d* spectrum, also signifies the presence of HMI onto NCs surface, confirming its adsorption, without any redox process (Fig. 2a, b). Adsorption kinetics for HMI adsorption over NCs surface as a significant parameter of an adsorbent for the practical uses, determining the rate of adsorption and equilibrium time of the metal ion uptake, is measured as HMI uptake capacity of NCs as a function of time, keeping other physical parameters like pH and temperature constant. Adsorption may or may not be fast and it depends on the time taken for equilibrium achievement. Kinetics of adsorption mechanism, may be pseudo-first-order (p1o) or pseudo-second-order (p2o) and usually both are applied to choose the more appropriate model for kinetic study as per the interpretation of kinetics data, in accordance with the Eqs. 19.1, 19.2 (Wang et al. 2019):

$$(\theta_t)_{p1o} = \theta_{eq}\left(1 - e^{-k_1 t}\right) \tag{19.1}$$

$$(\theta_t)_{p2o} = \frac{k_2\theta_{eq}^2 t}{1 + k_2\theta_{eq} t} \tag{19.2}$$

where, $\theta_t, \theta_{eq}, k_1, k_2$, are the adsorption capacity (mg g^{-1}) of an NCs at time *t*, during equilibrium, and rate constants for p1o (min^{-1}) and p2o (g mg^{-1} min^{-1}) reactions. Quite frequently, the p2o model is preferred over the pseudo first order model to describe M_yO_x NCs-HMI adsorption kinetics, as confirmed by the higher correlation coefficient (R^2) value and close agreement of experimental and theoretical adsorption capacity as compared to that of the latter. The process is governed by chemisorption *i.e.*, charge transfer between the surface hydroxyl groups of M_yO_x NCs NCs and

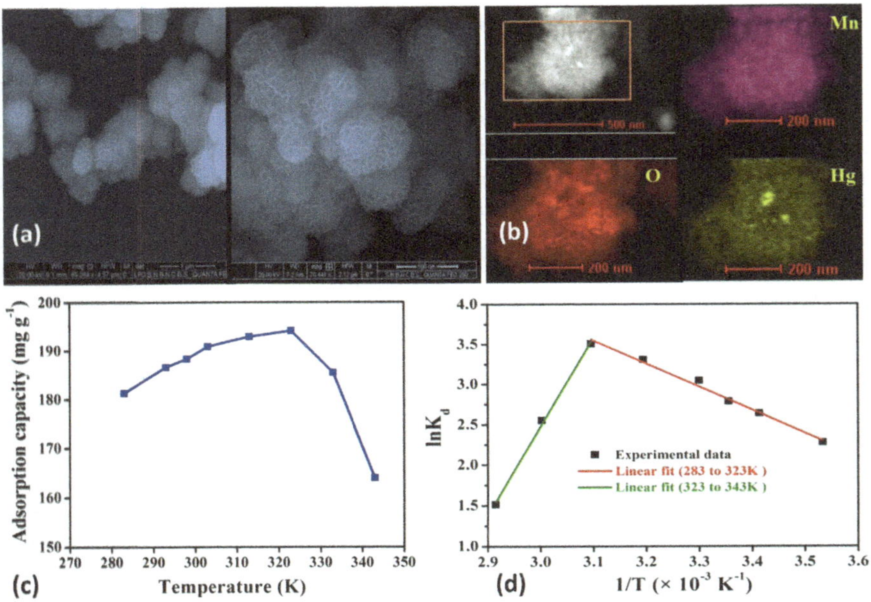

Fig. 19.2 **a** MnO_2 NCs FESEM images **b** MnO_2 NCs STEM images after Hg^{+2} uptake and corresponding elemental mapping the presence of Hg^{+2} ions (yellow zone) uniformly over the NCs, consisting of Mn(pink zone) and O(red zone), **c** Adsorption capacity of MnO_2 NCs versus temperature, **d** Plot of ln K_d versus 1/T. Reproduced with permission from Antochshuk et al. (2003) (https://doi.org/10.1039/D0DT01054E)

HMIs. However, at times, initial abruptly fast adsorption kinetics is attributed to high concentration of HMIs at the very early stage during the adsorption process. Adsorption isotherm depicting adsorbent-adsorbate interaction, are estimated by placing a constant amount of adsorbent in a series of flasks containing aqueous HMI solutions of equal volume and pH, with varied concentrations at constant temperatures. The sorption capability of the NCs generally increases with the increasing HMI concentration in the aqueous media. Langmuir model (LAM) is a monolayer adsorption of the adsorbate over homogeneous surface of the adsorbent that offers energetically equivalent adsorption sites having no inter adsorbate interactions. Unlike this, Freundlich model (FAM) which shows the multilayer adsorption of adsorbate over the heterogeneous surface of adsorbent with adsorption sites possessing distinct energies, with no inter adsorbate interactions, discussed by Eqs. 19.3, 19.4 (Antochshuk et al. 2003):

$$\left(\frac{\theta_{eq}}{C_{eq}}\right)_{LAM} = \frac{\theta_{max} K_L}{1 + K_L C_{eq}} \tag{19.3}$$

$$\left(\frac{\theta_{eq}}{C_{eq}}\right)_{FAM} = K_F \left(C_{eq}\right)^{\frac{1-n}{n}} \tag{19.4}$$

where, θ_{eq} and θ_{max},C_{eq}, K_L, K_F and n corresponds to equilibrium adsorption capacity of adsorbent, maximum adsorption capacity (mg g^{-1}), equilibrium concentration of adsorbate (mg L^{-1}), Langmuir equilibrium constant (L mg^{-1}) related to the free energy of adsorption, Freundlich equilibrium constant (mg g^{-1})(L $mg^{-1})^{1/n}$ and FAM equilibrium constant associated with system heterogeneity. Most of the reported M_yO_x NCs show higher correlation coefficient (R^2) acquired for LAM, illustrating monolayer coverage of HMI without any significant inter HMI interactions, owing to the even distribution of hydroxyl groups over its surface. The pH of the reaction mixture significantly controls the HMI adsorption both via adsorbate and adsorbate. Usually, M_yO_x NCs show an increase in adsorption with increasing pH up to 7.0, then a decrease in adsorption as the solution pH turns basic, particularly due to competitive adsorption between Hg^{+2} and H^+ ions over the surface of M_yO_x NCs along with excessive protonation of its own surface at lower pH, ensuing in weak binding ability for HMI due to electrostatic repulsion, affecting its uptake capacity and increases adsorption up to pH 7.0. The results from zero-point-charge pH (pH_{pzc}) of M_yO_x NCs usually lie in the neutral range. At pH above this surface charge of M_yO_x NCs must be negative facilitating binding of cationic species. Under alkaline condition, adsorption capacity is immensely decreased as HMIs formed hydroxyl complex of type $[(HMI)(OH)_z]^-$ because of its increased concentration, in solution, causing electrostatic repulsion between this complex and M_yO_x NCs, lowering its HMI capture capacity at higher pH (Kabiri et al. 2015).

Temperature Dependent Adsorption. Temperature dependent adsorption capacity of these NCs over a temperature range, with constant secondary experimental conditions, represents change in aqueous phase HMI adsorption capacity onto the M_xO_y NCs surface, which usually increase on increasing temperature up to a certain inversion point representing maximum adsorption capacity, after which it gradually decrease with further increase in temperature (Fig. 2c). The plot of ln K_d versus 1/T (Fig. 2d) is developed and change in free energy (ΔG°), enthalpy (ΔH°) and entropy (ΔS°) are estimated during the adsorption process as per Langmuir adsorption model using Eq. 19.5:

$$\ln K_d = \ln\left(\frac{\theta_{eq}}{C_{eq}}\right) = \ln K_L + \ln\left(\frac{\theta_{\max}}{\theta_{eq}}\right) = -\frac{\Delta G^o}{RT} = -\frac{\Delta H^o}{RT} + \frac{\Delta S^o}{R} \tag{19.5}$$

where, T, R and K_d refers to absolute temperature in Kelvin (K), universal gas constant (8.314 J mol^{-1} K^{-1}), and is equilibrium constant, calculated from $K_d = Q_e/C_e$ for different temperatures. The magnitudes of negative valued ΔG°, positive valued ΔH° and ΔS° increased with temperature, up to the inversion point, indicating good feasibility, endothermic nature and significant randomness enhancement because of release of chelate water molecules from the solvated HMI, during HMI adsorption process. However, as the temperature is increased further the sorption process being exothermic in nature, decreases the uptake capacity of the adsorbent, hence weakening the interaction between the active sites in M_xO_y NCs and HMIs. The combined kinetics and thermodynamics of HMI adsorption over M_yO_x NCs surface typically rely on the temperature dependent size of HMI-M_yO_x NCs complex formed during the adsorption process whose overall reaction rate, r_d depends on rate of the total HMI amount to adsorb over M_yO_x NCs surface (r_s) and rate of the adsorption of the HMI amount necessary due to the size alternation on every previous adsorption over M_yO_x NCs surface (r_c). This critical size of HMI-occupied hydrated M_yO_x NCs complex adjusts decrease of thermal activation energy barrier in adsorption process. Collision theory, expresses this overall r_d as Eq. 19.6:

$$r_d = f\left(\frac{k_B T}{h}\right) r_C^o r_S^o e^{(\Delta G_S^* - \Delta G_C^*)/RT} \tag{19.6}$$

where, ΔG_S^*, ΔG_C^*, and f denotes change in free energy of activation during adsorption process, interaction energy of hydrated HMI-M_yO_x NCs complex, collision theory-based temperature independent proportionality constant of the frequency factor, respectively. Furthermore, r_C^o and r_S^o depends on the number and size of HMI-M_yO_x NCs complex on the M_xO_y NCs. The thermally activated dissociation of this complex, corresponds to negative interaction energy for production of critical size of this complex, making the overall process to summarily look like Eq. 19.7:

$$\text{HMI}^{+n} + \text{M}_x\text{O}_y\text{NCs} \leftrightarrow \text{HMI}^{+n} - \text{M}_x\text{O}_y\text{NCs complex};\ k_d = \frac{k_f}{k_b} \tag{19.7}$$

where, k_f and k_b corresponds to rate of forward and backward reactions respectively, out of which the latter is unaffected by interaction energy of complexation, making K_d to be controlled only by ΔG_C^* and ΔG^o, owing to which, the free energy of activation for formation of HMI-M_xO_y NCs complex can be shown in Eq. 19.8:

$$\Delta G^o = \Delta G_f^* - \Delta G_b^* = \Delta G_{ads}^* - \Delta G_C^* \tag{19.8}$$

where, ΔG_{ads}^* is the amount of free energy change corresponding to HMI surface adsorption process. At temperature below inversion point, $\Delta G_C^* > \Delta G_S^*$, $\Delta G_C^* \approx 0$, and as $\left(\Delta G_S^* - \Delta G_C^*\right) < 0$, so ln K_d varies linearly with 1/T, and the dependence of temperature for HMI-M_xO_y NCs de-complexation or size-decrease with increase temperature is impractical. At temperatures above inversion point, positive, that indicates a positive ΔH^o, or the exothermic nature of HMI adsorption. A negative weakening of M_xO_y NCs-HMI interaction and HMI uptake capacity. Below inversion temperature, the slope of ln K_d versus 1/T, plot is negative, that indicates a positive ΔH^o or endothermic nature of HMI-M_xO_y NCs complexation process. A is positive ΔS^o under such a situation may lead to increase in randomness via de-complexation process. HMI selectivity often referred to as anti-interfere-ability as depicted by the NCs, is attributed to various aspects of HMI and NCs. It may be because of different electric charges, ionic radius and hydration energies of the HMIs (Das et al. 2020). NCs often prefer adsorption of HMIs that have lower hydration enthalpies as compared to the one having higher enthalpies, because the prior shows reduced solvation effects in aqueous solution, hence facilitating them to approach much closer to adsorbent surfaces.

19.2.2 Colorimetric NC Sensors

Optical sensors owing to its simplicity of naked eye detection without the use of any sophisticated instrumentation, play an essential part in fields such as pollution monitoring and medical diagnostics. Molecular sensing via unique quenching characteristics of MoS_2 NSs can be applied for both DNA biomolecule sensing. MoS_2 as a graphene analogue has two main advantages over graphene *i.e.,* being capable to offering an electron transfer path (as well as an activity site), and a platform for part of electronic activity such as metal NPs, enzymes, antigens, etc. (Ahmed and Gerischer 1979).

Surface Plasmon based Colorimetric NC sensors. SPR-based NP NCs as colorimetric sensors show color change on both aggregation and decomposition of metallic nanostructures. Liu et al. (2004) demonstrated DNAzyme-induced self-assembly of gold NPs used as colorimetric Pb^{+2} biosensor (Liu and Lu 2004). Guo et al. (2011) reported a dual colorimetric biosensor for glucose and H_2O_2 detection with the help of enzyme mimicking hemin-functionalized graphene NSs (Guo et al. 2011). Chen et al. (2011) reported a novel colorimetric polydiacetylenes based biosensor exploiting the distinctive chromatic transition property of used polymers (West et al. 2009). For detection of HMIs Priyadarshini et al. (2017) emphasized over AuNPs-based colorimetric sensors (Priyadarshini and Pradhan 2017). However, Vilela et al. (2012) showed this on the basis of aggregation of Au and Ag NPs (Vilela et al. 2012). Kim et al. (2012) reported receptor for colorimetric and t fluoresce detection of Pb^{+2}, Cd^{+2}, and Hg^{+2} ions (Kim et al. 2012). They basically are of following types:

Aggregation based Colorimetric NC sensors. Metal NPs of Au, Ag and Cu have localized surface plasma resonance (LSPR) that depends on the distance between the aggregated NPs causing change in color at difference distances. The aggregation caused bathochromic deviation of SPR peak and variation in optical properties of these NCs facilitate a naked-eye detection of color changes in NP solution, hence behaving as high-performance chemo/biosensors (Liu et al. 2020). For instance, Zhang et al. (2020) reported the self-assembly based synthesis of Hg^{+2} fluorescent peptide nanofiber (PNF) functionalized AuNP NCs, whereby the motif-designed peptide facilitated the self-assembly and stabilization of AuNPs, that finally show a bright red color. The specific interactions between self-assembled PNFs and Hg^{2+} causes the color change to purple owing to the aggregation AuNPs, hence bringing a bathochromic shift in its SPR peak. DNA aptamer and AuNPs-based Colorimetric sensor whereby DNA aptamer was developed by Li et al. (2016) and was applied to mediate the specific binding and AuNPs acts as the sensing probes for selective determination of microcystin-LR (MCLR), with LOD of 0.37 nm thus behaving as simple and sensitive probe. And linear detection range (LDR) of 0.5 nM to 7.5 μM

(Li et al. 2016). High salt concentration induced aggregation of AuNPs and wine-red to blue color transformation due to the interparticle plasmon coupling gets inhibited on the presence of DNA aptamer. However, on addition of MC-LR into the aptamer-functionalized AuNP NCs, it changed the structure of aptamer from the random coil to the regulated structure by aptamer-MC-LR complexation, triggering the aggregation of AuNPs and the change in color. This strategy could be applied to develop other types of aptamer-based colorimetric sensors. The thiocyanate (SCN^-) ion induced anti-aggregation property in the presence of aggregation inducing sulfuric acid in citrate-capped AuNP NCs developed by Deng et al. (2014) caused the color change from red to blue, hence functioning as a naked-eye detector of SCN^- ion with an LOD of 1 μM (Deng et al. 2014). Various other AuNPs acts as colorimetric sensors for naked-eye detection of proteins, bacteria, organophosphate pesticides, nitro-amine compounds, multiple HMIs, and clenbuterol have been fabricated successfully. Patel et al. (2015) reported specific colorimetric detection of carbendazim fungicide using color change mediated aggregation of 4-aminobenzenethiol (ABT) functionalized AgNP via interactions such as ion pair and π-π interactions, resulting in formation of conjugated network-like aggregation of AgNPs, with LOD of 1.04 μM and LDR of 10–100 μM (Patel et al. 2015). Elavarasi et al. (2014) found that citrate-capped AgNPs prepared by an optimal amount of sodium borohydride were capable of behaving as selective and sensitive colorimetric receptor for recognition of Cr^{+3} and Cr^{+4} ions, as these HMIs binds with the citrate groups to induce the aggregation of AuNPs (Elavarasi et al. 2014).

Decomposition based Colorimetric NC sensors. Metallic NPs display strong SPR properties in the visible area, but their decomposition or oxidation becomes the basis of fabrication of colorimetric sensors. Kumar et al. (2014) reported *ortho*-phenylenediamine functionalized AgNPs colorimetric sensor for detection of NO_2^- ions with LOD of 0.1 pM in real samples, via AgNPs to Ag^+ decomposition decolorizing its brownish yellow color complete absence of its SPR peak at AgNPs at 426 nm (Kumar and Anthony 2014). Chaiyo et al. (2015) reported thiosulfate catalyzed etching of Ag nanoplates and their pink violet color change, for fabrication of paper-based colorimetric sensor of Cu^{+2} ions, with LOD of 1.0 ng/mL (Chaiyo et al. 2015). Zhou et al. (2018) demonstrated fabrication of Ag nanoprism NCs as a multichannel sensor that got etched by dissolved oxygen at alkaline pH depicted by a blue-shift in its LSPR band. Different thiols showed high affinity towards these NCs at different pH values, with distinct LSPR peaks hence encouraging the use of these NCs for effective detection of glutathione, cysteine, dimercapto succinic acid, 3-mercaptopropionic acid, and dithiothreitol at nanomolar concentrations, by employing the data matric generated by principal component analysis (Zhou et al. 2018). Although Au and Ag NPs gave better results and were biocompatible, the high cost of these noble metals limited their wider applications, as an alternative, Cu NPs with much lower price and visible range LSPR have shown capability to somehow replace the noble metals. Hatamie et al. (2014) reported its fabrication and sensitivity as a potent LSPR colorimetric sensor for naked-eye S^{-2} ion detection, with LOD of 5 μM (Hatamie et al. 2014). However, unlike Au and Ag NPs, further applications of CuNPs gets upheld due to its tendency of getting oxidized in water. Soomro et al. (2014) reported a method to overcome this oxidation, by utilizing L-cysteine as a capping agent to prepare water-stabilized CuNPs that were later used for colorimetric sensing of Hg^{+2} ions with LOD of 43 nM, which could remove this layer, allowing CuNPs to oxidize, and aggregate together, changing its LSPR (Soomro et al. 2014).

Electron transfer-based sensors. PL of chemically exfoliated MoS_2 can also be controlled by regulating its bandgap. Another common optical sensor is the surface plasmon resonance (SPR) sensor, whose novel configuration depends on differential phase measurement by hybrid structure of MoS_2-graphene, that under optical excitation, transfer electrons from metal film to MoS_2-Graphene layer, leading to greater electric field enhancement on the sensing surface. The 3-layers of MoS_2 NSs improved the light absorption, while the single-layered graphene acted as a biorecognition element. Ionic sensors based on MoS_2 as reported by Yang et al. (2015) have recently, shown an unexpected phenomenon of MoS_2 NSs adsorption over rhodamine B isothiocyanate, later utilized for selective detection of Ag^+ ions even under bacteria solution with LOD of 1 nM under natural water, whereby Ag^+ ions were reduced to Ag^0, finally resulting in the separation of rhodamine B isothiocynate and MoS_2 NSs and fluorescence recovery (Mao et al. 2015).

Fluorescence based Colorimetric NC sensors. Organic dyes, fluorescent polymers and NMs (such as nanoclusters, QDs, carbon dots, etc.), have been employed to prepare colorimetric sensors that shows Turn ON/OFF behavior on interaction with analyte that could be transformed from "on" to "off" mode by addition of quenching analytes, or using the aggregation-induced emission behavior, or by addition of controlling analyte to react with the fluorescent probe bringing "on–off" change and addition of restoring analyte to bring it back. Many organic of probes showing fluorescence "on–off" behavior of HMIs such as Cu are synthesized. For instance, 4,4-difluorine-1,3,5,5–4-tetramethyl-4-bora-3a,4a-diaza-s-indacene (BODIPY) derivatives finds application as HMI-chemosensor, had an obvious practical disadvantage of less selectivity in fluorescence quenching and unsatisfactory trace-level detection. Zhang et al. (2014) developed derivative based on dye platform for example BODIPY to overcome these disadvantages by developing a new BODIPY derivative as "turn-on" fluorescence and colorimetric sensors that exhibited fluorescent turn-on behaviour towards Cu^{+2} ions with LOD of 25 nM and LDR of 0.1–8.0 μM (Wu et al. 2012). Beside fluorescent dyes, nanoscale fluorescent NCs have also been extensively used in chemo/biosensor technology. Owing to their small size, energy-level dispersion and unique physicochemical properties fluorescent Au and Ag NCs have been used in biochemical analysis. Li et al. (2017) prepared fluorescent bimetallic Au/Ag-DNA template based NCs, utilized as receptor for colorimetric detection of I^- anions, that quenched the fluorescence of NCs, and showed a color change from colorless transparent to purple red (Nasir et al. 2015).

Bio-mimicking Colorimetric NC sensors. Bio-enzyme-mimicking NCs apart from being equally efficient and regionally selective in catalysis to enzymes are also thermally stable, bio-selective, and cheap making them a better candidate for sensor technology as compared to their enzyme counterparts. Colorimetric detections are mostly done by employing enzyme- and biomimetic enzyme-mediated catalysis of 3,3′,5′5′-tetramethyl biphenylamine (TMB). Many biochemical and environmental processes involve Glucose and Hydrogen peroxide (H_2O_2). Their role is crucial as an accurate analyte in the arenas of bioanalysis, food and environmental science. Hou et al. (2015) reported immobilization of glucose oxidase ($GluO_x$) within magnetic zeolitic imidazolate framework 8 (mZIF-8@$GluO_x$) NCs that showed peroxidase-like colorimetric detection of glucose with LOD of 1.9 μM and a linear range from 0.005–0.15 mM, attributed to porous structure, high surface area and high glucose affinity of mZIF-8@$GluO_x$. On introducing TMB to the mimetic nanozyme systems and via its nanozyme-catalyzed oxidization, highly selective and sensitive colorimetric sensors with could be fabricated. Huang et al. (2018) developed unique MnO_2 NSs-TMB NCs as colorimetric sensor system for determination/separation of antioxidants that could react with MnO_2, and competitively inhibit the TMB-MnO_2 reaction (Huang et al. 2015). Environmental endocrine disruptors like bisphenol A (BPA) being widely used in plastic food packaging products results in pollution of food and environment. The molecular imprint polymer (MIP) membranes played significant role of providing multidimensional space for analyte adsorption and high affinity specific target-analyte combination. Kong et al. (2017) combined the adsorption active MIP membranes with peroxidase mimicking $ZnFe_2O_4$ to form microfluidic paper-based BPA detector with LOD of 6.18 nM and a LDR of 10–1000 nM, whereby, $ZnFe_2O_4$ NPs and paper cellulose fibers wrapped over MIP membranes, could generate •OH via H_2O_2-$ZnFe_2O_4$ reaction, promoting TMB oxidation and formation of blue color products in absence of BPA, but in its presence the H_2O_2–$ZnFe_2O_4$ NPs reaction was inhibited causing no color change (Kong et al. 2015). Peroxidase mimicking activity of NPs have also been improved by conjugating porphyrin-based stabilizers onto them. Liu et al. (2014) prepared 5,10,15,20-tetrakis(4-carboxyphenyl)-porphyrin (H_2-TCPP) functionalized ZnS NP NCs via reaction of Zn^{+2}-TCPP complex with thioacetamide, that showed a peroxidase mimetic, colorimetric TMB-H_2O_2 reaction catalyzing and glucose and H_2O_2 detection activity, owing to TCPP-ZnS NPs synergism (Fig. 3a) (Walia et al. 2013). Intrinsic peroxidase-mimics like GO, N-doped graphene, and $(BN)_x$ NSs, have been utilized as mimetic nanozymes for the development of colorimetric sensors. Zhang et al. (2014) reported an in-situ synthesis of PtNPs/GO NCs for colorimetric cancer cell detection, whereby the NCs served as peroxidase-like nanozymes (Zhang et al. 2014). Wu et al. (2017) prepared Fe_3CNPs encapsulated in N-rich graphene for fabrication of nanozyme-based colorimetric glucose detection (Wu et al. 2017). Zhang et al. (2017) reported $(BN)_x$ NSs/CuS NCs, for their peroxidase based colorimetric cholesterol determination ability with LOD of 2.9 μM and linear range of 10–100 μM (Zhang et al. 2017). Based on the fluorescence quenching activity of MoS_2 under certain circumstances may show peroxidase type (Lin et al. 2014) in accordance to Michaelis–Menten mechanism, catalyzing the peroxidase substrate 3,30,5,50-tetramethylbenzidine (TMB) to HO^-

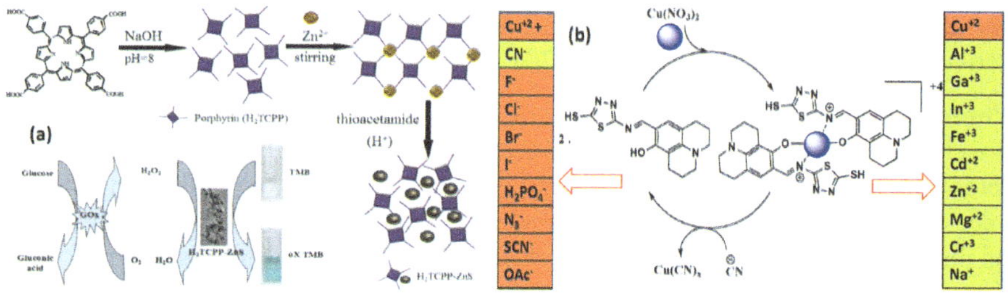

Fig. 19.3 **a** The illustration depicting the evolution of ZnS-Porphyrin (ZnS-H_2TCPP) NCs and their ability to colorimetrically sense glucose and H_2O_2, **Reproduced with permission from the reference in footnote,**[1] **b** The Illustration depicting the interaction of reporter ion and Cu^{+2} ion and its colorimetric sensing, followed by the decomposition this reporter-Cu^{+2} complex via CN^- showing a distinct colorimetric sensing ability of CN^-. Adapted from reference (Vilela et al., 2012)

reaction producing a blue color, visible to the unaided eye, over wide pH range(2.0–7.5) and small sample size, whereby within the MoS_2 NSs–TMB–H_2O_2 system, MoS_2 enables electron transfer between TMB and H_2O_2, facilitating the decomposition of H_2O_2 into OH^- ions. Moreover, the advantages of this technique can also be extended to the portable and convenient process for glucose detection. Many oxidative biological reactions have H_2O_2 as a by-product thus MoS_2-based H_2O_2 sensor are in great demand. Wang et al. (2013) reported ultrasonically produced pristine MoS_2 NPs (with unavoidable self-agglomerations) capable of detecting H_2O_2 at LOD of 2.5 nM, whereby the current response grows rapidly with as the size of MoS_2 NPs decrease, indicating that surface active sites are greatly influencing the catalytic activity (Zhang et al. 2015). Recently, MoS_2 NSs and MoS_2-NCs are widely used to detect HMIs, organic pollutants, pesticides etc. Chen et al. (2016) developed FET sensors for Hg^{+2} recognition based on DNA functionalized MoS_2 NSs/AuNP NCs that exhibited a rapid response within 1–2 s, a wide linear range of 0.1–10 nM and an ultralow LOD of 0.1 nM for Hg^{+2} recognition via the formation of [Thymine–Hg–Thymine]$^{+2}$ complex, with no interference from ions like Cd^{+2}, Pb^{+2}, Fe^{+2} and As^{+5} ions (Zhou et al. 2016). Song et al. (2018) developed AgNP-decorated nitrogen–fluorine codoped MoS_2 NCs (AgNPs–N/F–MoS_2) result in sensing system coupled with the inhibition of acetylcholinesterase to sense organophosphate pesticides like monocrotophos and chlorpyrifos with LODs of 0.2 pM and 3 pM respectively, mere on basis of superior conductivity and high electro-active surface area of the MoS_2-NCs, (Song et al. 2018). Su et al. (2016) decorated MoS_2 NSs with AuNPs, PtNPs and AuPtNPs, to improve its electrochemical catechol detection ability, with best detection performance offered MoS_2–AuPtNP NCs because of synergistic effects. Under optimal conditions, MoS_2–AuPtNP modified electrode offered LOD of 0.44 mM catechol (Su et al. 2016). Jia et al. (2017) reported MoS_2–Au/polyethylimine–hemin NCs as an electrochemical non-enzyme, anti-dopant against stimulants like clenbuterol, with LODs as low as 1.92 ng mL^{-1}, due to synergistic effects of the greater

[1] This article is published in Sensors and Actuators: B, 251, Liu et al. (2017).

surface area of the MoS_2 NSs, good conductivity of AuNP and electrocatalytic activity of hemin protein (Yang et al. 2017).

Photonic Structure based Colorimetric NC sensors. Colorimetric sensors of this novel class comprise structural colors, and their diverse photonic structures can interact with light through the phenomena of diffraction or scattering, interference showing different color change on different stimuli. Sensing systems based on photonic crystal materials owing to their distinct merit in synthesis and easy functionalization have gained much attention in rennet times. Encouraged by the photonic structure in the butterfly, Ye et al. (2013) demonstrated a structural photonic crystal based colorimetric sensor created by with inverse opal photonic film formed of target-responsive aptamer hydrogel, by replicating the template arrays of isotropic photonic crystal beads, featuring angle independence, which on immersion of the created film into a target solution, changed the conformation of the aptamer hydrogel causing the shrinkage of the hydrogel, resulting in angle-independent color change of the sensor film, with an LOD of 1 nM and LDR of 1 nM to 10 μM towards Hg^{+2} (Ye et al. 2013). Lu et al. (2016) fabricated a hierarchy of stimuli-responsive photonic structured materials as high-performance colorimetric sensors via chemical coating of butterfly wings with polyacrylamide that offered high water-absorption properties with developed photonic structure and excellent humidity-responsive properties (Lu et al. 2016). Yetisen et al. (2015) reported fabrication of a photonic nanosensors prepared via incorporating AgBr photonic crystals (as a tunable wavelength filter) in poly(acrylamide-cocarboxylic acid) (PAM) matrix (for dynamic volume modulation), that could diffract the illuminated white light, through visible range on addition of suitable HMIs, but after pendant 8-hydroxyquinoline is incorporated in PAM matrix, chelation of divalent HMIs result in the hypsochromic shift of the Bragg peak owing to the decreased Donnan osmotic pressure, and the system could quantitatively perceive Pb^{+2} and Cu^{+2} with LOD of 11.4 and 18.6 μM and LCR of 0.1–10.0 mM, respectively (Yetisen et al. 2015).

Interaction mediated Colorimetric NC Sensors. Ligands/receptors interaction based colorimetric sensors, depict presence and absence of analytes, by changing its emission spectra, resulting colorimetric response to analytes over various concentrations. Schiff base-based ligands have been exploited for the fabrication of such sensors. Wang et al. (2014) reported that on addition of Al^{+3} into CH_3CN/H_2O binary solution of his Schiff base derivative, 2-hydroxy naphthaldehyde isonicotinoyl hydrazine (HINH), based sensor, the solution rapidly changed from colorless to yellow due to HINH and Al^{+3} mediated specific interactions (Wang et al. 2014). Kumar et al. (2015) developed a fluorescent ON/OFF and colorimetric sensor for sensing of Cu^{+2} with detection limit of 10 pM exploiting Coordinative interactions amongst Schiff base and $Cu^{+2($Kumar et al. 2015). Multifunctional Schiff bases have been employed for the colorimetric recognition of other HMIs and anions. Apart from, Schiff bases, You et al. (2015) demonstrated the synthesis of a novel reporter ligands via chemical reaction between 5-amino-1,3,4-thiadiazole-2-thiol and 8-hydroxyjulolidine-9-carboxaldehyde, which on addition of these Cu^{+2} ions (LOD: 0.9 μM) reporter formed a Cu^{+2}-reporter complex, bringing yellow to orange color change, that however reverted back on addition of CN^- ions (LOD: 210 μM) as the complex got decomposed to reporter ligand (Fig. 3b) (You et al. 2015). Paper immobilized reporter ligands showed quick, simple, and real-time detection of Cu^{+2} ions. The high sensitivity, and repeatability enable its practical applications involved in protection and monitoring of environment. Colorimetric sensors for recognition of HMIs and anions have been reported by combining organic indicators with NMs, such as AgNPs and MOFs. Kumar et al. (2015) reported the reducing and surface functionalizing nature of organic ligands, to that with metal NPs like AgNPs, developing NCs that could be utilized as colorimetric sensor for the detection of Cd^{+2} and Pb^{+2} ions. Yu et al. (2014) initiated the era of MOFs-based NCs by preparing CuI and 1-benzimidazolyl-3,5,-bis(4-pyridyl) benzene-based Cu(I)-MOF NCs that were could detect humidity and HCHO with high sensitivity.

Instrument Mediated Colorimetric NC Sensors. Instruments based data analysis of colorimetric sensors employing scanner, computer, mobile phone etc. improve the sensitivity and result in high-throughput. Chen et al. (2014) developed a silica-gel plate printed, novel colorimetric sensor array using fifteen chemically responsive dyes for selective recognition of spoilage bacteria, and their interrelationships and VOC exudates, by bringing a scanned readout of colorific change, calibrated by a computer via hierarchical cluster analysis. However, the limitation of detecting cultured bacteria strains from real samples (Chen et al. 2014). An artificial olfaction system based colorimetric sensor array for measuring total viable count (TVC) in chicken were also developed. Jia et al. (2015) demonstrated cellphone camera-based calibration of colorimetric sensor array for the glucose determination in urine, via Whatman filter paper printed with 9 dyes (via Xerox Phaser 8560DB printer), which on addition of urine samples showed spots with different colors due to dye-glucose reactions, imaged under ambient lighting, using cellphones containing dye-contained sensor arrays. Microscope assisted colorimetric sensing of toxic substances has also been recently reported. Li et al. (2014) described a sensitive approach to sense adenosine and Hg^{+2} using atomic force microscopy (AFM)-based mapping, involving a single stranded DNA (ssDNA) monolayer modified silica and complimentary ssDNA functionalized aptamer AFM tip under closed proximity produced a strong AFM detectable force, and depict multiple colors based on force intensity, readout via AFM software with LOD of 0.1 nM and 10 pM for adenosine and Hg^{+2}, respectively.

Advanced Colorimetric NC Sensors. Recently, colorimetric sensing products like paptode and strips prepared for fast detection of various analytes prose practical advantages, simplicity, and visual quantification. Paptode sensors developed by Abbaspour et al. (2006) enabling easy and rapid determination of analytes. Thin layer chromatography (TLC) based paptode strips for colorimetric detection of ascorbic acid in vitamin C based on Fe^{+3} reduction by ascorbic acid and formation of red complex over TLC strips with LOD of 1 ppm and LDR of 20–200 ppm. Similarly, a colorimetric sensor prepared by immobilizing p-(dimethylamino)benzaldehyde onto TLC strip, exhibiting fast sensing of hydrazine with a LOD of 0.1 μg/mL and a LDR of 10–300 μg/mL (Abbaspour et al. 2010). However, TLC-based paptode sensors being simplicity, high scanning, low cost and time-savy, the performance affected TLC membrane uniformity. As a solution, Lapresta-Fernandez et al. (2013) prepared colorimetric disposable paper sensors for sensing K^+ ions whereby the sensing membranes were back-illuminated and the luminosity coming from a delimited zone were used to build analytical parameter (Lapresta-Fernández et al. 2013). Castaneda-Loaiza et al. (2019) demonstrated disposable colorimetric paper based Al^{+3} sensors employing poly(styrene–divinylbenzene) membrane disks modified with pyridoxal salicyloylhydrazone as colorimetric reagent, exhibiting LOD of 0.18 mg/L and LDR of 0.18 to 2 mg/L (Castañeda-Loaiza et al. 2019). Electrospinning as a mature technique for continuous polymer nanofiber production, have been utilized for fabrication of colorimetric sensor strips. Ding et al. (2011) prepared multicomponent colorimetric sensor system by coating nanofibres (NFs) membrane (PANI-EBNF) over glass substrate. PANI-EBNF were prepared by electrospinned half-oxidized polyaniline emeraldine base (PANI-EB), polyvinylbutyral (PVB) and polyamide-6 (PA-6) immersed into aqueous hydrazine solution to create fully reduced PANI leuco-emeraldine base membrane for Hg^{+2} detection, based on Hg^{+2}-PANI complexation that induced n–p* transition of the azobenzene moiety showing color change and LOD of 5 nM (Si et al. 2014). Similarly, De Almeida et al. (2015) used water soluble and chemically stable, polyvinyl alcohol (PVA) surface matrix, proposing first generation NFs colorimetric detector for glyphosate in water, which is a worldwide herbicide and a potential environmental pollutant, via two-step strategy of including glyphosate-CS_2 reaction forming dithiocarbamate and applying analyte sample over electro spun Cu-doped PVA NFs sensing strip, causing change in the glyphosate color in presence of dithiocarbamate (Almeida et al. 2015). Addition of an indicator onto NFs strips improves its colorimetric sensing by showing change of color on analyte addition. The sensing performance also depends on temperature, pH, the loading amount of indicator, strip size, etc. that must be optimized carefully for near-accurate measurements.

19.2.3 Fluorescence Quenching Based NC Sensors

Variety and Strategy. Solution phase fluorescence quenching mechanism involved in the finding of substances of environmental concern, involves electron/energy transfer via dynamic quenching whereby excited state fluorophore is deactivated by transmitting its energy to the nearby quencher without chemical structure alterations, and the emission intensity or life-time of luminescent NCs decreases together as illustrated by Stern–Volmer equation (Eq. 19.9):

$$\left(\frac{F_o}{F}\right) = 1 + K[Q] = 1 + k_q \tau_o [Q] \tag{19.9}$$

where, F_o, F, K, k_q, τ_o, and [Q] refers to fluorescence intensities in absence and presence of quencher molecule, Stern–Volmer quenching constant, bimolecular quenching constant, unquenched life-time and quencher concentration respectively. K is directly proportional to sensitivity of CT between fluorophore and quencher. More exposed fluorophore facilitates greater collisional interactions to quencher resulting in greater value of K. Dielectric constant/ solvent polarity and fluorophores-analyte supramolecular interactions also govern the quenching efficiency. Electronic properties/energy levels of the involved MOs of fluorophores and quenchers is also governed by electron transfer. These studies require must be conducted in oxygen free medium. Apart from dynamic quenching, in static quenching fluorophores form non-fluorescent complex with quencher in the ground state, and the fluorescence decay lifetime of any substance does not depend on the concentration of quencher. When both the processes takes place at the same time in fluorophores-quencher entity, change in fluorescence intensities is depicted in Eq. 19.10:

$$\left(\frac{F_o}{F}\right) = (1 + K_D[Q])(1 + K_S[Q]) = 1 + (K_D + K_S)[Q] + K_D K_S[Q]^2 \tag{19.10}$$

where, K_D and K_S refers to Stern–Volmer quenching constants for dynamic and static quenching respectively. At lower concentration, second order term is neglected and the plot of (F_o/F) versus [Q] gives a linear plot with slope are equal to the sum of both the quenching constants. At higher concentrations, the plot shows an upward curvature, owing to the consideration of squared term. At higher temperatures, diffusion, energy transfer and quenching rate increases. Viscous solvents usually lead to non fluoresensing fluorophore-analyte complex (Dutta et al. 2020).

Advancements. NPs functionalized MOF NCs (NPs-MOF NCs) offer intriguing properties owing to its greater porosity having ordered crystalline porous network and high SA/V as a result adsorption efficiency of MOF NCs is improved which enhances catalytic efficacy, size exclusive growth, agglomeration with selective material transport, selective detection of NACs, HMIs, small molecules etc. Owing to its low toxicity and high fluorescence quantum yield CQDs is widely applicable in developing materials with diverse applications. Li et al. (2012) with this motivation, reported fluorescent CQDs-(5′-(4-carboxyphenyl)-[1,1′:3′,1″-terphenyl]-4,4″-dicarboxylic acid) based NCs (CQDs-UMCM NCs) with particle dimensions nearly 1/12th of the size of particles as compared to the one synthesized via solvothermal approach (Li et al. 2015). Polar CQDs were incorporated with in NP–Zn–MOF that increases the hydrogen storage capacity compare to rest pristine MOFs and also better prompt sensing of NAC sensing serendipitously. Bare Zn–MOF showed no emission upon excitation at 420 nm but green emission was observed (λ_{em}: 500 nm) with CQDs on excitation at 420 nm. Solvent dependent emission was maximum with diethyl formamide (DEF) and minimum with nitrobenzene, providing an idea to explore NAC sensing behaviour leading to the observation that fluorescence intensity gradually got completely quenched on addition of 1000 ppm of 2,4,6-trinitrophenol (TNP) in DEF suspension of NP@Zn- MOF NCs, probably due to speedy electron transfer from amino groups which are electron rich present on the surface of CQDs, to the electron deficit NACs. Wanzhi et al. prepared rod like luminescent Rhodamine-Eu(1,3,5-benzenetricarboxylate) MOF NCs (RBH@Eu(BTC)MOF NCs) with dual optical detection of TNP, including colorimetric and ratiometric fluorescent sensing, whereby rhodamine based dye worked as "*turn-on*" probe for TNP sensing while, Eu(BTC) unit was deployed as "*turn-off*" receptor and supporting substrate for Rhodamine (Dhiman et al. 2020). Excitation spectrum of RBH-Eu(BTC) framework showed strong bands at 281 nm and two weak bands at 223 and 530 nm, corresponding to ligand $\pi \rightarrow \pi^*$ excitation, and inefficient rhodamine $\rightarrow Eu^{+3}$ energy transfer, with absence of 530 nm band representing absence of open ring rhodamine form *i.e.,* existence of thermally stable spirolactam structure. On excitation at 281 nm, RBH@Eu(BTC) displayed metal emission bands corresponding to $^5D_0 \rightarrow {}^7F_1$, $^5D_0 \rightarrow {}^7F_2$, $^5D_0 \rightarrow {}^7F_3$ and $^5D_0 \rightarrow {}^7F_4$ transitions, the most intense being $^5D_0 \rightarrow {}^7F_2$ (red) transition. When, 100 μM solution of TNP was added to the ethanolic suspension of these NCs, four absorption bands at 238, 300, 353 and 530 nm assigned to ligand $\pi^* \leftarrow \pi$ transition, open ring structure of rhodamine and absorption from TNP was observed. NAC triggered Rhodamine transition of non-fluorescent spirolactam to fluorescent xanthene configuration owing to supply of acidic –OH proton to oxygen atom of the spiro-ring. Visible range orange absorption is accounted for emission from TNP-RBH-Eu (BTC) adduct. Ratiometric fluorescent sensing mechanism by this NCs was attributed to dual proton-augmented structural transformation of rhodamine-dye followed by nitrophenyl-triggered quenching of Eu emission successively. Explosive vapours needs urgent detection in the field of security screening. Qui et al. (2012) reported luminescent NTs –Cd-centred MOF NCs with capacity to detect ppb level of NAC vapours, via employing Cd(BTC), Cd metal and BZT ligand as metal, connector and bridging ligand (Li et al. 2012). On excitation at 315 nm at room temperature, Cd-MOF-NT NCs showed emission band at 406 nm. NACs like nitrobenzene (NB), 2-nitrotoluene (2-NT), 4-nitrotoluene (4-NT), and 2,4-dinitrotoluene (DNT) showed strong quenching, with quenching of emission intensity of 72.5% within 10 s under DNT vapour, with LOD of 18.1 ppb, even after heating the sample at 80 °C for about 2 min even on emission intensity of Cd-MOF-NT NCs after ten consecutive cycles. Recently, Guo et al. (2017) Ag-MOF-embedded NCs (Ag-MOF-NCs) showed selective recognition of TNP over a concentration range of 1–29 μM in aqueous phase, fluorescence quantum yield of ~ 10.6%. AgNPs supported rod-like morphology of MOF surface, under DMF reduced Ag^+ to metallic Ag and embedded it on their surface. On excitation at 345 nm, these NCs showed blue emission at 440 nm which enhanced scope of luminescence-based NAC sensor against *p*-nitrophenol (PNP), 2,4-dinitrophenol (DNP), 2,4-dinitrotoluene (DNT), *o*-nitrophenol (ONP), 2,4,6-trinitrophenol (TNP). On addition of these NACs to NCs aqueous solution, maximum quenching was observed for TNP, particularly because of hydrogen-bonding interactions and RET between NCs and NACs (Zhang et al. 2018).

19.2.4 Field-Effect Transistor NC Sensors

Varieties and Strategy. Field-effect transistor NC Sensors are devoid of a physical gate having functionalized dielectric layer with specific receptors for selectively capturing the desired analyte, which then generates a concentration signal transducing into a readable format, such as a drain-to-source current or channel conductance. Atomically thin rapheme and MoS_2 show significant sensitivity as they offer high SA/V ratio enhancing the dielectric properties, and provide active sites for analyte adsorption. 2D MoS_2 based FET devices due to their excellent charge carrier mobility, sensitivity, and a high on/off ratio, are promising candidates for phototransistors, memory devices, and gas sensors. Single-layered MoS_2 has a direct bandgap of 1.8–1.9 eV, which is limited by the surface effects in case of its bulk counterpart used in FET channel as response generators on small changes in analytes. Analytes like NO (from transport effluents) behave as electron acceptor leading to p-type doping, while NH_3 (from bioactive effluents) as electron donor leading to n-type doping. Metal doped MoS_2 NCs, offers a higher sensitivity and detection diversity of the analyte, as observed for MoS_2–Pd NCs based FET device for H_2 detection as Pd could absorb H_2 and show changes in resistance leading to change in current response. Thiolated ligands conjugated on MoS_2 metal NCs, can be used for bio-relevant protein and biomolecule sensing. The natural direct bandgap of MoS_2, allows its FET sensors to sense pH up to 74 fold higher sensitivity than the rapheme-based pH sensor. Single/few layered MoS_2-based FET sensors could detect NO with limit of detection (LOD) of 0.8 ppm, particularly with scotch tape n-type semiconducting material, indicating rapid but unstable response by single-layered as compared to few layered counterparts.

Advancements. Late et al. (2013) worked on sensing performances for NH_3, NO_2, and humidity were investigated by with single to multi-layered crystalline MoS_2 sheet-based FET sensors to understand gas–solid interactions in layered structures, and the results revealed that 5-layered MoS_2 had better sensitivity than others. The energy level of the conduction and valence bands of can be tuned by the size and thickness of MoS_2 stating that can affect the energy levels of, bringing change in its electronic structure. Out of the p- and n-type MoS_2 nanoflakes based FET gas detectors, the n-type FET sensor showed improved performances for sensing NO_2 in 50% humidity than the GO or CNT-based FET. The sensitivity ratio was found to improve in case of room temperature H_2 detection due to the large SA/V ratio of few-layered MoS_2 and the sub-threshold region of the bias. Volatile organic compound (VOC) can also be detected by MoS_2-FET sensors, as confirmed by Kim et al. (2014) and his tunable thiolated (mercapto-undecanoic acid) MoS_2 NCs that exhibit distinct and sensitive VOC responses up to 1 ppm. The direct bandgap of MoS_2 makes it a unique candidate with electrical and optical properties, as when it shows photo-excitation, its mono-layered form allows efficient electron–hole (e/h) pair generation with high absorption rate, encouraging scientists to employ them as photodetectors (PDs) [Lopez-Sanchez et al. (2013) reported its photoresponsivity of 880 A W^{-1} and response wavelength range of 400–680 nm, that later turned to 1.8 AW^{-1} and with photodetectivity of 5×10^8 Jones for a wavelength of 850 nm, due to excellent surface controlled obtained by magnetron sputtering by Ling et al. (2015)]. Tentatively, mono-layered MoS_2-based PDs have overcome the intrinsic shortcomings of the traditional PDs through MoS_2/substrate designing.

19.3 Novel Matrices and Their Role in Sensing

Conventional matrices of graphite and boron nitride have been replaced by novel matrices that not only provide proper stability to NCs but also assist their sensing capabilities. Quite recently, there has been demand of such matrices that could contribute in the sensing properties of reinforcing materials. Materials like transition metal dichalcogenide have been proposed to provide wide coordination capability with greater stability. However, there has always been an inclination towards polymer or nanotube based matrices that are environmental friendly and are surface active. Some of the novel and recently employed matrices have been discussed in this section.

19.3.1 Halloysite Clay Nanotubes

Quite commonly, amino silane-functionalized halloysite clay nanotubes (HNTs) with abundant surface hydroxyl groups, chemically active inner-outer surfaces, cheap, environment friendly, and tunable surface have been exploited as support material and nucleation center for growing NCs, via common methods like light-assisted decomposition technique (Das et al. 2016). The amine functionalized HNTs with metal precursor solution ($KMnO_4$ for MnO_2) prepared with Millipore water into the glass vial, are added with NaOH solution (3.0 wt%, 3.0 mL), and is vortexed, followed by the resting of this properly closed glass vial having the reaction mixture under visible light for 6 h. For instance, MnO_2 NCs synthesized over the outer surface of clay NTs prepared by light-assisted hydrolytic decomposition of $KMnO_4$ under alkaline condition, produced negatively charged MnO_2 NPs in accordance to the following equation:

$$MnO_2(H_2O)(s) \rightarrow MnO_2(OH)^-(s) + H^+$$

These M_xO_y NCs gets immobilized over the outer surfaces of HNTs having positive surface charge through electrostatic interaction. These NPs may later grow to various morphologies, like the nanoflower morphology shown by monoclinic δ-MnO_2 NPs in case of MnO_2 NCs. These were made of intersected and wrinkled nanosheets (NSs), with thickness of 4 nm, lattice fringe spacing of 0.71 nm owing to (001) plane of MnO_2 NCs, and specific surface area of the NCs of 32 m^2 g^{-1} at 77 K.

19.3.2 Transition Metal Dichalcogenide NSs

Carbon based 2D system like of olyacry paved way for development of diverse 2D NMs like transition metal dichalcogenides (TMDCs), hexagonal boron nitride (h-BN), black phosphorus, and metal organic frameworks (MOFs), that have been extensively used in as sensors of HMIs and other organic moieties and molecules of environmental concern either in their native form or as a composite with metal ions, metal NPs or polymeric frameworks, because of their exceptional physical and chemical properties. Molybdenum disulfide (MoS_2), a distinctive member of this family, show structural and applicational similarities with olyacry-like NMs in terms of its layered structure, large surface area, easy functionalization, and a composition of S–Mo–S atomic layers interacting through weak van der Waals force. Unlike olyacry, it offers indirect bandgap to direct bandgap transformation on exfoliation to a single-layer NSs, and this ability offers bandgap tunability over the range of 1.2–1.9 eV, by controlling layer numbers of NSs. With rapid developments, scientist concluded that the practical applicability of pure MoS_2 NSs is not sufficient to meet the expected outcomes, and hence they felt a necessity of developing MoS_2-based NCs (MoS_2 NCs) and since then just like olyacry MoS_2 NSs were utilized as a powerful substrate to produce hybrids with olyacry, M_xO_y, metal nanostructures and organic compound resulting in the formation of NCs with better properties because of their synergistic effects. These MoS_2-based NCs offered much more choices and applications in the field of energy, food safety, biochemical analysis, disease diagnosis monitoring of environment and so on. Among varieties of MoS_2 NCs developed so far, MoS_2-noble metal NCs (MoS_2–NMNCs) have gained much attention owing to their extraordinary properties and applications, particularly due to synergism between the merits of MoS_2 NSs and noble metal nanostructures (Su et al. 2018).

Preparation MoS_2. Mono-layered MoS_2 have been prepared by either milder physical methods like low yielding, micromechanical exfoliation or laser thinning, or by chemical methods like chemo-repulsive force mediated liquid exfoliation, metal (Wang et al. (2013) via Li^+, Na^+ and K^+)/organo-metal (Ambrosi et al. (2015) via alkyl(Methyl, n-and t-butyl)lithiums, with $^2H \rightarrow {}^1T$ transition)/organo-solvent (Varrla et al. (2015) developed NSs, via n-methyl-pyrrolidone/isopropanol on a controllable scale and layer number) insertion, shear exfoliation based sonication or much quantitative aqueous phase electrochemical lithiation whereby non-bonding interactions, generated H_2 and hydroxides of Li^+ (You et al. demonstrated it with SO_4^{-2} as Li compounds are inflammable, demanding inert atmosphere), like other methods push MoS_2 layers apart, in a less scalable/controllable manner. Facile chemical reaction of Mo-S precursor later lead to involvement of post-synthetic methods like much abstract thermal annealing, high-yielding hydro/solvothermal (mostly via molybdic oxide + KSCN) reactions with modifications like high temperature annealing of $(NH_4)_2MoS_4$ layer over S (Liu et al. 2014), or low temperature autoclaving (180 °C, 24 h) to avoid excessive toxic gas/energy release, polydispersity and finally the morphology optimizable, chemical vapour deposition (CVD) with multiple precursor options (like Mo, MoO_3, MoO_2 etc.) and reported modifications like temperature varying inert Mo-film sulfurization on SiO_2 (Zhan et al. 2012) or high temperature S vapour annealing (annealing period to layer number correlation) over MoO_2 microplates (Wang et al. 2013), hence to form triangularly shaped continuous single layers with more stable S edges (Li and Zhu 2015).

Synthetic Protocols for MoS_2 NCs Highly sensitive, selective, stable and reliable MoS_2-NC sensors have also been synthesized. Low temperature and energy intensive self-assembly has been done to place noble metal NPs over MoS_2 sheet. High temperature, low yielding but effectively controlled CVD followed by NS-immersion into metal precursor solution via reduction followed by NP decoration over MoS_2 NS defects, to produce MoS_2-NP NCs (Sreeprasad et al. (2013) used hydroxylamine as a reducing agent, Hong et al. (2014) reported universal method whereby Pt and Ag precursors, benzyl alcohol and polyvinylpyrrolidone (PVP) as templating agents fomed belt-like nanoribbons, Yuwen et al. (2014) used carboxymethyl cellulose was used as stabilizer, Shi et al. (2014) reported MoS_2-Au NCs, Cui et al. (2013) reported lithium exfoliation based synthesis of MoS_2–SnO_2 NCs), or MoS_2-2D material NCs (MoS_2-2DM NCs) particularly with the much similar olyacry at lower temperatures (400 °C) to avoid 2D surface-wrinkling and restacking (Shi et al. (2012) presented MoS_2–olyacry NCs), dip-casting 3D olyacry network(3DGN) followed by a series of annealing treatments to enhance electrical conductivity between collector-depositor MoS_2 and generate MoS_2-3DGN NCs, or with carbon as reported by Zhou et al. (2015) to fabricate 2D MoS_2-C NCs using few layered MoS_2 ($\leq$5 layers), porous carbon NSs and cubic NaCl as the template to avoid the intimate interfacial contact, MoS_2 restacking and electrodic-ion transport resistance. Hydro/solvothermal reaction has been done with a mixture of $(NH_4)_2MoS_4$, N_2H_4, and olyacry oxide (GO)/ multifunctional L-cysteine polymer/olyacry sheets/CNTs in dimethyl formamide (DMF) at 200 °C to generate MoS_2 NSs over reduced substrate, (Wang et al. (2014) reported inert phase refluxing of $(NH_4)_2MoS_4$, $N_2H_4.xH_2O$, and olyacry sheets to develop MoS_2-Graphene NCs, annealing of MoS_2 NSs under N_2 for N-doped MoS_2 NSs that showed excellent photoluminescence response towards Hg^{2+}, Shi et al. (2013) reported, MoS_x–CNT NCs ($2 \leq x \leq 3$)). Simple reduction of metal and MoS_2 precursors synchronously to generate (Zhang et al. (2015) prepared MoS_2–AuNP NCs using N_2H_4 as reductant for $(NH_4)_2MoS_4$ and $HauCl_4$, generating S–Au bond anchoring Au NPs onto the MoS_2 NSs). Electrospinning, which facilitates large SA/V ratio and high yield, has also been applied for preparation of MoS_2-nanofibre composite (MoS_2-CNF NCs) by heat treatment embedding (Zhu et al. (2015) embedded ultra-small single-layered MoS_2 NPIs in CNF of diameter 50 nm), or by stabilizing the electro-spun solution (olyacrylonitrile and $(NH_4)_2MoS_4$ in DMF) followed by its inert phase high temperature carbonization (Xiong et al. (2015) reported large-scale flexible MoS_2–CNF NCs as free-standing and binder-free electrodes by electrospinning, solution stabilized at 400 °C for 2 h and carbonized at 800 °C for 1 h in Ar(95 vol%)/H_2(5 vol%), with content of approximately 83.2%). Microwave beam assisted synthesis with high local temperatures could generate NPs that can now be attached onto the vacant sites of MoS_2 NSs with ensured high embed-density of NPs. Lu et al. (2015) used focused laser beam to activate the

MoS_2 surface domains with unbound sulfur, promoting "on-demand" assembly of Au NPs, via immersion of MoS_2 film with micro-pattern active nucleation sites into metal precursor solution (Au^{+3}, Pt^{+2}, Pd^{+2}, Ag^+ and Cu^{+2} ions), allowing the selective anchoring of Au NPs in its modified domains, hence developing a promising protocol for MoS_2–NP NCs preparation for sensor applications (Wang et al. 2017).

Advancements. The need decides the type of synthetic protocol to be employed. There is a demand of precise structures, atomic arrangement in case of field-effect transistor (FET) sensors and high SA/V ratios, homogeneous NPs location on NSs in case of electrochemical sensors and optical sensors. CVD being an excellent candidate for creating hybrids with atom-layered materials is unable to prepare coaxial structures that can only can be prepared by the electrospinning method. Metallic NPs hybrids can be achieved by self-assembly or hydro/solvothermal methods with high yields and load efficiency, suitable for electrochemical sensors, and by laser beam assisted self-assembly with location controllability. Chemically exfoliated MoS_2 NSs contain small quantities of ^{1} T type structures, whose metallic crystal structure affect the quenching efficiency, as observed with ion-intercalated MoS_2 nanoflakes constructed by coexistent semi-conductive ^{2}H and metallic ^{1} T structures, resulting in significant quenching of photoluminescence in bio-systems. Shi et al. (2015) developed MoS_2-AuNP NCs whereby MoS_2 NSs prepared by CVD method were decorated AuNP by a spontaneous redox reaction deprived of any reducing agent, selectively on the edge sites or defective sites. The sizes and the densities of AuNPs supported on the surface of MoS_2 NSs were tuned by varying the concentration of the metal precursor ($HauCl_4$) and it was observed that AuNPs could easily cover the entire surface of MoS_2 NSs, but excess concentrations of $HauCl_4$ could damage these NSs. Similarly, MoS_2–PdNP NCs were reported by Kwon et al. (2016). However, NM NPs can also be supported on the surface of MoS_2 NSs by employing reducing agents like ascorbic acid, hydrated hydrazine, hydroxyl amine, sodium borohydride or sodium citrate. Zhang et al. (2016) prepared MoS_2–AuNP NCs by using L-ascorbic acid, and Hassanzadeh et al. (2018) prepared MoS_2-Ag nanocluster NCs by using an excess of $NaBH_4$ as a reducing agents. Self-assembly based synthetic routes are employed for synthesis of ordered structures or NCs by non-covalent interactions particularly because its low temperature and energy-saving set up. Through this technique, high quality MoS_2-Au nanorod NCs were prepared by Xia et al. (2014) by tuning the ratio of gold nanorods: MoS_2 NSs (v/v) from 1: 5 to 4: 5, finally choosing 2: 5 (v/v) for better structural stability and hydrogen evolution reaction. Electrodeposition being a simple, fast, and environmentally friendly method to synthesize MoS_2–NMNCs, is also a possible synthetic route. MoS_2 NSs solution is dropped onto a cleaned glass carbon (GCE) or other common electrodes and is dried overnight in an ambient environment, forming a MoS_2 film modified electrode (MoS_2/electrode) that is immersed in a metal precursor solution to get MoS2–NMNCs/electrodes using various electrochemical parameters. For instance, Su et al. (2014) reported the preparation of MoS_2-AuNP/GCE by electrodeposition method by immersing GCE in 0.5 mM $HauCl_4$ solution for AuNP decoration on MoS_2 film at a scan rate of 10 mV s^{-1}, within 60 s of deposition time and 1.1 to -0.2 V pulse potential. Later, its Pt analogue was also prepared by using the same method. Laser beam-assisted and microwave-assisted methods, have also been commonly used for preparation of MoS_2-NMNCs whereby, the size and the number of anchored AuNPs could be modified by the experimental conditions, such as laser powers, reaction times and thickness of the MoS_2 film (Yi et al. 2020).

19.4 Concluding Remarks and Future Prospects

Since the dawn of metal NC systems, their synthesis, characterization, and applications have developed a lot with constant ongoing progress. Much has been achieved in this course of time, but with advancements both in the quantity and toxicity of pollutants or contaminants, these NCs have also coevolved with time. Systems like mesoporous, negatively charged M_xO_y both in their naked or NC forms have been employed as a solid adsorbent for substances of environmental concerns like HMIs, via pseudo-second-order kinetic monolayer adsorption, increasing up to near neutral and then gradually decreasing at higher pH, suggesting a spontaneous, endothermic adsorption till inversion temperature followed by a nonspontaneous exothermic adsorption. These systems have proved their nature friendly and environmental remediation approach by their improved recyclability, affordability, and efficiency, encouraging commercial industries to utilize them for treatment of their chemical effluents without compromising with their profit. Systems like colorimetric chemo/bio NC sensors and their recent advancements like sensor arrays and colorimetric strips have transformed sensing process to Lyman's play. Different toxicities have been judged via NM based, bio-mimicking systems, probe-based photonic structure-based NCs on the basis of their type, structure, and function. Colorimetric sensing mechanisms of these NCs have been more or less concerned with aggregation, decomposition, neighboring interactions, nanozymes catalysis, and fluorescence response. These capabilities although making them an easy, quick, low-cost, and high performance entities for analytical, environmental, and food adulterant sensing, but have also exposed the vulnerability of their results *w.r.t.* physical, chemical and biological conditions, which still need to be addressed. Recent advancements in 2D NMs that mimic nanozymes for colorimetric sensing of biologically relevant molecules via conjugation of quantum confinement with unique optical and electronic properties have developed them as a suitable candidate for employment in 2D sensor array systems for instrument and software-based data analysis, making the design process and economy run parallel to one another for real-time and practical applications.

As a substrate the relevance of MoS_2 has been reported in numerous literature, but MoS_2-based NCs, also not neglected. Sensing applications of these systems are based on their optical, FET, and electrochemical capabilities, with the added advantage of being planar structure and large SA/V ratio. MoS_2-based FET with excellent gas and biomolecule sensing ability, MoS_2 optical sensors as a fluorescence quencher for fluorescent NCS type molecules, Electrochemical MoS_2 based NCs with ultrahigh H_2O_2 and glucose sensitivity has inspired their exploitation.

Although there are many achieved advancements, limitations like the absence of any defect-guided, layer-number controlled, effective, low-cost, easy-to-operate preparation methods for mono-layered MoS_2 NSs, absence of any protocol to study MoS_2 NSs-graphene/metal NPs/biomolecules/polymers interactions for electrical and optically tunability, and a foggy knowledge of MoS_2 electron transfer mechanism Sensing applications of all such NCs can still be improved by enhancing their sensitivity and stability, with a knowledge of their sensing mechanism, revealing properties still left to be explored. Traditional electronic devices have almost overcome the limitations of mono-layered MoS_2. With incorporation of biologically compatible components, a non-toxic and fast in vivo detection can also be attained. Hence, developments of NC sensors have come through greater advancements and is still a hot topic of research in inorganic chemistry and material science. These systems are still and will remain to be progressing to overcome newer environmental challenges and pollutant.

Acknowledgements Authors would like to acknowledge all those scientists, researchers, and reviewers referred in this chapter, for their notable works that helped us to compose the manuscript presented here. Authors are thankful to the Royal Chemical Society and Elsevier for grant of permission to use illustrations from reference [41] and [95] respectively.

References

Abbaspour A, Mirahmadi E, Khajehzadeh A (2010) Disposable sensor for quantitative determination of hydrazine in water and biological sample. Anal Methods 2(4):349–353. https://doi.org/10.1039/B9AY00291J

Ahmed SM, Gerischer H (1979) Influence of crystal surface orientation on redox reactions at semiconducting MoS_2. Electrochim Acta 24(6):705–711. https://doi.org/10.1016/0013-4686(79)87055-3

Aldewachi H, Chalati T, Woodroofe MN, Bricklebank N, Sharrack B, Gardiner P (2018) Gold nanoparticle-based colorimetric biosensors. Nanoscale 10(1):18–33. https://doi.org/10.1039/C7NR06367A

Antochshuk V, Olkhovyk O, Jaroniec M, Park IS, Ryoo R (2003) Benzoylthiourea-modified mesoporous silica for mercury (II) removal. Langmuir 19(7):3031–3034. https://doi.org/10.1021/la026739z

Arora R (2019) Adsorption of heavy metals–a review. Mater Today: Proc 18:4745–4750. https://doi.org/10.1016/j.matpr.2019.07.462

Banerjee D, Hu Z, Li J (2014) Luminescent metal–organic frameworks as explosive sensors. Dalton Trans 43(28):10668–10685. https://doi.org/10.1039/C4DT01196A

Boening DW (2000) Ecological effects, transport, and fate of mercury: a general review. Chemosphere 40(12):1335–1351. https://doi.org/10.1016/s0045-6535(99)00283-0

Carbone M, Gorton L, Antiochia R (2015) An overview of the latest graphene-based sensors for glucose detection: the effects of graphene defects. Electroanalysis 27(1):16–31. https://doi.org/10.1002/elan.201400409

Castañeda-Loaiza V, Díaz-de-Alba MI, Granado-Castro MD, Galindo-Riaño MD, Casanueva-Marenco MJ (2019) Disposable optical sensor for Al(III) ions determination by coupled colorimetric solid-phase extraction-reflectance spectroscopy in leachates from cookware, antacids and hygienic care products. Talanta 205:120102. https://doi.org/10.1016/j.talanta.2019.06.102

Chaiyo S, Siangproh W, Apilux A, Chailapakul O (2015) Highly selective and sensitive paper-based colorimetric sensor using thiosulfate catalytic etching of silver nanoplates for trace determination of copper ions. Anal Chim Acta 866:75–83. https://doi.org/10.1016/j.aca.2015.01.042
Chen Q, Li H, Ouyang Q, Zhao J (2014) Identification of spoilage bacteria using a simple colorimetric sensor array. Sens Actuators B 205:1–8. https://doi.org/10.1016/j.snb.2014.08.025
Chen Y, Tan C, Zhang H, Wang L (2015) Two-dimensional graphene analogues for biomedical applications. Chem Soc Rev 44(9):2681–2701. https://doi.org/10.1039/C4CS00300D
Cui Y, Zhang J, He H, Qian G (2018) Photonic functional metal–organic frameworks. Chem Soc Rev 47(15):5740–5785. https://doi.org/10.1039/C7CS00879A
Das S, Ghosh C, Jana S (2016) Moisture induced isotopic carbon dioxide trapping from ambient air. J Mater Chem A 4(20):7632–7640. https://doi.org/10.1039/C6TA00378H
Das S, Samanta A, Kole K, Gangopadhyay G, Jana S (2020) MnO2 flowery nanocomposites for efficient and fast removal of mercury(ii) from aqueous solution: a facile strategy and mechanistic interpretation. Dalton Trans 49(20):6790–6800. https://doi.org/10.1039/D0DT01054E
De Almeida LKS, Chigome S, Torto N, Frost CL, Pletschke BI (2015) A novel colorimetric sensor strip for the detection of glyphosate in water. Sens Actuators B 206:357–363. https://doi.org/10.1016/j.snb.2014.09.039
Deng HH, Wu CL, Liu AL, Li GW, Chen W, Lin XH (2014) Colorimetric sensor for thiocyanate based on anti-aggregation of citrate-capped gold nanoparticles. Sens Actuators B 191:479–484. https://doi.org/10.1016/j.snb.2013.10.058
Dhiman S, Kumar G, Luxami V, Singh P, Kumar S (2020) A stilbazolium dye-based chromogenic and red-fluorescent probe for recognition of 2,4,6-trinitrophenol in water. New J Chem 44(26):10870–10877. https://doi.org/10.1039/D0NJ00489H
Dutta A, Singh A, Wang X, Kumar A, Liu J (2020) Luminescent sensing of nitro aromatics by crystalline porous materials. Cryst Eng Comm 22(45):7736–7781. https://doi.org/10.1039/D0CE01087A
Elavarasi M, Rajeshwari A, Alex SA, Nanda Kumar D, Chandrasekaran N, Mukherjee A (2014) Simple colorimetric sensor for Cr(III) and Cr(VI) speciation using silver nanoparticles as a probe. Anal Methods 6(14):5161–5167. https://doi.org/10.1039/C4AY00877D
Fischer C, Oschatz M, Nickel W, Leistenschneider D, Kaskel S, Brunner E (2017) Bioinspired carbide-derived carbons with hierarchical pore structure for the adsorptive removal of mercury from aqueous solution. Chem Commun (camb) 53(35):4845–4848. https://doi.org/10.1039/c6cc08041c
Guo Y, Li J, Dong S (2011) Hemin functionalized graphene nanosheets-based dual biosensor platforms for hydrogen peroxide and glucose. Sens Actuators B 160(1):295–300. https://doi.org/10.1016/j.snb.2011.07.050
Hassan T, Salam A, Khan A et al (2021) Functional nanocomposites and their potential applications: a review. J Polym Res 28:36. https://doi.org/10.1007/s10965-021-02408-1
Hatamie A, Zargar B, Jalali A (2014) Copper nanoparticles: a new colorimetric probe for quick, naked-eye detection of sulfide ions in water samples. Talanta 121:234–238. https://doi.org/10.1016/j.talanta.2014.01.008
Henglein A (1989) Small-particle research: physicochemical properties of extremely small colloidal metal and semiconductor particles. Chem Rev 89(8):1861–1873. https://doi.org/10.1021/cr00098a010
Huang Y, Shi Y, Yang HY, Ai Y (2015) A novel single-layered MoS_2 nanosheet based microfluidic biosensor for ultrasensitive detection of DNA. Nanoscale 7(6):2245–2249. https://doi.org/10.1039/C4NR07162J
Jana S, Praharaj S, Panigrahi S, Basu S, Pande S, Chang CH, Pal T (2007) Light-induced hydrolysis of nitriles by photo produced α-MnO_2 nano rods on polystyrene beads. Org Lett 9(11):2191–2193. https://doi.org/10.1021/ol070750e
Jana S, Pande S, Sinha AK, Pal T (2008) Synthesis of super paramagnetic beta-MnO_2 organosol: a photocatalyst for the oxidative phenol coupling reaction. Inorg Chem 47(13):5558–5560. https://doi.org/10.1021/ic800499q

Jin Y, Wu S, Zeng Z, Fu Z (2017) Effects of environmental pollutants on gut microbiota. Environ Pollut 222:1–9. https://doi.org/10.1016/j.envpol.2016.11.045

Kabiri S, Tran DNH, Azari S, Losic D (2015) Graphene-diatom silica aerogels for efficient removal of mercury ions from water. ACS Appl Mater Interfaces 7(22):11815–11823. https://doi.org/10.1021/acsami.5b01159

Kim HN, Ren WX, Kim JS, Yoon J (2012) Fluorescent and colorimetric sensors for detection of lead, cadmium, and mercury ions. Chem Soc Rev 41(8):3210–3244. https://doi.org/10.1039/C1CS15245A

Kitagawa S, Uemura K (2005) Dynamic porous properties of coordination polymers inspired by hydrogen bonds. Chem Soc Rev 34(2):109–119. https://doi.org/10.1039/B313997M

Kong RM, Ding L, Wang Z, You J, Qu F (2015) A novel aptamer-functionalized MoS_2 nanosheet fluorescent biosensor for sensitive detection of prostate specific antigen. Anal Bioanal Chem 407(2):369–377. https://doi.org/10.1007/s00216-014-8267-9

Krabbenhoft DP, Sunderland EM (2013) Environmental science. Global change and mercury. Science 341(6153):1457–1458. https://doi.org/10.1126/science.1242838

Kumar VV, Anthony SP (2014) Highly selective silver nanoparticles-based label free colorimetric sensor for nitrite anions. Anal Chim Acta 842:57–62. https://doi.org/10.1016/j.aca.2014.06.028

Kumar J, Bhattacharyya PK, Das DK (2015) New duel fluorescent "on–off" and colorimetric sensor for copper(II): copper(II) binds through N coordination and pi cation interaction to sensor. Spectrochim Acta Part A 138:99–104. https://doi.org/10.1016/j.saa.2014.11.030

Kumar S, Bukkitgar SD, Singh S, Singh P, Reddy NS (2019) Electrochemical sensors and biosensors based on graphene functionalized with metal oxide nanostructures for healthcare applications. ChemistrySelect 4(18):5322–5337. https://doi.org/10.1002/slct.201803871

Lapresta-Fernández A, Gómez-Sánchez J, Capitan-Vallvey LF (2013) Transmission-mode scanner for potassium determination using colorimetric disposable sensors. Sensor Lett 11:368–376. https://doi.org/10.1166/sl.2013.2741

Li X, Zhu H (2015) Two-dimensional MoS_2: properties, preparation, and applications. J Materiomics 1(1):33–44. https://doi.org/10.1016/j.jmat.2015.03.003

Li R, Yuan YP, Qiu LG, Zhang W, Zhu JF (2012) A rational self-sacrificing template route to metal-organic framework nanotubes and reversible vapor-phase detection of nitro aromatic explosives. Small 8(2):225–230. https://doi.org/10.1002/smll.201101699

Li JS, Tang YJ, Li SL, Zhang SR, Dai ZH, Si L, Lan YQ (2015) Carbon nanodots functional MOFs composites by a stepwise synthetic approach: enhanced H_2 storage and fluorescent sensing. Cryst Eng Comm 17(5):1080–1085. https://doi.org/10.1039/C4CE02020K

Li X, Cheng R, Shi H, Tang B, Xiao H, Zhao G (2016) A simple highly sensitive and selective aptamer-based colorimetric sensor for environmental toxins microcystin-LR in water samples. J Hazard Mater 304:474–480. https://doi.org/10.1016/j.jhazmat.2015.11.016

Liang J, Li X, Yu Z, Zeng G, Luo Y, Jiang L, Wu H (2017) Amorphous MnO_2 modified biochar derived from aerobically composted swine manure for adsorption of Pb(II) and Cd(II). ACS Sustain Chem Eng 5(6):5049–5058. https://doi.org/10.1021/acssuschemeng.7b00434

Liu J, Lu Y (2004) Colorimetric biosensors based on DNAzyme-assembled gold nanoparticles. J Fluoresc 14(4):343–354. https://doi.org/10.1023/B:JOFL.0000031816.06134.d3

Liu B, Yang F, Zou Y, Peng Y (2014) Adsorption of phenol and p-nitrophenol from aqueous solutions on metal-organic frameworks: effect of hydrogen bonding. J Chem Eng Data 59(5):1476–1482. https://doi.org/10.1021/je4010239

Liu Z, Lau SP, Yan F (2015) Functionalized graphene and other two-dimensional materials for photovoltaic devices: device design and processing. Chem Soc Rev 44(15):5638–5679. https://doi.org/10.1039/C4CS00455H

Liu QY, Chen PP, Xu Z, Chen MM, Ding YA, Yue K (2017) A facile strategy to prepare porphyrin functionalized ZnS nanoparticles and their peroxidase-like catalytic activity for colorimetric sensor of hydrogen peroxide and glucose. Copyright Elsevier, pp 339–348

Liu D, Zhang Z, Wu J (2019) Elemental mercury removal by MnO_2 nanoparticle-decorated carbon nitride nanosheet. Energy Fuels 33(4):3089–3097. https://doi.org/10.1021/acs.energyfuels.9b00149

Liu B, Zhuang J, Wei G (2020) Recent advances in the design of colorimetric sensors for environmental monitoring. Environ Sci Nano 7(8):2195–2213. https://doi.org/10.1039/D0EN00449A

Lu Z, Yin Y (2012) Colloidal nanoparticle clusters: functional materials by design. Chem Soc Rev 41(21):6874–6887. https://doi.org/10.1039/C2CS35197H

Lu T, Zhu S, Chen Z, Wang W, Zhang W, Zhang D (2016) Hierarchical photonic structured stimuli-responsive materials as high-performance colorimetric sensors. Nanoscale 8(19):10316–10322. https://doi.org/10.1039/C6NR01875K

Lustig WP, Mukherjee S, Rudd ND, Desai AV, Li J, Ghosh SK (2017) Metal–organic frameworks: functional luminescent and photonic materials for sensing applications. Chem Soc Rev 46(11):3242–3285. https://doi.org/10.1039/C6CS00930A

Mao K, Wu Z, Chen Y, Zhou X, Shen A, Hu J (2015) A novel biosensor based on single-layer MoS_2 nanosheets for detection of Ag^+. Talanta 132:658–663. https://doi.org/10.1016/j.talanta.2014.10.026

Moghaddam HK, Pakizeh M (2015) Experimental study on mercury ions removal from aqueous solution by MnO_2/CNTs nanocomposite adsorbent. J Ind Eng Chem (washington, D C) 21:221–229. https://doi.org/10.1016/j.jiec.2014.02.028

Morsi MA, Asnag GM, Rajeh A, Awwad NS (2021) Nd:YAG nanosecond laser induced growth of Au nanoparticles within CMC/PVA matrix: multifunctional nanocomposites with tunable optical and electrical properties. Compos Commun 24:100662. https://doi.org/10.1016/j.coco.2021.100662.

Nasir MZM, Sofer Z, Ambrosi A, Pumera M (2015) A limited anodic and cathodic potential window of MoS_2: limitations in electrochemical applications. Nanoscale 7(7):3126–3129. https://doi.org/10.1039/C4NR06899H

Pan B, Pan B, Zhang W, Lv L, Zhang Q, Zheng S (2009) Development of polymeric and polymer-based hybrid adsorbents for pollutants removal from waters. Chem Eng J (Loughborough, Engl) 151(1):19–29. https://doi.org/10.1016/j.cej.2009.02.036

Patel GM, Rohit JV, Singhal RK, Kailasa SK (2015) Recognition of carbendazim fungicide in environmental samples by using 4-aminobenzenethiol functionalized silver nanoparticles as a colorimetric sensor. Sens Actuators B 206:684–691. https://doi.org/10.1016/j.snb.2014.09.095

Pourreza N, Rastegarzadeh S, Larki A (2014) Simultaneous preconcentration of Cd(II), Cu(II) and Pb(II) on Nano-TiO_2 modified with 2-mercaptobenzothiazole prior to flame atomic absorption spectrometric determination. J Ind Eng Chem (Amsterdam, Neth.) 20(5):2680–2686. https://doi.org/10.1016/j.jiec.2013.10.055

Priyadarshini E, Pradhan N (2017) Gold nanoparticles as efficient sensors in colorimetric detection of toxic metal ions: a review. Sens Actuators B 238:888–902. https://doi.org/10.1016/j.snb.2016.06.081

Qiao S, Chen J, Li J, Qu Z, Liu P, Yan N, Jia J (2009) Adsorption and catalytic oxidation of gaseous elemental mercury in flue gas over MnOx/alumina. Ind Eng Chem Res 48(7):3317–3322. https://doi.org/10.1021/ie801478w

Rao CNR, Ramakrishna Matte HSS, Maitra U (2013) Graphene analogues of inorganic layered materials. Angew Chem Int Ed 52(50):13162–13185. https://doi.org/10.1002/anie.201301548

Si Y, Wang X, Li Y, Chen K, Wang J, Yu J, Ding B (2014) Optimized colorimetric sensor strip for mercury(II) assay using hierarchical nanostructured conjugated polymers. J Mater Chem A 2(3):645–652. https://doi.org/10.1039/C3TA13867D

Song D, Wang Y, Lu X, Gao Y, Li Y, Gao F (2018) Ag nanoparticles-decorated nitrogen-fluorine co-doped monolayer MoS_2 nanosheet for highly sensitive electrochemical sensing of organophosphorus pesticides. Sens Actuators B 267:5–13. https://doi.org/10.1016/j.snb.2018.04.016

Soomro RA, Nafady A, Sirajuddin Memon N, Sherazi TH, Kalwar NH (2014) L-cysteine protected copper nanoparticles as colorimetric sensor for mercuric ions. Talanta 130:415–422. https://doi.org/10.1016/j.talanta.2014.07.023

Su S, Cao W, Zhang C, Han X, Yu H, Zhu D, Wang L (2016) Improving performance of MoS_2-based electrochemical sensors by decorating noble metallic nanoparticles on the surface of MoS_2 nanosheet. RSC Adv 6(80):76614–76620. https://doi.org/10.1039/C6RA12401A

Su S, Xu Y, Sun Q, Gu X, Weng L, Wang L (2018) Noble metal nanostructure-decorated molybdenum disulfide nanocomposites: synthesis and applications. J Mater Chem B 6(33):5323–5334. https://doi.org/10.1039/C8TB01659C

Tan C, Zhang H (2015) Two-dimensional transition metal dichalcogenide nanosheet-based composites. Chem Soc Rev 44(9):2713–2731. https://doi.org/10.1039/C4CS00182F

Tiwari JN, Vij V, Kemp KC, Kim KS (2016) Engineered carbon-nanomaterial-based electrochemical sensors for biomolecules. ACS Nano 10(1):46–80. https://doi.org/10.1021/acsnano.5b05690

Vilela D, González MC, Escarpa A (2012) Sensing colorimetric approaches based on gold and silver nanoparticles aggregation: chemical creativity behind the assay A Review. Anal Chim Acta 751:24–43. https://doi.org/10.1016/j.aca.2012.08.043

Walia S, Balendhran S, Wang Y, Kadir RA, Zoolfakar AS, Atkin P, Bhaskaran M (2013) Characterization of metal contacts for two-dimensional MoS_2 nanoflakes. Appl Phy Lett 103(23):232105. https://doi.org/10.1063/1.4840317

Wang Z, Cohen SM (2009) Postsynthetic modification of metal–organic frameworks. Chem Soc Rev 38(5):1315–1329. https://doi.org/10.1039/B802258P

Wang DF, Ke YC, Guo HX, Chen J, Weng W (2014) A novel highly selective colorimetric sensor for aluminum(III) ion using Schiff base derivative. Spectrochim Acta Part A 122:268–272. https://doi.org/10.1016/j.saa.2013.11.063

Wang X, Xing W, Feng X, Song L, Hu Y (2017) MoS_2/polymer nanocomposites: preparation, properties, and applications. Polymer Rev 57(3):440–466. https://doi.org/10.1080/15583724.2017.1309662

Wang H, Shen H, Shen C, Li YN, Ying Z, Duan Y (2019) Kinetics and mechanism study of mercury adsorption by activated carbon in wet oxy-fuel conditions. Energy Fuels 33(2):1344–1353. https://doi.org/10.1021/acs.energyfuels.8b03610

West MR, Hanks TW, Watson RT (2009) Polydiacetylene-based liposomes: an "Optical Tongue" for bacteria detection and identification. J Chem Educ 86(3):373. https://doi.org/10.1021/ed086p373

Wu S, Zeng Z, He Q, Wang Z, Wang SJ, Du Y, Zhang H (2012) Electrochemically reduced single-layer MoS_2 nanosheets: characterization, properties, and sensing applications. Small 8(14):2264–2270. https://doi.org/10.1002/smll.201200044

Wu S, Huang H, Feng X, Du C, Song W (2017) Facile visual colorimetric sensor based on iron carbide nanoparticles encapsulated in porous nitrogen-rich graphene. Talanta 167:385–391. https://doi.org/10.1016/j.talanta.2017.02.003

Wypych F, Schöllhorn R (1992) 1T-MoS_2, a new metallic modification of molybdenum disulfide. J Chem Soc Chem Commun 19:1386–1388. https://doi.org/10.1039/C39920001386

Xiong R, Keffer DJ, Fuentes-Cabrera M, Nicholson DM, Michalkova A, Petrova T, Lewis JP (2010) Effect of charge distribution on RDX adsorption in IRMOF-10. Langmuir 26(8):5942–5950. https://doi.org/10.1021/la9039013

Yang Y, Zhang H, Huang C, Yang D, Jia N (2017) Electrochemical non-enzyme sensor for detecting clenbuterol (CLB) based on MoS_2-Au-PEI-hemin layered nanocomposites. Biosens Bioelectron 89:461–467. https://doi.org/10.1016/j.bios.2016.04.019

Ye B, Rong F, Gu H, Xie Z, Cheng Y, Zhao Y, Gu Z (2013) Bioinspired angle-independent photonic crystal colorimetric sensing. Chem Commun 49(46):5331–5333. https://doi.org/10.1039/C3CC42122H

Yetisen AK, Montelongo Y, Qasim MM, Butt H, Wilkinson TD, Monteiro MJ, Yun SH (2015) Photonic nano sensor for colorimetric detection of metal ions. Anal Chem 87(10):5101–5108. https://doi.org/10.1021/ac504274q

Yi N, Cheng Z, Li H, Yang L, Zhu J, Zheng X, Cheng H (2020) Stretchable, ultrasensitive, and low-temperature NO_2 sensors based on MoS_2@rGO nanocomposites. Mater Today Phys 15:100265. https://doi.org/10.1016/j.mtphys.2020.100265

You GR, Park GJ, Lee JJ, Kim C (2015) A colorimetric sensor for the sequential detection of Cu^{2+} and CN^- in fully aqueous media: practical performance of Cu^{2+}. Dalton Trans 44(19):9120–9129. https://doi.org/10.1039/C5DT00772K

Zhang J, Xu S, Kumacheva E (2004) Polymer microgels: reactors for semiconductor, metal, and magnetic nanoparticles. J Am Chem Soc 126(25):7908–7914. https://doi.org/10.1021/ja031523k

Zhang LN, Deng HH, Lin FL, Xu XW, Weng SH, Liu AL, Chen W (2014) In situ growth of porous platinum nanoparticles on graphene oxide for colorimetric detection of cancer cells. Anal Chem 86(5):2711–2718. https://doi.org/10.1021/ac404104j

Zhang W, Zhang P, Su Z, Wei G (2015) Synthesis and sensor applications of MoS_2-based nanocomposites. Nanoscale 7(44):18364–18378. https://doi.org/10.1039/C5NR06121K

Zhang Y, Wang YN, Sun XT, Chen L, Xu ZR (2017) Boron nitride nanosheet/CuS nanocomposites as mimetic peroxidase for sensitive colorimetric detection of cholesterol. Sens Actuators B 246:118–126. https://doi.org/10.1016/j.snb.2017.02.059

Zhang E, Ju P, Guo P, Hou X, Hou X, Lv H, Zhang Y (2018) A FRET-based fluorescent and colorimetric probe for the specific detection of picric acid. RSC Adv 8(55):31658–31665. https://doi.org/10.1039/C8RA05468A

Zhao Z, Liu J, Cui F, Feng H, Zhang L (2012) One pot synthesis of tunable Fe_3O_4–MnO_2 core–shell nanoplates and their applications for water purification. J Mater Chem 22(18):9052–9057. https://doi.org/10.1039/C2JM00153E

Zhao T, Yao Y, Wang M, Chen R, Yu Y, Wu F, Zhang C (2017) Preparation of MnO_2-modified graphite sorbents from spent Li-Ion batteries for the treatment of water contaminated by lead, cadmium, and silver. ACS Appl Mater Interfaces 9(30):25369–25376. https://doi.org/10.1021/acsami.7b07882

Zhou W, Zou X, Najmaei S, Liu Z, Shi Y, Kong J, Idrobo JC (2013) Intrinsic structural defects in monolayer molybdenum disulfide. Nano Lett 13(6):2615–2622. https://doi.org/10.1021/nl4007479

Zhou X, Torabi M, Lu J, Shen R, Zhang K (2014) Nanostructured energetic composites: synthesis, ignition/combustion modeling, and applications. ACS Appl Mater Interfaces 6(5):3058–3074. https://doi.org/10.1021/am4058138

Zhou G, Chang J, Pu H, Shi K, Mao S, Sui X, Chen J (2016) Ultrasensitive mercury ion detection using DNA-functionalized molybdenum disulfide nanosheet/gold nanoparticle hybrid field-effect transistor device. ACS Sensors 1(3):295–302. https://doi.org/10.1021/acssensors.5b00241

Zhou Y, Huang W, He Y (2018) pH-Induced silver nanoprism etching-based multichannel colorimetric sensor array for ultrasensitive discrimination of thiols. Sens Actuators B 270:187–191. https://doi.org/10.1016/j.snb.2018.05.025

Zhu Q, Zhang B, Wang T, Feng X (2019) Synthesis and properties of porous δ-MnO_2/polymer millimeter-sized beads for Ni(II) removal. Microporous Mesoporous Mater 273:90–98. https://doi.org/10.1016/j.micromeso.2018.07.001

Chapter 20
Transition Metal and Conducting Polymers Nanocomposite for Sensing of Environmental Gases

Chandra Shekhar Kushwaha, Pratibha Singh, and Saroj Kr Shukla

Abstract The present book chapter describes the synthesis of transition metal and conducting polymer nanocomposites with their noble properties for sensing different environmental gases like humidity, ammonia, and other gaseous pollutants. The basic principle of sensing techniques like optical, electrical, and mechanical has been discussed along with their technical details and potentials. The composite dependents sensitivity of different gases is compared for selections and use of appropriate techniques for commercialization and scientific use. Although tremendous success has been reported the multifaceted entry of atmospheric pollutants from natural as well anthropogenic sources still indicated many challenges like remote monitoring are need to be encountered for effective monitoring of the environment. This, existing relation between success and challenges has been presented in the present review with the suitable significant illustration.

Keywords Conducting polymer nanocomposite · Environmental gases · Gaseous pollutants · Gas sensor

20.1 Introduction

Nano-confined transition metals, their oxides (TMOs), and conducting polymer composite (CPNCs) exhibits great potential in the preparation of optical and electronic devices, physical and chemical sensing along with different biomedical applications (Guo et al. 2015). In general, CPNCs are hybrid materials comprised of at least one constituent in the nano range i.e. 1–100 nm. The presence of heterogeneous structure along with nanoconfinement yields synergistic properties of both inorganic material and processability of polymers i.e. organic. $CPNC_S$ has exhibits several advantageous features like easy preparation, processability, surface reactivity, tunable electrical and optical properties. These properties are depending on constituents'

C. S. Kushwaha · P. Singh · S. K. Shukla (✉)
Department of Polymer Science, Bhaskaracharya College of Applied Sciences, University of Delhi, Delhi 110075, India

S. P. Singh et al. (eds.), *Metal Nanocomposites for Energy and Environmental Applications*, Energy, Environment, and Sustainability,
https://doi.org/10.1007/978-981-16-8599-6_20

oxides, conducting and designed molecular engineering during composite formations. In this regards both direct and indirect processes are adopted along with the use of nano confining tools such as hard and soft templates for precise control of shape and size (Singh and Shukla 2020). Thus, obtained nanostructured CPNCs are have been also for efficient sensing of different environmental gases like ammonia, humidity, volatile organic compounds, sulfur, and nitrogen gases with efficient sensing parameters (Shukla et al. 2017). The advantageous properties of CPNCs have been used for sensing different gases using different transduces viz optical, electrical, and mechanical. Although many significant achievements have been reported by scientists continuously still many limitations like their use in point of care need the attention of researchers. In the line above discussion present chapter discloses the studies of TMs and conducting polymer-based nanocomposites materials, which will include briefly the synthetic methods, properties of transition metal oxides containing conducting polymers, and their application as the sensor for monitoring of environmental gases along with sensing mechanism and potential applications.

20.2 Overview on TM/CPNCs

CPNCs is a multi-component hybrid structure in which at least one component in the nano range i.e. 1–100 nm, is also referred as fillers. The incorporation nano particle improves the surface area to incorporate large interface area between fillers and neat macromolecules for better properties like mechanical strength, thermal conductivity, electrical conductivity, corrosion resistance and interactivity. Further, nano fillers are classified as particulate, flake, and fibre, which exhibits individual reinforcing properties in conducting polymer matrix. Another, current aspects in CPNCs is dimensional manipulations like zero-dimensional, one-dimensional, two-dimensional, and three-dimensional to further modify the efficient use of CPNCs. In example the nanotubes, nanowires, nanofibers and nanorods are examples of one-dimensional nanomaterials, which exhibits the large aspect ratio and the optimises charge transfer rate in a one direction for advanced applications (Kondawar and Patil 2017). The heterogeneous structures with nano confinements yield high surface area, aligned crystallinity, super magnetism, and wide range of conductivity for different applications including sensors. The presence of metal oxides also produce catalytic behaviours for effective sensing of several organic compounds like organic dyes, agrochemicals, and pharmaceuticals (Zhan et al. 2017). The most frequently used conducting is polyaniline, polypyrrole, polythiophene, poly(3,4-ethylenedioxythiophene), and poly (paraphenylene) along with inorganic and organic fillers like metals, metal oxides, carbon nanostructures before or after polymerization. The illustration about structure formation during composite formation is shown in Fig. 20.1.

Thus, obtained CPNCs can be simple classified into two categories, (a) organic in an inorganic matrix and (b) inorganic in an organic matrix. In the first categories, the organic materials are embedded in the inorganic template, however in the second category inorganic constituents dispersed into the organic matrix. Both types of

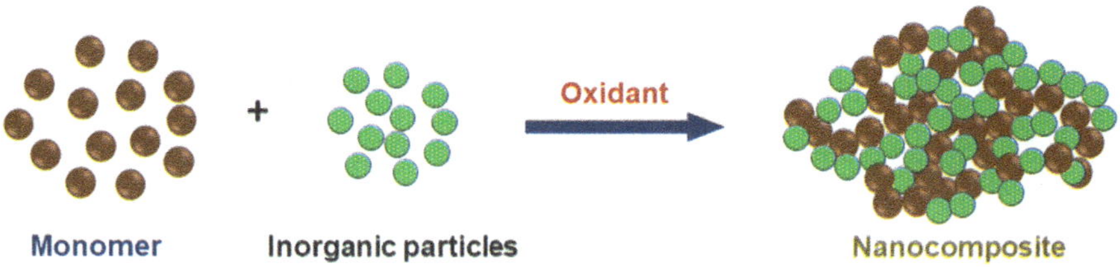

Fig. 20.1 Simple illustration for the formation of CPNCs

CPNCs are prepared after using different preparation methods like mixing and polymerization after employing a different forms of energy i.e. chemical, electrical, heat, and photo-radiation.

20.3 Synthetic Methods

The methods used for synthesis of transition metal oxide TMO-CPs nanocomposites are (i) ex-situ synthesis and (ii) in-situ synthesis along with different sub categories as represented in Fig. 20.2 (Janáky and Rajeshwar 2015). The main characteristics of all these methods are given in Table 20.1 along with advantages and disadvantages.

In *ex-situ* method both components were prepared separately and after that the both components were composed in the form of hybrid after applying solvation, heat, mechanical and light energy (Wang et al. 2013; Guzelturk et al. 2016). Some other advanced modifications like *ligand exchange, lamination and layer-by-layer* deposition were also used to achieve the highly ordered nanostructures in composites (We et al. 2014; Rawolle et al. 2013).

Further the in-situ procedure deals with the polymerisation of monomer was performed in the presence of filler or other component. The composite formed by

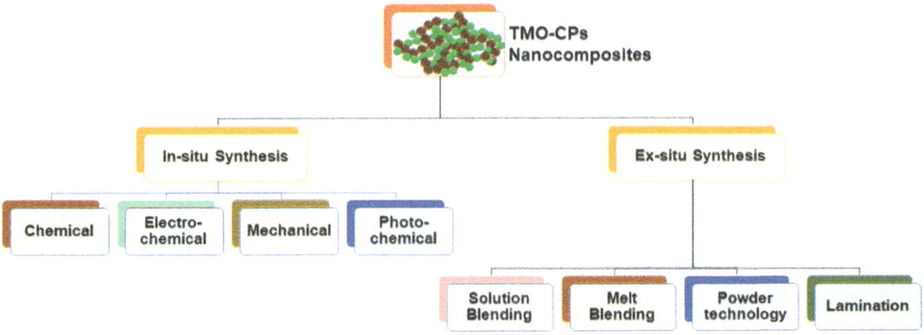

Fig. 20.2 Synthesis methods of TMO-based CPs nanocomposites

Table 20.1 Comparison of synthesis methods along with advantage and disadvantage

Synthesis methods	Advantages	Disadvantages	Process involved	Example
Ex-situ method	Easy to perform and scalable	Agglomeration and heterogenous distribution	Mechanical, solution, thermal and mixing	Wang et al. (2013)
In-situ method	Chemical functionality and molecular lever arrangements	Mostly possible at lab scale	Chemical and electrical polymerisation	Gao et al. (2016), Bora et al. (2014)
One-pot synthesis	Low cost	Uncontrolled structure and morphology	Chemical polymerisation and redox reaction Co-deposition	Bogdanović et al. (2015)

this method yield more arranged composite structure along with chemical functionalisation at molecular levels. The functionalisation of monomers also regulates the polymerisation methods in term of yield and polymerisation times. The polymerisation was performed by different routes like chemical electro chemical, mechanical and photochemical with due condition and precaution. The selection of polymerising agents are also significant impact on the properties like crystallinity, strength and conductance (Xia et al. 2013; Singh et al. 2013a). The simplification of method is another steps, in this context *one-pot* synthesis and use of greener solvents are preferred (Feng et al. 2014).

20.4 Properties

The TM and TMOs nanostructures containing conducting polymers nanohybrids exhibit properties of metal sensing are enhanced electrical conducting, interactivity, responsiveness, switch ability, and processability. These properties depend on the nature and size of inorganic components and polymer matrixes and chemical arrangements. Depending upon the synthesis techniques and the characteristics of the inorganic materials, the ultimate properties of the resulting composite are also controlled. The brief properties of CPNCs nanohybrids are shown in Fig. 20.3

These properties are synergized effect of constituent materials like metallic sites in polymer matrixes incorporate catalysis and luminesces for different applications. Further, the heterogeneity generated due to the presence of metal oxides creates p–n junction, catalytic centers, and optically active sites for use in different types of sensing (Sarma et al. 2002). For example, Shukla et al. has prepared the composites comprised of Fe_2O_3, chitosan, and polyaniline with catalytic properties. The catalytic property of composites was used for sensing paracetamol in residual water with efficient sensing parameters (Kushwaha and Shukla 2020). Functional groups

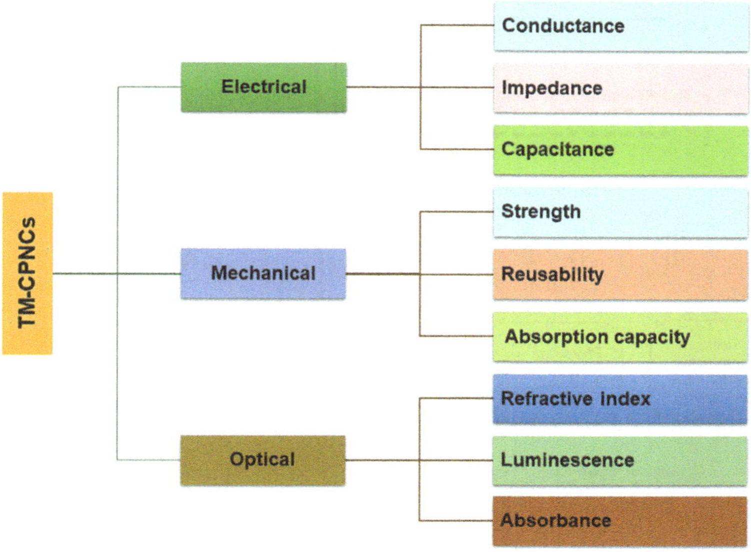

Fig. 20.3 Properties of TM and TMOs/CPs nanocomposites

viz. $-NH_2$ and –COOH have also been added to the composite particles and all these combinations and modifications have improved the applicability of conducting polymers in sensing applications. In an example, the composite comprised of polyaniline, chitosan, and silver nanoparticle exhibits multifold catalytic functionality towards different metal ions present in the milk samples. The enhanced functional catalytic activities are due to a conformational change in the secondary structure of composite due to interaction of metal ions (Khan and Husain 2019). Surface-induced optical responsiveness is another important feature of CPNCs for chemical and biosensing. The optical activities of polypyrrole/prussian blue/TiO_2 coated on plastic optical fiber have been demonstrated for effective sensing of ammonia than acetone and ethanol in the concentration range of 0–500 ppm at room temperature (Muthusamy et al. 2021).

20.5 CPNCs Based Environmental Sensor

The different evolved mechanical, optical, and electrical properties are explored for sensing different atmospheric gases. The basic mechanism lying in sensing different gases is illustrated in Fig. 20.3. In general, the sensor consists of sensing materials i.e. CPNCs along with monitoring tools like photometers in optical, multi-meter in electrical types. The processing process and optimum thickness of sensing materials

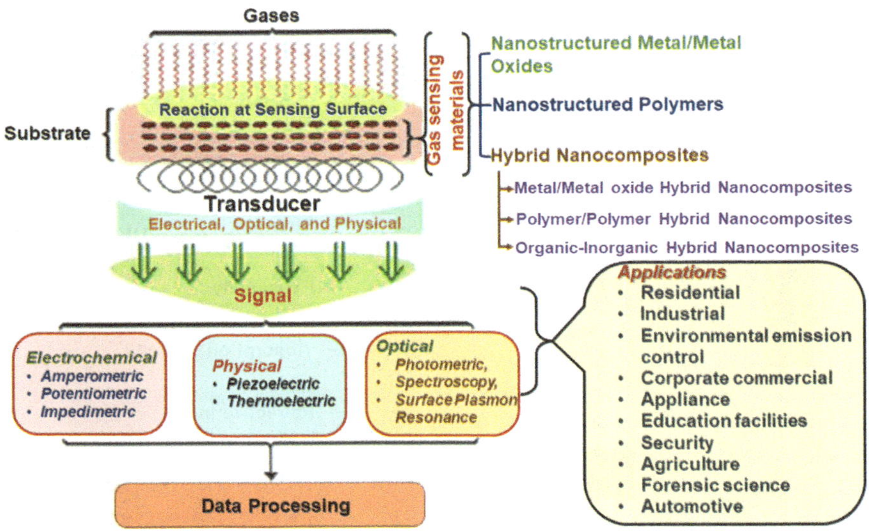

Fig. 20.4 Schematic of gas sensor fabrication techniques (Kaushik et al. 2015) (Figs. 20.4, 20.6 and 20.7 each have individual format)

are other features for the development of efficient gas sensing (Kaushik et al. 2015). The representative gas sensor fabrication technique is illustrated in Fig. 20.4.

In general, environmental gases are classified into two categories, (a) constituents gases, and (b) polluting gases.

20.5.1 Constituent Gases

The basic composition of different gases is illustrated in Fig. 20.5, however its minor composition changes with time and position.

The minor change in composition also disturbs the balance between them in term of their cumulative impact on atmospheric events and existing creature. Hence, their composition monitoring is essential for the prediction and welfare of human beings using different sensors. The sensors used for sensing constituent gases of the environment are oxygen, carbon dioxide, and humidity.

Sensing of oxygen is performed for gaseous as well as the dissolved states using different $CPNC_S$ based sensors. The presence of oxygen in air and water in adequate amounts is of utmost importance for the survival of organisms along with a presence in body fluid is equally important for the functioning of body organ physiological activities. The basic mechanism during oxygen sensing is interactive surface oxidation of sensing substrate, which yields a change in current, potential, resistance, and luminescence. The fluorescence quenching and optical fiber guide-based optical sensor other significant oxygen sensors with significant properties. The quenching is

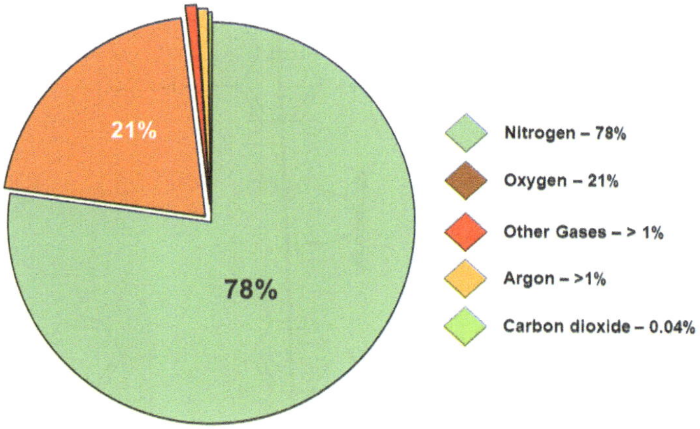

Fig. 20.5 Gaseous composition of the atmosphere

based on energy transfer from oxygen to fluorescent sensitive materials. The oxygen was also reduced into the peroxide during the interaction of oxygen on the surface of polyaniline and polypyrrole under optimized conditions. These oxygen reduction properties over CPNCs are the simple principle for oxygen sensing (Wang et al. 2019; Rabl et al. 2020).

Carbon dioxide is another important constituent gas of the atmosphere for use in photosynthesis and maintaining the oxygen balance. Further, the presence of carbon dioxide beyond the limit is responsible for the greenhouse effect, thus, its monitoring is important and several CPNCs sensors are used in this regard. The mechanism of CO_2 sensing is reducing and coordinating behavior with sensing materials. Functionalisation of conducting polymer like PANI develops interacting sites for effective carbon dioxide sensing. The sensing mechanism of the sodium salt of sulfonated PANI is shown in Fig. 20.6 (Doan et al. 2012).

However, the use of pH-sensitive dyes is another strategy to design a better optical sensor for carbon dioxides. The pH-sensitive dye interacts with CO_2 and exhibits induced optical response, the change is used for quantitative sensing applications.

Humidity is an important constituent of the atmosphere and is important for monitoring atmospheric processes like rain. Currently, humidity sensors are also important in human comforts, packaging, and medical fields (Shukla et al. 2016). A combination of nanostructure TM/TMO and conducting polymers such as PANI, PPy, PEDOT, and PTh, have been widely used for humidity sensors after using electrical, optical, and mechanical techniques. The basic process in humidity sensing is dissociative adsorption of water molecules over CPNCs, the generated ions change different electrical properties like resistance and conductance for humidity sensing. The simple resistive type of humidity sensing over CuO and PANI-based nanocomposite is shown in Fig. 20.7.

The scheme is illustrating that initial adsorption is non-ionising, while the subsequent layer is the ionising and explored for electrical type sensing. The role of

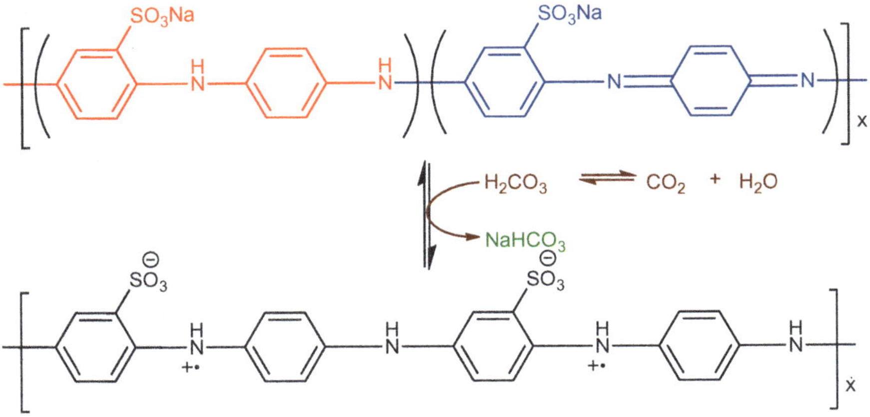

Fig. 20.6 Reversible change from an insulating sulfonated polyaniline (SPAN-Na) to conducting SPAN (self-doped) due to interaction with carbonic acid and subsequent protonation

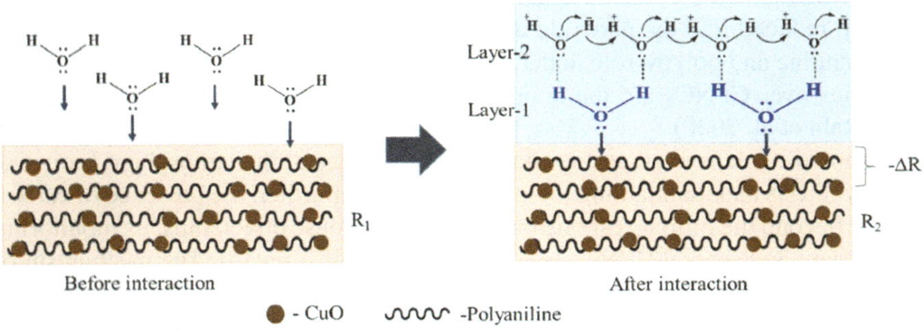

Fig. 20.7 The sensing mechanism of humidity

constituent composite is in optimizing the hydrophilicity, porosity, and conductivity. Shukla et al. have reported cellulose grafted polyaniline and polypyrrole composite with improved hydrophilicity and conductivity. The composites were used effective wide range humidity sensing after monitoring the change in resistance (Shukla 2012, 2013). The adsorption of water on the hydrophilic surface also changes optical properties like the refractive index. The change in the refractive index also changes the light out power and surface emissivity. Both induced properties were used for humidity sensing and a simple setup for opto-chemical sensing of humidity present in the atmosphere is shown in Fig. 20.8, using the optical fiber approach (Shukla et al. 2004).

Some other important CPNCs based materials for constituent gas sensing along with their properties are listed in Table 20.2.

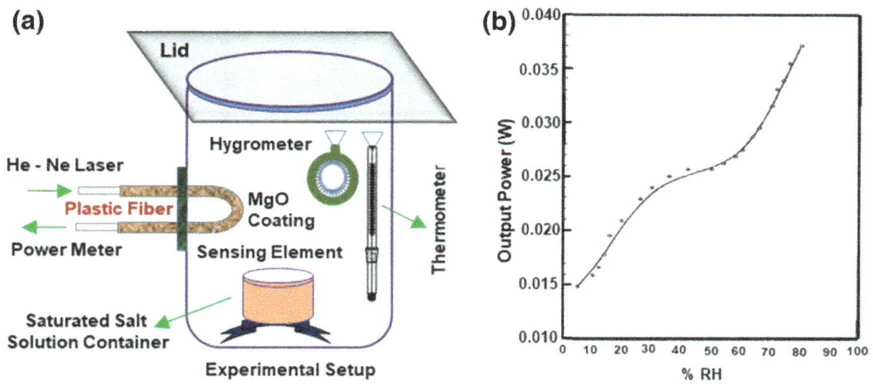

Fig. 20.8 Opto-chemical sensing of humidity using optical fiber approach (Shukla et al. 2004)

20.6 Pollutants

The most problematic environmental pollutants are added to the atmosphere through different natural and anthropogenic agencies. The major pollutants are nitrogen and sulfur compounds, volatile organic compounds, and hydrocarbon like chlorofluorohydrocarbon. Some of the nitrogen-based pollutants are ammonia and nitrogen-based oxides. In this regard, ammonia (NH_3) is a useful inorganic compound the fertilizers, household products, medicines, and explosives but its presence beyond the prescribed limit is harmful to humans, animals, and other creatures. The ammonia is monitored after using most of the sensing methods. Shukla et.al. have demonstrated sensing of efficient volatile ammonia from the soil over zinc oxide encapsulated polypyrrole prepared by in-situ chemical polymerization techniques. The sensing ammonia mechanism is illustrated in Fig. 20.9, along with the role of zinc oxide (Singh et al. 2021).

Different nitrogen oxides like nitrous and nitric oxide are also sensed to grade the pollution level and suitability of the atmosphere for human health using different TMOs and CPs composites. The facile properties of CPNCs have been also explored in NO_x sensing after exploring its reducing and acidic nature. In an example, tungsten oxide encapsulated chitosan grafted polyaniline has been explored for amperometric sensing of NO_2 in concentration range 100–500 ppb with sensing parameter of response time 8 s (Tiwari and Gong 2008). The presence of metal oxide in the polymer matrix facilitates the structural transition for effective NOx sensing. Yadav et al. have demonstrated NO_2 sensing over Fe_2O_3 and PANI nanocomposites using electrical transducers. The basic sensing mechanisms are based on electron transfer from NO_2 to Fe_2O_3 after adsorption on the composite surface (Sonker and Yadav 2017).

In recent years, various TM/TMO-based conducting polymer nanocomposites materials sensors were reported for the gaseous pollutants, and a wide variety of volatile organic compounds (VOCs) such as $CHCl_3$, alcohols, acetones, pyridine,

Table 20.2 Representative CPNCs based constituent gas sensors

S. No	CPNCs	Transducer	Analyte	Sensing range	Response and recovery time	References
1	TiO_2/PPy	Resistive	H_2O	30–90% RH	40 and 20 s	Su and Huang (2007)
2	Co-PANI	Optical	H_2O	20–95% RH	8 and 60 s	Vijayan et al. (2008)
3	PANI/α-Fe_2O_3	Resistive	H_2O	10–95% RH	70 and 90 s	Singh et al. (2018)
4	Ag-PANI	Optical	H_2O	5–95% RH	30 and 90 s	Fuke et al. (2010)
5	PPy/TaS_2	Resistive	H_2O	10–97% RH	10 and 20 s	Sunilkumar et al. (2019)
6	PANI/TiO_2	Chemiresistive	CO_2	70–80 ppm	80 s	Nimkar et al. (2015)
7	PPy/Cu phthalocyanine	Capacitive-type	CO_2	0.1–1%	Response time (34 s)	Riyazi and Azim Araghi (2020)
8	Bi_2O_3–loaded-PPy	Chemiresistive	CO_2	0.065 for 50 ppm	40 and 60 s	Choudhary and Waghuley (2019)
9	ZnO-d-PANI	Resistive	CO_2	0–100 ppm	59 and 101 s	Mude et al. 2017)
10	Ethyl cellulose-PPy	Resistive	CO_2	0.1–950 ppm	210 and 1560 s	Waghuley et al. (2008)
11	NiO-en-PPy	Resistive and electrochemical	H_2O_2	10–675 ppm in gas phase 10–720 ppm in liquid state	30 and 70 s in gas phase 150 and 20 s in liquid state	Singh et al. (2020)
12	BP-PS-PPy/Ag	Electrochemical (CV)	H_2O_2	1–20 μM	Response time – 2 s	Jlassi et al. (2019)
13	PPy@CaO/NiO-NTs	UV–VIS Spectro-photometer	H_2O_2	0.004–10 mM	–	Song et al. (2020)
14	Poly(azo-BBY)-rGO	EIS	Dissolved O_2	0.018–1.28 mmol L^{-1}	15 and 25 s	Olean-Oliveira and Teixeira (2018)

(continued)

Table 20.2 (continued)

S. No	CPNCs	Transducer	Analyte	Sensing range	Response and recovery time	References
15	Ag NPs-d-Ru(DPP)$_3$Cl$_2$/PMMA	0–13 mg/L	Dissolved O_2	0–13 mg/L	2.0 and 104 s	Jiang et al. (2017)

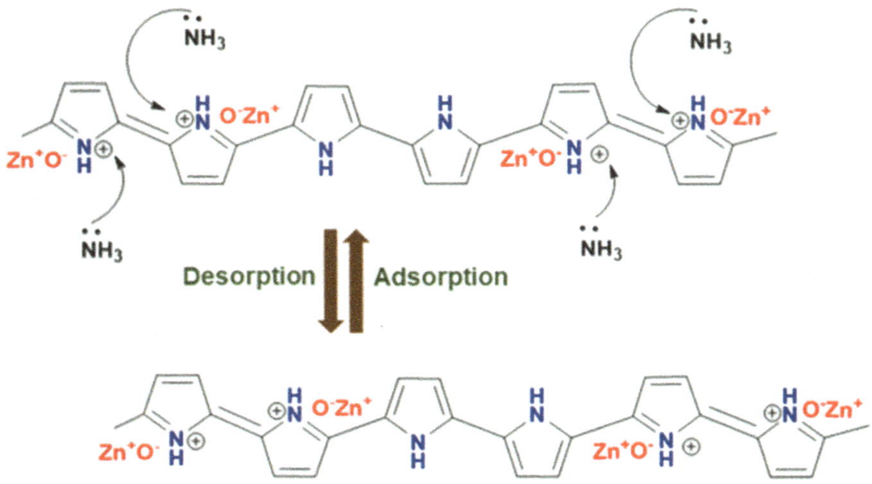

Fig. 20.9 The schematic of the ammonia sensing mechanism

dimethyl sulphoxide, and LPG (Gangopadhyay and De 2001). The important sensing mechanism in VOCs sensing is electro active interaction between organic molecules and transition metal and their compounds. These interactions transfer the charge as well as exhibits redox reactions. However, the presence of conducting polymer along with transition metal supports efficient response communication. The preparation method, as well as the processing technique, also controls several properties in VOCs sensing. For example, the acetone sensing in the gas phase was compared for PPy and PANI-based electrodes prepared by chemical oxidation, chemical vapor deposition, and impregnated oxidation technique (Do and Wang 2013). The sensitivity of PANI towards acetone was found greater than PPy prepared by chemical vapor deposition along with sensing parameters i.e. the sensitivity, response time, and recovery time of 5.908×10^{-4} ppm^{-1}, 1.0–3.0, and 3.0–10.0 min, respectively. Patil et al. have reported an LPG sensor using CPNCs comprised of polyaniline and Cu_2ZnSnS_4 composite prepared by electrodeposition of PANI over spray deposited film of Cu_2ZnSnS_4 (Patil et al. 2016). The sensor exhibited a 79% response for 780 ppm LPG concentration along with a response time of 120 s and recovery time of 125 s. The chemically polymerized CPNCs of PANI and In_2O_3 were reported for sensing hydrogen, carbon monoxide, and NO_2 at ambient temperature (Sadek et al. 2006). Similarly, the again synthesized PANI/WO_3 nanocomposite(Sadek et al. 2008) and employed in the sensor for detection of H_2 gas. Some important gaseous pollutants and VOCs sensors with their sensing parameters are shown in Table 20.3.

SO_x are other serious types of atmosphere polluting gases and produced through automotive engines, power, and industrial plant. Its presence in the atmosphere is responsible for acid rain and smog. The sensors reported for onsite monitoring of SO_x are semiconducting, polymer electrolyte, and piezoelectric crystal. In this regards several transition metals and conducting polymer composites like Polyaniline-WO_3

Table 20.3 TM/TMO and conducting polymer nanocomposites based gaseous pollutants and VOCs sensors and their sensing parameters

S. No	CPNCs	Transducer	Analyte	Sensing range	Response and recovery time	References
1	TiO_2/PANI	Resistive	NH_3	23–141 ppm	18 and 58 s	Tai et al. (2007)
2	Ag/PPy thin film on flexible PET	Chemiresistive	NH_3	5–50 ppm	6 s and ~ 80 min	Singh et al. (2013b)
3	α-Fe_2O_3/c-PANI	Electrochemical	Hydrazine	0. 2–40 μM	Less than 12 s	Harraz et al. (2016)
4	PANI-Polyethylene	Optical	NH_3	180–1800 ppm	15 s and 2 min	Jin et al. (2001)
5	PANI/TiO_2	QCM	Trimethylamine	10–200 ppm	58 and 280 s	Zheng et al. (2008)
6	Ag/PPy	Chemiresistive	NO_2	5–120 ppm	148 and 500 s	Karmakar et al. (2017)
7	Ag/PANI	Chemiresistive	H_2S	1–25 ppm	6 min, NA	Mekki et al. (2014)
8	PANI/$ZnMoO_4$	Resistive	LPG	800–1800 ppm	600 and 840 s	Bhanvase et al. (2015)
9	PPy-WO_3	Resistive	H_2S	100–1000 ppb	6 and 210 min	Su and Peng (2014)
10	PANI/CdS	Resistive	H_2S	10–100 ppm	32–413 s	Raut et al. (2012)
11	PANI/ γ-Fe_2O_3	Resistive	LPG	50–200 ppm	60 s	Sen et al. (2014)
12	PPy/TiO_2	QCM	NH_3, H_2S, C_2H_5OH and Me_3N	10–250 ppm	100 and 200 s	Cui et al. (2016)
13	PANI–WO_3	Resistive	SO_2	5–80 ppm	200 and 180 s	Chaudhary and Kaur (2015)
14	PANI/Cu	Resistive	$CHCl_3$	10–100 ppm	5 min	Athawale et al. (2002)
15	PANI/TiO_2	QCM	Trimethylamine	10–200 ppm	58 and 280 s	Zheng et al. (2008)
16	PANI/TiO_2	Resistive	Acetone	100–500 ppm	20 and 700 s	Bahru et al. (2020)

and Au/PANI are sued in SO_2 sensing with effective sensing parameters (Khan et al. 2019).

20.7 Conclusion and Future Challenges

The chapter presents an overview of transition metal and conducting polymer nanocomposite along with preparation and properties. Further, the use of properties optimized conducting polymer composite in sensing of different atmospheric gases is explained with properties and applications. The role of CPNCs towards sensing mechanisms has been explained based on adsorptive and catalytic interaction between atmospheric gases and CPNCS along with suitable schemes and illustrations. The existing challenges in the prospects of the sensor in point of care application and extrapolated application like humidity sensing in-breath monitoring are highlighted for newer researchers.

Acknowledgements Authors, CSK, and PS are thankful to CSIR New Delhi [Ref. No. 08/642(0002)/2016-EMR-I], and University Grants Commission, New Delhi, for granting fellowship. Authors are also obliged to Prof. Balaram Pani, Principal, Bhaskaracharya College of Applied Science, for maintaining academic culture in college.

References

Athawale AA, Pethkar S, Sharma S, Nirkhe C (2002) Chloroform vapour sensor based on copper/polyaniline nanocomposite. Sens Actua B Chem 85:131–136

Bahru R, Zamri MFMA, Shamsuddin AH, Mohamed MA (2020) Structural analyses of polyaniline–titanium oxide composite for acetone detection. J Mater Sci Mater Electron 31:1574–1584

Bhanvase BA, Darda NS, Veerkar NC, Shende AS, Satpute SR, Sonawane SH (2015) Ultrasound assisted synthesis of PANI/$ZnMoO_4$ nanocomposite for simultaneous improvement in anticorrosion, physico-chemical properties and its application in gas sensing. Ultras Sonochem 24:87–97

Bogdanović U, Vodnik V, Mitrić M, Dimitrijević S, Škapin SD, Žunič V, Budimir M, Stoiljković M (2015) Nanomaterial with high antimicrobial efficacycopper/polyaniline nanocomposite. ACS Appl Mater Interface 7:1955–1966

Bora C, Sharma J, Dolui S (2014) Polypyrrole/sulfonated graphene composite as electrode material for supercapacitor. J Phys Chem C 118:29688–29694

Chaudhary V, Kaur A (2015) Enhanced room temperature sulfur dioxide sensing behaviour of in situ polymerized polyaniline-tungsten oxide nanocomposite possessing honeycomb morphology. RSC Adv 5:73535–73544

Choudhary AR, Waghuley SA (2019) In situ Bi_2O_3-loaded polypyrrole nanocomposites for carbon dioxide gas sensing. Iran Polym J 28:933–941

Cui S, Yang L, Wang J, Wang X (2016) Fabrication of a sensitive gas sensor based on PPy/TiO_2 nanocomposites films by layer-by-layer self-assembly and its application in food storage. Sens Actua B Chem 233:337–346

Do JS, Wang SH (2013) On the sensitivity of conductimetric acetone gas sensor based on polypyrrole and polyaniline conducting polymers. Sens Actua B Chem 185:39–46

Doan TCD, Ramaneti R, Baggerman J, Van Der Bent JF, Marcelis ATM, Tong HD, Van Rijn CJM (2012) Carbon dioxide sensing with sulfonated polyaniline. Sens Actua B Chem 168:123–130
Feng ZL, Yao YY, Xu JK, Zhang L, Wang ZF, Wen YP (2014) One-step co-electrodeposition of graphene oxide doped poly(hydroxymethylated-3,4-ethylenedioxythiophene) film and its electrochemical studies of indole-3-acetic acid. Chin Chem Let 25:511–516
Fuke MV, Kanitkar P, Kulkarni M, Kale BB, Aiyer RC (2010) Effect of particle size variation of Ag nanoparticles in polyaniline composite on humidity sensing. Talanta 81:320–326
Gangopadhyay R, De A (2001) Conducting polymer composites: novel materials for gas sensing. Sens Actua B Chem 77:326–329
Gao S, Zang P, Dang L, Xu H, Shi F, Liu Z, Lei Z (2016) Extraordinarily high-rate capability of polyaniline nanorod arrays on graphene nanomesh. J Power Source 304:111–118
Guo T, Yao MS, Lin YH, Nan CW (2015) A comprehensive review on synthesis methods for transition-metal oxide nanostructures. CrystEngComm 17:3551–3585. https://doi.org/10.1039/c5ce00034c
Guzelturk B, Menk F, Philipps K, Kelestemur Y, Olutas M, Zentel R, Demir HV (2016) Colloidal nanoplatelet/conducting polymer hybrids: excitonic and material properties. J Phys Chem C 120:3573–3582
Harraz FA, Ismail AA, Al-Sayari SA, Al-Hajry A, Al-Assiri MS (2016) Highly sensitive amperometric hydrazine sensor based on novel α-Fe_2O_3/crosslinked polyaniline nanocomposite modified glassy carbon electrode. Sens Actua B Chem 234:573–582
Janáky C, Rajeshwar K (2015) The role of (photo)electrochemistry in the rational design of hybrid conducting polymer/semiconductor assemblies: from fundamental concepts to practical applications. Prog Polym Sci 43:96–135
Jiang Z, Yu X, Zhai S, Hao Y (2017) Ratiometric dissolved oxygen sensors based on ruthenium complex doped with silver nanoparticles. Sensors 17:548
Jin Z, Su Y, Duan Y (2001) Development of a polyaniline-based optical ammonia sensor. Sens Actua B Chem 72:75–79
Jlassi K, Sliem MH, Eid K, Krupa I, Chehimi MM, Abdullah AM (2019) Novel enzyme-free multifunctional bentonite/polypyrrole/silver nanocomposite sensor for hydrogen peroxide detection over a wide pH range. Sensors 19:4442
Karmakar N, Fernandes R, Jain S, Patil UV, Shimpi NG, Bhat NV, Kothari DC (2017) Room temperature NO_2 gas sensing properties of p-toluenesulfonic acid doped silver-polypyrrole nanocomposite. Sens Actua B Chem 242:118–126
Kaushik A, Kumar R, Arya SK, Nair M, Malhotra BD, Bhansali S (2015) Organic-inorganic hybrid nanocomposite-based gas sensors for environmental monitoring. Chem Rev 115:4571–4606
Khan M, Husain Q (2019) Effect of metal ions present in milk on the structure and functional integrity of native and polyaniline chitosan nanocomposites bound β-galactosidase: a multi-spectroscopic approach. Food Chem 279:312–320
Khan MAH, Rao MV, Li Q (2019) Recent advances in electrochemical sensors for detecting toxic gases: NO_2, SO_2 and H_2S. Sensors 19:905
Kondawar SB, Patil PT (2017) Conducting polymer nanocomposites for sensor applications. Conducting polymer hybrids. Springer, Cham, pp 223–267
Kushwaha CS, Shukla SK (2020) Electrochemical sensing of paracetamol using iron oxide encapsulated in chitosan-grafted-polyaniline. ACS Appl Polym Mater 2:2252–2259
Mekki A, Joshi N, Singh A, Salmi Z, Jha P, Decorse P, Lau-Truong S, Mahmoud R, Chehimi MM, Aswal DK, Gupta SK (2014) H2S sensing using in situ photo-polymerized polyaniline-silver nanocomposite films on flexible substrates. Org Electron 15:71–81
Mude KM, Mude BM, Raulkar KB, Yawale SS, Yawale SP (2017) Conducting polymer polyaniline as CO_2 gas sensor. Int J Chem Tech Res 10:494–500
Muthusamy S, Charles J, Renganathan B, Ganesan AR (2021) Ternary polypyrrole/prussian blue/TiO_2 nanocomposite wrapped poly-methyl methacrylate fiber optic gas sensor to detect volatile gas analytes. Optik 230:166289

Nimkar S, Kondawar S, More P (2015) Polyaniline/TiO_2 nanocomposite thin film based carbon dioxide gas sensor. Int J Res Biosci Agric Technol 12:e18

Olean-Oliveira A, Teixeira MFS (2018) Development of a nanocomposite chemiresistor sensor based on π-conjugated azo polymer and graphene blend for detection of dissolved oxygen. Sens Actua B Chem 271:353–357

Patil SJ, Lokhande AC, Yadav AA, Lokhande CD (2016) Polyaniline/Cu_2ZnSnS_4 heterojunction based room temperature LPG sensor. J Mater Sci Mater Electron 27:7505–7508

Rabl H, Wielend D, Tekoglu S, Seelajaroen H, Neugebauer H, Heitzmann N, Apaydin DH, Scharber MC, Sariciftci NS (2020) Are polyaniline and polypyrrole electrocatalysts for oxygen (O_2) reduction to hydrogen peroxide (H_2O_2)? ACS Appl Energy Mater 3:10611–10618

Raut BT, Chougule MA, Nalage SR, Dalavi DS, Mali S, Patil PS, Patil VB (2012) CSA doped polyaniline/CdS organic-inorganic nanohybrid: physical and gas sensing properties. Ceram Int 38:5501–5506, 4

Rawolle M, Sarkar K, Niedermeier MA, Schindler M, Lellig P, Gutmann JS, Moulin JF, Haese-Seiller M, Wochnik AS, Scheu C, Müller-Buschbaum P (2013) Infiltration of polymer hole-conductor into mesoporous titania structures for solid-state dye-sensitized solar cells. ACS Appl Mater Interface 5:719–729

Riyazi S, Azim Araghi ME (2020) Performance of interdigitated capacitive-type CO_2 sensor based on polypyrrole/copper phthalocyanine nanocomposite. J Mater Sci Mater Electron 31:3539–3548

Sadek AZ, Wlodarski W, Shin K, Kaner RB, Kalantar-Zadeh K (2006) A layered surface acoustic wave gas sensor based on a polyaniline/In_2O_3 nanofibre composite. Nanotechnol 17:4488–4492

Sadek AZ, Wlodarski W, Shin K, Kaner RB, Kalantar-zadeh K (2008) A polyaniline/WO3 nanofiber composite-based ZnO/64° YX $LiNbO_3$ SAW hydrogen gas sensor. Synth Met 158:29–32

Sarma TK, Chowdhury D, Paul A, Chattopadhyay A (2002) Synthesis of Au nanoparticle-conductive polyaniline composite using H_2O_2 as oxidising as well as reducing agent. Chem Comm 2:1048–1049

Sen T, Shimpi NG, Mishra S, Sharma R (2014) Polyaniline/γ-Fe_2O_3 nanocomposite for room temperature LPG sensing. Sens Actua B Chem 190:120–126

Shukla SK (2012) Synthesis of polyaniline grafted cellulose suitable for humidity sensing. Ind J Eng Mater Sci 19:417–420

Shukla SK (2013) Synthesis and characterization of polypyrrole grafted cellulose for humidity sensing. Int J Biol Macromol 62:531–536

Shukla SK, Parashar GK, Mishra AP, Misra P, Yadav BC, Shukla RK, Bali LM, Dubey GC (2004) Nano-like magnesium oxide films and its significance in optical fiber humidity sensor. Sens Actua B Chem 98:5–11

Shukla SK, Shukla SK, Govender PP, Agorku ES (2016) A resistive type humidity sensor based on crystalline tin oxide nanoparticles encapsulated in polyaniline matrix. Microchim Acta 183:573–580

Shukla SK, Kushwaha CS, Singh NB (2017) Recent developments in conducting polymer based composites for sensing devices. Mater Today Proc 4:5672–5681

Singh P, Shukla SK (2020) Advances in polyaniline-based nanocomposites. J Mater Sci 55:1331–1365

Singh K, Ohlan A, Pham VH, Balasubramaniyan RB, Varshney S, Jang J, Hur SH, Choi WM, Kumar M, Dhawan SK, Kong BS, Chung JS (2013a) Nanostructured graphene/Fe_3O_4 incorporated polyaniline as a high performance shield against electromagnetic pollution. Nanoscale 5:2411–2420

Singh A, Salmi Z, Joshi N, Jha P, Kumar A, Lecoq H, Lau S, Chehimi MM, Aswal DK, Gupta SK (2013b) Photo-induced synthesis of polypyrrole-silver nanocomposite films on N-(3-trimethoxysilylpropyl)pyrrole-modified biaxially oriented polyethylene terephthalate flexible substrates. RSC Adv 3:5506–5523

Singh P, Singh P, Shukla SK (2020) A structurally aligned nickel oxide encapsulated polypyrrole nanocomposite for hydrogen peroxide sensing. Dalton Trans 49:8744–8754

Singh P, Shekhar Kushwaha C, Shukla SK, Dubey GC (2018) Synthesis and humidity sensing property of α-Fe_2O_3 and polyaniline composite. Mater Today Proc 5:9118–9125

Singh P, Kushwaha CS, Singh VK, Dubey GC, Shukla SK (2021) Chemiresistive sensing of volatile ammonia over zinc oxide encapsulated polypyrrole based nanocomposite. Sens Actua B Chem 342:130042

Song N, Wang C, Lu X (2020) Rational design of hierarchical CoO/NiO nanosheets on conductive polypyrrole nanotubes for peroxidase mimicking and sensing application. ACS Sust Chem Eng 8:11069–11078

Sonker RK, Yadav BC (2017) Development of Fe_2O_3–PANI nanocomposite thin film based sensor for NO_2 detection. J Taiwan Inst Chem Eng 77:276–281

Su PG, Huang LN (2007) Humidity sensors based on TiO_2 nanoparticles/polypyrrole composite thin films. Sens Actua B Chem 123:501–507

Su PG, Peng YT (2014) Fabrication of a room-temperature H_2S gas sensor based on PPy/WO_3 nanocomposite films by in-situ photopolymerization. Sens Actua B Chem 193:637–643

Sunilkumar A, Manjunatha S, Chethan B, Ravikiran YT, Machappa T (2019) Polypyrrole–tantalum disulfide composite: an efficient material for fabrication of room temperature operable humidity sensor. Sens Actua A Phys 298:111593

Tai H, Jiang Y, Xie G, Yu J, Chen X (2007) Fabrication and gas sensitivity of polyaniline-titanium dioxide nanocomposite thin film. Sens Actua B Chem 125:644–650

Tiwari A, Gong S (2008) Electrochemical synthesis of chitosan-co-polyaniline/$WO_3 \cdot nH_2O$ composite electrode for amperometric detection of NO_2 gas. Electroanal 20:1775–1781

Vijayan A, Fuke M, Hawaldar R, Kulkarni M, Amalnerkar D, Aiyer RC (2008) Optical fibre based humidity sensor using co-polyaniline clad. Sens Actua B Chem 129:106–112

Waghuley SA, Yenorkar SM, Yawale SS, Yawale SP (2008) Application of chemically synthesized conducting polymer-polypyrrole as a carbon dioxide gas sensor. Sens Actua B Chem 128:366–373

Wang AL, Xu H, Feng JX, Ding LX, Tong YX, Li GR (2013) Design of Pd/PANI/Pd sandwich-structured nanotube array catalysts with special shape effects and synergistic effects for ethanol electro oxidation. J Am Chem Soc 135:10703–10709

Wang Q, Zhang J-M, Li S (2019) Minreview: recent advances in the development of gaseous and dissolved oxygen sensors. Instrument Sci Technol 47:19–50

We JH, Kim SJ, Cho BJ (2014) Hybrid composite of screen-printed inorganic thermoelectric film and organic conducting polymer for flexible thermoelectric power generator. Energy 73:506–512

Xia X, Chao D, Qi X, Xiong Q, Zhang Y, Tu J, Zhang H, Fan HJ (2013) Controllable growth of conducting polymers shell for constructing high-quality organic/inorganic core/shell nanostructures and their optical-electrochemical properties. Nano Let 13:4562–4568

Zhan C, Yu G, Lu Y, Wang L, Wujcik E, Wei S (2017) Conductive polymer nanocomposites: a critical review of modern advanced devices. J Mater Chem C 5:1569–1585

Zheng J, Li G, Ma X, Wang Y, Wu G, Cheng Y (2008) Polyaniline-TiO_2 nano-composite-based trimethylamine QCM sensor and its thermal behavior studies. Sens Actua B Chem 133:374–380

Chapter 21
Copper-Based Polymer Nanocomposites: Application as Sensors

Rama Kanwar Khangarot, Manisha Khandelwal, and Ravindra Singh

Abstract Metal-based nanoparticles (MNPs) like gold, silver, zinc and palladium etc., are frequently used for sensing applications. Among the MNPs, copper-based nanoparticles (CuNPs) are widely used for sensing applications due to their low cost, easy availability, and high electrical conductivity. The primary synthesis methods of CuNPs are direct mixing, in-situ polymerization, sol–gel method, and intercalation. Furthermore, the incorporation of copper-based nanoparticles (CuNPs) in polymeric matrices to generate copper-based polymer nanocomposites (CuPNCs) have revealed excellent sensing applications due to the low preparation cost, controllable pore size and surface chemistry, high stability and processability, and the possibility of achieving high tunable selectivity and sensitivity. The copper-based polymer nanocomposites (CuPNCs) act as different sensors, like gas, humidity, chemical, optical, and biosensors, especially glucose sensors, etc. Therefore, this chapter explores the promising aspects of copper-based polymer nanocomposites (CuPNCs) for various sensing applications. This chapter gives a clear overview of synthesis technologies of copper-based polymer nanocomposites (CuPNCs) utilized as sensing elements in various sensing applications. These include but are not limited to gas sensors, chemical sensors, optical sensors, humidity sensors, and biosensors.

Keywords Sensors · Polymers · Copper-based nanoparticles (CuNPs) · Nanocomposites · Synthesis

---

R. K. Khangarot (✉) · M. Khandelwal
Department of Chemistry, University College of Science, Mohanlal Sukhadia University, Udaipur, Rajasthan 313001, India
e-mail: ramakanwar@mlsu.ac.in

R. Singh
Department of Chemistry, Maharani Shri Jaya Government Post-Graduate College, Bharatpur, Rajasthan 321001, India

S. P. Singh et al. (eds.), *Metal Nanocomposites for Energy and Environmental Applications*, Energy, Environment, and Sustainability,
https://doi.org/10.1007/978-981-16-8599-6_21

21.1 Introduction

Polymers have been known as functional materials in various disciplines, including physics, chemistry, chemical engineering, and biomedicine. Despite the successful applications of polymers as a core material, some properties of polymers such as corrosion resistance, mechanical strength, temperature dependency, stability, specificity, and conductivity still require further perfections for advanced sensing applications. Numerous studies disclosed that the synergic effect of metal-based nanoparticles (MNPs) and polymer matrices well suppressed unadulterated polymer matrices' drawbacks (Hanemann and Szabó 2010). The synergism of metal-based nanoparticles (MNPs) and a wide range of polymers developed a new area of research for scientists (Fu et al. 2019; Xu et al. 2014). Metal-based polymer nanocomposites (MPNCs) have several interesting properties: tensile strength, elasticity, thermal stability, longevity, fast dynamic response, natural p-n characteristics, broad light absorption, selectivity, and sensitivity (Poyraz et al. 2013). Furthermore, the production of nanocomposites with controllable size, shape, and surface properties are essential for superior materials, including optoelectronic materials (Sengwa and Dhatarwal 2021), piezoelectric devices (Zhang et al. 2021), photocatalyst (Kazancioglu et al. 2021), and sensing devices (Sowmya et al. 2021), etc.

Meanwhile, the recent research in versatile areas like robotics engineering, artificial intelligence, biomedical, electronics, and food technology entails the advanced sensing technologies that coalesce miniaturization with high tactile sensitivity and low power consumption. Polymers such as carboxymethyl cellulose (CMC), chitosan (CS), polyaniline (PANI), polyindole (PIN), polyvinyl pyrrolidone (PVP), polypyrrole (Ppy), polyethylene oxide (PEO), phenothiazine (PTZ), polyolefin and, polyvinyl alcohol (PVA), etc., due to their conducting properties mainly used in sensors (Kaur et al. 2021).

Alternatively, metal-based nanoparticles (MNPs), for example, gold, silver, zinc and palladium etc., are frequently used for sensing applications. Among the metals used in nanomaterials, copper is a low-cost metal that is abundant on the earth. Besides metallic copper, its other forms, such as copper oxide, copper sulfide, and other copper-containing compounds, also have excellent properties. Due to their low cost, easy availability, and high electrical conductivity, copper-based nanoparticles (CuNPs) are widely used for sensing applications (Guo et al. 2016; Din and Rehan 2017). Copper oxide nanoparticles (CuONPs) are a p-type semiconductor and provide better analytical characteristics like low bandgap and wide linear range (Anik et al. 2019). Thus, the researchers showed a great interest in copper-based polymer nanocomposites (CuPNCs) obtained by copper-based nanoparticles (CuNPs) combined with semiconducting polymers. These nanocomposites are utilizing in sensors because of their better mechanical stability, sensitivity, portability, size compactness, and quick response time. They are simple and inexpensive to fabricate. They can function with low sample concentration at room temperature and low electricity consumption (Jawaid and Khan 2018).

This chapter provides a brief overview of the synthesis methods of copper-based polymer nanocomposites (CuPNCs). This chapter also enlightened the copper-based polymer nanocomposites (CuPNCs) applications as sensors, including gas sensors, humidity sensors, chemical sensors, optical sensors, and biosensors.

21.2 Synthesis of Copper-Based Polymer Nanocomposites (CuPNCs)

There are various methods to synthesize copper-based polymer nanocomposites (CuPNCs). The main problem in the fabrication of copper-based polymer nanocomposites (CuPNCs) is to get uniform and homogenous dispersion of nanoparticles in the polymer matrix. There are various techniques viz. spin coating (Jundale et al. 2013), electrochemical deposition (Ashokkumar et al. 2020), electrochemical polymerization (Liu et al. 2013), sol–gel method (Siddiqui et al. 2019), hydrothermal method (Gong et al. 2006), electrospinning (Li et al. 2006), solvent method (Liff et al. 2007), melt compounding (Lin et al. 2008), green synthesis (Pérez-Alvarez et al. 2021), co-precipitation method (Albokheet et al. 2021), pulsed laser ablation (Tommalieh 2021), mechano-chemical method (Maruthi et al. 2021), etc. have been utilized to synthesize copper-based polymer nanocomposites (CuPNCs). In short, we can classify these synthesis approaches into four broad domains, i.e., direct mixing, in-situ polymerization, sol–gel method, and Intercalation. Here, we will discuss these four techniques in detail.

21.2.1 Direct Mixing

A direct mixing method is a top-down approach to nanocomposites synthesis. In this method, nanocomposites are synthesized by direct mixing of nanofillers and polymer matrices. Polymer matrix and nanofiller are prepared distinctly and then assembled through the solution mixing method (in the presence of solvent) and melt mixing method (in the absence of solvent). For the solution mixing method, copper-based nanoparticles (CuNPs) and polymer are initially distinctly dispersed and dissolved in an appropriate solvent; later, the suspension of copper-based nanoparticles (CuNPs) is added into the polymer solution, followed by precipitation or casting of the mixture. This method is not facile for massive production due to the considerable amount of organic solvents is required for the reaction. Compared to the solution mixing method, the melt mixing method is a more practical and promising technique for the massive production of nanocomposites due to the absence of harmful organic solvents. The melt mixing method applies the shear force to the polymer melt, which plays a vital role in the dispersed phase size. Many scientists synthesized copper-based polymer nanocomposites (CuPNCs) by utilizing the direct mixing method. A

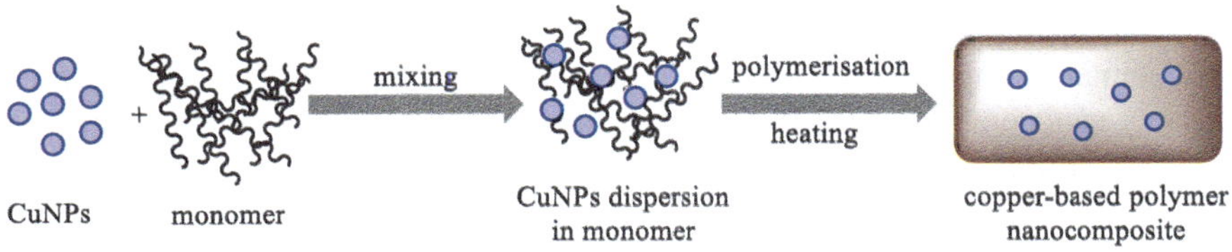

Fig. 21.1 Schematic diagram of direct mixing method of copper-based polymer nanocomposite (CuPNC) synthesis

schematic diagram of the direct mixing method of copper-based polymer nanocomposite (CuPNC) synthesis is shown in Fig. 21.1. For example, Bikiaris et al. prepared the high-density polyethylene (HDPE) nanocomposite with Cu-nanofillers using the melt mixing method (Bikiaris and Triantafyllidis 2013). Khil and co-workers synthesized the copper oxide nanoparticles (CuONPs) using the wet chemical method and hydroxyapatite nanoparticles-garlanded poly(DL-lactide-co-glycolide) ultrafine fibres using electrospinning technique. They then incorporated them together by the solution mixing method (Amina et al. 2014). CuNPs/poly-vinyl-methyl-ketone (PVMK) nanocomposites were synthesized via electrosynthesis of CuNPs followed by colloidal mixing of poly-vinyl-methyl-ketone solution (Cioffi et al. 2005).

Raza et al. synthesized the CuNPs nanofiller doped-polyvinyl alcohol nanocomposites using the solution casting technique (Aslam and Kalyar 2019). Later, they also reported a synthesis of tricomponent nanocomposite using a simple solvent evaporation technique. Initially, copper oxide/chitosan (CuO/CS) nanocomposite was synthesized then the resultant nanocomposite was incorporated into polyvinyl alcohol (PVA) matrix to manufacture a tricomponent composite (Aslam et al. 2021). The direct mixing method has an advantage over other methods owing to its ease of operation, suitability for massive production, and comparatively low cost.

21.2.2 *In-Situ Polymerization*

The in-situ polymerization method involves the one-step synthesis of nanocomposites, wherein the nanoparticles and polymers are produced in the same reaction vessel. The nanoparticles are initially dispersed in the monomer solution, and then the resultant mixture is polymerized via heat, radiation, or initiator. Nanocomposites' stability and size control can be improved by in-situ nanoparticle production in a polymer matrix. Nanocomposites' thermodynamic stability is prompted by the high surface energy of nanoparticles and strong interaction between nanoparticles and polymer matrix. A schematic diagram of in-situ polymerization of copper-based polymer nanocomposite (CuPNC) synthesis is shown in Fig. 21.2. For instance, Khodaei and Karegar synthesized copper nanoparticles/polyindole nanocomposites via a facile in-situ polymerization of indole using ammonium persulfate ($(NH_4)_2S_2O_8$) as the

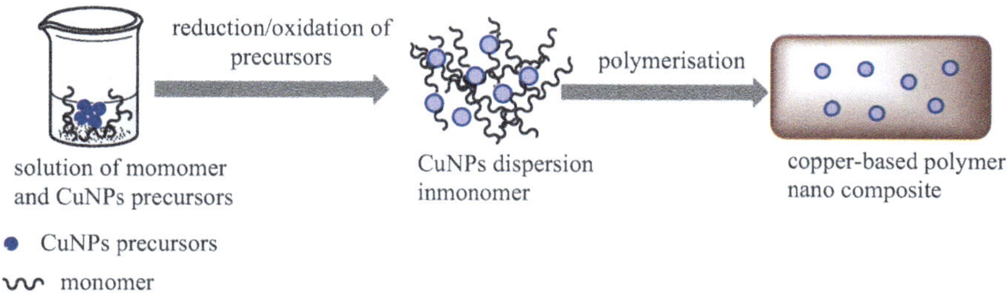

Fig. 21.2 Schematic diagram of in-situ polymerization of copper-based polymer nanocomposite (CuPNC) synthesis

oxidizing agent in an aqueous solution of copper nanoparticles (CuNPs) (Khodaei and Karegar 2021).

Consequently, Singh and Taunk prepared a CuS and polypyrrole (Ppy) hybrid nanocomposites. The polymerization was begun with the addition of $(NH_4)_2S_2O_8$ as an oxidizing agent in pyrrole monomer solution containing CuS nanoparticles (Singh and Taunk 2021). Gedanken and co-workers designed an MXene/poly(norepinephrine)/copper hybrid nanocomposite via the in-situ photopolymerization approach utilizing carbon dots as initiator (Das et al. 2021). Choudhury and colleagues synthesized the copper oxide/polyaniline nanofiber nanocomposite by the in-situ chemical deposition method (Ahmad et al. 2021). The primary key of the in-situ polymerization is the proper distribution of nanofillers into monomers without any agglomeration. At the same time, the drawback of the method is that the unreacted copper nanoparticles (CuNPs) may alter the properties of the nanocomposites (CuPNCs).

21.2.3 Sol–Gel Method

A sol–gel method is a bottom-up approach associated with two terms, sol and gel. The colloidal solution of inorganic nanoparticles is termed sol, and the polymer building block is termed gel. In this method, the polymer acts as a nucleating agent, allowing the inorganic nanoparticles to develop. With the growth of the nanoparticles, trapping of nanoparticles in polymer layers occurs, and nanocomposites obtain. The most significant advantage of this technology is the ability to tailor the microstructure of the hybrid material based on its target application. A schematic diagram of sol-gel method of copper-based polymer nanocomposite (CuPNC) synthesis is shown in Fig. 21.3. Using the sol–gel method, Guzman and co-workers synthesized copper nanoparticles (CuNPs) in polyvinyl chloride (PVC) polymer. A PVC plastisol solution is formed, followed by colloidal suspension of CuNPs (sol) and heating of PVC plastisol (gel) at 180 °C (Guzman et al. 2014). Siddique and co-workers synthesized the novel sheet-like copper oxide/sodium alginate bio-nano composite by the

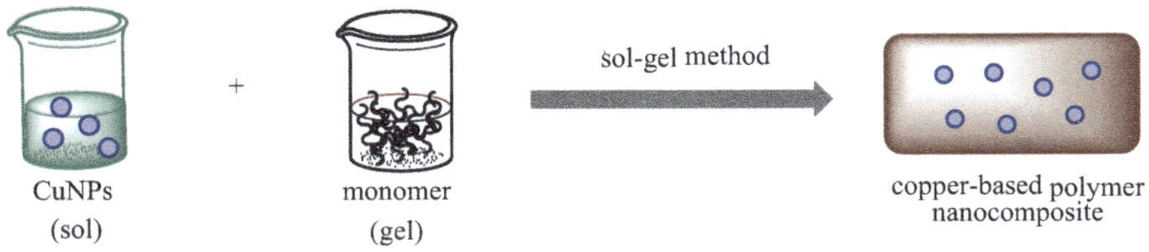

Fig. 21.3 Schematic diagram sol–gel method of copper-based polymer nanocomposite (CuPNC) synthesis

sol–gel method. The authors also illustrated the change in physical properties (dielectric constant, annealing temperature, conductivity) at the different concentrations of CuO and alginate (Siddiqui et al. 2019). Hammad et al. developed alumino-copper silicate nanocomposite using the sol–gel method (Hammad et al. 2021). The only shortcoming of the sol–gel process is the use of high temperature that damages the polymer matrix.

21.2.4 Intercalation

The dispersion of nanomaterials into the polymer matrix is a standard intercalation process. Nanomaterial incorporation in polymer matrices improves bulk properties such as shrinkage, flammability, and stiffness (George and Bhowmick 2008; Liu et al. 2021). Homogenous distribution of nanomaterial into the polymer matrix is essential for this method, and this can be achieved by using chemical and mechanical techniques. In chemical approach, nanoparticles are mixed in polymer, then the polymerization process occurs (Zeng et al. 2002; Koski et al. 2012). While, in the mechanical method, the different solutions of polymer and nanoparticles are mixed through mechanical devices. Over the time, this method is evolved as melt intercalation, in which the organic solvents are avoided. The melt intercalation method (Emam et al. 2021) follows a top-down approach and needs surface modification for homogenous distribution (Li et al. 2021). This method entails the annealing of polymer matrix at high temperature, then adding the filler and kneading the nanocomposite to ensure equal distribution (Xu et al. 2005). This method can also be used for those polymeric compounds which can't be obtained by in-situ polymerization. A schematic diagram of melt intercalation method of copper-based polymer nanocomposite (CuPNC) synthesis is shown in Fig. 21.4. Guo et al. synthesized graphene/polyaniline/copper oxide (PANI/CuO ternary) nanocomposites based on in-situ intercalation polymerization of PANI (Guo et al. 2021). Madhu and co-authors fabricated copper oxide/polyvinyl alcohol (CuO/PVA) nanocomposite via solution intercalation film casting method. Initially, the hydrothermal reaction occurred at 160 °C then heat treatment was given at 280 °C (Rao et al. 2015). However, the high temperature used for melting can damage the surface modification of the nanofillers.

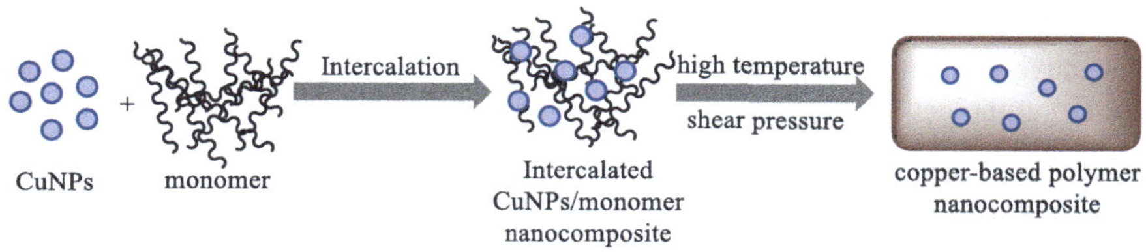

Fig. 21.4 Schematic diagram melt intercalation method of copper-based polymer nanocomposite (CuPNC) synthesis

21.3 Application of Copper-Based Polymer Nanocomposites (CuPNCs) as Sensors

Copper-based polymer nanocomposites (CuPNCs) have diverse applications in various areas, for example, environmental remediation (Chauhan et al. 2020), energy storage devices (Siwal et al. 2020), defense system (Peng et al. 2020), food industry (Shankar and Rhim 2016; Kausar 2020), catalytical chemistry (Zuraev et al. 2018), and sensors (Tong et al. 2017). A sensor is a device that senses some inputs (pressure, heat, moisture, light, sound, etc.) from its surroundings. It discovers the changes and induces the corresponding electrical signals into human-readable signals. (Chen et al. 2021). Due to advancements in fabrication technologies, affordable sensors with higher sensitivity and reliability have been made possible. Here we have discussed the sensing behaviour of copper-based polymer nanocomposites (CuPNCs) as different sensors, particularly in gas, humidity, chemical, optical, and biosensors, especially glucose sensors, etc.

21.3.1 Gas Sensors

During the past decades, different gas sensors have been fabricated involving diverse sensing methods and analytes. Gas sensors based on copper-based polymer nanocomposites (CuPNCs) have enhanced performance due to their unique electrical and optical properties, large surface-to-volume ratio, short response time, and high sensitivity (Wang et al. 2020a). Their exclusive properties allow the development of various types of gas sensors that can detect different gases such as ammonia (NH_3), hydrogen sulfide (H_2S), carbon dioxide (CO_2), chloroform ($CHCl_3$), and nitrogen dioxide (NO_2), etc. Many gases used in the industries are very harmful to human health that even their small leakage can damage civil life to havoc range (Naseem and King 2018; Yarandi et al. 2021; Wu et al. 2019; Lewis and Copley 2015; Chen et al. 2020). The performance of gas sensors can be evaluated by different parameters like selectivity, sensitivity, response time, reversibility or recovery time, fabrication cost, and stability (Yuan et al. 2019). For instance, we are discussing here some gas sensors fabricated by copper-based polymer nanocomposites (CuPNCs).

H_2S is a well-known poisonous gas in air pollution, and even a trace amount of it negatively affects the neurological system of human beings. This gas is an important target molecule for most gas sensors. In this context, Sarfraz et al. synthesized a printed H_2S gas sensor on paper substrate based on polyaniline/copper chloride (PANI/$CuCl_2$) nanocomposite. The fabricated sensor showed enhanced sensitivity towards H_2S gas (Sarfraz et al. 2013). Ayesh and co-workers developed a gas sensor using copper oxide nanoparticles (CuONPs), polyvinyl alcohol (PVA), and glycerol ionic liquid. At a very low temperature, the fabricated sensor showed high sensitivity towards H_2S gas at a relatively low concentration of 10 ppm. The sensitivity of the sensors depended on copper oxide nanoparticles (CuONPs) quantity and reaction temperature. The sensitivity is directly proportional to copper oxide nanoparticles (CuONPs) concentration, and the sensing behaviour was reversible, which enabled repeatable tests (Ayesh et al. 2016). Chen et al. studied the effect of polystyrene sulfonate/penicillamine/copper nanocomposites on H_2S gas. They incorporated this nanocomposites into a paper-based analytical device and found that this sensor could detect H_2S gas at very low sample volume (5 μL) (Chen et al. 2016). Cai and co-authors designed a DNA-Cu/Ag nanocomposite based novel fluorescence sensor. The proposed sensor showed high efficacy and sensitivity for H_2S containing blood samples (Ding et al. 2017). Lee's group studied another fluorescence sensor derived from cationic polyvinylpyrrolidone and copper. They observed that copper and sulfur had strong affinity towards each other. Selectivity of H_2S gas detection has remained a significant issue to date, owing to the greater amounts of interfering species from bio thiols. The sensor solved this problem and displayed high selectivity for H_2S gas over other anions and bithiols (Abd-Elaal et al. 2017). Du and colleagues designed a unique sensor that eliminated the interfering species and provided extraordinary results with a high affinity towards H_2S gas. This sensor was worked on two phenomena; (i) electrochemical (ii) photochemical. Due to combination of both phenomena fabricated polypyrrole (Ppy) and copper sulfide nanocrystallites (Cu_xS_y) sensors could be effectively used at the industrial level for the detection of H_2S gas with good precision and superior selectivity (Shang et al. 2019). Later, Greish and co-authors studied the sensing properties of chemoresistive gas sensors obtained by chitosan/copper oxide (CS/CuO) hybrid nanocomposite film. The authors discovered that the properties of sensor were improved due to a change in copper oxide nanoparticles (CuONPs) concentration. Large surface-to-volume ratio of the copper oxide nanoparticles (CuONPs) provided more reactive sites for the desorption of oxygen molecules. The fabricated sensor illustrated high repeatability up to 21 days with good sensitivity, quick response, and moderate reaction conditions (Ali et al. 2020). Another main shortcoming of the gas sensors is humidity dependency; humidity creates detrimental impacts on target gas molecules. To overcome this, Mahmoud et al. developed a low humidity dependent H_2S gas sensor that permitted its use for hands-on applications in a highly humid atmosphere. The fabricated carboxymethyl cellulose/copper oxide (CMC/CuO) nanocomposites illustrated quick response and high sensitivity towards H_2S gas at a low gas concentration of 15 ppm and at low working temperature (40 °C) (Hittini et al. 2020). In the same year, Kabel and colleagues synthesized an electrochemical sensor based on polypyrrole (Ppy) and copper oxide (CuO) that potentially detected H_2S gas at varying concentrations (Kabel et al. 2020).

Furthermore, NH_3 gas detection at a trace level is particularly critical, as it is widely employed in various industries, including chemical technologies, fertilizers, power plants, medical diagnosis, food processing, and environment protection. Ramgir and colleagues developed copper nanoparticles (CuNPs) functionalized polyaniline (PANI) nanocomposite, which had better response kinetics and superior sensitive response towards NH_3 gas at room temperature. The nanocomposite film detected NH_3 gas at less than 1 ppm concentration in a reversible manner. The doublet of nitrogen in the polymer backbone released the electrons when a reaction occurred between PANI and NH_3; the electron transfer increased the resistance of the gas sensor (Patil et al. 2015). The mechanism of the reaction is as follows:

$$PAH^+ + NH_3 \overset{\text{CuNPs}}{\rightleftharpoons} PA + NH_4^+$$

$$PA = \text{undoped repeated block of the PANI chains}$$

$$PAH^+ = \text{proton-doped repeated block of the PANI chains.}$$

Further, Naidu and co-workers synthesized copper complexed clay/poly-acrylic acid nanocomposites and examined their sensing properties towards NH_3 gas. The sensing properties were observed due to retention of NH_3 on fabricated nanocomposites by Lewis acid and Lewis base reaction between NH_3 and Cu ions (Liu et al. 2016). Liu and colleagues designed a colorimetric NH_3 gas sensor based on copper(II)-doped cellulose nanocrystals (Dai et al. 2017). Later, Ramesan and Dilsha fabricated a sensor based on polyaniline/phenothiazine/copper sulfide (PANI/PTZ/CuS) nanocomposites. Copper sulfide nanoparticles (CuSNPs) significantly improved the NH_3 sensing behaviour of the hybrid nanocomposites compared to pure PANI/PTZ hybrids. They illustrated that 10 wt% CuS nanocomposites showed good gas sensitivity and selectivity (Ramesan and Dilsha 2019). Ramesan and co-workers also fabricated another electrochemical sensor, Cu–Al_2O_3 reinforced polyindole (PIN) nanocomposites for NH_3 gas detection. In this case, they measured NH_3 gas in the 50–400 ppm concentration range and observed low response time. The Cu–Al_2O_3 composition strongly influenced the gas sensitivity of the nanocomposites film. It was inferred that PIN/Cu–Al_2O_3 nanocomposites were a suitable replacement as a sensor device owing to high thermal stability, dielectric constant and excellent sensitivity (Sankar et al. 2020). Ramesan's group further investigated the NH_3 gas sensing properties of Ppy/Cu–Al_2O_3 nanocomposites and discovered that sensitivity was increased with nanoparticles loading. It was also determined that the composite containing 5 wt% of the sample had the highest gas sensitivity (Menon et al. 2020). Zhang et al. employed an in-situ polymerization approach to obtain PANI/$CuFe_2O_4$ sensor for NH_3 gas detection. At 5 ppm NH_3, the response of PANI/$CuFe_2O_4$ sensor was very fast as compared to pure PANI and $CuFe_2O_4$ films. Good repeatability, high sensitivity, and fast recovery of the nanocomposites were attributed to the interface between p-n heterojunctions and cooperative effects (Wang et al. 2020a). Tariq and co-workers studied an NH_3 gas sensor made up of polyvinylpyrrolidone (PVP) and copper oxide nanoparticles (CuONPs). They observed that PVP/CuO nanocomposite 8 wt% content showed maximum selectivity and reproducibility towards NH_3 gas (Khan et al. 2020).

Riyazi and Araghi developed an interdigitated capacitive-type CO_2 gas sensor by in-situ chemical oxidative polymerization method with or without cationic surfactants. Under the exposure of CO_2 gas in the range of 0.1–1% at room temperature, the fabricated polypyrrole/copper phthalocyanine (Ppy/CuPc) nanocomposite exhibited good sensing performance. They found that fabricated nanocomposite had a vast interrelated network of Ppy nanofiber with large surface-to-volume ratio and porous structure (Riyazi and Araghi 2020).

Furthermore, monitoring volatile ethanol is crucial for both traffic safety and the ethanol production industry. Miyama and co-workers developed a CuS/polyolefine nanocomposites film as a p–n junction device used to detect ethanol. To prepare CuS, copper was electroless plated into porous polyolefine film and then sulfurized. However, this method is out-of-date nowadays because environmentally harmful chemicals were used in the reaction. They found that CuS/polyolefine nanocomposites only conduct high electric current when the CuS content is greater than 35 wt% (Miyama et al. 1990). Muradov et al. developed a sensor with copper sulfide (Cu_2S) in gelatin matrix and examined its sensitivity on vapours of different solvents such as acetone, methylethylketone, dichloroethane, and ethyl alcohol. It was observed that the fabricated sensor had the highest gas sensitivity and better percolation threshold towards ethyl alcohol which had the lowest molecular weight among the tested solvents (Muradov et al. 2007).

Athawale et al. synthesized a chloroform gas sensor with Cu/PANI hybrid nanocomposite thin films, which could detect chloroform vapours at < 100 ppm concentration range. The study showed that the sensor could discriminate between pure hexane and hexane/chloroform mixtures efficiently and selectively (Sharma et al. 2002).

Wang's group developed a NO_2 gas sensor prepared by copper phthalocyanine (CuPc) and polyvinyl alcohol (PVA) nanofibers. They found out that the average sensitivity was 829%/ppm, the detection limit was very low, less than 1 ppm, and the response time was less than10 min for detecting of NO_2 gas (Wang et al. 2020b).

21.3.2 Humidity Sensors

Humidity regulation is becoming extremely important for promoting human health (Guan et al. 2021) and boosting industrial processes (Tai et al. 2020). In the medical and chemical industries, measuring ambient humidity is critical. Humidity sensing is helpful in a various fields like civil engineering, aerospace engineering, corrosion analysis, food and agriculture industries, etc. (Sikarwar and Yadav 2015; Lee and Lee 2005). Humidity sensors are based on different inorganic and organic materials having other fabrication procedures and properties (Farahani et al. 2014).

Among them, copper-based polymer nanocomposites (CuPNCs) can provide all of the appealing features, including low response time, low power consumption, low cost, repeatability, reversibility, high sensitivity, and portability required for humidity sensors (Cichosz et al. 2018).

In humidity sensors, the device sensitivity is mainly determined by host polymer matrices. In 1993, Godovski et al. synthesized a humidity sensor derived from CuS nanoparticles embedded in polyvinyl alcohol (PVA). In this fabricated sensor, PVA matrix behaved as a water vapour receiver, while CuS was a conductor. They explained that on exposure of water vapours, the alteration in electrical conductivity of sensor occurred due to change in the dielectric permittivity of the system or due to swelling of PVA matrix. The PVA/CuS polymer nanocomposite percolation threshold was detected in the concentration range of 16–22 wt%, with a conductivity change from 10^{-9} to 10^{-2} $\Omega\ cm^{-1}$ (Godovski et al. 1993). Further, the in-situ polymer-inorganic solid-state reaction developed another humidity sensor based on polyphenylene sulfide (PPS), Cu_2S and Cu/Cu_2O. To test the humidity response, change in the resistance was measured as a function of relative humidity (RH). The sensitivity was changed by varying the polymer loading. The highest sensitivity was observed with the 1:1 molar ratio, zero sensitivity at 1:5 molar ratio of copper acetate and PPS, respectively, and then increased further by increased polymer loading (Adkar et al. 2011). Karimov et al. studied the humidity sensing property of CuO/poly-*N*-epoxypropylcarbazole. The authors examined the effect of humidity on the electrical properties of the nanocomposite films by calculating the capacitance and dissipation of the water vapour samples at two different frequencies of the applied voltage. They observed that the resistance and capacitance of the samples decreased by an increase in voltage frequency (Karimov et al. 2012). In 2017, Hashim's group successfully synthesized PVA/starch/CuO nanocomposites for sensing humidity at diverse temperatures and various copper oxide nanoparticles (CuONPs) concentrations. They found that electrical conductivity and absorbance increased with the concentration of copper oxide nanoparticles (CuONPs) while the band gap decreased. The electrical resistance of fabricated nanocomposites decreased with an increase in humidity concentration and temperature (Hashim et al. 2017). Later, Hadi and Hashim investigated an efficient humidity sensor based on carboxymethyl cellulose-starch-CuO ternary nanocomposite, exhibited outstanding sensing properties in the range of 60–90% RH (Hadi and Hashim 2017). Chani et al. synthesized chitosan/$CuMn_2O_4$ nanocomposites via an eco-friendly mechanical mixing technique. The developed nanocomposite's temperature and humidity sensing mechanism was based on impedance change. The fabricated temperature sensor had a sensing range of 21–80 °C, whereas the humidity sensor had 22–94% RH (Chani et al. 2019). Consequently, Yakubu et al. (2018), Singh and Shukla (2020) studied the humidity sensing properties of copper oxide/polyaniline (CuO/PANI) nanocomposites. They both discovered that fabricated nanocomposites had good humidity sensing ability. Singh and Shukla demonstrated the change in electrical resistance with humidity range 10–95%. The fabricated nanocomposites had a sensitivity 4.5 Ω/RH, recovery time 55 s, and response time 40 s. The resistance of the PANI and its CuO nanocomposites decreased with the increase in % RH. Further, Hashim and co-workers developed a humidity sensor based on polyvinyl alcohol/polyethylene oxide/copper oxide (PVA/PEO/CuO) nanocomposites. The prepared nanocomposites are flexible, inexpensive, sensitive and lightweight. At room temperature, the sensitivity values were ranged from 60.2% for (PVA/PEO) blend to 92.4% for 15 wt% copper oxide nanoparticles (CuONPs). They discovered that the electrical resistance decreased by increasing relative humidity and copper oxide nanoparticles (CuONPs) concentration (Hashim et al. 2019).

21.3.3 Chemical Sensors

In addition to humidity sensing and gas sensing, copper-based polymer nanocomposites (CuPNCs) are also utilized for sensing various chemicals. Khan et al. synthesized a nanohybrid by simple intercalation of CuO nano-sheets into polypropylene carbonate (PPC). The chemical sensing performance of PPC and CuO-PPC nanohybrid was examined using the current–voltage (I–V) technique utilizing nitrophenol as an organic pollutant. Nanohybrid showed 11.25 times higher sensitivity (4.50 μA cm^{-2} mM^{-1}) than pure PCC (0.40 μA cm^2 mM^{-1}) (Khan et al. 2012). Baraket and the authors reported a susceptible chemical sensor based on copper phthalocyanine-acrylate-polymer to detect phosphate ions. The developed chemical sensor showed good results for phosphate ion detection within the range of 10^{-10}–10^{-5} with a Nernstian sensitivity of 27.7 mV/decade compared to other interfering ions like chloride, sulfate, carbonate, and perchlorate (Barhoumi et al. 2017). A graphene/copper phthalocyanine/polyaniline nanocomposite based amperometric sensor was examined for detecting ascorbic acid. The effect of applied electric potential on the sensor was investigated in a neutral medium at various potentials ranging from 0 to + 0.6 V versus Ag/AgCl to obtain a high current response and avoid interference. Graphene and polyaniline (PANI) improved the electrocatalytic properties, conductivity, repeatability, and stability of nanocomposite by promoting electron transport of ascorbic acid at the electrode. Furthermore, the combination of PANI and copper phthalocyanine promoted the sensing properties. The efficiency of the fabricated sensor demonstrated a linear range from 5×10^{-7} M to 1.2×10^{-5} M, a low detection limit of 6.3×10^{-8} M (S/N = 3) and excellent sensitivity of 24.46 μA mM^{-1}. At the commercial level, this sensor gives good accuracy examining ascorbic acid in vitamin C tablets (Pakapongpan et al. 2014). Barhoumi and co-workers designed a novel, susceptible chemical sensor derived from ion-imprinted polyaniline matrix and copper nanoparticles (CuNPs). They investigated the sensing properties of fabricated nanocomposite on nitrate ions (NO_3^-). They observed that the linear range was from 1 M to 100 mM and a low detection limit for NO_3^- determination (Essousi et al. 2019).

21.3.4 Optical Sensors

Copper-based polymer nanocomposites (CuPNCs) are prospective materials for optical sensing applications owing to their unique optical characteristics and configurable band gap. Since it is difficult for chemical sensors to detect phenolic group selectively in the presence of other chemicals, Wang and Zheng proposed a fluorescent sensor based on chitosan (CS) and copper sulfide nanoparticles (CuSNPs) for phenolic group detection. The composite film exhibits high sensitivity towards the phenolic group, and a very low amount of phenolic group intensely increased the emission. The fluorescence enhancement of 1-naphthaphenol was the highest, followed by 2-isonaphthol and 2-amino phenol, respectively (Wang and Zheng 2016). Later, Belal et al. demonstrated the fluorescence sensor for kanamycin (aminoglycoside bactericidal antibiotic) using two splits of DNA aptamer and copper sulfide nanoparticles (CuSNPs). The kanamycin concentration in the range of 0.04–20 nM was shown to have a good linear relationship with fluorescence signal amplification (Belal et al. 2018). Moreover, Feng and colleagues reported a fluorescent "turn-on" conjugated polymer (poly(2,5-bis(4-(bis(pyridin-2-ylmethyl)amino)butoxyl)-1,4-phenylethynylene-1,4-poly(phenylene ethynylene)) (PPE-DPA) and Cu^{2+} based sensor to analyte hydrazine in aqueous solution. They concluded that the sensor detected hydrazine solution under optimized conditions with a low detection limit of 5.4×10^{-10} M (S/N = 3) and with a wide linear range from 0 to 70 nM (Hu et al. 2019).

21.3.5 Biosensors

Biosensors are analytical devices made up of biological sensing elements that can generate sensitive and selective analytical signals. The nanocomposites based bio-transducer converts the physiochemical changes caused by the interaction of nanocomposites with their host biological molecules, such as carbohydrates, proteins, enzymes, and antibodies, into readable or quantified electronic signals (Naresh and Lee 2021). In the current scenario, among various biological molecules sensing, glucose sensing is very crucial. We all know that diabetes is a long-term illness that causes metabolic problems (Maniruzzaman et al. 2020). The accurate determination of human glucose levels is helpful in preventing and treating diabetes. The number of diabetic patients is continuously increasing day by day. Therefore, it is essential to develop affordable and reliable technologies for the detection of glucose levels. Copper-based polymer nanocomposites (CuPNCs) are a promising material for non-enzymatic glucose sensing due to their properties like low-cost, high catalytic activity, and quickly modifiable behaviour (Chen et al. 2019). Non-enzymatic glucose sensors depend on oxidation and reduction of glucose. Wang and co-workers developed a non-enzymatic glucose sensor based on copper nanoparticles/poly(o-phenylenediamine) nanocomposites. They discovered that fabricated nanocomposites had response time

< 1 s, a low detection limit 0.25 μM and a linear range from 5.0 μM to 1.6 mM, which was an extremely high sensitivity for blood glucose detection (Liu et al. 2014). Nickel and copper oxide nanoparticles coated on PANI are attributed as a non-enzymatic glucose sensor with good sensitivity, a wide range of detection concentrations from 20 to 2500 μM, and a low detection limit of 2.0 μM (S/N = 3) at 0.6 V in alkaline medium (Ghanbari and Babaei 2016). Lu and colleagues synthesized the copper/poly(tetrafluoroethylene) nanocomposites based glucose sensor, revealing very high sensitivity up to 1417.1 $\mu A\ cm^{-2}\ mM^{-1}$ and a wide linear range from 0.01 to 1 mM of glucose concentration change (Tong et al. 2017). Bilal et al. reported the amperometric monitoring of glucose at hospitals with excellent selectivity via polyaniline/CuNi modified electrode (Bilal et al. 2018). Konakov et al. studied the formation of copper iodide/polypyrrole nanocomposites (CuI/Ppy) via facile one-pot chemical synthesis and found that fabricated nanocomposites were an excellent component for innovative biosensors (Konakov et al. 2021). Moreover, Radhakrishnan et al. designed a voltametric enzyme-free glucose sensor using Ppy/CuS/SiO_2 modified electrodes. They discovered that the synthesized sensor exhibited a speedy response time (< 0.5 s) and high durability (Radhakrishnan et al. 2020).

21.4 Conclusions and Outlook for the Future

Copper-based polymer nanocomposites (CuPNCs) have unique physiochemical characteristics by synergistic integration of the features of polymer matrix and copper-based nanofiller that open new avenues to improve the existing sensing technologies. In this chapter, the synthesis methods of copper-based polymer nanocomposites (CuPNCs) and their application as sensors have been comprehensively discussed. The nanocomposites possess many applications in different forms such as gas, humidity, chemical, optical, and biosensors.

Although substantial progress has been made in recent years; yet there are still numerous obstacles to overcome. Apart from the sensing applications discussed above, tremendous efforts have been made and continue to be made to design and fabricate new copper-based polymer nanocomposite (CuPNC) sensors. More steps are still required to investigate their potential as pressure sensors, biometric sensors, optical image sensors, chip sensors, and water pollution sensors, etc. We can expect even more remarkable sensor breakthroughs in the future.

References

Abd-Elaal AA, Tawfik SM, Lee Y-I (2017) Highly selective fluorescent probe based on new coordinated cationic polyvinylpyrrolidone for hydrogen sulfide sensing in aqueous solution. J Mol Liq 247:35–42

Adkar D, Hake A, Jadkar S, Adhyapak P, Mulik U, Amalnerkar D (2011) Humidity sensing performance of in-situ fabricated $Cu/Cu_2O/Cu_2S$-polymer nanocomposite via polyphenylene sulphide cyclisation route. J Nanosci Nanotechnol 11:6815–6822
Ahmad MW, Anand S, Fatima A, Yang D-J, Choudhury A (2021) Facile synthesis of copper oxide nanoparticles-decorated polyaniline nanofibers with enhanced electrochemical performance as supercapacitor electrode. Polym Adv Technol n/a
Albokheet W, Gouda M, Alfaiyz Y (2021) Removal of Methylene blue dye using copper oxide/carboxymethyl cellulose nanocomposite: kinetic, equilibrium and thermodynamic studies. Asian J Adv Res 1–18
Ali FIM, Mahmoud ST, Awwad F, Greish YE, Abu-Hani AFS (2020) Low power consumption and fast response H_2S gas sensor based on a chitosan-CuO hybrid nanocomposite thin film. Carbohydr Polym 236:116064
Amina M, Hassan MS, Al Musayeib NM, Amna T, Khil M-S (2014) Improved antibacterial activity of HAP garlanded PLGA ultrafine fibers incorporated with CuO: synthesis and characterization. J Sol-Gel Sci Technol 71:43–49
Anik Ü, Timur S, Dursun Z (2019) Metal organic frameworks in electrochemical and optical sensing platforms: a review. Microchim Acta 186:196
Ashokkumar SP, Vijeth H, Yesappa L, Niranjana M, Vandana M, Devendrappa H (2020) Electrochemically synthesized polyaniline/copper oxide nano composites: to study optical band gap and electrochemical performance for energy storage devices. Inorg Chem Commun 115:107865
Aslam M, Raza ZA, Siddique A (2021) Fabrication and chemo-physical characterization of CuO/chitosan nanocomposite-mediated tricomponent PVA films. Polym Bull 78:1955–1965
Aslam M, Kalyar M (2019) Investigation of structural and thermal properties of distinct nanofillers-doped PVA composite films. Polym Bull 76
Ayesh AI, Abu-Hani AFS, Mahmoud ST, Haik Y (2016) Selective H_2S sensor based on CuO nanoparticles embedded in organic membranes. Sens Actuators B Chem 231:593–600
Barhoumi L, Baraket A, Nooredeen NM, Ali MB, Abbas MN, Bausells J, Errachid A (2017) Silicon nitride capacitive chemical sensor for phosphate ion detection based on copper phthalocyanine–acrylate-polymer. Electroanalysis 29:1586–1595
Belal ASF, Ismail A, Elnaggar MM, Belal TS (2018) Click chemistry inspired copper sulphide nanoparticle-based fluorescence assay of kanamycin using DNA aptamer. Spectrochim Acta A Mol Biomol Spectrosc 205:48–54
Bikiaris D, Triantafyllidis K (2013) HDPE/Cu-nanofiber nanocomposites with enhanced antibacterial and oxygen barrier properties appropriate for food packaging applications. Mater Lett 93:1–4
Bilal S, Ullah W, Ali Shah A-H (2018) Polyaniline@CuNi nanocomposite: A highly selective, stable and efficient electrode material for binder free non-enzymatic glucose sensor. Electrochim Acta 284:382–391
Chani T, Karimov K, Khan S, Fatima N, Asiri AM (2016) Impedimetric humidity and temperature sensing properties of chitosan-$CuMn_2O_4$ spinel nanocomposite. Ceram Int 45
Chauhan D, Afreen S, Talreja N, Ashfaq M (2020) Multifunctional copper polymer-based nanocomposite for environmental and agricultural applications, pp 189–211
Chen P-C, Li Y-C, Ma J-Y, Huang J-Y, Chen C, Chang H-T (2016) Size-tunable copper nanocluster aggregates and their application in hydrogen sulfide sensing on paper-based devices. Sci Rep 6:24882
Chen H, Fan G, Zhao J, Qiu M, Sun P, Fu Y, Han D, Cui G (2019) A portable micro glucose sensor based on copper-based nanocomposite structure. New J Chem 43:7806–7813
Chen Q, Liu D, Cao B, Wang Z, Wang T, Wang B, Wang Z, Yang Y, Xiong X, Wang Y (2020) Enhanced photo degradation activity of organic pollutants contained in sewage through construction of a CuO/Ag composite nanostructure. J Electron Mater 49:2032–2039
Chen J, Zhu Y, Huang J, Zhang J, Pan D, Zhou J, Ryu JE, Umar A, Guo Z (2021) Advances in responsively conductive polymer composites and sensing applications. Polym Rev 61:157–193
Cichosz S, Masek A, Zaborski M (2018) Polymer-based sensors: a review. Polym Test 67:342–348

Cioffi N, Ditaranto N, Torsi L, Picca RA, De Giglio E, Sabbatini L, Novello L, Tantillo G, Bleve-Zacheo T, Zambonin PG (2005) Synthesis, analytical characterization and bioactivity of Ag and Cu nanoparticles embedded in poly-vinyl-methyl-ketone films. Anal Bioanal Chem 382:1912–1918

Dai S, Prempeh N, Liu D, Fan Y, Gu M, Chang Y (2017) Cholesteric film of Cu(II)-doped cellulose nanocrystals for colorimetric sensing of ammonia gas. Carbohydr Polym 174:531–539

Das P, Ganguly S, Saha A, Noked M, Margel S, Gedanken A (2021) Carbon-dots-initiated photopolymerization: an in situ synthetic approach for MXene/poly(norepinephrine)/copper hybrid and its application for mitigating water pollution. ACS Appl Mater Interfaces

Din MI, Rehan R (2017) Synthesis, characterization, and applications of copper nanoparticles. Anal Lett 50:50–62

Ding Y, Li X, Chen C, Ling J, Li W, Guo Y, Yan J, Zha L, Cai J (2017) A rapid evaluation of acute hydrogen sulfide poisoning in blood based on DNA-Cu/Ag nanocluster fluorescence probe. Sci Rep 7:9638

Emam HE, Ahmed HB, Abdelhameed RM (2021) Melt intercalation technique for synthesis of hetero-metallic@chitin bio-composite as recyclable catalyst for prothiofos hydrolysis. Carbohydr Polym 266:118163

Essousi H, Barhoumi H, Bibani M, Ktari N, Wendler F, Al-Hamry A, Kanoun O (2019) Ion-imprinted electrochemical sensor based on copper nanoparticles-polyaniline matrix for nitrate detection. J Sens 2019:1–14

Farahani H, Wagiran R, Hamidon MN (2014) Humidity sensors principle, mechanism, and fabrication technologies: a comprehensive review. Sensors 14

Fu S, Sun Z, Huang P, Li Y, Hu N (2019) Some basic aspects of polymer nanocomposites: a critical review. Nano Mater Sci 1:2–30

George JJ, Bhowmick AK (2008) Ethylene vinyl acetate/expanded graphite nanocomposites by solution intercalation: preparation, characterization and properties. J Mater Sci 43:702–708

Ghanbari K, Babaei Z (2016) Fabrication and characterization of non-enzymatic glucose sensor based on ternary NiO/CuO/polyaniline nanocomposite. Anal Biochem 498:37–46

Godovski DY, Sukharev VY, Volkov AV, Moskvina MA (1993) Absorption induced response of electrophysical characteristics of filled polymer-composite. J Phys Chem Solids 54:1613–1620

Gong J, Yu S-H, Qian H, Fei L (2006) Synthesis of copper/cross-linked poly(vinyl alcohol) (PVA) nanocables via a simple hydrothermal route. J Mater Chem 16

Guan X, Hou Z, Wu K, Zhao H, Liu S, Fei T, Zhang T (2021) Flexible humidity sensor based on modified cellulose paper. Sens Actuators B Chem 339:129879

Guo Y, Cao F, Lei X, Mang L, Cheng S, Song J (2016) Fluorescent copper nanoparticles: recent advances in synthesis and applications for sensing metal ions. Nanoscale 8:4852–4863

Guo S, Chen G, Wen J, Liu S, Qu J, He W, Song L, Zheng B (2021) In-situ intercalation polymerization synthesis and electromagnetic wave absorption property of graphene/PANI/CuO ternary composites. J Mater Sci Mater Electron 1–11

Guzman A, Arroyo J, Verde L, Rengifo J (2014) Synthesis and characterization of copper nanoparticles/polyvinyl chloride (Cu NPs/PVC) nanocomposites. Int Congr Sci Technol Metall Mater Sam–Conamet 2014 9:298–304

Hadi A, Hashim A (2017) Development of a new humidity sensor based on (Carboxymethyl cellulose–starch) blend with copper oxide nanoparticles. Ukr J Phys 62:1044–1049

Hammad A, Youssef A, Nahrawy A, Ali A, Hemdan B (2021) Sol-gel preparation of bioactive nanoporous (Al_2O_3:CuO:SiO_2): dielectric properties and wastewater decontamination. Int J Mater Eng Innov 12:37

Hanemann T, Szabó DV (2010) Polymer-nanoparticle composites: from synthesis to modern applications. Materials 3:3468–3517

Hashim A, Habeeb MA, Hadi A (2017) Synthesis of novel polyvinyl alcohol–starch-copper oxide nanocomposites for humidity sensors applications with different temperatures. Sens Lett 15:758–761

Hashim A, Al-Khafaji Y, Hadi A (2019) Synthesis and characterization of flexible resistive humidity sensors based on PVA/PEO/CuO nanocomposites. Trans Electr Electron Mater 20:530–536
Hittini W, Abu-Hani AF, Reddy N, Mahmoud ST (2020) Cellulose-copper oxide hybrid nanocomposites membranes for H_2S gas detection at low temperatures. Sci Rep 10:1–9
Hu Z, Yang T, Liu J, Zhang Z, Feng G (2019) Preparation and application of a highly sensitive conjugated polymer-copper (II) composite fluorescent sensor or detecting hydrazine in aqueous solution. Talanta 207:120203
Jawaid M, Khan MM (2018) Polymer-based nanocomposites for energy and environmental applications. Woodhead Publishing
Jundale DM, Navale ST, Khuspe GD, Dalavi DS, Patil PS, Patil VB (2013) Polyaniline–CuO hybrid nanocomposites: synthesis, structural, morphological, optical and electrical transport studies. J Mater Sci Mater Electron 24:3526–3535
Kabel K, Zaki T, Shabaan W (2020) Fabrication of H_2S gas sensor based on Ppy/CuO and Ppy/SnO_2 nanocomposites at room temperature. Egypt J Chem 63:2763–2774
Karimov S, Saleem M, Karieva Z, Mateen A, Chani T, Zafar Q (2012) Humidity sensing properties of Cu_2O-PEPC nanocomposite films. J Semicond 33
Kaur G, Kaur A, Kaur H (2021) Review on nanomaterials/conducting polymer based nanocomposites for the development of biosensors and electrochemical sensors. Polym-Plast Technol Mater. 60:502–519
Kausar A (2020) A review of high performance polymer nanocomposites for packaging applications in electronics and food industries. J Plast Film Sheeting 36:94–112
Kazancioglu EO, Aydin M, Arsu N (2021) Photochemical synthesis of nanocomposite thin films containing silver and gold nanoparticles with 2-thioxanthone thioacetic acid-dioxide and their role in photocatalytic degradation of methylene blue. Surf Interfaces 22:100793
Khan S, Rahman M, Akhtar K, Asiri AM, Alamry K, Seo J, Han H (2012) Copper oxide based polymer nanohybrid for chemical sensor applications. Int J Electrochem Sci 7
Khan HU, Tariq M, Shah M, Iqbal M, Jan MT (2020) Inquest of highly sensitive, selective and stable ammonia (NH_3) gas sensor: structural, morphological and gas sensing properties of polyvinylpyrrolidone (PVP)/CuO nanocomposite. Synth Met 268:116482
Khodaei MM, Karegar M (2021) Synthesis and characterization of copper nanoparticles stabilized with polyvinyl pyrrolidone and its performance on the conductivity and stability of polyindole. J Iran Chem Soc 18:863–872
Konakov AO, Dremova NN, Khodos II, Koch M, Zolotukhina EV, Silina YE (2021) One-pot synthesis of copper iodide-polypyrrole nanocomposites. Chemosensors 9
Koski K, Cha J, Reed B, Wessells C, Kong D, Cui Y (2012) High-density chemical intercalation of zero-valent copper into Bi_2Se_3 nanoribbons. J Am Chem Soc 134:7584–7587
Lee C-Y, Lee G-B (2005) Humidity sensors: a review. Sens Lett 3:1–15
Lewis RJ, Copley GB (2015) Chronic low-level hydrogen sulfide exposure and potential effects on human health: a review of the epidemiological evidence. Crit Rev Toxicol 45:93–123
Li Z, Huang H, Wang C (2006) Electrostatic forces induce poly(vinyl alcohol)-protected copper nanoparticles to form copper/poly(viny alcohol) nanocables via electrospinning. Macromol Rapid Commun 27:152–155
Li K, Bian S, Zhen W, Li H, Zhao L (2021) Performance, crystallization and rheological behavior of poly(lactic acid)/N-(2-hydroxyl) propyl-3-trimethyl ammonium chitosan chloride intercalated vermiculite grafted poly(acrylamide) nanocomposites. React Funct Polym 158:104791
Liff SM, Kumar N, McKinley GH (2007) High-performance elastomeric nanocomposites via solvent-exchange processing. Nat Mater 6:76–83
Lin B, Gelves GA, Haber JA, Pötschke P, Sundararaj U (2008) Electrical, morphological and rheological study of melt-mixed polystyrene/copper nanowire nanocomposites. Macromol Mater Eng 293:631–640
Liu A, Luong B, Kim JS, Kim B-K, Kim JS (2013) Synthesis and characterization of conducting polyaniline-copper composites. J Nanosci Nanotechnol 13:7728–7733

Liu L, Chen Y, Lv H, Wang G, Hu X, Wang C (2014) Construction of a non-enzymatic glucose sensor based on copper nanoparticles/poly(o-phenylenediamine) nanocomposites. J Solid State Electrochem 19:731–738

Liu E, Sarkar B, Wang L, Naidu R (2016) Copper-complexed clay/poly-acrylic acid composites: Extremely efficient adsorbents of ammonia gas. Appl Clay Sci 121–122:154–161

Liu K, Zhang X, Miao F, Wang Z, Zhang S, Zhang Y, Zhang P, Shao G (2021) In Situ electrochemical intercalation-induced phase transition to enhance catalytic performance for lithium-sulfur battery. Small 17:2100065

Maniruzzaman M, Rahman MJ, Ahammed B, Abedin MM (2020) Classification and prediction of diabetes disease using machine learning paradigm. Health Inf Sci Syst 8:7

Maruthi N, Faisal M, Raghavendra N, Prasanna BP, Manohara SR, Revanasiddappa M (2021) Anti-corrosive polyaniline-coated copper oxide (PANI/CuO) nanocomposites with tunable electrical properties for broadband electromagnetic interference shielding. Colloids Surf Physicochem Eng Asp 621:126611

Menon SS, Parvathi K, Ramesan MT (2020) Structural characterization, electrical properties and gas sensing applications of polypyrrole/Cu-Al_2O_3 hybrid nanocomposites. High Perform Polym 32:719–728

Miyama H, Yokoyama H, Nosaka Y (1990) Preparation and application of semiconductor–polymer composites. J Appl Polym Sci 41:2273–2283

Muradov MB, Nuriev MA, Eivazova GM (2007) Sensitivity of composites based on gelatin and nanoparticles Cu_2S and CdS to vapors of some organic compounds. Surf Eng Appl Electrochem 43:512–515

Naresh V, Lee N (2021) A review on biosensors and recent development of nanostructured materials-enabled biosensors. Sensors 21

Naseem S, King AJ (2018) Ammonia production in poultry houses can affect health of humans, birds, and the environment—techniques for its reduction during poultry production. Environ Sci Pollut Res 25:15269–15293

Pakapongpan S, Mensing J, Ditsayut P, Lomas T, Tuantranont A (2014) Highly selective electrochemical sensor for ascorbic acid based on a novel hybrid graphene-copper phthalocyanine-polyaniline nanocomposites. Electrochim Acta 133:294–301

Patil UV, Ramgir N, Karmakar N, Bhogale A, Debnath A, Aswal D, Gupta SK, Kothari DC (2015) Room temperature ammonia sensor based on copper nanoparticle intercalated polyaniline nanocomposite thin films. Appl Surf Sci 339

Peng T, Xiao R, Rong Z, Liu H, Hu Q, Wang S, Li X, Zhang J (2020) Polymer nanocomposite-based coatings for corrosion protection. Chem Asian J 15:3915–3941

Pérez-Alvarez M, Cadenas-Pliego G, Pérez-Camacho O, Comparán-Padilla VE, Cabello-Alvarado CJ, Saucedo-Salazar E (2021) Green synthesis of copper nanoparticles using cotton. Polymers 13

Poyraz S, Liu Z, Zhang L, Zhang X (2013) Conducting polymer—metal nanocomposites synthesis and their sensory applications. Curr Org Chem 17:2256–2267

Radhakrishnan S, Ganesan V, Kim J (2020) Voltammetric nonenzymatic sensing of glucose by using a porous nanohybrid composed of CuS@SiO_2 spheres and polypyrrole. Microchim Acta 187:260

Ramesan MT, Dilsha K (2019) Structural properties, conductivity, dielectric behaviour and gas sensing application of polyaniline/phenothiazine/copper sulphide blend nanocomposites. Mater Res Express 6:105328

Rao JK, Raizada A, Ganguly D, Mankad MM, Satayanarayana SV, Madhu GM (2015) Investigation of structural and electrical properties of novel CuO–PVA nanocomposite films. J Mater Sci 50:7064–7074

Riyazi S, Araghi M (2020) Performance of interdigitated capacitive-type CO_2 sensor based on polypyrrole/copper phthalocyanine nanocomposite. J Mater Sci Mater Electron 31

Sankar S, Naik AA, Anilkumar T, Ramesan MT (2020) Characterization, conductivity studies, dielectric properties, and gas sensing performance of in situ polymerized polyindole/copper alumina nanocomposites. J Appl Polym Sci 137:49145

Sarfraz J, Ihalainen P, Määttänen A, Peltonen J, Lindén M (2013) Printed hydrogen sulfide gas sensor on paper substrate based on polyaniline composite. Thin Solid Films 534:621–628

Sengwa RJ, Dhatarwal P (2021) PVA/MMT and (PVA/PVP)/MMT hybrid nanocomposites for broad-range radio frequency tunable nanodielectric applications. Mater Lett 299:130081

Shang H, Xu H, Jin L, Chen C, Song T, Wang C, Du Y (2019) Electrochemical-photoelectrochemical dual-mode sensing platform based on advanced Cu_9S_8/polypyrrole/ZIF-67 heterojunction nanohybrid for the robust and selective detection of hydrogen sulfide. Sens Actuators B Chem 301:127060

Shankar S, Rhim J-W (2016) Polymer nanocomposites for food packaging applications. In: Functional and physical properties of polymer nanocomposites, pp 29–55

Sharma S, Nirkhe C, Pethkar S, Athawale A (2002) Chloroform vapour sensor based on copper/polyaniline nanocomposite. Sens Actuators B Chem 85:131–136

Siddiqui V, Ansari A, Khan I, Akram M, Siddiqui W (2019) Sol-gel synthesis of copper (II) oxide/alginate (CuO/Alg) bio-nanocomposite and effects of rapid thermal annealing on its properties and structure. Mater Res Express 6

Sikarwar S, Yadav BC (2015) Opto-electronic humidity sensor: a review. Sens Actuators Phys 233:54–70. https://doi.org/10.1016/j.sna.2015.05.007

Singh P, Shukla SK (2020) Structurally optimized cupric oxide/polyaniline nanocomposites for efficient humidity sensing. Surf Interfaces 18:100410

Singh N, Taunk M (2021) In-situ chemical synthesis, microstructural, morphological and charge transport studies of polypyrrole-CuS hybrid nanocomposites. J Inorg Organomet Polym Mater 31:437–445

Siwal SS, Zhang Q, Devi N, Thakur VK (2020) Carbon-based polymer nanocomposite for high-performance energy storage applications. Polymers 12:505

Sowmya B, John A, Panda PK (2021) A review on metal-oxide based p–n and n–n heterostructured nano-materials for gas sensing applications. Sens Int 2:100085

Tai H, Duan Z, Wang Y, Wang S, Jiang Y (2020) Paper-based sensors for gas, humidity, and strain detections: a review. ACS Appl Mater Interfaces 12:31037–31053

Tommalieh MJ (2021) Gamma radiation assisted modification on electrical properties of polyvinyl pyrrolidone/polyethylene oxide blend doped by copper oxide nanoparticles. Radiat Phys Chem 179:109236

Tong Y, Xu J, Jiang H, Gao F, Lu Q (2017) One-step synthesis of novel Cu@polymer nanocomposites through a self-activated route and their application as nonenzymatic glucose sensors. Dalton Trans 46:9918–9924

Wang S, Zheng M (2016) In situ synthesis of chitosan-capped CuS nanoparticals film as an friendly fluorescence sensor for phenolic. Mater Technol 32:1–5

Wang L, Wang L, Yang G, Xie Q, Zhong S, Su X, Hou Y, Zhang B (2020b) Improvement of sensing properties for copper phthalocyanine sensors based on polymer nanofibers scaffolds. Langmuir 36:4532–4539

Wang X, Gong L, Zhang D, Fan X, Jin Y, Guo L (2002a) Room temperature ammonia gas sensor based on polyaniline/copper ferrite binary nanocomposites. Sens Actuators B Chem 322:128615

Wu S, Zhang L, Fan J, Zhou Y (2019) Dynamic risk analysis of hydrogen sulfide leakage for offshore natural gas wells in MPD phases. Process Saf Environ Prot 122:339–351

Xu J, Li RKY, Xu Y, Li L, Meng YZ (2005) Preparation of poly(propylene carbonate)/organo-vermiculite nanocomposites via direct melt intercalation. Eur Polym J 41:881–888

Xu P, Han X, Zhang B, Du Y, Wang H-L (2014) Multifunctional polymer–metal nanocomposites via direct chemical reduction by conjugated polymers. Chem Soc Rev 43:1349–1360

Yakubu IS, Muhammad U, A'isha AM (2018) Humidity sensing study of polyaniline/copper oxide nanocomposites. Int J Adv Acad Res 4

Yarandi MS, Mahdinia M, Barazandeh J, Soltanzadeh A (2021) Evaluation of the toxic effects of ammonia dispersion: consequence analysis of ammonia leakage in an industrial slaughterhouse. Med Gas Res 11:24–29

Yuan Z, Li R, Meng F, Zhang J, Zuo K, Han E (2019) Approaches to enhancing gas sensing properties: a review. Sensors 19:1495

Zeng Q, Wang D, Yu A, Lu M (2022) Synthesis of polymer-montmorillonite nanocomposites by in situ intercalative polymerization. Nanotechnology 13

Zhang C, Fan W, Wang S, Wang Q, Zhang Y, Dong K (2021) Recent progress of wearable Piezoelectric nanogenerators. ACS Appl Electron Mater

Zuraev AV, Grigoriev YV, Budevich VA, Ivashkevich OA (2018) Copper-polymer nanocomposite: an efficient catalyst for green Huisgen click synthesis. Tetrahedron Lett 59:1583–1586

Chapter 22
Stacked Stainless Steel Mesh with Iron Oxide Nanostructures as a Substrate for NO_x Emission Control of Diesel Engines

Sandeep Yadav, Piyush Avasthi, Viswanath Balakrishnan, and Atul Dhar

Abstract NO_x emissions caused by diesel engines are a major source of overall NO_x emissions. Selective Catalytic Reduction (SCR) is a well-known technology for controlling the emissions from clean-burning diesel engines. A substrate is the most important part of SCR and in this work, 1D nanostructures of iron oxide are grown over stainless steel (SS) mesh by thermal oxidation which is used as a catalyst support material on SS mesh-based substrate for SCR. Scanning electron microscopy (SEM) and X-ray diffraction (XRD) were used to confirm the growth of various phases of iron oxides on the surface of the SS mesh substrate. A noble metal and metal oxide-based catalyst; $NiFePdO_4$ prepared through a sol–gel auto-ignition method is considered as a catalyst for NO reduction. A new method has been developed for uniform catalyst coating along with the sample holder to hold the samples and to eliminate the gas bypassing during the NO reduction experiments. A circular disc-type stacked substrate structure is developed to achieve the maximum NO reduction efficiency with a minimum amount of catalyst. All the NO reduction experiments are performed in a lab-scale single zone heating furnace at 350 °C. For the catalyst loading of 3% catalyst, experiments were conducted with 4, 8, 16, and 24 stack mesh frames. The high conversion efficiency of 93% is achieved with 24 stacks which indicates the influence of simple design involving stacks in place of individual frames in enhancing the performance. This work investigates the feasibility of low-cost iron oxides as a catalyst support material for SCR of NO_x in diesel engines by making it compact, efficient, and affordable.

Keywords Selective catalytic reduction · Substrate · NOx conversion · Nanomaterials · Emission control technology

S. Yadav · P. Avasthi · V. Balakrishnan · A. Dhar (✉)
School of Engineering, Indian Institute of Technology Mandi, Mandi, Himachal Pradesh 175005, India
e-mail: add@iitmandi.ac.in

S. P. Singh et al. (eds.), *Metal Nanocomposites for Energy and Environmental Applications*, Energy, Environment, and Sustainability,
https://doi.org/10.1007/978-981-16-8599-6_22

Nomenclature

GHSV	Gas hourly space velocity
IOs	Iron oxides
NO_x	Oxides of nitrogen
PPM	Parts per million
SCR	Selective catalytic reduction
SEM	Scanning electron microscopy
SO_x	Oxides of sulphur
SS	Stainless steel
XRD	X-ray diffraction

22.1 Introduction

Environmental pollution produced by diesel engines is harmful to human health and can cause respiratory diseases. According to a report by the world health organization, nine persons out of every ten persons are breathing polluted air (World Health Organization 2019). NO_2 is a major environmental pollutant and component of smog and if the concentration of NO_2 is high in the air, it can irritate airways in the respiratory system of humans (Clean Air Technology Center 1999; Kagawa 2002; Nakomcic-Smaragdakis et al. 2014). Nitrogen oxides are mainly emitted into the atmosphere because of combustion processes in air-rich environments particularly through transport and industrial activities. Road transport is the major cause of mobile NO_x emissions and globally contributes to 40 to 70% of the NO_x emission. In mobiles sources, 85% of NO_x emissions are caused by diesel engines, mainly in the form of NO (Reşitoılu et al. 2015). NO, and NO_2 both are considered toxic, but NO_2 is five times more toxic as compared to NO. Moreover, NO_2 is not only an important air pollutant by itself but also reacts in the atmosphere to form ozone (O_3) and acid rain. Here the ozone is referred to as the tropospheric ozone that is, ozone in the ambient air that we breathe (Clean Air Technology Center 1999).

In the light of the above aspects, Selective Catalytic Reduction (SCR) is a proven active NOx emissions control technology system in which NOx emissions generated in the exhaust gas are adsorbed and then converted into environment-friendly natural components by the use of a reducing agent (Clean Air Technology Center 1999; Paridah et al. 2016; Guan et al. 2014; Masoudi et al. 2019; Wen et al. 2019). The reducing agent chemically reacts with exhaust and converts nitrogen oxides (NO_x) into nitrogen (N_2), water (H_2O), and small amounts of carbon dioxide (CO_2) and natural components of the air.

Strict emission norms are already enforced in major countries of the world (Vassallo 2018; European Union 2019; United Nations Environment Programme 2017; Sharif et al. 2019). In India, these emission norms are called Bharat stages (BS) which are equivalent to Euro norms. BS-VI is standardized with Euro 6 standards and

Fig. 22.1 NO_x emission levels of NO_x for different Bharat stages and its reduction from BS-1 to BS-6

has seen its implementation in the region of Delhi from 2018 and implemented nation-wide since April 2020. BS-VI requires a deterioration factor of 1.3, 1.3, and 1.15 for CO, HC, and NO_x, respectively compared to current Bharat Stage IV standards (Urdhwareshe 2018). NO_x emissions limits of each BS norm are shown in Fig. 22.1. Such stringent reduction in emissions requires radical technologies in combustion and post-treatment for both transport and power sector ICEs. This requires the additional cost for auto manufactures and simultaneously to the customers. In this regard, this work targets the optimization of expensive noble metal-based catalyst used for reduction of NO_x in SCR technology which can reduce the cost of overall SCR systems.

Currently, a single monolith has been founded as the best choice for the catalyst support used in commercial engines. It constitutes three different layers; substrate, washcoat, and catalyst materials and all play an important role along with the reducing agent in the NO_x reduction process. Most substrates that are being used for commercial engines are iron-chromium-aluminum alloys shaped in a cylindrical porous monolith (Robinson 2016; Stewart et al. 2018). The commonly used SCR catalysts are copper-zeolite (Cu-Z) but it needs further requirement of post-treatment in Diesel Oxidation Catalyst (DOC) (Cavataio et al. 2009; Kwak et al. 2010; Fedeyko 2009; Wang et al. 2020). DOC adds further cost to the system, but vanadium-based SCR solves this problem. In this study, we have used NH_3 as a reducing agent as it is most suitable in the NO_x adsorption in the SCR process (Guan et al. 2014; Theis and Lambert 2019). Urea-SCR over Cu-Z SCR has a relatively wide temperature range, better durability, and low NH_3 slip (Guan et al. 2014). Vanadium and titanium-based catalysts have solved most of the problems of Cu-Z-based catalysts, but the high cost associated with it is still the main concern in the field of SCR-based NO_x reduction (Cavataio et al. 2009; Vargas et al. 2007; Sagar et al. 2011; Maunula et al. 2013a, b; Choung et al. 2006; Rammelt et al. 2019; Christensen et al. 2019). Precious and noble metal-based catalysts like Pd, Rh, Pt, and Ti, etc. are used with cheap and non-noble metal like Ni, Fe, Co, and Cu, etc. (Xu et al. 2016; Heikens et al. 2019; Vedyagin et al. 2019; Chen et al. 2017; Resitoglu et al. 2019; Hamidzadeh et al. 2018). With nanofibers, the substrate in the form of the membrane can act as a catalyst if the substrate is catalytically active and porous enough (Dankeaw et al. 2019; Fujiwara et al. 2017). There is a need for research for the development of an affordable catalyst for NO_x reduction process without compromising on its durability and effectiveness. In this work Ni, Fe, and Pd based catalyst ($NiFePdO_4$) is used which was made with sol–gel auto-ignition method with ethylene glycol which is coated over high surface area washcoat material.

Washcoat material is used to increase the surface area and reduces the back pressure leading to high catalytic activity. Washcoat is sometimes the less considered layer of a substrate. It neither has any appealing factor of the noble metals nor the clearly shown part of the substrate. Nevertheless, washcoat plays a vital role in the efficient function of catalysts. The vital role of washcoat is to deliver more surface area for catalyst deposition. Coating bare substrate with catalytic active materials would create a catalyst that would have relatively low NO_x reduction activity. If the NO_x reduction efficiency needs to be controlled at a high level (such as BS-6) then it would take a larger size substrate for that application (Rice et al. 2010; Hirose et al. 2012; Pless et al. 2014; Rajadurai et al. 2006). Washcoating the substrate increases the surface area available for catalyst coating drastically over the bare substrate (Robinson 2016). This study shows the effect of IO washcoated substrate on the catalytic activity of noble metal and metal oxide-based catalysts. Catalyst is coated on IOs based high surface area washcoat which is grown by thermal oxidation of stainless steel 304-grade mesh. The performance of developed novel circular stacked SS substrate with IO washout for reduction of NO_x emissions for NH_3-SCR has been investigated. A $NiFePdO_4$ catalyst is tested for bare SS substrate and IO/SS substrate and NO conversion efficiency of 93% is achieved in 24 stacks of IO/SS mesh with 3% of catalyst loading. Outcomes of this work can be used for the development of cost-effective substrate for SCR systems providing high efficiency (~90%).

22.2 Materials and Method

22.2.1 Experimental Setup

Field emission scanning electron microscope (FESEM- FEI-Nova NanoSEM 450) has been used for characterizing the microstructures of catalyst coated IO nanostructures on SS mesh. X-ray diffraction has been carried out using the Rigaku Smart Lab diffractometer system. Experimental reactor setup consists of homemade CVD system using Thermo Scientific single-zone make tube furnace (Model Lindberg Blue M). Gas cylinders of NH_3 (5000 PPM), N_2 (99.999% pure), and NO (2500 PPM) were used for all the experiments and are coupled with a rotameter for the

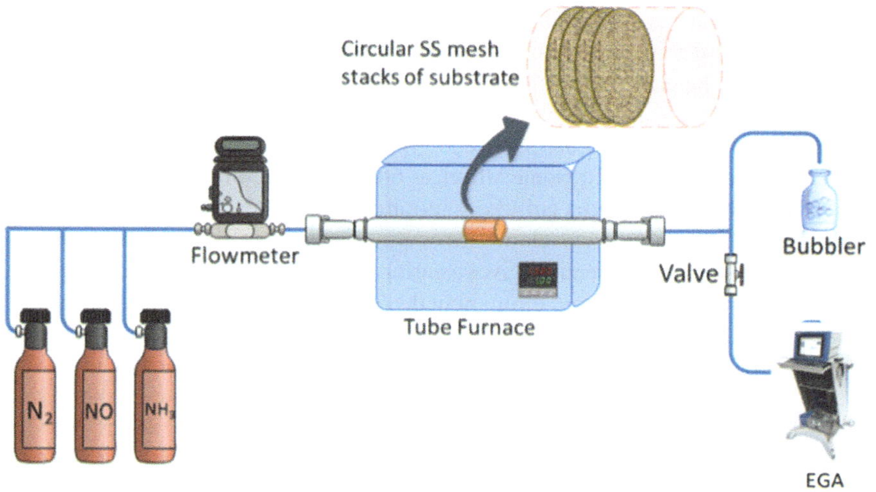

Fig. 22.2 Schematic diagram of reactor setup for NO_x reduction by SCR

flow control of the respective gases. All the experiments were done at 350 °C. A removable Tedlar gas sampling bag of a capacity of 10 L is used to collect the reduced gas after the catalytic converter. Uncollected gas was bypassed with the help of a bubbler connected to an exhaust fan. An exhaust Gas Analyzer (EGA) is used for NO_x characterization after the catalytic reduction. A schematic diagram of the reactor setup is shown in Fig. 22.2.

22.2.2 *Growth of 1-D Nanostructure Support*

Growth of iron oxide 1D nanostructure was carried out by controlled thermal oxidation of acid-treated SS mesh. In brief, SS mesh was etched in HCL (35%) for 30 min followed by acetone cleaning. The cleaned mesh was loaded into the furnace for heating up to 850 °C and then the temperature ramped down to 700 °C and maintained for 30 min in the presence of Argon gas with a flow rate of 200 ml/min. Scanning electron microscope (SEM) analysis was performed using FEI-Nova NanoSEM 450 equipped with energy-dispersive X-ray spectroscopy (EDS, AMETEK).

22.2.3 *Preparation of $NiFePdO_4$ Catalyst*

In this work noble metal and metal oxide-based catalyst is used for the NO_x reduction experiments. $NiFe_{1.93}Pd_{0.07}O_4$ was prepared through a sol–gel auto-ignition method with Nickel (II) Nitrate Hexahydrate, Ferric Nitrate Nonahydrate & Palladium (II)

Chloride as solutes and Ethylene Glycol as the solvent according to the procedure reported by Yang et al. (Xu et al. 2016). The number of ingredients required for the preparation of the catalyst was calculated using the basic principles of chemistry. The ingredients were accurately weighed using an electronic weighing machine and were added into a beaker. Later ethylene glycol was added to it. The mixture was thoroughly stirred in an ultrasonic stirrer at 60 °C till the solutes were completely dissolved in the solvent and a gel-like solution is prepared.

A new coating method is developed for coating the catalyst solution on the metallic substrates (samples). This method gives a uniform coating of catalyst on the substrates as the catalyst solution is stirred throughout the process. Metallic catalyst particles are not able to sit at the bottom of the beaker during the coating process. Agglomeration of catalyst particles over the substrate can be avoided by this method. The catalyst solution was heated at 60 °C for one hour and then circular substrates were dipped into it. Substrates are dried in a hot air oven for 15 min. This dipping and drying continue till the required amount of catalyst coating is achieved and once it is achieved, a sample is left for 2 h in a hot oven at 120 °C. Finally, the substrates are calcined in a muffle furnace at 600 °C for 4 h.

22.2.4 NO Reduction Testing Method

Placing circular stacked samples directly in the quartz tube is not feasible as there is some gap through which gas may bypass without having any contact with a substrate as shown in Figs. 22.3a, b. To overcome this gas bypassing, we have developed a stainless-steel non-reacting sample holder to place the substrate without having any gap as shown in Fig. 22.3c. The process of experimenting with a prepared sample of catalyst coated substrate starts with loading the circular stacks coated with catalyst within the sample holder in the quartz tube of the furnace as shown in Fig. 22.3d, e. The quartz tube is placed in the furnace and the sample is placed in the tube just above the thermocouple in the furnace. Temperature is set to heat at 350 °C. As the heating is turned ON, the flow of N_2 is turned ON for the purging, till the temperature of the furnace is reached 350 °C. When the temperature of the furnace reaches 350 °C, the flow of N_2 is turned OFF and the flow of NO and NH_3 is turned ON with the flow rate of 500 ml/min and 250 ml/min, respectively. The reduction of NO and NH_3 mixture takes place as it passes through the catalyst coated substrate gas and then this mixture is collected in the Tedlar bag for the NO characterization after the reduction in the EGA.

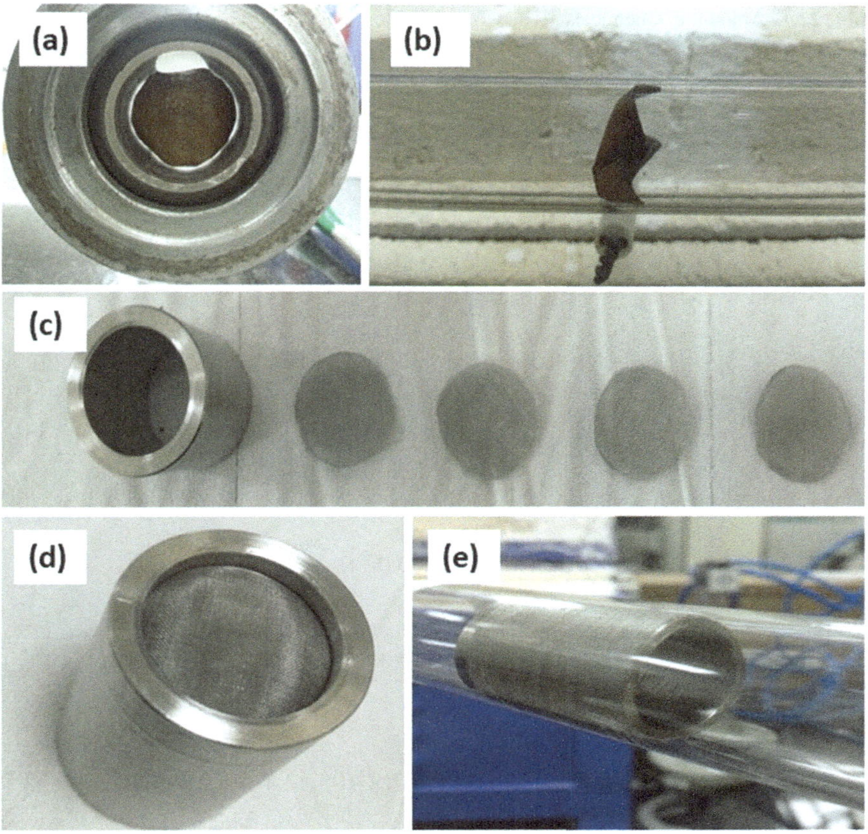

Fig. 22.3 Sample holder for testing samples in the reactor setup. **a** and **b** placing of samples in the quartz tube before sample holder, **c** sample holder with samples, **d** sample holder containing samples, **e** sample holder containing samples placed in the quartz tube

22.2.5 Calculation of the Number of Ingredients for the Preparation of Catalyst

The parent formula of the catalyst selected for this work is of the form $NiFe_{2-x}Pd_xO_4$. By substituting X = 0.07 in the formula, we get $\mathbf{NiFe_{1.93}Pd_{0.07}O_4}$. In 1 mol of the $\mathbf{NiFe_{1.93}Pd_{0.07}O_4}$ catalyst, there is 1 mol of Ni, 1.93 mol of Fe, 0.07 mol of Pd, and 4 mol of O

where,

$$\text{Ni} = \text{Nickel (Atomic Wt.} = 58.6934\text{ u)} \tag{22.1}$$

Table 22.1 Amount of Chemicals needed to prepare 15gm of catalyst

S. No.	Name of chemical	Chemical formula	Requirement
1	Nickel (II) Nitrate Hexahydrate	$Ni(NO_3)_2 \cdot 6H_2O$	18.70 g
	Ferric Nitrate Nonahydrate	$Fe(NO_3)_3 \cdot 9H_2O$	49.13 g
3	Palladium (II) Chloride	$PdCl_2$	0.781 g
4	Ethylene Glycol	$(CH_2OH)_2$	100 ml

$$\text{Fe} = \text{Iron (Atomic Wt.} = 55.845\ \text{u)} \tag{22.2}$$

$$\text{Pd} = \text{Palladium (Atomic Wt.} = 106.42\ \text{u} \tag{22.3}$$

$$\text{O} = \text{Oxygen (Atomic Wt.} = 15.994\ \text{u)} \tag{22.4}$$

The Molecular Wt. (M) of the catalyst is thus given by:

$$\text{M} = \text{A(Ni)} + 1.93 \times \text{A(Fe)} + 0.07 \times \text{A(Pd)} + 4 \times \text{A(O)}$$

where,

A (Ni), A (Fe), A (Pd), A (O) respectively correspond to the Atomic Wt. of Ni, Fe, Pd & O.

So,

$$M = 58.6934 + 1.93 \times 55.845 + 0.07 \times 106.42 + 4 \times 15.994 \tag{22.5}$$

After solving Eqs. (1) to (5)

$$\text{M} = 237.921\ \text{u}$$

Now, for 15 g of the $NiFe_{1.93}Pd_{0.07}O_4$ catalyst, we have to first calculate the number of ingredients (Ni, Fe, and Pd) required for making 15 g of the catalyst. The amount of oxygen is intentionally ignored because the weight of oxygen is self-adjusting. Table 22.1 shows the required chemicals for making 15gm of catalyst for the NO reduction through SCR.

22.2.6 *Estimation of Catalytic NO_x Reduction Efficiency*

All the catalytic reduction experiments are performed under a constant temperature condition of 350 °C. Catalytic NO reduction is defined as the ratio of reduced NO (in ppm) to the original NO (in ppm) over 100 scales as defined by Eq. (22.1) (Chen et al. 2017; Zha et al. 2018).

$$\text{Catalytic NO reduction } (\%) = \frac{(X - Y)}{X} \times 100$$

where,

X NO at the input (in ppm)
Y NO after catalytic reduction (in ppm).

22.2.7 *Estimation of Catalytic Coating*

Catalytic Coating is defined as the ratio of the weight of the catalyst coated on a substrate to the weight of the uncoated substrate.

$$\text{Catalytic coating } (\%) = \frac{(W1 - W2)}{W1} \times 100$$

where,

W1 Weight of coated substrate
W2 Weight of uncoated substrate.

Initially, the sample is directly put into the quartz tube as shown in Fig. 22.3a, b. But there is a gap between the periphery of the sample and the internal diameter of the quartz tube as can be seen from Fig. 22.3a. Due to this gap, gases are bypassing without passing through the sample which is increasing the NO_x levels in the output. This issue has been solved using a sample holder. As shown in Fig. 22.3c, d in which circular-shaped stacks can easily be put, and then the whole sample holder is put into the quartz tube (figure e). This solves the problem of gas bypassing. Substrate samples are made in the shape of circular stacks as shown in Fig. 22.3c. These types of stacks of substrate samples are placed in the sample holder with having some gap in between each stack. This gap increases the residence time of the gas for the catalytic activity which subsequently increases the NO_x conversion.

22.3 Results and Discussions

22.3.1 *Characterization of Support and Catalyst*

Characterization of support and catalyst was performed to confirm the growth and presence of a catalyst. Scanning electron microscopy (SEM) confirms the good amount of growth of iron oxide on the SS mesh. Scanning Electron Microscopy (SEM) images as shown in Fig. 22.4a, b are the images before having growth of the iron oxides that is bare mesh. SEM images Fig. 22.4c, d are the images after the growth of the iron oxides over SS mesh. Figure 22.4d clearly shows the wire-like

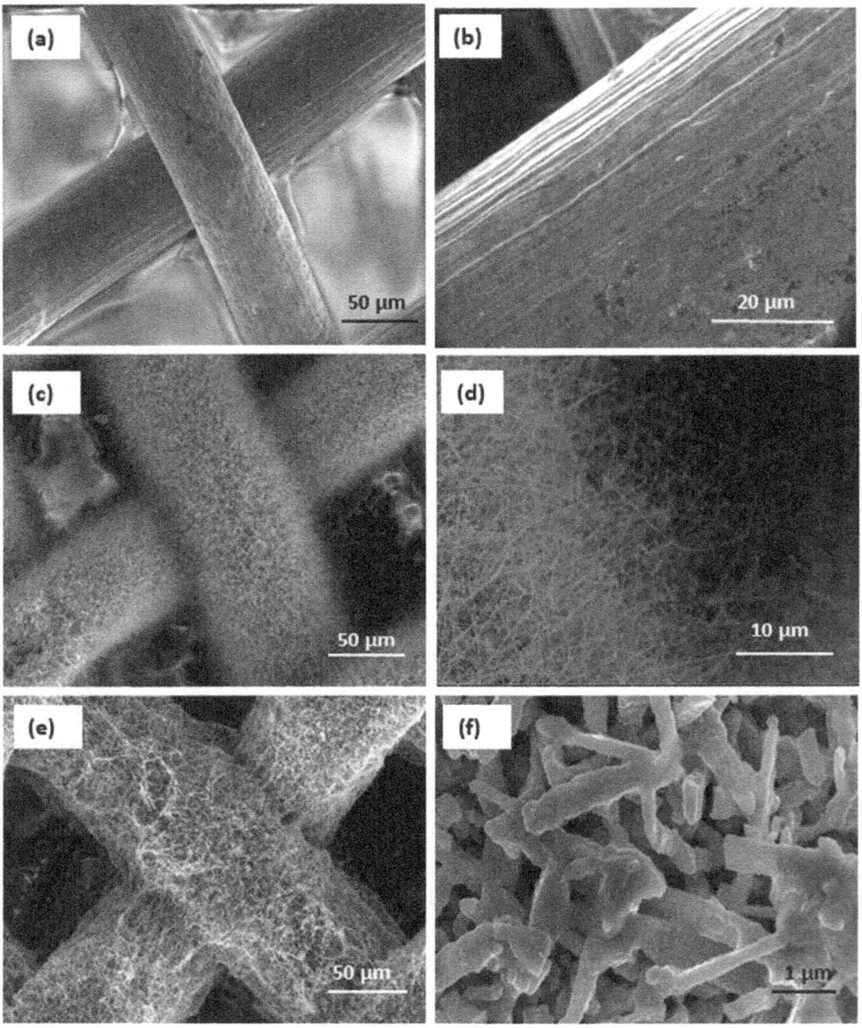

Fig. 22.4 SEM images of samples: **a** and **b** Bare SS mesh without the growth of iron oxides, **c**, **d** SS mesh with iron oxides grown over it, and **e**, **f** SS mesh with iron oxide grown over it coated with a catalyst

growth of iron oxides over the SS mesh. This confirms the growth of iron oxides over SS mesh. Figures 22.4e, f are the SEM images of iron oxide grown over SS mesh with a catalyst coated over it. This sample was used for the NO reduction characterization experiments. Powder XRD was carried out for finding the phases present in the prepared catalyst sample. Nickel, Palladium, Iron, and their oxides in the catalyst sample of $NiFePdO_4$ and the patterns were collected on X-ray diffraction, which was recorded using Rigaku Smartlab X-ray diffractometer with Cu Kα radiation at a scan speed of 1°/min. The prepared catalyst exhibited similar XRD peaks as reported by Yang et al. The Pd^{2+} doping level did not cause any significant change in the crystallinity of the samples, and the strong XRD reflection confirmed the good crystallization of the samples. A sharp peak at $2\Theta = 35.9°$ was also reported which corresponds to the NiO phase.

XRD measurements are performed for the confirmation of the presence of catalyst over the substrate. Figure 22.5a shows they are peaks corresponding to Palladium (Pd), Iron (Fe), Nickel (Ni), and their oxides, which confirms the mixed phase of catalyst. For further confirmation, XRD characterization for substrate grown with iron oxide is performed to find the phases of iron oxide present in it. In Fig. 22.5b, peaks of iron oxides shows that the mixed phase of iron oxides is found on the SS mesh in the form of Fe_2O_3 and Fe_3O_4. Majority of the peaks are of Fe_3O_4 in mixed phase of iron oxide. This confirms the proper growth of the iron oxides on the SS mesh.

22.3.2 Analysis and Optimization of NO Reduction Results

Initially, the experiments are done for optimizing the catalyst loading with different amounts of loadings (0.2%, 1%, and 3%). It is found that the sample with 3% catalyst loading was effective in reducing the NO. Three samples (all 8 stacks mesh frame each and 3% of catalyst loading) are tested i.e. bare SS mesh (bare SS), SS mesh with catalyst $NiFePdO_4$ coated over it (SS-$NiFePdO_4$) and SS mesh with IOs grown over it and catalyst $NiFePdO_4$ coated over it (SS-IO-$NiFePdO_4$) as shown in Fig. 22.6. It is found that 32%, 47%, and 58% of catalytic NO reduction are achieved, respectively. It is observed that SS-IO-$NiFePdO_4$ gives more NO reduction than the SS-$NiFePdO_4$, which shows that the IOs are playing a positive role as a substrate. NO reduction efficiency is increased because of an increase in total surface area due to the growth of IOs over SS mesh. This increase in surface area provides more reaction surface (as compared to bare mesh) sites for NO reduction process. This shows the significance of newly developed low-cost and high surface washcoat material as a good alternative for the substrate.

Further experiments are done for the optimization of a number of stacks with the optimized catalyst loading of 3% by measuring NO reduction with SS-IO-$NiFePdO_4$ catalyst at 0.2%, 1%, and 3% catalyst coating and four different numbers of stacks i.e. 4, 8, 16 and 24. At 1% of catalyst loading SS-IO-$NiFePdO_4$ shows higher catalytic NO reduction in all the 4 stacks, 8 stacks, and 16 stacks samples i.e. 38%, 43%, and 51%, respectively as compared to 35%, 41%, and 48% respectively which is shown by SS-$NiFePdO_4$. This is due to more surface area and uniform catalyst coating ability of iron oxides grown over SS. Most importantly it shows higher NO reduction efficiency at this low loading of 1% because of the better catalyst grasping ability of iron oxides. Figure 22.7 shows the comparison of NO reduction efficiencies of 4 stacks, 8 stacks, and 16 stacks samples of SS-$NiFePdO_4$ and SS-IO-$NiFePdO_4$ at 1% catalyst loading. This confirms that iron oxide-based catalyst support is effective in enhancing NO

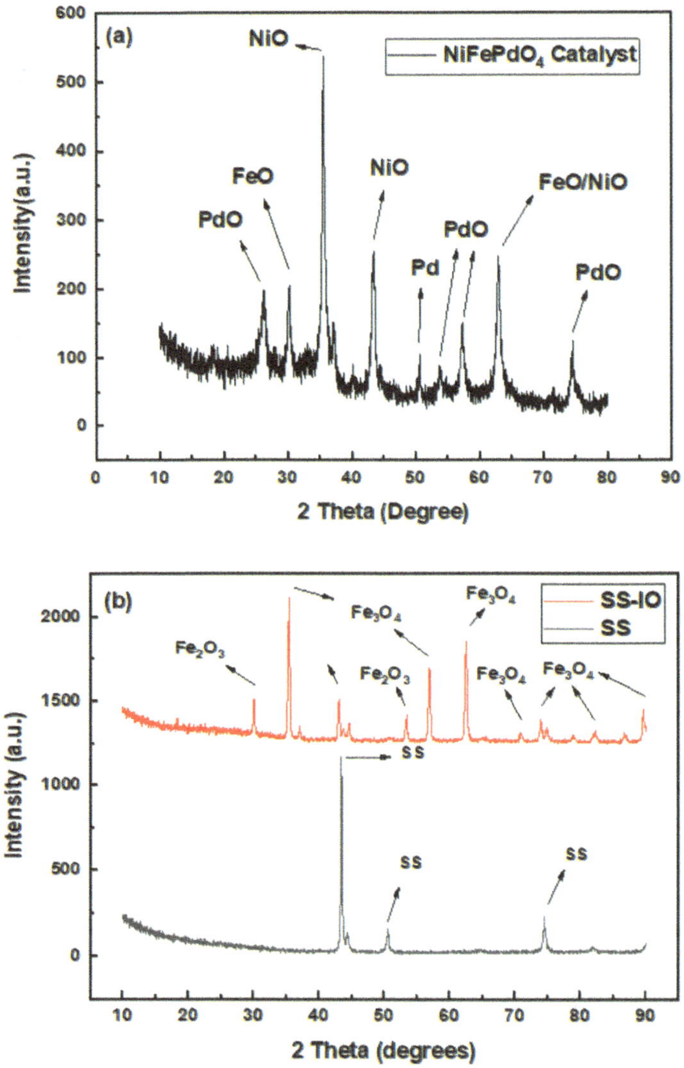

Fig. 22.5 X-ray diffraction pattern of **a** patterns of catalyst $NiFePdO_4$ and **b** bare SS (Black) and iron oxides grown over SS mesh (Red)

reduction. Now catalyst loading is increased to 3% and a number of stacks to 24 to get the target result of more than 90% NO reduction efficiency. Substrates with 3% of catalyst loading with the samples of SS-IO-$NiFePdO_4$ with 4, 8, 16, and 24 stacks achieved the catalytic NO reduction of 51%, 58%, 82%, and 93% respectively. Here, we have adopted the circular stacked substrate structure to maintain uniform gap between the stacks. This increases the residence time during the NO reduction

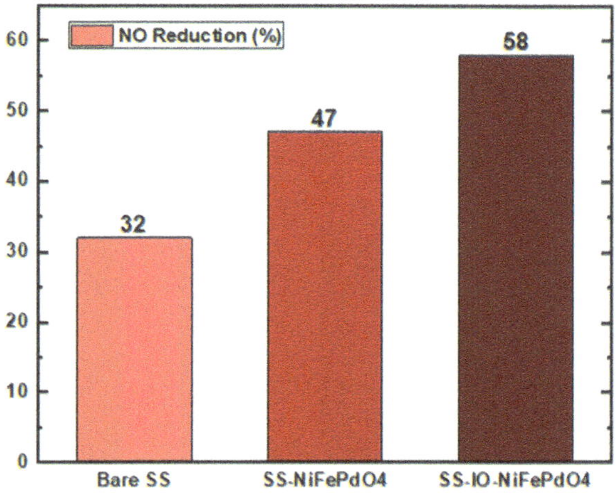

Fig. 22.6 Bar graph of NO reduction efficiency for various samples tested for 8 stacks mesh frame with 3% catalyst loading

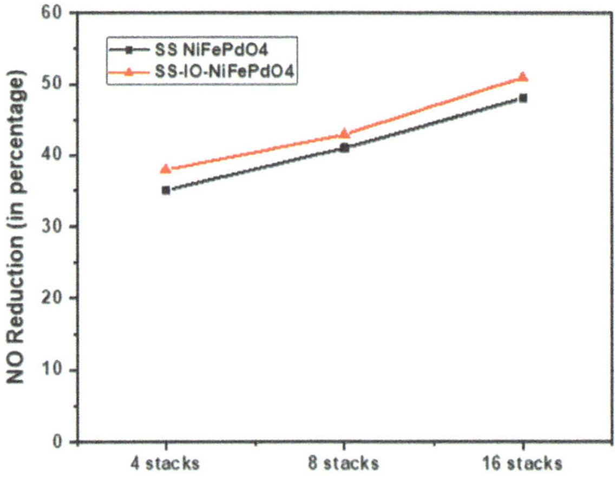

Fig. 22.7 Comparison of NO reduction efficiencies of 4 stacks, 8 stacks, and 16 stacks samples of $SS\text{-}NiFePdO_4$ and $SS\text{-}IO\text{-}NiFePdO_4$ at 1% of catalyst loading

process and provides more contact time with a catalyst which subsequently increases the NO reduction. In a honeycomb substrate structure, it is not possible to increase the residence time by adjusting the separation between stacks. The possibility to control the residence time is the main advantage of this low-cost flexible substrate investigated in this study.

Optimization for the number of stacks has been done to minimize the required catalyst without affecting the catalytic efficiency. Figure 22.8 shows the comparison of NO reduction efficiencies of 4 stacks, 8 stacks, 16 stacks, and 24 stacks samples

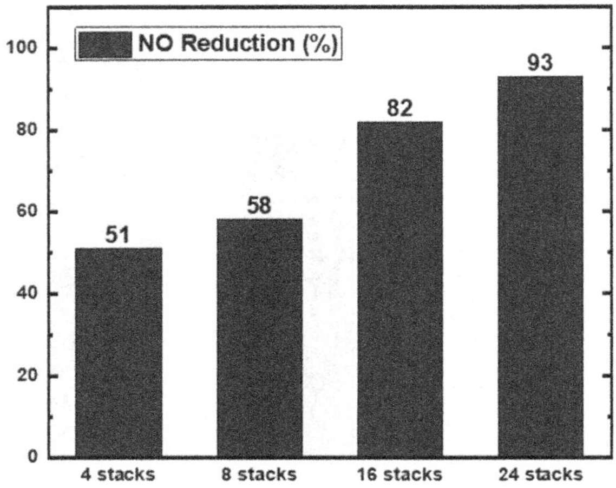

Fig. 22.8 Comparison of NO reduction efficiencies of 4 stacks, 8 stacks, 16 stacks, and 24 stacks mesh frame SS-IO-$NiFePdO_4$ at 3% of catalyst loading

of SS-IO-$NiFePdO_4$ at 3% of catalyst loading. Also, the number of stacks is a very important criterion as increasing the number of stacks from 4 to 24 stacks increases the NO reduction from 51 to 93%. As the number of stacks increased from 4 to 24, residence time also increased which provides higher NO catalytic reduction efficiency.

All the results shown in Figs. 22.6 and 22.7 may have an error of $\pm$ 5% based on the accuracy of EGA. This increase in NO reduction becomes more effective without the need of increasing the amount of catalyst coating beyond 3% for all the stacks. Although the total amount of catalyst would increase with the increase in the number of SS mesh stacks, the use of a low catalyst coating of 3% ensures better distribution of particles without losing their active surface area by aggregation. The developed method involving catalyst coating on SS mesh with IO nanostructures and their assembly into stacks demonstrates high NO reduction efficiency for diesel engines. This works investigated that it is more effective to have washcoat material grown over the substrate rather than the substrate without any washcoat material. Washcoat material and circular stacked design decrease the need for more catalysts which eventually decreases the overall cost of the SCR system.

22.4 Conclusions

Low-cost thermal oxidation of SS mesh is used for the growth of IOs which is used as a washcoat material in this work of NO reduction. $NiFePdO_4$ catalyst is coated on IO nanostructures with a high surface area may ensure uniform distribution of catalysts leading to high NO reduction efficiency. A circular stacked substrate structure is used for increasing the residence time over catalyst for effective NO reduction. Experimental investigation shows that 24 stacks of SS-IO-$NiFePdO_4$, with 3% of catalyst loading resulted in 93% NO reduction efficiency. This study highlights the importance of modification in the substrate structure in reducing the cost of SCR substrate to increase the NO reduction efficiency. Further work can be done to minimize the amount of catalyst needed for maximum NO reduction.

Acknowledgements We acknowledge the financial support from DST (ECR/2015/000135) and also thank Advanced Materials Research Centre, IIT Mandi (H.P.) for materials characterization facilities.

References

Cavataio G, Jen HW, Warner JR, Girard JW, Kim JY, Lambert CK (2009) Enhanced durability of a Cu/Zeolite based SCR catalyst. SAE Int J Fuels Lubr 1(1):477–548

Clean Air Technology Center (1999) Nitrogen oxides (NO_x), why and how they are controlled. Epa-456/F-99–006R, no. November, p. 48.

Chen X et al (2017) Design strategies for SCR catalysts with improved N_2 selectivity: the significance of nano-confining effects by titanate nanotubes. Environ Sci Nano 4(2):437–447

Choung JW, Nam IS, Ham SW (2006) Effect of promoters including tungsten and barium on the thermal stability of V_2O_5/sulfated TiO_2 catalyst for NO reduction by NH_3. Catal Today 111(3–4):242–247

Christensen SR, Hansen BB, Pedersen KH, Thøgersen JR, Jensen AD (2019) Selective catalytic reduction of NO_x over V_2O_5-WO_3-TiO_2 SCR catalysts—a study at elevated pressure for maritime pre-turbine SCR configuration. Emiss Control Sci Technol 5(3):263–278

Dankeaw A et al (2019) Highly porous Ce-W-TiO_2 free-standing electrospun catalytic membranes for efficient de-NO_x via ammonia selective catalytic reduction. Environ Sci Nano 6(1):94–104

European Union (2019) Worldwide emission standards and related regulations—passenger cars/light and medium duty vehicles

Fedeyko JM et al (2009) Development of thermally durable Cu/SCR catalysts. SAE Tech Pap

Fujiwara K, Okuyama K, Pratsinis SE (2017) Metal-support interactions in catalysts for environmental remediation. Environ Sci Nano 4(11):2076–2092

Guan B, Zhan R, Lin H, Huang Z (2014) Review of state of the art technologies of selective catalytic reduction of NO_x from diesel engine exhaust. Appl Therm Eng 66(1–2):395–414

Hamidzadeh M, Ghassemzadeh M, Tarlani A, Sahebdelfar S (2018) The effect of hydrothermal impregnation of Ni Co, and Cu on HZSM-5 in the nitrogen oxide removal. Int J Environ Sci Technol 15(1):93–104

Heikens S, Ulrich V, Joshi PS, Saruhan-Brings B, Grünert W (2019) Three-way catalysis with noble metal-substituted La(Fe, Co)O_3 perovskites—the role of noble and base metal components. Emiss Control Sci Technol 5(4):353–362

Hirose S et al (2012) Newly developed cordierite honeycomb substrate for SCR coating realizing system compactness and low backpressure. SAE Tech. Pap. Ser., vol. 1.

Kagawa J (2002) Health effects of diesel exhaust emissions a mixture of air pollutants of worldwide concern. Toxicology 182:349–353

Kwak JH, Tonkyn RG, Kim DH, Szanyi J, Peden CHF (2010) Excellent activity and selectivity of Cu-SSZ-13 in the selective catalytic reduction of NO_x with NH_3. J Catal 275(2):187–190

Masoudi M, Hensel J, Tegeler E (2019) A review of the 2018 U.S.-DOE CLEERS conference: trends and deeper insights in reduction of NO_x and particulate in diesel and gasoline engines and

advances in catalyst materials, mechanisms, and emission control technologies. Emiss Control Sci Technol
Maunula T, Viitanen A, Kinnunen T, Kanniainen K (2013) Design of durable vanadium-SCR catalyst systems for heavy-duty diesel applications. SAE Tech Pap 5(2)
Maunula T, Kinnunen T, Kanniainen K, Viitanen A, Savimaki A (2013) Thermally durable vanadium-SCR catalysts for diesel applications. SAE Tech Pap 2(4)
Nakomcic-Smaragdakis B, Cepic Z, Cepic M, Stajic T (2014) Data analysis of the flue gas emissions in the thermal-power plant firing fuel oil and natural gas. Int J Environ Sci Technol 11(2):269–280
Paridah M, Moradbak A, Mohamed A, Abdulwahab Taiwo Owolabi F, Asniza M, Abdul Khalid SH (2016) We are IntechOpen , the world ' s leading publisher of Open Access books Built by scientists, for scientists TOP 1 %. Intech i(tourism), 13
Pless JD, Naseri M, Klink W, Spreitzer G, Chatterjee S, Markatou P (2014) Development of SCR on high porosity substrates for heavy duty and off-road applications. SAE Int J Commer Veh 7(1):177–185
Rajadurai S, Jacob S, Serrell C, Morin R, Kircanski Z (2006) Wiremesh substrates for oxidation, TWC and SCR converters. Adv Propuls Emiss 2:97–105
Rammelt T, Kuc AB, Böhm J, Heine T, Gläser R (2019) Nature and surface interactions of sulfur-containing deposits on V_2O_5-WO_3/TiO_2 catalysts for SCR-DeNOx. Emiss Control Sci Technol 5(4):297–306
Resitoglu IA, Keskin A, Özarslan H, Bulut H (2019) Selective catalytic reduction of NO_x emissions by hydrocarbons over Ag–Pt/Al_2O_3 catalyst in diesel engine. Int J Environ Sci Technol 16(11):6959–6966
Reşitoḷlu IA, Altinişik K, Keskin A (2015) The pollutant emissions from diesel-engine vehicles and exhaust aftertreatment systems. Clean Technol Environ Policy 17(1):15–27
Rice M, Kramer J, Mueller-Haas K, Mueller R (2010) Innovative substrate technology for high performance heavy duty truck SCR catalyst systems. SAE Tech Pap Ser 1
Robinson JW (2016) Understanding catalysts a handbook for the user Part 2—Catalysts all around us. Handbook
Sagar A, Trovarelli A, Casanova M, Schermanz K (2011) A new class of environmental friendly vanadate based NH_3 SCR catalysts exhibiting good low temperature activity and high temperature stability. SAE Int J Engines 4(1):1839–1849
Sharif PM et al (2019) International regulation of vehicle emissions control rules and its influence on academic engine development experimental study and vehicle manufacturing. IOP Conf Ser Mater Sci Eng 469(1)
Stewart ML, Kamp CJ, Gao F, Wang Y, Engelhard MH (2018) Coating distribution in a commercial SCR filter. Emiss. Control Sci. Technol. 4(4):260–270
Theis JR, Lambert CK (2019) Effect of reductants on the NO_x storage performance of a Pd/CZO low temperature NO_x adsorber. Emiss. Control Sci. Technol., 215–224
United Nations Environment Programme (2017) Vehicle emissions standards world map 2017
Urdhwareshe R (2018) The Automotive Research Association of India, "2 Wheeler Vehicles : BS VI"
Vargas MAL, Casanova M, Trovarelli A, Busca G (2007) An IR study of thermally stable V_2O_5-WO_3-TiO_2 SCR catalysts modified with silica and rare-earths (Ce, Tb, Er). Appl Catal B Environ 75(3–4):303–311
Vassallo JE (2018) Status of emission control science and technology in Argentina. Emiss. Control Sci. Technol. 4(2):73–77
Vedyagin AA et al (2019) Synthesis and study of bimetallic Pd-Rh system supported on zirconia-doped alumina as a component of three-way catalysts. Emiss Control Sci Technol 5(4):363–377
Wang H et al. (2020) "Selective catalytic oxidation of ammonia over nano Cu/zeolites with different topologies. Environ Sci Nano
Wen M, Li G, Liu H, Chen J, An T, Yamashita H (2019) Metal-organic framework-based nanomaterials for adsorption and photocatalytic degradation of gaseous pollutants: recent progress and challenges. Environ Sci Nano 6(4):1006–1025

World Health Organization, 9 out of 10 people worldwide breathe polluted air, but more countries are taking action. [Online]. Available: https://www.who.int/news-room/detail/02-05-2018-9-out-of-10-people-worldwide-breathe-polluted-air-but-more-countries-are-taking-action. Accessed: 10 Dec 2019

Xu C, Sun W, Cao L, Yang J (2016) Selective catalytic reduction of nitric oxide by hydrogen over $NiFe_2$-$xPdxO_4$ catalysts at low temperature. Chem Eng J 283:1137–1144

Zha K et al (2018) Improved NO:X reduction in the presence of alkali metals by using hollandite Mn-Ti oxide promoted Cu-SAPO-34 catalysts. Environ Sci Nano 5(6):1408–1419

The manufacturer's authorised representative in the EU is Springer Nature Customer Service Centre GmbH, Europaplatz 3, 69115 Heidelberg, Germany. If you have any concerns regarding our products, please contact ProductSafety@springernature.com

Printed and bound by CPI Group (UK) Ltd, Croydon, CR0 4YY
30/09/2026
02225293-0020